T0180508

Advances in Intelligent Systems and Computing

Volume 1000

The series "Advances in Intelligent Systems and Computing" contains publications on theory, applications, and design methods of Intelligent Systems and Intelligent Computing. Virtually all disciplines such as engineering, natural sciences, computer and information science, ICT, economics, business, e-commerce, environment, healthcare, life science are covered. The list of topics spans all the areas of modern intelligent systems and computing such as: computational intelligence, soft computing including neural networks, fuzzy systems, evolutionary computing and the fusion of these paradigms, social intelligence, ambient intelligence, computational neuroscience, artificial life, virtual worlds and society, cognitive science and systems, Perception and Vision, DNA and immune based systems, self-organizing and adaptive systems, e-Learning and teaching, human-centered and human-centric computing, recommender systems, intelligent control, robotics and mechatronics including human-machine teaming, knowledge-based paradigms, learning paradigms, machine ethics, intelligent data analysis, knowledge management, intelligent agents, intelligent decision making and support, intelligent network security, trust management, interactive entertainment, Web intelligence and multimedia.

The publications within "Advances in Intelligent Systems and Computing" are primarily proceedings of important conferences, symposia and congresses. They cover significant recent developments in the field, both of a foundational and applicable character. An important characteristic feature of the series is the short publication time and world-wide distribution. This permits a rapid and broad dissemination of research results.

** Indexing: The books of this series are submitted to ISI Proceedings, EI-Compendex, DBLP, SCOPUS, Google Scholar and Springerlink **

More information about this series at http://www.springer.com/series/11156

Ralph Baker Kearfott · Ildar Batyrshin ·
Marek Reformat · Martine Ceberio ·
Vladik Kreinovich
Editors

Fuzzy Techniques: Theory and Applications

Proceedings of the 2019 Joint World Congress
of the International Fuzzy Systems
Association and the Annual Conference
of the North American Fuzzy Information
Processing Society IFSA/NAFIPS'2019
(Lafayette, Louisiana, USA, June 18–21, 2019)

 Springer

Editors
Ralph Baker Kearfott
Department of Mathematics
University of Louisiana at Lafayette
Lafayette, LA, USA

Ildar Batyrshin
Center for Computing Research (CIC)
Instituto Politecnico Nacional
Ciudad de Mexico, Mexico

Marek Reformat
Department of Electrical
and Computer Engineering
University of Alberta
Edmonton, AB, Canada

Martine Ceberio
Department of Computer Science
University of Texas at El Paso
El Paso, TX, USA

Vladik Kreinovich
Department of Computer Science
University of Texas at El Paso
El Paso, TX, USA

ISSN 2194-5357   ISSN 2194-5365   (electronic)
Advances in Intelligent Systems and Computing
ISBN 978-3-030-21919-2   ISBN 978-3-030-21920-8   (eBook)
https://doi.org/10.1007/978-3-030-21920-8

This Springer imprint is published by the registered company Springer Nature Switzerland AG
The registered company address is: Gewerbestrasse 11, 6330 Cham, Switzerland

Preface

This book contains papers from the 2019 World Congress of the International Fuzzy Systems Association IFSA/NAFIPS'2019, which will be held jointly with the Annual Conference of the North American Fuzzy Information Processing Society in Lafayette, Louisiana, USA, on June 18–21, 2019. These papers cover many interesting and important theoretical issues related to fuzzy techniques, as well as many exciting and successful applications of the fuzzy methodology.

A special section of this book is devoted to papers from the affiliated 12th International Workshop on Constraint Programming and Decision Making CoProd'2019 (Lafayette, Louisiana, June 17, 2019).

Our many thanks to all the authors for their excellent papers, to all the anonymous referees for their important work, to IFSA and NAFIPS leadership, especially to IFSA President Javier Montero, for their support, to Janusz Kacprzyk and to all the Springer staff for helping us to publish this volume, and of course to the readers for their interest—this is for you, the readers, that these books are published. We hope that in this book, you will find new ideas that can be directly applied, as well as ideas and challenges that will inspire you to make further progress.

Thanks again and enjoy!

R. Baker Kearfott
Ildar Batyrshin
Marek Reformat
Martine Ceberio
Vladik Kreinovich

Contents

**Constraint Programming And Decision Making Papers
from the 12th International Workshop CoProd'2019**

Aggregation and Pre-aggregation Functions. Theory And Applications

A New Axiomatic Approach
to Interval-Valued Entropy

Humberto Bustince[1]($\boxtimes$), Javier Fernandez[1], Iosu Rodriguez[1], Borja de la Osa[1],
Cédric Marco-Detchart[1], Jose Antonio Sanz Delgado[1], and Zdenko Takáč[2]

[1] Departamento de Estadistica, Informatica y Matematicas,
Universidad Publica de Navarra, 31006 Pamplona, Spain
{bustince,fcojavier.fernandez,iosu.rodriguez,delaosa.47954,
cedric.marco,joseantonio.sanz}@unavarra.es
[2] Institute of Information Engineering, Automation and Mathematics,
The Slovak University of Technology, Bratislava, Slovakia
zdenko.takac@stuba.sk

Abstract. In this work we propose a new definition of interval-valued
entropy taking into account the width of the considered membership
intervals. We build these new entropies by aggregating normal E_N functions.

1 Introduction

Most of the recent developments in the interval-valued fuzzy sets theory [17] only
consider a partial order [1–6,8,9,12,14–16,20,21] and do not take into account
the widths of the involved intervals.

As the width of the intervals can be understood as a measure of the level
of uncertainty involved in the construction of the corresponding interval-valued
fuzzy set, the objective of this work is to define axiomatically a new interval-
valued entropy measure for interval-valued fuzzy sets which takes into account
the width of the involved membership intervals.

To build this new entropy, we extend to the interval-valued fuzzy setting the
notion of normal E_N-function, taking also into account the use of admissible
(total) orders.

The paper is organized as follows. We start with some preliminaries. In
Sect. 3, we present the definition of interval-valued normal E_N functions, and,
in Sect. 4, we discuss the new definition of width-based interval-valued entropy.
We finish with some conclusions and references.

2 Preliminaries

We recall now some notions relevant for this work. We denote

$$L([0,1]) = \{[\underline{X}, \overline{X}] \mid 0 \le \underline{X} \le \overline{X} \le 1\}.$$

© Springer Nature Switzerland AG 2019
R. B. Kearfott et al. (Eds.): IFSA 2019/NAFIPS 2019, AISC 1000, pp. 3–12, 2019.
https://doi.org/10.1007/978-3-030-21920-8_1

Capital letters will be used to denote elements in $L([0,1])$ as well as the bounds of intervals. The width of the interval $X \in L([0,1])$ will be denoted by $w(X)$, clearly $w(X) = \overline{X} - \underline{X}$. An interval function $f : (L([0,1]))^n \to L([0,1])$ is called w-preserving if $w(X_1) = \ldots = w(X_n)$ implies $w(f(X_1, \ldots, X_n)) = w(X_1)$.

A fuzzy set in an universe U is a mapping $A : U \to [0,1]$, an interval-valued fuzzy set is a mapping $A : U \to L([0,1])$. The class of all fuzzy sets in U is denoted by $FS(U)$ and the class of all interval-valued fuzzy sets in U by $IVFS(U)$.

Definition 1. An *order relation* on $L([0,1])$ is a binary relation $\leq$ on $L([0,1])$ such that, for all $X, Y, Z \in L([0,1])$,

(i) $X \leq X$, (*reflexivity*),
(ii) $X \leq Y$ and $Y \leq X$ imply $X = Y$, (*antisymmetry*),
(iii) $X \leq Y$ and $Y \leq Z$ imply $X \leq Z$, (*transitivity*).

An order relation on $L([0,1])$ is *linear* if any two elements of $L([0,1])$ are comparable, i.e., if for every $X, Y \in L([0,1])$, $X \leq Y$ or $Y \leq X$. An order relation on $L([0,1])$ is *partial* if it is not linear.

We will denote by $\precsim_L$ the partial order relation on $L([0,1])$ induced by the usual partial order on $\mathbb{R}^2$, that is:

$$[\underline{X}, \overline{X}] \precsim_L [\underline{Y}, \overline{Y}] \qquad \text{iff} \qquad \underline{X} \leq \underline{Y} \text{ and } \overline{X} \leq \overline{Y}. \tag{1}$$

This is the order relation most widely used in the literature [13].

We denote by $\leq_L$ any order in $L([0,1])$ (which can be partial or total) with $0_L = [0,0]$ as its minimal element and $1_L = [1,1]$ as its maximal element. To denote a total order in $L([0,1])$ with these minimal and maximal elements, we use the notation $\leq_{TL}$.

Example 1. Xu and Yager's order (see [22]) is an example of total order on $L([0,1])$: $[\underline{X}, \overline{X}] \leq_{XY} [\underline{Y}, \overline{Y}]$ if

$$\begin{cases} \underline{X} + \overline{X} < \underline{Y} + \overline{Y} \text{ or} \\ \underline{X} + \overline{X} = \underline{Y} + \overline{Y} \text{ and } \overline{X} - \underline{X} \leq \overline{Y} - \underline{Y}. \end{cases} \tag{2}$$

This definition of Xu and Yager's order was originally provided for Atanassov intuitionistic fuzzy pairs.

An *admissible* order on $L([0,1])$ is a total order $\leq_{TL}$ that refines the partial order $\precsim_L$. These admissible orders can be built using aggregation functions.

Proposition 1 ([10]). *Let $M_1, M_2 : [0,1]^2 \to [0,1]$ be two aggregation functions such that for all $X, Y \in L([0,1])$, the equalities $M_1(\underline{X}, \overline{X}) = M_1(\underline{Y}, \overline{Y})$ and $M_2(\underline{X}, \overline{X}) = M_2(\underline{Y}, \overline{Y})$ can only hold simultaneously if $X = Y$. The order $\leq_{M_1, M_2}$ on $L([0,1])$ given by*

$$X \leq_{M_1, M_2} Y \text{ if } \begin{cases} M_1(\underline{X}, \overline{X}) < M_1(\underline{Y}, \overline{Y}) \text{ or} \\ M_1(\underline{X}, \overline{X}) = M_1(\underline{Y}, \overline{Y}) \text{ and } M_2(\underline{X}, \overline{X}) \leq M_2(\underline{Y}, \overline{Y}) \end{cases}$$

is an admissible order on $L([0,1])$.

Example 2. Take $\alpha \in [0, 1]$ and consider the aggregation function

$$K_\alpha(x, y) = (1 - \alpha)x + \alpha y.$$

If $\alpha, \beta \in [0, 1]$ with $\alpha \neq \beta$, we can obtain an admissible order $\leq_{\alpha, \beta}$ just taking $M_1(x, y) = K_\alpha(x, y)$ and $M_2(x, y) = K_\beta(x, y)$. See [10] for more details.

We also need the definition of interval-valued negation.

Definition 2. Let $\leq_L$ be an order relation in $L([0, 1])$. A function $N \colon L([0, 1]) \to L([0, 1])$ is an *interval-valued negation function* (IV negation) if it is a decreasing function with respect to the order $\leq_L$ such that $N(0_L) = 1_L$ and $N(1_L) = 0_L$. A negation N is called a *strong negation* if $N(N(X)) = X$ for every $X \in L([0, 1])$. An interval $\varepsilon \in L([0, 1])$ is called an *equilibrium point* of the IV negation if $N(\varepsilon) = \varepsilon$.

We bring here the definition of interval-valued aggregation function.

Definition 3. Let $n \geq 2$. An (n-dimensional) *interval-valued (IV) aggregation function* in $L([0, 1])$ with respect to $\leq_L$ is a mapping $M_{IV} \colon (L([0, 1]))^n \to L([0, 1])$ which verifies:

(i) $M_{IV}(0_L, \cdots, 0_L) = 0_L$.
(ii) $M_{IV}(1_L, \cdots, 1_L) = 1_L$.
(iii) M_{IV} is a non-decreasing function with respect to $\leq_L$.

We say that $M_{IV} \colon (L([0, 1]))^n \to L([0, 1])$ is a *decomposable n-dimensional IV aggregation function* associated with M_L and M_U, if there exist n-dimensional aggregation functions $M_L, M_U \colon [0, 1]^n \to [0, 1]$ such that $M_L \leq M_U$ and

$$M_{IV}(X_1, \ldots, X_n) = \left[M_L \left(\underline{X_1}, \ldots, \underline{X_n} \right), M_U \left(\overline{X_1}, \ldots, \overline{X_n} \right) \right] \tag{3}$$

for all $X_1, \ldots, X_n \in L([0, 1])$.

Finally we recover here the notion of E_N function which is going to be crucial for our work.

Definition 4 [7]. Let $n \colon [0, 1] \to [0, 1]$ be a strong negation with the equilibrium point e (i.e., an involutive decreasing function such that $n(e) = e$). A function $E_N \colon [0, 1] \to [0, 1]$ is called a *normal E_N-function* w.r.t. n if it satisfies the following conditions:

1. $E_N(e) = 1$;
2. $E_N(x) = 0$ if and only if $x = 0$ or $x = 1$;
3. If $y \leq x \leq e$ or $y \geq x \geq e$, then $E_N(x) \geq E_N(y)$.

Remark 1. For any $p \in [0, \infty]$, the function $E_N^p(x) = 1 - |2x - 1|^p$ is a normal E_N-function regardless which strong negation n we consider as long as the latter has equilibrium point $e = 1/2$.

3   Interval-Valued Normal E_N-functions

As a first step to build entropy functions, we extend the definition of E_N function to the interval-valued setting. Later we will collect together E_N-functions to get the entropy.

Definition 5. Let $N : L([0,1]) \to L([0,1])$ be an interval-valued strong negation w.r.t. a total order $\leq_{TL}$ with the equilibrium point ε (i.e., a decreasing involutive function such that $N(\varepsilon) = \varepsilon$). A function $EN_{IV} : L([0,1]) \to L([0,1])$ is called an *interval-valued normal E_N-function* w.r.t. N if it satisfies the following conditions:

1. $EN_{IV}(\varepsilon) = [1 - w(\varepsilon), 1]$;
2. $EN_{IV}(X) = 0_L$ if and only if $X = 0_L$ or $X = 1_L$;
3. If $Y \leq_{TL} X \leq_{TL} \varepsilon$ or $Y \geq_{TL} X \geq_{TL} \varepsilon$ where $w(X) = w(Y)$, then $EN_{IV}(X) \geq_{TL} EN_{IV}(Y)$.

Example 3. Given $X_0 \in L([0,1])$ where $X_0 \leq_{TL} [0,1]$ and $X_0 \neq 0_L$, we can obtain the following trivial example of IV normal E_N function w.r.t any IV strong negation and any admissible order.

$$EN_{IV}(X) = \begin{cases} 0_L, & \text{if } X = 0_L \text{ or } X = 1_L, \\ [1 - w(X), 1], & \text{if } X = \varepsilon, \\ X_0, & \text{otherwise,} \end{cases}$$

Now we provide a method for constructing interval-valued E_N functions which preserve the width of the intervals.

Proposition 2. *Let $n : [0,1] \to [0,1]$ be a strong negation with equilibrium point e. Let $\alpha, \beta \in]0,1[$, $\beta \neq \alpha$ and $N : L([0,1]) \to L([0,1])$ be a strong IV negation w.r.t. $\leq_{\alpha,\beta}$ with equilibrium point ε such that $K_\alpha(\varepsilon) = e$. Let $E_N : [0,1]^2 \to [0,1]$ be a normal E_N-function w.r.t. n. Then, the function $EN_{IV} : L([0,1]) \to L([0,1])$ given by*

$$EN_{IV}(X) = \Big[\max\big(0, E_N\left(K_\alpha(X)\right) - w(X)\big),$$
$$\max\big(E_N\left(K_\alpha(X)\right), w(X)\big)\Big] \tag{4}$$

is an IV normal E_N-function w.r.t. N. Moreover, EN_{IV} is w-preserving.

Proof. Equation (4) can be simplified:

$$EN_{IV}(X) = \begin{cases} [E_N\left(K_\alpha(X)\right) - w(X), E_N\left(K_\alpha(X)\right)], & \text{if } E_N\left(K_\alpha(X)\right) \geq w(X), \\ [0, w(X)], & \text{otherwise.} \end{cases} \tag{5}$$

Clearly, EN_{IV} is well-defined and w-preserving, so it remains to prove the three conditions in Definition 5:

1. Since $E_N(K_\alpha(\varepsilon)) = E_N(e) = 1 \geq w(\varepsilon)$, we have

$$EN_{IV}(\varepsilon) = [E_N(K_\alpha(\varepsilon)) - w(\varepsilon), E_N(K_\alpha(\varepsilon))] = [1 - w(\varepsilon), 1].$$

2. $EN_{IV}(X) = 0_L$ if and only if $E_N(K_\alpha(X)) = 0$ and $w(X) = 0$ if and only if $X \in \{0_L, 1_L\}$.

3. Let $w(X) = w(Y)$ and $Y \leq_{\alpha,\beta} X \leq_{\alpha,\beta} \varepsilon$. Then $K_\alpha(Y) \leq K_\alpha(X) \leq K_\alpha(\varepsilon) = e$, hence $E_N(K_\alpha(Y)) \leq E_N(K_\alpha(X))$ and consequently $EN_{IV}(Y) \leq_{\alpha,\beta} EN_{IV}(X)$. The same conclusion can be drawn for $Y \geq_{\alpha,\beta} X \geq_{\alpha,\beta} \varepsilon$.

Corollary 1. *Let $\leq_{XY}$ be the Xu and Yager order and $N : L([0,1]) \to L([0,1])$ be an IV strong negation w.r.t. $\leq_{XY}$ with the equilibrium point ε such that $\underline{\varepsilon} + \overline{\varepsilon} = 1$. Let $E_N : [0,1]^2 \to [0,1]$ be a normal E_N-function w.r.t. a strong negation n with the equilibrium point $e = 1/2$ such that $E_N(x) \geq 1 - |2x - 1|$ for all $x \in [0,1]$. Then, the function $EN_{IV} : L([0,1]) \to L([0,1])$ given by*

$$EN_{IV}(X) = \left[E_N\left(\frac{\underline{X} + \overline{X}}{2}\right) - w(X), E_N\left(\frac{\underline{X} + \overline{X}}{2}\right) \right] \tag{6}$$

is an IV normal E_N-function w.r.t. N. Moreover, EN_{IV} is w-preserving.

Proof. Since (6) is a special case of (4), we only need to show that $E_N\left(\frac{\underline{X}+\overline{X}}{2}\right) - w(X) \geq 0$. Let $\frac{\underline{X}+\overline{X}}{2} \geq 0.5$. Then

$$E_N\left(\frac{\underline{X} + \overline{X}}{2}\right) - w(X) \geq 1 - |\underline{X} + \overline{X} - 1| - w(X)$$

$$= 1 - \underline{X} - \overline{X} + 1 - \overline{X} + \underline{X} = 2 - 2\overline{X} \geq 0.$$

Similarly, $E_N\left(\frac{\underline{X}+\overline{X}}{2}\right) - w(X) \geq 2\underline{X} \geq 0$ for $\frac{\underline{X}+\overline{X}}{2} < 0.5$, hence the the proof is completed.

From now on, $d_\alpha(c)$ denotes the maximal possible length of an interval $X \in L([0,1])$ such that $K_\alpha(X) = c$ where $c \in [0,1]$ and $\alpha \in [0,1]$. Then (see Proposition 7 in [1]):

$$d_\alpha(K_\alpha(X)) = \wedge\left(\frac{K_\alpha(X)}{\alpha}, \frac{1 - K_\alpha(X)}{1 - \alpha}\right).$$

Proposition 3. *Let $\alpha \in [0,1]$. For any $X \in L([0,1])$ let*

$$\lambda_\alpha(X) = \frac{\overline{X} - \underline{X}}{d_\alpha(K_\alpha(X))} = \begin{cases} (\overline{X} - \underline{X})\frac{\alpha}{K_\alpha(X)} & \text{if } K_\alpha(X) \leq \alpha, \\ (\overline{X} - \underline{X})\frac{1-\alpha}{1-K_\alpha(X)} & \text{if } K_\alpha(X) \geq \alpha. \end{cases} \tag{7}$$

If $n : [0,1] \to [0,1]$ is a strong negation, then the mapping $N_{\alpha,n} : L([0,1]) \to L([0,1])$ given by

$$\begin{cases} N_{\alpha,n}(0_L) = 1_L, \\ N_{\alpha,n}(1_L) = 0_L, \\ N_{\alpha,n}(X) = Y, & \text{if } X \in L([0,1])\backslash\{0_L, 1_L\}, \end{cases} \tag{8}$$

where

$$\begin{cases} K_\alpha(Y) = n(K_\alpha(X)), \\ \lambda_\alpha(Y) = 1 - \lambda_\alpha(X), \end{cases}$$

is a strong IV negation on $L([0,1])$ with respect to the order $\leq_{\alpha,\beta}$ for any $\beta \neq \alpha$.

Moreover, $N_{\alpha,n}$ has the unique equilibrium point ε that is given by $K_\alpha(\varepsilon) = e$ and $\lambda_\alpha(\varepsilon) = \frac{1}{2}$, where e is the equilibrium point of the strong negation n.

Proof. See Theorem 4 and Proposition 8 in [1].

Example 4. (i) Let us consider the construction of IV normal E_N-function given by Corollary 1. Consider the normal E_N-function $E_N^p(x) = 1 - |2x - 1|^p$ where $p \in [1, \infty[$. Let $\leq_{XY}$ be the Xu and Yager order. Then (see Theorem 2 in [1]) $N(X) = [c' - r', c' + r']$ where $c = \frac{X + \overline{X}}{2}$, $r = \frac{\overline{X} - \underline{X}}{2}$, $a = \min(c, 1 - c)$, $c' = 1 - c$ and $r' = a - r$, is a strong IV negation w.r.t. $\leq_{XY}$ with the unique equilibrium point $[1/4, 3/4]$. Then

$$EN_{IV}^p(X) = \left[E_N^p \left(\frac{\underline{X} + \overline{X}}{2} \right) - w(X), E_N^p \left(\frac{\underline{X} + \overline{X}}{2} \right) \right]$$

is a class of IV normal E_N-functions w.r.t. N.

(ii) Now let us consider the construction of IV normal E_N-function given by Proposition 2. Consider the normal E_N-function $E_N^p(x) = 1 - |2x - 1|^p$ where $p \in]0, \infty[$. Let $N_{\alpha,n}$ be the strong IV negation given in Proposition 3. Then

$$EN_{IV}^{p,\alpha}(X) = \left[\max\left(0, E_N^p \left(K_\alpha(X) \right) - w(X) \right), \max\left(E_N^p \left(K_\alpha(X) \right), w(X) \right) \right],$$

for $\alpha \in]0, 1[$, is a class of IV normal E_N-functions w.r.t. $N_{\alpha,n}$.

4   Aggregation of IV E_N-functions for Getting IV Entropies

In this section an IV entropy is defined for interval-valued fuzzy sets and it is built aggregating IV normal E_N-functions.

First we summarize the main results on w-preserving IV aggregation functions from [11] that will be needed in this section.

We start with a construction method of IV aggregation functions w.r.t. $\leq_{\alpha,\beta}$.

Theorem 1 [11]. *Let $\alpha, \beta \in [0,1]$, $\beta \neq \alpha$. Let $M_1, M_2 : [0,1]^n \rightarrow [0,1]$ be aggregation functions where M_1 is strictly increasing. Then $M_{IV} : (L([0,1]))^n \rightarrow L([0,1])$ defined by:*

$$M_{IV}(X_1, \ldots, X_n) = Y,$$

where

$$\begin{cases} K_\alpha(Y) = M_1 \left(K_\alpha(X_1), \ldots, K_\alpha(X_n) \right), \\ \lambda_\alpha(Y) = M_2 \left(\lambda_\alpha(X_1), \ldots, \lambda_\alpha(X_2) \right), \end{cases}$$

for all $X_1, \ldots, X_n \in L([0,1])$, is an IV aggregation function with respect to $\leq_{\alpha,\beta}$.

And next we provide a construction method of IV aggregation functions w.r.t. $\leq_{\alpha,\beta}$ which would preserve the width of input intervals is given.

(P1) $M(cx_1, \ldots, cx_n) \geq cM(x_1, \ldots, x_n)$ for all $c \in [0,1]$, $x_1, \ldots, x_n \in [0,1]$.
(P2) $M(x_1, \ldots, x_n) \leq 1 - M(1 - x_1, \ldots, 1 - x_n)$ for all $x_1, \ldots, x_n \in [0,1]$.

Theorem 2 [11]. *Let $\alpha, \beta \in [0,1]$, $\beta \neq \alpha$. Let $M_1, M_2 : [0,1]^n \to [0,1]$ be aggregation functions such that M_1 is strictly increasing, $M_1(x_1, \ldots, x_n) \geq M_2(x_1, \ldots, x_n)$ for all $x_1, \ldots, x_n \in [0,1]$, M_1 or M_2 satisfies property (P1) and M_1 or M_2 satisfies property (P2). Then $M_{IV} : (L([0,1]))^n \to L([0,1])$ defined by:*

$$M_{IV}(X_1, \ldots, X_n) = Y,$$

where

$$\begin{cases} K_\alpha(Y) = M_1\left(K_\alpha(X_1), \ldots, K_\alpha(X_n)\right), \\ w(Y) = M_2\left(w(X_1), \ldots, w(X_n)\right), \end{cases}$$

for all $X_1, \ldots, X_n \in L([0,1])$, is an IV aggregation function with respect to $\leq_{\alpha,\beta}$. Moreover, if M_2 is idempotent, then M_{IV} is w-preserving.

Lemma 1 [11]. *Let $M_{IV} : (L([0,1]))^n \to L([0,1])$ be defined as in Theorem 2.*

(i) *If*
  - $M_1(x_1, \ldots, x_n) = 0$ *if and only if* $x_1 = \ldots = x_n = 0$ *and*
  - $M_2(x_1, \ldots, x_n) = 0$ *if and only if* $x_1 = \ldots = x_n = 0$,

  then $M_{IV}(X_1, \ldots, X_n) = 0_L$ if and only if $X_1 = \ldots = X_n = 0_L$. Moreover, if $\alpha \neq 0$, then the restriction on M_2 can be skipped.

(ii) *If*
  - $M_1(x_1, \ldots, x_n) = 1$ *if and only if* $x_1 = \ldots = x_n = 1$ *and*
  - $M_2(x_1, \ldots, x_n) = 0$ *if and only if* $x_1 = \ldots = x_n = 0$,

  then $M_{IV}(X_1, \ldots, X_n) = 1_L$ if and only if $X_1 = \ldots = X_n = 1_L$. Moreover, if $\alpha \neq 1$, then the restriction on M_2 can be skipped.

(iii) *M_{IV} is idempotent if and only if M_1 and M_2 are idempotent.*

4.1 Width-Based IV Entropies

From now on, $\widetilde{X}$ denotes the IVFS A in U such that $A(u) = X$ for all $u \in U$ where $X \in L([0,1])$.

Definition 6. Let $\leq_{TL}$ be a total order in $L([0,1])$. Let N be a strong IV negation with respect to $\leq_{TL}$ with an equilibrium point $\varepsilon \in L([0,1])$. A function $E : IVFS(U) \to L([0,1])$ is an *IV entropy* on $IVFS(U)$ with respect to the strong IV negation N if for all $A, B \in IVFS(U)$:

(E1) $E(A) = 0_L$ if and only if A is crisp;
(E2) $E(\widetilde{\varepsilon}) = [1 - w(\varepsilon), 1]$;
(E3) $E(A) \leq_{TL} E(B)$, if $w(A(u)) = w(B(u))$ and $A(u) \leq_{TL} B(u) \leq_{TL} \varepsilon$ or $A(u) \geq_{TL} B(u) \geq_{TL} \varepsilon$ for all $u \in U$.

The definition is taken from [7], however, the third axiom is relaxed to take into account uncertainty linked to intervals. Now, a construction method of IV entropy by aggregation of normal E_N-functions is given.

Proposition 4. *Let* $U = \{u_1, \ldots, u_n\}$ *and let* $N : L([0,1]) \to L([0,1])$ *be a strong IV negation w.r.t. a total order* $\leq_{TL}$. *Let* $M_{IV} : (L([0,1]))^n \to L([0,1])$ *be an idempotent IV aggregation function w.r.t.* $\leq_{TL}$ *satisfying* $M_{IV}(X_1, \ldots, X_n) = 0_L$ *if and only if* $X_1 = \ldots = X_n = 0_L$. *Let* $EN_{IV} : L([0,1]) \to L([0,1])$ *be an IV normal* E_N-*function w.r.t.* N *(given by Definition 5). Then the function* $E : IVFS(U) \to L([0,1])$ *defined by:*

$$E(A) = M_{IV}\big(EN_{IV}\left(A(u_1)\right), \ldots, EN_{IV}\left(A(u_n)\right)\big)$$

for all $A \in IVFS(U)$ *is an IV entropy on* $IVFS(U)$ *with respect to the strong IV negation* N.

Proof. The proof is straightforward.

We study the conditions under which the function EN_{IV} given by Eq. (4) can be applied in the previous proposition to obtain an IV entropy.

Corollary 2. *Let* $U = \{u_1, \ldots, u_n\}$ *and* $\alpha, \beta \in]0,1[$ *where* $\beta \neq \alpha$. *Let* $M_{IV} : (L([0,1]))^n \to L([0,1])$ *be an IV aggregation function w.r.t.* $\leq_{\alpha,\beta}$ *given by aggregation functions* M_1, M_2 *as in Proposition 2. Let* $EN_{IV} : L([0,1]) \to L([0,1])$ *be an IV normal* E_N-*function given by a normal* E_N-*function* E_N *as in Proposition 2 where* N *is a strong IV negation w.r.t.* $\leq_{\alpha,\beta}$ *with an equilibrium point* ε. *Let* $E : IVFS(U) \to L([0,1])$ *be a function defined by:*

$$E(A) = M_{IV}\big(EN_{IV}\left(A(u_1)\right), \ldots, EN_{IV}\left(A(u_n)\right)\big)$$

for all $A \in IVFS(U)$. *Then*

(i) *E satisfies axiom (E1), if*
    $M_1(x_1, \ldots, x_n) = 0$ *if and only if* $x_1 = \ldots = x_n = 0$.
(ii) *E satisfies axiom (E2), if* M_1 *and* M_2 *are idempotent.*
(iii) *E satisfies axiom (E3) w.r.t.* $\leq_{\alpha,\beta}$.
(iv) *Let* M_2 *be idempotent. Then, for all* $A \in IVFS(U)$, $w(A(u_1)) = \ldots = w(A(u_n))$ *implies* $w(E(A)) = w(A(u_1))$.

Proof. The proof of (i) follows from Lemma 1 (i), that of (ii) from Lemma 1 (iii), and the proof of (iii) and (iv) is straightforward.

Example 5. Let $\alpha, \beta \in]0,1[$ where $\beta \neq \alpha$. A function $E : IVFS(U) \to L([0,1])$ defined as in Corollary 2, is an entropy for IVFSs w.r.t. $\leq_{\alpha,\beta}$, if, for instance, $M_1(x_1, \ldots, x_n) = M_2(x_1, \ldots, x_n) = \frac{x_1 + \ldots + x_n}{n}$ for all $x_1, \ldots, x_n \in [0,1]$; and E_N be any E_N-normal function.

5   Conclusions

In this work we have proposed a new definition of width-based interval-valued entropy and we have shown how these new entropies can be built aggregating normal IV E_N functions.

In future works we intend to develop possible applications of these new concepts in fields such as image processing [2,16] or classification [18–20].

Acknowledgments. This work has been suported by research project TIN2016-77356-P (AEI/UE,FEDER) of the Spanish Government and by Project VEGA 1/0614/18.

References

1. Asiain, M.J., Bustince, H., Mesiar, R., Kolesárová, A., Takáč, Z.: Negations with respect to admissible orders in the interval-valued fuzzy set theory. IEEE Trans. Fuzzy Syst. **26**, 556–568 (2018)
2. Barrenechea, E., Bustince, H., De Baets, B., Lopez-Molina, C.: Construction of interval-valued fuzzy relations with application to the generation of fuzzy edge images. IEEE Trans. Fuzzy Syst. **19**(5), 819–830 (2011)
3. Barrenechea, E., Fernandez, J., Pagola, M., Chiclana, F., Bustince, H.: Construction of interval-valued fuzzy preference relations from ignorance functions and fuzzy preference relations. Appl. Decis. Making Knowl.-Based Syst. **58**, 33–44 (2014)
4. Bentkowska, U., Bustince, H., Jurio, A., Pagola, M., Pekala, B.: Decision making with an interval-valued fuzzy preference relation and admissible orders. Appl. Soft Comput. **35**, 792–801 (2015)
5. Burillo, P., Bustince, H.: Construction theorems for intuitionistic fuzzy sets. Fuzzy Sets Syst. **84**, 271–281 (1996)
6. Bustince, H.: Indicator of inclusion grade for interval-valued fuzzy sets. Application to approximate reasoning based on interval-valued fuzzy sets. Int. J. Approximate Reasoning **23**(3), 137–209 (2000)
7. Bustince, H., Barrenechea, E., Pagola, M.: Relationship between restricted dissimilarity functions, restricted equivalence functions and normal E_N-functions: image thresholding invariant. Pattern Recogn. Lett. **29**(4), 525–536 (2008)
8. Bustince, H., Barrenechea, E., Pagola, M., Fernández, J.: Interval-valued fuzzy sets constructed from matrices: application to edge detection. Fuzzy Sets Syst. **160**, 1819–1840 (2009)
9. Bustince, H., Barrenechea, E., Pagola, M., Fernández, J., Xu, Z., Bedregal, B., Montero, J., Hagras, H., Herrera, F., De Baets, B.: A historical account of types of fuzzy sets and their relationship. IEEE Trans. Fuzzy Syst. **24**(1), 179–194 (2016)
10. Bustince, H., Fernandez, J., Kolesárová, A., Mesiar, R.: Generation of linear orders for intervals by means of aggregation functions. Fuzzy Sets Syst. **220**, 69–77 (2013)
11. Bustince, H., Marco-Detchart, C., Fernandez, J., Wagner, C., Garibaldi, J., Takáč, Z.: Similarity between interval-valued fuzzy sets taking into account the width of the intervals and admissible orders. Fuzzy Sets Syst. (Submitted)
12. Castillo, O., Melin, P.: A review on interval type-2 fuzzy logic applications in intelligent control. Inf. Sci. **279**, 615–631 (2014)

13. Cornelis, C., Deschrijver, G., Kerre, E.E.: Implication in intuitionistic fuzzy and interval-valued fuzzy set theory: construction, classification, application. Int. J. Approximate Reasoning **35**(1), 55–95 (2004)
14. Choi, H.M., Mun, G.S., Ahn, J.Y.: A medical diagnosis based on interval-valued fuzzy sets. Biomed. Eng.-Appl. Basis Commun. **24**(4), 349–354 (2012)
15. Couto, P., Jurio, A., Varejao, A., Pagola, M., Bustince, H., Melo-Pinto, P.: An IVFS-based image segmentation methodology for rat gait analysis. Soft Comput. **15**(10), 1937–1944 (2011)
16. Jurio, A., Pagola, M., Mesiar, R., Beliakov, G., Bustince, H.: Image magnification using interval information. IEEE Trans. Image Process. **20**(11), 3112–3123 (2011)
17. Sambuc, R.: Function phi-flous application a l'aide au diagnostic en pathologie thyroidienne. Ph.D. thesis, University of Marseille (1975)
18. Sanz, J.A., Fernández, A., Bustince, H., Herrera, F.: A genetic tuning to improve the performance of fuzzy rule-based classification systems with interval-valued fuzzy sets: degree of ignorance and lateral position. Int. J. Approximate Reasoning **52**(6), 751–766 (2011)
19. Sanz, J.A., Fernández, A., Bustince, H., Herrera, F.: Improving the performance of fuzzy rule-based classification systems with interval-valued fuzzy sets and genetic amplitude tuning. Inf. Sci. **180**(19), 3674–3685 (2010)
20. Sanz, J.A., Fernandez, A., Bustince, H., Herrera, F.: IVTURS: a linguistic fuzzy rule-based classification system based on a new interval-valued fuzzy reasoning method with tuning and rule selection. IEEE Trans. Fuzzy Syst. **21**(3), 399–411 (2013)
21. Wang, J., Guo, Q.: Ensemble interval-valued fuzzy cognitive maps. IEEE Access **6**, 38356–38366 (2018)
22. Xu, Z.S., Yager, R.R.: Some geometric aggregation operators based on intuitionistic fuzzy sets. Int. J. Gen. Syst. **35**, 417–433 (2006)

Aggregation Operators to Evaluate the Relevance of Classes in a Fuzzy Partition

Fabián Castiblanco[1]([✉]) [iD], Camilo Franco[2] [iD], Javier Montero[3] [iD],
and J. Tinguaro Rodríguez[3] [iD]

[1] Faculty of Economic, Administrative and Accounting Sciences,
Gran Colombia University, Bogotá, Colombia
fabianalberto.castiblanco@ugc.edu.co
[2] Department of Industrial Engineering, Andes University, Bogotá, Colombia
[3] Department of Statistics, Complutense University, Madrid, Spain

Abstract. In this paper we propose a comparison process that allows evaluating the relevance of a class in a fuzzy partition. This process starts by forming a commutative group structure based on a *fuzzy classification system*. On this structure, proximity relations are proposed to establish the comparison process. As our proposal is based on a fuzzy classification system, i.e., recursive De Morgan triples, we study the properties that the operators of such triples must satisfy. Therefore, we propose the study of a property that we have called *weakly self-dual*.

Keywords: Fuzzy classification systems · Relevance · Fuzzy partition · Weakly Self-dual

1 Introduction

Fuzzy classification systems, as proposed in [1] and [2], were conceived as a structure allowing the treatment of complex classification problems following two key ideas. On the one hand, under the theoretical framework proposed by Dombi [3, 4] about aggregation operators, fuzzy classification systems were proposed as an alternative approach for non-associative connectives through the use of the concept of recursiveness. On the other hand, fuzzy classification systems were proposed as a structure allowing the evaluation of the established classification from De Morgan triples i.e., by using recursive rules (satisfying the De Morgan laws) to evaluate three key characteristics of the family of classes that are obtained in a fuzzy partition: redundancy, coverage and relevance.

Regarding the redundancy and coverage property, several studies have been developed allowing the emergence of the well-known overlap functions and grouping functions, as proposed in [5] in the case of overlap functions (see also [6] for the multidimensional case), as well as in [7] and [8] in the case of grouping functions.

Concerning the relevance property, there are some approaches and first proposals about their study applied to unsupervised classification problems. For instance, in [2] a study based on statistical tools is proposed. In [9], relevance is addressed as a problem

R. B. Kearfott et al. (Eds.): IFSA 2019/NAFIPS 2019, AISC 1000, pp. 13–21, 2019.
https://doi.org/10.1007/978-3-030-21920-8_2

of dimensionality reduction, and in [10] a first approach from a commutative group structure formed by a set of operators is proposed.

However, there is no general characterization of the property of relevance or a general consensus on the kind of functions or parameters that allow evaluating and measuring such property. In [10] several elements allowing the characterization of the relevance property are established. In the particular case of fuzzy classification, the relevance of a class in a fuzzy partition depends on two fundamental aspects: on the one hand, a set of information about the changes in the partition once a class is included or excluded and, on the other hand, a measure of the intensity of the changes.

According to the above, for the study of relevance property two processes are required. First, a comparison process in which all the available information about the classes is compared with the information obtained after eliminating or adding a class. Second, a process of evaluation (classification) in which each class is labeled on a given scale bounded by total relevance and total irrelevance.

The two processes described above are performed on information inferred from the classes, in particular, on aggregate information of the classes. In general, the information can refer to the overlap, the coverage, the relevance and other indices related to relevant points, dispersion and shape of the classes, as happens in statistics. Such information, in principle, is obtained from aggregation operators with the objective of capturing the essential complexity of a fuzzy partition performed.

For each property studied a particular kind of aggregation operators is required, according to the nature and conception of the specific property. In general, aggregation operators are classified according to the properties they fulfill and depending on the context and purposes, certain properties are required. A general study of such properties can be found in [11], and more specific studies can be found on migrativity [12], consistency and stability [13], semiautoduality [14] or quasiconvexity [15].

Therefore, based on the elements given in [10] on the characterization of the property of relevance we propose: (1) A comparison process to measure the relevance of a class in a fuzzy partition, (2) A study on the properties that must be fulfilled by the operators of aggregation according to our methodology.

2   Comparison Process

In [1] and [2] a fuzzy classification system is a finite family C of fuzzy sets or classes (each $c \in C$ with its associated membership function $\mu_c(x) : X \to [0, 1]$), together with a recursive triplet[1] (φ, ϕ, N), where:

1. ϕ is a standard recursive rule such that $\phi_2(0, 1) = \phi_2(1, 0) = 0$
2. $N : [0, 1] \to [0, 1]$ is a strong negation function, i.e., a bijective strictly decreasing function such that $N \circ N(\mu(x)) = \mu(x)$ for all $\mu(x) \in [0, 1]$
3. φ is a standard recursive rule such that $\varphi_n(\mu_1(x), \ldots, \mu_n(x)) = N[\phi_n(N(\mu_1(x)), \ldots, N(\mu_n(x)))], \forall n > 1$.

---

[1] Recursiveness is a property of a sequence of operators $\{\phi_n\}_{n>2}$ allowing the aggregation of any number of items from two-dimensional operators, see [16].

Notice that, ϕ_n is a conjunctive recursive rule in the sense that $\phi_n(\mu_1(x),$ $\ldots, \mu_n(x)) = 0$ whenever there is j such that $\mu_j(x) = 0$, while φ_n is a disjunctive recursive rule, in the sense that $\varphi_n(\mu_1(x), \ldots, \mu_n(x)) = 1$ whenever there is j such that $\mu_j(x) = 1$.

From the fuzzy classification system (C, φ, ϕ, N), we consider two new mappings $\sigma_n : [0, 1]^n \to [0, 1]$ and $\delta_n : [0, 1]^n \to [0, 1]$, defined as:

$$\sigma_n(\mu_1(x), \ldots, \mu_n(x)) = N(\varphi_n(\mu_1(x), \ldots, \mu_n(x))) \tag{1}$$

$$\delta_n(\mu_1(x), \ldots, \mu_n(x)) = N(\phi_n(\mu_1(x), \ldots, \mu_n(x))) \tag{2}$$

In particular, if $\varphi_n(\mu_1(x), \ldots, \mu_n(x))$ represents the degree of coverage of the classes, then $\sigma_n(\mu_1(x), \ldots, \mu_n(x))$ represents the degree of non-coverage of the classes, understanding φ_n as a proposition and $N(\varphi_n)$ as the negation of such proposition. In a similar way, if $\phi_n(\mu_1(x), \ldots, \mu_n(x))$ represents the degree of redundancy of the classes, then $\delta_n(\mu_1(x), \ldots, \mu_n(x))$ represents the degree of non-redundancy of the classes.

Notice that as $\varphi_n(\mu_1(x), \ldots, \mu_n(x)) = N[\phi_n(N(\mu_1(x)), \ldots, N(\mu_n(x)))]$, from (1) and (2), we have,

$$\sigma_n(\mu_1(x), \ldots, \mu_n(x)) = \phi_n(N(\mu_1(x)), \ldots, N(\mu_n(x))) \tag{3}$$

And therefore, from (1), (2) and (3) we can write φ_n, σ_n and δ_n in terms of ϕ_n. We denote:

$$\phi_n^* = N[\phi_n(N(\mu_1(x)), \ldots, N(\mu_n(x)))] = \varphi_n(\mu_1(x), \ldots, \mu_n(x)) \tag{4}$$

$$\phi_n^\wedge = \phi_n(N(\mu_1(x)), \ldots, N(\mu_n(x))) = \sigma_n(\mu_1(x), \ldots, \mu_n(x)) \tag{5}$$

$$\phi_n^\sim = N(\phi_n(\mu_1(x), \ldots, \mu_n(x))) = \delta_n(\mu_1(x), \ldots, \mu_n(x)) \tag{6}$$

Thus, "$*$", "$\wedge$" and "$\sim$" are operations on ϕ_n forming φ_n, σ_n and δ_n, i.e., each one is a unary operation. In this sense, we have another operation called identity and denoted by i, such that $\phi_n^i = \phi_n(\mu_1(x), \ldots, \mu_n(x))$. Keeping this in mind, $G_o = \{\phi_n^i, \phi_n^*, \phi_n^\wedge, \phi_n^\sim\} = \{\phi_n, \varphi_n, \sigma_n, \delta_n\}$ and let "$\odot$" denote the composition operation on $B = \{i, *, \wedge, \sim\}$. Therefore, Table 1 summarize the results of such operation,

From Table 1 it is easy to determine that $(G_o, \odot)$ has a commutative group structure.

Table 1. Operation $\odot$

$\odot$	ϕ_n	σ_n	δ_n	φ_n
ϕ_n	ϕ_n	σ_n	δ_n	φ_n
σ_2	σ_n	ϕ_n	φ_n	δ_n
δ_n	δ_n	φ_n	ϕ_n	σ_n
φ_n	φ_n	δ_n	σ_n	ϕ_n

Remember that for each set of classes of a fuzzy partition, the previous group is constituted by the degree of covering, the degree of overlap, the degree of non-overlap and the degree of non-covering. Thus, we are interested in establishing a relation that allows identifying the set of classes that corresponds with the best partition. In general, a partition is a *high-quality partition* when it has high degrees of covering and low degrees of overlap. However, when we include the degree of non-coverage and the degree of non-overlap, the classification system acquires more values to be compared.

Since our purpose is to establish a comparison process between the classes of a fuzzy partition, we follow the proposal in [17] and establish the global degree for each operator. For instance, given an universe X, and a family of fuzzy classes C over this universe, if $\varphi_n(\mu_1(x), \ldots, \mu_n(x))$ represents the degree of coverage of the partition for the element $x \in X$, then φ_n^T represents the degree of global covering of X and is formed by the aggregation of the degrees of covering for all items $x \in X$. Such aggregation function can be of very different nature (conjunctive, disjunctive or average).

Keeping this in mind, we propose to establish a comparison process based on proximity relations (sometimes called compatibility relations) between the elements of $(G_o, \odot)$. For this reason, first we compute the global degree of the values obtained through a certain relation that allows in some sense to compare the elements of the group. Finally, on such global degrees, we establish a proximity relation. We show the idea of this process through Example 1.

Example 1. *Given the commutative group $G_o = \{\varphi_n, \phi_n, \sigma_n, \delta_n\}$ and considering that for all $\theta_n, \lambda_n \in G_o$, $\theta_{kn} = \theta_n(\mu_1(x_k), \ldots, \mu_n(x_k))$ is the aggregation of the membership functions μ_n of the element $x_k \in X$ to the n classes, then we define the operator $\rho^T : G_o \times G_o \to [0, 1]$ for all $\theta_n, \lambda_n \in G_o$ such that $\rho^T(\theta_n, \lambda_n) = \frac{\sum_{k=1}^m |\theta_{kn} - \lambda_{kn}|}{m}$ with $m = |X|$. In this way, ρ^T is defined as the distance for each pair (θ_n, λ_n), and thus, the relation $w(\theta_n, \lambda_n) = 1 - \rho^T(\theta_n, \lambda_n)$ is defined as a proximity relation.*

The values obtained by such a relation allow, in principle, to establish an order about the preferences of the decision maker. For instance a decision maker can establish that, *it is desirable that the degree of proximity between the global degree of covering and the global degree of overlapping to be low and lower than the global degree of proximity between the global degree of overlapping and the global degree of non-covering.*

Once the proximity relation between the elements of the group is established, we propose to eliminate (or add) classes from the family of initial classes and compute again the values of the relation w. Changes in such values allow decision makers to determine the number of classes according to their preferences.

3   Aggregation Operators for Comparison Process

Our purpose in this section is to first delve into a property that certain aggregation operators satisfy and which is called *weakly self-dual*. Subsequently, under the framework of our proposal, we present the properties that an aggregation operator must satisfy to properly measure the relevance of a class in a fuzzy partition.

3.1   Weakly Self-dual Operators

Following [19] (see also, [3] and [11]) there is a set of aggregation operators satisfying the functional equation, $\phi_2(x,y) + \phi_2(1-x, 1-y) = 1$. Such operators are characterized through the so-called *symmetric sum* and are called also, self-dual aggregation operators. In the framework presented in [19], the self-duality corresponds to that $\phi_2(x,y) = \varphi_2(x,y)$ with $\varphi_2(x,y) = N(\phi_2(N(x), N(y)))$ and $N(x) = 1 - x$. Keeping this in mind, we propose to consider a weak version of the functional equation such that $\phi_2(x,y) - \phi_2(N(x), N(y)) = 0$, with N a strong negation and ϕ correspond to the conjunctive aggregation operator of our fuzzy classification system.

The idea of considering this weak version of the functional equation is motivated by the following fact. If we consider a fuzzy classification, where a Ruspini's partition is obtained, i.e., the total sum of the membership degrees of an object to all classes must be equal to 1, then in a partition with two classes it holds that $\mu_1(x) + \mu_2(x) = 1$. In this case, if the employed operators satisfy the weak version of the functional equation under the strong negation $N(\mu(x)) = 1 - \mu(x)$ and considering the elements of our commutative group, then we have that,

$$\phi_2(\mu_1(x), \mu_2(x)) = \phi_2(1 - \mu_1(x), 1 - \mu_2(x))$$

or, equivalently $\phi_2(\mu_1(x), \mu_2(x)) = \sigma_2(\mu_1(x), \mu_2(x))$. Therefore, we would be losing information for our comparison process, as the degree of redundancy would be equivalent to the degree of non-coverage.

According to the above, we present the following result:

Proposition 1. *Given a fuzzy classification system* (C, ϕ, φ, N) *where* $N(\mu(x)) = 1 - \mu(x)$ *and* $\phi_2(\mu_1(x), \mu_2(x)) = \phi_2(\mu_2(x), \mu_1(x))$ *(i.e.,* ϕ_2 *is symmetric) it holds that* $\phi_2(\mu_1(x), \mu_2(x)) + \varphi_2(\mu_1(x), \mu_2(x)) = 1$ *if and only if* $\mu_1(x) + \mu_2(x) = 1$.

Proof. (Necessity)   If $\phi_2(\mu_1(x), \mu_2(x)) + \varphi_2(\mu_1(x), \mu_2(x)) = 1$ then, $\phi_2(\mu_1(x), \mu_2(x)) + 1 - \phi_2(1 - \mu_1(x), 1 - \mu_2(x)) = 1$. Thus, $\phi_2(\mu_1(x), \mu_2(x)) = \phi_2(1 - \mu_1(x), 1 - \mu_2(x))$. Suppose that $\mu_1(x) + \mu_2(x) \neq 1$. Therefore, if we consider $\mu_1(x) = 0$ we have that $\mu_2(x) \neq 1$, then $0 = \phi_2(0, \mu_2(x)) = \phi_2(1, 1 - \mu_2(x))$, which is impossible. Therefore $\mu_1(x) + \mu_2(x) = 1$.

(Sufficiency)   If $\mu_1(x) + \mu_2(x) = 1$ then $\mu_1(x) = 1 - \mu_2(x)$ and as $\phi_2(\mu_1(x), \mu_2(x)) = \phi_2(\mu_2(x), \mu_1(x))$, we have that $\phi_2(\mu_1(x), 1 - \mu_1(x)) = \phi_2(1 - \mu_1(x), \mu_1(x))$ and thus $1 - \phi_2(\mu_1(x), 1 - \mu_1(x)) = \varphi_2(\mu_1(x), 1 - \mu_1(x))$ then $\phi_2(\mu_1(x), \mu_2(x)) + \varphi_2(\mu_1(x), \mu_2(x)) = 1$. ∎

Under our framework, Proposition 1 establishes that for any De Morgan triple (ϕ, φ, N), with $N(\mu(x)) = 1 - \mu(x)$ and ϕ_2 symmetric, the weak version of the functional equation is always fulfilled for a Ruspini partition with two classes.

According to the above, under certain conditions we have a set of De Morgan triple (ϕ, φ, N) such that $\phi_2(\mu_1(x), \mu_2(x)) = \sigma_2(\mu_1(x), \mu_2(x))$, or equivalently, such that $\phi_2(\mu_1(x), \mu_2(x)) = \phi_2(N(\mu_1(x)), N(\mu_2(x)))$. Keeping this in mind, we propose the following definition:

Definition 1. *An aggregation operator* $\psi_2 : [0, 1]^2 \to [0, 1]$ *is called weakly self-dual if* $\forall x, y$, *such that* $x + y = 1$, *then* $\psi_2(x, y) = \psi_2(N(x), N(y))$ *with* N *a strong negation.*

18   F. Castiblanco et al.

There are many possible examples of weakly self-dual aggregation operator. In general, as can easily be deduced from Proposition 1, all symmetric two-dimensional operators together with the strong negation $N(x) = 1 - x$ fulfills the property, for instance,

1. $\psi_2(x,y) = \min(x,y)$
2. $\psi_2(x,y) = xy$,
3. $\psi_2(x,y) = \frac{3xy}{1+2xy}$

An example of operators that do not satisfy the property are,

4. $\psi_2(x,y) = \frac{xy}{\frac{2}{3}+\frac{1}{3}(x+y-xy)}$ with $N(x) = \frac{1-x}{1+2x}$,
5. $\psi_2(x,y) = xy^2$ with $N(x) = 1 - x$.

3.2 Properties of Operators that Measure Relevance

Taking into account our proposal about a comparison process that allows studying the relevance of the classes, it is necessary to determine the properties that the aggregation operators must satisfy. For instance, it is not desirable that the degree of overlap of two classes depends on the order in which the classes are aggregated.

Remember that our proposal is based on a fuzzy classification system (C, ϕ, φ, N), where ϕ is a recursive rule, i.e., a family of aggregation functions and (ϕ, φ, N) forms a De Morgan triple. As the operators φ, σ and δ can be written in terms of ϕ, in principle we will only refer to properties on the conjunctive operator.

According to the fuzzy classification system definition, we have three basic properties for ϕ:

1. Boundary conditions, $\phi_2(0,0) = 0$ and $\phi_2(1,1) = 1$.
2. Monotonicity, if $x_1 \geq x_2$ and $y_1 \geq y_2$ then $\phi_2(x_1,y_1) \geq \phi_2(x_2,y_2)$.
3. ϕ_2 is a conjunctive recursive rule in the sense that $\phi_2(x,y) = 0$ whenever that $x = 0$ or $y = 0$.

Properties 1 and 2 are referred to the indispensable property of any aggregation operator according with [18].

From Property 3, we have that ϕ_2 is not stable for the strong negation[2] N, i.e., in general, $\phi_2(N(x),N(y)) \neq N(\phi_2(x,y))$. In others words, $\phi_2(x,y) = \varphi_2(x,y)$ is not true for all $x,y \in [0,1]$. If ϕ_2 is stable for the strong negation N, then $0 = \phi_2(0,1) = N(\phi_2(N(0),N(1))) = N(0) = 1$, which is impossible. Notice that this condition is fundamental in a comparison process. We can obtain more information about the quality of the classes when this property on the operators is not fulfilled.

Taking into account the above, we also require that, in principle, $\phi_2(x,y) \neq \sigma_2(x,y)$. Notice that if equality holds then $\phi_2((x,y)) = N(\varphi_2(x,y))$ and thus, $\phi_2(x,y) = \phi_2(N(x),N(y))$. Therefore, in the framework of the proposal of this paper, we establish as additional basic properties, the followings:

---

[2] An operator ϕ_2 is stable for the strong negation N if $\phi_2(N(x),N(y)) = N(\phi_2(x,y))$. [20]

4. Not weakly self-dual. $\forall x, y \in (0, 1)$ $\phi_2(x, y) \neq \phi_2(N(x), N(y))$.

Similarly, since our proposed comparison of classes implies the elimination or addition of classes, it is necessary that the employed aggregation operator does not have a neutral element different from 1 (in the conjunctive case) i.e., different from trivial idempotent elements. Remember that the neutral element can be omitted from aggregation inputs without influencing the final output. Specifically, $e \in [0, 1]$ is called a neutral element of ϕ, if $\phi_2(e, x) = x$, $\forall x$. For instance, the operator $\phi_2(x, y) = \min (x, y)$ has as neutral element $e = 1$, while $\phi_2(x, y) = \frac{xy}{xy + (1-x)(1-y)}$ has as neutral element $e = \frac{1}{2}$.

5. Neutral element. The only neutral elements of ϕ is 1 or 0.

Some other properties may be desirable according to the specific application. For example, symmetry and continuity are desirable properties in some image segmentation problems. However, in the framework of the methodology proposed in this paper, only the described properties are necessary to measure the relevance of a class in a fuzzy partition.

With the properties 1–5 we seek to give the first elements for the characterization of the aggregation operators that allow carrying out a comparison process between the classes of a fuzzy partition. Such a process, in principle, allows determining the relevance of a class by examining the changes produced by eliminating the class. In general, a class should be added to the partition when the values obtained by the comparison process in the whole class set are not satisfactory for the decision maker.

4   Final Comments

The main idea of this paper is that the relevance of a class or a set of classes in a fuzzy partition can be evaluated through a comparison process, taking into account the variation of different indexes when such class or classes are removed or added.

In this way, a commutative group structure has been established from four aggregation operators that measure the degree of overlap, the degree of coverage, the degree of non-coverage and the degree of non-overlap. On such a group, a proximity relation has been proposed.

The comparison is carried out through degrees of proximity for pairs of characteristics, such as the proximity between the coverage degree and the overlap degree, or the proximity between the degree of grouping and non-overlap.

In general, we establish that a class will be relevant in a fuzzy partition if by eliminating it, the values obtained in the comparison process change according to the decision-maker preferences.

Regarding the comparison process, we identify some properties necessary to carry out an effective process. Therefore, we propose to study the two-dimensional aggregation operators that satisfy the functional equation $\phi_2(x, y) - \phi_2(N(x), N(y)) = 0$. From such study, we propose a definition for those operators such that, $\phi_2(x, y) = \phi_2(N(x), N(y))$ and we call them *weakly self-dual*.

As future work, we propose to apply our proposal in unsupervised image segmentation in order to determine the optimal number of classes on fuzzy partitions. Other properties may be required on this kind of problems and therefore, may be included in our proposal.

Similarly, it is necessary to carry out a study about how to choose or establish the proximity relation and the possibility of choosing a *similarity relation*. On the other hand, considering the definition of weakly self-dual operators and the other properties studied, it is necessary to establish a characterization of the operators that allow carrying out the proposed comparison process and therefore, measure the relevance of the classes in a fuzzy partition.

Acknowledgements. This research has been partially supported by the Government of Spain (grant TIN2015-66471-P), the Government of Madrid (grant S2013/ICCE-2845), Complutense University (UCM Research Group 910149) and Gran Colombia University (grant JCG2018-CEAC-03).

References

1. Amo, A., Montero, J., Biging, G., Cutello, V.: Fuzzy classification systems. Eur. J. Oper. Res. **156**, 495–507 (2004)
2. Del Amo, A., Gómez, D., Montero, J., Biging, G.: Relevance and redundancy in fuzzy classification systems. Mathware Soft Comput. **8**, 203–216 (2001)
3. Dombi, J.: Basic concepts for a theory of evaluation: the aggregative operator. Eur. J. Oper. Res. **10**, 282–293 (1982)
4. Dombi, J.: A general class of fuzzy operators, the Demorgan class of fuzzy operators and fuzziness measures induced by fuzzy operators. Fuzzy Sets Syst. **8**, 149–163 (1982)
5. Bustince, H., Fernandez, J., Mesiar, R., Montero, J., Orduna, R.: Overlap functions. Nonlinear Anal. Theory Methods Appl. **72**, 1488–1499 (2010)
6. Gómez, D., Rodríguez, J.T., Montero, J., Bustince, H., Barrenechea, E.: N-Dimensional overlap functions. Fuzzy Sets Syst. **287**, 57–75 (2016)
7. Bustince, H., Pagola, M., Mesiar, R., Hüllermeier, E., Herrera, F.: Grouping, overlap, and generalized bientropic functions for fuzzy modeling of pairwise comparisons. IEEE Trans. Fuzzy Syst. **20**, 405–415 (2012)
8. Bedregal, B., Dimuro, G.P., Bustince, H., Barrenechea, E.: New results on overlap and grouping functions. Inf. Sci. (Ny) **249**, 148–170 (2013)
9. Castiblanco, F., Montero, J., Rodríguez, J.T., Gómez, D.: Quality assessment of fuzzy classification: an application to solvency analysis. Fuzzy Econ. Rev. **22**, 19–31 (2017)
10. Castiblanco, F., Franco, C., Montero, J., Rodríguez, J.T.: Relevance of classes in a fuzzy partition. A study from a group of aggregation operators. In: Barreto, G.A., Coelho, R. (eds.) Fuzzy Information Processing, pp. 96–107. Springer, Cham (2018)
11. Calvo, T., Kolesarova, A., Komornikova, M., Mesiar, R.: Aggregation operators, properties, classes and construction methods. In: Calvo, T., et al. (eds.) Aggregation Operators. New Trends and Applications, pp. 3–104. Physica-Verlag, Heidelberg (2002)
12. Bustince, H., De Baets, B., Fernandez, J., Mesiar, R., Montero, J.: A generalization of the migrativity property of aggregation functions. Inf. Sci. (Ny) **191**, 76–85 (2012)

13. Gómez, D., Rojas, K., Montero, J., Rodríguez, J.T., Beliakov, G.: Consistency and stability in aggregation operators: an application to missing data problems. Int. J. Comput. Intell. Syst 7(3), 595–604 (2014)
14. Bustince, H., Montero, J., Barrenechea, E., Pagola, M.: Semiautoduality in a restricted family of aggregation operators. Fuzzy Sets Syst. **158**, 1360–1377 (2007)
15. Janiš, V., Král, P., Renčová, M.: Aggregation operators preserving quasiconvexity. Inf. Sci. (Ny) **228**, 37–44 (2013)
16. Cutello, V., Montero, J.: Recursive connective rules. Int. J. Intell. Syst. **14**, 3–20 (1999)
17. Castiblanco, F., Gómez, D., Montero, J., Rodríguez, J.T.: Aggregation tools for the evaluation of classifications. In: Fuzzy Systems Association and 9th International Conference on Soft Computing and Intelligent Systems (IFSA-SCIS), 2017 Joint 17th World Congress of International, pp. 1–5, Otsu, Japan. IEEE (2017)
18. Klir, G.J., Folger, T.A.: Fuzzy Sets, Uncertainty, and Information. Prentice Hall, Upper Saddle River (1988)
19. Silvert, W.: Symmetric summation: a class of operations on fuzzy sets. IEEE Trans. Man Cybern. **9**, 657–659 (1979)
20. Fodor, J., Roubens, M.: Fuzzy Preference Modelling and Multicriteria Decision Support Theory and Decision Library Series D: System Theory, Knowledge Engineering and Problem Solving (1994)

The Influence of Induced OWA Operators in a Clustering Method

Aranzazu Jurio$^{(\boxtimes)}$, Mikel Sesma-Sara, Jose Antonio Sanz Delgado,
and Humberto Bustince

Departamento de Estadistica, Informatica y Matematicas,
Universidad Publica de Navarra, Pamplona, Spain
`aranzazu.jurio@unavarra.es`

Abstract. In this work we present an adaptation of the well known k-means algorithm for clustering. The proposal increases the flexibility of the algorithm to calculate the representative value of each cluster. To do so, we work with Induced Ordered Weighting Averaging operators. These instances of aggregation functions are able to increase or decrease the influence of the data in the final result depending on the specific values of the weights. We present an experimental study to show how these operators are able to modify the representatives of the clusters. We also compare our results over some standard datasets.

1 Introduction

Clustering is an unsupervised classification problem where the purpose is to find the natural groups that exist in a dataset. It is based on the idea that the data belonging to the same group must have similar characteristics whereas the data that belong to different groups must be different in the same characteristics [7].

Clustering methods can be generally divided into two types: hierarchical methods and partitional ones. Hierarchical methods construct a tree based on the similarities between data [8, 14]. On the other side, partitional methods divide the whole dataset into a fixed number of clusters, each of them represented by a single point. Usually this point is the one whose sum of distances to all the data in the cluster is minimum [9, 10, 12].

In this work we focus in one of the most used partitional methods: the k-means [5, 12]. It selects as representative of each cluster the average of all the data that belong to it, i.e. the centroid. Moreover, to determine the cluster that every datum belongs to, it looks for the nearest centroid. The solution of this problem is made iteratively.

We believe that the centroid of a dataset may not be the best representation of all the data. For example, if there exist some outliers among the data, the centroid will be deviated from the corresponding place. This idea was also pointed out by some authors in the literature. One of the proposed solutions is to use the median instead of the mean to calculate the representatives (k-medoids) [9].

© Springer Nature Switzerland AG 2019
R. B. Kearfott et al. (Eds.): IFSA 2019/NAFIPS 2019, AISC 1000, pp. 22–32, 2019.
https://doi.org/10.1007/978-3-030-21920-8_3

We extend this idea to the use of other aggregation functions to calculate the representative of each cluster. In this work we focus on Induced Weighted Averaging (IOWA) [18] operators because of two main aspects: (i) they are simple enough in order to be quickly calculated and (ii) they offer enough variability due to the weighting vector. We need a quick calculation of the representatives because this step is made many times through the process of clustering.

The difference between IOWA operators and any other weighted averaging operators is that the weight associated with each value to aggregate depends on the position of that value when all of them are sorted based on another characteristic. We take advantage of this circumstance in order to calculate the representative of each cluster. To do so, we diminish the importance of the values that are far away from the others.

The remaining of this work is organized as follows: in Sect. 2 we recall some well known concepts that are used through the rest of the work. In Sect. 3 we propose our new algorithm for clustering that is based on IOWA operators. In Sects. 4 and 5 we show the performance of this algorithm. We start with some visual examples to focus on some specific characteristics and we continue with the performance over several standard datasets. Finally, in Sect. 6 we outline the conclusions of this work.

2   Preliminaries

In this section we recall some well known concepts that are used through the work. We start with the definitions of aggregations function, ordered weighted averaging operator and induced ordered weighted averaging operator.

Definition 1 [3,11]. An *n-ary aggregation function* is defined as a function

$$M : [0,1]^n \to [0,1]$$

such that

(i) $M(x_1, \ldots, x_n) \le M(y_1, \ldots, y_n)$ whenever $x_i \le y_i$ for all $i \in 1, \ldots, n$.
(ii) $M(0, \ldots, 0) = 0$ and $M(1, \ldots, 1) = 1$.

A particular instance of aggregation functions frequently used in many applications are OWA operators given by Yager [16].

Definition 2 [16]. Let ω be a weight vector, i.e., $\omega = (\omega_1, \ldots, \omega_n) \in [0,1]^n$ with $\omega_1 + \ldots + \omega_n = 1$. The *Ordered Weighted Averaging* operator associated with ω, OWA_ω, is a mapping $OWA_\omega : [0,1]^n \to [0,1]$ defined by

$$OWA_\omega(x_1, \ldots, x_n) = \sum_{i=1}^{n} \omega_i x_{(i)}$$

where $x_{(i)}$, $i = 1, \ldots, n$, denotes the i-th greatest component of the input $(x_1, \ldots, x_n)$.

A more general type of OWA operator is called the Induced Ordered Weighted Averaging (IOWA) [18] Operator. These operators take pairs to aggregate, called OWA pairs. The first component is used to induce an order (order inducing variable) while the second component is the value to aggregate (argument variable).

Given a set ot OWA pairs $X = \{(u_1, a_1), \ldots, (u_n, a_n)\}$, the aggregation of them is done in two steps. The first one is the creation of the ordered argument vector B_u so that b_j is the a value of the OWA pair having the j-th largest u value. The second step is the weighted aggregation of that vector B_u.

$$A((u_1, a_1), \ldots, (u_n, a_n)) = W^T B_u$$

One possible way of determining the weights for the IOWA operators is derived from fuzzy linguistic quantifiers [19]. Two examples of fuzzy linguistic quantifiers are *at least half* and *most*. When these quantifiers are proportional, they can be represented by fuzzy sets in the interval $[0, 1]$. The membership degree of each value represents the proportion in which that value is compatible with the meaning of the quantifier. If the quantifiers are increasing, the membership function can be calculated as

$$Q(x) = \begin{cases} 0 & \text{if } x \leq a \\ \dfrac{x - a}{b - a} & \text{if } a \leq x \leq b \\ 1 & \text{if } x > b \end{cases}$$

Some linguistic quantifiers that we use in this work are *at least half*, *as many as possible* and *most* (see Fig. 1).

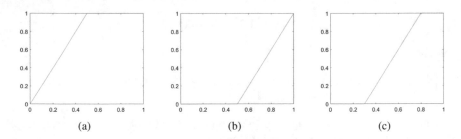

(a)   (b)   (c)

Fig. 1. Membership functions of the fuzzy linguistic quantifiers *at least half* (a), *as many as possible* (b) and *most* (c).

From the linguistic quantifiers it is possible to calculate the weights for the OWA operator [17]. If it is proportional increasing, as the ones we use in this work, the weights for the aggregation are given as follows:

$$\omega_i = Q\left(\frac{i}{n}\right) - Q\left(\frac{i-1}{n}\right), i = 1..n.$$

2.1 k-means Algorithm

K-means [5, 12] is one of the most well known and used clustering algorithms. It belongs to the methods called partitional, i.e. it divides all the existing data into a given number of clusters. The idea of this algorithm is that every cluster can be represented by a unique value. The goal is to minimize the sum of the distances between every datum and the representative of its corresponding cluster.

Let c be the given number of clusters, $x_1, \ldots, x_n$ the data to be partitioned and v_i the representative value of cluster i. Then, following the mentioned goal, the objective function to minimize is the following:

$$J = \sum_{i=1}^{c} \sum_{x_k \in cluster_i} ||x_k - v_i||_A^2$$

where $||x||_A = \sqrt{x^l A x}$ is any norm associated with an inner product.

The minimization of this function is done in an iterative process. Starting from a random initialization of the representatives of each cluster, then it is updated the cluster each datum belongs to. In this step, each datum is assigned to the cluster whose representative is closer to it. It is done using the following formula:

$$cluster_i = \{x_p : ||x_p - v_i||^2 \leq ||x_p - v_j||^2, \forall j, 1 \leq j \leq c\}$$

Once it is updated the data that belong to each cluster, it is time to update the representatives of the clusters. Using the function above, this representatives are calculated as the average value of all the data, it means, as the centroids.

$$v_i = \frac{1}{|cluster_i|} \sum_{x_j \in cluster_i} x_j$$

This process of updating in two steps is repeated until the centroids do not change in two consecutive iterations. At this moment, the objective function is minimized and that is the final solution of the problem.

3 Clustering Based on IOWA Operators

In this section we explain our proposed algorithm for clustering. The main idea is quite similar to the k-means algorithm: we want to assign every datum to the cluster whose representative is the closest one. However, the main novelty is the way to calculate the representatives of each cluster. We strongly believe that calculating them as the average of all the belonging data is not the best option. By doing this, all the data have the same importance, no matter whether they are close to the other data of the cluster or they are far away. In this procedure, the outliers can modify the representatives, when this is not desirable.

To avoid this situation, we propose to use an IOWA aggregation of the data for the calculation of the representative of each cluster. In our case, the pairs to be aggregated include the information about the data and how far they were from the previous representatives (representatives in the previous iteration). Then, the argument variable is the datum itself while the order inducing variable is the distance to the representative.

The clustering algorithm starts with a random initialization of the representatives. After this initialization, it starts an iterative process made of two steps: update of the cluster every datum belong to and update of the representatives.

Based on the representatives, we are able to assign every datum to a cluster. Specifically, we assign it to the one whose representative is closer to the datum. This step is exactly the same as in the k-means algorithm.

$$cluster_i = \{x_p : ||x_p - v_i||^2 \leq ||x_p - v_j||^2, \forall j, 1 \leq j \leq c\}$$

Once we know which data belong to which cluster, we update the representatives. To do so, we sort all of them taking into account the distance to the previous representative of the cluster they belong to. Then, we aggregate them using an specific weighting vector for each cluster.

$$A((u_1, a_1), \ldots, (u_n, a_n)) = W^T B_u$$

These vectors have the same distribution for all the clusters, but they change in the number of elements they are made of. Each one is adapted to the number of data that belong to that cluster.

The two steps of the iterative process are repeated until the system is stabilized and the centres do not change.

4   Illustrative Examples

In this section we use some examples with data in two dimensions in order to see the results of the algorithm. We show its performance in several scenarios and we analyse the results based on the weighting vectors.

In Fig. 2 we show the shapes of the weighting vectors that we use in our experimentations. The first three images, (a), (b) and (c), correspond to the weighting vectors derived from the three linguistic quantifiers shown in Fig. 1, that is, *at least half*, *as many as possible* and *most*. The fourth image corresponds to a weighting vector equal for all data. This last weighting vector makes our algorithm really similar to the k-means.

With these four vectors, we can observe the development of the clustering algorithm when we calculate the representatives of the clusters following four different ideas:

- Giving more importance to the data which are far from the representative.
- Giving more importance to the data which are close to the representative.
- Giving more importance to the data which are at a middle distance to the representative.
- Giving the same importance to all data.

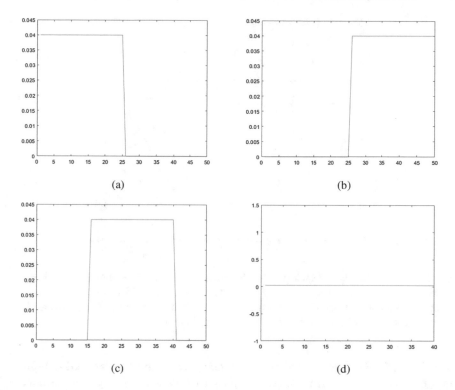

Fig. 2. Weighting vector. (a) Derived from *at least half*, (b) derived from *as many as possible*, (c) derived from *most* and (d) equal for all data.

We assume that it is better to make more significant the values which are close to the representatives, i.e., to reduce the weight assigned to the data which are far from the representatives.

We start with a simple example, in which there are two separated clusters. In Fig. 3 we show the data to classify and the representatives obtained using our proposal and the k-means algorithm. Looking at the figure, it is clear that if we select equal weights for all the data in the cluster, our final result is exactly the same as the one obtained by the k-means (magenta star and red square). Moreover, in this example the representatives obtained by all the weighting vector are quite similar.

However, if we add some outliers to the dataset, the representative values change considerably. In Fig. 4 we show this example. The cluster in the right includes the four outliers added to the dataset. As we can see, the representatives of all the versions but the one shown in green diamonds are clearly moved towards the outliers. This behaviour is undesired, as most of the data (50) have not changed and only 4 new data have been added.

This result confirms our idea that, in order to have a result not influenced by the outliers, it is better to calculate the representative giving more importance to

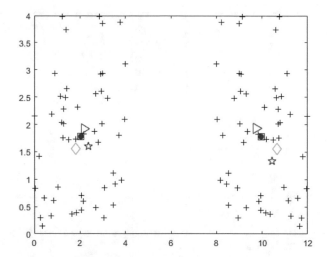

Fig. 3. Data to classify in black pluses. Representatives obtained using different weighting vectors: derived from *at least half* in red triangles, derived from *as many as possible* in green diamonds, derived from *most* in purple stars and equal for all data in blue squares. Centroids obtained using k-means in red circles.

the data which is close to that representative (calculated in our example by the weighting vector derived from *as many as possible*, shown in green diamonds).

This behaviour can lead, in extreme cases, even to wrong classification of the data. In Fig. 5 we show a dataset made of two clusters and one outlier. On the left we observe the final classification obtained by our proposal using the weighting vector derived from *as many as possible*. On the right we observe the final classification obtained by the kmeans. It is easy to see that, due to the outlier, the centroid of the right cluster in the k-means has moved too much to the right. Now, one of the data belonging to this cluster is too far from the centroid, so it is classified as the cluster on the left.

5 Experimental Study

In this section we use our proposal to classify some standard datasets. We have selected a benchmark of 12 real-world datasets selected from the KEEL dataset repository [1,2], which are publicly available on the corresponding webpage (http://www.keel.es/dataset.php). Table 1 summarizes the properties of the selected datasets, showing for each dataset the number of attributes (#Atts.), the number of examples (#Ex.) and the number of classes (#Class.).

These datasets are prepared to test supervised classification methods, so they have the real class which every datum belongs to. We use these real classes to test the performance of the clustering methods. We execute our algorithm using

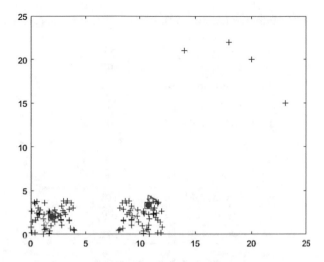

Fig. 4. Data to classify in black pluses. Representatives obtained using different weighting vectors: derived from *at least half* in red triangles, derived from *as many as possible* in green diamonds, derived from *most* in purple stars and equal for all data in blue squares. Centroids obtained using k-means in red circles.

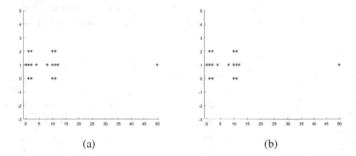

| | (a) | | | (b) | |

Fig. 5. Classification of a dataset made of two clusters and one outlier. (a) Our proposed method with the weighting vector derived from *as many as possible*. (b) K-means algorithm.

all the weighting vectors proposed in this work, and we also execute the k-means. Both algorithms have a random initialization, so we run any of them 25 times to select the best result. For each execution we show the number of examples correctly classified (see Table 2). In bold are highlighted the best result for each dataset.

We observe that the weighting vector derived from *as many as possible* is the one with the biggest average accuracy. Indeed, it gets the best results in 8 out of 12 datasets. On the other side, *at least half* and equal (as well as k-means) only win in 2 out of 12 datasets.

We want to complete this study with the use of a hypothesis validation technique in order to try to give statistical support to the analysis of the results [6, 13].

Table 1. Information about the datasets used in this section.

Id	Dataset	#Atts.	#Ex.	#Class
hab	Haberman	3	306	2
pho	Phoneme	5	5404	2
ban	Banana	2	5300	2
app	Appendicitis	7	106	2
tit	Titanic	3	2201	2
iri	Iris	4	150	3
bal	Balance	4	625	3
new	Newthyroid	5	215	3
hay	Hayes-roth	4	160	3
tae	Tae	5	151	3
gla	Glass	9	214	7
eco	Ecoli	7	336	8

Table 2. Percentage of examples correctly classified for each dataset and each weighting vector. First column: weighting vector derived from *at least half*; second column: weighting vector derived from *as many as possible*; third column: weighting vector derived from *most*; fourth column: equal weights for all data; fifth column: k-means algorithm.

Id	At least half	As many as possible	most	equal	k-means
hab	52.29	**55.88**	51.96	52.29	52.29
pho	**75.85**	68.62	69.25	66.78	66.78
ban	55.75	58.53	**58.96**	56.68	56.68
app	82.08	87.74	**88.68**	83.02	83.02
tit	**77.60**	**77.60**	**77.60**	**77.60**	**77.60**
iri	64.67	**66.67**	66.00	64.67	64.67
bal	53.76	**70.08**	68.00	62.72	62.72
new	83.26	60.47	64.19	**86.05**	**86.05**
hay	53.75	**54.37**	51.25	52.50	52.50
tae	41.06	**41.72**	**41.72**	40.40	40.40
gla	38.32	**41.12**	39.72	39.25	39.25
eco	55.36	**56.55**	54.46	56.25	56.25
Average	61.15	**61.61**	60.98	61.52	61.52
Wins/loses	2/10	8/4	4/8	2/10	2/10

We will use a non-parametric test because the initial conditions that guarantee the reliability of the parametric tests cannot be fulfilled, which imply that the statistical analysis loses credibility with these parametric tests [4]. Specifically, we employ the Wilcoxon rank test [15] as a non-parametric statistical procedure for making pairwise comparisons between two algorithms.

In Table 3 we show the results of the Wilcoxon rank test to compare any of our modifications with the original k-means algorithm. We use $\alpha = 0.1$.

Table 3. Wilcoxon test to compare any of our proposals (R+) against the k-means algorithm (R−), sorted by p-value. atLest is weighting vector derived from *at least half*. asMany is the weighting vector derived from *as many as possible*. most is the weighting vector derived from *most*. equal is the weighting vector with equal weights for all data.

Comparison	R+	R−	Hypothesis	p-value
asMany vs. k-means	65.5	12.5	Rejected for asMany	0.0537
most vs. k-means	52.5	25.5	Accepted	0.32
atLeast vs. k-means	28	50	Accepted	0.4961
equal vs. k-means	0	0	Accepted	1

The results from the statistical test confirm that the best version of our proposal is the one that gives more importance to the data which are close to a representative. Indeed it allows us to asseverate with a high level of confidence that it is better than the k-means. On the contrary, the other weighting vectors are not statistically better than the well known method.

We can then summarize that in all the examples we have studied, the weighting vectors more suitable to calculate the representative of a cluster are the ones that give more importance to the data which are close to it.

6   Conclusions

In this work we have studied the possibility of calculating the representative of each cluster using a different aggregation operator than the mean. In particular, we have worked with induced ordered weighted averaging operators. Taking advantage of the order of the data, we have experimentally studied different types of weighting vectors, giving more importance to the data which are close to the representative or to the data which are far from it. Under these circumstances, it is better to work with the data which are close to the representative.

From the results, we can conclude that it is promising to study different ways of calculating representative values of a cluster. We have just focused on IOWA operators, but many other aggregation operators can be studied.

Acknowledgments. This work is partially supported by the Public University of Navarra under the project PJUPNA13 and by the Spanish Government under the project TIN2016-77356-P (AEI/FEDER, UE).

References

1. Alcala-Fdez, J., Sanchez, L., Garcia, S., del Jesus, M.J., Ventura, S., Garrell, J., Otero, J., Romero, C., Bacardit, J., Rivas, V., Fernandez, J.C., Herrera, F.: KEEL: a software tool to assess evolutionary algorithms for data mining problems. Soft Comput. **13**, 307–318 (2009)
2. Alcala-Fdez, J., Fernandez, A., Luengo, J., Derrac, J., Garcia, S., Sanchez, L., Herrera, F.: KEEL data-mining software tool: data set repository, integration of algorithms and experimental analysis framework. J. Multiple-Valued Logic Soft Comput. **17**, 255–287 (2011)
3. Calvo, T., Mayor, G., Mesiar, R. (eds.).: Aggregation Operators: New Trends and Applications, pp. 3–104. Physica-Verlag, Heidelberg (2002)
4. Demšar, J.: Statistical comparisons of classifiers over multiple data sets. J. Mach. Learn. Res. **7**, 1–30 (2006)
5. Forgy, E.W.: Cluster analysis of multivariate data: efficiency versus interpretability of classifications. Biometrics **21**, 768–769 (1965)
6. Garcia, S., Fernandez, A., Luengo, J., Herrera, F.: A study of statistical techniques and performance measures for genetics-based machine learning: accuracy and interpretability. Soft Comput. **13**, 959–977 (2009)
7. Jain, A.K., Murty, M.N., Flynn, P.J.: Data clustering: a review. ACM Comput. Surv. **31**, 264–323 (1999)
8. Johnson, S.C.: Hierarchical clustering schemes. Psychometrika **32**, 241–254 (1967)
9. Kaufman, L., Rousseeuw, P.J.: Clustering by means of medoids. In: Y. Dodge (ed.) Statistical Data Analysis Based on the L_1–Norm and Related Methods, pp. 405–416. North-Holland (1987)
10. Lloyd, S.P.: Least square quantization in PCM. IEEE Trans. Inf. Theory **28**, 129–137 (1982)
11. Klir, G.J., Folger, T.A.: Fuzzy Sets, Uncertainty and Information. Prentice Hall, Englewood Cliffs (1988)
12. MacQueen, J.B.: Some methods for classification and analysis of multivariate observations. In: Proceedings of 5th Berkeley Symposium on Mathematical Statistics and Probability, vol. 1, pp. 281–297 (1967)
13. Sheskin, D.: Handbook of Parametric and Nonparametric Statistical Procedures, 2nd edn. Chapman & Hall/CRC, London (2006)
14. Ward Jr., J.H.: Hierarchical grouping to optimize an objective function. J. Am. Stat. Assoc. **58**, 236–244 (1963)
15. Wilcoxon, F.: Individual comparisons by ranking methods. Biometrics **1**, 80–83 (1945)
16. Yager, R.R.: On ordered weighted averaging aggregation operators in multicriteria decision making. IEEE Trans. Syst. Man Cybern. **18**(1), 183–190 (1988)
17. Yager, R.R.: Families of OWA operators. Fuzzy Sets Syst. **59**, 125–148 (1993)
18. Yager, R.R., Filev, D.: Induced ordered weighted averaging operators. IEEE Trans. Syst. Man Cybern. **29**, 141–150 (1999)
19. Zadeh, L.A.: A computational approach to fuzzy quantifiers in natural languages. Comput. Math. Appl. **9**, 149–184 (1983)

Local Properties of Strengthened Ordered Directional and Other Forms of Monotonicity

Mikel Sesma-Sara[1(✉)], Laura De Miguel[1], Radko Mesiar[2], Javier Fernandez[1], and Humberto Bustince[1]

[1] Public University of Navarra, Pamplona, Spain
{mikel.sesma,laura.demiguel,fcojavier.fernandez,bustince}@unavarra.es
[2] Slovak University of Technology, Bratislava, Slovakia
mesiar@math.sk

Abstract. In this study we discuss some of the recent generalized forms of monotonicity, introduced in the attempt of relaxing the monotonicity condition of aggregation functions. Specifically, we deal with weak, directional, ordered directional and strengthened ordered directional monotonicity. We present some of the most relevant properties of the functions that satisfy each of these monotonicity conditions and, using the concept of pointwise directional monotonicity, we carry out a local study of the discussed relaxations of monotonicity. This local study enables to highlight the differences between each notion of monotonicity. We illustrate such differences with an example of a restricted equivalence function.

1   Introduction

A function $f : [0,1]^n \to [0,1]$ satisfying the conditions $f(0,\ldots,0) = 0$, $f(1,\ldots,1) = 1$ and $f(x_1,\ldots,x_n) \leq f(y_1,\ldots,y_n)$ for $x_i \leq y_i \in [0,1]^n$, for all $i \in \{1,\ldots,n\}$, is said to be an aggregation function [2,12]. Aggregation functions are aimed at fusing data by representing with a single value the information coming from n values. There exist a great number of works that study aggregation functions, both from the theoretical [1,7,10] and the practical [9,16,21] perspectives.

On occasion, the monotonicity condition of aggregation functions is restrictive, in the sense that it prevents some, otherwise valid, non-monotone functions to enter the framework of data fusion functions [22]. This is the case of some averaging functions like Gini and Lehmer means [3] and other mixture functions [4].

In that respect, in the literature various proposals of weaker forms of monotonicity can be found. The first one was introduced by Wilkin and Beliakov with the name of weak monotonicity [22]. A function is said to be weakly increasing if whenever the value of all its arguments increase by the same amount, the value of the function increases. An analagous manner to look at this concept is as increasingness along the ray defined by the vector $(1,\ldots,1)$. This led to

© Springer Nature Switzerland AG 2019
R. B. Kearfott et al. (Eds.): IFSA 2019/NAFIPS 2019, AISC 1000, pp. 33–43, 2019.
https://doi.org/10.1007/978-3-030-21920-8_4

the generalization of weak monotonicity to directional monotonicity [8], which considers monotonicity along directions defined by any vector $\overrightarrow{r} \in \mathbb{R}^n \setminus \{\overrightarrow{0}\}$. Functions that are directionally monotone have been used in classification systems with sound results [13–15]. Weak and directional monotonicity have been extended to more general frameworks, such as lattices, intervals, intuitionistic fuzzy values, etc. in [20].

There also exist other generalizations of monotonicity, such as ordered directional (OD) monotonicity [6] and strengthened ordered directional (SOD) monotonicity [19]. These two concepts are based on directional monotonicity but the direction of increasingness varies from one point of the domain to another, depending on the relative sizes of the components of the specific point.

In this work, we study the cited notions of weak, directional, ordered directional and strengthened ordered directional monotonicity from a local point of view. All these are global properties of monotonicity. We make use of the concept of pointwise directional monotonicity, which is a local condition (studied at a specific point of the domain) of monotonicity along rays, to study the relations and differences between the mentioned relaxed forms of monotonicity. We characterize each of the forms in terms of pointwise directional monotonicity and we illustrate these findings with an example of a restricted equivalence function [5].

This paper is organized as follows. In Sect. 2 we fix the notation that is used in this work. In Sect. 3 we present the notions of weak, directional, ordered directional and strengthened ordered directional monotonicity, along with some of their most relevant properties and some examples. In Sect. 4, we present the definition of pointwise directional monotonicity and show how it can be use to characterize of each of the cited monotonicity forms. In Sect. 5 we expose some concluding remarks.

2   Preliminaries

Let us first set the notation for this work. We refer to points in the unit hypercube as $\mathbf{x} = (x_1, \ldots, x_n) \in [0,1]^n$ and we denote real vectors, which are used to express directions in the space, as $\overrightarrow{r} \in \mathbb{R}^n$. In particular, we use the notations $\mathbf{0} = (0, \ldots, 0)$ and $\mathbf{1} = (1, \ldots, 1)$. To order elements in $[0,1]^n$, we consider the product order $\leq_L$ inherited from the standard total order of $[0,1]$, i.e., for any $\mathbf{x}, \mathbf{y} \in [0,1]^n$ we say that $\mathbf{x} \leq_L \mathbf{y}$ if $x_i \leq y_i$ for all $i \in \{1, \ldots, n\}$.

Moreover, in this work, the order of the inputs of a function $f : [0,1]^n \to [0,1]$ has a great impact in the conditions of monotonicity that affect f at the point $\mathbf{x}$ formed by these inputs. Consequently, in this work we deal with permutations of the components of points $\mathbf{x} \in [0,1]^n$ and we use the following notation: let $\mathscr{S}_n$ be the set of all permutations of n elements, $\sigma \in \mathscr{S}_n$ and $\mathbf{x} \in [0,1]^n$, then $\mathbf{x}_\sigma$ denotes the point $(x_{\sigma(1)}, \ldots, x_{\sigma(n)}) \in [0,1]^n$. Similarly, given $\overrightarrow{r} \in \mathbb{R}^n$ we use the notation $\overrightarrow{r}_\sigma = (r_{\sigma(1)}, \ldots, r_{\sigma(n)}) \in \mathbb{R}^n$.

Once the basic notation is established, we recall the concept of an aggregation function.

Definition 1. Let $n \in \mathbb{N}$ and $A : [0,1]^n \to [0,1]$. We say that A is an *aggregation function* if

1. $A(\mathbf{0}) = 0$ and $A(\mathbf{1}) = 1$;
2. A is increasing with respect to all its arguments, i.e., if $\mathbf{x}, \mathbf{y} \in [0,1]^n$ such that $\mathbf{x} \leq_L \mathbf{y}$, then $A(\mathbf{x}) \leq A(\mathbf{y})$.

Note that, throughout the paper, we use the term *increasing* to refer to the property of non-decreasingness. We will explicitly state *strict increasingness* when needed.

3   Different Relaxations of Monotonicity

3.1   Definitions

In this section we gather some of the recently introduced relaxed forms of monotonicity. Among those that are object of study in this work, weak monotonicity was introduced first [22].

Definition 2. Let $n \in \mathbb{N}$ and $f : [0,1]^n \to [0,1]$. We say that f is *weakly increasing* if for all $\mathbf{x} \in [0,1]^n$ and $c > 0$ such that $\mathbf{x} + c\mathbf{1} \in [0,1]^n$, it holds that $f(\mathbf{x}) \leq f(\mathbf{x}+c\mathbf{1})$. Similarly, f is *weakly decreasing* if for all $\mathbf{x} \in [0,1]^n$ and $c > 0$ such that $\mathbf{x} + c\mathbf{1} \in [0,1]^n$, it holds that $f(\mathbf{x}) \geq f(\mathbf{x}+c\mathbf{1})$.

If a function f is both weakly increasing and weakly decreasing, then f is said to be weakly constant.

This concept was extended considering an arbitrary direction $\overrightarrow{r} \in \mathbb{R}^n$ instead of the vector $\mathbf{1}$, leading to the notion of directional monotonicity [8].

Definition 3. Let $n \in \mathbb{N}$, $\overrightarrow{r} \in \mathbb{R}^n \setminus \{\overrightarrow{0}\}$ and $f : [0,1]^n \to [0,1]$. We say that f is $\overrightarrow{r}$*-increasing*, if for all $c > 0$ and $\mathbf{x} \in [0,1]^n$ such that $\mathbf{x} + c\overrightarrow{r} \in [0,1]^n$, it holds that $f(\mathbf{x}) \leq f(\mathbf{x} + c\overrightarrow{r})$. Similarly, f is $\overrightarrow{r}$*-decreasing*, if for all $c > 0$ and $\mathbf{x} \in [0,1]^n$ such that $\mathbf{x} + c\overrightarrow{r} \in [0,1]^n$, it holds that $f(\mathbf{x}) \geq f(\mathbf{x} + c\overrightarrow{r})$.

If a function f is both $\overrightarrow{r}$-increasing and $\overrightarrow{r}$-decreasing, then f is said to be $\overrightarrow{r}$-constant.

The concepts given in Definitions 2 and 3 require that a function satisfies a property of monotonicity along a ray which remains the same for all $\mathbf{x} \in [0,1]^n$. The following two relaxed forms of monotonicity are related to directions in $\mathbb{R}^n$ as well, but the direction of the monotonicity property changes from certain points of the domain to others.

The next relaxed monotonicity form is known as ordered directional monotonicity, or OD monotonicity for short, [6].

Definition 4. Let $n \in \mathbb{N}$, $\overrightarrow{r} \in \mathbb{R}^n \setminus \{\overrightarrow{0}\}$ and $f : [0,1]^n \to [0,1]$. We say that f is *ordered directionally (OD)* $\overrightarrow{r}$*-increasing* if for all $c > 0$, $\sigma \in \mathscr{S}_n$ and $\mathbf{x} \in [0,1]^n$ with $x_{\sigma(1)} \geq \cdots \geq x_{\sigma(n)}$ such that

$$1 \geq x_{\sigma(1)} + cr_1 \geq \cdots \geq x_{\sigma(n)} + cr_n \geq 0, \tag{1}$$

it holds that

$$f(\mathbf{x}) \leq f(\mathbf{x} + c\overrightarrow{r}_{\sigma^{-1}}),$$

where σ^{-1} is the inverse permutation of σ. Similarly, we say that f is OD $\overrightarrow{r}$-*decreasing* if for all $c > 0$, $\sigma \in \mathscr{S}_n$ and $\mathbf{x} \in [0,1]^n$ with $x_{\sigma(1)} \geq \cdots \geq x_{\sigma(n)}$ such that

$$1 \geq x_{\sigma(1)} + cr_1 \geq \cdots \geq x_{\sigma(n)} + cr_n \geq 0,$$

it holds that

$$f(\mathbf{x}) \geq f(\mathbf{x} + c\overrightarrow{r}_{\sigma^{-1}}).$$

If a function f is both OD $\overrightarrow{r}$-increasing and OD $\overrightarrow{r}$-decreasing, then f is said to be OD $\overrightarrow{r}$-constant.

Based on the concept of OD monotonicity, in [19] the concept of strengthened ordered directional monotonicity was introduced.

Definition 5. Let $n \in \mathbb{N}$, $\overrightarrow{r} \in \mathbb{R}^n \setminus \{\overrightarrow{0}\}$ and $f : [0,1]^n \to [0,1]$. We say that f is *strengthened ordered directionally (SOD)* $\overrightarrow{r}$-*increasing* if for all $c > 0$, $\sigma \in \mathscr{S}_n$ and $\mathbf{x} \in [0,1]^n$ with $x_{\sigma(1)} \geq \cdots \geq x_{\sigma(n)}$ such that $\mathbf{x}_\sigma + c\overrightarrow{r} \in [0,1]^n$, it holds that

$$f(\mathbf{x}) \leq f(\mathbf{x} + c\overrightarrow{r}_{\sigma^{-1}}).$$

Similarly, we say that f is *SOD* $\overrightarrow{r}$-*decreasing* if for all $c > 0$, $\sigma \in \mathscr{S}_n$ and $\mathbf{x} \in [0,1]^n$ with $x_{\sigma(1)} \geq \cdots \geq x_{\sigma(n)}$ such that $\mathbf{x}_\sigma + c\overrightarrow{r} \in [0,1]^n$, it holds that

$$f(\mathbf{x}) \geq f(\mathbf{x} + c\overrightarrow{r}_{\sigma^{-1}}).$$

If a function f is both SOD $\overrightarrow{r}$-increasing and SOD $\overrightarrow{r}$-decreasing, then f is said to be SOD $\overrightarrow{r}$-constant.

Example 1. Let $L : [0,1]^2 \to [0,1]$ be the Lehmer mean, i.e., the function given by

$$L(x,y) = \frac{x^2 + y^2}{x + y},$$

with the convention $\frac{0}{0} = 0$. This function is weakly increasing. In fact, L only increases along the direction given by the vector $(1,1)$ [8].

Example 2. Let $f : [0,1]^2 \to [0,1]$ be the function given by

$$f(x,y) = |x - y|.$$

Function f is SOD $\overrightarrow{r}$-increasing for all $\overrightarrow{r} = (r_1, r_2) \in \mathbb{R}^2 \setminus \{(0,0)\}$ such that $r_1 \geq r_2$.

Indeed, let $\mathbf{x} \in [0,1]^2$, $\sigma \in \mathscr{S}_2$ and $c > 0$ such that $x_{\sigma(1)} \geq x_{\sigma(2)}$ and $\mathbf{x} + c\overrightarrow{r}_{\sigma^{-1}} \in [0,1]^2$. For the case $x_1 \geq x_2$, clearly $\sigma = (1\ 2)$ and since $r_1 \geq r_2$, it holds that

$$f(x,y) = |x - y| \leq |x - y| + cr_1 - cr_2 = |x + cr_1 - y - cr_2| = f(x_1 + cr_1, x_2 + cr_2).$$

The case in which $x_1 \leq x_2$ is similar taking into account that $\sigma = (2\ 1)$.

3.2   Properties

The direction given by a vector $\overrightarrow{r} \in \mathbb{R}^n \setminus \{\overrightarrow{0}\}$ and the one given by any other vector $\alpha \overrightarrow{r}$ for any $\alpha > 0$ coincide. In consequence, for all the monotonicity conditions with which we deal, it is equivalent to increase along $\overrightarrow{r}$ and along $\alpha \overrightarrow{r}$ for any $\alpha > 0$.

Proposition 1. *Let $\overrightarrow{r} \in \mathbb{R}^n \setminus \{\overrightarrow{0}\}$, $f\colon [0,1]^n \to [0,1]$ and $\alpha > 0$. Then,*

1. *f is weakly increasing if and only if f is $(\alpha \overrightarrow{1})$-increasing;*
2. *f is $\overrightarrow{r}$-increasing if and only if f is $(\alpha \overrightarrow{r})$-increasing;*
3. *f is OD $\overrightarrow{r}$-increasing if and only if f is OD $(\alpha \overrightarrow{r})$-increasing;*
4. *f is SOD $\overrightarrow{r}$-increasing if and only if f is SOD $(\alpha \overrightarrow{r})$-increasing.*

By Proposition 1, the norm of a vector $\overrightarrow{r}$ has no effect in any of the conditions of monotonicity. Therefore, we can find a unique representative vector for each direction by requiring that $\|\overrightarrow{r}\| = 1$.

Moreover, from the definitions, it is clear that weak monotonicity is a particular case of directional monotonicity and, due to the restriction (1), it is also clear that if a function f is SOD $\overrightarrow{r}$-monotone for some $\overrightarrow{r} \in \mathbb{R}^n \setminus \{\overrightarrow{0}\}$, then f is also OD $\overrightarrow{r}$-monotone. The converse implication is not true [19]. However, if the components of $\overrightarrow{r}$ are decreasingly ordered, then the concepts of OD and SOD monotonicity are equivalent.

Proposition 2. *Let $f : [0,1]^n \to [0,1]$ and let $\overrightarrow{r} \in \mathbb{R}^n \setminus \{\overrightarrow{0}\}$ such that $r_1 \geq \ldots \geq r_n$. Function f is OD $\overrightarrow{r}$-increasing if and only if it is SOD $\overrightarrow{r}$-increasing.*

Obviously, Propositions 1 and 2 can be equivalently stated in terms of OD and SOD $\overrightarrow{r}$-decreasingness and $\overrightarrow{r}$-constantness.

The next result shows another difference between OD monotone functions and SOD monotone functions.

Proposition 3. *Let $\overrightarrow{r} \in \mathbb{R}^n \setminus \{\overrightarrow{0}\}$. Then,*

1. *A function $f : [0,1]^n \to [0,1]$ is $\overrightarrow{r}$-increasing if and only if it is $(-\overrightarrow{r})$-decreasing;*
2. *A function $f : [0,1]^n \to [0,1]$ is OD $\overrightarrow{r}$-increasing if and only if it is OD $(-\overrightarrow{r})$-decreasing.*

Proposition 3 does not hold for SOD monotone functions. In fact, in Sect. 3.1 we present an example (Example 2) of a function that is SOD $\overrightarrow{r}$-increasing but not SOD $(-\overrightarrow{r})$-decreasing.

A relevant property that is satisfied by any function that meets one of the discussed monotonicity criteria is that if it increases along two directions, then it increases along the positive convex combination of those directions. The following three results state this fact for each concept of monotonicity.

Theorem 1 ([8]). *Let $\overrightarrow{r}, \overrightarrow{s} \in \mathbb{R}^n \setminus \{\overrightarrow{0}\}$ and $a, b > 0$. Let $\mathbf{x} \in [0,1]^n$ and $c > 0$ such that if $\mathbf{x}$ and $\mathbf{x} + c(a\overrightarrow{r} + b\overrightarrow{s}) \in [0,1]^n$, then $\mathbf{x} + ca\overrightarrow{r} \in [0,1]^n$ or $\mathbf{x} + cb\overrightarrow{s} \in [0,1]^n$. Then, if a function $f : [0,1]^n \to [0,1]$ is $\overrightarrow{r}$-increasing and $\overrightarrow{s}$-increasing, it is also $(a\overrightarrow{r} + b\overrightarrow{s})$-increasing.*

Theorem 2 ([6]). *Let $\overrightarrow{r}, \overrightarrow{s} \in \mathbb{R}^n \setminus \{\overrightarrow{0}\}$ and $a, b > 0$. Let $\mathbf{x} \in [0,1]^n$, $c > 0$ and $\sigma \in \mathscr{S}_n$ such that if $1 \geq x_{\sigma(1)} \geq \ldots \geq x_{\sigma(n)} \geq 0$ and*

$$1 \geq x_{\sigma(1)} + c(ar_1 + bs_1) \geq \ldots \geq x_{\sigma(n)} + c(ar_n + bs_n) \geq 0,$$

then either

$$1 \geq x_{\sigma(1)} + car_1 \geq \ldots \geq x_{\sigma(n)} + car_n \geq 0,$$

or

$$1 \geq x_{\sigma(1)} + cbs_1 \geq \ldots \geq x_{\sigma(n)} + cbs_n \geq 0.$$

Then, if a function $f : [0,1]^n \to [0,1]$ is OD $\overrightarrow{r}$-increasing and OD $\overrightarrow{s}$-increasing, it is also OD $(a\overrightarrow{r} + b\overrightarrow{s})$-increasing.

Theorem 3 ([19]). *Let $\overrightarrow{r}, \overrightarrow{s} \in \mathbb{R}^n$ and $a, b > 0$. Let $\mathbf{x} \in [0,1]^n$, $c > 0$ and $\sigma \in \mathscr{S}_n$ such that if $1 \geq x_{\sigma(1)} \geq \ldots \geq x_{\sigma(n)} \geq 0$ and $\mathbf{x}_\sigma + c(a\overrightarrow{r} + b\overrightarrow{s}) \in [0,1]^n$, then either $\mathbf{x} + ca\overrightarrow{r} \in [0,1]^n$ or $\mathbf{x} + cb\overrightarrow{s} \in [0,1]^n$. Then, if a function $f : [0,1]^n \to [0,1]$ is SOD $\overrightarrow{r}$-increasing and SOD $\overrightarrow{s}$-increasing, it is also SOD $(a\overrightarrow{r} + b\overrightarrow{s})$-increasing.*

As a consequence of Theorems 1, 2 and 3, it is equivalent for a function to increase, in the standard sense, and to increase, in the sense of the three notions of monotonicity, with respect a set of vectors that span the set of all positive vectors.

Theorem 4 ([19]). *Let $f : [0,1]^n \to [0,1]$ and $(\overrightarrow{e}_1, \ldots, \overrightarrow{e}_n)$ be the canonical basis of $\mathbb{R}^n$, i.e., the set of vectors such that $\overrightarrow{e}_i = (0, \ldots, 0, \underset{i}{1}, 0, \ldots, 0) \in \mathbb{R}^n$ for each $i \in \{1, \ldots, n\}$. Then, the following statements are equivalent:*

1. *f is increasing;*
2. *f is $\overrightarrow{e}_i$-increasing for all $i \in \{1, \ldots, n\}$;*
3. *f is OD $\overrightarrow{e}_i$-increasing for all $i \in \{1, \ldots, n\}$;*
4. *f is SOD $\overrightarrow{e}_i$-increasing for all $i \in \{1, \ldots, n\}$.*

Note that for all $\sigma \in \mathscr{S}_n$ it holds that $\overrightarrow{1}_{\sigma^{-1}} = \overrightarrow{1}$ and, hence, it is straight that for a function $f : [0,1]^n \to [0,1]$ it is equivalent to be weakly increasing, OD $\overrightarrow{1}$-increasing and SOD $\overrightarrow{1}$-increasing.

4   Local Study of the Different Notions of Monotonicity

The aforementioned conditions of monotonicity are global properties, in the sense that they require to be fulfilled for all the points in the domain $[0,1]^n$. In [17], the local notion of pointwise directional monotonicity was introduced.

Definition 6. Let $\overrightarrow{r} \in \mathbb{R}^n \setminus \{\overrightarrow{0}\}$ and $f : [0,1]^n \to [0,1]$. We say that f is $\overrightarrow{r}$-*increasing* at $\mathbf{x} \in [0,1]^n$ if for all $c > 0$ such that $\mathbf{x} + c\overrightarrow{r} \in [0,1]^n$, it holds that $f(\mathbf{x}) \leq f(\mathbf{x} + c\overrightarrow{r})$. Similarly, f is $\overrightarrow{r}$-*decreasing* at $\mathbf{x} \in [0,1]^n$ if for all $c > 0$ such that $\mathbf{x} + c\overrightarrow{r} \in [0,1]^n$, it holds that $f(\mathbf{x}) \geq f(\mathbf{x} + c\overrightarrow{r})$.

Thus, weak and directional monotonicity can be characterized by means of pointwise directional monotonicity. Clearly, given $\overrightarrow{r} \in \mathbb{R}^n \setminus \{\overrightarrow{0}\}$, it is equivalent for a function $f : [0,1]^n \to [0,1]$ to be $\overrightarrow{r}$-increasing and to be $\overrightarrow{r}$-increasing at $\mathbf{x}$, for all $\mathbf{x} \in [0,1]^n$. The case of weak monotonicity follows similarly considering the vector $\overrightarrow{r} = \overrightarrow{1}$.

Remark 1. Note that the property of directional monotonicity stated in Proposition 3 does not hold for pointwise directional monotonicity at a specific point. Namely, if a function $f : [0,1]^n \to [0,1]$ is $\overrightarrow{r}$-increasing at $\mathbf{x} \in [0,1]^n$ for some vector $\overrightarrow{r} \in \mathbb{R}^n \setminus \{\overrightarrow{0}\}$, it does not necessarily hold that f is $(-\overrightarrow{r})$-decreasing at $\mathbf{x}$.

Indeed, let $f : [0,1]^2 \to [0,1]$ be the function given by

$$f(x,y) = 2x^2 + 2y^2 - 2x - 2y + 1.$$

One can easily verify that function f has a global minimum at the point $(0.5, 0.5)$, in which $f(0.5, 0.5) = 0$, and, also, that $f(x,y) > 0$ for all $(x,y) \in [0,1]^2 \setminus \{(0,0)\}$. Hence, f is $\overrightarrow{r}$-increasing at $(0.5, 0.5)$ for every possible $\overrightarrow{r} \in \mathbb{R}^n \setminus \{\overrightarrow{0}\}$. In particular, given $\overrightarrow{r} \in \mathbb{R}^n \setminus \{\overrightarrow{0}\}$, f is $\overrightarrow{r}$-increasing and $(-\overrightarrow{r})$-increasing. However, f is not $\overrightarrow{r}$-decreasing with respect to any direction $\overrightarrow{r} \in \mathbb{R}^n \setminus \{\overrightarrow{0}\}$.

As for weak monotonicity and directional monotonicity, it is possible to characterize the concepts of OD monotonicity and SOD monotonicity in terms of pointwise directional monotonicity. For that, we need to introduce first some specific subsets of $[0,1]^n$.

Let $\sigma \in \mathscr{S}_n$ and let us set $\Omega_\sigma \subset [0,1]^n$ as follows:

$$\Omega_\sigma = \{\mathbf{x} \in [0,1]^n \mid x_{\sigma(1)} \geq x_{\sigma(2)} \geq \ldots \geq x_{\sigma(n)}\}.$$

Thus, Ω_σ is the set of points $\mathbf{x} \in [0,1]^n$ such that $\mathbf{x}_\sigma$ is decreasingly ordered.

Let us first start with the characterization of SOD monotonicity in terms of pointwise directional monotonicity.

Theorem 5. *Let* $\overrightarrow{r} \in \mathbb{R}^n \setminus \{\overrightarrow{0}\}$. *A function* $f : [0,1]^n \to [0,1]$ *is SOD* $\overrightarrow{r}$-*increasing if and only if, for every* $\sigma \in \mathscr{S}_n$, f *is* $\overrightarrow{r}_{\sigma^{-1}}$-*increasing at* $\mathbf{x}$ *for every* $\mathbf{x} \in \Omega_\sigma$.

Proof. Let f be SOD $\overrightarrow{r}$-increasing. If $\mathbf{x} \in [0,1]^n$ and $\sigma \in \mathscr{S}_n$ such that $x_{\sigma(1)} \geq \ldots \geq x_{\sigma(n)}$, then $\mathbf{x} \in \Omega_\sigma$. Since f is SOD $\overrightarrow{r}$-increasing, then by Definition 5, it holds that

$$f(\mathbf{x}) \leq f(\mathbf{x} + c\overrightarrow{r}_{\sigma^{-1}}),$$

for some $c > 0$. Hence, f is $\overrightarrow{r}_{\sigma^{-1}}$-increasing at $\mathbf{x}$.

Conversely, let f be $\overrightarrow{r}_{\sigma^{-1}}$-increasing at $\mathbf{x}$ for every $\mathbf{x} \in \Omega_\sigma$, for every $\sigma \in \mathscr{S}_n$. Thus, if $\mathbf{x} \in [0,1]^n$ and $\sigma \in \mathscr{S}_n$ such that $x_{\sigma(1)} \geq \cdots \geq x_{\sigma(n)}$, by the $\overrightarrow{r}_{\sigma^{-1}}$-increasingness of f at $\mathbf{x}$, it is clear that $f(\mathbf{x}) \leq f(\mathbf{x} + c\overrightarrow{r}_{\sigma^{-1}})$, and, therefore, f is SOD $\overrightarrow{r}$-increasing.

The preceding result, Theorem 5, cannot be straightforwardly translated to OD monotonicity. In fact, in Example 3 we can find an OD $\overrightarrow{r}$-increasing function that is not $\overrightarrow{r}_{\sigma^{-1}}$-increasing at $\mathbf{x}$ for all $\mathbf{x} \in \Omega_\sigma$. It suffices to consider a vector $\overrightarrow{r} \in \mathbb{R}^2 \setminus \{\overrightarrow{0}\}$ such that $r_1 < r_2$.

Nevertheless, it is true that the converse implication holds naturally for OD monotone functions.

Corollary 1. *Let $\overrightarrow{r} \in \mathbb{R}^n \setminus \{\overrightarrow{0}\}$. If, for every $\sigma \in \mathscr{S}_n$, a function $f : [0,1]^n \to [0,1]$ is $\overrightarrow{r}_{\sigma^{-1}}$-increasing at $\mathbf{x}$ for every $\mathbf{x} \in \Omega_\sigma$, then f is OD $\overrightarrow{r}$-increasing.*

Proof. Let f be $\overrightarrow{r}_{\sigma^{-1}}$-increasing at $\mathbf{x}$ for every $\mathbf{x} \in \Omega_\sigma$, for every $\sigma \in \mathscr{S}_n$. By Theorem 5, f is SOD $\overrightarrow{r}$-increasing and, since SOD monotonicity implies OD monotonicity, f is OD $\overrightarrow{r}$-increasing.

Corollary 1 fails to be a characterization because OD monotonicity does not require monotonicity conditions for points $\mathbf{x} \in \Omega_\sigma$ such that $\mathbf{x} + c\overrightarrow{r}_{\sigma^{-1}} \notin \Omega_\sigma$, as it means that condition (1) in Definition 4 is not satisfied.

The next result shows the modifications that need to be done in order to characterize OD monotonicity in terms of pointwise directional monotonicity.

Theorem 6. *Let $\overrightarrow{r} \in \mathbb{R}^n \setminus \{\overrightarrow{0}\}$. A function $f : [0,1]^n \to [0,1]$ is OD $\overrightarrow{r}$-increasing if and only if, for every $\sigma \in \mathscr{S}_n$, the restricted function $f|_{\Omega_\sigma} : \Omega_\sigma \to [0,1]$ is $\overrightarrow{r}_{\sigma^{-1}}$-increasing at $\mathbf{x}$ for every $\mathbf{x} \in \Omega_\sigma$.*

Proof. Let f be OD $\overrightarrow{r}$-increasing and let $c > 0$, $\sigma \in \mathscr{S}_n$ and $\mathbf{x} \in [0,1]^n$ such that $x_{\sigma(1)} \geq \cdots \geq x_{\sigma(n)}$ and

$$1 \geq x_{\sigma(1)} + cr_1 \geq \cdots \geq x_{\sigma(n)} + cr_n \geq 0.$$

Then, $\mathbf{x} \in \Omega_\sigma$ and $\mathbf{x} + c\overrightarrow{r}_{\sigma^{-1}} \in \Omega_\sigma$. Therefore, $f(\mathbf{x}) = f|_{\Omega_\sigma}(\mathbf{x})$ and $f(\mathbf{x} + c\overrightarrow{r}_{\sigma^{-1}}) = f|_{\Omega_\sigma}(\mathbf{x} + c\overrightarrow{r}_{\sigma^{-1}})$. Thus, since f is OD $\overrightarrow{r}$-increasing, it holds that

$$f|_{\Omega_\sigma}(\mathbf{x}) = f(\mathbf{x}) \leq f(\mathbf{x} + c\overrightarrow{r}_{\sigma^{-1}}) = f|_{\Omega_\sigma}(\mathbf{x} + c\overrightarrow{r}_{\sigma^{-1}}).$$

The converse implication follows analogously.

The next is an example that remarks the differences between SOD and OD monotonicity from a local point of view. The function in it is a restricted equivalence function (REF) [5], which have been used for diverse applications [11,18].

Example 3. If we consider the function $REF : [0,1]^2 \to [0,1]$ given by

$$REF(x,y) = 1 - |x - y|.$$

To study the pointwise directional monotonicity of REF, let us first note that for a vector $\vec{r} = (r, r) \in \mathbb{R}^2 \setminus \{(0,0)\}$ it holds that $REF(x, y) = REF(x + cr, y + cr)$ for every $(x, y) \in [0, 1]^2$. Therefore, REF is (r, r)-constant for all $0 \neq r \in \mathbb{R}$. Now, if $\vec{r} = (r_1, r_2) \in \mathbb{R}^2 \setminus \{(0,0)\}$ is such that $r_1 \neq r_2$, note that

$$REF(0.8, 0.8) = 1 > 1 - |cr_1 - cr_2| = REF(0.8 + cr_1, 0.8 + cr_2).$$

Thus, REF is not $\vec{r}$-increasing at $(0.8, 0.8)$ for any $\vec{r}$ such that $r_1 \neq r_2$. Consequently, by the characterization of global directional montonicity in terms of pointwise directional monotonicity, REF is (r_1, r_2)-increasing if and only if $r_1 = r_2 \neq 0$.

Let us proceed to study OD and SOD monotonicity of REF. In the 2-dimensional case, there exist two possible permutations $\sigma_1 = (1\ 2)$ and $\sigma_2 = (2\ 1) \in \mathscr{S}_2$, whose respective inverse coincides with themselves and which define the following subsets:

$$\Omega_{\sigma_1} = \{\mathbf{x} \in [0, 1]^2 \mid x_1 \geq x_2\}, \text{ and}$$
$$\Omega_{\sigma_2} = \{\mathbf{x} \in [0, 1]^2 \mid x_1 \leq x_2\}.$$

Clearly, by the same argument as before, REF is OD and SOD (r, r)-increasing for all $0 \neq r \in \mathbb{R}$. Hence, we can focus on the case of $\vec{r} = (r_1, r_2) \in \mathbb{R}^2 \setminus \{(0,0)\}$ such that $r_1 \neq r_2$.

On the one hand, if $r_1 > r_2$, let $(0.8, 0.2) \in [0, 1]^2$. It holds that $(0.8, 0.2) \in \Omega_{\sigma_1}$ and there exists $c > 0$ such that $(0.8 + cr_1, 0.2 + cr_2) \in \Omega_{\sigma_1}$ and, hence,

$$REF(0.8, 0.2) = 0.4 > 1 - |0.6 + c(r_1 - r_2)| = REF(0.8 + cr_1, 0.2 + cr_2).$$

Therefore, by Theorems 5 and 6, REF is neither OD, nor SOD (r_1, r_2)-increasing for $\vec{r} \in \mathbb{R}^2 \setminus \{(0,0)\}$ such that $r_1 > r_2$.

On the other hand, if $r_1 < r_2$, first let $(0.8, 0.8) \in \Omega_{\sigma_1}$. Thus, there exists $c > 0$ such that $(0.8 + cr_1, 0.8 + cr_2) \in \Omega_2$ and

$$REF(0.8, 0.8) = 1 > 1 - c(r_2 - r_1) = REF(0.8 + cr_1, 0.8 + cr_2).$$

Consequently, by Theorem 5, REF is not SOD $\vec{r}$-increasing.

However, it is easy to check that in the cases that $(x, y), (x, y) + c(r_1, r_2) \in \Omega_{\sigma_1}$ and $(x, y), (x, y) + c(r_2, r_1) \in \Omega_{\sigma_2}$, the conditions $REF(x, y) \leq REF(x + cr_1, y + cr_2)$ and $REF(x, y) \leq REF(x + cr_2, y + cr_1)$, respectively, hold. Therefore, REF is OD $\vec{r}$-increasing.

In conclusion, REF is SOD $\vec{r}$-increasing if and only if $\vec{r} = (r, r) \in \mathbb{R}^2 \setminus \{(0,0)\}$ and REF is OD $\vec{r}$-increasing if and only if $\vec{r} = (r_1, r_2) \in \mathbb{R}^2 \setminus \{(0,0)\}$ such that $r_1 \leq r_2$.

5   Conclusions

We have gone over some of the recently introduced relaxed forms of monotonicity. Particularly, we have discussed the notions of weak, directional, OD and

SOD monotonicity. We have also discussed some of the properties that functions verifying the cited monotonicity conditions satisfy and we have carried out a study of the local effect of this monotonicity conditions at specific points of the domain. In that attempt, we have recalled the notion of pointwise directional monotonicity and we have used it to characterize these relaxations. Additionally, we have pointed out some differences between OD and SOD monotonicity by means of an instance of a restricted equivalence function.

Acknowledgments. This work is supported by the project TIN2016-77356-P (AEI/FEDER, UE), by the Public University of Navarra under the project PJUPNA13 and by Slovak grant APVV-14-0013.

References

1. Bedregal, B., Reiser, R., Bustince, H., Lopez-Molina, C., Torra, V.: Aggregation functions for typical hesitant fuzzy elements and the action of automorphisms. Inf. Sci. **255**, 82–99 (2014)
2. Beliakov, G., Bustince, H., Calvo, T.: A Practical Guide to Averaging Functions. Studies in Fuzziness and Soft Computing. Springer, Heidelberg (2016)
3. Beliakov, G., Calvo, T., Wilkin, T.: Three types of monotonicity of averaging functions. Knowl.-Based Syst. **72**, 114–122 (2014). https://doi.org/10.1016/j.knosys. 2014.08.028
4. Beliakov, G., Calvo, T., Wilkin, T.: On the weak monotonicity of Gini means and other mixture functions. Inf. Sci. **300**, 70–84 (2015)
5. Bustince, H., Barrenechea, E., Pagola, M.: Restricted equivalence functions. Fuzzy Sets Syst. **157**(17), 2333–2346 (2006). https://doi.org/10.1016/j.fss.2006.03.018
6. Bustince, H., Barrenechea, E., Sesma-Sara, M., Lafuente, J., Dimuro, G.P., Mesiar, R., Kolesárová, A.: Ordered directionally monotone functions. Justification and application. IEEE Trans. Fuzzy Syst. **26**(4), 2237–2250 (2018). https://doi.org/ 10.1109/TFUZZ.2017.2769486
7. Bustince, H., Fernández, J., Kolesárová, A., Mesiar, R.: Generation of linear orders for intervals by means of aggregation functions. Fuzzy Sets Syst. **220**, 69–77 (2013)
8. Bustince, H., Fernandez, J., Kolesárová, A., Mesiar, R.: Directional monotonicity of fusion functions. Eur. J. Oper. Res. **244**(1), 300–308 (2015). https://doi.org/10. 1016/j.ejor.2015.01.018
9. De Miguel, L., Sesma-Sara, M., Elkano, M., Asiain, M., Bustince, H.: An algorithm for group decision making using n-dimensional fuzzy sets, admissible orders and OWA operators. Inf. Fusion **37**, 126–131 (2017). https://doi.org/10.1016/j.inffus. 2017.01.007
10. Deckỳ, M., Mesiar, R., Stupňanová, A.: Deviation-based aggregation functions. Fuzzy Sets Syst. **332**, 29–36 (2018)
11. Galar, M., Fernandez, J., Beliakov, G., Bustince, H.: Interval-valued fuzzy sets applied to stereo matching of color images. IEEE Trans. Image Process. **20**(7), 1949–1961 (2011)
12. Grabisch, M., Marichal, J., Mesiar, R., Pap, E.: Aggregation Functions. Cambridge University Press, Cambridge (2009)
13. Lucca, G., Sanz, J., Dimuro, G., Bedregal, B., Asiain, M.J., Elkano, M., Bustince, H.: CC-integrals: Choquet-like Copula-based aggregation functions and its application in fuzzy rule-based classification systems. Knowl.-Based Syst. **119**, 32–43 (2017). https://doi.org/10.1016/j.knosys.2016.12.004

14. Lucca, G., Sanz, J.A., Dimuro, G.P., Bedregal, B., Bustince, H., Mesiar, R.: CF-integrals: a new family of pre-aggregation functions with application to fuzzy rule-based classification systems. Inf. Sci. **435**, 94–110 (2018)

15. Lucca, G., Sanz, J.A., Dimuro, G.P., Bedregal, B., Mesiar, R., Kolesárová, A., Bustince, H.: Preaggregation functions: construction and an application. IEEE Trans. Fuzzy Syst. **24**(2), 260–272 (2016). https://doi.org/10.1109/TFUZZ.2015.2453020

16. Paternain, D., Fernandez, J., Bustince, H., Mesiar, R., Beliakov, G.: Construction of image reduction operators using averaging aggregation functions. Fuzzy Sets Syst. **261**, 87–111 (2015). https://doi.org/10.1016/j.fss.2014.03.008

17. Sesma-Sara, M., De Miguel, L., Roldán López de Hierro, A.F., Lafuente, J., Mesiar, R., Bustince, H.: Pointwise directional increasingness and geometric interpretation of directionally monotone functions. Information Sciences (Submitted)

18. Sesma-Sara, M., De Miguel, L., Pagola, M., Burusco, A., Mesiar, R., Bustince, H.: New measures for comparing matrices and their application to image processing. Appl. Math. Model. **61**, 498–520 (2018)

19. Sesma-Sara, M., Lafuente, J., Roldán, A., Mesiar, R., Bustince, H.: Strengthened ordered directionally monotone functions. Links between the different notions of monotonicity. Fuzzy Sets Syst. **357**, 151–172 (2019). https://doi.org/10.1016/j.fss.2018.07.007

20. Sesma-Sara, M., Mesiar, R., Bustince, H.: Weak and directional monotonicity of functions on Riesz spaces to fuse uncertain data. Fuzzy Sets Syst. (In press). https://doi.org/10.1016/j.fss.2019.01.019

21. Wei, G.: Some geometric aggregation functions and their application to dynamic multiple attribute decision making in the intuitionistic fuzzy setting. Int. J. Uncertainty Fuzziness Knowl.-Based Syst. **17**(02), 179–196 (2009)

22. Wilkin, T., Beliakov, G.: Weakly monotonic averaging functions. Int. J. Intell. Syst. **30**(2), 144–169 (2015). https://doi.org/10.1002/int.21692

On the Influence of Interval Normalization in IVOVO Fuzzy Multi-class Classifier

Mikel Uriz[1,2(✉)], Daniel Paternain[1,2], Humberto Bustince[1,2], and Mikel Galar[1,2]

[1] Department of Statistics, Computer Science and Mathematics,
Public University of Navarre, Campus Arrosadia s/n, 31006 Pamplona, Spain
[2] Institute of Smart Cities, Public University of Navarre,
Campus Arrosadia s/n, 31006 Pamplona, Spain
{mikelxabier.uriz,daniel.paternain,bustince,mikel.galar}@unavarra.es

Abstract. IVOVO stands for Inverval-Valued One-Vs-One and is the combination of IVTURS fuzzy classifier and the One-Vs-One strategy. This method is designed to improve the performance of IVTURS in multi-class problems, by dividing the original problem into simpler binary ones. The key issue with IVTURS is that interval-valued confidence degrees for each class are returned and, consequently, they have to be normalized for applying a One-Vs-One strategy. However, there is no consensus on which normalization method should be used with intervals. In IVOVO, the normalization method based on the upper bounds was considered as it maintains the admissible order between intervals and also the proportion of ignorance, but no further study was developed. In this work, we aim to extend this analysis considering several normalizations in the literature. We will study both their main theoretical properties and empirical performance in the final results of IVOVO.

1 Introduction

In Machine Learning, classification problems consists in learning a model from a set of labelled training examples capable of predicting the class of new, previously unseen, examples. Classification problems are divided into two major groups depending on the number of classes that the learning algorithm should deal with: two-class (binary) and multi-class problems. The latter are usually considered to be more difficult due to the greater overlapping between decision boundaries. Fortunately, multi-class problems can be reduced to binary ones using decomposition strategies [1]. Among them, the One-Vs-One (OVO) [2] strategy is widely used for this purpose. In OVO, the original problem is divided into as many binary problems as pairs of classes, which are solved by independent *base classifiers*. Consequently, new examples are classified by querying all base classifiers and aggregating their outputs. These kinds of strategies are not only useful for classifiers without inherent multi-class support, but also for those capable of managing multiple classes [2].

© Springer Nature Switzerland AG 2019
R. B. Kearfott et al. (Eds.): IFSA 2019/NAFIPS 2019, AISC 1000, pp. 44–57, 2019.
https://doi.org/10.1007/978-3-030-21920-8_5

Fuzzy Rule-Based Classification Systems (FRBCSs) are state-of-the-art classifiers. Their main characteristic is that the model obtained is expressed by a number of rules using human-readable linguistic labels [3]. The OVO strategy has also shown to improve the accuracy of these models when addressing multiclass problems [4–6]. In [4] and [5], FARC-HD [7] was extended using OVO and proposing the usage of overlap functions to improve the final performance of the model. Afterwards, in [6], OVO was combined with IVTURS [8], handling interval-valued confidence outputs in the OVO aggregation phase for the first time. The new classifier was named as IVOVO and is center of attention in this work.

In [6], we addressed the two main obstacles when designing a OVO system dealing with interval-valued outputs: the usage of a normalization strategy for intervals and the adaptation of OVO aggregations for intervals. The interval normalization consisted in making the upper bounds sum to one, which preserves the order and ignorance. However, in the literature there is still no consensus on how intervals should be normalized and hence, other methods could have been considered. In this work, we aim to complement our previous work studying the influence of different interval normalization methods in the performance of IVOVO. We want to study whether this could be a key issue to achieve the best performance or if any of the studied alternatives is valid so as to achieve a competitive performance.

To do so, we will consider different ways for normalizing intervals [9] and study their effects in the resulting intervals both from the theoretical and applied point of view. That is, we will not only evaluate their performance in IVOVO, but we will also study whether the different normalizations are able to maintain the order established between intervals as well as other properties that may be expected after normalization.

The experimental study with IVOVO will consider twenty-two numerical datasets from the KEEL dataset repository [10]. The analysis will be supported by the usage of non-parametric statistical tests, as suggested in the specialized literature [11]. As aggregations in OVO, the voting [12] and Win Weighted Voting (WinWV) [4] will be considered.

The structure of this work is as follows. Section 2 describes preliminary concepts of FRBCSs, OVO and IVOVO required to understand the rest of the work. Section 3 details the different interval normalizations analyzed and studies their main properties. In Sect. 4 both the experimental framework and the corresponding experimental results are presented. Finally, in Sect. 5 we draw the conclusions.

2 IVOVO: Interval-Valued One-Vs-One

IVOVO stands for Interval-Valued One-Vs-One, and is based on the application of the OVO strategy to IVTURS fuzzy classifier, which outputs interval-valued confidence degrees instead of real-valued ones. For this reason, in this section we recall IVOVO and its main components: IVTURS and OVO.

2.1   Fuzzy Rule-Based Classification Systems: IVTURS

FRBCSs create models consisting of human-readable rules based on the usage of linguistic labels [3]. To generate a rule base, a learning algorithm is applied to a training set $\mathscr{D}_T$ having P labeled examples $x_p = (x_{p1}, \ldots, x_{pn}), p = \{1, \ldots, P\}$, where x_{pi} is the value of the i-th attribute ($i = \{1, 2, \ldots, n\}$) of the p-th training example. Each example belongs to a class $y_p \in \mathbb{C} = \{C_1, C_2, \ldots, C_m\}$, where m is the number of classes of the problem.

IVTURS algorithm [8] is based on FARC-HD (Fuzzy Association Rule-based Classification model for High-Dimensional problems) [7]. Both use rules with the following structure:

$$\text{Rule } R_j : \text{ If } x_1 \text{ is } A_{j1} \text{ and } \ldots \text{ and } x_{n_j} \text{ is } A_{jn_j} \text{ then Class} = C_j \text{ with } RW_j$$
$$(1)$$

where R_j is the label of the j-th rule, $x = (x_1, \ldots, x_n)$ is a vector representing the example, $A_{ji} \in \mathbb{X}_i$ is a linguistic label modeled by a triangular membership function (where $\mathbb{X}_i = \{X_{i1}, \ldots, X_{il}\}$ is the set of linguistic labels for the i-th antecedent, being l the number of linguistic labels in this set), C_j is the class label and RW_j is the rule weight computed using the certainty factor defined in [13].

The main difference in the rule representation between FARC-HD and IVTURS is that the latter take advantage of Interval-Valued Fuzzy Sets (IVFSs) to model the uncertainty under the definition of the linguistic labels, and hence its membership functions are defined by IVFSs instead of FSs. Accordingly, the whole Fuzzy Reasoning Method (FRM) needs to be adapted to work with interval along all its steps. As a consequence, the confidence (association) degree for each class obtained in the final step is also an interval. Therefore, the final class is taken as the one with the largest confidence degree (according to an admissible order, see Sect. 2.2).

With respect to the rule learning algorithm, FARC-HD was composed of three steps (see [7] for more details): a fuzzy association rule extraction, a candidate rule pre-screening, and a genetic rule selection and lateral tuning. IVTURS makes use of FARC-HD for carrying out the rule extraction, but without performing the last step. Then, it introduces IVFSs and finally uses a genetic algorithm to tune the interval FRM and carry out a rule selection.

2.2   Admissible Orders Between Intervals

The problem when dealing with intervals is the *a priori* non existence of a total order. Then, we are not always able to compare any pair of intervals, and therefore we cannot establish which is the greatest (or lowest) from a set of intervals. We recall that in IVTURS we must select the largest interval-valued confidence degree.

To solve this problem, we base ourselves on the concept of admissible orders, i.e. linear (total) orders with certain properties that can be defined in the interval setting (see, for example [14,15]). Specifically, in this work we focus on the

admissible order defined by Xu and Yager in [16], which is given as follows. Let
$\mathbb{L}$ the set of all positive closed subintervals, i.e.

$$\mathbb{L} = \{x = [\underline{x}, \overline{x}] | \underline{x}, \overline{x} \in \mathbb{R} \text{ with } 0 \leq \underline{x} \leq \overline{x}\}.$$

Then, for each $x, y \in \mathbb{L}$, we have that $x \leq_{XY} y$ if and only if $\frac{\underline{x}+\overline{x}}{2} < \frac{\underline{y}+\overline{y}}{2}$ or $(\frac{\underline{x}+\overline{x}}{2} = \frac{\underline{y}+\overline{y}}{2}$ and $\overline{x} - \underline{x} \geq \overline{y} - \underline{y})$. For the sake of completeness, will also denote by $L([0,1])$ the set of all closed subintervals of the unit interval. Clearly, $L([0,1]) \subset \mathbb{L}$.

2.3 One-Versus-One (OVO)

In OVO the original m class problem is transformed into a $m(m-1)/2$ sub-problems (all possible pair of classes). Therefore, each base classifier will learn to distinguish a pair of classes $\{C_i, C_j\}$. To predict the class of a new examples, each classifier is expected to provide a pair confidence degrees $r_{ij}, r_{ji} \in [0,1]$ in favor of classes C_i and C_j, respectively. For simplicity, these outputs are stored in a *score-matrix* R. In the case of fuzzy classifiers, these pairs are rarely normalized [4,5]. This fact requires a normalization step so that the outputs of all the base classifiers are in the same scale. Normalization with real-valued confidence degrees is direct, but it is not so straightforward with intervals.

2.4 IVOVO: Interval-Valued One-Vs-One

IVOVO [6] refers to the combination of IVTURS and OVO to enhance the performance of the former in multi-class problems. Nevertheless, there are two main issues when using OVO with IVTURS because the score-matrix is filled by interval confidence scores: (1) there is no consensus on which normalization strategy should be applied; (2) the aggregations needs to be adapted to work with intervals.

Hereafter we recall how these issues were addressed in [6]. Recall that the score-matrix is formed of intervals (R):

$$R = \begin{pmatrix} - & r_{12} & \cdots & r_{1m} \\ r_{21} & - & \cdots & r_{2m} \\ \vdots & & & \vdots \\ r_{m1} & r_{m2} & \cdots & - \end{pmatrix} \tag{2}$$

$r_{ij}, r_{ji} \in \mathbb{L}$ corresponding to the confidence degrees for classes C_i, C_j, respectively.

In IVOVO, the score-matrix R was normalized to a new score-matrix R^u in such a way that all the elements are closed sub-intervals in $[0,1]$, that is, $r_{ij}^u \in L([0,1])$ for every $i, j, i \neq j$ (according to the theory described in [8]). This was done by normalizing them according to the upper bounds:

$$
r_{ij}^u = \begin{cases} \left[\dfrac{r_{ij}}{\overline{r}_{ij} + \overline{r}_{ji}}, \dfrac{\overline{r}_{ij}}{\overline{r}_{ij} + \overline{r}_{ji}} \right] & \text{if } \overline{r}_{ij} \neq 0 \quad \text{or} \quad \overline{r}_{ji} \neq 0 \\[2mm] [0.5, 0.5] & \text{otherwise} \end{cases}
\tag{3}
$$

Interestingly, this normalization allows one to maintain the proportion of ignorance and the order between intervals. After normalizing, $\overline{r}_{ij}^u + \overline{r}_{ji}^u = 1$ holds. However, although this normalization presented good experimental results, no further analysis was developed on its suitability and influence with respect to other normalization strategies. This is why we will elaborate on this aspect.

Regarding the adaptation of the aggregations methods for OVO, they mainly consisted in using the interval arithmetic. We recall the voting strategy and the WinWV strategy as they will be the ones considered in the experimental study (notice that WV was shown to perform worse than WinWV when considering fuzzy classifiers and IVOVO).

- *Voting strategy (Vote)*: $Class = arg \max\limits_{i=1,\ldots,m} \sum\limits_{1 \le j \neq i \le m} s_{ij}$, where s_{ij} is 1 if $r_{ij}^u > r_{ji}^u$ and 0 otherwise.
- *WinWV*: $Class = arg \max\limits_{i=1,\ldots,m} \sum\limits_{1 \le j \neq i \le m} s_{ij}$, where s_{ij} is r_{ij}^u if $r_{ij}^u > r_{ji}^u$ and 0 otherwise.

3   Different Approaches for the Normalization of Intervals

As we have stated in the introduction, one of the main problems when facing an interval-valued OVO decomposition strategy is how to perform a normalization of the interval-valued confidences. On the one hand, it is interesting to know how each normalization "transforms" the original intervals. For example, we want to know whether the resulting interval belongs to $L([0,1])$ (is bounded). On the other hand, and more interesting, we want to know if the normalization is able to keep the original ordinal structure of the interval confidences. That is, if $r_{ij} \le_{XY} r_{ji}$, we wonder if this order is kept by the new normalized intervals.

As we have recalled in Sect. 2.4, the normalization adopted in the original IVOVO algorithm was done according to the upper bonds of the original intervals, i.e., making $\overline{r}_{ij}^u + \overline{r}_{ji}^u = 1$. One of the most interesting property of this method is the fact that the normalized intervals are always elements of $L([0,1])$. Moreover, it was also mentioned that the admissible order $\le_{XY}$ between the original intervals is kept in the normalized ones, i.e., if $x \le_{XY} y$, then $x^u \le_{XY} y^u$.

3.1   Normalization by the Lower Bound and the Middle Point

While the original IVOVO normalization considered the upper bounds, in this subsection we analyze the usage of other points within the interval to perform the normalization. Specifically, we explore the normalization according to the middle points and to the lower bounds. Formally, if $x, y \in \mathbb{L}$, the normalization based on the middle points and the lower bounds are given, respectively, by

$$
x^m = \begin{cases} \left[\dfrac{\underline{x}}{(\underline{x}+\overline{x}+\underline{y}+\overline{y})/2}, \dfrac{\overline{x}}{(\underline{x}+\overline{x}+\underline{y}+\overline{y})/2} \right] & \text{if } \underline{x}+\underline{y}+\overline{x}+\overline{y} \neq 0 \\ [0.5, 0.5] & \text{otherwise} \end{cases} \tag{4}
$$

$$
x^l = \begin{cases} \left[\dfrac{\underline{x}}{\underline{x}+\underline{y}}, \dfrac{\overline{x}}{\underline{x}+\underline{y}} \right] & \text{if } \underline{x}+\underline{y} \neq 0 \\ [0.5, 0.5] & \text{otherwise} \end{cases} \tag{5}
$$

It is worth noting that if in the normalization by the upper bound we had that $\overline{x}^u + \overline{y}^u = 1$, now we have that $\frac{\underline{x}^m + \overline{x}^m}{2} + \frac{\underline{y}^m + \overline{y}^m}{2} = 1$ and that $\underline{x}^l + \underline{y}^l = 1$. Another important differences between x^m, x^l and x^u is the fact that x^m, x^l need not belong to $L([0, 1])$, even if x, y do.

Proposition 1. *Let $x, y \in \mathbb{L}$. The following items hold:*

1. $x^m(y^m), x^l(y^l) \in \mathbb{L}$;
2. $\underline{x}^m(\underline{y}^m), \underline{x}^l(\underline{y}^l) \leq 1$ *for every* $x, y \in \mathbb{L}$;
3. $\overline{x}^m(\overline{y}^m) > 1$ *whenever* $\overline{x}(\overline{y}) > \underline{x} + \underline{y} + \overline{y}(\underline{y} + \underline{x} + \overline{x})$;
4. $\overline{x}^l(\overline{y}^l) > 1$ *whenever* $\overline{x}(\overline{y}) > \underline{x} + \underline{y}$.

However, even if the normalized intervals exceeds $L([0, 1])$, the order relation between x and y is kept under these transformations.

Proposition 2. *Let $x, y \in \mathbb{L}$. If $x \leq_{XY} y$, then $x^m \leq_{XY} y^m$ and $x^l \leq_{XY} y_l$, with x^m, x^l being the normalized intervals given by Eqs. 4 and 5, respectively.*

Finally, we must notice that the normalization according to the lower bounds present an undesirable behavior when $\underline{x} + \underline{y} = 0$, since we will always have $x^l = y^l = [0.5, 0.5]$, discarding the information provided by the upper bounds of x and y.

Example 1. Let $x = [0, 0.6]$, $y = [0, 0.9]$. The normalized intervals according to the three normalization based on the upper bound, middle point and lower bound are given, respectively, by $x^u = [0.0, 0.4]$, $y^u = [0.0, 0.6]$, $x^m = [0.0, 0.8]$, $y^m = [0.0, 1.2]$, $x^l = [0.5, 0.5]$ and $y^l = [0.5, 0.5]$. Observe that, up to some extent, both x^u, y^u and x^m, y^m keep original information of x and y. This loss of information in x^l may be problematic in certain applications, specially if the property $\underline{x} + \underline{y} = 0$ frequently appears.

3.2   Other Normalization Methods

Apart from the normalizations based on values within the interval, such as the lower, middle and upper bound, in the literature one can find other approaches which were originally given for normalizing interval weighting vectors (see, for example [9]).

The first method (Other1) we recall here is the one based on interval arithmetic [17], which is the natural extension of the normalization of numbers. However, it is known to produce too wide intervals. The normalization is as follows: given $x, y \in \mathbb{L}$,

$$
x^{o1} = \begin{cases} \left[\dfrac{\underline{x}}{\overline{x} + \overline{y}}, \dfrac{\overline{x}}{\underline{x} + \underline{y}} \right] & \text{if } \underline{x} + \underline{y} \neq 0 \text{ and } \overline{x} + \overline{y} \neq 0 \\ [0.5, 0.5] & \text{otherwise} \end{cases} \tag{6}
$$

The normalized intervals following this methodology need not belong to $L([0,1])$, since $\overline{x}^{o1} > 1$ whenever $\overline{x} > \underline{x} + \underline{y}$. Moreover, it does not keep the ordinal structure of x and y under $\leq_{XY}$ and it shares the undesirable loss of information when $\underline{x} + \underline{y} = 0$.

Example 2. Let $x = [0.0, 0.3]$ and $y = [0.2, 0.2]$, where clearly $x <_{XY} y$. After applying the normalization, we have that $x^{o1} = [0.0, 1.5]$, $y^{o1} = [0.4, 1.0]$ and $y^{o1} <_{XY} x^{o1}$, so the order relation has been inverted.

The last two methods (Other2 and Other3) we analyze in this section are based on the "interval extended zero" method proposed in [18]. The normalization is performed by multiplying each original interval by an interval "weight", that is given by the following two formulae:

$$
w^{o2} = \left[\frac{1}{\overline{x} + \overline{y}}, \frac{2}{\overline{x} + \overline{y}} - \frac{\underline{x} + \underline{y}}{(\overline{x} + \overline{y})^2} \right], \tag{7}
$$

$$
w^{o3} = \left[\frac{1}{\underline{x} + \underline{y}} - \frac{y_{max} + y_{min}}{2(\underline{x} + \underline{y})}, \frac{1}{\overline{x} + \overline{y}} + \frac{y_{max} + y_{min}}{2(\overline{x} + \overline{y})} \right] \tag{8}
$$

where

$$
y_{max} = 1 - \frac{\underline{x} + \underline{y}}{\overline{x} + \overline{y}}, \quad y_{min} = \frac{\overline{x} + \overline{y} - \underline{x} - \underline{y}}{\underline{x} + \underline{y} + \overline{x} + \overline{y}}.
$$

Finally, the normalized intervals are given by

$$
x^{o2} = \begin{cases} w^{o2} x = [\underline{w}^{o2}\underline{x}, \overline{w}^{o2}\overline{x}], & \text{if } \overline{x} + \overline{y} \neq 0 \\ [0.5, 0.5] & \text{otherwise} \end{cases} \tag{9}
$$

$$
x^{o3} = \begin{cases} w^{o3} x = [\underline{w}^{o3}\underline{x}, \overline{w}^{o3}\overline{x}] & \text{if } \underline{x} + \underline{y} \neq 0 \text{ and } \overline{x} + \overline{y} \neq 0 \\ [0.5, 0.5] & \text{otherwise} \end{cases} \tag{10}
$$

Here again, we cannot assure the belonging of x^{o2}, x^{o3} to $L([0,1])$, especially if x or y are very small intervals. However, they differ on how the special case $x^{o2(o3)} = [0.5, 0.5]$ is obtained. Observe that in Eq. 9, we must assure $\overline{x} + \overline{y} \neq 0$, which means $x, y \neq [0,0]$, while in Eq. 10 we can obtain $[0.5, 0.5]$ as long as $\underline{x} + \underline{y} = 0$, with the consequent loss of information.

If we analyze the order according to $\leq_{XY}$, we have that its maintenance is violated by both approaches under certain conditions. Although in this work we have not fully analyze the conditions under which the order is kept, we have an interesting partial result that make us glimpse that it is mostly respected.

Proposition 3. *Let* $x, y \in \mathbb{L}$ *such that* $x <_{XY} y$. *If* $\underline{x} > \underline{y}$, *then* $x^{o2(o3)} <_{XY}$ $y^{o2(o3)}$.

Example 3. Let $x - [0.1, 0.7]$ and $y - [0.4, 0.5]$, having $x <_{XY} y$. Applying Eq. 9 we have that $w^{o2} = [0.83, 1.32]$ and $x^{o2} = [0.083, 0.924]$, $y^{o2} = [0.332, 0.66]$ with the relation $y^{o2} <_{XY} x^{o2}$.

Now, let $x = [0.03, 0.44]$, $y = [0.14, 0.35]$. Applying Eq. 10 we have that $w^{o3} = [1, 6747, 3.3696]$ and $x^{o3} = [0.05, 1.483]$, $y^{o3} = [0.23, 1.18]$ and the relation $y^{o3} <_{XY} x^{o3}$.

4   Experimental Study

The main goal of this experimental study is to empirically analyze the influence of the normalization in IVOVO. We want to assess the performance of the different normalizations using the same experimental framework as the one in [6]. To do so, as explained earlier, we will study the results using two OVO aggregations, Vote and WinWV. This is an interesting issue because the results in Vote will serve as a measure of how many times the order relation between intervals has been broken and how much this affects the results. Otherwise, WinWV will allow us to measure the quality of the final normalized intervals as they are directly used in the aggregation. Moreover, notice that in previous experiments [4–6], Vote always performed better than WinWV when dealing with fuzzy classifiers. One could expect that a better normalization could lead to better performance in WinWV, closing the gap between both aggregations.

4.1   Experimental Framework

To carry out the experimental study we have considered twenty-two datasets from the KEEL dataset repository [10]. These are the same datasets as those considered in previous works [4–6]. In Table 1, we present a summary of all the datasets, indicating the number of examples (#Ex.), the number of attributes (#Atts.), the number of numerical (#Num.) and nominal (#Nom.) attributes, and the number of classes (#Class.).

We have used a *5-fold stratified cross-validation model* following the *Distribution Optimally Balanced Cross Validation* procedure [19]. Non-parametric

Table 1. Summary description of the datasets.

Id.	Dataset	#Ex.	#Atts.	#Num.	#Nom.	#Class.	Id.	Dataset	#Ex.	#Atts.	#Num.	#Nom.	#Class.
aut	autos	159	25	15	10	6	bal	balance	625	4	4	0	3
cle	cleveland	297	13	13	0	5	con	contraceptive	1473	9	6	3	3
der	dermatology	358	34	1	33	6	eco	ecoli	336	7	7	0	8
gla	glass	214	9	9	0	7	hay	hayes-roth	132	4	4	0	3
iri	iris	150	4	4	0	3	lym	lymphography	148	18	3	15	4
new	newthyroid	215	5	5	0	3	pag	pageblocks	548	10	10	0	5
pen	penbased	1100	16	16	0	10	sat	satimage	643	36	36	0	7
seg	segment	2310	19	19	0	7	shu	shuttle	2175	9	9	0	7
tae	tae	151	5	3	2	3	thy	thyroid	720	21	21	0	3
veh	vehicle	846	18	18	0	4	vow	vowel	990	13	13	0	11
win	wine	178	13	13	0	3	yea	yeast	1484	8	8	0	10

statistical tests are used to support our conclusions as suggested in the specialized literature [11]. More specifically, we use the Wilcoxon rank test to carry out pairwise comparisons and Aligned Friedman test to carry out multiple method comparison.

For IVTURS we used the configuration recommended by the authors: 5 fuzzy labels for each variable, 3 as maximum depth of the tree, a minimum support of 0.05, a minimum confidence of 0.8, 50 individuals as population size, 30 bits per gene for the Gray codification and a maximum of 20000 evaluations.

4.2   Influence of Normalization Strategies in IVOVO

Tables 2 and 3 show the classification accuracy obtained by each normalization method using both Vote and WinWV aggregations methods, respectively.

Attending at these results, there are several points to be highlighted:

- In Vote, NoNorm, Upper and Middle perform exactly the same. This was expected as Upper and Middle do not alter the order relation between intervals. Although the same could be expected by Lower, it needs to go through the else part (see Eq. 5) in many more cases, making a lot of intervals to become $[0.5, 0.5]$, causing a decrease in accuracy.
- Also in Vote, the rest of the normalizations provides different results for different datasets, but looking at the overall performance Other2 and Other3 seems to behave better than Other1. There are datasets such as satimage, shuttle or vehicle, where differences are clear. Notice the three of them break the order relation in some cases. We should point out that, in general, the greater the number of times the order relation is broken, the greater the loss of performance is. Other1 and Other3 also suffer the same problem as Lower with the else part.
- With respect to WinWV, Middle is the best performer. This result may suggest that Upper (used in IVOVO) is not the most adequate for this purpose. However, we cannot make such a claim without carrying out the proper statistical analysis.

Table 2. Classification accuracy obtained by IVOVO in testing (Vote).

Dat	NoNorm	Upper	Lower	Middle	Other1	Other2	Other3
aut	0.7713	0.7713	0.7652	0.7713	0.7592	0.7652	0.7652
bal	0.8512	0.8512	0.8030	0.8512	0.8045	0.8608	0.8094
cle	0.5457	0.5457	0.5457	0.5457	**0.5524**	0.5492	0.5457
con	0.5364	0.5364	0.5364	0.5364	0.5323	0.5330	0.5337
der	0.9529	0.9529	0.9529	0.9529	0.9529	0.9529	0.9529
eco	0.8167	0.8167	0.8137	0.8167	0.8286	0.8286	0.8195
gla	0.7098	0.7098	0.7148	0.7098	0.6911	**0.7202**	0.7188
hay	0.7445	0.7445	0.7445	0.7445	0.7445	0.7445	0.7445
iri	**0.9533**	**0.9533**	**0.9533**	**0.9533**	**0.9533**	**0.9533**	**0.9533**
lym	0.8052	0.8052	0.7983	0.8052	0.7983	0.8052	0.7983
new	**0.9488**	**0.9488**	**0.9488**	**0.9488**	0.9116	**0.9488**	0.9442
pag	0.9435	0.9435	0.9435	0.9435	0.9435	0.9435	0.9435
pen	0.9519	0.9519	0.9410	0.9519	0.9282	0.9473	0.9391
sat	0.8198	0.8198	0.8198	0.8198	0.7169	0.7481	0.8028
seg	**0.9216**	**0.9216**	0.9178	**0.9216**	0.9056	0.9173	0.9156
shu	**0.9435**	**0.9435**	**0.9435**	**0.9435**	0.8634	0.8721	0.9055
tae	0.5677	0.5677	0.5677	0.5677	0.5613	0.5742	0.5742
thy	**0.9417**	**0.9417**	**0.9417**	**0.9417**	0.9403	0.9403	0.9403
veh	0.7091	0.7091	0.7091	0.7091	0.6558	0.6913	0.7032
vow	**0.8990**	**0.8990**	0.8949	**0.8990**	0.8768	0.8980	0.8919
win	0.9663	0.9663	0.9663	0.9663	0.9609	0.9609	0.9663
yea	0.5957	0.5957	0.5944	0.5957	0.5829	0.5951	**0.5977**
AVG	**0.8134**	**0.8134**	0.8098	**0.8134**	0.7938	0.8068	0.8075

- Anyway, WinWV shows the importance of a good normalization. Lower and Other1 achieve the worst results, with performances far from the rest. Other2 and Other3 are able to overcome Upper in terms of overall accuracy and looking at NoNorm the need for normalization can be observed.

These statements needs to be validated performing the proper statistical analysis. First, Table 4 shows the results of the Aligned Friedman Ranks tests, one for each OVO aggregation to focus on the differences among normalizations.

According to the tests, NoNorm, Middle and Upper are equally effective with Vote. Other2, Lower and Other3 get lower ranks, although only statistical differences are found with Other1. In WinWV, Middle is the best performer in terms of ranks, and in this case significant differences are found against NoNorm, Lower and Other1. The others give a p-value of 1.0 due to the much greater differences with the rest in the comparison. For this reason, we carry out pairwise Wilcoxon tests to compare the best four alternatives in this case (Table 5).

Table 3. Classification accuracy obtained by IVOVO in testing (WinWV).

	NoNorm	Upper	Lower	Middle	Other1	Other2	Other3
aut	0.7195	0.7585	0.6400	0.7652	0.6402	0.7592	**0.7719**
bal	**0.8625**	0.8254	0.8033	0.8447	0.8064	0.8543	0.8157
cle	0.3974	0.5458	0.4785	0.5389	0.4885	0.5457	0.5422
con	0.5161	**0.5371**	0.5350	0.5364	0.5303	0.5344	0.5343
der	0.9274	**0.9669**	**0.9669**	**0.9669**	**0.9669**	**0.9669**	**0.9669**
eco	0.7839	0.8137	0.4296	0.8316	0.4443	**0.8373**	0.8283
gla	0.5456	0.6453	0.5261	0.6966	0.5257	0.7064	0.7055
hay	0.7440	**0.7522**	**0.7522**	**0.7522**	**0.7522**	**0.7522**	**0.7522**
iri	**0.9533**	**0.9533**	0.8067	**0.9533**	0.8067	**0.9533**	**0.9533**
lym	0.7838	**0.8123**	0.8054	**0.8123**	0.8054	**0.8123**	0.8054
new	0.9163	0.9395	0.9395	**0.9488**	0.9023	**0.9488**	0.9442
pag	0.5444	0.8801	0.8404	**0.9471**	0.8387	0.9453	0.9453
pen	0.8375	0.9393	0.4514	**0.9538**	0.4551	0.9474	0.9529
sat	0.6717	0.8183	0.5465	**0.8245**	0.4810	0.7762	0.8106
seg	0.6723	0.8931	0.5979	0.9164	0.5931	0.9117	0.9130
shu	0.7872	0.8946	0.3941	0.9426	0.3771	0.8717	0.9050
tae	0.5279	0.5679	0.5364	**0.5744**	0.5297	0.5742	**0.5744**
thy	0.9251	0.9348	0.9389	**0.9417**	0.9250	0.9403	0.9403
veh	0.6250	0.7103	0.6618	**0.7161**	0.6203	0.6972	0.7079
vow	0.7202	0.8424	0.3545	0.8707	0.3354	0.8667	0.8687
win	0.9609	**0.9717**	0.8942	0.9663	0.8996	0.9609	0.9663
yea	0.4509	0.5803	0.1986	0.5937	0.1986	0.5910	0.5971
AVG	0.7215	0.7992	0.6408	**0.8134**	0.6328	0.8070	0.8092

Table 4. Aligned Friedman test

Method	Vote Rank (p-value)	WinWV Rank (p-value)
NoNorm	59.00 (−)	98.16 (0.0003+)
Middle	59.00 (1.0)	45.25 (−)
Upper	59.00 (1.0)	53.70 (1.0)
Other2	77.18 (0.5604)	48.00 (1.0)
Lower	78.84 (0.5604)	121.61 (0.0000+)
Other3	89.55 (0.1156)	49.91 (1.0)
Other1	119.93 (0.0000+)	125.86 (0.0000+)

+ near the p-value means that statastical differences are found at 95% confidence.

Table 5. Wilcoxon test for WINWV aggregation method

		Middle	Other2	Other3	Upper
Middle	(W/T/L)	-	13/5/4	13/5/4	15/4/3
	p-value	-	0.0973*	0.0531*	0.0012+
Other2	(W/T/L)		-	10/5/7	12/4/6
	p-value		-	0.3801	0.1309
Other3	(W/T/L)			-	12/3/7
	p-value			-	0.0001+

* and + near the p-value mean that statistical differences are found at 90% and 95% confidence, respectively. (W/T/L) stands for (Wins/Ties/Losses)

Table 6. Wilcoxon test for best VOTE method and best WINWV method

Comparison	R+	R−	Hypothesis	p-value
NoNorm_Vote (IVOVO) vs Upper_WinWV	194.5	58.5	Rejected for NoNorm_Vote	0.0273
NoNorm_Vote vs Middle_WinWV	113.5	139.5	Not rejected	0.6726

From these tests we can conclude that Middle outperforms all the other contenders. Therefore, with WinWV using the Middle point as normalization factor seems to be beneficial.

Our last comparison will compare the results between Vote and WinWV considering both the normalization used in the original IVOVO [6] (Upper) and the best performer in this work (Middle). We will compare the best WinWV alternative versus the Vote with NoNorm (which is the same as any normalization not altering the order relation). The results of the comparison are presented in Table 6.

The outputs of the Wilcoxon tests allows us to conclude that normalization is crucial to achieve the best performance. In our previous work [6], although WinWV allowed us to increase the performance of WV it did not allowed to overcome simple Vote (first test). However, with a better normalization (in this case using Middle), statistical differences in favour of Vote are transformed into a comparison won by WinWV (although without statistical differences).

5   Conclusions

In this paper we have focused on analyzing the influence of different normalization methods for intervals in IVOVO. To do so, we have considered five ways of normalizing interval and we have analyzed some of their main properties, such as whether they maintain the order relation between intervals (considering Xu and Yager's admissible order). Then, we have carried out an experimental study were the high influence of normalization has been shown. Overall, the normalization based on the middle point has shown to perform well with both Vote and WinWV aggregations. More interestingly, the usage of this normalization has allowed us for the first time to improve the performance of Vote strategy using the confidences given by a fuzzy classifier.

For future work we aim to carry out a deeper study including more normalization methods. From a theoretical point of view, we are interested in analyzing whether all the normalization methods based on internal points (lower, middle and upper) satisfy the usual properties demanded to normalized interval-valued vector. Moreover, we want to extend the maintenance of not only the Xu and Yager's order, but many other admissible orders. From an applied point of view, we will check if new methods allow us to outperform the results presented in this work, specially when considering WinWV.

Acknowledgment. This work has been partially supported by the Spanish Ministry of Science and Technology under the project TIN2016-77356-P and the Public University of Navarre under the project PJUPNA13.

References

1. Lorena, A., Carvalho, A., Gama, J.: A review on the combination of binary classifiers in multiclass problems. Artif. Intell. Rev. **30**(1–4), 19–37 (2008)
2. Galar, M., Fernández, A., Barrenechea, E., Bustince, H., Herrera, F.: An overview of ensemble methods for binary classifiers in multi-class problems: experimental study on one-vs-one and one-vs-all schemes. Pattern Recogn. **44**(8), 1761–1776 (2011)
3. Ishibuchi, H., Nakashima, T., Nii, M.: Classification and Modeling with Linguistic Information Granules: Advanced Approaches to Linguistic Data Mining. Springer, Heidelberg (2004). https://doi.org/10.1007/b138232
4. Elkano, M., et al.: Enhancing multiclass classification in FARC-HD fuzzy classifier: on the synergy between n-dimensional overlap functions and decomposition strategies. IEEE Trans. Fuzzy Syst. **23**(5), 1562–1580 (2015)
5. Elkano, M., Galar, M., Sanz, J., Bustince, H.: Fuzzy rule-based classification systems for multi-class problems using binary decomposition strategies: on the influence of n-dimensional overlap functions in the fuzzy reasoning method. Inf. Sci. **332**, 94–114 (2016)
6. Elkano, M., Galar, M., Sanz, J., Lucca, G., Bustince, H.: IVOVO: a new interval-valued one-vs-one approach for multi-class classification problems. In: 17th International Fuzzy Systems Association (IFSA), pp. 1–6 (2017)
7. Alcalá-Fdez, J., Alcalá, R., Herrera, F.: A fuzzy association rule-based classification model for high-dimensional problems with genetic rule selection and lateral tuning. IEEE Trans. Fuzzy Syst. **19**(5), 857–872 (2011)
8. Sanz, J., Fernández, A., Bustince, H., Herrera, F.: IVTURS: a linguistic fuzzy rule-based classification system based on a new interval-valued fuzzy reasoning method with tuning and rule selection. IEEE Trans. Fuzzy Syst. **21**(3), 399–411 (2013)
9. Pavlačka, O.: On various approaches to normalization of interval and fuzzy weights. Fuzzy Sets Syst. **243**, 110–130 (2014)
10. Alcalá-Fdez, J., et al.: KEEL data-mining software tool: data set repository, integration of algorithms and experimental analysis framework. J. Multiple-Valued Logic Soft Comput. **17**(2–3), 255–287 (2011)

11. García, S., Fernández, A., Luengo, J., Herrera, F.: A study of statistical techniques and performance measures for genetics-based machine learning: accuracy and interpretability. Soft Comput. **13**(10), 959–977 (2009)
12. Friedman, J.: Another approach to polychotomous classification. Technical report, Department of Statistics, Stanford University (1996)
13. Ishibuchi, H., Yamamoto, T., Nakashima, T.: Hybridization of fuzzy GBML approaches for pattern classification problems. IEEE Trans. Syst. Man Cybern. B **35**(2), 359–365 (2005)
14. Bustince, H., Fernandez, J., Kolesárová, A., Mesiar, R.: Generation of linear orders for intervals by means of aggregation functions. Fuzzy Sets Syst. **220**, 69–77 (2013)
15. Paternain, D., Miguel, L.D., Ochoa, G., Lizasoain, I., Mesiar, R., Bustince, H.: The interval-valued choquet integral based on admissible permutations. IEEE Trans. Fuzzy Syst. (in press)
16. Xu, Z.S., Yager, R.R.: Some geometric aggregation operators based on intuitionistic fuzzy sets. Int. J. Gen. Syst. **35**(4), 417–433 (2006)
17. Xu, R.: Fuzzy least-squares priority method in the analytic hierarchy process. Fuzzy Sets Syst. **112**, 395–404 (2000)
18. Sevastjanov, P., Dymova, L., Bartosiewicz, P.: A new approach to normalization of interval and fuzzy weights. Fuzzy Sets Syst. **198**, 34–45 (2012)
19. Moreno-Torres, J., Saez, J., Herrera, F.: Study on the impact of partition-induced dataset shift on k-fold cross-validation. IEEE Trans. Neural Netw. Learn. Syst. **23**(8), 1304–1312 (2012)

Fuzzy Optimization and Decision Making: Theory, Algorithms and Applications

Group Assessment of Comparable Items from the Incomplete Judgments

Tomoe Entani[✉]

University of Hyogo, Kobe, Japan
entani@ai.u-hyogo.ac.jp

Abstract. In this study, we propose an approach to derive a group assessment of an item as its weight vector on multiple viewpoints. When there are a group of decision makers who give the judgments of the item as comparison matrices on the viewpoints, it is reasonable that the weight vector of the item is the core of those by all the decision makers. Each decision maker's weight vector basically includes his/her given comparison matrix, which represents only a part of his/her thinking. Namely, there is an inclusion relation between a comparison and a ratio of the corresponding weights. In addition, there are items other than the target one. A decision maker gives the comparison matrices of some of the other items if s/he knows them, as well as the target one. It is natural that there is a correlation between the judgment of the target item to those of the others. The correlation is taken into consideration from the aspect of consistency of his/her judgments. We define a fuzzy degree of the consistency with all the comparison matrices s/he gives. As the consistency degree for a comparison matrix increases, it may become unable to satisfy the inclusion relation between the comparison matrix and the weight vector. Hence, we introduce a fuzzy degree of inclusion relation in order to relax it. There is a trade-off between them. Therefore, by maximizing both degrees we obtain the weight vector of the target item from the comparison matrices of the target item considering the consistency of each decision maker's judgments. The proposed approach is applicable even in the case that a group of given comparison matrices is incomplete such that some comparisons in a comparison matrix are missing and/or the comparison matrices of some items are missing.

1 Introduction

We are concerned with assigning a number to a thing without the scale to measure among various kinds of decision problems, in this study. Once a crisp value is assigned to each of a group of items, we can rank them linearly. In order for more detail, the crisp value would be extended into multiple and/or interval values of some criteria. The focus is on representing each item with multiple interval values corresponding to the viewpoints for the better understanding of the characteristic of the item. Moreover, for a fair understanding from a broad perspective, we need multiple decision makers who evaluate the items. For instance, the issue of

© Springer Nature Switzerland AG 2019
R. B. Kearfott et al. (Eds.): IFSA 2019/NAFIPS 2019, AISC 1000, pp. 61–72, 2019.
https://doi.org/10.1007/978-3-030-21920-8_6

an allocation of employees may be the right person in right place based on their character rather than linearly rank them and each employee may be assessed by more than one manager. This study proposes an approach to obtain the interval values of multiple viewpoints of a target item from a group of judgments of multiple items including the target one by multiple decision makers.

The decision maker's judgment is given as a pairwise comparison matrix, which is commonly used in multi-criteria decision making. A decision maker compares each pair of viewpoints at a time and gives a comparison. It represents his/her intuitive judgment on how much better the item is from one viewpoint over the other one. It is sometimes difficult to give the comparisons of all the pairs of viewpoints so that the given comparison matrix could be incomplete with missing comparisons. The first method to estimate the missing comparisons were based on the connecting path of the comparisons [1] and some other methods have been proposed in succession. Once the comparison matrix is completed, the weight vector is obtained so as to approximate it. The closer the weight vector to the comparison matrix, the more efficient it becomes [2]. In Analytic Hierarchy Process (AHP), several methods have been proposed for this kind of weighting problems such as the eigenvector method [3] and the least square method [4], and these methods have been compared [2,5,6]. Although these methods assume a crisp weight vector, Interval AHP assumes an interval weight vector [7]. It is based on the idea that a decision maker does not perceive a precise weight vector but s/he has a range of the weight vector in his/her mind. The interval weight vector is obtained so as to include the comparison matrix as precise as possible. There is no need to complete the missing comparisons, but to focus on the inclusion relation of the given comparisons. From the viewpoint of the certainty of the obtained weight vector, the other objective function has been introduced [8]. Furthermore, the objective function has been relaxed to express the vagueness of the decision maker's evaluation more sufficiently [9].

In recent AHP papers, the consistency among the comparisons has been discussed, while an experiment showed that the level of inconsistency in the judgments provides no significant effect [6]. One of the ways to treat inconsistency is to introduce the (in)consistency index to distinguish whether the comparison matrix is not too inconsistent [3]. When the inconsistency stems from the large-scale of the comparison matrix, it can be decomposed into consistent small pieces [10]. Some axiomatic properties of inconsistency index have been shown [11]. It becomes complex to measure the consistency of a fuzzy comparison matrix, which is suitable for handling the vagueness of human judgments. One of the consistency indexes with fuzzy comparison matrix is based on geometric mean and it is applicable with a crisp comparison matrix [12]. The other is based on the inclusion relation [13]. Moreover, the inconsistent comparisons have been found by two measurements of congruence and dissonance and corrected [14]. In an additive sense, the consistency has been defined and the missing comparisons have been completed [15]. These (in)consistency indexes are used to complete the comparison matrix. Meanwhile, in Interval AHP since the consistency is measured by a certain weight of the interval weight vector which reflects the inconsistency, the interval weight vector with the assumed consistency can be obtained. Furthermore, the possibility of the interval weight vector motivates us to take the other comparison matrices into consideration as well as the comparison matrix of the target one for a fair evaluation from a broad perspective.

This study is on the basis of Interval AHP. Its goal is to obtain the weight vector of a target item mostly from the comparison matrices of the item given by all the decision makers and in addition from those of the other items to consider the decision makers' character. On one hand, there are multiple comparison matrices of the item. There are two basic approaches of aggregation: one is to aggregate the comparison matrices and the other is to aggregate the weight vectors by all the experts [16]. The former group comparison matrix is obtained based on the distance between the comparison matrices so as to maximize the consensus [17]. The latter collective weight vector is obtained with the expert's confidence degree based on the consistency in his/her comparison matrix [18]. It is pointed out that the group comparison matrix depends on the choices of inconstancy indexes [19]. Instead of these stepwise aggregation approaches, the inclusion relation of the group weight vector to each decision maker's weight vector is used [20]. The group and individual weight vectors are obtained from the comparison matrices without aggregation process. On the other hand, each decision maker gives the comparison matrices of some of the other items, as well as the target one, in a similar manner. Therefore, there is a correlation between that of the target item and those of the other items. In other words, the weight vector of the target item corresponds not only to that of the item but also to those of the other items to some extent. The items of which one decision maker gives the judgments may be different from the items of which the other decision maker does. In that sense a group of given comparison matrices is incomplete.

This paper's outline is as follows. In Sect. 2, we define the consistency of a decision maker's judgment from the possibilistic view. Section 3 is the case of a group of decision makers. We denote a group assessment as the core of those of all the decision makers. Therefore, there is no aggregation step, such as with the importance weights of decision makers, which some conventional combining methods have. Moreover, in Sect. 4, we formulate the problem to obtain the weight vector of the target item considering the consistency of judgments and the inclusion relation of the comparison matrix and weight vector. Instead of weighting decision makers, the consistency of their judgments is considered. Then, we show the numerical example in Sect. 5 to illustrate the proposed method and draw the conclusion in Sect. 6.

2   Consistency of Judgments

2.1   Single Evaluated Item

Assume multiple viewpoints for an assessment of an item. It is denoted as interval weight vector of n viewpoints: $\boldsymbol{W} = (W_1, \ldots, W_n)^t$, where each element is $W_i = [\underline{w}_i, \overline{w}_i]$. The interval weight vector is normalized as follows.

$$\sum_{j \neq i} \underline{w}_j + \overline{w}_i \leq 1, \ \sum_{j \neq i} \overline{w}_j + \underline{w}_i \geq 1, \forall i, \tag{1}$$

where redundancy in each interval weight to make the sum be one is excluded. In the case of $\underline{w}_i = \overline{w}_i = w_i, \forall i$, two kinds of inequalities are reduced into $\sum_i w_i = 1$, which is the usual crisp normalization.

The interval weight vector $\boldsymbol{W}$ mentions that the weight of the item on viewpoint i is surely more than its lower bound $\underline{w}_i$. The certain weight in the interval weight vector is surely assigned to one of the viewpoints denoted as follows.

$$s = \sum_i \underline{w}_i, \tag{2}$$

where $s \leq 1$ because of (1). The remainder, $1-s$, represents the uncertain weight because of being assigned to more than two viewpoints [8]. Therefore, the greater the certain weight, s, is, the more certain the interval weight vector becomes. We prefer a more certain weight with a higher s.

The assessment of an item is obtained from the following pairwise comparison matrix of the item on n viewpoints. A decision maker assesses the item by comparing the superiority from two viewpoints, such that the item is much better from this viewpoint than the other one and repeats evaluating from all the pairs of viewpoints.

$$\boldsymbol{A} = [a_{ij}] = \begin{bmatrix} 1 & \cdots & a_{1n} \\ \vdots & a_{ij} & \vdots \\ a_{n1} & \cdots & 1 \end{bmatrix}, \tag{3}$$

where each element a_{ij} represents how much better the item is from viewpoint i over viewpoint j. These elements are reciprocal and transitive: $a_{ii} = 1, a_{ij}a_{ji} = 1 \forall i, j$, so that in order to fulfill the comparison matrix the decision maker has to give $n(n-1)/2$ comparisons. However, in a real situation, s/he skips some of them since s/he may be unfamiliar with them and/or get tired of the repeat. Then, we often have an incomplete comparison matrix in the sense of missing comparisons.

The other kind of incompleteness in the comparison matrix is inconsistency. The given comparisons in $\boldsymbol{A}$ do not always satisfy the following transitivity, since a viewpoint is compared to the other $n-1$ viewpoints by a decision maker intuitively.

$$a_{ij} = a_{il}a_{lj}, \ \forall i, j, l \in \{1, \ldots, n\} \text{ such that } i \neq j, \ j \neq l, \ l \neq i. \tag{4}$$

If and only if the comparison matrix satisfies (4), it is perfectly consistent.

Various methods to derive the weight vector from such an incomplete comparison matrix have been proposed. In Interval AHP, the interval weight vector satisfies the following inclusion relation with the given comparison matrix [7].

$$a_{ij} \in \frac{W_i}{W_j}, \forall i, j, i > j \leftrightarrow \frac{\underline{w}_i}{\overline{w}_j} \leq a_{ij} \leq \frac{\overline{w}_i}{\underline{w}_j}, \ \varepsilon \leq \underline{w}_i, \forall i, j, i > j$$
$$\leftrightarrow 0 \leq a_{ij}\overline{w}_j - \underline{w}_i, \ 0 \leq \overline{w}_i - a_{ij}\underline{w}_j, \ \varepsilon \leq \underline{w}_i, \forall i, j, i > j, \tag{5}$$

where ε is a small positive number and a fraction of intervals are defined as the maximum range. It is based on the idea that the comparison matrix represents only a part of the decision maker's thinking. The inclusion relation can be satisfied as the lower and/or upper bounds of the interval weights become smaller and/or greater, respectively. Such a wide interval weight vector with a small certain weight seems to be a result of taking the possibility of the given comparisons as much as possible. It is reasonable from the possibilistic viewpoint, however, too much possibility often makes the result be useless in a real situation. Hence, the unnecessary possibility should be reduced.

One of the approaches is to maximize the certain weight (8) so that the obtained weight vector approaches to the given comparison matrix [8].

$$\max \sum_i \underline{w}_i - \varepsilon \sum_i \overline{w}_i,$$
$$s.t. \sum_{j \neq i} \underline{w}_j + \overline{w}_i \leq 1, \ \sum_{j \neq i} \overline{w}_j + \underline{w}_i \geq 1, \forall i, \qquad (6)$$
$$0 \leq a_{ij}\overline{w}_j - \underline{w}_i, \ 0 \leq \overline{w}_i - a_{ij}\underline{w}_j, \ \varepsilon \leq \underline{w}_i, \forall i, j, i > j,$$

where the variables are the interval weight $[\underline{w}_i, \overline{w}_i], \forall i$ and the upper bounds are secondarily minimized to reduce redundancy. In the other approach, both are done simultaneously: $\min \sum_i (\overline{w}_i - \underline{w}_i)$, [7].

The interval weight vector is obtained from the given comparison matrix by (6). It should be noted that if some of the comparisons are missing but there is at least one comparison for each viewpoint, the problem (6) is solvable. We do not need the procedure to estimate the missing comparisons. The decision maker needs not to give all the comparisons but to give some of them with confidence. This kind of the easiness for the decision maker confirms the reliability of the given data and is an advantage of (6).

The inconsistency of the given comparison matrix is discussed a lot in AHP. The well-known consistency index is based on eigenvalue [3] and one has been proposed in additive sense [15]. In addition, many methods transform an inconsistent comparison matrix to be consistent. We define the consistency in the comparison matrix $\boldsymbol{A}$ as the certain weight of the obtained interval weight vector from the possibilistic viewpoint as follows.

$$s^* = \sum_i \underline{w}_i^*, \qquad (7)$$

where the more s^*, the more consistent $\boldsymbol{A}$. We have $s^* = 1$ with the crisp weight vector, instead of interval one, since the comparison matrix is perfectly consistent (4). Moreover, we discuss the consistency of a comparison matrix by comparing it with the other comparison matrices given by the decision maker in the next section.

2.2 Multiple Comparable Items

Assume multiple items, $k = 1, \ldots, m$, each of which a decision maker assesses on n viewpoints. S/he gives at most m pairwise comparison matrices: $\boldsymbol{A}_k, \forall k$ as in (3). We cannot have the comparison matrix of the item which s/he is not in the mood to evaluate and/or does not know well. In the sense of missing comparison matrices, a group of the comparison matrices is incomplete. The consistency in each comparison matrix A_k is obtained by (7) and we denote it as s_k^*. Then, we find that the consistency degrees of the decision maker varies from $\underline{s}$ to $\overline{s}$ as follows.

$$\underline{s} = \min_k \; s_k^*, \; \overline{s} = \max_k \; s_k^*. \tag{8}$$

This possible range of consistency represents the character of the judgments by the decision maker. For instance, when $[\underline{s}, \overline{s}]$ is almost 1 with a small width, it mentions that the decision maker always gives consistent judgments, while small $[\underline{s}, \overline{s}]$ with a wide width mentions that s/he may be unfamiliar with these items.

Since it is possible for the decision maker to give the comparison matrix with the consistency degree $\overline{s}$, we reconsider the weight vector from A_k as if its consistency degree is $\overline{s}$. In order to ensure the highest consistency, the inclusion relation in (5) cannot be satisfied. Hence, it is relaxed by $c_k \geq 0$. By minimizing the relaxation the interval weight vector W_k is obtained by the following problem.

$$\begin{aligned}
\min \; & c_k - \varepsilon \sum_i \overline{w}_{ki}, \\
s.t. \; & \sum_i \underline{w}_{ki} \geq \overline{s}, \\
& \sum_{j \neq i} \overline{w}_{kj} + \overline{w}_{ki} \leq 1, \; \sum_{j \neq i} \overline{w}_{kj} + \underline{w}_{ki} \geq 1, \forall i, \\
& -c_k \leq a_{kij} \overline{w}_{kj} - \underline{w}_{ki}, \; -c_k \leq \overline{w}_{ki} - a_{kij} \underline{w}_{kj}, \; \varepsilon \leq \underline{w}_{ki}, \forall i, j, i > j,
\end{aligned} \tag{9}$$

where c_k is added to the variables and a replacement for 0 in (6).

The inclusion relation (5) is based on the idea that the given comparison matrix is a part of the decision maker's judgment. It considers an independent comparison matrix but not the other comparison matrices which the decision maker gives. In (9), the inclusion relation for the comparison matrix is extended by the influence of the other comparison matrices. Because of the first constraint of the others' influence, the condition for the weight vector is extended from strictly including the comparisons to roughly including them by the third kinds of constraints. When the degree of consistency of A_k is much lower than the highest one among the comparison matrices by the decision maker: $s_k^* \leq \overline{s}$, it needs the more relaxation with the greater c_k. Contrary, A_k is the most consistent comparison matrix among those by the decision maker, there is no relaxation $c_k = 0$.

3   Group Assessment

In the former section, there are multiple items $k = 1, \ldots, m$. In this section, assume multiple decision makers, $d = 1, \ldots, p$, in addition. Each decision maker d gives the comparison matrix of each item k: A_k^d as in (3). If all the decision makers give the judgments of all the items, there are mp comparison matrices as shown in Table 1, where W_k^d is the weight vector corresponding to A_k^d. The goal of this study is to obtain a group assessment of the target item $k' \in \{1, \ldots, m\}$ as interval weight vector of n viewpoints: $W_{k'} = (W_{k'1}, \ldots, W_{k'n})^t$, where each element is $W_{k'i} = [\underline{w}_{k'i}, \overline{w}_{k'i}]$ and normalized by (1).

As for the target item k', there are p comparison matrices by p decision makers so that there could be p weight vectors of the target item at the row in

Table 1. A group of comparison matrices of multiple items

item	decision maker comparison matrix → weight vector					goal
	1	$\cdots$	d	$\cdots$	p	
target k'	$A_{k'}^1 \to W_{k'}^1$	$\cdots$	$A_{k'}^d \to W_{k'}^d$	$\cdots$	$A_{k'}^p \to W_{k'}^p$	$\to W_{k'}$
1	$A_1^1 \to W_1^1$	$\cdots$	$A_1^d \to W_1^d$	$\cdots$	$A_1^p \to W_1^p$	
$\vdots$	$\vdots$	$\vdots$	$\vdots$	$\vdots$	$\vdots$	
m	$A_m^1 \to W_m^1$	$\cdots$	$A_m^d \to W_m^d$	$\cdots$	$A_m^p \to W_m^p$	

Table 1. In this sense, it is reasonable that the weight vector of the target item is the core of those by all the decision makers. Hence, the core relation is denoted as follows.

$$W_{k'} \subseteq W_{k'}^d, \forall d, \leftrightarrow \underline{w}_{k'i}^d \leq \underline{w}_{k'i}, \overline{w}_{k'i} \leq \overline{w}_{k'i}^d, \forall i, d,$$
$$\leftrightarrow 0 \leq \underline{w}_{k'i} - \underline{w}_{k'i}^d, 0 \leq \overline{w}_{k'i}^d - \overline{w}_{k'i}, \forall i, d, \tag{10}$$

where the core weight vector should be completely included in each weight vector.

We remind that there could be various $W_{k'}^d$ since the given $A_{k'}^p$ represents only a part of decision maker's thinking. Furthermore, the decision maker just gives vague intuitive judgments. Because of this kind of incompleteness of a comparison matrix, the comparison matrices of the other items are taken into consideration as well as that of the target one. Then, both $W_{k'i}$ and $W_{k'i}^d, \forall d$ are obtained simultaneously from $A_{k'}^d, \forall d$.

Our approach is not a stepwise one such as first obtaining $W_{k'}^d$ from $A_{k'}^d$ for each d and then deriving $W_{k'}$ from them. We combine (9) for all the decision makers and add the core condition (10). Then, the problem to obtain the minimum relaxations of all the decision makers, $c_{k'}^d, \forall d$, is formulated as follows.

$$\begin{aligned}
\min \ & c_{k'}^d - \varepsilon \sum_i \overline{w}_{k'i}, \\
s.t. \ & \sum_{j \neq i} \underline{w}_{k'j} + \overline{w}_{k'i} \leq 1, \ \sum_{j \neq i} \overline{w}_{k'j} + \underline{w}_{k'i} \geq 1, \forall i, \\
& 0 \leq \underline{w}_{k'i} - \underline{w}_{k'i}^d, 0 \leq \overline{w}_{k'i}^d - \overline{w}_{k'i}, \forall d, i, \\
& \sum_i \underline{w}_{k'i}^d \geq \overline{s}^d, \forall d, \\
& \sum_{j \neq i} \underline{w}_{k'j}^d + \overline{w}_{k'i}^d \leq 1, \ \sum_{j \neq i} \overline{w}_{k'j}^d + \underline{w}_{k'i}^d \geq 1, \forall d, i, \\
& -c_{k'}^d \leq a_{k'ij}^d \overline{w}_{k'j}^d - \underline{w}_{k'i}^d, \ -c_{k'}^d \leq \overline{w}_{k'i}^d - a_{k'ij}^d \underline{w}_{k'j}^d, \ \varepsilon \leq \underline{w}_{k'i}^d, \forall d, i, j, i > j,
\end{aligned}$$
$$\tag{11}$$

where the variables are the interval weights $[\underline{w}_{k'i}^d, \overline{w}_{k'i}^d], [\underline{w}_{k'i}, \overline{w}_{k'i}], \forall d, i$ and the relaxation of inclusion relation $c_k^d, \forall d$. The first and second kinds of constraints are the normalization and core conditions, respectively. The optimal solution of the weight vector of a decision maker may not be the most precise to his/her comparison matrix but it is certain as required and contents with the others because of the variables $W_{k'}^{d'}$ and $W_{k'}$. Denote the optimal solutions of the relaxation for decision maker d by (11) as $c_{k'}^{d*}$. It is a minimum relaxation of the inclusion relation (5) needed for the highest degree of consistency for the decision maker (8) and the existence of the core of all the weight vectors (10).

4   Trade-Off Between Consistency of Judgment and Inclusion Relation

In (11), the degree of consistency for each comparison matrix is assumed to be the highest among those of all the comparison matrices by decision maker d. The highest degree is replaced into his/her fuzzy consistency degree, $\tilde{s}^d$, whose membership function in Fig. 1(a) is as follows.

$$
\mu^d(s) = \begin{cases} 0, & s \le \underline{s}^d, \\ (s - \underline{s}^d)/(\overline{s}^d - \underline{s}^d), & \underline{s}^d \le s \le \overline{s}^d, \\ 1, & \overline{s}^d \le s, \end{cases} \tag{12}
$$

where $\underline{s}^d$ and $\overline{s}^d$ are the highest and lowest degrees of consistency of each decision maker d by (8). In order for the consistency degree to be more than α, the certain weight in the interval weight vector is $s = \sum_i \underline{w}^d_{ki} \ge \alpha(\overline{s}^d - \underline{s}^d) + \underline{s}^d$.

In order to ensure the highest degree of the consistency $\overline{s}^d$, the inclusion relation should be relaxed by $c^{d*}_{k'}$ by (11). In the case that the degree is less than the highest, the relaxation can be smaller. We define fuzzy inclusion degree for target item k' of decision maker d, whose membership function in Fig. 1(b) as follows.

$$
\mu^d_{k'}(c) = \begin{cases} 0, & c \le -c^{d*}_{k'}, \\ 1/c, & -c^{d*}_{k'} \le c \le 0 \\ 1, & 0 \le c, \end{cases} \tag{13}
$$

which can be considered as fuzzy number $\tilde{0}$. In order for the inclusion degree to be more than α, 0 is replaced into $c \le c^{d*}_{k'}(\alpha - 1)$. There is a trade-off between the degrees of the consistency and the inclusion relation. Specifically, the higher consistency of the comparison matrix makes it the more difficult for the weight vector to satisfy the inclusion relation (5).

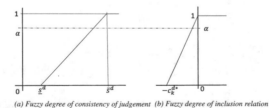

(a) Fuzzy degree of consistency of judgement (b) Fuzzy degree of inclusion relation

Fig. 1. Fuzzy degrees of consistency and inclusion relation

By maximizing both degrees of consistency and inclusion relation, the problem to obtain the weight vector of target item k' is formulated as follows.

$$\max\ \alpha + \varepsilon(\overline{w}_{k'i} - \underline{w}_{k'i})$$

$$s.t.\ \sum_{j\neq i} \underline{w}_{k'j} + \overline{w}_{k'i} \leq 1,\ \sum_{j\neq i} \overline{w}_{k'j} + \underline{w}_{k'i} \geq 1, \forall i,$$

$$0 \leq \underline{w}_{k'i} - \underline{w}^d_{k'i}, 0 \leq \overline{w}^d_{k'i} - \overline{w}_{k'i}, \forall d,$$

$$\sum_i \underline{w}^d_{k'i} \geq \alpha(\overline{s}^d - \underline{s}^d) + \underline{s}^d, \forall d,$$

$$\sum_{j\neq i} \underline{w}^d_{k'j} + \overline{w}^d_{k'i} \leq 1,\ \sum_{j\neq i} \overline{w}^d_{k'j} + \underline{w}^d_{k'i} \geq 1, \forall d, i,$$

$$c^{d*}_{k'}(\alpha - 1) \leq a^d_{k'ij}\overline{w}^d_{k'j} - \underline{w}^d_{k'i},\ c^{d*}_{k'}(\alpha - 1) \leq \overline{w}^d_{k'i} - a^d_{k'ij}\underline{w}^d_{k'j},\ \varepsilon \leq \underline{w}^d_{k'i}, \forall d, i, j, i > j,$$

$$(14)$$

where the variables are α and $\boldsymbol{W}^d_{k'}, \forall d$ and $\boldsymbol{W}_{k'}$.

There are two kinds of incompleteness in the given data in Table 1. One is that a decision maker d does not give some comparison matrices on some items, although it is that a group of comparison matrices is incomplete. We remove those who do not give the comparison matrices of target item k' from the group. The other is that a decision maker d does not give some comparisons on item k, which is that the comparison matrix is incomplete. Moreover, the given comparisons are often inconsistent. We focus on the inclusion relation of the given comparisons. In this way, the proposed model (14) is solvable and derives the weight vector of the target, even when Table 1 is incomplete in both senses. Therefore, the decision makers are not forced to give all the comparisons on all the items but can give some with confidence.

5   Numerical Example

Three decision makers: DM ($d = 1, 2, 3$) gave the judgments of five items ($k = 1, \ldots, 5$) on five viewpoints ($i, j = 1, \ldots, 5$). The decision maker compared two viewpoints of an item and gave the comparisons on which viewpoint and how many times the item is better than on the other viewpoint. If all the decision makers compared all the pairs of viewpoints of all the items, we could have 15 pairwise comparison matrices each of which consists of 10 comparisons. However, some are missing in our example such as the given incomplete data in Table 2. DM 1 and DM 2 did not give the comparison matrices of items 1 and 3, respectively, and as for item 5, DM 3 did not give 2 comparisons of viewpoints 1 and 3, and viewpoints 3 and 4.

Table 2. Given comparison matrices, and highest and lowest degrees of consistency

	DM 1	DM 2	DM 3
item 1	$-$	A^2_1	A^3_1
item 2	A^1_2	A^2_2	A^3_2
item 3	A^1_3	$-$	A^3_3
item 4	A^1_4	A^2_4	A^3_4
item 5	A^1_5	A^2_5	A^3_5
$\overline{s}$	0.977	0.907	0.958
$\underline{s}$	0.902	0.829	0.850

$$\text{where } A^3_5 = \begin{bmatrix} 1 & 1/4 & - & 4 & 1/2 \\ - & 1 & 6 & 5 & 3 \\ - & - & 1 & - & 1/5 \\ - & - & - & 1 & 1/4 \\ - & - & - & - & 1 \end{bmatrix}$$

The goal is to obtain the weight vectors of five items: $W_1, \ldots, W_5$ from the comparison matrices in Table 2. The consistency degrees $s^d_k, \forall k, d$ are obtained by (6) for 13 comparison matrices. Although DM 3 gives incomplete A^3_5, where two comparisons are missing, the problem (6) is solvable by removing their inclusion relation constraints (5). For each decision maker, the highest and lowest consistency degrees by (8) are shown at the bottom two rows of Table 2. Then, the fuzzy consistency degree of each decision maker is defined by (12). We find that DM 1, DM 3, and DM 2, in order, tend to give consistent judgments. The higher consistency degree for the decision maker is more preferable because of a more certain weight vector, although there is a trade-off between the consistency and the inclusion relation (5). Hence, the minimum relaxation for the inclusion relation in the case of the highest consistency is obtained by (11). We have $c_1^{2*} = 0.166$ for DM 2 and $c_1^{3*} = 0.075$ for DM 3. It is reasonable that DM 2 of the lower consistency degree is required the more relaxation for a consensus.

First, we set the target item as $k' = 1$. Its weight vector is obtained from its judgments by DM 2 and DM 3 and those of the other items by them at the right two columns of Table 2. By (14), the weight vector of item 1 is obtained as $W_1 = ([0.125, 0.130], [0.085, 0.091], [0.209, 0.215], [0.355, 0.361], [0.220, 0.226])$, whose lower bounds are illustrated in Fig. 2. It represents the group assessment of item 1 and is a better consensus in the sense that the consistency of each decision maker's judgments is considered and there is no additional step to aggregate their judgments. These are the advantages over the other methods which need an aggregation step and a threshold for acceptable consistency level. We find that item 1 is more superior from viewpoint 4 than the other viewpoints.

Figure 2 also shows the obtained weight vectors of the others in the same way by setting them as a target one by one. In the cases of items 2, 4, and 5, their weight vectors are obtained from the judgments of three decision makers. Comparing these five weight vectors, we easily find out the character of the items. Item 4 is balanced on all the viewpoints and item 3 is specialized in viewpoint 2. They are not only from the judgments of the target items but also those of the other items from the aspect of the consistency of the decision makers' judgments. Although the weights of item 1 are intervals, those of the other items are crisp. This is because of the core condition (10) and this often happens that the decision makers judgments are diverse. The core condition could be relaxed in a similar manner as the inclusion relation by introducing fuzzy inclusion degrees. It will be our future work.

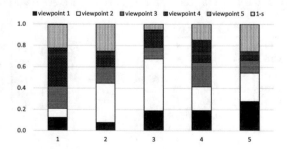

Fig. 2. Weights of 5 items on five viewpoints

6   Conclusion

This paper proposed the approach to obtain the weight vector of a target item from a group of judgments of multiple items by multiple decision makers. The weight vector is the core of those by all the decision makers. It is obtained mostly from the comparison matrices of the item and in addition from those of the other items to take their character into consideration. We introduced two kinds of fuzzy degrees of consistency of the comparison matrix and inclusion relation between the comparison matrix and weight vector. There is a trade-off between them such that the higher the consistency degree, the lower the inclusion degree. The former fuzzy consistency degree is defined with the highest and lowest degrees of the consistency among those of all the items of which the decision maker gives the judgments. The latter fuzzy inclusion degree is denoted as fuzzy number 0 and defined with the minimum relaxation to ensure the highest consistency degree. The obtained weight vector represents a group assessment. It could be the advantage of the proposed approach to consider the given judgments and their consistency simultaneously since it is not easy for us to take two different aspects of the comparison matrix into consideration. Another advantage is the reliability of the given judgments because of the easiness for the decision makers. They need to give neither all the comparisons in each comparison matrix nor all the comparison matrices of the items but can give some of them with confidence. The proposed method can be applied without estimating the missing data.

References

1. Haker, P.T.: Incomplete pairwise comparisons in the analytic hierarchy process. Math. Model. **9**(11), 837–848 (1987)
2. Bozóki, S., Fülöp, J.: Efficient weight vectors from pairwise comparison matrices. Eur. J. Oper. Res. **264**(2), 419–427 (2018)
3. Saaty, T.L.: The Analytic Hierarchy Process. McGraw-Hill, New York (1980)
4. Jensen, R.E.: An alternative scaling method for priorities in hierarchical structures. J. Math. Psychol. **28**(3), 317–332 (1984)
5. Saaty, T.L., Vargas, L.G.: Comparison of eigenvalue, logarithmic least squares and least squares methods in estimating ratios. Math. Model. **5**(5), 309–324 (1984)
6. Ishizaka, A., Siraj, S.: Are multi-criteria decision-making tools useful? An experimental comparative study of three methods. Eur. J. Oper. Res. **264**(2), 462–471 (2018)
7. Sugihara, K., Tanaka, H.: Interval evaluations in the analytic hierarchy process by possibilistic analysis. Comput. Intell. **17**(3), 567–579 (2001)
8. Entani, T., Sugihara, K.: Uncertainty index based interval assignment by interval AHP. Eur. J. of Oper. Res. **219**(2), 379–385 (2012)
9. Inuiguchi, M., Innan, S.: Improving interval weight estimations in interval AHP by relaxations. J. Adv. Comput. Intell. Intell. Inform. **21**, 1135–1143 (2017)
10. Jalao, E.R., Wu, T., Shunk, D.: An intelligent decomposition of pairwise comparison matrices for large-scale decisions. Eur. J. Oper. Res. **238**(1), 270–280 (2014)
11. Brunelli, M., Fedrizzi, M.: Axiomatic properties of inconsistency indices for pairwise comparisons. J. Oper. Res. Soc. **66**, 1–15 (2015)

12. Ramík, J., Korviny, P.: Inconsistency of pair-wise comparison matrix with fuzzy elements based on geometric mean. Fuzzy Sets Syst. **161**(11), 1604–1613 (2010)
13. Kubler, S., Derigent, W., Voisin, A., Robert, J., Traon, Y.L., Viedma, E.H.: Measuring inconsistency and deriving priorities from fuzzy pairwise comparison matrices using the knowledge-based consistency index. Knowl.-Based Syst. **162**, 147–160 (2018). Special Issue on Intelligent Decision-Making and Consensus Under Uncertainty in Inconsistent and Dynamic Environments
14. Siraj, S., Mikhailov, L., Keane, J.A.: Contribution of individual judgments toward inconsistency in pairwise comparisons. Eur. J. Oper. Res. **242**(2), 557–567 (2015)
15. Fedrizzi, M., Giove, S.: Incomplete pairwise comparison and consistency optimization. Eur. J. Oper. Res. **183**(1), 303–313 (2007)
16. Forman, E., Peniwati, K.: Aggregating individual judgments and priorities with the Analytic hierarchy process. Eur. J. Oper. Res. **108**(1), 165–169 (1998)
17. Wan, S., Wang, F., Dong, J.: Additive consistent interval-valued atanassov intuitionistic fuzzy preference relation and likelihood comparison algorithm based group decision making. Eur. J. Oper. Res. **263**(2), 571–582 (2017)
18. Wan, S., Wang, F., Dong, J.: A group decision making method with interval valued fuzzy preference relations based on the geometric consistency. Inf. Fusion **40**, 87–100 (2018)
19. Brunelli, M., Fedrizzi, M.: Boundary properties of the inconsistency of pairwise comparisons in group decisionsy. Eur. J. Oper. Res. **240**, 765–773 (2015)
20. Entani, T., Inuiguchi, M.: Pairwise comparison based interval analysis for group decision aiding with multiple criteria. Fuzzy Sets Syst. **274**(1), 79–96 (2015)

Interval Weight Estimation Methods Satisfying Desirable Properties in Interval AHP

Masahiro Inuiguchi[✉]

Graduate School of Engineering Science, Osaka University,
1-3 Machikaneyama-cho, Toyonaka, Osaka, Japan
inuiguti@sys.es.osaka-u.ac.jp

Abstract. From the viewpoint that the inconsistency of a pairwise comparison matrix comes from the vagueness of decision maker's evaluation, interval AHP representing the vagueness as interval weights was proposed. It has been shown that the interval weight vector estimated by the conventional estimation method does not reflect the vagueness well. Several alternative estimation methods have been proposed. However, those methods do not always satisfy the following desirable properties while the conventional method does: (0) the method is parameter-free, (i) the given pairwise comparison matrix is realizable under the estimated interval weight vector, (ii) the unique accurate crisp weight vector is estimated under a consistent crisp pairwise comparison matrix, (iii) the estimated interval weight vector satisfies the normality condition and (iv) a proper interval weight vector is estimated from a consistent interval pairwise comparison matrix. In this paper, we show that the previous alternative estimation methods do not satisfy some of those properties and propose novel interval weight estimation methods satisfying those five requirements. The non-uniqueness of optimal normalized interval weight vector is also taken care in this paper.

1 Introduction

In multiple criteria decision making [1], the aggregation of criterion-wise evaluations is necessary to obtain the overall evaluation. The weighted average is used most frequently and is the simplest model for the aggregation. When the weighted average model is adopted, the estimation of proper weights representing the decision maker's preference becomes one of main topics. There are many methods for weight estimations [1]. The types of preference information obtained from the decision maker is different among the methods. Analytic Hierarchy Process (AHP) [4] is one of the most used methods for weight estimation because of its simplicity of the process. It assumes a pairwise comparison matrix (PCM) whose (i, j)-component shows the relative importance of the i-th criterion to the j-th criterion. We focus on weight estimation from a pairwise comparison matrix. Given a PCM, weights are estimated by the maximum eigenvalue method or the geometric mean method [5].

© Springer Nature Switzerland AG 2019
R. B. Kearfott et al. (Eds.): IFSA 2019/NAFIPS 2019, AISC 1000, pp. 73–84, 2019.
https://doi.org/10.1007/978-3-030-21920-8_7

A PCM obtained from the decision maker is often inconsistent. Then to measure the inconsistency, we use consistency index (C.I.) and reject the PCM whose C.I. is more than a certain threshold. If the pairwise comparison matrix is accepted, priority weights with minimum errors has been calculated by the maximum eigenvalue method, the geometric mean method or other methods, and used for the decision making without deep consideration of the errors existing in the given PCM.

On the other hand, we may assume that the inconsistency comes from the vagueness of human evaluation. Namely, human perception of the weights would be imprecise so that human can accept any priority weights in a certain range. From this point of view, interval AHP [7] is proposed to estimate the range of admissible priority weights as intervals. Moreover, this method is extended to the case where the interval pairwise comparison matrix is given by the decision maker (see [6]). Those methods are advantageous in keeping the source of inconsistency as interval weights during the decision analysis so that the confidence of preference evaluation for each pair of alternatives is analyzed while the conventional AHP analysis. Some other approaches [9, 10] have been proposed in the case where an interval PCM is given. Similar approaches [11] have been proposed when an intuitionistic fuzzy preference relation is given, where an intuitionistic fuzzy preference relation corresponds to an interval PCM.

The author and his colleagues showed that the interval weights estimated by the conventional method in interval AHP [7] under a crisp PCM do not reflect the vagueness of the decision maker. To improve the refection of the vagueness of human evaluation in the estimated interval weights, β-relaxation method [3], γ-relaxation method [3], maximizing minimum range (MMR) method [3] as well as maximizing minimum range method with weights λ (MMR$^\lambda$ method) [3] and error-based methods [8] have been proposed. We showed that the error-based methods satisfied better the following desirable properties: (i) the given pairwise comparison matrix is realizable under the estimated interval weight vector, (ii) the unique accurate crisp weight vector is estimated under a consistent crisp pairwise comparison matrix, (iii) the estimated interval weight vector satisfies the normality condition and (iv') the quality of estimated interval priority weights should be improved as the number of observations of the PCM increases. To treat property (iv'), we extend each of the previous interval weight estimation methods under a crisp PCM to the corresponding interval weight estimation methods under an interval PCM whose (i, j) component is an interval defined by the minimum and maximum among the (i, j) components of the observed PCMs.

In this paper, we consider more desirable properties: (0) the method is parameter-free, (i) the given pairwise comparison matrix is realizable under the estimated interval weight vector, (ii) the unique accurate crisp weight vector is estimated under a consistent crisp pairwise comparison matrix, (iii) the estimated interval weight vector satisfies the normality condition and (iv) a proper interval weight vector is estimated from a consistent interval pairwise comparison matrix. Namely, property (0) is added and property (iv) is a stronger requirement than property (iv'). The consistency of interval pairwise comparison matrix implies that the given interval pairwise comparison matrix is the same as the interval pairwise comparison matrix obtained from a normalized interval weight vector. We show that the previous parameter-free estimation methods improving the conventional estimation method in interval AHP do not satisfy (iv). We propose modifications of the previous methods to satisfy property (iv). Moreover, as the author [2] revealed recently, the normalized interval weight vector corresponding to a consistent interval pairwise comparison matrix is not unique. The non-uniqueness of the estimated normalized interval weight vector is also taken care properly.

This paper is organized as follows. In next section, we briefly review previous estimation methods for interval weights and preference evaluation under interval weights. In Sect. 3, we describe the desirable properties of interval weight estimation method and the unsatisfaction of property (iv) by some of previous methods. The revised methods satisfying all desirable properties are given in Sect. 4. In Sect. 5, some concluding remarks are given.

2   Interval AHP

2.1   Estimation by Minimizing Widths of Interval Weights

In the classical AHP, a weight vector $w = (w_1, w_2, \ldots, w_n)^{\mathrm{T}}$ for criteria or for alternatives is estimated from a PCM A.

$$A = \begin{pmatrix} 1 & \cdots & a_{1n} \\ \vdots & a_{ij} & \vdots \\ a_{n1} & \cdots & 1 \end{pmatrix}, \tag{1}$$

where, $a_{ij} = 1/a_{ji}$, $a_{ii} = 1$, $i, j \in N - \{1, 2, \ldots, n\}$. The (i, j) component a_{ij} represents the relative importance of the i-th criterion/alternative to the j-th criterion/alternative. In other words, a_{ij} shows the evaluation of w_i/w_j. However, the human evaluation is not very precise, so that the given A does not always satisfy the transitivity $a_{ij} \cdot a_{jk} = a_{ik}$, $i, j, k \in N = \{1, 2, \ldots, n\}$. In the classical AHP, w_i, $i \in N$ are estimated so as to minimize the differences between a_{ij} and w_i/w_j, $i, j \in N$.

On the other hand, in interval AHP, the inconsistency of A is assumed to come from the vagueness of human perception of a priority weight so that the weight would be represented by an interval rather than by a real number. Therefore, interval weights $W_i = [w_i^{\mathrm{L}}, w_i^{\mathrm{R}}]$, $i \in N$ are estimated from the given PCM. In the conventional estimation method, the interval weights $W_i = [w_i^{\mathrm{L}}, w_i^{\mathrm{R}}]$, $i \in N$ are estimated as an optimal solution to the following linear programming (LP) problem (see [7]):

$$\text{minimize} \sum_{i \in N} (w_i^{\mathrm{R}} - w_i^{\mathrm{L}}), \text{ sub. to } a_{ij} w_j^{\mathrm{L}} \leq w_i^{\mathrm{R}}, \ a_{ij} w_j^{\mathrm{R}} \geq w_i^{\mathrm{L}}, \ i, j \in N \ (i < j),$$

$$\sum_{i \in N \backslash j} w_i^{\mathrm{R}} + w_j^{\mathrm{L}} \geq 1, \ \sum_{i \in N \backslash j} w_i^{\mathrm{L}} + w_j^{\mathrm{R}} \leq 1, \ j \in N,$$

$$w_i^{\mathrm{R}} \geq w_i^{\mathrm{L}} \geq \varepsilon, \ i \in N, \tag{2}$$

where ε is a very small positive number introduced to treat positivity of interval weights. The first line of Problem (2) shows the evaluations function, i.e., minimization of the sum of widths of interval priority weights. The constraints on the second line represent that the given pairwise comparison matrix A is realizable from $W_i = [w_i^{\mathrm{L}}, w_i^{\mathrm{R}}]$, $i \in N$. The constraints on the third line are the normalization condition of $W_i = [w_i^{\mathrm{L}}, w_i^{\mathrm{R}}]$, $i \in N$. The constraint on the forth line shows the natural condition and the positivity of intervals $W_i = [w_i^{\mathrm{L}}, w_i^{\mathrm{R}}]$, $i \in N$.

It was demonstrated by numerical experiments that the estimated interval priority weights do not reflect well the vagueness of human evaluation (see [3]). Then β-relaxation method [3], γ-relaxation method [3], maximizing minimum range (MMR) method [3] and MMR$^\lambda$ method [3] have been proposed. The performances of β-relaxation method and γ-relaxation method depend on the selections of parameters β and γ, and the parameter tuning may require a certain effort. Then MMR and MMR$^\lambda$ methods are proposed as parameter-free estimation methods. We describe the procedure of MMR method because the proposed approach is a modification of MMR method.

MMR method [3] is based on the minimal range concept. When we obtain a set of realizations of a variable z as $\{z_1, z_2, \ldots, z_q\}$, we know that the range of z includes $[\min_{i=1,2,\ldots,q} z_i, \max_{i=1,2,\ldots,q} z_i]$. Therefore, we call this range as minimal range of z. In the case when a crisp PCM $A = (a_{ij})$ is given, we estimate the minimal range of the priority weight w_k as $[\min_{i \in N \setminus k} a_{ki} w_i, \max_{i \in N \setminus k} a_{ki} w_i]$. However, w_i, $i \in N \setminus k$ are not given. Therefore, we estimate first w_i, $i \in N \setminus k$ and then the minimal range of w_k. For the estimation of w_i, $i \in N \setminus k$, we apply the conventional interval weight estimation method. As we obtain $W_i = [w_i^{\mathrm{L}}, w_i^{\mathrm{R}}]$, $i \in N \setminus k$, the minimal range of w_k can be estimated by $[\min_{i \in N \setminus k} a_{ki} w_i^{\mathrm{L}}, \max_{i \in N \setminus k} a_{ki} w_i^{\mathrm{R}}]$. However, we may have multiple $W_i = [w_i^{\mathrm{L}}, w_i^{\mathrm{R}}]$, $i \in N \setminus k$ for the optimal estimation, we estimate W_k including all possible variations. Accordingly, first we estimate W_i, $i \in N \setminus k$ by solving Problem (2) replacing $\sum_{i \in N}(w_i^{\mathrm{R}} - w_i^{\mathrm{L}})$ with $\sum_{i \in N \setminus k}(w_i^{\mathrm{R}} - w_i^{\mathrm{L}})$. Let $\hat{d}_{\bar{k}}$ be the optimal value to this problem. Then we solve the following two kinds of LP problems for each $k \in N$:

$$\text{maximize } w_k^{\mathrm{R}}, \text{ sub. to } a_{ij} w_j^{\mathrm{L}} \leq w_i^{\mathrm{R}}, \ a_{ij} w_j^{\mathrm{R}} \geq w_i^{\mathrm{L}}, \ i,j \in N \ (i < j),$$
$$\sum_{i \in N \setminus j} w_i^{\mathrm{R}} + w_j^{\mathrm{L}} \geq 1, \ \sum_{i \in N \setminus j} w_i^{\mathrm{L}} + w_j^{\mathrm{R}} \leq 1, \ j \in N, \tag{3}$$
$$\sum_{i \in N \setminus k} (w_i^{\mathrm{R}} - w_i^{\mathrm{L}}) = \hat{d}_{\bar{k}}, \ w_i^{\mathrm{R}} \geq w_i^{\mathrm{L}} \geq \varepsilon, \ i \in N,$$

$$\text{minimize } w_k^{\mathrm{L}}, \text{ sub. to } a_{ij} w_j^{\mathrm{L}} \leq w_i^{\mathrm{R}}, \ a_{ij} w_j^{\mathrm{R}} \geq w_i^{\mathrm{L}}, \ i,j \in N \ (i < j),$$
$$\sum_{i \in N \setminus j} w_i^{\mathrm{R}} + w_j^{\mathrm{L}} \geq 1, \ \sum_{i \in N \setminus j} w_i^{\mathrm{L}} + w_j^{\mathrm{R}} \leq 1, \ j \in N, \tag{4}$$
$$\sum_{i \in N \setminus k} (w_i^{\mathrm{R}} - w_i^{\mathrm{L}}) = \hat{d}_{\bar{k}}, \ w_i^{\mathrm{R}} \geq w_i^{\mathrm{L}} \geq \varepsilon, \ i \in N,$$

We define $\hat{w}_i^{\mathrm{L}}(k)$ and $\hat{w}_i^{\mathrm{R}}(k)$, $i \in N$ by w_i^{L} and w_i^{R}, $i \in N$ at an optimal solution to Problem (3), respectively, and $\breve{w}_i^{\mathrm{L}}(k)$ and $\breve{w}_i^{\mathrm{R}}(k)$, $i \in N$ by w_i^{L} and w_i^{R}, $i \in N$

at an optimal solution to Problem (4), respectively. Then we obtain estimated interval priority weights $\tilde{W}_i = [\tilde{w}_i^{\mathrm{L}}, \tilde{w}_i^{\mathrm{R}}]$, $i \in N$ defined by

$$\tilde{w}_j^{\mathrm{R}} = \max_{k \in N} \max \left(\hat{w}_j^{\mathrm{R}}(k), \check{w}_j^{\mathrm{R}}(k) \right), \quad \tilde{w}_j^{\mathrm{L}} = \min_{k \in N} \min \left(\hat{w}_j^{\mathrm{L}}(k), \check{w}_j^{\mathrm{L}}(k) \right). \qquad (5)$$

Interval weights are estimated also in the case an interval PCM is given [6]. In the case of interval PCM $A = ([a_{ij}^{\mathrm{L}}, a_{ij}^{\mathrm{R}}])$, we assume $a_{ij}^{\mathrm{L}} = 1/a_{ji}^{\mathrm{R}}$ for any $i, j \in N$ ($i \neq j$). Sugihara et al. [6] proposed two method, i.e., lower approximation problem and upper approximation problem assuming that the given interval components of the interval PCM show the approximate fluctuation range of the components. However, in this paper, we assume that the interval components of interval PCM show minimally possible range of the components, we can utilize upper approximation problem only. Let $[a_{ij}^{\mathrm{L}}, a_{ij}^{\mathrm{R}}]$ be the (i, j)-component of the given interval PCM. Then upper approximation problem is formulated as

$$\text{minimize} \sum_{i \in N} (w_i^{\mathrm{R}} - w_i^{\mathrm{L}}), \text{ sub. to } a_{ij}^{\mathrm{R}} w_j^{\mathrm{L}} \leq w_i^{\mathrm{R}}, \ a_{ij}^{\mathrm{L}} w_j^{\mathrm{R}} \geq w_i^{\mathrm{L}}, \ i, j \in N \ (i < j),$$

$$\sum_{i \in N \backslash j} w_i^{\mathrm{R}} + w_j^{\mathrm{L}} \geq 1, \ \sum_{i \in N \backslash j} w_i^{\mathrm{L}} + w_j^{\mathrm{R}} \leq 1, \ j \in N,$$

$$w_i^{\mathrm{R}} \geq w_i^{\mathrm{L}} \geq \varepsilon, \ i \in N. \qquad (6)$$

When $a_{ij}^{\mathrm{L}} = a_{ij}^{\mathrm{R}}$, $i, j \in N$, Problem (6) is reduced to Problem (2). Namely, Problem (6) is a generalization of Problem (2).

MMR method can also be generalized to the case where an interval PCM is given. To this end, we replace the constraints shown on the second line of Problem (2) with the constraints shown on the second line of Problem (6) in all linear programming problems appeared in the procedure of MMR method. When $a_{ij}^{\mathrm{L}} = a_{ij}^{\mathrm{R}}$, $i, j \in N$ ($i \neq j$), the estimation method under an interval PCM is reduced to that under a PCM. Therefore, we describe interval estimation methods only under a given interval PCM.

2.2 Estimation by Minimizing Deviation from the Given PCM

In the estimation methods described in the previous subsection, the sum of widths are minimized. This implies less imprecise interval weights are originally preferred. However, because given preference information is a PCM, it may be more natural to estimate the interval weights based on the deviation from the given PCM. From this point of view, Torisu and Inuiguchi [8] modified previously proposed estimation methods. In this paper, we briefly introduce two interval weight estimation methods based on the deviation from the given PCM corresponding to the conventional estimation method and MMR method. Corresponding to Problem (2), we obtain the following interval weight estimation methods based on the deviation from the given interval PCM:

$$\text{minimize} \sum_{i \in N} \sum_{j \in N \backslash i} \delta_{ij}, \text{ sub. to } \sqrt{a_{ij}^{\mathrm{R}}} w_j^{\mathrm{L}} + \delta_{ij} = \sqrt{a_{ji}^{\mathrm{L}}} w_i^{\mathrm{R}}, \ i, j \in N (i \neq j),$$

$$\sum_{i \in N \backslash j} w_i^{\mathrm{R}} + w_j^{\mathrm{L}} \geq 1, \ \sum_{i \in N \backslash j} w_i^{\mathrm{L}} + w_j^{\mathrm{R}} \leq 1, \ j \in N, \qquad (7)$$

$$w_i^{\mathrm{R}} \geq w_i^{\mathrm{L}} \geq \varepsilon, \ \delta_{ij} \geq 0, \ i, j \in N (i \neq j),$$

where δ_{ij} shows a deviation between a_{ij} and $w_i^{\mathrm{R}}/w_j^{\mathrm{L}}$. Because $\delta_{ij} \geq 0$ and $\delta_{ji} \geq 0$, $\sqrt{a_{ij}^{\mathrm{R}} w_j^{\mathrm{L}}} + \delta_{ij} = \sqrt{a_{ji}^{\mathrm{L}} w_i^{\mathrm{R}}}$ and $\sqrt{a_{ji}^{\mathrm{R}} w_i^{\mathrm{L}}} + \delta_{ji} = \sqrt{a_{ij}^{\mathrm{R}} w_i^{\mathrm{L}}}$ are equivalent to $a_{ij}^{\mathrm{R}} w_j^{\mathrm{L}} \leq w_i^{\mathrm{R}}$ and $a_{ij}^{\mathrm{L}} w_j^{\mathrm{R}} \geq w_i^{\mathrm{L}}$, respectively.

The idea of minimizing the deviation from the given PCM is applied also to MMR method. It is composed of the following three steps: first we obtain the minimum sum of widths of interval priority weights W_i, $i \in N \setminus k$ by solving Problem (2) replacing $\sum_{i \in N} \sum_{j \in N \setminus i} \delta_{ij}$ with $\sum_{i \in N \setminus k} \sum_{j \in N \setminus \{i,k\}} \delta_{ij}$. Let $\hat{\Delta}_k$ be the optimal value to this problem. Then we solve the following two kinds of LP problems for each $(k,l) \in N \times (N \setminus k)$:

$$
\begin{aligned}
&\text{maximize } a_{kl}^{\mathrm{R}} w_l^{\mathrm{L}}, \text{ sub. to } \sum_{i \in N \setminus k} \sum_{j \in N \setminus \{i,k\}} \delta_{ij} \leq \hat{\Delta}_k, \\
&\qquad \sqrt{a_{ji}^{\mathrm{R}} w_i^{\mathrm{L}}} + \delta_{ij} = \sqrt{a_{ij}^{\mathrm{L}} w_j^{\mathrm{R}}}, \ i,j \in N \ (i \neq j), \\
&\qquad \sum_{i \in N \setminus j} w_i^{\mathrm{R}} + w_j^{\mathrm{L}} \geq 1, \ \sum_{i \in N \setminus j} w_i^{\mathrm{L}} + w_j^{\mathrm{R}} \leq 1, j \in N, \\
&\qquad w_i^{\mathrm{R}} \geq w_i^{\mathrm{L}} \geq \varepsilon, \ i \in N, \ \delta_{ij} \geq 0, \ i,j \in N(i \neq j),
\end{aligned}
\tag{8}
$$

$$
\begin{aligned}
&\text{minimize } w_k^{\mathrm{R}}, \text{ sub. to } \sum_{i \in N \setminus k} \sum_{j \in N \setminus \{i,k\}} \delta_{ij} \leq \hat{\Delta}_k, \ a_{kl}^{\mathrm{R}} w_l^{\mathrm{L}} = \breve{w}_k^{\mathrm{R}}(l), \\
&\qquad \sqrt{a_{ji}^{\mathrm{R}} w_i^{\mathrm{L}}} + \delta_{ij} = \sqrt{a_{ij}^{\mathrm{L}} w_j^{\mathrm{R}}}, \ i,j \ (i \neq j), \\
&\qquad \sum_{i \in N \setminus j} w_i^{\mathrm{R}} + w_j^{\mathrm{L}} \geq 1, \ \sum_{i \in N \setminus j} w_i^{\mathrm{L}} + w_j^{\mathrm{R}} \leq 1, \ j \in N, \\
&\qquad w_i^{\mathrm{R}} \geq w_i^{\mathrm{L}} \geq \varepsilon, \ i \in N, \ \delta_{ij} \geq 0, \ i,j \in N(i \neq j),
\end{aligned}
\tag{9}
$$

where $\breve{w}_k^{\mathrm{R}}(l)$ is the optimal value of Problem (8). Let $\hat{w}_k^{\mathrm{R}}(l)$ be the optimum value of equation (9). Moreover, we solve the following two LP problems sequentially for each $(k,l) \in N \times (N \setminus k)$:

$$
\begin{aligned}
&\text{minimize } a_{kl}^{\mathrm{L}} w_l^{\mathrm{R}}, \text{ sub. to } \sum_{i \in N \setminus k} \sum_{j \in N \setminus \{i,k\}} \delta_{ij} \leq \hat{\Delta}_k, \\
&\qquad \sqrt{a_{ji}^{\mathrm{R}} w_i^{\mathrm{L}}} + \delta_{ij} = \sqrt{a_{ij}^{\mathrm{L}} w_j^{\mathrm{R}}}, \ i,j \ (i \neq j), \\
&\qquad \sum_{i \in N \setminus j} w_i^{\mathrm{R}} + w_j^{\mathrm{L}} \geq 1, \ \sum_{i \in N \setminus j} w_i^{\mathrm{L}} + w_j^{\mathrm{R}} \leq 1, \ j \in N, \\
&\qquad w_i^{\mathrm{R}} \geq w_i^{\mathrm{L}} \geq \varepsilon, \ i \in N, \ \varepsilon_{ij} \geq 0, \ i,j \in N(i \neq j),
\end{aligned}
\tag{10}
$$

$$
\begin{aligned}
&\text{maximize } w_k^{\mathrm{L}}, \text{ sub. to } \sum_{i \in N \setminus k} \sum_{j \in N \setminus \{i,k\}} \delta_{ij} \leq \hat{\Delta}_k, \ a_{kl}^{\mathrm{R}} w_l^{\mathrm{L}} = \breve{w}_k^{\mathrm{L}}(l), \\
&\qquad \sqrt{a_{ji}^{\mathrm{R}} w_i^{\mathrm{L}}} + \varepsilon_{ij} = \sqrt{a_{ij}^{\mathrm{L}} w_j^{\mathrm{R}}}, \ i,j \ (i \neq j), \\
&\qquad \sum_{i \in N \setminus j} w_i^{\mathrm{R}} + w_j^{\mathrm{L}} \geq 1, \ \sum_{i \in N \setminus j} w_i^{\mathrm{L}} + w_j^{\mathrm{R}} \leq 1, \ j \in N, \\
&\qquad w_i^{\mathrm{R}} \geq w_i^{\mathrm{L}} \geq \varepsilon, \ i \in N, \ \delta_{ij} \geq 0, \ i,j \in N(i \neq j),
\end{aligned}
\tag{11}
$$

where $\breve{w}_k^{\mathrm{L}}(l)$ is the optimal value of Problem (8). Let $\hat{w}_k^{\mathrm{L}}(l)$ be the optimum value of Eq. (11). Finally, the estimated interval priority weights $\hat{W}_k = [\hat{w}_k^{\mathrm{L}}, \hat{w}_k^{\mathrm{R}}]$,

$k \in N$ are obtained by

$$\widetilde{w}_k^{R} = \max_{l \in N} \widehat{w}_k^{R}(l), \qquad \widetilde{w}_k^{L} = \min_{l \in N} \widehat{w}_k^{L}(l). \tag{12}$$

3 Unsatisfaction of Desirable Properties

3.1 Desirable Properties

In this paper, a parameter-free estimation method of a normalized interval weight estimation vector satisfying (i) the given pairwise comparison matrix is realizable under the estimated interval weight vector, (ii) the unique accurate crisp weight vector is estimated under a consistent crisp pairwise comparison matrix, (iii) the estimated interval weight vector satisfies the normality condition and (iv) a proper interval weight vector is estimated from a consistent interval pairwise comparison matrix. Properties (i) and (iii) are represented in the constraints of all estimation problems for a normalized interval weight vector. Property (ii) is obviously satisfied. Property (iv) is satisfied with the conventional method (2) and the corresponding deviation minimization method (7). But not satisfied with MMR method and MMR method with dissimilarity minimization. In this paper, we revise the MMR method to satisfy property (iv) as well as properties (i)–(iii).

3.2 Non-unique Solutions Under a Consistent interval PCM

Inuiguchi [2] found that there are multiple normalized interval weight vectors associated with a consistent interval PCM. For example, consider a normalized interval weight vector W composed of $W_1 = [w_1^{L}, w_1^{R}] = [0.2, 0.4]$, $W_2 = [w_2^{L}, w_2^{R}] = [0.3, 0.5]$ and $W_3 = [w_3^{L}, w_3^{R}] = [0.3, 0.4]$, and another normalized interval weight vector V composed of $V_1 = [v_1^{L}, v_1^{R}] = [2/11, 4/11]$, $V_2 = [v_2^{L}, v_2^{R}] = [3/11, 5/11]$ and $V_3 = [v_3^{L}, v_3^{R}] = [3/11, 4/11]$. For those, we have $[w_i^{L}/w_j^{R}, w_i^{L}/w_j^{R}] = [v_i^{L}/v_j^{R}, v_i^{L}/v_j^{R}], \forall i, j \in \{1, 2, 3\}$ $(i \neq j)$. Therefore, we have

$$\left(\left[\frac{w_i^{L}}{w_j^{R}}, \frac{w_i^{R}}{w_j^{L}} \right] \right) = \left(\left[\frac{v_i^{L}}{v_j^{R}}, \frac{v_i^{R}}{v_j^{L}} \right] \right) = \begin{pmatrix} 1 & [\frac{2}{5}, \frac{4}{3}] & [\frac{1}{2}, \frac{4}{3}] \\ [\frac{3}{5}, \frac{5}{2}] & 1 & [\frac{3}{4}, \frac{5}{3}] \\ [\frac{3}{4}, \frac{2}{1}] & [\frac{3}{5}, \frac{4}{3}] & 1 \end{pmatrix}. \tag{13}$$

Namely, more than two normalized interval weight vectors have an exactly same interval PCM. This fact implies that a normalized interval weight vector minimizing the deviation from a given interval PCM is quite often non-unique. We note that there may exist also multiple normalized interval weight vectors minimizing the deviation from a given interval PCM whose associated interval PCMs are different. However, the existence of multiple optimal solutions with different interval PCMs is less frequent. We take care mainly on the multiplicity of optimal solutions having a same interval PCM because of its high frequency.

The non-uniqueness of the solution to the estimation problem is one of reasons of the unsatisfaction of property (iv) with MMR methods. Before describing the idea to cope with the non-uniqueness, we show the following property of a normalized interval weight vector.

Theorem 1. *For any normalized interval weight vector* $W = (W_1, W_2, \ldots, W_n)^{\mathrm{T}}$ *with* $W_i = [w_i^{\mathrm{L}}, w_i^{\mathrm{R}}]$, $i \in N$, *there exists a normalized interval weight vector* $\bar{W} = (\bar{W}_1, \bar{W}_2, \ldots, \bar{W}_n)^{\mathrm{T}}$ *with* $\bar{W}_i = [\bar{w}_i^{\mathrm{L}}, \bar{w}_i^{\mathrm{R}}]$, $i \in N$ *satisfying*

$$\frac{\bar{w}_i^{\mathrm{L}}}{\bar{w}_j^{\mathrm{R}}} = \frac{w_i^{\mathrm{L}}}{w_j^{\mathrm{R}}}, \ i, j \in N (i \neq j), \quad \sum_{i \in N} (\bar{w}_i^{\mathrm{L}} + \bar{w}_i^{\mathrm{R}}) = 2. \tag{14}$$

Proof. Let $\bar{w}_i^{\mathrm{L}} = 2w_i^{\mathrm{L}} / \sum_{j \in N}(w_j^{\mathrm{L}} + w_j^{\mathrm{R}})$ and $\bar{w}_i^{\mathrm{R}} = 2w_i^{\mathrm{R}} / \sum_{j \in N}(w_j^{\mathrm{L}} + w_j^{\mathrm{R}})$, $i \in N$. We have $\sum_{i \in N}(\bar{w}_i^{\mathrm{L}} + \bar{w}_i^{\mathrm{R}}) = 2$. Moreover, $w_i^{\mathrm{L}} + \sum_{j \in N \setminus i} w_i^{\mathrm{R}} \geq 1 \geq w_i^{\mathrm{R}} + \sum_{j \in N \setminus i} w_i^{\mathrm{L}}$, $i \in N$ implies $\bar{w}_i^{\mathrm{L}} + \sum_{j \in N \setminus i} \bar{w}_i^{\mathrm{R}} \geq 1 \geq \bar{w}_i^{\mathrm{R}} + \sum_{j \in N \setminus i} \bar{w}_i^{\mathrm{L}}$, $i \in N$. □

Because $w_i^{\mathrm{L}} / w_j^{\mathrm{R}} = \bar{w}_i^{\mathrm{L}} / \bar{w}_j^{\mathrm{R}}$, $i, j \in N$ $(i \neq j)$ hold for W and $\bar{W}$, if W is an optimal solution to interval weight estimation problem by minimization of the deviation from the given interval PCM, $\bar{W}$ is also an optimal solution. Then Theorem 1 implies that the addition of constraint $\sum_{i \in N}(w_i^{\mathrm{L}} + w_i^{\mathrm{R}}) = 2$ to any interval weight estimation problems has no influence on the feasibility.

To cope with this non-uniqueness, we consider all normalized interval weight vectors optimal for the estimation problems. Let $\bar{W} = (\bar{W}_1, \bar{W}_2, \ldots, \bar{W}_n)^{\mathrm{T}}$ be an estimated normalized interval weight vector, where $\bar{W}_i = [\bar{w}_i^{\mathrm{L}}, \bar{w}_i^{\mathrm{R}}]$, $i \in N$ and $\bar{W}$ does not always satisfy (14). We define

$$t^{\mathrm{L}} = \min_{i \in N} \frac{1}{\bar{w}_i^{\mathrm{L}} + \sum_{j \in N \setminus i} \bar{w}_i^{\mathrm{R}}}, \quad t^{\mathrm{R}} = \max_{i \in N} \frac{1}{\bar{w}_i^{\mathrm{R}} + \sum_{j \in N \setminus i} \bar{w}_i^{\mathrm{L}}}. \tag{15}$$

Then any normalized interval weight vector $W = (W_1, W_2, \ldots, W_n)^{\mathrm{T}}$ defined by

$$W = t\bar{W}, \text{ or equivalently, } W_i = [w_i^{\mathrm{L}}, w_i^{\mathrm{R}}], \ w_i^{\mathrm{L}} = t\bar{w}_i^{\mathrm{L}}, \ w_i^{\mathrm{R}} = t\bar{w}_i^{\mathrm{R}}, i \in N \tag{16}$$

with some $t \in [t^{\mathrm{L}}, t^{\mathrm{R}}]$ has the same quality in the sense of the deviations between $w_i^{\mathrm{L}} / w_j^{\mathrm{R}}$ and a_{ij}^{L}, $i, j \in N$ $(i \neq j)$. Accordingly, using (16), for the estimation problem, it suffices to obtain a normalized interval vector.

3.3   Unsatisfaction of Property (iv)

In this subsection, we show that the MMR methods described in previous section do not always satisfy property (iv) and approach to the reasons. To this end, we assume a consistent interval PCM $A = ([a_{ij}^{\mathrm{L}}, a_{ij}^{\mathrm{R}}])$ is given, where the consistency is defined by

$$a_{ik}^{\mathrm{L}} a_{ik}^{\mathrm{R}} = a_{ij}^{\mathrm{L}} a_{ij}^{\mathrm{R}} a_{jk}^{\mathrm{L}} a_{jk}^{\mathrm{R}}, \quad a_{ij}^{\mathrm{R}} a_{jk}^{\mathrm{L}} = a_{il}^{\mathrm{R}} a_{lk}^{\mathrm{L}}, \tag{17}$$
$$\forall i, j, k, l \in N \text{ such that } i, j, k \text{ and } l \text{ are different one another.}$$

Let assume that there are many normalized interval weight vector minimizing the deviation from $A = ([a_{ij}^{\mathrm{L}}, a_{ij}^{\mathrm{R}}])$. This assumption is satisfied frequently in real world problems. Let $\bar{w}_i^{\mathrm{L}}$ and $\bar{w}_i^{\mathrm{R}}$, $i \in N$ compose a normalized interval weight vector $\bar{W}$ associated with the given interval PCM. Namely, we have $\bar{w}_i^{\mathrm{R}}/\bar{w}_j^{\mathrm{L}} = a_{ij}^{\mathrm{R}}$, $i \in N$. For this normalized interval weight vector we calculate t^{L} and t^{R} of (15). We note that for any selection of $\bar{W}$, $t^{\mathrm{L}}\bar{W}$ and $t^{\mathrm{R}}\bar{W}$ are constant.

First let us take MMR method with minimizing widths of interval weights in consideration. We have the following theorem.

Theorem 2. *Assume (17). At the first stage of MMR method with minimizing widths of interval weights, we minimize $\sum_{i \in N \setminus k}(w_i^{\mathrm{R}} - w_i^{\mathrm{L}})$ under constraints of (6). The optimal solution to this problem satisfies $w_i^{\mathrm{L}} = t^{\mathrm{L}}\bar{w}_i^{\mathrm{L}}$, $i \in N$ and $w_i^{\mathrm{R}} = t^{\mathrm{L}}\bar{w}_i^{\mathrm{R}}$, $i \in N \setminus k$. However, at an optimal solution to Problem (3) under interval PCM $A = ([a_{ij}^{\mathrm{L}}, a_{ij}^{\mathrm{R}}])$, we have $\hat{w}_k^{\mathrm{R}}(k) = 1 - \sum_{i \in N \setminus k} t^{\mathrm{L}}\bar{w}_i^{\mathrm{L}}$.*

Proof. The first part of the theorem is obtained from the fact that $\sum_{i \in N \setminus k} t^{\mathrm{L}}\bar{w}_i^{\mathrm{R}} + t^{\mathrm{L}}\bar{w}_k^{\mathrm{L}} \geq 1$ and $t^{\mathrm{L}}\bar{w}_i^{\mathrm{R}} \leq a_{ij}^{\mathrm{L}}(t^{\mathrm{L}}\bar{w}_j^{\mathrm{R}})$, $i, j \in N$ $(i \neq j)$ are satisfied with equalities. On the other hand, the upper bound of w_k^{R} has no restriction except $\sum_{i \in N \setminus k} w_i^{\mathrm{L}} + w_k^{\mathrm{R}} \leq 1$. $\square$

As we obtain $\hat{w}_k^{\mathrm{R}}(k) = 1 - \sum_{i \in N \setminus k} t^{\mathrm{L}}\bar{w}_i^{\mathrm{L}} > t^{\mathrm{L}}\bar{w}_k^{\mathrm{R}}$, property (iv) is not satisfied.

Now, let us consider MMR method with minimizing the deviation from a given interval PCM. We have the following theorem.

Theorem 3. *Assume (17). At the first stage of MMR method with minimizing the deviation, we minimize $\sum_{i \in N \setminus k} \sum_{j \in N \setminus \{i,k\}} \delta_{ij}$ under constraints of (7). At the optimal solution, we have $\delta_{ij} = 0$, $i, j \in N \setminus k$ $(i \neq j)$ and $w_i^{\mathrm{L}} = t\bar{w}_i^{\mathrm{L}}$ and $w_i^{\mathrm{R}} = t\bar{w}_i^{\mathrm{R}}$ with any $t \in [t^{\mathrm{L}}, t^{\mathrm{R}}])$, $i \neq k$. Moreover, at the optimal solutions of Problems (8) and (9), we have $w_i^{\mathrm{L}} = t^{\mathrm{R}}\bar{w}_i^{\mathrm{L}}$, $i \in N \setminus k$ and $w_i^{\mathrm{R}} = t^{\mathrm{R}}\bar{w}_i^{\mathrm{R}}$, $i \in N$. However, w_k^{L} can take the value $1 - \sum_{i \in N \setminus k} t^{\mathrm{R}}\bar{w}_i^{\mathrm{R}} < t^{\mathrm{R}}\bar{w}_k^{\mathrm{L}}$. Similarly, at the optimal solutions of Problems (10) and (11), we have $w_i^{\mathrm{L}} = t^{\mathrm{L}}\bar{w}_i^{\mathrm{L}}$, $i \in N$ and $w_i^{\mathrm{R}} = t^{\mathrm{L}}\bar{w}_i^{\mathrm{R}}$, $i \in N \setminus k$. However, w_k^{R} can take the value $1 - \sum_{i \in N \setminus k} t^{\mathrm{L}}\bar{w}_i^{\mathrm{L}} > t^{\mathrm{L}}\bar{w}_k^{\mathrm{R}}$.*

Proof. We can prove in a similar way to the proof of Theorem 2. $\square$

As shown in Theorem 3, there is no guarantee that the solution obtained by (12) satisfy (iv). Moreover, Problems (9) and (11) have no influence on solutions.

3.4 Consideration of Deviations Between $w_i^{\mathrm{L}}/w_j^{\mathrm{R}}$ and a_{ij}^{L}

In our estimation, if a normalized interval weight vector $\bar{W}$ is an optimal estimation, any normalized interval weight vector W is also an optimal estimation if deviations between $w_i^{\mathrm{L}}/w_j^{\mathrm{R}}$ and a_{ij}^{L}, $i \in N$ are same as those between $\bar{w}_i^{\mathrm{L}}/\bar{w}_j^{\mathrm{R}}$ and a_{ij}^{L}, $i \in N$. Indeed, any normalized interval weight vector W defined by (16) satisfies the requirement. Namely, the evaluation function should be invariant the change of the solution from $\bar{W}$ to $t\bar{W}$. However, the evaluation functions $\sum_{i \in N \setminus k}(w_i^{\mathrm{R}} - w_i^{\mathrm{L}})$ and $\sum_{i \in N \setminus k} \sum_{j \in N setminus \{i,k\}} \delta_{ij}$ are both variant. Therefore, we replace those evaluation functions with evaluation functions

$\sum_{i\in N\setminus k}(w_i^{\mathrm{R}}-w_i^{\mathrm{L}})/\sum_{i\in N}(w_i^{\mathrm{L}}+w_i^{\mathrm{R}})$ and $\sum_{i\in N\setminus k}\sum_{j\in N\setminus\{i,k\}}\delta_{ij}/\sum_{i\in N}(w_i^{\mathrm{L}}+w_i^{\mathrm{R}})$ which are invariant.

By these replacement, the estimation problems become linear fractional programming problems. However, fortunately, we can prove that the linear fractional programming problems are equivalent to the original linear programming problems with an additional linear constraint $\sum_{i\in N}(w_i^{\mathrm{L}}+w_i^{\mathrm{R}})=2$. Moreover, by the addition of constraint $\sum_{i\in N}(w_i^{\mathrm{L}}+w_i^{\mathrm{R}})=2$ does not have the influence on the assertions in Theorems 2 and 3 except the last assertions of those theorems.

4   The Improved MMR Methods

We propose the revised MMR methods considering the properties shown in Theorems 2 and 3. Based on the first assertions in Theorems 2 and 3, we propose the following revised MMR methods:

Revised MMR Method with Minimizing Widths: First, for each $k\in N$, we minimize $\sum_{i\in N\setminus k}(w_i^{\mathrm{R}}-w_i^{\mathrm{L}})$ under constraints of (6) with an additional constraint $\sum_{i\in N}(w_i^{\mathrm{L}}+w_i^{\mathrm{R}})=2$ and we define $\hat{w}_j^{\mathrm{L}}(k)$ and $\hat{w}_j^{\mathrm{R}}(k)$, $j\in N$ by an optimal solution. Then, we define $\check{w}_i^{\mathrm{L}}$, $\check{w}_i^{\mathrm{R}}$, $i\in N$ by

$$\check{w}_i^{\mathrm{L}}=\min_{k\in N}\hat{w}_i^{\mathrm{L}}(k),\quad \check{w}_i^{\mathrm{R}}=\max_{k\in N\setminus i}\hat{w}_i^{\mathrm{R}}(k),\quad i\in N. \tag{18}$$

Because the interval weights $[\check{w}_i^{\mathrm{L}},\check{w}_i^{\mathrm{R}}]$, $i\in N$ does not always satisfy the normalization condition, we obtain $\bar{w}_i^{\mathrm{L}}$, $\bar{w}_i^{\mathrm{R}}$, $i\in N$ by solving the following problem:

$$\text{minimize}\sum_{i\in N}(w_i^{\mathrm{R}}-w_i^{\mathrm{L}}),\text{ subject to }\sum_{i\in N\setminus j}w_i^{\mathrm{R}}+w_j^{\mathrm{L}}\ge 1,\ \sum_{i\in N\setminus j}w_i^{\mathrm{L}}+w_j^{\mathrm{R}}\le 1,\ j\in N,$$
$$w_j^{\mathrm{R}}\ge\check{w}_j^{\mathrm{R}},\ \check{w}_j^{\mathrm{L}}\ge w_j^{\mathrm{L}}\ge\varepsilon,\ j\in N. \tag{19}$$

Then whole solution is obtained from $\bar{w}_i^{\mathrm{L}}$, $\bar{w}_i^{\mathrm{R}}$, $i\in N$.

Revised MMR Method with Minimizing Deviations (1): For each k, we tentatively define $\check{w}_i^{\mathrm{L}}(k)$, $i\in N$ by the optimal solution to a problem minimizing $\sum_{i\in N\setminus k}\sum_{j\in N\setminus\{i,k\}}\delta_{ij}$ under constraints of Problem (7) with an additional constraint $\sum_{i\in N}(w_i^{\mathrm{L}}+w_i^{\mathrm{R}})=2$. Define $l=\arg\max_{i\in N\setminus k}a_{ki}^{\mathrm{R}}\check{w}_i^{\mathrm{L}}(k)$. For l, we solve Problem (8) and we redefine $\check{w}_i^{\mathrm{L}}(k)$, $i\in N$, and $\check{w}_i^{\mathrm{R}}(k)$, $i\in N$ by the optimal solution to Problem (8). Then, we define $\check{w}_i^{\mathrm{L}}$, $\check{w}_i^{\mathrm{R}}$ $i\in N$ by

$$\check{w}_i^{\mathrm{L}}=\min_{k\in N\setminus i}\check{w}_i^{\mathrm{L}}(k),\quad \check{w}_i^{\mathrm{R}}=\max_{k\in N}\check{w}_i^{\mathrm{R}}(k),\quad i\in N. \tag{20}$$

After obtaining $\check{w}_i^{\mathrm{L}}$ and $\check{w}_i^{\mathrm{R}}$, $i\in N$, we apply the same process of Revised MMR method with minimizing widths.

Revised MMR Method with Minimizing Deviations (2): For each k, we tentatively define $\check{w}_i^{\mathrm{R}}(k)$, $i\in N$ by the optimal solution to a problem minimizing

$\sum_{i \in N \setminus k} \sum_{j \in N \setminus \{i,k\}} \delta_{ij}$ under constraints of Problem (7) with an additional constraint $\sum_{i \in N} (w_i^L + w_i^R) = 2$. Define $l = \arg\min_{i \in N \setminus k} a_{ki}^L \breve{w}_i^R(k)$. For l, we solve Problem (10) and we redefine $\breve{w}_i^L(k)$, $i \in N$, and $\breve{w}_i^R(k)$, $i \in N$ by the optimal solution to Problem (10). Then, we define $\breve{w}_i^L$, $\breve{w}_i^R$ $i \in N$ by

$$\breve{w}_i^L = \min_{k \in N} \breve{w}_i^L(k), \quad \breve{w}_i^R = \max_{k \in N \setminus i} \breve{w}_i^R(k), \quad i \in N. \tag{21}$$

After obtaining $\breve{w}_i^L$ and $\breve{w}_i^R$, $i \in N$, we apply the same process of Revised MMR method with minimizing widths.

Solutions of those MMR methods satisfy property (iv) by Theorems 2 and 3.

5 Concluding Remarks

In this paper, we investigated the interval weight estimation methods satisfying desirable properties: (0) the method is parameter-free, (i) the given pairwise comparison matrix is realizable under the estimated interval weight vector, (ii) the unique accurate crisp weight vector is estimated under a consistent crisp pairwise comparison matrix, (iii) the estimated interval weight vector satisfies the normality condition and (iv) a proper interval weight vector is estimated from a consistent interval pairwise comparison matrix. The previous MMR methods do not satisfy (iv) but (0)–(iii) while the conventional method and the corresponding deviation minimization method satisfy all. Solutions of MMR methods reflect the human vague evaluation to a certain extent although solutions of the conventional method and the corresponding deviation minimization method do not reflect it well. We analyzed the solutions obtained by MMR methods to discover the reasons why previous MMR methods do not satisfy (iv). Based on the results of the analysis, we propose the revised MMR methods satisfying all desirable properties. The investigation on the performances of the revised MMR methods is one of future topics.

Acknowledgement. This work was supposed by JSPS KAKENHI Grant Number 17K18952.

References

1. Greco, S., Ehrgott, M., Figueira, J.R. (eds.): Multiple Criteria Decision Analysis: State of the Art Surveys, vol. 2, 2nd edn. Springer, New York (2016)
2. Inuiguchi, M.: Non-uniqueness of interval weight vector to consistent interval pairwise comparison matrix and logarithmic estimation methods. In: Huynh, V.-N., et al. (eds.) Integrated Uncertainty in Knowledge Modelling and Decision Making: 5th International Symposium, IUKM 2016. LNAI, vol. 9978, pp. 39–50 (2016)
3. Inuiguchi, M., Innan, S.: Comparison among several parameter-free interval weight estimation methods from a crisp pairwise comparison matrix. In: CD-Proceedings of 14th MDAI, pp. 61–76 (2017)
4. Saaty, T.L.: The Analytic Hierarchy Process. McGraw-Hill, New York (1980)

5. Saaty, T.L., Vargas, C.G.: Comparison of eigenvalue, logarithmic least squares and least squares methods in estimating ratios. Math. Model. **5**(5), 309–324 (1984)
6. Sugihara, K., Ishii, H., Tanaka, H.: Interval priorities in AHP by interval regression analysis. Eur. J. Oper. Res. **58**, 45–754 (2004)
7. Sugihara, K., Tanaka, H.: Interval evaluations in the analytic hierarchy process by possibility analysis. Comput. Intell. **17**(3), 567–579 (2001)
8. Torisu, I., Inuiguchi, M.: Increasing convergence of the quality of estimated interval weight vector in interval AHP. In: Proceedings of 2018 Joint 10th International Conference on Soft Computing and Intelligent Systems and 19th International Symposium on Advanced Intelligent Systems (SCIS & ISIS 2018), pp. 1400–1405 (2018)
9. Wang, Y.-M., Elhag, T.M.S.: A goal programming method for obtaining interval weights from an interval comparison matrix. Eur. J. Oper. Res. **177**, 458–471 (2007)
10. Wang, Z.-J.: Uncertain index based consistency measurement and priority generation with interval priorities in the analytic hierarchy process. Comput. Industr. Eng. **83**, 252–260 (2015)
11. Xia, M., Xu, Z.: Interval weight generation approaches for reciprocal relations. Appl. Math. Model. **38**, 828–838 (2014)

Fuzzy Systems in Economics and Management

Weighted Averages in the Ordered Weighted Average Inflation

Ernesto León-Castro[1](✉), Fabio Blanco-Mesa[2],
and José M. Merigó[3,4]

[1] Universidad Católica de la Santísima Concepción,
Av. Alonso de Ribera 2850, Concepción, Región del Bío Bío, Chile
ernesto134@hotmail.com
[2] Universidad Pedagógica y Tecnológica de Colombia,
Av. Central del Norte, 39-115, 150001 Tunja, Colombia
fabio.blanco01@uptc.edu.co
[3] School of Systems, Management and Leadership,
Faculty of Engineering and Information Technology,
University of Technology Sydney, Ultimo, NSW 2007, Australia
[4] Department of Management Control and Information Systems,
School of Economics and Business, University of Chile,
Av. Diagonal Paraguay 257, 8330015 Santiago, Chile
jmerigo@fen.uchile.cl

Abstract. This paper presents the ordered weighted average weighted average inflation (OWAWAI). The OWAWAI operator is a new formulation for calculating inflation that provides different criteria for the association between the arguments and weights. OWAWAI presents the possibility to generate new approaches that under- or overestimate the results according to the knowledge and expertise of the decision maker. The works present an approach in Chile inflation.

Keywords: OWA operator · Inflation · Decision making

1 Introduction

Inflation has become an important financial indicator not only for the monetary policies but also for individuals' and enterprises' financial decisions (Malmendier and Nagel 2015). In the case of policy makers, it is important for improving their forecasting and policy choices, and in the case of the enterprises and individuals, their choices in the housing market, real expenditure decisions and macroeconomic outcomes are influenced by expected inflation (Woodford and Walsh 2005).

Many models have been developed in order to find the relations in inflation, such as the relationship between openness and inflation (Romer 1993), between inflation and unemployment (Ruge-Murcia 2004) and between the behaviour of the central bank and inflation (Hayat *et al.* 2018). An important aspect of the central bank of each country is that these banks use different techniques in order to calculate inflation. A common element is that these banks divide the different elements of the economy, assign a

© Springer Nature Switzerland AG 2019
R. B. Kearfott et al. (Eds.): IFSA 2019/NAFIPS 2019, AISC 1000, pp. 87–95, 2019.
https://doi.org/10.1007/978-3-030-21920-8_8

specific weight to each with respect to the total inflation and calculate the change in the consumer price index monthly.

To accomplish this task, this paper proposes using aggregation operators. The specific aggregation operator that was used is the ordered weighted average (OWA) operator developed by Yager (1988). This operator was selected because it can provide different scenarios between the maximum and the minimum results, and this information is helpful for the enterprises that seek to make decisions about different aspects of the enterprise, such as income, expenses and profits.

2   Preliminaries

2.1   Inflation Formulas for Latin America Countries

2.1.1   Chile
In Chile, the calculation of the inflation is based on the determination of the Consumer Price Index of 12 different divisions that have different associated weights. This is presented in Table 1.

Table 1. Divisions and weights used to calculate inflation in Chile

Division	Weight
Food and non-alcoholic beverages	19.05855
Alcoholic beverages and tobacco	3.31194
Clothing and footwear	4.48204
Housing and basic services	13.82810
Equipment and maintenance of the home	7.02041
Health	6.44131
Transportation	14.47381
Communications	5.00064
Recreation and culture	6.76121
Education	8.08996
Restaurants and hotels	4.37454
Miscellaneous goods and services	7.15749

See: http://www.ine.cl/estad%C3%ADsticas/precios/ipc

To calculate the inflation, each of the divisions is compared with their previous month's value with the formula $\left(\frac{CPI_n - CPI_{n-1}}{CPI_{n-1}}\right)(100)$ and, at the end, each value is multiplied by its weight and summed in order to obtain the inflation rate.

2.2   Basics and Extensions of the OWAWA Operator

To better understand the operator that will be used to improve inflation, it is important to understand the aggregation operator that is being used in the traditional formulation and the weighted average (WA) operator (Beliakov *et al.* 2007; Torra and Narukawa 2007). They are defined as follows.

Definition 1. A WA operator of dimension n is a mapping $WA : R^n \to R$ that has an associated weighting vector V, with $v_j \in [0,1]$ and $\sum_{i=1}^{n} v_i = 1$, such that the following exists:

$$WA(a_1, \ldots, a_n) = \sum_{i=1}^{n} v_i a_i, \qquad (1)$$

where a_i represents the argument variable.

This idea was developed further by Yager (1988) by adding a reordering step in the WA operator, this new operator was called ordered weighted average (OWA) operator and one of its main characteristic is that it is possible to obtain the maximum and minimum operator. The formulation is.

Definition 2. An OWA operator of dimension n is a mapping $F : R^n \to R$ with a weight vector $w = [w_1, w_2, \ldots, w_n]^T$, where $w_j \in [0,1]$, $1 \le i \le n$ and $\sum_{j=1}^{n} w_j = 1$, such that:

$$OWA(a_1, a_2, \ldots, a_n) = \sum_{j=1}^{n} w_j b_j, \qquad (2)$$

where b_j is the jth element that is the largest of the collection a_n.

The OWA operator was developed further by Merigó (2011) with the inclusion of another weighting average vector into the formulation (Merigó *et al.* 2017). By doing this, it is possible to include two different weighting vectors with some degree of importance for each that can help us better understand a problem or a situation. The formulation is as follows.

Definition 3. An OWAWA operator of dimension n is a mapping $OWAWA : R^n \to R$ that has an associated weighing vector W of dimension n such that $w_j \in [0,1]$ and $\sum_{j=1}^{n} w_j = 1$ and is calculated according to the following formula:

$$OWAWA(a_1, \ldots, a_n) = \sum_{j=1}^{n} \widehat{v}_j b_j, \qquad (3)$$

where b_j is the jth largest of the a_i, each argument a_i has an associated weight (WA) v_i with $\sum_{i=1}^{n} v_i = 1$ where $v_i \in [0,1]$, $\widehat{v}_j = \beta w_j + (1-\beta) v_j$ where $\beta \in [0,1]$ and v_j is ordered according to b_j, that is, according to the jth largest a_i. Also note that if the reordering step is omitted, then the OWAWA operator becomes the weighted average weighted average (WAWA) operator.

3 The Ordered Weighted Average Weighted Average Inflation

3.1 The OWAWA Inflation

A particular case of the inflation is that each country has different ways to provide a result based on different factors that are important to each one. Another characteristic is that not all the country uses the same weights for the same factor, in this sense, a way to provide a better understand of the phenomenon of the inflation is by adding another weighting vector that can be specific to the characteristic of the enterprise or that market that will be evaluated and by doing that it is possible to take into the same formulation the idea of the country inflation and the specific needs of the decision maker.

This new operator is called the ordered weighted average weighted average inflation (OWAWAI) and its definition is as follows

Definition 4. The OWAWAI operator of dimension n is a mapping $F : R^n \rightarrow R$ with a weight vector $w = [w_1, \ldots, w_n]^T$, where $w_j \in [0, 1]$, $1 \leq i \leq n$ and $\sum_{i=1}^{n} w_j = w_1 + \ldots + w_n = 1$ and can be defined as:

$$OWAWAI(i_1, i_2, \ldots, i_n) = \sum_{k=1}^{n} \widehat{v}_j h_j, \qquad (4)$$

where h_j is the jth element, which is the largest of the collection $i_1, i_2, \ldots, i_n$. Each element of the collection represents the factors that are considered and used in order to obtain the average inflation. Each argument i_i has an associated weight (WA) v_i with $\sum_{i=1}^{n} v_i = 1$ where $v_i \in [0, 1]$, $\widehat{v}_j = \beta w_j + (1 - \beta)v_j$, where $\beta \in [0, 1]$ and v_j is ordered according to h_j, that is, according to the jth largest i_i.

An extension can be obtained if an induced reordering step is done. This operator is the induced ordered weighted average weighted average inflation (IOWAWAI) and is defined as follows.

Definition 5. The IOWAWAI operator of dimension n is a mapping $IOWAWAI : R^n \times R^n \rightarrow R$ that has an associated weighting vector W of dimension n, where the sum of the weights is 1 and $w_j \in [0, 1]$, where an induced set of ordering variables is included (u_i) so the formula is

$$IOWAWAI(\langle u_1, i_1 \rangle, \ldots, \langle u_n, i_n \rangle) = \sum_{k=1}^{n} \widehat{v}_j h_j, \qquad (5)$$

where h_j is the i_i value of the OWA pair $<u_i, i_i>$ that has the jth largest u_i. u_i is the order-inducing variable, and i_i are the inflation factors. Each argument i_i has an associated weight (WA) v_i with $\sum_{i=1}^{n} v_i = 1$ where $v_i \in [0, 1]$, $\widehat{v}_j = \beta w_j + (1 - \beta)v_j$, where $\beta \in [0, 1]$ and v_j is ordered according to h_j, that is, according to the jth largest i_i

It is important to note that another extension can be made if the weighting vector is unbounded in this sense that the heavy ordered weighted average weighted average inflation (HOWAWAI) is obtained and can be defined as follows.

Definition 6. The HOWAWAI operator is a map $R^n \rightarrow R$ that is associated with a weight vector w, where $w_j \in [0,1]$ and $1 \leq \sum_{j=1}^{n} w_j \leq n$, such that

$$HOWAWAI(i_1, i_2, \ldots, i_n) = \sum_{k=1}^{n} \widehat{v}_j h_j, \tag{6}$$

where h_j is the jth largest element of the collection $i_1, i_2, \ldots, i_n$, each argument i_i has an associated weight (WA) v_i with $1 < \sum_{i=1}^{n} v_i < n$ or even $-\infty < \sum_{i=1}^{n} v_i < \infty$ and $v_i \in [0,1]$, $\widehat{v}_j = \beta w_j + (1-\beta)v_j$ and $\beta \in [0,1]$ and v_j ordered according to h_j, that is, according to the jth largest of the i_i

Finally, if both of the main characteristic of the Definition 8 and 9 are included in one formulation the induced heavy ordered weighted average weighted average inflation (IHOWAWAI) operator is done. Its definition is.

Definition 7. The IHOWAWAI operator of dimension n is a mapping $IOWAWAI$: $R^n \times R^n \rightarrow R$ that has an associated weighting vector W of dimension n, with $w_j \in [0,1]$ and $1 \leq \sum_{j=1}^{n} w_j \leq n$ and an induced set of ordering variables is included (u_i) so the formula is

$$IHOWAWAI(u_1, i_1, \ldots, u_n, i_n) = \sum_{k=1}^{n} \widehat{v}_j h_j, \tag{7}$$

where h_j is the i_i value of the OWA pair $< u_i, i_i >$ having the jth largest u_i. u_i is the order-inducing variable, and i_i are the inflation factors. Each argument i_i has an associated weight (WA) v_i with $1 < \sum_{i=1}^{n} v_i < n$ or even $-\infty < \sum_{i=1}^{n} v_i < \infty$ where $v_i \in [0,1]$, $\widehat{v}_j = \beta w_j + (1-\beta)v_j$, where $\beta \in [0,1]$ and v_j is ordered according to h_j, that is, according to the jth largest i_i.

4 Numerical Example

To explain the new inflation formulations, the same information that was presented in Table 2 will be used, but in this case another weighting vector, a heavy vector and an induced vector will be used. This information is presented in Table 2.

Table 2. Information to calculate the inflation aggregation operator for Chile in August 2018

Division	Inflation	Official weighting vector	Expert weighting vector
Food and non-alcoholic beverages	0.47	0.19059	0.20
Alcoholic beverages and tobacco	0.18	0.03312	0.05
Clothing and footwear	0.56	0.04482	0.10
Housing and basic services	0.63	0.13828	0.10
Equipment and maintenance of the home	0.13	0.07020	0.05
Health	−0.05	0.06441	0.10
Transportation	−0.54	0.14474	0.15
Communications	0.35	0.05001	0.03
Recreation and culture	−0.52	0.06761	0.05
Education	–	0.08090	0.10
Restaurants and hotels	0.13	0.04375	0.03
Miscellaneous goods and services	0.48	0.07157	0.04

With the information in Table 2, the different results are presented in Tables 3, 4, 5 and 6. (Note that $\beta = 40\%$ for the original weighting vector and $\beta = 60\%$ for the expert weighting vector.)

Table 3. OWAWAI operator results

Inflation	Official weighting vector	Expert weighting vector	Weighted inflation
−0.54	0.03312	0.03	−0.017
−0.52	0.04375	0.03	−0.018
−0.05	0.04482	0.04	−0.002
0	0.05001	0.2	0
0.13	0.06441	0.05	0.007
0.13	0.06761	0.05	0.007
0.18	0.0702	0.05	0.010
0.35	0.07157	0.10	0.031
0.47	0.0809	0.10	0.043
0.48	0.13828	0.10	0.055
0.56	0.14474	0.10	0.066
0.63	0.19059	0.15	0.105
		OWAWAI result	**0.288**

Table 4. IOWAWAI operator results

Inflation	Official weighting vector	Expert weighting vector	Weighted inflation
0.48	0.07157	0.04	0.025
0.18	0.03312	0.05	0.008
0.13	0.04375	0.03	0.005
−0.52	0.06761	0.05	−0.030
0.13	0.0702	0.05	0.008
0.35	0.05001	0.03	0.013
0.63	0.13828	0.10	0.073
0	0.0809	0.10	0
−0.54	0.14474	0.15	−0.080
−0.05	0.06441	0.10	−0.004
0.56	0.04482	0.10	0.044
0.47	0.19059	0.20	0.092
		IOWAWAI result	**0.153**

Table 5. HOWAWAI operator results

Inflation	Official weighting vector	Expert heavy weighting vector	Weighted inflation
−0.54	0.03312	0.04	−0.020
−0.52	0.04375	0.04	−0.022
−0.05	0.04482	0.05	−0.002
0	0.05001	0.05	0
0.13	0.06441	0.05	0.007
0.13	0.06761	0.07	0.009
0.18	0.0702	0.10	0.016
0.35	0.07157	0.10	0.031
0.47	0.0809	0.12	0.049
0.48	0.13828	0.15	0.070
0.56	0.14474	0.15	0.083
0.63	0.19059	0.22	0.131
		HOWAWAI result	**0.352**

Table 6. IHOWAWAI operator results

Inflation	Official weighting vector	Expert heavy weighting vector	Weighted inflation
0.48	0.07157	0.04	0.025
0.18	0.03312	0.07	0.010
0.13	0.04375	0.04	0.005
−0.52	0.06761	0.05	−0.030
0.13	0.0702	0.05	0.008
0.35	0.05001	0.05	0.018
0.63	0.13828	0.10	0.073
0	0.0809	0.10	0
−0.54	0.14474	0.15	−0.080
−0.05	0.06441	0.15	−0.006
0.56	0.04482	0.12	0.050
0.47	0.19059	0.22	0.098
		IHOWAWAI result	**0.171**

As seen in the numerical example, the inflation was originally 0.158 using only the official weighting vector, but with the use of the different operators, the different scenarios increased the inflation from 0.153 to 0.352, which was more than double the original estimate for the month. This is important when the experts make decisions because it is possible to generate specific results depending on the area of the enterprise.

Finally, it is important to note that with the use of the aggregation operators an important amount of data have been included in the decision making process, in this sense, analyzing each of the scenarios generated is necessary to understand the inflation better and how it will impact on the finance of the company. Also, when the information between the different aggregation operators and the traditional inflation are in conflict we suggest that the most complex operator (in this case the IHOWAWAI operator) should be taken into account, this because is the operator that includes more information about the problem in the result and must be the closest to the reality. Another important thing to consider, is that when the decision must be made quickly and not all the information needed for the use of the IHOWAWAI operator cannot be obtained, the analysis between the traditional formula and the most simple aggregation operator (in this case the OWAWAI) must be done to visualize how much the difference can be and determine if the decision must be done hasty or not.

5   Conclusions

The main purpose of the paper is to provide a new aggregation operator called the ordered weighted average weighted average inflation (OWAWAI) operator. The main characteristics of these new formulations is that they can provide new inflation scenarios that can be calculated using the expectations, knowledge and characteristics of the market of the enterprise and, depending on the complexity of the situation or the problem, different formulations can be calculated

For future research, new extensions of the OWA operator and applications can be derived by using the Bonferroni, moving averages and its application in other areas of engineering, business, economics and finance.

References

Beliakov, G., Pradera, A., Calvo, T.: Aggregation Functions: A Guide for practiTioners, vol. 221. Springer, Heidelberg (2007)

Hayat, Z., Balli, F., Rehman, M.: Does inflation bias stabilize real growth? Evidence from Pakistan. J. Policy Model. **40**(6), 1083–1103 (2018)

Malmendier, U., Nagel, S.: Learning from inflation experiences. Q. J. Econ. **131**(1), 53–87 (2015)

Merigó, J.M.: A unified model between the weighted average and the induced OWA operator. Expert Syst. Appl. **38**(9), 11560–11572 (2011)

Merigó, J.M., Palacios-Marqués, D., Soto-Acosta, P.: Distance measures, weighted averages, OWA operators and Bonferroni means. Appl. Soft Comput. **50**, 356–366 (2017)

Romer, D.: Openness and inflation: theory and evidence. Q. J. Econ. **108**(4), 869–903 (1993)

Ruge-Murcia, F.J.: The inflation bias when the central bank targets the natural rate of unemployment. Eur. Econ. Rev. **48**(1), 91–107 (2004)

Torra, V., Narukawa, Y.: Modeling Decisions: Information Fusion and Aggregation Operators. Springer, Heidelberg (2007)

Woodford, M., Walsh, C.E.: Interest and prices: foundations of a theory of monetary policy. Macroecon. Dyn. **9**(3), 462–468 (2005)

Yager, R.R.: On ordered weighted averaging aggregation operators in multicriteria decision-making. IEEE Trans. Syst. Man Cybern. **18**(1), 183–190 (1988)

Inter-relation Between Interval and Fuzzy Techniques

A Note on the Centroid, Yager Index and Sample Mean for Fuzzy Numbers

Juan Carlos Figueroa-García[1]([⊠]), Jairo Soriano-Mendez[2],
and Miguel Alberto Melgarejo-Rey[2]

[1] Universidad Distrital Francisco José de Caldas, Bogotá, Colombia
`jcfigueroag@udistrital.edu.co`
[2] Engineering Department, Universidad Distrital Francisco José de Caldas,
Bogotá, Colombia
`{josoriano,mmelgarejo}@udistrital.edu.co`

Abstract. This paper compares three well known defuzzifications methods with the sample mean for fuzzy numbers, namely: Yager index, centroid and possibilistic mean. Fuzzy random variable generation was performed to carry out the comparison over the necessity of statistical independence. Our experimental evidence suggests the four approaches exhibits interesting differences among them.

Keywords: Yager index · Centroid · Fuzzy numbers ·
Random variables

1 Introduction and Motivation

Computation of expected values for fuzzy sets/numbers has an important role in fuzzy theory, and it has been widely applied to fuzzy logic systems, optimization, data analysis etc. Some important methods for defuzzification of fuzzy sets/numbers have been proposed by Yager [12], Carlsson and Fullér [1], and Wu and Mendel [11], and they have been compared from different perspectives.

Kolmogorov [5] proposed a generalized theory of random variables in which the sample mean is (in the limit) the expected value of any random variable, and its relationship to probability measures was proven later on. His main results leads us to think that his theory can be also applied to fuzzy measures which is the main focus of this paper.

Therefore, we compare the Yager index, the centroid, and the possibilistic mean of a fuzzy number to the sample mean of its α-levels considered as random variables. We provide some numerical experiments for all methods and some considerations about its results are given.

This paper is organized as follows: Sect. 2 provides some introductory material to fuzzy sets/numbers. Section 3 describes the Yager index, the possibilistic mean and the centroid of a fuzzy number. Sample mean and random variables are presented in Sect. 4. Numerical experiments are described and analyzed in Sect. 5. Finally, we draw some conclusions in Sect. 6.

© Springer Nature Switzerland AG 2019
R. B. Kearfott et al. (Eds.): IFSA 2019/NAFIPS 2019, AISC 1000, pp. 99–105, 2019.
https://doi.org/10.1007/978-3-030-21920-8_9

2   Basics of Fuzzy Sets/Numbers

Let $\mathcal{P}(X)$ be the class of all crisp sets, $\mathcal{F}(X)$ is the class of all fuzzy sets, $\mathcal{F}_1(X)$ is the class of all convex fuzzy sets and $I = [0,1]$ is the set of values in the unit interval. A fuzzy set namely A is characterized by a membership function $\mu_A : X \to I$ defined over a universe of discourse $x \in X$. Thus, a fuzzy set A is the set of ordered pairs $x \in X$ and its membership degree, $\mu_A(x)$, i.e.,

$$A = \{(x, \mu_A(x)) \mid x \in X\}. \tag{1}$$

Let us denote $\mathcal{F}_1(\mathbb{R})$ as the class of all fuzzy numbers, then a fuzzy number is defined as follows:

Definition 1 Let $A : \mathbb{R} \to I$ be a fuzzy subset of the reals. Then $A \in \mathcal{F}_1(\mathbb{R})$ is a Fuzzy Number (FN) iff there exists a closed interval $[x_l, x_r] \neq \emptyset$ with a membership function $\mu_A(x)$ such that:

$$\mu_A(x) = \begin{cases} c(x) & for \ x \in [c_l, c_r], \\ l(x) & for \ x \in [-\infty, x_l], \\ r(x) & for \ x \in [x_r, \infty], \end{cases} \tag{2}$$

where $c(x) = 1$ for $x \in [c_l, c_r]$, $l : (-\infty, x_l) \to I$ is monotonic non-decreasing, continuous from the right, i.e. $l(x) = 0$ for $x < x_l$; $l : (x_r, \infty) \to I$ is monotonic non-increasing, continuous from the left, i.e. $r(x) = 0$ for $x > x_r$.

The α-cut of a set $A \in \mathcal{F}_1(\mathbb{R})$, $^\alpha A$ is the set of values with a membership degree equal or greatest than α, this is:

$$^\alpha A = \{x \mid \mu_A(x) \geqslant \alpha\} \ \forall \ x \in X, \tag{3}$$

$$^\alpha A = \left[\inf_x {}^\alpha \mu_A(x), \ \sup_x {}^\alpha \mu_A(x) \right] = \left[\check{a}_\alpha, \hat{a}_\alpha \right]. \tag{4}$$

And the α-*level* of $A \in \mathcal{F}_1(\mathbb{R})$ namely A_α are the endpoints of $^\alpha A$ i.e.

$$A_\alpha = \{x \mid \mu_A(x) = \alpha \text{ for some } x \in X\}, \tag{5}$$

$$A_\alpha = \left\{ \inf_x {}^\alpha \mu_A(x), \ \sup_x {}^\alpha \mu_A(x) \right\} = \{ \check{a}_\alpha, \hat{a}_\alpha \}. \tag{6}$$

3   Yager Index, Possibilistic Mean and Centroid of a Fuzzy Numbers

3.1   Yager Index of a Fuzzy Number

Yager [12] has proposed one of the most important and widely applied ranking methods for convex fuzzy sets. As Chaudhuri and Rosenfeld [2] and Hung and Yang [4] defined regarding α levels for computing distances, the Yager Index for convex fuzzy sets comes from the idea of integration of the arithmetic mean for every $^{\alpha}A$, namely $I(A)$:

$$I(A) = \frac{1}{2} \int_{[0,1]} (\check{a}_{\alpha} + \hat{a}_{\alpha}) \, d\alpha \tag{7}$$

For discrete $x \in X$ and $x_i < x_{i+1}$ we have:

$$I(A) = \frac{1}{2} \sum_{i=1}^{n} (\check{a}_{\alpha_i} + \hat{a}_{\alpha_i}) \Delta_{\alpha_i} \tag{8}$$

where $[\check{a}_{\alpha}, \hat{a}_{\alpha}]$, $\alpha \in [0,1]$ is the α-level of a fuzzy number A, and Δ_{α_i} is a delta step on α_i.

Roughly speaking, $I(A)$ is a Lebesgue integral (a Riemann integral in the discrete case) computed over α instead of $x \in X$, since its domain is $I = [0,1]$ and it is easier to measure in the case of fuzzy numbers. $I(A)$ has some desirable properties such as reflexivity, symmetry, and transitivity which are important properties since these allow to use it as a reliable measure of A.

3.2   Carlsson-Fullér Possbilistic Mean of a Fuzzy Number

Carlsson and Fullér [1] proposed a method for computing the possibilistic mean of a fuzzy set. Thus, the possibilistic mean of A, $M(A)$ is computed as follows:

$$M(A) = \int_{[0,1]} \alpha(\check{a}_{\alpha} + \hat{a}_{\alpha}) \, d\alpha. \tag{9}$$

This method is also based on α-cuts and can be seen as an less weighted α-based centroid of A.

3.3   Centroid of a Fuzzy Number

The well known centroid of a fuzzy set A, $C(A)$ is computed as follows:

$$C(A) = \frac{\int_{x} x \cdot \mu_A(x) \, dx}{\int_{x} \mu_A(x) \, dx} \tag{10}$$

$$= \frac{1}{|A|} \int_{x} x \cdot \mu_A(x) \, dx \tag{11}$$

whereas its discrete version is expressed as:

$$C(A) = \frac{\sum_{i=1}^{n} x_i \cdot \mu_A(x_i)}{\sum_{i=1}^{n} \mu_A(x_i)} \tag{12}$$

4   Sample Mean and Random Variables

We alternatively can consider α-levels as random variables $X(\omega)$ defined over a sample space Ω (analogous to the universe of discourse of A) in which the set of probable events $\omega \in \Omega$ is a δ-algebra to then compute its *sample mean*. We can also use random variable generation from α-cuts to compare them to $I(A), M(A)$ and $C(A)$.

4.1   Sample Mean

The sample mean of $\breve{a}_\alpha, \hat{a}_\alpha$ given an amount of α-cuts namely $\bar{x}_A$ is as follows:

$$\bar{x}_A = \frac{1}{2\alpha} \sum_i^n \breve{a}_{\alpha_i} + \hat{a}_{\alpha_i} \tag{13}$$

where n is the selected amount of α-cuts.

4.2   Fuzzy Random Variable Generation

Another approach that will be used in this paper is to generate random samples using $\breve{a}_\alpha, \hat{a}_\alpha$ and two uniform random numbers $U_1 \in [0,1]$ and $U_2 \in [0,1]$ as proposed by Varón-Gaviria [10] and Pulido-López [9]. Their proposal is summarized in Method 1.

Method 1. α-cut simulation

Require: $\mu_A \in \mathcal{F}_1(\mathbb{R})$ (see Eq. (4)) and $n \in \mathbb{N}_+$
    for $i : 1 \to n$ **do**
        Compute $U_{i,1} \in [0,1]$ and $U_{i,2} \in [0,1]$
        Set $\alpha_i = U_{i,1}$
        Compute $^\alpha A_i = [\breve{a}_{\alpha_i}, \hat{a}_{\alpha_i}]$
        If $U_{i,2} \leqslant 0.5$ then $x_i = \breve{a}_{\alpha_i}$, otherwise set $x_i = \hat{a}_{\alpha_i}$
    end for
    return $\bar{x}_A = \sum_1^n x_i / n$ as the sample mean of $X(\omega)$

It is worth to note that every x_i is a realization of $X(\omega)$ with membership α. The main idea behind is to generate statistically independent fuzzy random variables to then compare $\bar{x}_A$ to $I(A), M(A)$ and $C(A)$.

5   Numerical Experiments

In order to compare all four methods, we selected four shapes: Gaussian $G(c, \delta)$, triangular $T(a, c, b)$, exponential $e(c, \theta_l, \theta_r)$ and quadratic $Q(c, \beta_l, \beta_r)$, then we set three different combinations of parameters per shape which consists on a symmetric and two asymmetric membership functions. We use 1000 α-cuts i.e. 2000 samples and all equations presented in Sects. 3 and 4.

The sample mean is computed assuming $\{\check{a}_\alpha, \hat{a}_\alpha\} \in \mathbb{R}$ as random variables which means that they are considered just as samples without a particular membership degree. For the fuzzy random variable method, we generate 2000 realizations per shape using Method 1. The results are shown in Table 1.

Table 1. Obtained results for different fuzzy numbers

Shape	Gaussian fuzzy sets $G(c,\delta)$			Triangular fuzzy sets $T(a,c,b)$		
Measure	$G(50,7)$	$G(10,5)$	$G(5,1)$	$T(10,20,30)$	$T(0,20,25)$	$T(15,20,50)$
$\bar{x}_A^*$	50	10	5	20	16.25	26.25
$\bar{x}_A^{**}$	49.98	10.01	5.02	19.98	16.26	26.24
$I(A)$	50	10	5	20	16.25	26.25
$C(A)$	50	10	5	20	15	28.33
$M(A)$	50	10	5	20	17.5	24.17
Shape	Exponential fuzzy sets $e(c,\theta_l,\theta_r)$			Quadratic fuzzy sets $Q(c,\beta_l,\beta_r)$		
Measure	$e(50,5,5)$	$e(50,2,7)$	$e(50,8,1)$	$Q(10,10,10)$	$Q(10,2,20)$	$Q(10,10,2)$
$\bar{x}_A^*$	50	47.5	53.5	20	16.25	26.25
$\bar{x}_A^{**}$	50.01	47.47	53.52	19.98	16.26	26.24
$I(A)$	50	47.5	53.5	20	16.25	26.25
$C(A)$	50	45	56.99	20	15	28.33
$M(A)$	50	48.76	51.75	20	17.5	24.17

5.1 Obtained Results

Table 1 shows the obtained results for the selected combinations per fuzzy set in which $\bar{x}_A^*$ is the sample mean of the 1000 α-cuts used to obtain $I(A)$ and $M(A)$ and $\bar{x}_A^{**}$ is the sample mean of the 2000 fuzzy random variables generated using Method 1.

It is worth to note that all methods obtain the same results for symmetric fuzzy numbers e.g. Gaussian and symmetric shapes since symmetry leads to center all expected values at c. This is the same property shown by probability distributions and it is an expected result.

Asymmetric fuzzy numbers show a different behavior since all of them lead to $\bar{x}_A^* \approx \bar{x}_A^{**} \approx I(A)$. Moreover, $\bar{x}_A^* \neq C(A) \neq M(A)$ which is very interesting and it leads us to infer the following:

1. The Yager index $I(A)$ seems to agree to the classical expectation of a random variable (see Kolmogorov [5]) since $I(A) \approx \bar{x}_A^* \approx \bar{x}_A^{**}$.

2. The centroid and Carlsson-Fullér approaches do not converge to the Yager index since $I(A)$ is closer to the parameter $c \in \mathbb{R}$ of all asymmetric fuzzy sets. On the other hand, $I(A) = C(A) = M(A) = c$ for symmetric fuzzy numbers.
3. The centroid $C(A)$ seems to be the farthest measure from c while $I(A)$ is the closest.
4. The possibilistic mean $M(A)$ is always somewhere between $I(A)$ and $C(A)$ (except for symmetric shapes).

Therefore, it is clear the relationship between $I(A)$ and the expectation of random variables in the Kolmogorov's sense. On the other hand, $C(A)$ and $M(A)$ still give an idea of the expected value of A although they are farther from the parameter c.

6   Concluding Remarks

We have compared three well known location measures for fuzzy numbers to the sample mean $\bar{x}$ of their α-cuts: the Yager index $I(A)$, the centroid $C(A)$ and the possibilistic mean $M(A)$. There is clear evidence of the differences among them where $I(A) = \bar{x}$ only.

Whereas $I(A) = \bar{x}$ we can see that $\bar{x}_A^* \neq C(A) \neq M(A)$, and also that $I(A)$ is closer to c than both $C(A)$ and $M(A)$. This means that $C(A)$ and $M(A)$ are more sensible to asymmetry than $I(A)$ and $\bar{x}_A$.

There is evidence to think that Kolmogorov's random variables theory also applies to fuzzy sets/numbers in the sense that the Yager index seems to be the sample mean counterpart. Further definitions about fuzzy random variables have been provided by Kruse [6] and Kwarkernaak [7,8].

Future Work

Similar approaches to Type-2 fuzzy sets (see Figueroa-García et al. [3]) are a natural extension of the presented analysis. A theoretical analysis and a formal framework of the experimental evidence is needed as well.

References

1. Carlsson, C., Fullér, R.: On possibilistic mean value and variance of fuzzy numbers. Fuzzy Sets Syst. **122**(1), 315–326 (2001)
2. Chaudhuri, B., RosenFeld, A.: A modified Hausdorff distance between fuzzy sets. Inf. Sci. **118**, 159–171 (1999)
3. Figueroa-García, J.C., Chalco-Cano, Y., Román-Flores, H.: Yager index and ranking for interval type-2 fuzzy numbers. IEEE Trans. Fuzzy Syst. **81**(1), 93–102 (2018)
4. Hung, W.L., Yang, M.S.: Similarity measures between type-2 fuzzy sets. Int. J. Uncertainty, Fuzziness Knowl.-Based Syst. **12**(6), 827–841 (2004)
5. Kolmogorov, A.N.: Foundations of the Theory of Probability. Chelsea Publishing, New York (1956)

6. Kruse, R.: The strong law of large numbers for fuzzy random variables. Inf. Sci. **28**(1), 233–241 (1982)
7. Kwarkernaak, H.: Fuzzy random variables I. Inf. Sci. **15**(1), 1–29 (1978)
8. Kwarkernaak, H.: Fuzzy random variables II. Inf. Sci. **17**(1), 153–178 (1979)
9. Pulido-López, D.G., García, M., Figueroa-García, J.C.: Fuzzy uncertainty in random variable generation: a cumulative membership function approach. Communications in Computer and Information Science, vol. 742, no. 1, pp. 398–407 (2017). https://doi.org/10.1007/978-3-319-66963-2_36
10. Varón-Gaviria, C.A., Barbosa-Fontecha, J.L., Figueroa-García, J.C.: Fuzzy uncertainty in random variable generation: an α-cut approach. Lecture Notes in Artificial Intelligence, vol. 10363, no. 1, pp. 1–10 (2017)
11. Wu, D., Mendel, J.M.: A comparative study of ranking methods, similarity measures and uncertainty measures for interval type-2 fuzzy sets. Inf. Sci. **179**(1), 1169–1192 (2009)
12. Yager, R.: A procedure for ordering fuzzy subsets of the unit interval. Inf. Sci. **24**(1), 143–161 (1981)

Why Grade Distribution Is Often Multi-modal: An Uncertainty-Based Explanation

Olga Kosheleva[1], Christian Servin[2], and Vladik Kreinovich[3]([✉])

[1] Department of Teacher Education, University of Texas at El Paso, El Paso, TX 79968, USA
olgak@utep.edu

[2] Computer Science and Information Technology Systems Department, El Paso Community College, 919 Hunter, El Paso, TX 79915, USA
cservin@gmail.com

[3] Department of Computer Science, University of Texas at El Paso, El Paso, TX 79968, USA
vladik@utep.edu

Abstract. There are many different independent factors that affect student grades. There are many physical situations like this, in which many different independent factors affect a phenomenon, and in most such situations, we encounter normal distribution – in full accordance with the Central Limit Theorem, which explains that in such situations, distribution should be close to normal. However, the grade distribution is definitely not normal – it is multi-modal. In this paper, we explain this strange phenomenon, and, moreover, we explain several observed features of this multi-modal distribution.

1 Formulation of the Problem

Many Different Factors Affect the Student Grades. Many different independent factors affect the student's grade in a class. The grade can be affected by a student's preparedness for different sections of the material, by the student's degree of involvement in other classes, by how well the professor's teaching style matches the student's learning style, by possible personal problems – the list can go on and on.

Based on This, One Would Expect Normal Distribution of the Grades. Situations when the result comes from the joint effect of a large number of independent factors are ubiquitous in real life. From the mathematical viewpoint, such situations have been well analyzed. It is known that the distribution of the sum of a large number of relatively small independent random variables is close to Gaussian (normal)—this is the gist of the so-called Central Limit Theorem; see, e.g., [16]. And indeed, normal distributions are encountered in many such situations.

© Springer Nature Switzerland AG 2019
R. B. Kearfott et al. (Eds.): IFSA 2019/NAFIPS 2019, AISC 1000, pp. 106–112, 2019.
https://doi.org/10.1007/978-3-030-21920-8_10

Based on the above explanation, one would expect that in a large class, grades would also be normally distributed. But they are not.

A Puzzling Fact: Grade Distribution Is Multi-modal. Even for a relatively large class, we very rarely see the bell-shaped curve of a normal distribution. In reality, the distribution is multi-modal.

This multi-modality is a well-known phenomenon: so well-known that many professors use it for grading. To avoid a natural student's complaint about grading fairness, that this student with 89.9 got a B but someone with a practically indistinguishable grade of 90.1 gets an A, experienced teachers recommend to use, as an A-or-B threshold, not some arbitrary number like 90, but the largest gap between the grades which is close to 90. This way, there is a significant gap between B and A students, and thus, the grades are viewed as more fair than before.

Such a gap can always be found – exactly because the distribution is multi-modal, because between the modes, the probability density gets very low, and thus, gaps between neighboring grades become much larger than in the vicinity of each mode.

But why? It is great that we can use multi-modality, but the question remains: why? In this paper, we provide a possible uncertainty-based explanation for this unexpected phenomenon.

Additional Observations. Another interesting phenomenon is that the number of modes does not stay the same throughout the students' studies: for under-graduate students, we have more modes, while for graduate students, we observe fewer modes. This seems to be in perfect accordance with the fact that in under-graduate studies, we usually use more different grades: A, B, and C, while for graduate students, C is practically a failure grade, so, in effect, we only use As and Bs. Convenient, but why?

Yet another convenient-but-why observation is that modes are almost equidistant, so the corresponding clusters are indeed close to the usual groupings of 90–100, 80–90, 70–80, etc.

In this paper, we try to explain these additional observations as well.

2   Analysis of the Problem and the Resulting Qualitative Explanation

Important Phenomenon: Students Help Each Other. At first glance, the situation with grades is the same as with other cases when Central Limit Theorem works: e.g., in situations like Brownian motion where the motion of a particle is caused by a joint effect of many different phenomena.

At first glance, the situation is the same, and if the students were simply randomly affected by all the factors mentioned above, we probably would have observed exactly the same normal distribution as in many physical situations.

But there is a big difference between students and particles: students help each other. This help may not always be a huge contribution to the student's success, but, as everyone who has ever studied knows well, it does provide an important help. Students ask questions to each other, students exchange ideas – and often even form study groups to study together, and it helps.

How does this helping phenomenon affect the resulting grade distribution?

What Happens When Two Students Study Together: Ideal Case. In the ideal case, when two students study with each other, they exchange knowledge, and at the end, both get the exact same amount of knowledge. To be more precise, each student knows exactly what he knew before + what the other student knew. As a result, if at this moment, we give them a test, they will get the exact same grade – reflecting their exact same state of knowledge.

What Happens in Practice. In practice, this ideal exchange of information only happens when students are at approximately the same level of knowledge. If we try to bring together two students with a big gap between them – e.g., a straight A student and an almost-failing student – this rarely helps, because most students lack the ability to clearly explain things to those who know much less.

For example, in the department where one of the authors (VK) works, when we started hiring undergraduate instructional assistants for classes, it turned out that students who in their time got B for the corresponding classes were much better in helping new students than those who got A – the A students knew material much better, but they could not as convincingly explain it to the new students.

Not only students, starting professors (and even some should-have-been-experienced professors) have the same limitation.

As a result, this exchange of knowledge happens only when the difference between the students' levels of knowledge – i.e., the difference between their grades – is small. The larger the difference, the less probable it is that the knowledge exchange will happen.

Eventually, the students get better in this knowledge exchange skills: as they progress from undergraduate students to graduate ones, their ability improves, and the threshold beyond which they cannot effectively exchange knowledge increases.

What Happens as a Result: A Qualitative Description. How does the existence of this collaboration gap affect distribution of student grades? To understand the effect on the qualitative level, let us consider a simplified model of student grade distribution.

Suppose that originally, students' knowledge levels are uniformly distributed – at least on some segment of the grades interval. This can be simplified into saying that the students' grades are initially distributed with the same step h. In other words, these grades, when sorted in the increasing order, form the following sequence:

$$g_0 < g_1 = g_0 + h < g_2 = g_0 + 2h < \ldots < g_k = g_0 + k \cdot h < \ldots < g_n = g_0 + n \cdot h.$$

In the beginning, the least-well performing student is the one who is the most desperate for help. So, it is reasonable to expect that first, the student whose grade is g_0 will reach for help. The person most appropriate for helping this g_0-level student is the person whose grade is the closest to g_0 – i.e., the student with the grade g_1. They start actively collaborating, and, as the result of this collaboration, the reach the same grade level – which is close to g_1. For simplicity, let us assume that their grade level is now exactly g_1.

The student whose grade is g_2 also needs help—so he/she contacts the closest better student, the one with the grade level g_3. As a result of their collaboration, they both reach the same level g_3.

Similarly, the first yet-unpaired student g_4 teams us with g_5, so their grade level is now g_5, etc. As a result of this first round of exchanges, we have pairs of students whose grades are

$$g_1 = g_0 + h, g_3 = g_0 + 3h, g_5 = g_0 + 5h, \ldots ;$$

note that the gap between different levels has doubled, from h to $2h$.

Now, the same process starts again: students at level g_1 are the most eager for help, so they contact students at the next level g_3 to form a study group. As a result of their joint study, all four of them reach the level g_3.

Students at the lowest not-yet-involved level g_5 contact students from the level g_7 and all get to the level g_7, etc. Now, we have a new list of grades:

$$g_3 = g_0 + 3h < g_7 + 7h < g_{11} = g_0 + 11h < \ldots$$

The gap has doubled again, to $4h$. At the next iteration of this process, the gap will double again – until it reaches the threshold after which the mutual exchange of knowledge becomes difficult.

As a result, instead of the original uniform distribution, we have big groups with approximately the same level of knowledge – separated by gaps in which there are no students with this particular grade.

From Simplified Model to Real Life Situations. Of course, the above description is oversimplified. In reality, the original distances $g_i - g_{i-1}$ are not exactly equal, and the effects are also not always the same. As a result, what we get is not the above simplified picture, but rather a smoothed version of it: instead of groups of students with identical grades, we have groups with close grades – i.e., in effect, we will have a multi-modal distribution.

So, the mutual help indeed explains why grade distribution is multi-modal.

Why There Are Fewer Modes for Grades of Graduate Students? The same phenomenon explains why for graduate students, we usually have fewer modes than for undergraduate ones: graduate students have already learned how to exchange knowledge, so for them, the threshold above which they cannot productively collaborate is much higher. As a result, they continue merging into a single cluster even when at the undergraduate level, we would have reached the original merging threshold and stopped. As a result, for graduate students, we have fewer clusters – i.e., fewer modes of the resulting distribution.

3   Towards a Quantitative Analysis

Analogy with Physics. To looks for a quantitative analysis of the situation, let us look for other situations when a similar phenomenon occurs.

What is the above phenomenon? We started with a distribution which was perfectly uniform. This distribution was symmetric, in the sense that it does not change – at least locally – if we simply shift all the grades by the same number h. Then what happens if two nearby students with grades g_i and g_{i+1} start collaborating? As a result, the knowledge of both students reaches the same level g_{i+1}: $g'_i = g'_{i+1} = g_{i+1}$. Now, we get a gap of width $2h$ between the levels $g'_{i-1} = g_{i-1}$ and $g'_i = g_{i+1}$.

The distribution is no longer invariant with respect to a shift by h – even if many pairs exchange their knowledge. The original symmetry is broken.

This phenomenon of spontaneous symmetry breaking is ubiquitous in physics; see, e.g., [2,17]. We can easily observe this phenomenon: e.g., if we drop a breakable rotationally symmetric vase, it will not break into rotationally symmetric pieces: it will break into irregular ones.

This phenomenon is very important: without it, our Universe would remain the same highly homogenous and isotropic blurb that it was close to the Big Bang. Luckily, gravity acts as the spontaneous symmetry breaking mechanism. Specifically, if a small fluctuation appears and at some location, the density at this location becomes slightly larger than at other locations, then this heavier location will start attracting other particles. As it attracts them, its mass increases and it attracts more and more – until the whole original homogeneous cloud disintegrates into what we call proto-galaxies [3,4,8].

In Physics, Researchers Go Beyond Qualitative Explanations. Not only this mechanism explains symmetry breaking, it explains all the observed shapes of celestial bodies, such as spiral galaxies and planetary systems like ours in which distances of the planets to the central star form a geometric progression; see, e.g., [3,4,8]. (This mechanics also explains relative frequencies of different shapes.)

To understand the corresponding explanations, we need to know the basic ideas of statistical physics, according to which it is not very probable to go from a completely symmetric state to a state with no symmetries at all: it is much more probable that – at least at first – some symmetries will be preserved, and the more symmetries will be preserved, the more probable the corresponding transition. For example, a solid body (i.e., matter in highly symmetric – usually crystal – state), when heated, usually does not immediately gets transformed into a completely asymmetric state of gas, it first gets transformed into the state of the liquid in which some symmetries are preserved [2,17].

What Are the Symmetries in the Grades Case? How can we apply the above idea in our example – of grade distribution?

In the gravity case, the original symmetries were easy to find: rotations, shifts, probably scalings. To apply a similar approach to grade distribution, we need to understand what are the natural symmetries here. Let us brainstorm.

How are grades formed? Usually, by simply adding the grades corresponding to different assignments – and these grades, in their turn, are obtained by simply adding grades on different problems or parts or aspects of each assignment. Some assignments are very tough, some are much easier. There have to be easier assignments: we are talking mass education, not training students to win at an international student olympiad in computer science.

What does it mean that the assignment is relatively easy? That on this particular assignment (or part of the assignment), practically all the students will get a very good grade. One professor may give a certain number of such assignments, another professor may give one more such relatively simple task. The difference between the grades given by these two professors will be exactly the grade e on this extra assignment.

So, depending on who teaches a class, for the same level of knowledge, students may get grades g_i from one professor and grades $g_i + e$ from another one. This shift $g_i \to g_i + e$ is therefore a reasonable symmetry here. In other words, the original situation is invariant under all possible shifts $g \to g + e$.

Now, spontaneous symmetry breaking occurs, and the situation is no longer fully symmetric. However, in line with the general ideas from statistical physics, the most probable situation is that *some* of the original symmetries will remain. In other words, there remains some value e_0 so that – at least locally – the resulting distribution will not change if we simply add e_0 to all the grades. In particular, this means that if we add e_0 to one mode (i.e., to one local maximum of the corresponding probability distribution), then we should again encounter a mode—i.e., yet another local maximum. So, in the first approximation, local maxima (modes) are almost equidistant – and, as we have mentioned, this is exactly what we observe!

Thus, this equidistance distribution can also be explained by our analysis.

Future Work. To make the conclusions more quantitative, we need to provide a formal explanation of the threshold. In the above analysis, we viewed the threshold as, in effect, an interval beyond which collaboration is not productive – in line with the interval uncertainty (see, e.g., [5,7,9,11,12]). However, in practice, this threshold is not precise, it is imprecise – so we believe that the use of fuzzy techniques (see, e.g., [1,6,10,13–15,18]) will lead to an even better description of this phenomenon.

Acknowledgments. This work was supported in part by the US National Science Foundation via grant HRD-1242122 (Cyber-ShARE Center of Excellence).

The authors are thankful to the anonymous referees for valuable suggestions.

References

1. Belohlavek, R., Dauben, J.W., Klir, G.J.: Fuzzy Logic and Mathematics: A Historical Perspective. Oxford University Press, New York (2017)
2. Feynman, R., Leighton, R., Sands, M.: The Feynman Lectures on Physics. Addison Wesley, Boston (2005)
3. Finkelstein, A., Kosheleva, O., Kreinovich, V.: Astrogeometry: towards mathematical foundations. Int. J. Theor. Phys. **36**(4), 1009–1020 (1997)
4. Finkelstein, A., Kosheleva, O., Kreinovich, V.: Astrogeometry: geometry explains shapes of celestial bodies. Geombinatorics **VI**(4), 125–139 (1997)
5. Jaulin, L., Kiefer, M., Didrit, O., Walter, E.: Applied Interval Analysis, with Examples in Parameter and State Estimation, Robust Control, and Robotics. Springer, London (2001)
6. Klir, G., Yuan, B.: Fuzzy Sets and Fuzzy Logic. Prentice Hall, Upper Saddle River (1995)
7. Kreinovich, V., Lakeyev, A., Rohn, J., Kahl, P.: Computational Complexity and Feasibility of Data Processing and Interval Computations. Kluwer, Dordrecht (1998)
8. Li, S., Ogura, Y., Kreinovich, V.: Limit Theorems and Applications of Set Valued and Fuzzy Valued Random Variables. Kluwer Academic Publishers, Dordrecht (2002)
9. Mayer, G.: Interval Analysis and Automatic Result Verification. de Gruyter, Berlin (2017)
10. Mendel, J.M.: Uncertain Rule-Based Fuzzy Systems: Introduction and New Directions. Springer, Cham (2017)
11. Moore, R.E., Kearfott, R.B., Cloud, M.J.: Introduction to Interval Analysis. SIAM, Philadelphia (2009)
12. Rabinovich, S.G.: Measurement Errors and Uncertainties: Theory and Practice. Springer, New York (2005)
13. Nguyen, H.T., Kreinovich, V.: Nested intervals and sets: concepts, relations to fuzzy sets, and applications. In: Kearfott, R.B., Kreinovich, V. (eds.) Applications of Interval Computations, pp. 245–290. Kluwer, Dordrecht (1996)
14. Nguyen, H.T., Walker, C., Walker, E.A.: A First Course in Fuzzy Logic. Chapman and Hall/CRC, Boca Raton (2019)
15. Novák, V., Perfilieva, I., Močkoř, J.: Mathematical Principles of Fuzzy Logic. Kluwer, Boston, Dordrecht (1999)
16. Sheskin, D.J.: Handbook of Parametric and Nonparametric Statistical Procedures. Chapman and Hall/CRC, Boca Raton (2011)
17. Thorne, K.S., Blandford, R.D.: Modern Classical Physics: Optics, Fluids, Plasmas, Elasticity, Relativity, and Statistical Physics. Princeton University Press, Princeton (2017)
18. Zadeh, L.A.: Fuzzy sets. Inf. Control **8**, 338–353 (1965)

How to Fuse Expert Knowledge: Not Always "And" but a Fuzzy Combination of "And" and "Or"

Christian Servin[1], Olga Kosheleva[2], and Vladik Kreinovich[2(✉)]

[1] Computer Science and Information Technology Systems Department, El Paso Community College, 919 Hunter, El Paso, TX 79915, USA
cservin@gmail.com
[2] University of Texas at El Paso, El Paso, TX 79968, USA
{olgak,vladik}@utep.edu

Abstract. In the non-fuzzy (e.g., interval) case, if two expert's opinions are consistent, then, as the result of fusing the knowledge of these two experts, we take the intersection of the two sets (e.g., intervals) describing the expert's opinions. In the experts are inconsistent, i.e., if the intersection is empty, then a reasonable idea is to assume that at least one of these experts is right, and thus, to take the union of the two corresponding sets. In practice, expert opinions are often imprecise; this imprecision can be naturally described in terms of fuzzy logic – a technique specifically designed to describe such imprecision. In the fuzzy case, expert opinions are not always absolutely consistent or absolutely inconsistent, they may be consistent to a certain degree. In this case, we show how the above natural idea of fusing expert opinions can be extended to the fuzzy case. As a result, we, in general, get not "and" (which would correspond to the intersection), not "or" (which would correspond to the union), but rather an appropriate fuzzy combination of "and"- and "or"-operations.

1 Fusing Expert Knowledge: Formulation of the Problem

Need to Fuse Knowledge of Different Experts. Expert estimates of different quantities are usually not very accurate – e.g., in situations when measurements are also possible, measurement results are usually much more accurate than expert estimates.

When we can perform measurements, we can further increase the measurement accuracy if we use several different measuring instruments and then combine ("fuse") their results. It is known that such combinations are usually more accurate than all original measurement results.

In many situations, measurements are not realistically possible, so we have to rely on expert estimates only. In such situations, we can increase the accuracy of the resulting estimates the same way we increase the accuracy of measurement results: by combining (fusing) estimates of several experts.

R. B. Kearfott et al. (Eds.): IFSA 2019/NAFIPS 2019, AISC 1000, pp. 113–120, 2019.
https://doi.org/10.1007/978-3-030-21920-8_11

Examples. To estimate the temperature, we can ask two experts. Suppose that:

- one expert states that the temperature is between 22 and 25 °C, and
- another expert states the temperature is in the low seventies, i.e., between 70 and 75 °F – which corresponds to between 21 and 24 °C.

Then we can conclude that the actual temperature is larger than 22 °C and smaller than 24 °C – i.e., the actual temperature is between 22 and 24 °C.

In this case, if we only asked one expert, we would have an interval of width 3 that contains the actual (unknown) temperature value. But by fusing the opinions of the two experts, we get a narrower interval [22, 24] of width 2 – i.e., we have indeed increased the accuracy.

Fusion is also possible on a non-quantitative level. For example, we can ask experts whether the wind is weak, moderate, or strong. Suppose that:

- one expert says that the wind is not weak, while
- another expert says that the wind is not strong.

By combining the opinions of both experts, we can conclude that the wind is moderate.

On the other hand, if we only asked one of the experts, we would not be able to come to this conclusion:

- if we only took into account the opinion of the first expert, then we would only be able to conclude that the wind is either moderate or strong;
- similarly, if we only took into account the opinion of the first expert, then we would only be able to conclude that the wind is either moderate or weak.

Fusing Expert Knowledge: Non-fuzzy Case. To understand how to best combine expert estimates, let us start with the case when expert estimates are crisp (non-fuzzy), i.e., when for each possible value of the estimated quantity, the expert is either absolutely sure that this value is possible or is absolutely sure that the given value is not possible. In this case, each expert estimate provides us with a set of possible values of the corresponding quantity. In most practical cases, this set is an interval $[\underline{x}, \overline{x}]$.

In these terms, when we have estimates of two different experts, this means that:

- based on the opinions of the first expert, we form a set S_1 of numbers which are, according to this expert, possible values of the estimates quantity;
- also, based on the opinions of the second expert, we form a set S_2 of numbers which are, according to this expert, possible values of the estimates quantity.

In general, different experts take into account different aspects of the situation. For example, the first expert may know the upper bound $\overline{x}$ on the corresponding quantity. In this case, the set S_1 consists of all the numbers which are smaller than or equal to $\overline{x}$, i.e., $S_1 = (-\infty, \overline{x}]$. The second expert may know the lower bound $\underline{x}$, in which case $S_2 = [\underline{x}, \infty)$. In such situations, a natural way to fuse the knowledge is to consider numbers which are possible according to both experts, i.e., in mathematical terms, to consider the intersection $S_1 \cap S_2$ of the two sets S_1 and S_2.

A problem occurs when this intersection is empty, i.e., when the opinions of two experts are inconsistent. This happens: experts are human and can thus make mistakes. In this case, an extreme option is to say that since experts are not consistent with each other, this means that we do not trust what each of them says, so we can as well ignore both opinions; the result of fusion is then the whole real line.

A more reasonable option is to conclude that, yes, both experts cannot be true, but we cannot conclude that both are wrong; they are experts after all, so it is reasonable to assume that one of them is right; in this case, the result of the fusion is the union $S_1 \cup S_2$ of the two sets.

In other words, the fusion $S_1 f S_2$ of the sets S_1 and S_2 has the following form;

- if $S_1 \cap S_2 \neq \emptyset$, then $S_1 f S_2 = S_1 \cap S_2$;
- otherwise, if $S_1 \cap S_2 = \emptyset$, then $S_1 f S_2 = S_1 \cup S_2$.

Example. Suppose that:

- one expert says that the temperature is between 22 and 25, and
- another one claims that it is between 18 and 21.

In this case, the intersection of the corresponding intervals $[22, 25]$ and $[18, 21]$ is empty – which means that the experts cannot be both right. What we can conclude – if we still believe that one of them is right – is that the temperature is either between 22 and 25 or between 18 and 21.

Need to Consider the Fuzzy Case. In practice, experts are rarely absolutely confident about their opinions. Usually, they are only confident to a certain degree. As a result, to adequately describe expert knowledge, we need to describe, for each number x, the degree to which, according to this expert, the number x is possible. This is the *fuzzy logic* approach; see, e.g., [1–7].

In the computer, "true" (="absolute certain") is usually represented as 1, and "false" (="absolutely certain this is false") is represented as 0. It is therefore reasonable to describe intermediate degrees of confidence by numbers intermediate between 0 and 1. Thus, to describe an expert's estimate, we need to have a function $\mu(x)$ that assigns, to each value x of the corresponding quantity, a number $\mu(x) \in [0, 1]$ that describes to what extent the value x is possible. Such a function is known as a *membership function*, or, alternatively, a *fuzzy set*. From this viewpoint, to be able to fuse expert estimates, we need to be able to fuse fuzzy sets.

A traditional approach to fusing fuzzy knowledge simply takes the intersection – which is then usually normalized, i.e., multiplied by a constant so that the maximum value is 1. However, this does not work if the expert opinions are inconsistent. We should therefore take into account that the expert opinions can be inconsistent – or, more generally, consistent to a certain degree. How can take this into account?

What We Do in This Paper. In this paper, we show how to extend the above fusion operation to the fuzzy case.

2   Analysis of the Problem

In General, How Notions are Generalized to the Fuzzy Case. The usual way to generalize different notions to the fuzzy case is as follows:

- First, we describe the original notion in logical terms, by using "and', "or", and quantifiers "for all" (which is, in effect, infinite "and") and "exists" (which is, in effect, infinite "or").
- Then, we replace each "and" operation with the fuzzy "and"-operation $f_\&(a, b)$ (also known as t-norm) and every "or"-operation with the fuzzy "or"-operation $f_\vee(a, b)$ (also known as t-conorm).

In selecting the t-norms and t-conorms, we need to be careful, in the following sense.

- If we have a universal quantifier – i.e., an infinite "and" – and we use, e.g., a product t-norm $f_\&(a, b) = a \cdot b$, then the product of infinitely many values smaller than 0 will be most probably simply 0. So, if we have an infinite "and" (=universal quantifier), the only t-norm that leads to meaningful results is the minimum

$$f_\&(a, b) = \min(a, b).$$

- Similarly, if we have an existential quantifier – i.e., an infinite "or" – and we use, e.g., an algebraic sum t-conorm $f_\vee(a, b) = a + b - a \cdot b$, then the result of applying this operation to infinitely many values larger than 0 will be most probably simply 1. So, if we have an infinite "or" (=existential quantifier), the only t-conorm that leads to meaningful results is the maximum

$$f_\vee(a, b) = \max(a, b).$$

How to Define Degree of Consistency. Let us use the above-described general approach to define the degree of consistency. In the non-fuzzy case, two expert opinions are consistent if there exists a value x for which both the first expert and the second expert agree that this value is possible.

For each real number x representing a possible value of the quantity of interest:

- let $\mu_1(x)$ denote the degree to which the first expert believes the value x to be possible, and

- let $\mu_2(x)$ denote the degree to which the second experts believes the value x to be possible.

Then, for each value x, the degree to which both experts consider the value x to be possible is equal to $f_\&(\mu_1(x), \mu_2(x))$, where $f_\&(a, b)$ is an appropriate fuzzy "and"-operation (t-norm).

In line with the above general scheme for generalizing notions into fuzzy, the existential quantifier over x is translated into maximum over x (which corresponds to the use of the maximum "or"-operation $f_\vee(a, b) = \max(a, b)$). Thus, we get the following formula for the degree $d(\mu_1, \mu_2)$ for which two membership functions are consistent:

$$d(\mu_1, \mu_2) = \max_x f_\&(\mu_1(x), \mu_2(x)). \qquad (1)$$

Accordingly, in line with a general description of negation in fuzzy logic, the degree to which the expert opinions are *inconsistent* can be computed as

$$1 - d(\mu_1, \mu_2).$$

Resulting Definition of Fusion. According to the definition given in the previous section, in the non-fuzzy case, the value x belongs to the fused set if:

- either the two sets describing expert opinions are consistent, and x belongs to the intersection of the two sets,
- or the two sets describing expert opinions are inconsistent, and x belongs to the union of these two sets.

Let us use the general methodology to generalize the above description to the fuzzy case. For each x:

- we know the degree $d(\mu_1, \mu_2)$ to which the experts are consistent, and
- we know the degree $f_\&(\mu_1(x), \mu_2(x))$ to which x belongs to the intersection.

Thus, the degree to which the expert opinions are consistent *and* x belongs to the intersection can be obtained by applying the "and"-operation $f_\&(a, b)$ to these two degrees. Thus, we get the value

$$f_\&(d(\mu_1, \mu_2), f_\&(\mu_1(x), \mu_2(x))) = f_\&(d(\mu_1, \mu_2), \mu_1(x), \mu_2(x)). \qquad (2)$$

Similarly, for each x:

- we know the degree $1 - d(\mu_1, \mu_2)$ to which the experts are inconsistent, and
- we know the degree $f_\vee(\mu_1(x), \mu_2(x)) = \max(\mu_1(x), \mu_2(x))$ to which x belongs to the union.

Thus, the degree to which the expert opinions are inconsistent *and* x belongs to the union can be obtained by applying the "and"-operation $f_\&(a, b)$ to these two degrees. Thus, we get the value

$$f_\&(1 - d(\mu_1, \mu_2), \max(\mu_1(x), \mu_2(x))). \qquad (3)$$

To find the degree $\mu(x)$ to which the value x belongs to the fused set, we need to apply the "or"-operation $f_\vee(a, b) = \max(a, b)$ to the degrees (2) and (3). As a result, we get the following formula.

3   Resulting Formula: Formulation and Example

Resulting Formula: General Case. If we know the functions $\mu_1(x)$ and $\mu_2(x)$ that describe the opinions of the two experts, then, to describe the fused opinion, we should take the function

$$\mu(x) = \max(d_1(x), d_2(x)), \tag{4}$$

where

$$d_1(x) \stackrel{\text{def}}{=} f_{\&}(d(\mu_1, \mu_2), \mu_1(x), \mu_2(x)), \tag{5}$$

$$d_2(x) \stackrel{\text{def}}{=} f_{\&}(1 - d(\mu_1, \mu_2), \max(\mu_1(x), \mu_2(x)))), \tag{6}$$

$f_{\&}(a, b)$ is an "and"-operation (t-norm) and the degree $d(\mu_1, \mu_2)$ is determined by the formula

$$d(\mu_1, \mu_2) = \max_x f_{\&}(\mu_1(x), \mu_2(x)). \tag{1}$$

Discussion. We can see that this fused fuzzy set is not exactly "and", it is not exactly "or" – it is a fuzzy combination of "and" and "or".

Case when $f_{\&}(a, b) = \min(a, b)$. In the case when, as the "and"-operation, we select the simplest possible "and"-operation $f_{\&}(a, b) = \min(a, b)$, the above formulas (4)–(6) can be further simplified. Namely, by definition of the degree of consistency $d(\mu_1, \mu_2)$, this degree is the largest of the values $f_{\&}(\mu_1(x), \mu_2(x))$. Thus, for every x, we have $f_{\&}(\mu_1(x), \mu_2(x)) \le d(\mu_1, \mu_2)$. Therefore, since our "and"-operation is minimum, we get a simplified expression for the formula (2):

$$f_{\&}(d(\mu_1, \mu_2), f_{\&}(\mu_1(x), \mu_2(x)) = f_{\&}(\mu_1(x), \mu_2(x)) = \min(\mu_1(x), \mu_2(x)).$$

Thus, for the minimum "and"-operation, the formulas (4)–(6) take the following simplified form:

$$\mu(x) = \max(\min(\mu_1(x), \mu_2(x)), \min(1 - d(\mu_1, \mu_2), \max(\mu_1(x), \mu_2(x)))). \tag{7}$$

Example. To illustrate the above formula, let us consider a simple case when the "and"-operation is minimum, and the membership functions are triangular functions of the same width. To make the computations even easier, let us select, as a starting point for measuring x, the arithmetic average between the most probable values corresponding to the two experts, and let us select the measuring unit so that the half-width of each membership function is 1.

In this case, the triangular membership functions are described by the formulas $\mu_1(x) = \max(0, 1 - |x - a|)$ and $\mu_2(x) = \max(0, 1 - |x + a|)$, for some $a > 0$. This value a is the half of the difference between the most probable value (a) according to the first expert and the most probable value according to the second expert ($-a$): $a = \dfrac{a - (-a)}{2}$.

When $a \geq 1$, the two membership functions have no intersection at all, so $d(\mu_1, \mu_2) = 0$, and the fused set is simply their union $\max(\mu_1(x), \mu_2(x))$, a bi-modal set whose graph consists of the two original triangles.

The more interesting case is when $a < 1$. In this case, the two sets have some degree of intersection. For such values a, as one can easily check, the intersection $f_\&(\mu_1(x), \mu_2(x))$ is also a triangular function $\max(0, 1 - a - |x|)$. The maximum $d(\mu_1, \mu_2)$ of this function is attained when $x = 0$ and is equal to $1 - a$.

Correspondingly, the degree to which the two expert opinions are inconsistent is equal to $1 - d(\mu_1, \mu_2) = 1 - (1 - a) = a$. By applying the formula (7), we can now conclude the following.

When $a \leq 0.5$, the fused expression is still a *fuzzy number*, i.e., a membership function which first increases and then decreases. Specifically:

- The fused function $\mu(x)$ starts being non-zero at the value $x = -1 - a$; between the value $-1 - a$ and -1, it grows as $\mu(x) = x - (-1 - a) = 1 + a - x$.
- Between the values $x = -1$ and $x = -(1 - 2a)$, the fused function remains constant $\mu(x) = a$.
- Between $x = -(1 - 2a)$ and $x = 0$, it grows as $\mu(x) = 1 - a + x$, until it reaches the value $1 - a$.
- Then, for x from 0 to $1 - 2a$, it decreases as $\mu(x) = 1 - a - x$ until it reaches the value a for $x = 1 - 2a$.
- Then, the value stays constant $\mu(x) = a$ until we reach $x = 1$.
- Finally, for x between 1 and $1 + a$, the values decreases as $\mu(x) = (1 + a) - x$, until it reaches 0 for $x = 1 + a$ – and stays 0 after that.

We can normalize the resulting function, by dividing it by its largest possible value $1 - a$. Then, the constant levels increase to $\dfrac{a}{1 - a}$.

When $a > 0.5$, we simply get the union cut-off at level $1 - a$, i.e.,

$$\mu(x) = \min(1 - a, \max(\mu_1(x), \mu_2(x))).$$

Acknowledgments. This work was supported in part by the US National Science Foundation via grant HRD-1242122 (Cyber-ShARE Center of Excellence).

The authors are thankful to the anonymous referees for valuable suggestions.

References

1. Belohlavek, R., Dauben, J.W., Klir, G.J.: Fuzzy Logic and Mathematics: A Historical Perspective. Oxford University Press, New York (2017)
2. Klir, G., Yuan, B.: Fuzzy Sets and Fuzzy Logic. Prentice Hall, Upper Saddle River (1995)
3. Mendel, J.M.: Uncertain Rule-Based Fuzzy Systems: Introduction and New Directions. Springer, Cham (2017)
4. Nguyen, H.T., Kreinovich, V.: Nested intervals and sets: concepts, relations to fuzzy sets, and applications. In: Kearfott, R.B., Kreinovich, V. (eds.) Applications of Interval Computations, pp. 245–290. Kluwer, Dordrecht (1996)

5. Nguyen, H.T., Walker, C., Walker, E.A.: A First Course in Fuzzy Logic. Chapman and Hall/CRC, Boca Raton (2019)
6. Novák, V., Perfilieva, I., Močkoř, J.: Mathematical Principles of Fuzzy Logic. Kluwer, Boston (1999)
7. Zadeh, L.A.: Fuzzy sets. Inf. Control **8**, 338–353 (1965)

General Papers

Intuitionistic Fuzzy Model of Traffic Jam Regions and Rush Hours for the Time Dependent Traveling Salesman Problem

Ruba Almahasneh[1(✉)], Boldizsar Tuu-Szabo[2], Peter Foldesi[3], and Laszlo T. Koczy[1,2]

[1] Department of Telecommunications and Media Informatics,
Budapest University of Technology and Economics Informatics,
Budapest, Hungary
{mahasnehr, koczy}@tmit.bme.hu

[2] Department of Information Technology, Széchenyi István University,
Gyor, Hungary
tuu.szabo.boldizsar@sze.hu

[3] Department of Logistics Technology, Széchenyi István University,
Gyor, Hungary
foldesi@sze.hu

Abstract. The Traveling Salesman Problem (TSP) is one of the most extensively studied NP-hard graph search problems. Many researchers published numerous approaches for quality solutions, applying various techniques in order to find the optimum (least cost) or semi optimum solution. Moreover, there are many different extensions and modifications of the original problem, The Time Dependent Traveling Salesman Problem (TD TSP) is a prime example. TD TSP indeed was one of the most realistic extensions of the original TSP towards assessment of traffic conditions [1]. Where the edges between nodes are assigned different cost (weight), considering whether they are traveled during the rush hour periods or they cross the traffic jam regions. In such conditions edges are assigned higher costs [1]. In this paper we introduce an even more realistic approach, the IFTD TSP (Intuitionistic Fuzzy Time Dependent Traveling Salesman Problem); which is an extension of the classic TD TSP with the additional notion of intuitionistic fuzzy sets. Our core concept is to employ intuitionistic fuzzy sets of the cost between nodes to quantify traffic jam regions, and the rush hour periods. Since the intuitionistic fuzzy sets are generalizations of the original fuzzy sets [2], then our approach is a usefully extended, alternative model of the original abstract problem. By demonstrating the addition of intuitionistic fuzzy elements to quantify the intangible jam factors and rush hours, and creating an inference system that approximates the tour cost in a more realistic way [3]. Since our motivation is to give a useful and practical alternative (extension) of the basic TD TSP problem, the DBMEA (Discrete Bacterial Memetic Evolutionary Algorithm) was used in order to calculate the (quasi-)optimum or semi optimum solution. DBMEA has been proven to be effective and efficient in a wide segment of NP-hard problems, including the original TSP and the TD TSP as well [4]. The results from the runs based on the extensions of the family of benchmarks generated from the original TD TSP benchmark data set showed rather good and credible initial results.

© Springer Nature Switzerland AG 2019
R. B. Kearfott et al. (Eds.): IFSA 2019/NAFIPS 2019, AISC 1000, pp. 123–134, 2019.
https://doi.org/10.1007/978-3-030-21920-8_12

Keywords: Intuitionistic fuzzy sets · Traveling Salesman Problem ·
Time Dependent Traveling Salesman Problem · Fuzzy costs · Jam region ·
Rush hour period · Discrete Bacterial Memetic

1   Introduction

The Traveling Salesman Problem (TSP) originally attempts to find the optimal (shortest) route starting from the company headquarters so that all cities, shops or other locations on the agenda are visited exactly once and then the salesman returns to the starting point [5]. Although TD TSP (Time Dependent Travel Salesman Problem) proposed a unique concept using crisp numbers to quantify traffic jam factors which is considered a step forward towards cost determination [1]. TD TSP proposes that fixed parts of the graph (the traffic jam region) have time dependent costs. Thus, those edges are assigned specific lower cost (weight) in the non-rush hour period, and another (higher) one in the rush hour time. The aim is to find the tour with the lowest total cost in the same way as with the TSP. However, realizing the limitation and impreciseness of such representation was our driving motivation in this research paper. In particular, the impracticality of actualizing such model on real-life scenarios, poses huge challenge. Due to the unrealistic simplification of the rush hours and the jam regions calculation, this was overcome in the IFTD TSP approach. In addition, there are several generalizations of fuzzy set theory for various objectives; the notion of intuitionistic fuzzy sets (IFSs) introduced by Atanassov [2] is an interesting and useful one. Although fuzzy sets are (special) IFSs, the converse is not accurately true [3]. In fact, there are situations where IFS theory is more convenient to deal with [6]; such as in the TD TSP with jam factor. The literature shows several successful IFSs-based models where uncertainty factors were effectively represented [7]. Moreover, IFS theory has been applied in different areas that have to do with decision making under high hesitation and vagueness degrees and it proved being successful [8]. In the present paper we study the extended TD TSP problem using the notions of IFS theory by introducing the IFTD TSP model. In our proposed model we apply the intuitionistic approach on jam regions (with the assumption some edges cross city centers) on the one hand, and on rush hours on the other hand (assuming some edges will be traveled during rush hours).

2   Traditional TSP Cost Calculation

The original TSP was first formulated in 1930, and till present, is one of the most intensively studied combinatorial optimization problems. The original problem involves a salesman who starts the journey from the company headquarters, visits each destination city or shop only once, and then returns to the original starting point. The ultimate goal is to find the route that allows the salesman to visit all destination points with the minimum overall travelled distance. TSP can be defined as a graph search problem with edge weights as per (1). To formulate the symmetric case with n nodes (cities) $C_{ij} = C_{ji}$, so a graph can be considered where there is only one arc between every two nodes. Let $X_{ij} = \{0, 1\}$ be the decision variable $(i = 1,2, \ldots, n$ and $j = 1,2, \ldots, n)$, and $X_{ij} = 1$, means that the arc connecting node i to node j is an element of the tour. Let

$$x_{ij} = 0 (i = 1, 2, \ldots, n)$$

$$G_{TSP} = \left(V_{cities}, E_{conn} \right)$$

$$V_{cities} = \{v1, v2, \ldots, vn\}, E_{conn} \subseteq \{(vi, vj) | i \neq j\} \tag{1}$$

$$C : V_{cities} \times V_{cities} \to R, C = \left(C_{ij} \right)_{n \times n}$$

C is called cost matrix, where (C_{ij}) represents the cost of going from city i to city j. The goal is to find the directed Hamiltonian cycle with minimal total length. Formulated in another way the goal is to find vertices that produces the least total cost.

$$\left(\sum_{i=1}^{n-1} C_{pi,Pi+1} \right) + C_{pn,p1} \tag{2}$$

Depending on the properties of the cost matrix TSPs are divided into two classes; symmetric and asymmetric. If the cost matrix is symmetric $\left(C_{ij} = C_{ji} \right)$ for all i and j then the TSP is called symmetric, else it is asymmetric.

The calculation of time required to cover the distance between cities is vital. Since the cost elements are time dependent, then the actual cost between two cities can be determined only if the total time elapsed is precisely resolved. Furthermore, if a cost element is growing then the required time is growing as well, since the velocity of the travel salesman is held constant.

Analyzing actual salesman tours, particularly in city centers, the topography and rush hours in variant locations are factors that must be looked at as uncertain values, more precisely as fuzzy numbers. Hence, relevant data for estimated tour distance between two nodes is not constant. On the contrary, it can be more appropriate to represent this imprecision using the intuitionistic fuzzy model. This in turn will introduce more realistic trips measurements and ultimately optimized solution for TD TSP problem with traffic jam factors.

3   The Discrete Bacterial Memetic Evolutionary Algorithm

The DBMEA is a memetic algorithm, a combination of the bacterial evolutionary algorithm and 2-opt, 3-opt local search [9]. Memetic algorithms combine the global search evolutionary algorithms with local search methods; thus they are able to eliminate the disadvantages of both methods. Also, in many cases the addition of local search approaches usually can improve significantly the performance of the classical evolutionary algorithms, so the Memetic algorithms can be efficient tools for solving TSP and other NP-hard optimization problems [10]. Hence, significant improvement in the performance of the classical evolutionary algorithms was achieved. Since its run time is more predictable than in the case of Concorde algorithm, then DBMEA is considered efficient, especially for large-sized problems with a complicated structure [10]. Moreover, testing DBMEA for the biggest tested instance, it was capable of giving even a better solution than the best-known value and the average runtime was smaller than in the case of the state-of-the-art methods for the problem.

4 The Intuitionistic Fuzzy Time Dependent Traveling Salesman Problem (IFTD TSP)

In our previous work, we introduced model of Fuzzy Rule-base Algorithm for the Time Dependent Traveling Salesman Problem (3FTD TSP). The model completely redefined the TSP and transformed it to 3FTD TSP; were the costs between the nodes that depend on time represented by fuzzy numbers (costs). In addition, the model was able to express the costs influenced by the jam factors (such as jam regions and rush hours) and ultimately quantify the overall tour length [11]. Afterwards, we applied DBMEA simulation on the 3FTD TSP model, and the preliminary results confirm the effectiveness and predictability, and the universality of the proposed technique [11]. In this paper, we moved one step further and extended the model using intuitionistic fuzzy theory as will be explained in details in the coming sections.

A. Preliminary Definitions
Let us start with a short review of basic concepts and definitions related to intuitionistic fuzzy sets which are used in the upcoming sections in a more analytical formal model

- Definition 1

Let a universal set E be fixed. An intuitionistic fuzzy set or IFS A in E is an object having the form

$$A = \{\langle x, \mu_A(x), \nu_A(x)\rangle | x \in E\}$$
$$0 \leq \mu_A(x) + \nu_A(x) \leq 1 \tag{3}$$

The amount $\pi_A(x) = 1 - (\mu_A(x) + \nu_A)$ is called the hesitation part, which may cater to either the membership value or to the non-membership value, or to both

- Definition 2

If A and B are two IFSs of the set E, Then

$$A \subset B \, iff$$

$$\forall x \in E, \quad \left[\begin{array}{c} \mu_A(x) \leq \mu_B(x) \text{ and} \\ \nu_A(x) \geq \nu_B(x) \end{array} \right] \tag{4}$$

$$A \supset B \text{ iff } B \subset A$$

$$A = B \text{ iff } \forall x \in E, \quad \left[\begin{array}{c} \mu_A(x) = \mu_B(x) \text{ and} \\ \nu_A(x) = V_B(x) \end{array} \right] \tag{5}$$

$$A \cap B = \{ \langle x, min(\mu_A(x), \mu_B(x)), max(\nu_A(x), \nu_B(x)) \rangle | x \in E \} \tag{6}$$

classic fuzzy set has the form

$$A \cup B = \{ \langle x, max(\mu_A(x), \mu_B(x)), min(\nu(x), \nu_B(x)) \rangle | x \in E \} \{ \langle x, \mu_A(x), \mu_{A^c}(x) \rangle | x \in E \} \tag{7}$$

- Definition 3

Let X and Y be two fuzzy sets. An intuitionistic fuzzy relation (IFR) R from X to Y is an IFS of $X \times Y$, characterized by the membership function μ_R and the non-membership function ν_R. An IFR R from X to Y will be denoted by $R(X \rightarrow Y)$.

- Definition 4

If A is an IFS of X, the max-min-max composition of the IFR R $(X \rightarrow Y)$ with A is an IFS B of Y denoted by $(B = R \circ A)$ and is defined by the membership function

$$\mu_{R \circ A}(y) = \vee_x[\mu_A(x) \wedge \mu_R(x, y)] \tag{8}$$

and the non-membership function

$$\nu_{R \circ A}(y) = \wedge_x[\nu_A(x) \vee \nu_R(x, y)] \tag{9}$$

Previous formulas hold for all Y.

- Definition 5

Let $Q(X \rightarrow Y)$ and $R(Y \rightarrow Z)$ be two IFRs. The max-min-max composition $(R \circ Q)$; is the intuitionistic fuzzy relation from X to Z, defined by the membership function

$$\mu_{R \circ Q}(x, z) = \vee_y[\mu_Q(x, y) \wedge \mu_R(y, z)] \tag{10}$$

and the non-membership function given by

$$V_{R \circ Q}(x, z) = \wedge_y[\nu_Q(x, y) \vee \nu_R(y, z)] \tag{11}$$

$$\forall(x, z) \in X \times Z \text{ and } \forall y \in Y$$

Let A be an IFS of the set J, and R be an IFR from J to C. Then the Max-Min-Max composition (10) B of IFS A with the IFR R $(J \rightarrow C)$ denoted by $B = A \circ R$ denotes the cost of the edges as an IFS B of C with the membership function given by

$$\mu_B(c) = \vee_{j \in J} [\mu_A(j) \wedge \mu_R(j, c)] \tag{12}$$

and the non-membership function given by

$$v_B(c) = \wedge_{j \in J} [v_A(j) \vee v_R(j, c)] \tag{13}$$

$$\forall c \in C. \, (Here \wedge = Min \, and \vee Max)$$

If the state of the edge E is described in terms of an IFS A of J; then E is assumed to be the assigned cost in terms of IFSs B of C, through an IFR R from J to C, which is assumed to be given by a knowledge base directory (or by experts who are able to translate the jam degrees of association and non-association according to geographical areas) on the destination cities and the extent (membership) to which each one is included in the jam region. This will be translated to the degrees of association and non-association, respectively, between jam and cost.

Now, let us expand this concept to a finite number of edges E that form a whole tour for a salesman. Let there be n edges E_i; $= 1; \, 2; .... \, n$; in a trip (from starting point to final destination). Thus $e_i \in E$. Let R be an IFR $(J \rightarrow C)$ and construct an IFR Q from the set of edges E to the set of jam factors J. Clearly, the composition T of IFRs R and $(T = R \circ Q)$ give the cost for each edge from E to C given by the membership function

$$\mu_T(e_i, c) = \vee_{j \in J} [\mu_Q(e_i, j) \wedge \mu_R(j, c)] \tag{14}$$

and the non-membership function given by

$$v_T(e_i, c) = \wedge_{j \in J} [v_Q(e_i, j) \vee v_R(j, c)] \tag{15}$$

$$\forall e_i \in E \, and \, c \in C$$

For given R and Q, the relation $(T = R \circ Q)$ can be computed. From the knowledge of Q and T, an improved version of the IFR R can be computed, for which the following holds valid:

(i) $J_R = \mu_R - v_R \cdot \pi_R$ is greatest
(ii) The equality $T = R \circ Q$ is retained.

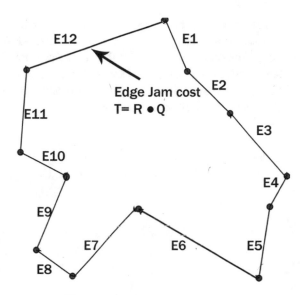

Fig. 1. Tour for a Simple Example

Clearly, this proposed concept improved version of R will be a more significant IFR translating the higher degrees of association and lower degrees of non-association of J as well as lower degrees of hesitation to the cost evaluation C. If almost equal values for different C in T are obtained, then we consider the case for which hesitation is least. From a refined version of R one may infer cost from jam factors in the sense of a paired value, one being the degree of association and other the degree of non-association.

B. IFTD TSP Application on a TD TSP Case

Let there be a simple tour that has only 12 edges ($E1$ .... $E12$) as shown in Fig. 1. Each edge connects two nodes. Thus, each edge eventually will have a jam cost, depending on the jam area(s) it crosses. Table 1 shows each edge and the jam factors associated. The ultimate goal is to be able to calculate the total tour jam factor which will be multiplied by the physical distance between two nodes. Hence, quantifying the jam cost for each edge that is part of that tour path. The intuitionistic fuzzy relation $Q(E \rightarrow J)$ is given as shown in Table 1. Let the set of jam costs be C as given in Table 2. The intuitionistic fuzzy relation $R(J \rightarrow C)$ as given in Table 2. Therefore the composition $(T = R \circ Q)$ as given in Table 3. We calculated jam region cost factors (c_{jam}) as given in Table 4 where the four cost factors are ($c_1 = 1.2$, $c_2 = 1.5$, $c_3 = 2$, $c_4 = 5$) with weighted average calculations:

$$c_{jam_e} = \frac{\sum_i j_i \times c_i}{\sum_i j_i} \tag{16}$$

The rush hour cost factors of each tour edge (c_{rush}) are determined in similar intu-itionistic model. The relations between time and the rush hour periods ($\hat{Q}$) are described with intuitionistic fuzzy functions Fig. 2. An intuitionistic fuzzy relation ($\hat{R}$) is given between the rush hour periods and the cost factors similarly, as was done for jam regions in Table 2. Then the composition ($\hat{T} = \hat{Q} \circ \hat{R}$) is calculated. Finally rush hour cost factors were calculated with weighted averaging.

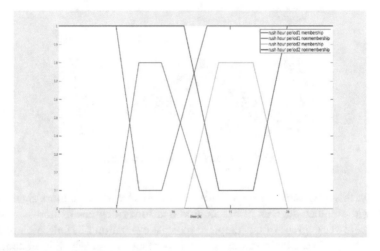

Fig. 2. Fuzzy membership and non-membership functions of the rush hour periods

The cost of the edges is calculated taking into account the two cost factors (jam region and rush hour cost factors):

- if $c_{jam_e} > 0$

  (the edge belongs to at least one of the jam regions)
  $AND c_{rush_e} > 0 (in\ rush\ hours)$

$$c_e = c_{jam_e} \times c_{rush_e} \times dist_e$$

- else

$$C_e = dist_e$$

where $dist_e$ is the Euclidean distance.

Table 1. Route1 = (edge1...edge12)

(Q)	Jam Region1	Jam Region2	Jam Region3	Jam Region4
E1	(0.8, 0.1)	(0.6, 0.1)	(0, 1)	(0, 1)
E2	(0, 1)	(0, 1)	(0.2, 0.8)	(0.6, 0.1)
E3	(0.8, 0.1)	(0.8, 0.1)	(0, 1)	(0, 1)
E4	(0, 1)	(0, 1)	(0, 0.6)	(0.2, 0.7)
E5	(0.8, 0.1)	(0.8, 0.1)	(0, 0.6)	(0.2, 0.7)
E6	(0, 0.8)	(0.4, 0.4)	(0, 1)	(0, 1)
E7	(0, 1)	(0, 1)	(0.6, 0.1)	(0.1, 0.7)
E8	(0, 0.8)	(0.4, 0.4)	(0.6, 0.1)	(0.1, 0.7)
E9	(0.6, 0.1)	(0.5, 0.4)	(0, 1)	(0, 1)
E10	(0, 1)	(0, 1)	(0.3, 0.4)	(0.7, 0.2)
E11	(0, 0.8)	(0.4, 0.4)	(0.6, 0.1)	(0.1, 0.7)
E12	(0.4, 0.4)	(0.6, 0.1)	(0, 1)	(0.2, 0.8)

Table 2. Jam costs

Jam area (R)	Cost factor 1 (c1)	Cost factor 2 (c2)	Cost factor 3 (c3)	Cost factor4 (c4)
Jam Region1	(0.4, 0)	(0.7, 0)	(0.3, 0.3)	(0.1, 0.7)
Jam Region2	(0.3, 0.5)	(0.2, 0.6)	(0.6, 0.1)	(0.2, 0.4)
Jam Region3	(0.1, 0.7)	(0, 0.9)	(0.2, 0.7)	(0.8, 0)
Jam Region4	(0.4, 0.3)	(0.4, 0.3)	(0.2, 0.6)	(0.2, 0.7)

Table 3. $T = R \circ Q$

Jam Cost (T)	Cost factor1	Cost factor2	Cost factor3	Cost factor4
E1	(0.4, 0.1)	(0.7, 0.1)	(0.6, 0.1)	(0.2, 0.4)
E2	(0.4, 0.3)	(0.4, 0.3)	(0.2, 0.6)	(0.2, 0.2)
E3	(0.4, 0.1)	(0.7, 0.1)	(0.6, 0.1)	(0.2, 0.4)
E4	(0.2, 0.7)	(0.2, 0.7)	(0.2, 0.7)	(0.2, 0.6)
E5	(0.3, 0.1)	(0.7, 0.1)	(0.6, 0.1)	(0.2, 0)
E6	(0.3, 0.5)	(0.2, 0.6)	(0.4, 0.4)	(0.2, 0.4)
E7	(0.1, 0.7)	(0.1, 0.7)	(0.2, 0.7)	(0.6, 0.1)
E8	(0.3, 0.5)	(0.2, 0.6)	(0.4, 0.4)	(0.6, 0.1)
E9	(0.4, 0.1)	(0.6, 0.1)	(0.5, 0.3)	(0.2, 0.4)
E10	(0.4, 0.3)	(0.2, 0.6)	(0.2, 0.6)	(0.3, 0.4)
E11	(0.3, 0.5)	(0.2, 0.6)	(0.4, 0.4)	(0.6, 0.1)
E12	(0.4, 0.4)	(0.4, 0.4)	(0.6, 0.1)	(0.2, 0.4)

Table 4. Inuitionistic jam region cost factors for edges

J_R	(J_R1)	(C 1)	(J_R2)	(C 2)	(J_R3)	(C 3)	(J_R4)	(C4)	Total Jam region cost factors
E1	0.35	1.2	0.68	1.5	0.57	2	0.04	5	1.695
E2	0.31	1.2	0.31	1.5	0.08	2	0.08	5	1.791
E3	0.35	1.2	0.68	1.5	0.57	2	0.04	5	1.695
E4	0.13	1.2	0.13	1.5	0.13	2	0.08	5	2.151
E5	0.24	1.2	0.68	1.5	0.57	2	0.2	5	2.04
E6	0.2	1.2	0.08	1.5	0.32	2	0.36	5	2.917
E7	0	1.2	0	1.5	0.13	2	0	5	2
E8	0.2	1.2	0.08	1.5	0.32	2	0.57	5	3.291
E9	0.35	1.2	0.57	1.5	0.44	2	0.04	5	1.682
E10	0.31	1.2	0.08	1.5	0.08	2	0.18	5	2.388
E11	0.2	1.2	0.08	1.5	0.32	2	0.57	5	3.291
E12	0.32	1.2	0.32	1.5	0.57	2	0.04	5	1.763

5 Computational Result

The DBMEA algorithm was modified and tested for the new IFTD TSP problem. The bier127 problem was selected for the test because we have some former results for TD TSP with this instance to validate the obtained results [12, 13].

These TSP benchmark instances were first examined by Schneider for the Time Dependent TSP [12]. He defined a traffic jam region (coordinate of the left corner point (7080, 7200), width is 6920, and the height is 9490). In our test this jam region was divided into four equal sized subdivisions. Velocity v is 6000 m/h in each test. Two rush hour periods were defined as in Fig. 2.

Our algorithm was tested on an Intel Core i7-7500U 2.7 GHz, 8 GB of RAM memory workstation under Linux Mint 18. The results were calculated by averaging five test runs Table 5. The Table contains the total time in hours required to visit each

Table 5. Computational results for IFTD TSP

Cost factors	DBMEA		
	Best elapsed time	Average elapsed time	Average runtime [s]
$c_1 = 1.01$ $c_2 = 1.05$ $c_3 = 1.1$ $c_4 = 1.2$	20.605	20.712	170.535
$c_1 = 1.05$ $c_2 = 11$ $c_3 = 1.2$ $c_4 = 1.5$	21.166	21.349	176.691

(continued)

Table 5. (*continued*)

Cost factors	DBMEA		
	Best elapsed time	Average elapsed time	Average runtime [s]
$c_1 = 1.1$ $c_2 = 1.3$ $c_3 = 1.5$ $c_4 = 2$	22.061	22.186	192.313
$c_1 = 1.2$ $c_2 = 1.5$ $c_3 = 2$ $c_4 = 5$	22.885	23.124	207.919
$c_1 = 1.5$ $c_2 = 2$ $c_3 = 5$ $c_4 = 10$	23.867	24.185	195.417
$c_1 = 2$ $c_2 = 5$ $c_3 = 10$ $c_4 = 15$	23.909	24.302	207.109
$c_1 = 5$ $c_2 = 10$ $c_3 = 20$ $c_4 = 50$	24.058	24.629	124.910

location with different cost factors. This new problem is a generalization of the original TSP (with $c_1 = 1.2, c_2 = 1.5, c_3 = 2, c_4 = 5$ the edge costs are the Euclidean distances, so this parameters result in the TSP problem). As it can be seen in Table 5. DBMEA also found high quality solutions with big cost factors.

6   Conclusions and Future Work

In this paper, we proposed a novel intuitionistic fuzzy set based model for the realistic extension of the TD TSP, namely, the IFTD TSP model, for considering jam factors and rush hours in the original TDT SP problem by applying the IFS theory. Obviously, this improved version will be a more promising approach in translating the higher degrees of association and lower degrees of non-association of the jam factor and rush hours as well as lower degrees of hesitation to any edge cost and, ultimately, more realistic calculation for the traveled routes. Our future work will focus on applying the DBMEA meta-heuristics [14] for determining the (quasi) optimal tours (as this algorithm has been repeatedly successfully applied for various NP-hard graph search problems with rather good accuracy, very good predictability and rather universality) to the model and test its efficiency on a larger number of benchmarks, with more complicated cases.

Acknowledgment. This work was supported by National Research, Development and Innovation Office (NKFIH) K124055. Supported by the ÚNKP-18-3 New National Excellence Program of the Ministry of Human Capacities.

References

1. Tüű-Szabó, B., Földesi, P., Kóczy, L.T.: The discrete bacterial memetic evolutionary algorithm for solving the one-commodity pickup-and-delivery traveling salesman problem (2018)
2. Atanassov, R.K.: Intuitionistic fuzzy sets. Fuzzy Sets Syst. **20**, 87–96 (1986)
3. Boran, F., Genç, S., Ku, M., Akay, D.: A multi-criteria intuitionistic fuzzy group decision making for supplier selection with TOPSIS method. Expert Syst. Appl. **36**, 11363–11368 (2009)
4. Földesi, P., Botzheim, J.: Modeling of loss aversion in solving fuzzy road transport travelling salesman problem using eugenic bacterial memetic algorithm. Memet. Comput. **2**(4), 259–271 (2010)
5. Applegate, D.L., Bixby, R.E., Chvátal, V., Cook, W.J.: The Traveling Salesman Problem: A Computational Study, pp. 1–81. Princeton University Press, Princeton (2006)
6. Biswas, R.: On fuzzy sets and intuitionistic fuzzy sets. Notes Intuitionistic Fuzzy Sets **3**, 3–11 (1997)
7. Gau, W.L., Buehrer, D.J.: Vague sets. IEEE Trans. Syst. Man Cybern. **23**(2), 610–614 (1993)
8. Szmidt, E., Kacprzyk, J.: Intuitionistic fuzzy sets in group decision making. NIFS **2**(1), 11–14 (1996)
9. Moscato, P.: On evolution, search, optimization, genetic algorithms and martial arts—towards memetic algorithms. Technical report Caltech Concurrent Computation Program, Report. 826, California Institute of Technology, Pasadena, USA (1989)
10. Kóczy, L.T., Földesi, P., Tüű-Szabó, B.: An effective discrete bacterial memetic evolutionary algorithm for the traveling salesman problem. Int. J. Intell. Syst. **32**(8), 862–876 (2017)
11. Kóczy, L.T., Földesi, P., Tüű-Szabó, B., Almahasneh, R.: Modeling of fuzzy rule-base algorithm for the time dependent traveling salesman problem. In: Proceedings of the IEEE International Conference on Fuzzy Systems (FUZZ-IEEE), (2019, under review)
12. Schneider, J.: The time-dependent traveling salesman problem. PhysicaA **314**, 151–155 (2002)
13. Tüű-Szabó, B., Földesi, P., Kóczy, T.L.: Discrete bacterial memetic evolutionary algorithm for the time dependent traveling salesman problem. In: International Conference on Information Processing and Management of Uncertainty in Knowledge-Based Systems (IPMU 2018), Cadíz, Spain, pp. 523–533. Springer, Cham (2018)
14. Kóczy, L.T., Földesi, P., Tüű-Szabó, B.: A discrete bacterial memetic evolutionary algorithm for the traveling salesman problem. In: IEEE World Congress on Computational Intelligence (WCCI 2016), Vancouver, Canada, pp. 3261–3267 (2016)

New Fuzzy Approaches to Cryptocurrencies Investment Recommendation Systems

Vinícius Luiz Amaral, Emmanuel Tavares F. Affonso,
Alisson Marques Silva$^{(\boxtimes)}$, Gray Farias Moita,
and Paulo Eduardo Maciel Almeida

Graduate Program in Mathematical and Computational Modeling,
Federal Center for Technological Education of Minas Gerais – CEFET-MG,
Belo Horizonte, Brazil
viniciusluiz.doamaral@gmail.com, emmanuelcomp@gmail.com,
{alisson,gray,pema}@cefetmg.br

Abstract. This work proposes the use of Computational Intelligence algorithms to predict cryptocurrencies values based on historical values. After predicting the value of the currencies for up to three days following the current one using an evolving algorithm, two approaches were presented to suggest the investment: the first one uses only the result of the forecast to provide a suggestion of investment; in contrast, the second approach, in addition to using the prediction data returned by the evolving system, also applies a Mamdani system, based on expert knowledge, to suggest to the users what to do with their invested value. After performing and processing the historical data of three cryptocurrencies, the suggestions offered by both approaches were compared to the actual quote. The comparison presented results with a total assertiveness rate of over 90% for the three cryptocurrencies evaluated, according to established criteria, for both the evolving approach and the hybrid approach. Computational experiments suggest that the two proposed approaches are promising and competitive with alternatives reported in the literature.

1 Introduction

Cryptocurrencies emerged in 2008 after a large drop in the *Dow Jones* index known as the Subprime Mortgage Crisis. In this scenario, between late 2008 and early 2009, one or more programmers under the pseudonym Satoshi Nakamoto launched Bitcoin, the first virtual and decentralized currency [1]. Since then, with the popularization of Bitcoin, several other coins of this nature were created giving the subject even more notoriety in the global financial market. For some, they are seen as a threat to the monopoly of central government power. For others, they mean a potential secondary currency [2].

Statistics show that the volume of transactions involving digital coins outweighs the volume of traditional currencies exchange, even the US dollar [2].

© Springer Nature Switzerland AG 2019
R. B. Kearfott et al. (Eds.): IFSA 2019/NAFIPS 2019, AISC 1000, pp. 135–147, 2019.
https://doi.org/10.1007/978-3-030-21920-8_13

Realtime information on the volume of dollar and Bitcoin transactions can be obtained in [3]. As they are not regulated/controlled by any official agency, variation in the cryptocurrency value are mainly a consequence of changes in its market demand/acceptance and the increase in the number of digital wallets. These factors make predicting their value a difficult task. One of the ways of forecasting is by analyzing its historical behavior. In this analysis, historical data is used to identify patterns of behavior and to detect the best time to buy or sell the money invested.

Currently, several researches present techniques and methodologies to predict the quotation and/or the number of cryptocurrency transactions using this information to guide investments. One of the techniques that have been used in this sense is the sentiment analysis, with extraction of comments from investors in dedicated communities or on social networks like Twitter and Facebook. For example, in [4] a model is proposed for forecasting the number of transactions and the quotation value of three cryptocurrencies extracting comments from investors in online communities. The purpose of this study is to identify the types of comments that most influence price fluctuations and the number of transactions for each cryptocurrency. Equally in [5], a similar correlation is established, this time with data extracted from Twitter. The results obtained in these two studies indicate the relationship between comments in social networks and oscillations in the price of cryptocurrencies.

On the other hand, several Computational Intelligence techniques have been used in crypto-prediction problems. Generally, these techniques use historical cryptocurrencies information to forecast. In [6], a free platform is offered for investors, which provides forecast for the value of several cryptocurrencies. The algorithm is based on an LSTM-Network (Long Short-Term Memory) and promises assertiveness ranging from 70% to 90%. In [7], Bitcoin quotation prediction experiments were performed, in which Recurrent Neural Networks (RNN) obtained better results than Vector Auto-Regression (VAR) models.

Artificial Neural Networks were used in the work of [8–10]. In [8], a Non-Linear Autoregressive Neural Network with Exogenous Input (NARX) is proposed to predict next-day prices for Bitcoin. In [9], a Bayesian Neural Network (BNN), which uses the structure of an MultiLayer Perceptron (MLP) with Bayesian statistics, was employed. This network was compared to a Support Vector Regression model and obtained the best results to predict the current exchange rate for the current day. Similarly, in [10], MLP networks were employed to predict the direction of Bitcoin variation, obtaining hit rates between 58% and 85%.

Deep learning networks were used in the work of [11] and [12]. In [11], deep learning is employed to train a Long Short-Term Memory (LSTM) model. Architectures based on MLP and RNN structures were also used to predict the direction of price changes for eight different cryptocurrencies. In this experiments, the LSTM model presented a better performance than other models to predict the direction of price variation for different cryptocurrencies. On the other hand, [12] proposed two different deep learning models for prediction, Bayesian optimised recurrent neural network and LSTM model. Both deep learning models are benchmarked with the popular model for time series forecasting know autoregressive integrated moving average (ARIMA). In the experiments the deep learning methods outperform the ARIMA forecast, where the LSTM presented better performance among all.

In this context, this paper proposes and evaluates two approaches for the recommendation of investments in cryptocurrencies using Evolving Fuzzy Systems (EFS). The first approach uses an EFS and the other one combines EFS with a Mamdani Fuzzy System.

The remaining part of the paper is organized as follows. Next section presents the proposed approaches to investment recommendation system. First, we present an approach based on an EFS. Afterwards, we detail an hybrid approach that uses an EFS and a traditional Mamdani model. Section 3 evaluates the performance of the proposed approaches considering historical data of three cryptocurrencies (BitCoin, Ethereum and LiteCoin). Finally, Sect. 4 concludes the paper summarizing its contributions and listing issues for future development.

2 Evolving Fuzzy Models

This section presents the evolving models used in this paper: eMG – Multivariable Gaussian Evolving Fuzzy System [13]; eNFN – Evolving Neo-Fuzzy Neuron [14]; eTS – evolving Takagi-Sugeno [15] e; xTS – eXtended Takagi-Sugeno [16].

2.1 eMG – Multivariable Gaussian Evolving Fuzzy System

Proposed by [13], the eMG uses multivariable Gaussian-type membership functions to represent its data clusters. This algorithm allows the inclusion of a new rule, the modification of the parameters of an existing rule, and the union of two redundant clusters. For each new sample presented to the eMG the cluster structure is updated by a compatibility measure and a calculated alert index for each of the clusters.

The measure of compatibility uses a calculation between the distance of the current observation and the centers of the clusters (M distance), which produces ellipsoidal clusters whose axes are not necessarily parallel to the input variable axes. The measure of compatibility is computed by $p_i^k = \exp[-\frac{1}{2}M(x^k, c_i^k)]$.

The M distance can be calculated by $M(x^k, c_i^k) = (x^k - c_i^k)(\sum_i^k)^{-1}(x_k - c_i^k)^T$, where $\sum_i^k$ is the dispersion matrix that represents the shape of the clusters and is estimated at each iteration by $\sum_i^{k+1} = (1 - \alpha(p_i^k)^{1-\lambda_i^k})(\sum_i^k - \alpha(p_i^k)^{1-\lambda_i^k}(x^k - c_i^k)(x^k - c_i^k)^T)$.

The membership functions characterize the clusters and are represented by a central vector and a dispersion matrix. The eMG has a mechanism for automatically adjusting the distance threshold value based on the size of the input space in order to avoid the so-called dimensionality curse. The compatibility threshold can be found by $T^p = exp[-1/2\chi^2_{m,\alpha}]$, where $\chi^2_{m,a}$ is α upper unilateral confidence interval of a Qui-Quadratic distribution with m degrees of freedom, and m is the size of the input space.

The alert index threshold is calculated by $T_a = 1 - \alpha/\omega$, where α significance level to calculate the compatibility and alert thresholds, ω is the size of the window to monitor the dynamics of the compatibility measure. New clusters are created when the compatibility measure is less than the compatibility threshold for all clusters, and the alert index is greater than the alert threshold for the cluster with the highest compatibility value.

2.2   eNFN – Evolving Neo-Fuzzy Neuron

eNFN [14] is an evolving model developed under the structure of Neo-Fuzzy-Neuron [17], which uses a set of Takagi-Sugeno [18] Zero Order models and complementary triangular membership functions. In this model, new membership functions can be added, deleted and/or have their parameters adjusted depending on the input data and the modeling error. The consequent parameters of the rules are updated using a gradient algorithm with optimal learning rate [19].

The creation of membership functions uses information about the global modeling error and the local error to decide on the creation of a new rule. For this, it is calculated recursively for the input x_t average value $\hat{\mu}_{g_t} = \hat{\mu}_{g_{t-1}} - \beta(\hat{\mu}_{g_{t-1}} - e_t)$ and the variance of the global error $\hat{\sigma}^2_{g_t} = (1 - \beta)(\hat{\sigma}^2_{g_{t-1}} + \beta(\hat{\mu}_{g_t} - e_t)^2)$, where $e_t = y_t - \hat{y}_t$ and β is the learning rate. The average value $\hat{\mu}_{b_{ti}^*}$ of the local error corresponding to the most active membership function (b_i^*) is recursively computed for the input x_{ti} by $\hat{\mu}_{b_{ti}^*} = \hat{\mu}_{b_{ti-1}^*} - \beta(\hat{\mu}_{b_{ti-1}^*} - e_t)$. If $\hat{\mu}_{b_{ti}^*} > \hat{\mu}_{g_t} + \hat{\sigma}^2_{g_t}$ and $dist > \tau$, then a new membership function is created and inserted, where $dist$ is the distance between the modal value of the function to be created and its adjacent functions, and τ is a threshold created to avoid complex models and overfitting.

The exclusion of functions from this model uses the age of the rule concept [16]. This age is determined by its inactivity time, that is, the time interval that a membership function is left unactivated and is obtained by $age_j = t - active_j$, where j is the index of the membership function, $active$ denotes the instance of time that the j-th rule has been activated, and t is the current time instance. If the age of the rule with the longest inactivity time is greater than a threshold defined as the parameter of the algorithm, the rule is excluded. Detailed information on the algorithm and on eNFN parameter settings can be found in [14].

2.3   eTS – Evolving Takagi-Sugeno

eTS [15] was a milestone in the development of evolutionary models. This is a model of the Takagi-Sugeno type that continually updates its rules and structure in a recursive way, adding and modifying existing rules. Its output is determined as the weighted average normalized by the activation of membership functions. The algorithm implements an unsupervised pool to update the rule base. This occurs with each new sample presented to the model. Each cluster defines an antecedent of the rules, so the algorithm updates its structure to each processed sample, being able to generate a new cluster or just update an existing cluster. The fuzzy sets of the antecedent of the rules use Gaussian pertinence functions.

Learning is done using an incremental subtractive clustering algorithm called eClustering [20]. In this clustering method, each point in space may be a possible candidate to be the center of the cluster. That choice is determined by a potential function. The potential function of a sample i is defined as a measure of the spatial proximity of point s_i to all others by $P_i = \frac{1}{N} \sum_{j=1}^{N} e^{-\frac{4}{r^2} ||s_i - s_j||^2}$, where P_i is the potential of the ith sample, N is the number of samples, r is the radius of influence of the cluster. The point that is surrounded by a greater number of other points will have a higher potential value, so it will be defined as a cluster center. The next cluster center is also defined as the point of greatest potential. The procedure repeats until the potential of all data points is reduced below a certain threshold. The potential of each new sample $P_t(s_t)$ in t is estimated recursively by $P_t(s_t) = \frac{t-1}{(t-1)(v_t+1)+\sigma_t - 2\nu_t}$, where $v_t = \sum_{j=1}^{n+1}(s_{tj})^2$, $\sigma_t = \sum_{l=1}^{t-1}\sum_{j=1}^{n+1}(s_{lj})^2$, $\nu_t = \sum_{j=1}^{n+1} s_{tj}\beta_{tj}$, and $\beta_{tj} = \sum_{l=1}^{t-1} s_{lj}$. The parameters v_t and ν_t are computed from s_t, while s_t and σ_t can be computed recursively. The center potential of each of the l clusters (rules) $P_t(s_l^*)$ on t is updated recursively by $P_t(s_l^*) = \frac{(t-1)P_{t-1}(s_l^*)}{t-2+P_{t-1}(s_l^*)+P_{t-1}(s_l^*)\sum_{j=1}^{n+1}(d_{l(t-1)j})^2}$.

At each iteration, the potential of the new sample is calculated. If the potential of this sample is greater than all the clusters and is close to an existing cluster, it will replace this center, otherwise this sample will be the center of a new cluster. On the other hand, if its potential is less than the potential of all cluster centers, the parameters of the rules of the cluster with the shortest Euclidean distance are updated. The learning of rules consequents is treated as a Least Squares Problem. The parameter vector of the linear models is found by minimizing a global objective function with RLS (Recursive Least Square) or minimizing a local objective function for each rule by wRLS (weighted Recursive Least Square).

2.4   xTS – eXtended Takagi-Sugeno

xTS [16] is an extended eTS template and was proposed by [16]. One of the new functions added in this model is the age index, a concept that is applied to measure the quality of clusters. Another factor also used to evaluate the quality of the cluster is the population, but the population is only a support and can be obtained by $S_l \leftarrow S_l + 1$; $l = \arg\min_l^R ||s_t - s_l^*||^2$, $l = 1...R$, where S_l is the support of the l-th cluster.

The support of a cluster indicates the amount of samples that are in its zone of influence and also the power of generalization of the rule. Support divided by the number of samples is one of the measures to evaluate the quality of the rule. This measure can be used to ignore/exclude or not a rule. In addition, another quality factor is the age index. This index is defined as already processed less the mean value of the temporal index of the samples represented by the cluster. A AGE is obtained by $AGE_j = t - \frac{2A_j}{t+1}$, where $A_j = \sum_{i=1}^{POP_j} idx_j$ is the cumulative time of arrival and idx_j the time index of j-th sample. The cumulative time of arrival is calculated similarly to the population of the cluster, but uses the temporal index of the sample obtained by $idx_j = idx_j + t; j = \arg\min_j ||s_t - s_j^*||^2$. This index takes values in the range $(0, t)$. For values close to 0 indicate that the cluster is new, values close to t suggest an old cluster.

The range radius of each cluster is estimated recursively based on the local spatial density by $r_{tlj} = \rho r_{(t-1)lj} + (1 - \rho)\sigma_{tlj}$; $r_{lj1} = 1$, $l = \arg\min_{i=1}^N ||s_t - s_l^*||$, where ρ is a constant that regulates the compatibility between the new information and the old one. σ_{tlj} is the local dispersion over the space of the input data obtained by $\sigma_{tij} = \sqrt{\frac{1}{S_{tl}} \sum_{l=1}^{S_{tl}} ||x_i^* - x_l||}$.

3   Evaluation Evolving Fuzzy Models

This section details the experiments carried out to evaluate the evolving fuzzy models. The model that achieves the best performance will be used to implement the proposed approaches. The models are evaluated by RMSE (Root Mean Square Error).

To evaluate the approaches we used a database with historical market data from different cryptocurrencies, available on the Kaggle platform. The historic of the cryptocurrencies of this database was taken from CoinMarketCap [21]. Among the 1000 coins available, 3 of the most relevant were selected. The selection criteria used were: sets with complete data; sufficient number of data; coins with high market value. The three cryptocurrencies chosen were Bitcoin (BTC), Ethereum (ETH) and Litecoin (LTC). From the data sets, daily data were collected between 01/01/2017 and 04/30/2018 totaling 1455 samples. It should be noted that the data were used without normalization.

The goal is to make the prediction of closing price of the cryptocurrencies for 1 (y_1), 2 (y_2) and 3 (y_3) days subsequent to the current date, considering inputs x_1 highest value reached in the day, x_2 lowest value reached on day, x_3 closing value, x_4 4 volume sold, x_5 market capitalization.

The experiments were executed with models provided by their respective authors, with eMG and eNFN versions being implemented in Matlab and eTS and xTS in Java. The models were evaluated by the $RMSE$ (Root Mean Squared Error) (1)

$$RMSE = \frac{1}{N}\left(\sum_{t=1}^{N}(y_t - \hat{y}_t)\right)^{\frac{1}{2}},\tag{1}$$

where N is the number of samples, $\hat{y}_t$ is the desired output, y_t is the estimated output. The experiments simulates on-line processing, i.e., the parameters and structure of the models evolve for all samples of the data set viewed as a data stream.

Table 1 presents the results by RMSE for 1 (y_1), 2 (y_2) and 3 (y_3) days subsequent to the date current. Considering the RMSE, it is noted that the best performance was obtained by eNFN. On the other hand, the results obtained by eMG, eTS and xTS are comparable. The eNFN obtained the best performance in predicting cryptocurrencies and will be used in the experiments of the system of recommendation of investments in cryptocurrencies.

Table 1. Forecasting the closing price of the cryptocurrencies – RMSE

Model	y_1	y_2	y_3
eNFN	**16.3086**	**15.9954**	**15.6831**
eMG	19.9694	19.9408	19.7170
eTS	19.9619	19.9821	20.0035
xTS	19.9633	19.9832	20.0033

4 Proposed Approaches

This paper proposes two approaches to cryptocurrencies investment recommendation. The first approach (Sect. 4.1) uses an evolving fuzzy system and the second (Sect. 4.2) combines an evolving fuzzy system and a Mamdani system.

4.1 Evolving Proposed Approach

In the proposed evolving approach the eNFN uses five input variables (x_1 the greatest value of the day, x_2 the lowest value of the day, x_3 the closing value, x_4 the trading volume, x_5 the market capitalization) and produces as an output the forecast of the closing value that can be 1 (y_1), 2 (y_2) or 3 (y_3) days after the current date. Compute the difference between the current value of the coin (y_0) and the one predicted for the next day (y_i), i.e., in which i corresponds the day predicted. Thus, it can reach the values $D_i = y_0 - y_i$, that are assessed:

- If the $|D_i|$ is lower than the Stability Rate (SR), it is suggested to maintain the investment.

- If the D_i is greater than the value set for RBS (minimum Rate of change for success in Buying or Selling suggestions), a purchase action is suggested;
- If the D_i is lower than the value set for RBS, a sales action is suggested.

where SR represents the maximum real variation for an investment maintenance suggested action to be considered correct and RBS represents the minimum real change to consider that a buy or sell action was effective.

4.2   Hybrid Approach

Hybrid approach to investment recommendation is composed of two steps: first one employs an evolving fuzzy algorithm to forecast the daily closing market value of a specific crypto-coin for the next three days; second step uses a fuzzy system of the Mamdani type [22], which receives as input the value forecasted in the first step and outputs the investment recommendation. A representation of the proposed model is shown in Fig. 1.

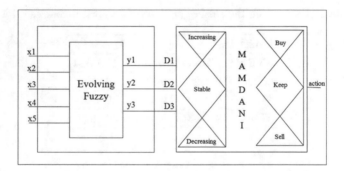

Fig. 1. Data flow in the forecasting steps and calculation of the best investment.

As it can be observed in Fig. 1, the evolving algorithm in the first step uses five input variables (x_1 the greatest value of the day, x_2 the lowest value of the day, x_3 the closing value, x_4 the trading volume, x_5 the market capitalization) and produces as an output the forecast of the closing value of the three following days (y_1, y_2, y_3).

The second part of the proposed solution consists of an expert system based on fuzzy sets that uses the Mamdani inference method. In this system, based on the forecast calculated for the next three days, a recommendation is made to the user about what to do with their invested amount. The system returns an action, which can be: buy, sell or keep the investment. The Mamdani model was created with three input variables D_1, D_2 e D_3. Each variable receives a discrete value that represents the expected change in the closing value for the next three days (positive values indicate that the forecast is increasing and, consequently, negative values indicate that the forecast is falling). This input value is calculated from the results presented by the eNFN algorithm, that is, $D_1 = y_1 - y_0$, $D_2 = y_2 - y_0$ and $D_3 = y_3 - y_0$, where y are the eNFN predictions. The variables were modeled to be composed of five linguistic terms that indicate the prediction of the behavior for the price of the cryptocurrency for a given day in relation to the current day, being: fall, slight fall, stable, slight increase and increase. For the linguistic terms "fall" and "increase" trapezoidal membership functions were used and triangular functions for the other terms.

The set of rules of the Mamdani model was constructed and their respective weights were adjusted based on expert knowledge. The composition of the rules portrays that if the forecast returned by the eNFN in D_1 is increasing, the recommended action is likely to be Buy. If the forecast is a slight increase, it is likely that the suggestion is also Buy, but this rule has less weight in view of the lower variation returned in the forecast. For the opposite direction the operation is the same: in case of a forecast of a decrease in D_1, the suggestion is for Sell and, in case of a slight fall, the suggested action will also be Sell, but with less weight. The predictions D_2 and D_3 were used only as aid to determine if the coin tended to remain stable. The rules of the inference model are described in Table 2.

Table 2. Mamdani's model rules.

	D_1	D_2	D_3	Action	Weight
1	Increasing	-	-	Buy	1
2	Slightly Increasing	-	-	Buy	0.8
3	Decreasing	-	-	Sell	1
4	Slightly Decreasing	-	-	Sell	0.8
5	Stable	-	-	Keep	0.4
6	Stable	Stable	Stable	Keep	1

The system has a unique output variable, named Action. This variable was modeled with three linguistic terms defined by trapezoidal functions: Buy, Sell and Keep. Centroid method was used for defuzzification. After defuzzification, if the system returns a value lower than 1, it is understood that the action to be taken is to Buy. If the returned value is greater than 2, the suggested action is to Sell. Finally, if the value is between 1 and 2, the maintenance of the current investment, according to the system, is the best choice.

4.3 Proposed Approaches Evaluation

This section summarizes the computational experiments performed to evaluate the proposed approaches. We used in these experiments the data set described in Sect. 3. For this, first the models perform the prediction of the closing value for the next day/days. Then, is calculate D (D_1 to Evolving Approach and D_1,

D_2 and D_3 to Hybrid Approach), i.e., the difference between the current value and the predicted value. The value of D is used to suggest the action (Buy, Keep or Sell). The validation if the answer given by each model was consistent with the actual price of the currency for the next day is based in the value of the SR and RBS parameters. In these experiments, the SR and RBS parameters were defined with values between 1 and 4. The best results for the two approaches were obtained with $SR = 2$ and $RBS = 1$.

The results obtained by the Evolving Approach and the Hybrid Approach are presented in Tables 3 and 4, respectively. For each cryptocurrency, the tables shows the accuracy (hit percentage) in the recommendations to Buy, Keep and Sell. Tables also present the number of samples, the number and the percentage of correct recommendations.

In the results obtained by the Evolving Approach (Table 3) we can see that the average accuracy for the three crypto-coins was 92.30%. The best results were obtained in the sell recommendations with a 99.20% and the worst results in keep recommendations with 85.69%. The accuracy in buy recommendations is 94.63%.

Table 3. Correctly submitted suggestions – Evolving Approach.

Correctly submitted suggestions (SR = 2 e RBS = 1)						
Evolving Approach						
	Samples	Hits	Buy(%)	Keep(%)	Sell(%)	Total(%)
BTC	485	436	95.20	80.86	99.08	89.90
ETH	485	459	95.95	90.22	99.22	94.64
LTC	485	448	92.63	86.67	99.30	92.37
Total	1455	1343	94.63	85.69	99.20	92.30

Table 4. Correctly submitted suggestions – Hybrid Approach.

Correctly submitted suggestions (SR = 2 e RBS = 1)						
Hybrid Approach						
	Samples	Hits	Buy(%)	Keep(%)	Sell(%)	Total(%)
BTC	485	441	93.62	83.24	83.24	90.93
ETH	485	460	91.88	94.93	98.67	94.85
LTC	485	454	91.57	90.00	99.36	93.61
Total	1455	1355	92.36	88.94	98.61	93.13

Table 4 presents the results obtained by the Hybrid Approach. The average accuracy for the three crypto-coins was 93.13%. The best results were obtained in the sell recommendations with a 98.61% and the worst results in keep recommendations with 88.94%. The accuracy in buy recommendations is 92.36%.

Comparing the results of the two approach, we noted that they are comparable. Computational experiments show that two proposed approach are promising and competitive with alternatives reported in the literature.

5 Conclusion

This paper presented two approaches for recommending investments in virtual currencies. The proposal consists in the use of an evolving fuzzy system applied in the forecasting of cryptocurrencies. The calculation of the change forecast for the next days in relation to the current value of the currency is used as a basis for suggestion of investment in two different approaches, the first being based only on evolving fuzzy model and the second uses an hybrid approach. The first approach considers only the output of the evolving algorithm in the suggestion of an investment action. The action returned may be one of three possible: buy, sell or keep the amount invested. In the second approach, in addition to the use of the result of the prediction returned by the evolving algorithm, a Mamdani system was also used, based on expert knowledge, which also returns a suggestion of investment.

Initially, experiments were performed with four evolving fuzzy models in the prediction of the quotation. The evolving models obtained good results. The best performance was obtained by eNFN. Therefore, this model was used for the recommendation system. In the evaluation of the investment recommendation, recent historical data from three different cryptocurrencies were used. For the three currencies evaluated, both approaches proved to be efficient, with the hybrid approach showing a little better in the overall result. The results obtained are considered satisfactory since the solution reached a high rate of assertiveness. Therefore, the proposed approach are promising alternative to construction of investment recommendation systems.

Future work shall the implementation of a loss control mechanism in the case of successive investment maintenance suggestions. The importance of this implementation is seen when considering a scenario where the SR value is equal to 2, and there are, for example, 10 successive declines in the quotation value, less than 2%. If the algorithm suggested in this interval, ten times, that the investment be maintained, the investor would record accumulated loss without being alerted. In addition, it is believed that results can be improved by creating rules and parameter adjustments of the Mamdani model based on data. Finally, it is suggested to perform a study and implementation of an algorithm that allows the model to return not only an action but also a value that quantifies how much to buy (for a purchase suggestion) or the quantity that must be sold (in the case of a sales suggestion).

Acknowledgements. This work was supported by Brazilian National Research Council (CNPq).

References

1. Nakamoto, S.: Bitcoin: a peer-to-peer electronic cash system (2008). Accessed 2 Mar 2018
2. Grinberg, R.: Bitcoin: an innovative alternative digital currency. Hastings Sci. Technol. Law J. **4**, 160 (2011). SSRN: https://ssrn.com/abstract=1817857. Accessed 2 Mar 2018
3. Investing.com. Investing (2018). https://br.investing.com/crypto/bitcoin/btc-usd-chart. Accessed 16 Mar 2018
4. Kim, Y.B., Kim, J.G., Kim, W., Im, J.H., Kim, T.H., Kang, S.J., Kim, C.H.: Predicting fluctuations in cryptocurrency transactions based on user comments and replies. PLoS ONE **11**, e0161197 (2016)
5. Kaminski, J., Gloor, P.A. Nowcasting the bitcoin market with Twitter signals. CoRR, vol. abs/1406.7577 (2014). http://arxiv.org/abs/1406.7577
6. NeuroBot. Neural Network Algorithm. https://neurobot.trading/. Accessed 17 Aug 2018
7. El-Abdelouarti Alouaret, Z.: Comparative study of vector autoregression and recurrent neural network applied to bitcoin forecasting, July 2017. http://oa.upm.es/47934/
8. Indera, N., Yassin, I., Zabidi, A., Rizman, Z.: Non-linear autoregressive with exogeneous input (NARX) Bitcoin price prediction model using PSO-optimized parameters and moving average technical indicators. J. Fundam. Appl. Sci. **9**(3S), 791–808 (2017)
9. Jang, H., Lee, J.: An empirical study on modeling and prediction of bitcoin prices with Bayesian neural networks based on blockchain information. IEEE Access **6**, 5427–5437 (2018)
10. Sin, E., Wang, L.: Bitcoin price prediction using ensembles of neural networks. In: International Conference Natural Computation, Fuzzy Systems and Knowledge Discovery, pp. 666–671 (2017)
11. Spilak, B.: Deep neural networks for cryptocurrencies price prediction. Ph.D. dissertation, Humboldt-Universitat zu Berlin, May 2018
12. McNally, S., Roche, J., Caton, S.: Predicting the price of Bitcoin using Machine Learning. In: International Conferene Parallel, Distributed and Network-Based Processing, pp. 339–343, March 2018
13. Lemos, A., Caminhas, W., Gomide, F.: Multivariable Gaussian evolving fuzzy modeling system. IEEE Trans. Fuzzy Syst. **19**(1), 91–104 (2011)
14. Silva, A.M., Caminhas, W., Lemos, A., Gomide, F.: A fast learning algorithm for evolving neo-fuzzy neuron. Appl. Soft Comput. **14**, 194–209 (2014)
15. Angelov, P.P., Filev, D.P.: An approach to online identification of Takagi-Sugeno fuzzy models. IEEE Trans. Syst. Man Cybern. **34**(1), 484–498 (2004)
16. Angelov, P., Zhou, X.: Evolving fuzzy systems from data streams in real-time. In: 2006 International Symposium on Evolving Fuzzy Systems, pp. 29–35, September 2006
17. Yamakawa, T.: Silicon implementation of a fuzzy neuron. IEEE Trans. Fuzzy Syst. **4**(4), 488–501 (1996)
18. Takagi, T., Sugeno, M.: Fuzzy identification of systems and its applications to modeling and control. IEEE Trans. Syst. Man Cybern. **15**(1), 116–132 (1985)
19. Caminhas, W., Gomide, F.: A fast learning algorithm for neofuzzy networks. In: Proceedings of the Information Processing and Management of Uncertainty in Knowledge Based Systems, pp. 1784–1790 (2000)

20. Angelov, P.: An approach for fuzzy rule-base adaptation using online clustering. Int. J. Approximate Reasoning **25**(3), 275–289 (2004)
21. CoinMarketCap. Cryptocurrency Market Capitalizations (2018). https://coinma rketcap.com/. Accessed 01 June 2018
22. Mamdani, E., Assilian, S.: An experiment in linguistic synthesis with a fuzzy logic controller. Int. J. Hum Comput Stud. **51**(2), 135–147 (1999)

On Fuzzy Optimization Foundation

Laécio C. Barros, Nilmara J. B. Pinto[(✉)], and Estevão Esmi

Institute of Mathematics, Statistics and Scientific Computing,
University of Campinas, Campinas, Brazil
laeciocb@ime.unicamp.br, nilmarabiscaia@gmail.com, eelaureano@gmail.com

Abstract. In this work we discuss the fuzzy optimization problem, in order to provide a mathematical approach to the foundation of optimization problem in the fuzzy context. By the Zadeh's extension principle we revisit the decision method stated by Bellman and Zadeh.

1 Introduction

Fuzzy optimization problems (FOPs) were firstly analyzed in the linear case, and after that in the non-linear case. Furthermore with advance of studies, the form to consider uncertainties of the fuzzy type has been changed. For more details the reader can refer to the surveys [8,17,20].

The FOPs were first introduced by Bellman and Zadeh in 1970 [3]. In that work the authors tried to express the elements of FOP in fuzzy terms, setting a real objective function with fuzzy constrains or the goal described by a fuzzy function with real or fuzzy evaluated constraints.

The solution for this problem was constructed by maximizing membership degree of all the possible solutions in the decision set. Subsequently several authors study these problems, such as, Zimmermann [29,30], Verdegay [22], Werners [24], Rommelfanger [18] and Mohamed [10]. Thereafter were considered multiple objective functions [31], and still type-2 fuzzy constraints [6,7].

Applications have been done in several areas, such as agronomic planning [9,15], financial market [12], environment [19], industrial process [21,23], among others.

The second approach arises with Wu [25–27], where the objective function is given by a non-linear d function. For that type of FOP were established optimality conditions similar to Karush-Kuhn-Tucker optimality conditions for classical optimization problem, by using fuzzy calculus [5,13,14]. Also, there is a version with constraints given by fuzzy-valued functions [16].

This paper focuses on a discussion of the most applied method, from to Bellman and Zadeh [3], in order to describe the method with more mathematical rigor. Our purpose is not to establish new methods, but instead to provide theoretical background according to the well-founded tools available in fuzzy set theory. This is made by means of Zadeh's extension principle.

This paper is organized as follows. In Sect. 2 we briefly recall the classical optimization problem. In Sect. 3 we present some basic definitions and results

© Springer Nature Switzerland AG 2019
R. B. Kearfott et al. (Eds.): IFSA 2019/NAFIPS 2019, AISC 1000, pp. 148–156, 2019.
https://doi.org/10.1007/978-3-030-21920-8_14

from fuzzy set theory. In Sect. 4, we develop the method of decision from Bellman and Zadeh in terms of Zadeh's extension principle.

2   Classical Optimization

This section presents some basic concepts of optimization problems.

Optimization problems (OPs), in general, consist in minimizing or maximizing a function f called objective function. The inputs of f must satisfy some properties described by *feasible set* or *feasible region* Ω. Mathematically we have

$$\min f \qquad \qquad \text{(OP)}$$
$$\text{subject to } x \in \Omega.$$

Usually such problems are considered in the form $f : \mathbb{R}^n \to \mathbb{R}$, and $\Omega \subseteq \mathbb{R}^n$. The problem is said unconstrained if the feasible set Ω is equal to $\mathbb{R}^n$; otherwise the problem is said to be constrained.

A point $x^* \in \Omega$ is said to be a local minimizer of (OP) if there is ε-neighbourhood around x^*, with $\varepsilon > 0$, denoted by $V_\varepsilon(x^*)$, such that $f(x^*) \leq f(x)$, for all $x \in V_\varepsilon(x^*) \cap \Omega$. Moreover, if there is no $\widehat{x} \in \Omega$ such that $f(\widehat{x}) < f(x^*)$, then x^* is called global minimizer of (OP).

The constraints can be given by inequalities or equalities. In the first case, the feasible set is $\Omega = \{x \in \mathbb{R}_+^n; f_i(x) \geq 0\}$, where $\mathbb{R}_+^n = \{(x_1, \ldots, x_n) \in \mathbb{R}^n; x_i \geq 0, i = 1, \ldots, n\}$. For the case where the constrains are given by equalities, one can consider that for any $a, b \in \mathbb{R}$, the equality $a = b$ can be rewrite as two inequalities $a \leq b$ and $b \leq a$. Henceforth we only consider the first case. The problem (OP) is well-defined if there exists at least one point $x^* \in \Omega$.

The simplest and most widely studied form of (OP) is the case in which the objective function and the constraints are linear, that is, $f(x) = c^t x$ and $\Omega = \{x \in \mathbb{R} : Ax \leq b\}$, where $A \in \mathbb{R}^{m \times n}$, $c \in \mathbb{R}^n$ and $b \in \mathbb{R}^m$. The vector c is said to be the cost vector, and the feasible region is a polytope.

In this work the feasible set will be considered as a fuzzy set and we will discuss how this affects the problem. In the next section we present some concepts on fuzzy set theory.

3   Fuzzy Set Theory

A fuzzy subset A of a universe X is characterized by its membership function $\mu_A : X \to [0, 1]$, where $\mu_A(x)$, or simply $A(x)$, represents the degree of the element x in the set A, for all $x \in X$. The class of fuzzy subsets of X is denoted by the symbol $\mathscr{F}(X)$. All classical subset A of X can be viewed as a particular fuzzy set considering its characteristic function as its membership function.

The operators t-norm and s-norm are defined by the functions $t, s : [0, 1] \times [0, 1] \to [0, 1]$ both associative, commutative and increasing. In addition, they satisfies $t(1, x) = 1tx = x$, and $s(0, x) = 0sx = x, \forall x \in [0, 1]$. The minimum and maximum operators, denoted by $t = \min = \wedge$ and $s = \max = \vee$, are examples of t-norm and s-norm respectively. These operators can be used to model the connectives "*and*" and "*or*". Also the classical set operations, such as union and intersection, can be extended by $t = \wedge$ and $s = \vee$, respectively [1].

A function $f : X \to Y$ can be extended, by the Zadeh's extension principle below, to a function $\widehat{f}$ where the arguments are given by fuzzy sets [28].

Definition 1. Let $f : X_1 \times \ldots \times X_n \to Y$. The Zadeh's extension of f at $(A_1, \ldots, A_n) \in \mathscr{F}(X_1 \times \ldots \times X_n)$ is the fuzzy set $\widehat{f}(A_1, \ldots, A_n) \in \mathscr{F}(Y)$ whose membership function is given by

$$\widehat{f}(A_1, \ldots, A_n)(y) = \sup_{(x_1, \ldots, x_n) \in f^{-1}(y)} A_1(x_1) \wedge \ldots \wedge A_n(x_n), \ \forall y \in Y, \quad (1)$$

where $f^{-1}(y) = \{(x_1, \ldots, x_n) \in X_1 \times \ldots \times X_n : f(x_1, \ldots, x_n) = y\}$ and by definition $\sup \emptyset = 0$.

For each $A \in \mathscr{F}(X)$, its α-cuts, $\forall \alpha \in (0, 1]$, are defined by the classical set $[A]_\alpha = \{x \in X; \ \mu_A(x) \geq \alpha\}$. If X is a topological space, then the 0-cut of A is given by $[A]_0 = cl\{x \in X; \ \mu_A(x) > 0\}$, where $cl \ Y$ denotes the closure of $Y \subseteq X$.

A fuzzy set $A \in \mathscr{F}(\mathbb{R})$ is said to be a fuzzy number if its α-cuts are bounded, closed and non-empty nested intervals [1]. Therefore each α-cut of a fuzzy number is an interval, and thus is written $[A]_\alpha = [a_\alpha^-, a_\alpha^+]$. The class of fuzzy numbers is denoted by $\mathbb{R}_\mathscr{F}$.

The most common fuzzy number is the trapezoidal fuzzy number, denoted by $(a; b; c; d)$, where $a \leq b \leq c \leq d$, whose α-cuts are given by $[A]_\alpha = [a + \alpha(b - a), d - \alpha(d - c)], \forall \alpha \in [0, 1]$. Every interval $[a, b]$ can be viewed as $(a; a; b; b)$, thus the classical intervals are particular cases of fuzzy numbers.

The Zadeh's extension of a function can be characterized by means of α-cuts as follows.

Theorem 1 [2,11]. *Let $f : \mathbb{R} \to \mathbb{R}$ a continuous function. The α-cuts of $\widehat{f}$ are given by*

$$[\widehat{f}(A)]_\alpha = f([A]_\alpha), \ \forall \alpha \in [0, 1]. \quad (2)$$

Using the Zadeh's extension principle and Theorem 1, we revisit fuzzy optimization problems in the next section.

4   Fuzzy Optimization Problems

Bellman and Zadeh [3] considered that a solution for an optimization problem in form of (OP) must satisfy at the same time the constraints and the search for the minimum. Figure 1 illustrates that is not enough minimize the objective function f, it is necessary analyze the intersection between the feasible set *and* seek the minimum, simultaneously.

The formulation of this problem is given as follows

$$\min \; f(x) = c^t x \qquad (3)$$
$$\text{subject to } Ax \preceq b,$$

where the inequality $\preceq$ can be seen as a relaxation of the constraints. Bector and Chandra [4] provided a fine survey for an analogous problem to (3), where the vector b contains fuzzy entries and for the minimum concept is fuzzy.

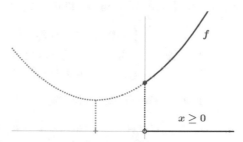

Fig. 1. The decision scheme for the case where objective function is $f(x) = (x + 2)^2$ and the constraint is $x \geq 0$.

Denoting the feasible set by $R = \{x \in \mathbb{R}; \; Ax \preceq b\}$, a point $x \in \mathbb{R}$ has membership degree $\mu_R(x)$ and x satisfies the objective function, denoted by O, with degree $\mu_O(x)$. Then, according to [3], we are looking for a point x^* that maximizes the membership function of the decision set given by

$$\mu_D(x) = \min[\mu_O(x), \mu_R(x)], \forall x \in \mathbb{R}. \qquad (4)$$

That is, x^* is the element in $O \cap R$ with bigger membership degree.

Nonetheless the membership function μ_O of objective function is found intuitively. Based on 0− and 1−cuts of R, Bellman and Zadeh [3] estimated the extreme values for the objective function, denoted by $z_{\min}$ and $z_{\max}$, respectively. From this an arbitrary membership function μ_O is established.

In order to provide a deep mathematical point of view for this methodology, we suggest an approach with more mathematical rigor than the aforementioned problem. This is done by firstly considering fuzzy constraints.

4.1   Fuzzy Constraints

First, let us comprehend the meaning of a fuzzy constraint $x \preceq B$, where $B \in \mathbb{R}_{\mathscr{F}}$ with α-cuts $[B]_\alpha = [b_\alpha^-, b_\alpha^+]$.

(1) Let $x \preceq B$ be the fuzzy set $L = \{x \in \mathbb{R} : x \preceq B\}$.
(2) Let $x \succeq B$ be the fuzzy set $U = \{x \in \mathbb{R} : x \preceq B\}$.

The membership function of L and U are given below.

First, if $x \leq y$, then the membership degree of x is, at least, equal to membership degree of y to L. The fuzzy set μ_L can be written in terms of μ_B:

$$\mu_L(x) = \sup_{x \leq y} \mu_B(y) = \begin{cases} 1, & \text{if } x \leq b_1^+ \\ \mu_B(x), & \text{if } b_1^+ < x \leq b_0^+ \\ 0, & \text{if } x \geq b_0^+ \end{cases}.$$

Analogously, if $x \geq y$, then the membership degree of x with respect to U must be at least equal to the degree of y in U. The fuzzy set μ_U is given by

$$\mu_U(x) = \sup_{x \geq y} \mu_B(y) = \begin{cases} 0, & \text{if } x \leq b_0^- \\ \mu_B(x), & \text{if } b_0^- < x \leq b_1^- \\ 1, & \text{if } x \geq b_1^- \end{cases}.$$

An example of these fuzzy sets is depicted in Fig. 2.

Remark 1. Note that $\min\{\mu_L(x), \mu_U(x)\} = \mu_B(x)$, that is, $L \cap U = B$. Hence this definition preserves the antisymmetry inherent to classical relation $\leq$ (or $\geq$).

Example 2. *Let $B = [b_1, b_2]$ be a real interval, where $b_1 \leq b_2$.*

The sets L and U have membership functions as follows

$$\mu_L(x) = \sup_{x \leq y} \mu_B(y) = \begin{cases} 1, & \text{if } x \leq b_2 \\ 0, & \text{if } x > b_2 \end{cases} \quad, \quad \text{and } \mu_U(y) = \begin{cases} 0, & \text{if } x \leq b_1 \\ 1, & \text{if } x > b_1 \end{cases}.$$

The intersection, $L \cap U$, is depicted in Fig. 3.

Following, we study the constraint $g(x) \preceq B$ (or $g(x) \succeq B$) for $g : \mathbb{R} \to \mathbb{R}$ a monotonic function. By Theorem 1, for a continuous function $g : \mathbb{R} \to \mathbb{R}$, the Zadeh's extension of g has its α-cuts given by

$$[\widehat{g}(A)]_\alpha = g([A]_\alpha) = g([a_\alpha^-, a_\alpha^+]) = \begin{cases} [g(a_\alpha^-), g(a_\alpha^+)], & \text{if } g \text{ is increasing} \\ [g(a_\alpha^+), g(a_\alpha^-)], & \text{if } g \text{ is decreasing.} \end{cases}$$

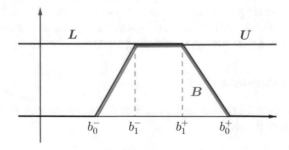

Fig. 2. Membership function of L in blue and U in red, and the intersection B between L and U in green.

Note that if g is a bijection, then $\mu_{\widehat{g}(A)}(y) = \mu_A(g^{-1}(y))$. In addition, if g is a bijection increasing we define

(3) For $g(x) \preceq B \Leftrightarrow x \preceq \widehat{g}^{-1}(B)$, we define the fuzzy set $L_g = \{x \in \mathbb{R} : x \preceq \widehat{g}^{-1}(B)\}$;

(4) For $g(x) \succeq B \Leftrightarrow x \succeq \widehat{g}^{-1}(B)$, we define the fuzzy set $U_g = \{x \in \mathbb{R} : x \succeq \widehat{g}^{-1}(B)\}$;

where $\widehat{g}^{-1}(B)$ is the Zadeh's extension of g^{-1}.

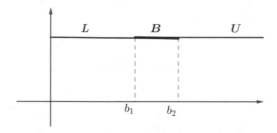

Fig. 3. The intersection between L and U results in the interval $B = \{x \in \mathbb{R}; b_1 \leq x \leq b_2\}$.

In this way, the membership function of L_g is

$$
\mu_{L_g}(x) = \begin{cases} 1, & \text{if } x \leq g^{-1}(b_1^+) \\ \mu_{g^{-1}(D)}(x), & \text{if } g^{-1}(b_1^+) \leq x \leq g^{-1}(b_0^+) \\ 0, & \text{if } x \geq g^{-1}(b_0^+) \end{cases} = \begin{cases} 1, & \text{if } g(x) \leq b_1^+ \\ \mu_B(y(x)), & \text{if } b_1^+ \leq g(x) \leq b_0^1 \\ 0, & \text{if } g(x) \geq b_0^+ \end{cases}
$$

(5)

and the membership function of U_g is

$$
\mu_{U_g}(x) = \begin{cases} 1, & \text{if } g(x) \geq b_1^- \\ \varphi_B(g(x)), & \text{if } b_0^- \leq g(x) \leq b_1^-. \\ 0, & \text{if } g(x) \geq b_0^- \end{cases}
$$

(6)

The simplest example arises when g is linear, that is, when g is given by $g(x) = Ax$ for all $x \in \mathbb{R}$, where $A \in \mathbb{R}$. In this, case it is easy to obtain an explicit formula for the membership functions of L_g and U_g. Moreover, note that the formulas (5) and (6) can naturally be extended for the case where g is a function from $\mathbb{R}^n$ to $\mathbb{R}$.

Using this membership functions, we can describe the term $\mu_O(x)$ in Eq. (4) taking into account the constraint L_g or U_g.

4.2   Fuzzification of Objective Function

In order to use the Bellman and Zadeh's decision (4), it is important to develop the term μ_O that describes the objective function, considering the fuzzy constraint in Eq. (7).

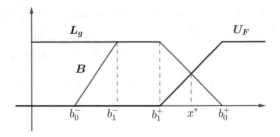

Fig. 4. The intersection between U_F and the constraint L_g, with g given by the identity.

$$\min \ f(x) \tag{7}$$
$$\text{subject to } g(x) \preceq B.$$

That is, the function f can be associated with a fuzzy set U_F that represents the degree in which an element x is a minimizer of problem (7). For example, if f is decreasing and g is increasing, one can defined U_F as follows:

$$\mu_{U_F}(x) = \begin{cases} 1, & \text{if } x \geq g(b_0^+) \\ \dfrac{f(x) - f(g(b_1^+))}{f(g(b_0^+)) - f(g(b_1^+))}, & \text{if } g(b_0^+) \leq x < g(b_1^+) \ . \\ 0, & \text{if } x < g(b_1^+) \end{cases} \tag{8}$$

We focus on this last case. A solution of (7) is an element x^* such that its degree of membership to $L_g \cap U_F$ is maximum. Figure 4 illustrates a case where x^* is unique and consists of the solution of the following equation:

$$\frac{f(b_1^+) - f(x)}{f(b_1^+) - f(b_0^+)} = \mu_B(g(x)). \tag{9}$$

In this way, the solution of (OP) is obtained from $L_g \cap U_F$, which has the form of (4). In the particular case where B is a trapezoidal fuzzy number and f and g are given by $f(x) = -cx$ and $g(x) = Ax$, with $c, A > 0$, we obtain from Eq. (9) that $x^* = \frac{b_0^+ + b_1^+}{1 + A}$. This case is illustrated in Fig. 4 for $A = 1$.

We observe that the formula (8) holds for nonlinear functions f. Moreover, for modelling questions, is common to use different values of $b_0^-, b_0^+, b_1^+, b_1^-$ as lower limits for U_F [17]. In our approach, $z_{\min} = f(b_1^+)$ and $z_{\max} = f(b_0^+)$.

5   Final Remarks

In this paper we studied the fuzzy optimization problem by revisiting the decision method stated by Bellman and Zadeh [3]. We focused on a specific type of FOP, where the constraints are modelled by fuzzy numbers.

We discussed different forms of dealing with FOPs. In particular, we focused on the case where the objective function is given by a real-valued-function and the fixed term in the constraint is given by a fuzzy number. Here we established a mathematical foundation for this well-known problem, first studied by Bellman and Zadeh [3].

We used the Zadeh's extension principle in order to describe the uncertainty of the constraints during the process of searching the minimum solution. The method proposed in [3] is equal to the method proposed here for constraints given by monotonic functions.

Acknowledgement. The authors would like to thank the financial support of CNPq under grant 306546/2017-5, CAPES under grant no. 1691227, and, FAPESP under grant no. 2016/26040-7.

References

1. Barros, L.C., Bassanezi, R.C., Lodwick, W.A.: A First Course in Fuzzy Logic, Fuzzy Dynamical Systems, and Biomathematics. Springer, Berlin, Heidelberg (2017)
2. Barros, L.C., Bassanezi, R.C., Tonelli, P.A.: On the continuity of the Zadeh's extension. In: Proceedings of the IFSA 1997 Congress, pp. 1–6 (1997)
3. Bellman, R.E., Zadeh, L.A.: Decision-making in a fuzzy environment. Manag. Sci. **17**(4), B141–B164 (1970)
4. Bector, C.R., Chandra, S.: Fuzzy Mathematical Programming and Fuzzy Matrix Games. Studies in Fuzziness and Soft Computing, vol. 169. Springer, Berlin, Heidelberg (2005)
5. Chalco-Cano, Y., Silva, G.N., Rufián-Lizana, A.: On the Newton method for solving fuzzy optimization problems. Fuzzy Sets Syst. **272**, 60–69 (2015)
6. Figueroa-García, J. C.: Linear programming with interval type-2 fuzzy right hand side parameters. In: NAFIPS 2008 - 2008 Annual Meeting of the North American Fuzzy Information Processing Society, pp. 1–6 (2008)
7. Figueroa-García, J.C., Hernandez, G.: Computing optimal solutions of a linear programming problem with interval type-2 fuzzy constraints. In: Hybrid Artificial Intelligent Systems, pp. 567–576 (2012)
8. Klir, G.J., Yuan, B.: Fuzzy Sets and Fuzzy Logic: Theory and Applications. Prentice-Hall, New York (1995)
9. Mjelde, K.M.: Fuzzy resource allocation. Fuzzy Sets Syst. **19**(3), 239–250 (1986)
10. Mohamed, R.H.: The relationship between goal programming and fuzzy programming. Fuzzy Sets Syst. **89**, 215–222 (1997)
11. Nguyen, H.T.: A note on the extension principle for fuzzy sets. J. Math. Anal. Appl. **64**, 369–380 (1978)
12. Östermark, R.: Fuzzy linear constraints in the capital asset pricing model. Fuzzy Sets Syst. **30**(2), 93–102 (1989)
13. Osuna-Gómez, R., Chalco-Cano, Y., Rufián-Lizana, A., Hernández-Jiménez, B.: Necessary and sufficient conditions for fuzzy optimality problems. Fuzzy Sets Syst. **296**, 112–123 (2016)
14. Osuna-Gómez, R., Hernández-Jiménez, B., Chalco-Cano, Y., Ruiz-Garzón, G.: Different optimum notions for fuzzy functions and optimality conditions associated. Fuzzy Optim. Decis. Making **17**, 177–193 (2018)

15. Owsiński, J.W., Zadrozny, S., Kacprzyk, J.: Analysis of water use and needs in agriculture through a fuzzy programming model. In: Kacprzyk, J., Orlovski, S.A. (eds.) Optimization Models Using Fuzzy Sets and Possibility Theory, pp. 377–395. Springer, Dordrecht (1987)
16. Pathak, V.D., Pirzada, U.M.: Necessary and sufficient optimality conditions for nonlinear fuzzy optimization problem. Int. J. Math. Sci. Educ. **4**(1), 1–16 (2011)
17. Pedrycz, W., Gomide, F.: An Introduction to Fuzzy Sets. The MIT Press, London (1998)
18. Rommelfanger, H.: Fuzzy linear programming and applications. Eur. J. Oper. Res. **92**, 512–527 (1996)
19. Sommer, G., Pollatschek, M.A.: A fuzzy programming approach to an air pollution regulation problem. Prog. Cybern. Syst. Res. **3**, 303–313 (1978)
20. Tang, J., Wang, D., Fung, R.Y.K., Yung, K.-L.: Understanding of fuzzy optimization: theories and methods. J. Syst. Sci. Complexity **17**(1), 117–136 (2004)
21. Trappey, J.F.C., Richard Liu, C., Chang, T.C.: Theory and application in manufacturing, fuzzy non-linear programming (1988)
22. Verdegay, J.L.: Fuzzy mathematical programming. In: Gupta, M.M., Sanchez, E. (eds.) Fuzzy Information and Decision Processes, North Holland, Amsterdam (1982)
23. Verdegay, J.L.: Applications of fuzzy optimization in operational research. Control Cybern. **13**(3), 229–239 (1984)
24. Werners, B.: Interactive multiple objective programming subject to flexible constraints. Eur. J. Oper. Res. **31**, 342–349 (1987)
25. Wu, H.C.: Duality theory in fuzzy linear programming problems with fuzzy coefficients. Fuzzy Optim. Decis. Making **2**, 61–73 (2003)
26. Wu, H.C.: Saddle point optimality conditions in fuzzy optimization. Fuzzy Optim. Decis. Making **2**, 261–273 (2003)
27. Wu, H.C.: The optimality conditions for optimization problems with fuzzy-valued objective functions. Optimization **57**(3), 476–489 (2008)
28. Zadeh, L.A.: The concept of a linguistic variable and its application to approximate reasoning-I, II, III. Inf. Sci. **8**, 301–357 (1975)
29. Zimmermann, H.-J.: Applications of fuzzy set theory to mathematical programming. Inf. Sci. **36**, 29–58 (1985)
30. Zimmermann, H.-J.: Description and optimization of fuzzy systems. Int. J. Gen. Syst. **2**, 209–215 (1976)
31. Zimmermann, H.-J.: Fuzzy programming and linear programming with several objective functions. Fuzzy Sets Syst. **1**, 45–55 (1978)

Fuzzy Systems with Sigmoid-Based Membership Functions as Interpretable Neural Networks

Barnábas Bede[✉]

DigiPen Institute of Technology, 9931 Willows Rd NE, Redmond, WA 98052, USA
bbede@digipen.edu

Abstract. In this paper new interpretable neural network architectures are proposed. A Neural Network with sigmoid activation function is converted into a sigmoid-based approximation operator, which, at its turn, can be approximated by a fuzzy system of Takagi-Sugeno type. Altogether this process shows that a neural network with sigmoid activation functions can be approximated by a Takagi-Sugeno fuzzy system. As the TS fuzzy system provides interpretability, while Neural Networks provide approximation capability, we obtain a novel interpretable Neural Network Architecture. Interpretability of the TS fuzzy system can provide insights into data in a new way, enhancing decision making.

1 Introduction

Fuzzy Sets, Fuzzy Systems and Neural Networks are connected as different paradigms within Soft Computing and Approximate Reasoning [14–16]. The strong connection between Fuzzy Systems and Neural Networks is hidden in their approximation properties. Both Fuzzy Systems and Neural Networks have universal approximation properties [1,4,12,13]. Also, there are many hybrid architectures, that combine Fuzzy Systems with Neural Networks [5,6,8], with the ANFIS architecture being the most well known such architecture [6]. Interpretability of Deep Learning of Neural Networks, has been a problem where intense ongoing research [2,9]. Radial Basis Function networks, can be seen as both fuzzy systems and interpretable neural networks [7]. In the present paper we propose a new architecture based on a difference of sigmoid functions. One single sigmoid function can model inequality relations, however we want to consider interpretation by localizing values. It is clear that the difference of two sigmoid functions, with the same slope parameter is a bell-shaped function, which can be interpreted as a fuzzy membership function.

Another idea that we would like to explore is inter-approximation between Fuzzy Systems of Takagi-Sugeno type, and Neural Networks. This idea comes from the fact that we know that both Fuzzy Systems and Neural Networks have universal approximation properties. As a consequence, naturally raises the idea of Fuzzy Systems approximating Neural Networks and vice-versa. We discuss inter-approximation between a TS fuzzy system with linguistic antecedents and piecewise constant output [10,11], and a neural networks with sigmoid type activation

© Springer Nature Switzerland AG 2019
R. B. Kearfott et al. (Eds.): IFSA 2019/NAFIPS 2019, AISC 1000, pp. 157–166, 2019.
https://doi.org/10.1007/978-3-030-21920-8_15

functions [3,4]. We also study suitable interpretable neural network architectures and deep learning architectures based on the above combined approach.

2   Preliminaries

A fuzzy system of Takagi-Sugeno type [10] with piece-wise constant outputs is based on fuzzy rules of the type

$$\text{if } \mathbf{x} \text{ is } A_i \text{ then } y \text{ is } w_i,$$

$i = 1, ..., n$, with antecedents A_i and crisp output w_i and input $\mathbf{x} \in \mathbf{R}^m$. A Takagi-Sugeno system can be seen as a function of the form

$$TS(\mathbf{x}) = \frac{\sum_{i=1}^{n} A_i(\mathbf{x}) w_i}{\sum_{i=1}^{n} A_i(\mathbf{x})},$$

An Adaptive Network-based Fuzzy Inference Systems (ANFIS) is a TS fuzzy system with added learning algorithms [6].

The interpretation of a TS fuzzy system is based on the fuzzy rule base that we consider above. It is easy to see that all the variants and fuzzy systems have the interpretation of an approximation. Namely, given an unknown function f, and knowing that $f(x_i) = y_i$ then in a neighborhood of x_i the value of the function should be near y_i. As a fuzzy rule we can formulate

$$\text{if } \mathbf{x} \text{ is about } x_i, \text{ then } y \text{ is about } y_i, \ i = 1, \ldots, n$$

Neural networks can also be seen as approximation operators. Let us consider a neural network with sigmoid activation function $\varphi(x) = \dfrac{1}{1 + e^{-x}}$

$$NN(x) = w_0 + \sum_{i=1}^{n} w_i \varphi(a_i(x - b_i)).$$

This type of neural network can be interpreted as a fuzzy system

$$\text{if } x \text{ is } A_i \text{ then } y+ = w_i, \ i = 0, \ldots, n.$$

3   Takagi-Sugeno Approximation of a Neural Network

Let us consider a neural network with one input, output, one hidden layer, and sigmoid activation function $\varphi(x) = \frac{1}{1+e^{-x}}$ (Figs. 1 and 2):

$$NN(x) = w_0 + \sum_{i=1}^{n} w_i \varphi(a_i(x - b_i)). \tag{1}$$

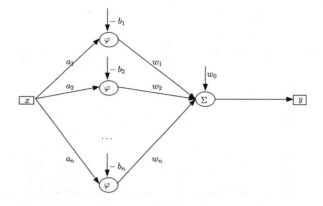

Fig. 1. Structure of a Neural Network

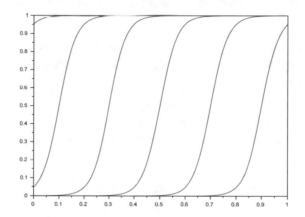

Fig. 2. Membership functions of sigmoid type.

We assume $b_1 < b_2 < ... < b_n$. Also, since $\varphi(-x) = 1 - \varphi(x)$, we may assume $a_1, ..., a_n > 0$. Indeed, if we have a term $a_i < 0$ we can write it as

$$\varphi(a_i(x - b_i)) = \varphi(-|a_i|(x - b_i)) = 1 - \varphi(|a_i|(x - b_i))$$

and after rearranging the terms we get a Neural Network with all $a_i > 0$.

We construct an approximation operator with sigmoid-based membership functions

$$A_0(x) = 1 - \varphi(a_1(x - b_1)),$$

$$A_i(x) = \varphi(a_i(x - b_i)) - \varphi(a_{i+1}(x - b_{i+1})), \quad i = 1, ..., n - 1$$

$$A_n(x) = \varphi(a_n(x - b_n)).$$

The shape of the sigmoid-based membership functions obtained as differences of sigmoids are shown in Fig. 3.

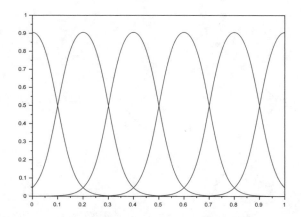

Fig. 3. Sigmoid-based membership functions of a Takagi Sugeno fuzzy system.

We can easily see that $A_i(x)$ are bell-shaped, but not necessarily positive for all $x \in \mathbf{R}$, so they are not proper fuzzy sets in the most general case. Further, we observe that

$$\sum_{i=0}^{n} A_i(x) = 1.$$

The approximation operator can be expressed as a sigmoid-based approximation (SA)

$$SA(x) = \sum_{i=0}^{n} A_i(x) \cdot y_i. \tag{2}$$

Theorem 1. *The neural network and the TS fuzzy system descried above are equivalent in the sense that*

(i) Given a TS fuzzy system as in (2), it can be written as a neural network as in (1), with weights

$$w_0 = y_0,$$

$$w_i = -y_{i-1} + y_i, \ i = 1, ..., n,$$

(ii) Given a Neural Network as in (1), it can be written as a sigmoid-based approximation as in (2), with weights

$$y_0 = w_0,$$

$$y_i = \sum_{j=0}^{i} w_j, \ i = 1, ..., n.$$

Proof. We observe that $SA(x)$ and $NN(x)$ are linear combinations of the same functions

$$\{1, \varphi(a_1(x - b_1)), \varphi(a_2(x - b_2)), ..., \varphi(a_n(x - b_n))\}.$$

Let us consider the equation

$$SA(x) = NN(x), \forall x \in \mathbf{R}.$$

We obtain

$$\sum_{i=0}^{n} A_i(x) \cdot y_i = w_0 + \sum_{i=1}^{n} w_i \varphi(a_i(x - b_i)).$$

Rewriting $A_i(x)$ in terms of the sigmoid functions we obtain

$$y_0(1 - \varphi(a_1(x - b_1))) + \sum_{i=1}^{n-1} y_i(\varphi(a_i(x - b_i)) - \varphi(a_{i+1}(x - b_{i+1}))) + y_n \varphi(a_n(x - b_n))$$

$$= w_0 + \sum_{i=1}^{n} w_i \varphi(a_i(x - b_i)).$$

As the system

$$\{1, \varphi(a_1(x - b_1)), \varphi(a_2(x - b_2)), ..., \varphi(a_n(x - b_n))\}$$

is linear independent we obtain

$$y_0 = w_0, \quad -y_0 + y_1 = w_1, ..., -y_n + y_n = w_n,$$

or equivalently,

$$w_0 = y_0,$$

$$w_i = -y_{i-1} + y_i, \quad i = 1, ..., n,$$

which proves (i).

The relations in (ii) follow immediately.

As the neural network was written as a sigmoid-based approximation, there was no loss of information in this initial step. Now we construct the Takagi-Sugeno approximation of the neural network.

We start from sigmoid-based approximation (SA)

$$SA(x) = \sum_{i=0}^{n} A_i(x) \cdot y_i$$

we construct the TS fuzzy rules with piece-wise constant consequences

$$\text{if } x \text{ is } A_i \text{ then } y = y_i, \quad i = 0, \ldots, n.$$

We assume $b_1 < b_2 < ... < b_n$. Also we assume $a_1, ..., a_n > 0$. Let

$$c_0 = a_0$$

$$c_i = \frac{a_i + a_{i+1}}{2}, i = 1, .., n - 1,$$

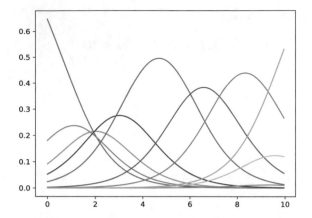

Fig. 4. Antecedents of a TS fuzzy approximation of a Neural Network.

and
$$c_n = a_n.$$

We define the fuzzy sets
$$B_0(x) = 1 - \varphi(c_1(x - b_1)),$$
$$B_i(x) = \varphi(c_i(x - b_i)) - \varphi(c_i(x - b_{i+1})), \quad i = 1, ..., n - 1$$
$$B_n(x) = \varphi(c_n(x - b_n)),$$

and we define the TS fuzzy system
$$TS(x) = \frac{\sum_{i=0}^{n} B_i(x) \cdot y_i}{\sum_{k=0}^{n} B_k(x)} \tag{3}$$

In Fig. 4 the shape of the sigmoid-based antecedents is illustrated.

As a first example let us consider the problem of function approximation. The function $f(x) = \sin x/(x + 0.25)$ is approximated by a Neural Network and the corresponding TS approximation. The results are illustrated in Fig. 5.

The idea to approximate a Neural Network by a TS fuzzy system allows us to "convert" a Neural Network into a TS fuzzy system and vice-versa. This provides a methodology for designing interpretable fuzzy systems. In Fig. 6 this procedure is illustrated. An expert can design and interact with the TS fuzzy system, while the Neural Network component of the system can learn efficiently from data. Altogether this system allows for an interpretable fuzzy system.

However the denominator is not constant we can further simplify our fuzzy system assuming an approximation of the form
$$TSL(x) = \sum_{i=0}^{n} B_i(x) \cdot z_i$$

which is a linear combination of the function $B_i(x)$ where z_i can be calculated for example, based on a least squares learning method.

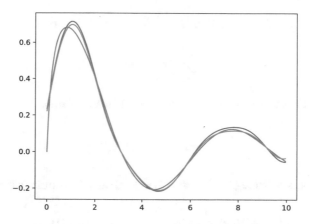

Fig. 5. Original function $f(x) = \sin x/(x+0.25)$ (orange), Neural Network (green) and TS fuzzy system (blue).

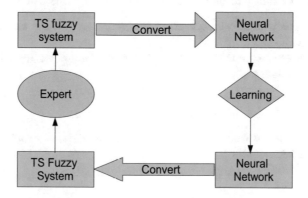

Fig. 6. TS approximation of a Neural Network as a proposal for an interpretable Neural Network

4 Multi-dimensional Case

In order to extend the above result to the multidimensional case we can adopt various approaches. In the approach we chose we allow multidimensional input, with multidimensional weights.

Let $f : \mathbf{R^n} \to \mathbf{R}$ be a function. Let $b_1, b_2, ..., b_n$ be real numbers. Let $\mathbf{u}_i \in \mathbf{R^n}$, $i = 1, ..., n$. We can define multi-dimensional functions

$$A_0(\mathbf{x}) = 1 - \varphi(\mathbf{u}_1 \cdot \mathbf{x} - b_1)$$

$$A_i(\mathbf{x}) = \varphi(\mathbf{u}_i \cdot \mathbf{x} - b_i) - \varphi(\mathbf{u}_{i+1} \cdot \mathbf{x} - b_{i+1}), i = 1, ..., n-1$$

$$A_n(\mathbf{x}) = \varphi(\mathbf{u}_n \cdot \mathbf{x} - b_n).$$

We consider the rules

$$\text{if } \mathbf{x} \text{ is } A_i \text{ then } y = y_i, \ i = 0, \dots, n.$$

Similar to the one-dimensional case we have a partition of 1, i.e.,

$$\sum_{i=0}^{n} A_i(\mathbf{x}) = 1$$

The Sigmoid-based approximation associated to these rules is

$$SA(\mathbf{x}) = \sum_{i=0}^{n} A_i(\mathbf{x}) \cdot y_i. \tag{4}$$

A neural network with sigmoidal activations an same parameters can be written as

$$NN(\mathbf{x}) = w_0 + \sum_{i=1}^{n} w_i \varphi(\mathbf{u}_i \cdot \mathbf{x} - b_i)). \tag{5}$$

The following theorem can be obtained similar to the one dimensional case.

Theorem 2. *The neural network and the approximation SA descried above are equivalent in the sense that*

(i) Given an approximation SA as in (4), it can be written as a neural network as in (5), with weights

$$w_0 = y_0,$$

$$w_i = -y_{i-1} + y_i, \ i = 1, ..., n,$$

(ii) Given a Neural Network as in (5), it can be written as an approximation SA as in (4), with weights

$$y_0 = w_0,$$

$$y_i = \sum_{j=0}^{i} w_j, \ i = 1, ..., n.$$

Similar to the one-dimensional case, the functions $A_i(\mathbf{x})$ given above are not necessarily proper fuzzy sets, as $A_i(\mathbf{x})$ may become negative.

Let $f : \mathbf{R^n} \to \mathbf{R}$ be a function. Let $b_1, b_2, ..., b_n$ be real numbers. Let $\mathbf{v}_i \in \mathbf{R^n}$, $i = 1, ..., n$. We can define multi-dimensional functions

$$B_0(\mathbf{x}) = 1 - \varphi(\mathbf{v_1} \cdot \mathbf{x} - b_1)$$

$$B_i(\mathbf{x}) = \varphi(\mathbf{v}_i \cdot \mathbf{x} - b_i) - \varphi(\mathbf{v}_i \cdot \mathbf{x} - b_{i+1}), i = 1, ..., n - 1$$

$$B_n(\mathbf{x}) = \varphi(\mathbf{v}_n \cdot \mathbf{x} - b_n).$$

Please note that having the same slope parameter $\mathbf{v_i}$ in the subtracted sigmiod functions in B_i will define it as a valid fuzzy set. However the property that they form a partition of 1 is lost in this case. We consider the fuzzy TS rules

$$\text{if } \mathbf{x} \text{ is } B_i \text{ then } y = y_i, \ \ i = 0, \ldots, n.$$

The Takagi-Sugeno fuzzy system associated to these rules is

$$TS(x) = \frac{\sum_{i=0}^{n} B_i(\mathbf{x}) \cdot y_i}{\sum_{i=0}^{n} B_i(\mathbf{x})}.$$

We observe that the system can be simplified if we do not normalize our system. In this case we can write the linearized TS system as

$$TSL(x) = \sum_{i=0}^{n} B_i(\mathbf{x}) \cdot z_i.$$

This system can be used as a simplified interpretable fuzzy system. The coefficients z_i are calculated using a learning algorithm.

5   Towards Deep Learning TS Fuzzy System with Sigmoid Antecedents

The multi-dimensional considerations in the previous section provide the opportunity to investigate deep learning architectures. We will construct the Layer L of the proposed neural network as a TS fuzzy system. Let us consider layer L with $n+1$ neurons having the activations

$$B_0(\mathbf{x}) = 1 - \varphi(\mathbf{v}_1 \cdot \mathbf{x} - b_1)$$

$$B_i(\mathbf{x}) = \varphi(\mathbf{v}_i \cdot \mathbf{x} - b_i) - \varphi(\mathbf{v}_i \cdot \mathbf{x} - b_{i+1}), i = 1, ..., n - 1$$

$$B_n(\mathbf{x}) = \varphi(\mathbf{v}_n \cdot \mathbf{x} - b_n).$$

As a layer of the proposed neural network we can consider the linearized TS system as

$$TSL(x) = \sum_{i=0}^{n} B_i(\mathbf{x}) \cdot \mathbf{z}_i.$$

In this setting $\mathbf{x} \in \mathbf{R^m}$, $\mathbf{z}_i \in \mathbf{R^p}$. Given that we have both input and output multi-dimensional, we can combine layers of this type into a neural network. In such an architecture, deep learning becomes possible.

6   Conclusions and Further Research

The present paper provides a system for interpretable learning based on a TS fuzzy system constructed using difference of sigmoid functions. The proposed system has promising interpretability and learning abilities. The proof of concept for the proposed system is given in a function approximation example. A deeper investigation using real data would be the next step to assess the effectiveness of the proposed system. An extension to the multi-dimensional case is given, together with directions for further investigation relative to deep learning.

References

1. Bede, B.: Mathematics of Fuzzy Sets and Fuzzy Logic. Springer, Heidelberg (2013)
2. Bonanno, D., Nock, K., Smith, L., Elmore, P., Petry, F.: An approach to explainable deep learning using fuzzy inference. In: Next-Generation Analyst V, SPIE Proceedings, vol. 10207, International Society for Optics and Photonics (2017)
3. Costarelli, D., Spigler, R.: Approximation results for neural network operators activated by sigmoidal functions. Neural Netw. **44**, 101–106 (2013)
4. Cybenko, G.: Approximation by superpositions of a sigmoidal function. Math. Control Signals Syst. **2**(4), 303–314 (1989)
5. Horikawa, S.-I., Furuhashi, T., Uchikawa, Y.: On fuzzy modeling using fuzzy neural networks with the back-propagation algorithm. IEEE Trans. Neural Netw. **3**(5), 801–806 (1992)
6. Jang, J.-S.R.: ANFIS: adaptive-network-based fuzzy inference system. IEEE Trans. Syst. Man Cybern. **23**(3), 665–685 (1993)
7. Jin, Y., Sendhoff, B.: Extracting interpretable fuzzy rules from RBF networks. Neural Process. Lett. **17**(2), 149–164 (2003)
8. Kasabov, N.K.: Learning fuzzy rules and approximate reasoning in fuzzy neural networks and hybrid systems. Fuzzy Sets Syst. **82**(2), 135–149 (1996)
9. Montavon, G., Samek, W., Muller, K.R.: Methods for interpreting and understanding deep neural networks. Digit. Signal Process. **73**, 1–15 (2018)
10. Sugeno, M.: An introductory survey of fuzzy control. Inf. Sci. **36**, 59–83 (1985)
11. Tanaka, K., Sugeno, M.: Stability analysis and design of fuzzy control systems. Fuzzy Sets Syst. **45**(2), 135–156 (1992)
12. Wang, L.-X., Mendel, J.M.: Fuzzy basis functions, universal approximation, and orthogonal least-squares learning. IEEE Trans. Neural Netw. **3**(5), 807–814 (1992)
13. Ying, H.: General SISO Takagi-Sugeno fuzzy systems with linear rule consequent are universal approximators. IEEE Trans. Fuzzy Syst. **6**(4), 582–587 (1998)
14. Zadeh, L.A.: Fuzzy Sets. Inf. Control **8**, 338–353 (1965)
15. Zadeh, L.A.: The concept of a linguistic variable and its application to approximate reasoning - I. Inf. Sci. **8**(3), 199–249 (1975)
16. Zadeh, L.A.: Fuzzy logic, neural networks, and soft computing. Fuzzy Sets, Fuzzy Logic, And Fuzzy Systems, pp. 775–782 (1996). Selected Papers by Zadeh, L.A

Type-2 Fuzzy Logic Augmentation of the Imperialist Competitive Algorithm with Dynamic Parameter Adaptation

Emer Bernal[(✉)], Oscar Castillo, José Soria, and Fevrier Valdez

Tijuana Institute of Technology, Tijuana, BC, Mexico
emerbernalj@gmail.com, ocastillo@tectijuana.mx

Abstract. In this paper we propose the utilization of type-2 fuzzy systems for the dynamic adjustment of parameters in the imperialist competitive algorithm (ICA), a type-1 fuzzy system was used as a basis, with decades as the input variable and the beta parameter as the output variable, then it was extended to interval type-2 fuzzy systems, and three variants with triangular, Gaussian and trapezoidal membership functions were performed. The imperialist competitive algorithm is based on the concept of imperialism, where the strongest countries absorb the weakest and make then their colonies. To measure the performance of the proposed method 10 mathematical functions with different number of decades are used and finally, a comparison was made between our variants and the results obtained with the type-1 fuzzy system to observe their behavior in the face of optimization problems.

1 Introduction

The dynamic adaptation of parameters utilizing fuzzy logic, in particular type-1 and type-2 fuzzy systems, in metaheuristic algorithms, has gained great importance in recent years because it has shown significant improvements in the performance of various metaheuristics algorithms found in the literature [14, 28].

Fuzzy logic and fuzzy sets were originally proposed by Zadeh, and this originated the creation of fuzzy systems with different applications such as in metaheuristic algorithms and modeling and control systems [7, 11, 16]. Initially type-1 fuzzy systems, that represent the imprecision with numerical values in a range of $[0, 1]$, came to replace traditional sets when it is difficult to establish or find an exact value in some type of measurement [17]. In addition, when the problem or situation contains a higher degree of uncertainty, type-2 fuzzy systems can be used because they work better with high levels of uncertainty or lack of information [15, 17].

We propose dynamic adjustment of the parameters in the imperialist competitive algorithm using fuzzy logic. Three distinct fuzzy systems were designed to dynamically adjust the beta parameter and to measure the performance of the proposed fuzzy imperialist competitive algorithm (FICAT2) utilizing interval type-2 fuzzy systems versus fuzzy imperialist competitive algorithm (FICA) using type-1 fuzzy systems.

© Springer Nature Switzerland AG 2019
R. B. Kearfott et al. (Eds.): IFSA 2019/NAFIPS 2019, AISC 1000, pp. 167–176, 2019.
https://doi.org/10.1007/978-3-030-21920-8_16

The rest of the paper is conformed as follows. Section 2 provides a re-view of imperialist competitive algorithm. Section 3 presents the methodology used in the proposed approach. Section 4 presents the mathematical benchmark functions. Section 5 summarizes the simulation of results obtained and finally the conclusions.

2   Imperialist Competitive Algorithm

Atashpaz-Gargari and Lucas proposed the imperialist competitive algorithm (ICA) in 2007 [3], this algorithm is inspired by imperialism, where all the most powerful countries aspire to make a colony the less powerful countries and thus absorb them. In the field of metaheuristic algorithms, the imperialist competitive algorithm takes as a basis the social political progress unlike other metaheuristics or evolutionary algorithms that are based on bio-inspired phenomena [1,11].

In the imperialist competitive algorithm, we start with a randomly generated population, where all individuals are called countries. The best positioned countries are considered the imperialist countries and the remaining countries are the colonies. All the colonies are divided among the imperialist countries according to their power, in this case their fitness function [3,4,10].

Once the initial population is generated and divided into imperialist countries and colonies, the colonies begin to move towards their imperialist country (known as assimilation process). The colonies move X distance toward their imperialist, where the movement X is a random number that is generated by a uniform distribution within the interval $(0, \beta, d)$, where β is a number in the range of 1 to 2 and d is the distance between the colony and the imperialist [2,6].

$$x \sim U(0, \beta, d) \tag{1}$$

The power of an empire is calculated based on the power of the imperialist country and the power of the colonies that are part of the empire. Each empire tries to take possession of other empires, which leads to an imperialist competition that diminishes the power of the weaker empires and increases the power of the strongest empires, causing the weak empires to collapse and their colonies become part of the most powerful empires [3,8].

After a period of time, all the empires will collapse periodically except the strongest empire, in this way all the colonies will belong and be controlled by a single imperialist country ending the imperialist competition [5,13].

3   Proposed Methodology

In this paper we propose a method based on interval type-2 fuzzy systems for the dynamic adjustment of parameters in the imperialist competitive algorithm in specific beta parameters. Initially, a type-1 fuzzy system presented in [7], where the decades represent the input variable and the beta parameter the output

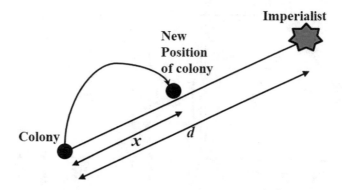

Fig. 1. Motion of a colony toward its imperialist.

variable with triangular membership functions labeled as "low", "medium" and "high" as shown in Fig. 2 [7].

To use the parameter of decades as input variable, we use a percentage, so when the algorithm starts the decades will be "low" and as time passes will be considered "high". The following mathematical expression is used to model the concept of the input variable decades [7,12] (Fig. 1).

$$Decades = \frac{Current\ Decade}{Total\ Number\ of\ Decades} \tag{2}$$

The proposed approach based on interval type-2 (IT2) fuzzy systems in the competitive imperialist algorithm for the dynamic adjustment of the beta parameter is illustrated in Fig. 3.

The first IT2 fuzzy system proposed can be found in Fig. 4, which was designed with triangular membership functions that are represented as follows [9].

$$\tilde{\mu}(x) = \left[\underline{\mu}(x), \overline{\mu}(x)\right] = itritype2(x, [a_1, b_1, c_1, a_2, b_2, c_2]),$$
$$where\ a_1 < a_2,\ b_1 < b_2,\ c_1 < c_2 \tag{3}$$

The second IT2 fuzzy system proposed is shown in Fig. 5, which was designed with Gaussian membership functions that are represented as follows [9].

$$\tilde{\mu}(x) = \left[\underline{\mu}(x), \overline{\mu}(x)\right] = igaussmtype2(x, [\sigma, m_1, m_1]),$$
$$where\ m_1 < m_2 \tag{4}$$

Finally the third IT2 fuzzy system proposed can be found in Fig. 6, which was designed with trapezoidal membership functions that are represented as follows [9].

$$\tilde{\mu}(x) = \left[\underline{\mu}(x), \overline{\mu}(x)\right] = itrapatype2(x, [a_1, b_1, c_1, d_1, a_2, b_2, c_2, d_2]),$$
$$where\ a_1 < a_2,\ b_1 < b_2,\ c_1 < c_2,\ d_1 < d_2 \tag{5}$$

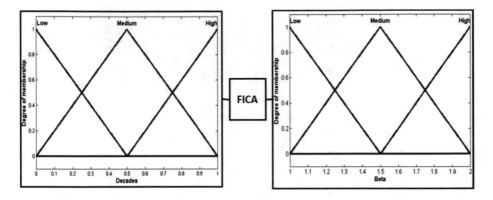

Fig. 2. Type-1 fuzzy system for Beta.

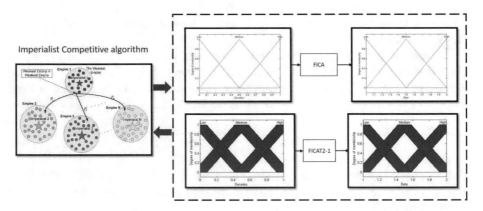

Fig. 3. General Proposal dynamic parameter adaptation.

4   Benchmark Functions

In this section we present the mathematical functions used to measure the performance of the imperialist competitive algorithm and our proposal using interval type-2 (IT2) fuzzy systems to the adaptation of parameters in the imperialist competitive algorithm.

In the area of metaheuristic algorithms it is common to use benchmark functions to measure their performance, and in this work we use 10 mathematical functions that are presented below with their mathematical expression, search space and their optimal value [7,14].

$$f_1(x) = \sum_{i=1}^{n} x_2^i \quad with \ x_j \ \varepsilon \ [-5.12, 5.12] \ and \ f(x^*) = 0 \tag{6}$$

$$f_2(x) = \sum_{i=1}^{n} |x_i| + \prod_{i=1}^{n} |x_i| \quad with \ x_j \ \varepsilon \ [-10, 10] \ and \ f(x^*) = 0 \tag{7}$$

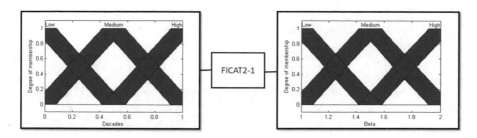

Fig. 4. IT2 fuzzy system FICAT2-1.

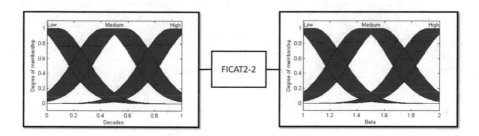

Fig. 5. IT2 fuzzy system FICAT2-2.

$$f_3(x) = \sum_{i=1}^{n} \left(\sum_{j-1}^{i} x_j \right)^2 \quad \text{with } x_j \; \varepsilon \, [-100, 100] \text{ and } f\,(x^*) = 0 \tag{8}$$

$$f_4(x) = max_i \, \{|x_i|, 1 \le i \le n\} \quad \text{with } x_j \; \varepsilon \, [-100, 100] \text{ and } f\,(x^*) = 0 \tag{9}$$

$$f_5(x) = \sum_{i=1}^{n-1} \left[100 \left(x_{i+1} - x_i^2 \right)^2 + (x_i - 1)^2 \right] \quad \text{with } x_j \; \varepsilon \, [-30, 30] \text{ and } f\,(x^*) = 0$$

$$\tag{10}$$

$$f_6(x) = \sum_{i=1}^{n} \left([x_i + 0.5] \right)^2 \quad \text{with } x_j \; \varepsilon \, [-30, 30] \text{ and } f\,(x^*) = 0 \tag{11}$$

$$f_7(x) = \sum_{i=1}^{n} i x_i^4 + random \, [0, 1] \quad \text{with } x_j \; \varepsilon \, [-1.28, 1.28] \text{ and } f\,(x^*) = 0 \tag{12}$$

$$f_8(x) = \sum_{i-1}^{n} -x_i sin \left(\sqrt{|x_i|} \right) \quad \text{with } x_j \; \varepsilon \, [-500, 500] \text{ and } f\,(x^*) = -418.9829$$

$$\tag{13}$$

$$f_9(x) = \sum_{i-1}^{n} \left[x_i^2 - 10cos\left(2\pi x_i\right) + 10 \right] \quad with \; x_j \; \varepsilon \left[-5.12, 5.12\right] \; and \; f\left(x^*\right) = 0$$

(14)

$$f_{10}(x) = -20exp\left(-0.2\sqrt{\frac{1}{n}\sum_{i=1}^{n} x_i^2}\right) - exp\left(\frac{1}{n}\sum_{i=1}^{n} cos\left(2\pi x_i\right)\right),$$

$$with \; x_j \; \varepsilon \left[-32, 32\right] \; and \; f\left(x^*\right) = 0 \quad (15)$$

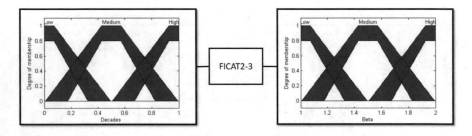

Fig. 6. IT2 fuzzy system FICAT2-3

5   Experimental Results

Imperialist competitive algorithm (ICA) and our proposal based on interval type 2 (IT2) fuzzy system for the adjustment of the beta parameter will be implemented for 10 mathematical functions in all cases for 30 dimensions, the results obtained by the competitive imperialist algorithm and the proposed method are shown in tables by number of decades. The parameters that were used to perform the experiments are as follows [7] (Table 1):

Table 1. Parameters for ICA and FICAT2 [7].

Parameters	Values
No. Dimensions	30
No. Countries	200
No. Imperialists	10
Revolution rate	0.2
Xi	0.02
Beta	Dynamic

Table 2. Simulation of results for 1000 decades.

Functions	ICA	FICA BETA	FICAT2-1	FICAT2-2	FICAT2-3
$f(1)$	1.12E−23	8.79E−25	1.86E−24	1.15E−24	2.95E−25
$f(2)$	3.39E−56	4.41E−52	4.46E−52	4.10E−52	5.63E−51
$f(3)$	5822.3589	5909.1780	5576.7949	7663.4388	4647.6901
$f(4)$	8.6002	0.8314	0.9575	1.0461	1.1748
$f(5)$	24.1293	171.5793	292.4097	18.6498	22.2757
$f(6)$	6.59E−17	1.06E−22	2.78E−21	9.94E−22	2.57E−22
$f(7)$	9.76E−41	11.6622	11.8203	11.9042	11.8252
$f(8)$	2535.7687	2195.7988	2270.0006	2217.7382	2018.7521
$f(9)$	123.8072	101.0117	104.0068	106.2300	103.5337
$f(10)$	13.0151	12.9013	12.3078	14.2735	13.3222

Table 2 shows the average obtained by implementing the imperialist competitive algorithm and our proposal using IT2 fuzzy systems to dynamically adjust the parameters of the ICA algorithm, the bold results represent the best obtained for each of the benchmark functions for 1000 decades.

Table 3. Simulation of results for 2000 decades.

Functions	ICA	FICA BETA	FICAT2-1	FICAT2-2	FICAT2-3
$f(1)$	9.26E−51	3.24E−51	5.16E−50	1.87E−52	4.38E−51
$f(2)$	2.22E−141	4.42E−116	1.81E−114	9.08E−117	4.30E−115
$f(3)$	7928.6501	3638.9330	3794.4171	4876.5903	4404.9118
$f(4)$	0.6159	0.0102	0.0106	0.0159	0.03324
$f(5)$	11.7993	8.0392	474.1157	7.1485	11.5097
$f(6)$	1.06E−17	2.47E−20	1.95E−16	2.97E−20	9.00E−23
$f(7)$	9.95E−92	11.2356	11.3735	11.2075	11.0072
$f(8)$	2349.0898	1889.5634	1837.3269	1934.4571	1750.6182
$f(9)$	114.7958	70.1576	57.4281	58.5647	36.0353
$f(10)$	12.8467	12.9715	13.9097	13.2210	11.5814

Table 3 shows the average obtained by implementing the imperialist competitive algorithm and our proposal using IT2 fuzzy systems to dynamically adjust the parameters of the ICA algorithm, the bold results represent the best obtained for each of the benchmark functions for 2000 decades.

Table 4. Simulation of results for 3000 decades.

Functions	ICA	FICA BETA	FICAT2-1	FICAT2-2	FICAT2-3
$f(1)$	1.18E−29	1.51E−75	4.06E−79	1.48E−77	2.11E−67
$f(2)$	7.50E−231	3.19E−183	7.32E−182	7.69E−183	1.10E−178
$f(3)$	6863.9257	7563.2744	5910.2325	6980.2326	4796.6941
$f(4)$	1.05E−02	1.39E−04	2.71E−04	1.42E−04	8.35E−04
$f(5)$	13.0497	9.9357	10.0435	7.8241	8.4234
$f(6)$	5.40E−19	1.86E−19	1.05E−20	1.81E−27	3.52E−15
$f(7)$	9.87E−144	10.9519	11.0305	11.1723	10.9067
$f(8)$	2345.0356	1804.9977	1965.0972	2020.0730	1935.3296
$f(9)$	114.2893	68.0383	67.0418	62.6640	43.2350
$f(10)$	15.3479	11.9827	13.6105	13.0194	12.2402

Table 4 shows the average obtained by implementing the imperialist competitive algorithm and our proposal using IT2 fuzzy systems to dynamically adjust the parameters of the ICA algorithm, the bold results represent the best obtained for each of the benchmark functions for 3000 decades.

6   Conclusions

In this paper, several type 1 and interval type 2 fuzzy systems were performed for the dynamic adjustment of the parameters in the imperialist competitive algorithm. A comparison was made of the results obtained that were divided by number of decades.

We can conclude that improvements have been obtained on average since in some cases our proposal manages to overcome the imperialist competitive algorithm and the method using type-1 fuzzy systems ac-cording to Tables 2, 3 and 4.

The use of interval type-2 fuzzy systems can be an alternative to achieve significant improvements by dynamically adapting the parameters in metaheuristic algorithms [18–21,23]. Also, we can consider other areas of application like in [22, 24–28].

Acknowledgements. We want to show our gratitude to CONACYT and Tijuana institute of technology for the resources provided for the development of our research.

References

1. Atashpaz-gargari, E., Hashemzadeh, F., Rajabioun, R., Lucas, C.: Colonial competitive algorithm: a novel approach for PID controller design in MIMO distillation column process. Int. J. Intell. Comput. Cybern. **1**(3), 337–355 (2008)
2. Atashpaz-gargari, E., Lucas, C.: Imperialist competitive algorithm for minimum bit error rate beamforming. Int. J. Bio Inspired Comput. **1**(2), 125–133 (2009)
3. Atashpaz-gargari, E., Lucas, C.: Imperialist competitive algorithm: an algorithm for optimization inspired by imperialistic competition. In: Evolutionary Computation, pp. 4661–4667 (2007)
4. Bernal, E., Castillo, O., Soria, J.: Imperialist competitive algorithm applied to the optimization of mathematical functions: a parameter variation study. In: Design of Intelligent Systems Based on Fuzzy Logic. Neural Networks and Nature-Inspired Optimization, vol. 601, pp. 219–232. Springer (2015)
5. Bernal, E., Castillo, O., Soria, J.: Imperialist competitive algorithm with dynamic parameter adaptation applied to the optimization of mathematical functions. In: Nature Inspired Design of Hybrid Intelligent Systems, vol. 667, pp. 329–341. Springer (2017)
6. Bernal, E., Castillo, O., Soria, J., Valdez, F.: Imperialist competitive algorithm with dynamic parameter adaptation using fuzzy logic applied to the optimization of mathematical functions. Algorithms **10**(1), 18 (2017)
7. Bernal, E., Castillo, O., Soria, J.: A fuzzy logic approach for dynamic adaptation of parameters in galactic swarm optimization. In: Annual Conference of the North American Fuzzy Information Processing Society (NAFIPS), pp. 1–6. IEEE (2017)
8. Bernal, E., Castillo, O., Soria, J.: Fuzzy logic for dynamic adaptation in the imperialist competitive algorithm. In: IEEE Symposium Series on Computational Intelligence (SSCI), pp. 1–7. IEEE (2017)
9. Castro, J.R., Castillo, O., Martinez, L.G.: Interval type-2 fuzzy logic toolbox. Eng. Lett. **15**(1), 89–98 (2007)
10. Duan, H., Huang, L.Z.: Imperialist competitive algorithm optimized artificial neural networks for UCAV global path planning. Neurocomputing **125**, 166–171 (2013)
11. Mahmoodabadi, M.J., Jahanshahi, H.: Multi objective optimized fuzzy-PID controllers for fourth order nonlinear systems. Eng. Sci. Technol. Int. J. **19**(2), 1084–1098 (2016)
12. Melin, P., Olivas, F., Castillo, O., Valdez, F., Soria, J., Valdez, M.: Optimal design of fuzzy classification systems using PSO with dynamic parameter adaptation through fuzzy logic. Expert Syst. Appl. **40**(8), 3196–3206 (2012)
13. Mitchell, M.: An Introduction to Genetic Algorithms. MIT Press, Cambridge (1999)
14. Mirjalili, S., Mirjalili, S.M., Lewis, A.: Grey Wolf optimizer. Adv. Eng. Softw. **68**, 46–61 (2014)
15. Ontiveros-robles, E., Melin, P., Castillo, O.: Comparative analysis of noise robustness of type 2 fuzzy logic controllers. Kybernetika **54**(1), 175–201 (2018)
16. Valdez, F., Melin, P., Castillo, O.: An improved evolutionary method with fuzzy logic for combining particle swarm optimization and genetic algorithms. Appl. Soft Comput. **11**(2), 2625–2632 (2011)
17. Zadeh, L.: Fuzzy logic = computing with words. IEEE Trans. Fuzzy Syst. **4**(2), 103–111 (1996)
18. Leal Ramírez, C., Castillo, O., Melin, P., Rodríguez Díaz, A.: Simulation of the bird agestructured population growth based on an interval type-2 fuzzy cellular structure. Inf. Sci. **181**(3), 519–535 (2011)

19. Cázarez-Castro, N.R., Aguilar, L.T., Castillo, O.: Designing type-1 and type-2 fuzzy logic controllers via fuzzy Lyapunov synthesis for nonsmooth mechanical systems. Eng. Appl. Artif. Intell. **25**(5), 971–979 (2012)
20. Castillo, O., Melin, P.: Intelligent systems with interval type-2 fuzzy logic. Int. J. Innov. Comput. Inf. Control **4**(4), 771–783 (2008)
21. Mendez, G.M., Castillo, O.: Interval type-2 TSK fuzzy logic systems using hybrid learning algorithm. In: IEEE International Conference on Fuzzy Systems, pp. 230–235 (2005)
22. Castillo, O., Melin, P.: Intelligent control of complex electrochemical systems with a neuro-fuzzy-genetic approach. IEEE Trans. Ind. Electron. **48**(5), 951–955 (2001)
23. Rubio, E., Castillo, O., Valdez, F., Melin, P., Gonzalez, C.I., Martinez, G.: An extension of the fuzzy possibilistic clustering algorithm using type-2 fuzzy logic techniques. Adv. Fuzzy Syst. **2017**, 23 (2017)
24. Aguilar, L., Melin, P., Castillo, O.: Intelligent control of a stepping motor drive using a hybrid neuro-fuzzy ANFIS approach. Appl. Soft Comput. **3**(3), 209–219 (2003)
25. Melin, P., Castillo, O.: Adaptive intelligent control of aircraft systems with a hybrid approach combining neural networks, fuzzy logic and fractal theory. Appl. Soft Comput. **3**(4), 353–362 (2003)
26. Melin, P., Amezcua, J., Valdez, F., Castillo, O.: A new neural network model based on the LVQ algorithm for multi-class classification of arrhythmias. Inf. Sci. **279**, 483–497 (2014)
27. Melin, P., Castillo, O.: Modelling, Simulation and Control of Non-Linear Dynamical Systems: An Intelligent Approach Using Soft Computing and Fractal Theory. CRC Press, Boca Raton (2001)
28. Melin, P., Sánchez, D., Castillo, O.: Genetic optimization of modular neural networks with fuzzy response integration for human recognition. Inf. Sci. **197**, 1–19 (2012)

Least Square Method with Quasi Linearly Interactive Fuzzy Data: Fitting an HIV Dataset

Nilmara J. Biscaia Pinto[✉], Estevão Esmi, Vinícius Francisco Wasques, and Laécio C. Barros

Institute of Mathematics, Statistics and Scientific Computing, University of Campinas, Campinas, Brazil
nilmarabiscaia@gmail.com, eelaureano@gmail.com, vwasques@outlook.com, laeciocb@ime.unicamp.br

Abstract. In this manuscript we propose a method to fit a dataset with uncertainty. These data are described by interactive fuzzy numbers. The relationship of interactivity is associated with the notion of joint possibility distribution. We focus on a specific type of interactivity namely linear interactivity. We use this concept to introduce a class of fuzzy numbers called quasi linearly interactive fuzzy numbers. We provide an application to fit a dataset of the HIV disease to illustrate the proposed method.

1 Introduction

The well-known least square method consists of producing a continuous function that best fit the data of a dataset [3]. The fuzzy set theory can be used to describe data with uncertainty [14]. The fuzzy least square method (FLSM) arises when the dataset is composed by fuzzy numbers.

The study of the FLSM have started with Tanaka *et al.* [18] analysing a fuzzy regression problem. Subsequently some authors, such as Celmins [5] and Diamond [6], continued to develop this subject, but always considering the input data as triangular fuzzy numbers. These approaches were used in several applications in [16,17,19,21].

Recently, Pinto *et al.* [15] proposed a method that takes into account a relationship between fuzzy numbers called interactivity. They provided fuzzy functions that approximates a dataset given by fuzzy numbers linearly interactive (or also called completely correlated).

Some phenomena may not be fully characterized by this type of relation among the data. Nevertheless they can be approximated by linear interactivity data. In order to include these cases, we introduce the notion of quasi linearly interactive fuzzy numbers, and thereafter incorporate in the method proposed by Pinto *et al.* [15].

© Springer Nature Switzerland AG 2019
R. B. Kearfott et al. (Eds.): IFSA 2019/NAFIPS 2019, AISC 1000, pp. 177–189, 2019.
https://doi.org/10.1007/978-3-030-21920-8_17

According to the annual report of Joint United Nations Program on HIV/AIDS [11], the estimated number of people newly infected with HIV, available on [20], is naturally uncertain due to barriers faced to assess the number of infected people. Jafelice *et al.* [10] described the dynamics of HIV via fuzzy differential equations and Laiate *et al.* [12] studied this dynamic using Choquet calculus. Here, we consider the data from 2010 to 2017, and we apply the FLSM to find a fuzzy function that fits these data using the concept of quasi linearly interactivity.

This paper is organized as follows. In Sect. 2 we briefly recall the classical least squares method and some basic definitions and results of fuzzy set theory. In Sect. 3, we develop the extension of the classical least squares method for the case where dataset is composed by linearly interactive fuzzy numbers. In Sect. 4 we establish the class of quasi linearly interactive fuzzy numbers, and finally, in Sect. 5, we apply the proposed method to forecast the number of HIV infected people.

2   Mathematical Background

This section presents some basic concepts and the main results of the least square method and the fuzzy set theory.

2.1   Least Square Method

Let $f : [c, d] \to \mathbb{R}$ be a continuous function. Given n functions $g_1, \ldots, g_n$, where $g_i : \mathbb{R} \to \mathbb{R}$ for $i = 1, \ldots, n$. The goal is to determine n coefficients $a_1, \ldots, a_n \in \mathbb{R}$ such that the function $\varphi : \mathbb{R} \to \mathbb{R}$ given by

$$\varphi(x) = a_1 g_1(x) + \ldots + a_n g_n(x) \tag{1}$$

is the best approximation of the function f, that is, $||\varphi - f|| < \varepsilon$, where $\varepsilon > 0$ for some $|| \cdot ||$. Hence the function φ is obtained by minimizing the distance between f and φ. More precisely, let $|| \cdot ||_2$ be the $\mathscr{L}^2$-norm defined on the class of the continuous functions from $[c, d]$ to $\mathbb{R}$ (denoted by $C([c, d])$) given by

$$||h||_2 = \left(\int_c^d |h(s)|^2 ds \right)^{1/2}, \quad \forall h \in C([c, d]).$$

When some values of f are known, say $D = \{f(x_1) = y_1, \ldots, f(x_m) = y_m\}$, the function φ have to satisfy $\varphi(x_i) \approx y_i$, for all $i = 1, \ldots, m$. Therefore the following minimization problem must be solved.

$$\min_{a_1, \ldots, a_n \in \mathbb{R}} {}^1/_2 ||(\varphi(x_1) - y_1, \ldots, \varphi(x_m) - y_m)||_2^2. \tag{2}$$

The real coefficients $a_1, \ldots, a_n$ that minimize (2) are obtained by solving the following matrix equation called normal equation:

$$Ma = b,$$

where

$$
M = \begin{bmatrix} \sum_{k=1}^{m} g_1(x_k)g_1(x_k) & \cdots & \sum_{k=1}^{m} g_1(x_k)g_n(x_k) \\ \vdots & \ddots & \vdots \\ \sum_{k=1}^{m} g_n(x_k)g_1(x_k) & \cdots & \sum_{k=1}^{m} g_n(x_k)g_n(x_k), \end{bmatrix}, a = \begin{bmatrix} a_1 \\ \vdots \\ a_n \end{bmatrix} \text{ and } b = \begin{bmatrix} \sum_{k=1}^{m} y_k g_1(x_k) \\ \vdots \\ \sum_{k=1}^{m} y_k g_n(x_k) \end{bmatrix}.
$$

If the matrix M is non-singular, say $P = M^{-1} = [p_{ij}]$, then a is obtained by

$$
a = Pb. \tag{3}
$$

Therefore, each coefficient a_i is given by

$$
a_i = p_{i1}\left(\sum_{k=1}^{m} y_k g_1(x_k)\right) + \ldots + p_{in}\left(\sum_{k=1}^{m} y_k g_n(x_k)\right)
$$
$$
= c_{i1}y_1 + \ldots + c_{im}y_m
$$

where $c_{ik} = \sum_{j=1}^{n} p_{ij}g_j(x_k)$, for $i = 1, \ldots, n$ and $k = 1, \ldots, m$.

If the matrix M is singular then the aforementioned method can be performed considering P as the pseudoinverse of M.

Since the parameters of the function φ can be obtained by (3), we can rewrite the function φ in terms of $y_1, \ldots, y_m$ as follows:

$$
\varphi(x) = a_1 g_1(x) + \ldots + a_n g_n(x)
$$
$$
= (c_{11}y_1 + \ldots + c_{1m}y_m)g_1(x) + \ldots + (c_{n1}y_1 + \ldots + c_{nm}y_m)g_n(x)
$$
$$
= \left(\sum_{j=1}^{n} g_j(x)c_{j1}\right)y_1 + \ldots + \left(\sum_{j=1}^{n} g_j(x)c_{jm}\right)y_m
$$
$$
= s_1(x)y_1 + \ldots + s_m(x)y_m
$$

where

$$
s_i = \left(\sum_{j=1}^{n} g_j(x)c_{ji}\right) \tag{4}
$$

for $i = 1, \ldots, n$.

2.2   Fuzzy Set Theory

A fuzzy subset A of a universe X is characterized by a function $\mu_A : X \to [0, 1]$, called membership function, where $\mu_A(x)$, or simply $A(x)$, represents the membership degree of x in A, for all $x \in X$. The class of fuzzy sets of X is denoted by the symbol $\mathscr{F}(X)$. Each classical subset A of X can be viewed as a particular fuzzy set whose membership function is given by its characteristic function.

The α−cuts of a fuzzy set $A \in \mathcal{F}(X)$, denoted by $[A]_\alpha$, are defined as $[A]_\alpha = \{x \in X : A(x) \geq \alpha\}$, $\forall \alpha \in (0, 1]$. In addition if X is also a topological space, then we can define the 0−cut of A by $[A]_0 = cl\{x \in X : A(x) > 0\}$, where $cl\, Y$, $Y \subseteq X$, denotes the closure of Y.

The Zadeh's extension principle [22] can be viewed as a mathematical method to extend a function $f : X \to Y$ to a function that have fuzzy sets as arguments.

Definition 1. Let $f : X_1 \times \ldots \times X_n \to Y$. The Zadeh's extension of f at $(A_1, \ldots, A_n) \in \mathcal{F}(X_1 \times \ldots \times X_n)$ is the fuzzy set $\hat{f}(A_1, \ldots, A_n) \in \mathcal{F}(Y)$ whose membership function is given by

$$\hat{f}(A_1, \ldots, A_n)(y) = \sup_{(x_1, \ldots, x_n) \in f^{-1}(y)} \min\{A_1(x_1), \ldots, A_n(x_n)\}, \ \forall \, y \in Y, \quad (5)$$

where $f^{-1}(y) = \{(x_1, \ldots, x_n) \in X_1 \times \ldots \times X_n : f(x_1, \ldots, x_n) = y\}$ and by definition $\sup \emptyset = 0$.

A fuzzy set $A \in \mathcal{F}(\mathbb{R})$ is called a fuzzy number if its α−cuts are closed, bounded and non-empty nested intervals for all $\alpha \in [0, 1]$ [1]. Since each α−cut of a fuzzy number A is an interval, then we can write $[A]_\alpha = [a_\alpha^-, a_\alpha^+]$. The class of fuzzy numbers is denoted by $\mathbb{R}_{\mathcal{F}}$.

The triangular fuzzy number, denoted by the triple $(a; b; c)$ where $a \leq b \leq c$, is an example of an element of $\mathbb{R}_{\mathcal{F}}$. The α−cuts of this fuzzy number are given by $[A]_\alpha = [a + \alpha(b - a), c - \alpha(c - b)]$, $\forall \alpha \in [0, 1]$. Note that a real number a is a particular case of triangular fuzzy number since we have $a \equiv (a; a; a)$.

Let be the following operator on $\mathbb{R}_{\mathcal{F}}$

$$\langle A, B \rangle = \int_0^1 a_\alpha^- b_\alpha^- d\alpha + \int_0^1 a_\alpha^+ b_\alpha^+ d\alpha \quad \forall A, B \in \mathbb{R}_{\mathcal{F}}.$$

Also consider the function $||.|| : \mathbb{R}_{\mathcal{F}} \to \mathbb{R}$ given as follows

$$||A|| = \sqrt{\langle A, A \rangle}. \quad (6)$$

The operator given by Eq. (6) satisfies the properties of norm and we use it as a norm for fuzzy numbers, regardless that the set of fuzzy numbers is not a vector space.

A fuzzy relation R over $X = X_1 \times \ldots \times X_n$ is defined as any fuzzy subset of X, whose membership function is given by $R : X_1 \times \ldots \times X_n \to [0, 1]$, where $R(x_1, \ldots, x_n) \in [0, 1]$ represents the degree of relationship among $x_1, \ldots, x_n$ with respect to R.

A fuzzy relation $J \in \mathcal{F}(\mathbb{R}^n)$ is said to be a *joint possibility distribution* of $A_1, \ldots, A_n \in \mathbb{R}_{\mathcal{F}}$ if

$$A_i(y) = \sup_{x \in X : x_i = y} J(x_1, \ldots, x_n),$$

for all $y \in \mathbb{R}$ and for all $i = 1, \ldots, n$.

One example of joint possibility distribution is provide as follows. Given a t-norm t, that is, a commutative, associative, and increasing operator $t : [0,1]^2 \to [0,1]$ satisfying $t(x,1) = x \, t \, 1 = x$ for all $x \in [0,1]$. A fuzzy relation J_t given by

$$J_t(x_1, \ldots, x_n) = A_1(x_1) \, t \, \ldots \, t \, A_n(x_n) \tag{7}$$

is said to be a t-norm-based joint possibility distribution of $A_1, \ldots, A_n \in \mathbb{R}_{\mathscr{F}}$ [7]. In particular, when J is given by (7) with the minimum t-norm $t = \min$, we say that $A_1, \ldots, A_n$ are *non-interactive*. Otherwise, we say that $A_1, \ldots, A_n$ are *J-interactive* or simply *interactive* [4,8,22]. Hence the notion of interactivity between fuzzy numbers is given by means of joint possibility distributions.

Carlsson *et al.* [4] introduced a possible type of interactivity relation between two fuzzy numbers called *completely correlation* or *linear interactivity*, which is not based on t-norms. Subsequently the authors of [15] generalized this concept for n $(n > 2)$ fuzzy numbers. Specifically, the fuzzy numbers $A_1, \ldots, A_n \in \mathbb{R}_{\mathscr{F}}$ are said to be linearly interactive (LI) (or completely correlated) if the joint possibility distribution $J = J_L$ is given by

$$J_L(x_1, \ldots, x_n) = \chi_L(x_1, \ldots, x_n) A_i(x_i), \quad \forall i = 1, \ldots, n$$

where χ_L represents the characteristic function of the set $L = \{(u, q_2 u + r_2, \ldots, q_n u + r_n) : u \in \mathbb{R}\}$, $q_i, r_i \in \mathbb{R}$, with $q_i \neq 0$, $\forall i = 1, \ldots, n$.

For each $\alpha \in [0,1]$, the $\alpha-$cut of J_L is given as follows

$$[J_L]_\alpha = \{(x, q_2 x + r_2, \ldots, q_n x + r_n) \; : \; x \in [A_1]_\alpha\}.$$

Moreover, for all $i = 2, \ldots, n$, we have that $[A_i]_\alpha = q_i [A_1]_\alpha + \{r_i\}$.

The next definition is a generalization of Zadeh's extension principle (see (5)) which is called *sup-J extension principle* [9].

Definition 2. Let $J \in \mathscr{F}(\mathbb{R}^n)$ be a joint possibility distribution of $A_1, \ldots, A_n \in \mathbb{R}_{\mathscr{F}}$ and let $f : \mathbb{R}^n \to \mathbb{R}$. The $\sup -J$ extension of f at $(A_1, \ldots, A_n)$ is defined by

$$\widehat{f}_J(A_1, \ldots, A_n)(y) = \widehat{f}(J)(y) = \sup_{(x_1, \ldots, x_n) \in f^{-1}(y)} J(x_1, \ldots, x_n),$$

where $f^{-1}(y) = \{(x_1, \ldots, x_n) \in \mathbb{R}^n : f(x_1, \ldots, x_n) = y\}$.

The next theorem characterizes the extension $\widehat{f}_J$ by means of its α-cuts.

Theorem 1 [2,13]. *Let $f : \mathbb{R}^n \to \mathbb{R}$ be a continuous function and $J \in \mathscr{F}(\mathbb{R}^n)$. We have that*

$$[\widehat{f}_J(A_1, \ldots, A_n)]_\alpha = f([J]_\alpha), \quad \forall \alpha \in [0,1].$$

By Theorem 1, if the sup-J extension of f at $(A_1, \ldots, A_n)$ is a fuzzy number, then the $\alpha-$cuts of $\widehat{f}_J(A_1, \ldots, A_n) = \widehat{f}(J)$ can be written as follows:

$$[\widehat{f}(J)]_\alpha = \left[\inf_{(x_1, \ldots, x_n) \in [J]_\alpha} f(x_1, \ldots, x_n) \; , \; \sup_{(x_1, \ldots, x_n) \in [J]_\alpha} f(x_1, \ldots, x_n) \right].$$

The next section presents the least square method proposed by the authors of [15] which consists in fitting a dataset given by interactive fuzzy numbers with respect to the joint possibility distribution $J = J_L$.

3   Least Squares Method for Interactive Fuzzy Data

Let us consider the uncertainty dataset $D = \{(x_1, Y_1), \ldots, (x_m, Y_m)\} \subset \mathbb{R} \times \mathbb{R}_{\mathscr{F}}$ where $Y_1, \ldots, Y_m$ are given by linearly interactive fuzzy numbers. The method, presented in [15], produces a fuzzy function $\Phi : \mathbb{R} \to \mathbb{R}_{\mathscr{F}}$, that approximates the data $Y_1, \ldots, Y_m$, which is given by means of the sup-J extension of a function $\varphi : \mathbb{R} \to \mathbb{R}$ of the form

$$\varphi(x) = a_1 g_1(x) + \ldots + a_n g_n(x),$$

where $a_1, \ldots, a_n \in \mathbb{R}$ and $g_1, \ldots, g_n$ are real-valued-functions.

The function Φ can be written by means of α-cuts as follows

$$[\Phi(x)]_\alpha = [\min\{\langle S(x), Y_\alpha^- \rangle, \langle S(x), Y_\alpha^+ \rangle\}, \max\{\langle S(x), Y_\alpha^- \rangle, \langle S(x), Y_\alpha^+ \rangle\}],$$

where $[Y]_1 = [y_{1\alpha}^-, y_{1\alpha}^+]$, $Y_\alpha^- = (y_{1\alpha}^-, q_2 y_{1\alpha}^- + r_2, \ldots, q_m y_{1\alpha}^- + r_m)$, $Y_\alpha^+ = (y_{1\alpha}^+, q_2 y_{1\alpha}^+ + r_2, \ldots, q_m y_{1\alpha}^+ + r_m)$, and $S(x) = (s_1(x), s_2(x), \ldots, s_m(x))$, $x \in \mathbb{R}$, is given by (4).

In this paper we propose a methodology, based on the aforementioned method, to produce a fuzzy function that fits a dataset given by *quasi completely correlated fuzzy numbers*, whose definition is given in the next section.

4   Quasi Linearly Interactive Fuzzy Numbers

The linearly interactive fuzzy numbers can be used in several problems, for example, in fitting a longitudinal dataset [15]. However, in general, the uncertainty data may not be precisely described by this type of interactivity. Our purpose is to relax the definition of linear interactivity. To this end, we introduce the following definition.

Definition 3. Let $A, B \in \mathbb{R}_{\mathscr{F}}$. The fuzzy numbers A and B are said to be quasi linearly interactive (QLI) if, for all $\varepsilon > 0$ there exist $q, r \in \mathbb{R}$, with $q \neq 0$, such that

$$\|B - (qA + r)\| < \varepsilon,$$

where the norm is given by (6) and the symbol "$-$" stands for the standard difference between fuzzy numbers.

The above definition stands a concept which consists in finding parameters $q, r \in \mathbb{R}$ such that $qA + r$ is the closest fuzzy number LI with B.

We adapted the method given in Sect. 3 for QLI fuzzy numbers. To this end, we consider the dataset $D = \{(x_1, Y_1), \ldots, (x_m, Y_m)\} \subset \mathbb{R} \times \mathbb{R}_{\mathscr{F}}$. In order to guarantee that the pairs (Y_1, Y_i), $i = 2, \ldots, m$, are QLI we are led to the following minimization problem:

$$\min_{q_i, r_i} \|(q_i Y_1 + r_i) - Y_i\|^2, \tag{8}$$

where $q_i \neq 0$ for all $i = 2, \ldots, m$.

From Eq. (6), we have $||(q_iY_1 + r_i) - Y_i||^2 = \langle (q_iY_1 + r_i) - Y_i, (q_iY_1 + r_i) - Y_i \rangle$. Writing the endpoints of Y_1 and Y_i by $f(\alpha)$ and $g(\alpha)$, respectively, the development of this problem can be given as follows. Let $f, g : [0,1] \to \mathbb{R}$ be real functions and the residual function $h : \mathbb{R}^* \times \mathbb{R} \to \mathbb{R}_+$ given by

$$h(q, r) = \int_0^1 ((qf(\alpha) - r) - g(\alpha))^2 d\alpha.$$

Since $\langle f, g \rangle = \int_0^1 f(\alpha)g(\alpha)d\alpha$, we have

$$h(q, r) = q^2||f||_2^2 + 2qr\langle f, 1 \rangle + r^2 - 2q\langle f, g \rangle - 2r\langle g, 1 \rangle + ||g||_2^2$$

where 1 is the constant function.

The values (q, r) that minimize the function h satisfy the following conditions:

$$\begin{cases} \dfrac{\partial h}{\partial q}(q, r) = -2\langle f, g \rangle + 2q||f||_2^2 + 2r\langle f, 1 \rangle = 0 \\ \dfrac{\partial h}{\partial r}(q, r) = -2\langle g, 1 \rangle + 2q\langle f, 1 \rangle + 2r = 0 \end{cases}$$

Thus q and r are obtained by solving the following system

$$\begin{pmatrix} ||f||_2^2 & \langle f, 1 \rangle \\ \langle f, 1 \rangle & 1 \end{pmatrix} \begin{pmatrix} q \\ r \end{pmatrix} = \begin{pmatrix} \langle f, g \rangle \\ \langle g, 1 \rangle \end{pmatrix}. \tag{9}$$

Let us consider the functions h_i^1, for $q > 0$, and h_i^2, for $q < 0$, given by

$$h_i^1(q, r) = \int_0^1 (qY_1^-(\alpha) + r - Y_i^+(\alpha))^2 d\alpha + \int_0^1 (qY_1^+(\alpha) + r - Y_i^-(\alpha))^2 d\alpha, \tag{10}$$

$$h_i^2(q, r) = \int_0^1 (qY_1^+(\alpha) + r - Y_i^+(\alpha))^2 d\alpha + \int_0^1 (qY_1^-(\alpha) + r - Y_i^-(\alpha))^2 d\alpha, \tag{11}$$

where $Y_i^-(\alpha) = (Y_i)_\alpha^-$ and $Y_i^+(\alpha) = (Y_i)_\alpha^+$, for $i = 1, \ldots, m$.

Note that $||(q_iY_1 + r_i) - Y_i||^2 = h_i^1(q, r)$, for $q > 0$ and $||(q_iY_1 + r_i) - Y_i||^2 = h_i^2(q, r)$, for $q < 0$. According to (9), the values q_i and r_i that minimize the functions (10) and (11) are given respectively by the systems

$$Nu_i = v_i^1, \tag{12}$$
$$Nu_i = v_i^2, \tag{13}$$

where

$$N = \begin{pmatrix} ||Y_1||_2^2 & \dfrac{\langle Y_i^-, 1 \rangle + \langle Y_i^+, 1 \rangle}{2} \\ \langle Y_i^-, 1 \rangle + \langle Y_i^+, 1 \rangle & \end{pmatrix}, \quad u_i = \begin{pmatrix} q_i \\ r_i \end{pmatrix}$$

and

$$v_i^1 = \begin{pmatrix} \dfrac{\langle Y_1^-, Y_i^+ \rangle + \langle Y_1^+, Y_i^- \rangle}{\langle Y_1^-, 1 \rangle + \langle Y_1^+, 1 \rangle} \end{pmatrix}, v_i^2 = \begin{pmatrix} \dfrac{\langle Y_1^-, Y_i^- \rangle + \langle Y_1^+, Y_i^+ \rangle}{\langle Y_1^-, 1 \rangle + \langle Y_1^+, 1 \rangle} \end{pmatrix}.$$

The vectors u_i^1 and u_i^2 are the solutions of systems (12) and (13), and will be chosen the pair that produces the minimum value of residual function h, that is,

$$(q_i, r_i) = argmin\{h_i^1(q_i^1, r_i^1), h_i^2(q_i^2, r_i^2)\}. \tag{14}$$

Hence $q_i Y_1 + r_i$ is the fuzzy number that best approximates Y_i.

Thus, to fit a dataset given by QLI fuzzy numbers, we produce a fuzzy function as in Eq. (1), where each coefficients q_i and r_i are given by the solution of (14).

In the next section we provide an example of this method.

5   Application to the HIV Dataset

In this section we present an application of the method given in Sect. 4 to fit a HIV dataset obtained from [20]. More precisely, we consult the data of the estimated number of people newly (from 2010 to 2017) infected with HIV and we may estimate the number of infected individuals for the year of 2018. The data which we consider have different behaviors and they are taking from three countries from different continents: Cambodia (Asia), Zimbabwe (Africa) and Venezuela (South America).

The data are modeled by the triangular fuzzy numbers of the form $Y = (a; b; c)$. For all simulations we consider the functions $g_1(x) = x^2$, $g_2(x) = x$, and, $g_3(x) = 1$, for all $x \in \mathbb{R}$. Here we only present the dataset of Cambodia (see Table 1). The dataset of the others countries can be consulted in [20].

Table 1. Dataset of the estimated number of people newly infected with HIV in Cambodia per year and the respectively parameters q and r

Year	Infected	(q, r)
2010	$(1400; 1600; 1800)$	$(1, 0)$
2011	$(1200; 1300; 1500)$	$(0.75, 125)$
2012	$(1000; 1100; 1300)$	$(0.75, -75)$
2013	$(880; 1000; 1100)$	$(0.55, 115)$
2014	$(770; 860; 950)$	$(0.45, 140)$
2015	$(670; 750; 830)$	$(0.40, 110)$
2016	$(610; 680; 750)$	$(0.35, 120)$
2017	$(530; 590; 650)$	$(0.30, 110)$

Figures 1, 2, 3, 4, 5 and 6 depict the fuzzy functions Φ_C, Φ_Z and Φ_V that approximates the dataset of Cambodia, Zimbabwe and Venezuela, respectively.

In the Table 2 the data from 2016 and 2017 was taken from [20] for Cambodia and Zimbabwe, and for Venezuela the data were provided until 2016.

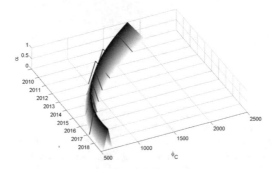

Fig. 1. The 3-dimensional view of the fuzzy function Φ_C. The red triangles represent the data given by QLI fuzzy numbers. The gray lines represent the α-cuts of the fuzzy solutions, where their endpoints for α varying from 0 to 1 are represented respectively from the gray-scale lines varying from white to black.

In according to Table 2 the newly infected individuals in Cambodia and Zimbabwe in 2018 decreases with respect to the year of 2017. For the case of Venezuela, we predicted that the newly infected individuals of 2017 increases with respect to 2016 and, in 2018, decreases with respect to 2017.

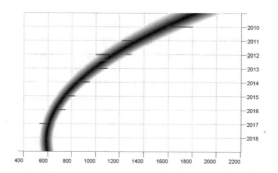

Fig. 2. The bidimensional view of the fuzzy function Φ_C. The red triangles represent the data given by QLI fuzzy numbers. The gray lines represent the α-cuts of the fuzzy solutions, where their endpoints for α varying from 0 to 1 are represented respectively from the gray-scale lines varying from white to black.

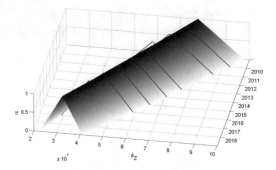

Fig. 3. The 3-dimensional view of the fuzzy function Φ_Z. The red triangles represent the data given by QLI fuzzy numbers. The gray lines represent the α-cuts of the fuzzy solutions, where their endpoints for α varying from 0 to 1 are represented respectively from the gray-scale lines varying from white to black.

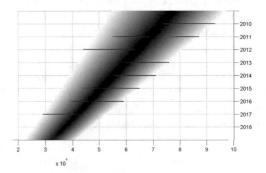

Fig. 4. The bidimensional view of the fuzzy function Φ_Z. The red triangles represent the data given by QLI fuzzy numbers. The gray lines represent the α-cuts of the fuzzy solutions, where their endpoints for α varying from 0 to 1 are represented respectively from the gray-scale lines varying from white to black.

Table 2. Dataset of the estimated number of people newly infected with HIV in Cambodia, Zimbabwe and Venezuela per years 2016, 2017 and 2018

Year	Cambodia	Zimbabwe	Venezuela
2016	$(610; 680; 750)$	$(33000; 46000; 59000)$	$(2060; 2483; 2920)$
2017	$(530; 590; 650)$	$(29000; 41000; 52000)$	$(2413; 2844; 3272)$
2018	$(530; 589; 647)$	$(25806; 37053; 46766)$	$(2401; 2799; 3194)$

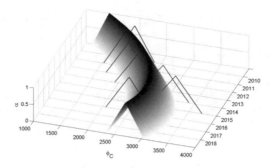

Fig. 5. The 3-dimensional view of the fuzzy function Φ_V. The red triangles represent the data given by QLI fuzzy numbers. The gray lines represent the α-cuts of the fuzzy solutions, where their endpoints for α varying from 0 to 1 are represented respectively from the gray scale lines varying from white to black.

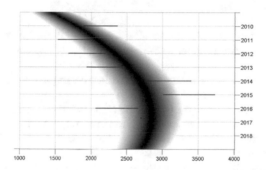

Fig. 6. The bidimensional view of the fuzzy function Φ_V. The red triangles represent the data given by QLI fuzzy numbers. The gray lines represent the α-cuts of the fuzzy solutions, where their endpoints for α varying from 0 to 1 are represented respectively from the gray-scale lines varying from white to black.

6   Final Remarks

In this paper we studied the fuzzy least square method to fit a dataset whose data are given by interactive fuzzy numbers. We focused on a specific type of interactivity namely linear interactivity.

We introduced a new subclass of fuzzy numbers called quasi linearly interactive fuzzy numbers, which incorporates the linearly interactive fuzzy numbers. Based on the method proposed by the authors of [15], we provided a least square method to fit a dataset given by QLI fuzzy numbers, which can be also applied to LI fuzzy numbers. This means that the methodology provided here is more general than the method given in [15].

We illustrated the methodology by fitting a HIV dataset of the number of infected individuals in Cambodia, Zimbabwe and Venezuela. It was possible to estimate the number of infected population in following years.

Recall that the data were modelled by triangular fuzzy numbers. However, we reinforce that the same method can be applied to trapezoidal, Gaussian and others types of fuzzy numbers.

Acknowledgements. The authors would like to thank the financial support of CAPES under grant no 1691227 and Finance Code 001, CNPq under grants 142414/2017-4 and 306546/2017-5, and, FAPESP under grant no 2016/26040-7.

References

1. Barros, L.C., Bassanezi, R.C., Lodwick, W.A.: A First Course in Fuzzy Logic, Fuzzy Dynamical Systems, and Biomathematics. Springer, Heidelberg (2017)
2. Barros, L.C., Bassanezi, R.C., Tonelli, P.A.: On the continuity of the Zadeh's extension. In: Proceedings of IFSA 1997 Congress, pp. 1–6 (1997)
3. Buckingham, R.A.: Numerical Methods. Sir Isaac Pitman & Sons Ltd., London (1966)
4. Carlsson, C., Fullér, R., Majlender, P.: Additions of completely correlated fuzzy numbers. In: IEEE International Conference on Fuzzy Systems, vol. 1, pp. 535–539 (2004)
5. Celmins, A.: Least squares model fitting to fuzzy vector data. Fuzzy Sets Syst. **22**, 245–269 (1987)
6. Diamond, P.: Fuzzy least squares. Inf. Sci. **46**, 141–157 (1988)
7. Dubois, D., Prade, H.: Additions of interactive fuzzy numbers. IEEE (1981)
8. Dubois, D., Prade, H.: Possibility Theory: An Approach to the Computerized Processing of Information. Springer, New York (1988)
9. Fullér, R., Majlender, P.: On interactive fuzzy numbers. Fuzzy Sets Syst. **143**, 355–369 (2004)
10. Jafelice, R.S.M., Barros, L.C.C., Bassanezi, R.C.: Study of the dynamics of HIV under treatment considering fuzzy delay. Comput. Appl. Math. **33**, 45–61 (2014)
11. Joint United Nations Programme on HIV/AIDS (UNAIDS): Knowledge is power - know your status, know your viral load (2018)
12. Laiate, B. Jafelice, R.M., Esmi, E., Barros, L.C.: Transference dynamics of HIV with or without treatment using choquet calculus. In: Proceedings of Fifth Brazilian Conference on Fuzzy Systems, pp. 323–334 (2018). (in Portuguese)
13. Nguyen, H.T.: A note on the extension principle for fuzzy sets. J. Math. Anal. Appl. **64**, 369–380 (1978)
14. Nguyen, H.T., Wu, B.: Fundamentals of Statistics with Fuzzy Data. Springer, Berlin (2006)
15. Pinto, N.J.B., Wasques, V.F., Esmi, E., Barros, L.C.: Least squares method with interactive fuzzy coefficient: application on longitudinal data. In: Fuzzy Information Processing, pp 132–143. Springer, Cham (2018)
16. Seng, K.-Y., Nestorov, I., Vicini, P.: Fuzzy least squares for identification of individual pharmacokinetic parameters. IEEE Trans. Biomed. Eng. **56**(12), 2796–2805 (2009)
17. Takemura, K.: Fuzzy least squares regression analysis for social judgment study. J. Adv. Computat. Intell. Intell. Inform. **9**(5), 461–466 (2005)
18. Tanaka, H., Uejima, S., Asai, K.: Linear regression analysis with fuzzy model. IEEE Trans. Syst. **6**, 903–907 (1982)

19. Torfi, F., Farahani, R.Z., Mahdavi, I.: Fuzzy least-squares linear regression app-roach to ascertain stochastic demand in the vehicle routing problem. Appl. Math. **2**, 64–73 (2011)
20. World Health Organization (WHO). https://www.who.int/hiv/data/en/. Accessed Dec 2018
21. Wu, B., Tseng, N.-F.: A new approach to fuzzy regression models with application to business cycle analysis. Fuzzy Sets Syst. **130**, 33–42 (2002)
22. Zadeh, L.A.: The concept of a linguistic variable and its application to approximate reasoning-i, ii, iii. Inf. Sci. **8**, 301–357 (1975)

Combining ANFIS and Digital Coaching for Good Decisions in Industrial Processes

Christer Carlsson[✉]

Institute for Advanced Management Systems Research,
Abo Akademi University, Auriga Business Center, 20100 Turku, Finland
christer.carlsson@abo.fi

Abstract. The context we address is the digitalization of industry and industrial processes. Digitalization brings enhanced logistics network and value chain integration, which are effective instruments to meet increasing competition and slimmer margins for productivity and profitability. Digitalization also brings pronounced requirements for effective planning, problem solving and decision-making. Decision analytics, including soft computing, will meet the challenges from growing global competition that major industrial corporations face and will help solve the problems of big data/fast data that digitalization is generating as a by-product. A new mantra is gaining support - *powerful, intelligent systems will be effective for the digitalization of industrial processes*. The discussion has paid less attention to the fact that users need advanced knowledge and skills to benefit from the intelligent systems. We need both an effective transfer of knowledge from developers, experts and researchers to users and support for daily use and operations as automated, intelligent industrial systems are complex to operate. We call this knowledge mobilization, and work out how ANFIS models and digital coaching contribute to good decisions in large, complex industrial processes.

Keywords: Industrial processes · ANFIS · Digital coaching

1 Introduction

The digitalization of industrial processes is mostly seen as progress and offering advantages. This will produce "some data processing" (as the discussion goes) but the digitalization enthusiasts do not always work out the actual volumes. In a classical example, the automation of diagnostic messages from the Frecciarossa-1000 train appears to be innovation and progress [8]. Each train beams 5000 messages per second, in a journey of 3 h this will be 54 M messages; if each message is 10 bytes, then a single journey produces about 540 MB; a train fleet is composed of several trains that normally operate every day, some trains make multiple journeys per day. The diagnostic messages are part of an automatic detection process to find faults that are shaping up before they can turn into full-scale problems and possible disasters. The fault detection relies on analytics to work out the process with optimal precision and effort, and the emerging faults prevented with effective maintenance programs. The volumes involved offer challenges to "big data analytics", tasks that are most often

R. B. Kearfott et al. (Eds.): IFSA 2019/NAFIPS 2019, AISC 1000, pp. 190–200, 2019.
https://doi.org/10.1007/978-3-030-21920-8_18

underestimated. The operations of the maintenance programs appear – quite surprisingly – to offer no challenges.

In a recent report called Competing in 2020: Winners and Losers in the Digital Economy [10] Harvard Business Review worked out the impact digitalization will have in a few key industrial sectors. The method was a survey aimed at 783 senior managers, executives and board members; all of them indicated that they are digital decision makers or influencers. Among the respondents 16% stated that their companies are digital (most products/operations depend on digital technology), 23% that they are non-digital (few if any products/operations depend on digital technology) and 61% that they are hybrid (some products/operations depend on digital technology).

The report found (among a number of results) significant productivity and profitability gaps between digital leaders ["digitals"] and the rest ["non-digitals"]. The digitals work with big data and analytics (84%), but only 34% of the non-digitals. The digitals use cognitive computing/AI (51%), but only 7% of the non-digitals. The digitals have data science and data engineering professionals on staff (62%), the non-digitals much fewer (20%); all professionals working for the digitals have the ability to work with and make sense of data and analytics (76%), which is not that common for the non-digitals (30%). A tempting conclusion is that strong analytics capability is key to digital business – companies that want to compete in the digital economy will have to invest in people, processes and technology that work with and make use of data and analytics.

The HBR study promotes "strong analytics capability" which is nice and well but misses a crucial point, the analytics capability is not readily available in most organizations (the "non-digitals").

We have found out in work with industrial partners [6, 7, 20] that there are three levels of knowledge and skills: (i) some knowledge of the key concepts and an intuitive understanding of what can be done with analytics; (ii) sufficient insight and knowledge to read (in written reports, from interactive information systems), absorb and make decisions based on analytics results; (iii) skills and knowledge to work with descriptive, predictive and prescriptive analytics tools and prescriptive, quantitative analysis when supported with intelligent systems. In a digital economy, the level (ii) is a minimum requirement but before long, level (iii) capabilities are necessary.

Digitalization brings in big data ("fast data" for streaming big data), which recent claims [19] state that it will be impossible to use analytics as huge amounts of data make the algorithms impossible or impractical to use. The claims are that it will take too much time as fast decision making in almost real-time is a necessity in the digital economy ("the fast eat the slow" as the slogan goes). The rejoinder is another slogan - "if there is time to make bad decisions it should also be time to make good decisions".

We have come across some common wisdom among managers. Senior, experienced managers claim –when problems are complex and difficult – that experience and intuition will always beat any algorithms. We sometimes give them a counter example developed by Kahneman [18]: *Consider an urn with 100 black and white balls where we know that the black and white balls appear in equal proportions, 50 black and 50 white balls. You pick a ball at random, what is the probability that this ball is black? Common wisdom shows that the probability is 0.50. Next, you get another urn, which contains black and white balls, but you get no knowledge of the number of black or white balls. Now, you pick a ball at random, what is the probability that it is black? A typical answer is 0.50 (because this was the right answer a moment ago).* Intuition is not so good and expert judgements are inferior to algorithms in most cases.

The rest of the paper has the following structure: Sect. 2 introduces and describes the handling of a complex industrial process. Section 3 works out adaptive neuro-fuzzy inference system (ANFIS) models. Section 4 introduces a short state-of-the-art of digital coaching and shows how digital coaches can support the use of ANFIS models. Section 5 gives a short summary and some conclusions.

2   Problem Solving and Complex Industrial Processes

A blast furnace is a good context for complex industrial processes. Blast furnaces are massive, expensive investments that most senior managers do not know in detail but about which they make business decisions, build strategic and tactical plans and for which they solve problems of various kinds. In the following we will work through a case, we have built from insight in the manufacturing industry (cf. [3–5, 8] for details; this paper is a summary of previously published, extensive and detailed papers [8]). We give a simplified process description of a blast furnace – the key part of the iron production process – in Fig. 1.

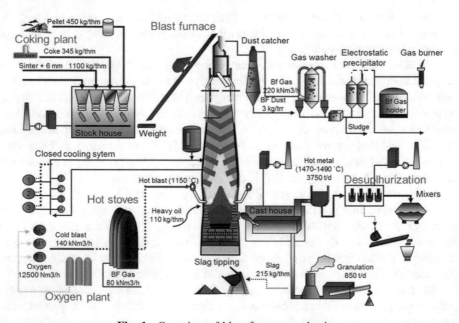

Fig. 1. Overview of blast furnace production

The blast furnace turned out to be very difficult to operate in terms of finding optimal settings. The quality of the raw material plays a role but significant parts are the effectiveness of monitoring technology and problem solving skills of the operators. The monitoring technology has some challenges as the temperature in the core of the blast furnace varies around 1600 C and no measuring equipment can survive.

There are 17 process factors in continuous monitoring from which experienced engineers decide the status of the blast furnace processes and take corrective actions when needed. There is some level of uncertainty on how effective these corrective actions are and have been.

Thus, we proposed to add some new technology and some new, possibly more advanced analytics tools. The engineers installed a sensor system with 16 sensor pairs on the outside and at the top of the blast furnace to collect data on the operations and to find out if the sensor data could add to the monitoring of the 17 processes factors and to the knowledge about the best corrective actions.

We collected data with the sensor system and over the 17 process factors over a 1-month period. This built on the assumption that there are some patterns in the data that would become visible and provide insight (on level (i) *some knowledge of the key concepts and an intuitive understanding of what can be done*). The patterns came from experiments with classical ARIMA time series modelling (the ARIMA typically runs several thousand iterations with the training data set and validates the results of each iteration against the original data; the model then generates a test data set and evaluates it against the original data). A typical ARIMA result is in Fig. 2.

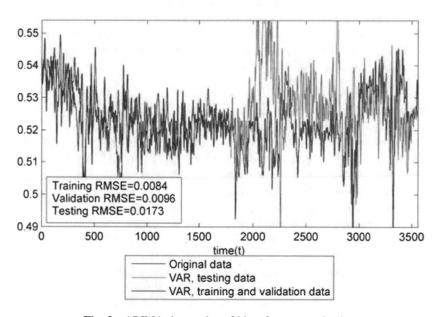

Fig. 2. ARIMA time series of blast furnace production

There were three typical reactions to the time series – (i) impressive visualization, there are clearly some factors we need to get hold of, please try to find them; (ii) this does not show much more than a number of variations for which we cannot find any reasonable explanations, please work out some other approach; (iii) time series do not show much of the key parts of the process, please work out some other algorithms with better predictive capabilities.

We soon realized that experienced engineers have a wealth of tacit knowledge on what factors are essential for the blast furnace operations. Semi-structured workshops with the engineers became knowledge mobilization processes, in which the engineers formulated their insight in response to statistical work we carried out with the collected, large datasets (cf. [19] for a similar approach).

We first realized that not all the data building up the big data set could be relevant – one processing phase in the blast furnace takes on average 8 h, the process is slow and data collected on factors with 1-s intervals is probably not relevant. The second insight – there is a subset of factors that should be essential and important, better find these. The third insight came with statistical analysis – because the process is slow, there is probably time lags between factor inputs and process outputs. We used a classical tool to find essential factors and to identify optimal time lags - principal component analysis (PCA). There is a bonus with PCA, big datasets reduce to quite reasonable size with redundant and irrelevant data cleaned out.

The blast furnace is a complex process and the interaction of key and core factors that decide a continuous and high iron quality are something of a mystery. Nevertheless, there is some consensus that the $etaCO = CO_2/(CO + CO_2)$ is a commonly accepted efficiency measure for a blast furnace. The etaCO is the gas utilization degree, the percentage of carbon monoxide transformed to carbon dioxide in the blast furnace burn process. The factors we found with the PCA all have an impact on the etaCO. The task is to work out models to describe, explain and predict the time series relations between etaCO and the identified factors.

3 Adaptive Neuro-Fuzzy Inference System Models

The choice of model was the Adaptive Neuro-Fuzzy Inference System (ANFIS) (first developed by Jang [16] and later enhanced by [1, 2, 14, 25]. ANFIS is a hybrid inference system that combines a neural network-type structure with fuzzy logic components. The ANFIS has some good features for the present case: (i) it is able to handle non-stationary data (changing variance and/or changing mean over the series); (ii) it does not require to transform input time series to remove trends etc. and (iii) it is able to find complex non-linear relations within data. The drawbacks of ANFIS relate to the "curse of dimensionality" – the duration of the runs and the volume of training data required increases exponentially with the addition of more inputs.

The ANFIS builds on *input-output* variables, a set of *if-then* rules and a fuzzy inference system (FIS); the rules build on *linguistic variables* for the entries (A_i, B_j) and the consequent of each rule is a linear combination of the input values and a constant term (r), in principle as follows:

> *Rule 1*: If x is A_1 and y is B_1 then $z_1 = p_1x + q_1y + r_1$
> *Rule 2*: If x is A_2 and y is B_2 then $z_2 = p_2x + q_2y + r_2$

The ANFIS inference builds on a neural network of five layers that adapt by means of the gradient descent and square minimum methods [16]. In ANFIS, layer 1 implements the membership functions for the linguistic variables; layer 2 calculates the firing strength of a rule; layer 3 determines the normalized firing strength, and layer 4 calculates the output of each rule, multiplying it by the respective level of normalized strength. Finally, layer 5 generates the FIS total output, for which x and y are the variables for the antecedent and consequent, respectively, while z is the output of the system [16, 17] (Fig. 3).

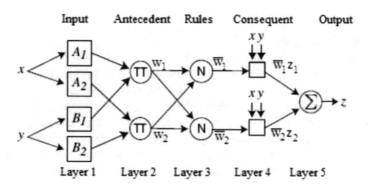

Fig. 3. The adaptive neuro-fuzzy inference system (ANFIS)

The results of the initial ANFIS prototype experiments proved promising as compared to the ARIMA models. This encouraged us to test the ANFIS on larger datasets, blast furnace data for 3 months. The best performing three ANFIS models were run on the new datasets and the results were recorded in repeated experiments (Fig. 4 shows the results with the best ANFIS model).

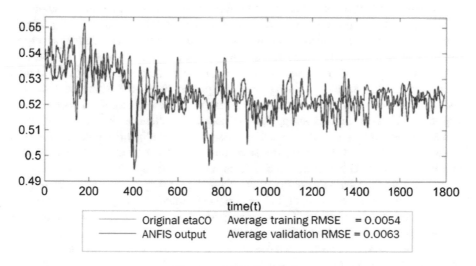

Fig. 4. ANFIS test results

Let us for a moment return to the user context and to the engineers and operators for whom we have recommended to use the ANFIS models. There are obviously some challenges besides the theory and the analytics included in the models. The training and validation of the models (with selections of combinations of input factors) with new datasets require some expertise before the etaCO recommendations are useful.

We propose to include digital coaches with the ANFIS models to guide and support the users.

4 Digital Coaching

A central challenge has appeared in digitalization, the human users of advanced automated systems are *"the weak links"*. Large, automated systems need advanced algorithms and large complex computational systems. System users have diverse backgrounds and different levels of experience. Some users understand everything and master the systems in a short time. They will quickly start to contribute to the use and development of the systems. On the other hand, there are some users, who are slow to learn and/or are not motivated. It will take time for them to reach even a minimum of acceptable levels.

The digital coaching systems got started a few years ago [15] as an answer to the demand on human operators to master advanced automated systems for monitor and control of complex, very large industrial process systems. Digital coaching will work with data that is collected from digital devices, instruments, tools, monitoring systems, sensor systems, software systems, data and knowledge bases, data warehouses, etc. and then processed to be usable for the digital systems that will guide and support users.

Digital coaching requires that we master the transition from data to information, and then on to knowledge, this is *digital fusion*. Data fusion collects and harmonizes data from a variety of sources with different formats and labels. Information fusion uses analytics tools to build syntheses of data to describe, explain and predict key features for problem solving. Knowledge fusion uses ontology to build and formalize insight from data and information fusion as a basis for computational intelligence methods, AI, machine learning, soft computing, approximate reasoning, etc. (cf. [21–24] for details).

Fern et al. [13] develop a theory base for personalized AI systems that work as personal assistants to support human users with tasks they do not fully know how to carry out. This type of technology has gained much attention in the last 10 years because of the growing use of automated systems with intelligent functions. Fern et al. [13] work out a model where the assistant (an AI system) observes the user (represented as a goal-oriented agent) and selects assistive actions (from a closed set of actions) to help the user achieve his goals. Digital coaching gets growing attention in modern medical research and digital health [11] where AI, big data analytics and computational intelligence combine to give patients better control of their diagnosis, care and recovery management from disease.

It appears that the present state-of-the-art is not sufficient to offer a reasonable conceptual framework for the digital coaching. The approach we have in mind is to reduce complexity and dynamics by identifying key performance factors and their interaction logic. The ANFIS models capture insight on interaction between key factors and the etaCO outcomes. To this, we need to add heuristics and experience. Our proposal is to use a combination of ontology (with imprecise elements and relations) to capture experience, and then to process the imprecise logic to get conclusions that can be worked out with families of ANFIS models.

We propose to use digital coaching to guide and support users through the operation of ANFIS models with explanations of each part of the models. The constructs we have in mind build on the technology of multi-agent systems [9, 11, 12], which represent a viable approach to include distributed artificial intelligence support in problem solving and decision making. Individual coaches specialize,

Coach CA_1 on exploration of problem area
Coach CA_2 on design of layer 1
Coach CA_3 on fusion and cleaning of data for layer 1
Coach CA_4 on design of layer 2 and selection of antecedents
Coach CA_5 on selections of inputs, layer 2
Coach CA_6 on design of inference rules of layer 3
Coach CA_7 on interpretations of error terms of layer 4
Coach CA_8 guides iterations on layer 4; stops if iteration errors at acceptable level
Coach CA_9 guides on etaCO to continue training or to produce predictions

The coaches integrate with the models (Fig. 5) and activate when needed through links with the labels *what*, *why*, *how* and *when*, which are typical opening questions when there is a need for guidance.

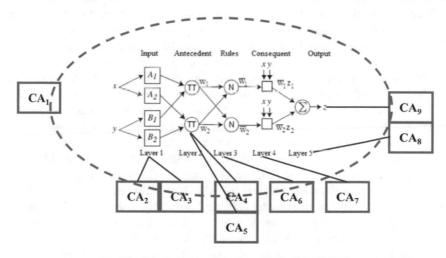

Fig. 5. Coaches CA_1–CA_9 integrated with ANFIS

The digital coaching offers a bridge between two contexts – the analytics representation of explicit knowledge built from data and information collected from the blast furnace processes and the fuzzy ontology representation from partly explicit, linguistic knowledge and tacit knowledge collected through work with experienced engineers.

The process to work out the fuzzy ontology needs also to be interactive with the engineers. Overall, digital coaching offers some good challenges for both analytics and technology development.

5 Summary and Conclusions

The digitalization gains footholds in industrial processes and drives a "fourth industrial revolution" [8] to improve productivity and cost effectiveness, which is one of the ways to secure economic growth in a modern society. We noted that digitalization also drives "some data processing", which is a consequence of sensor technology, information and communication technology, and the availability of sufficient and fast, cloud-service based data processing capacity. The availability of tools appears to drive the progress of digitalization.

Harvard Business Review carried out a study [10] with 783 senior executives and found that they have recognized that the fast growing volumes of data generated by digital technology will require "strong analytics capability" in the corporate world. The executives also admitted that they did not have the requisite competence nor personnel with the necessary levels of knowledge, skill and experience.

In work with industrial partners [6, 7, 20], we typically found three levels of knowledge and skills among senior managers: (i) an intuitive understanding of some analytics, (ii) sufficient knowledge to read, absorb and make decisions with analytics, and (iii) knowledge to work with analytics when supported with intelligent systems.

Here we used the blast furnace as a context and example of what managers need to tackle in order to manage complex industrial processes. The ARIMA time series (Fig. 2) is a typical "brute force" result, i.e. modelers feed a large data set into an analytics package and try to make sense of the results and explain them to managers. We pointed out that the use of classical statistics methods – like PCA – make much more sense and reduce big data sets to more reasonable sizes. Then reasonable size data sets invite work with analytics models and tools that better fit the requirements of the problems at hand. The ANFIS offers a number of advantages: it handles non-stationary data and works with trends, and it finds complex non-linear relations within data.

Advanced models – like the ANFIS – introduce another set of problems. The mathematics is advanced and unavailable to managers. The results (Fig. 4) are visually persuasive – they are clearly better than the ARIMA time series results – but very few managers can actually figure out why they are better. This is a dilemma we have with fuzzy analytics tools – we can prove that they are better than classical operational research and time series tools but we cannot explain why and how to managers.

Digital coaching is at least a partial answer. The digital coaches are another by-product of digitalization, the development of digital health care. The idea is to build a storyline of diagnostic findings and treatment decisions that make health care under-standable. Then digital coaches will support and guide patients to manage their own care without continuous presence of health care professionals. The digital coaches monitor the care processes, note, and correct any deviations from optimal care practice.

In much the same way, we propose that digital coaches could guide and support the use of ANFIS models. We propose that the coaching has a basis in fuzzy ontology [8] to offer a basis and conceptual framework for digital fusion. Multi-agent systems technology offers tools to build automatic coaching systems (this still requires experimental validation and verification) but there are some visions of joint human-machine intelligence [10] that we need to follow up.

Business processes – problem solving and decision making – in a digital economy require that managers process relevant data quickly and accurately, find key factors and their interrelations, and then make good (best, optimal) decisions that quickly turn into decisive actions. The processes happen "on-the-move" or in fast, interactive meetings coordinated with digital tools in multiple settings. Nevertheless, limited time will never be an excuse for bad-quality decisions. Analytics and digital coaching and support need to develop to meet the requirements of a digital economy.

The ANFIS is an example of advanced modelling that will drive digitalization; there are hundreds of families of similar models. Hence, it follows that digital coaching need to develop to offer support for the models, which offers opportunities for research in advanced fuzzy modelling, soft computing and fuzzy ontology.

References

1. Akkoc, S.: An empirical comparison of conventional techniques, neural networks and the three stage hybrid adaptive neuro fuzzy inference system (ANFIS) model for credit scoring analysis: the case of Turkish credit card data. Eur. J. Oper. Res. (2012). https://doi.org/10.1016/j.ejor.2012.04.009
2. Afshar-Kazemi, M.A., Toloie-Eshlaghy, A., Raze Motadel, M., Saremi, H.: Product lifecycle prediction using adaptive network-based fuzzy inference system. In: International Conference on Innovation, Management and Service IPEDR, Singapore, 14 September 2011, pp. 230–236 (2011)
3. Carlsson, C.: Soft computing in analytics: handling imprecision and uncertainty in strategic decisions. Fuzzy Econ. Rev. **XVII**(2), 3–21 (2012)
4. Carlsson, C., Mezei, J., Brunelli, M.: Decision making with a fuzzy ontology. Soft Comput. **16**(7), 1143–1152 (2012)
5. Carlsson, C., Mezei, J., Brunelli, M.: Fuzzy ontology used for knowledge mobilization. Int. J. Intell. Syst **28**(1), 52–71 (2013)
6. Carlsson, C., Brunelli, M., Mezei, J.: A soft computing approach to mastering paper machines. In: 2012 Proceedings of HICSS-46, HICSS.2013.61, pp. 1394–1401. IEEE (2013)
7. Carlsson, C., Heikkilä, M., Mezei, J.: Fuzzy entropy used for predictive analytics. In: Kahraman, C., Kaymak, U., Yazici, A. (eds.) Fuzzy Logic in its 50th Year. New Developments, Directions and Challenges. Studies in Fuzziness, vol. 341, pp. 187–210. Springer, Heidelberg (2016)
8. Carlsson, C.: Decision analytics - key to digitalization. Inf. Sci. **460–461**, 424–438 (2018)
9. Carlsson, C.: Decision support in virtual organizations: the case for multi-agent support. Group Decis. Negot. **11**(9), 185–221 (2002)
10. Competing in 2020: Winners and Losers in the Digital Economy (2017). A Harvard Business Review Analytic Services Report, 25 April 2017
11. Fagherazzi, G., Ravaud, P.: Digital Diabetes: Perspectives for Diabetes Prevention, Management and Research. Diabetes & Metabolism (2018, in press)

12. Ferber, J.: Multi-agent Systems, An Introduction to Distributed Artificial Intelligence. Addison-Wesley, Great Britain (1999)
13. Fern, A., Natarajan, S., Judah, K., Tadepalli, P.: A decision-theoretic model of assistance. J. Artif. Intell. Res. **49**, 71–104 (2014)
14. Firat, C.A., Cevik, A., Gokceoglu, C.: Some applications of adaptive neuro-fuzzy inference system (ANFIS) in geotechnical engineering. Comput. Geotech. **40**, 14–33 (2012)
15. Fricoteaux, L., Thouvenin, I., Mestre, D.: GULLIVER: a decision-making system based on user observation for an adaptive training in informed virtual environments. Eng. Appl. Artif. Intell. **33**(2014), 47–57 (2014)
16. Jang, J.R.: ANFIS: adaptive-network-based fuzzy inference system. IEEE Trans. Syst. Man Cybern. **23**(3), 665–685 (1993)
17. Jang, J.R., Sun, C.T.: Neuro-fuzzy modelling and control. I: Proceedings of the IEEE, pp. 378–406, March 1995
18. Kahneman, D.: Thinking Fast and Slow. Farrar, Straus and Giroux, New York (2011)
19. Manco, G., Ritaccoa, E., Rulloe, P., Galluccid, L., Astillc, W., Dianne Kimber, D., Marco Antonelli, M.: Fault detection and explanation through big data analysis on sensor streams. Expert Syst. Appl. **87**(2017), 141–156 (2017)
20. Mezei, J., Brunelli, M., Carlsson, C.: A fuzzy approach to using expert knowledge for tuning paper machines. J. Oper. Res. Soc. **68**(6), 605–616 (2017)
21. Morente-Molinera, J.A., Wikström, R., Carlsson, C., Viedma-Herrera, E.: A linguistic mobile decision support system based on fuzzy ontology to facilitate knowledge mobilization. Decis. Support Syst. **81**, 66–75 (2016)
22. Morente-Molinera, J.A., Mezei, J., Carlsson, C., Viedma-Herrera, E.: Improving supervised learning classification methods using multi-granular linguistic modelling and fuzzy entropy. Trans. Fuzzy Syst. **25**(5), 1078–1089 (2016)
23. Morente-Molinera, J.A., Wikström, R., Carlsson, C., Cabrerizo, F.J., Pérez, I.J., Herrera-Viedma, E.: A novel android application design based on fuzzy ontologies to carry out local based group decision making processes. In: 13th International Conference on Modeling Decisions for Artificial Intelligence, MDAI 2016. Springer, Andorra (2016)
24. Morente-Molinera, J.A., Mezei, J., Carlsson, C., Herrera-Viedma, E.: Using multi-granular fuzzy linguistic modelling methods for supervised classification learning purposes. In: Proceedings of FUZZ-IEEE 2017, Paper # 50. IEEE Computational Intelligence Society, Naples (2017)
25. Mullai, P., Arulselvi, S., Huu-Hao, N., Sabarathinam, P.L.: Experiments and ANFIS modelling for the biodegradation of penicillin-G wastewater using anaerobic hybrid reactor. Bioresour. Technol. **102**, 5492–5497 (2011)

On the Implementation of Evolving Dynamic Cognitive Maps

Joao Paulo Carvalho[✉]

INESC-ID/Instituto Superior Técnico, Universidade de Lisboa,
R. Alves Redol, 9, 1000-029 Lisbon, Portugal
joao.carvalho@inesc-id.pt

Abstract. Fuzzy Cognitive Maps (FCM) and other Dynamic Cognitive Maps (DCM) allow simulation of the evolution of complex qualitative dynamic systems through time. However, the DCM model is static by itself in the sense that its cognitive configuration, i.e., the concepts' definitions, the relations among the concepts and the structure of the map, do not change with time. This paper introduces DCM meta-states, a simple but versatile Finite State Machine based mechanism that can be used to implement Evolving FCM and generic Evolving Dynamic Cognitive Maps (Ev-DCM).

Keywords: Fuzzy Cognitive Maps · Rule-Based Fuzzy Cognitive Maps · Dynamic Cognitive Maps · Evolving temporal modeling

1 Introduction

In 1986 Bart Kosko introduced Fuzzy Cognitive Maps (FCM) as a tool to model the dynamics of qualitative systems such as Social, Economic or Political systems. Such systems are composed of a number of dynamic qualitative concepts interrelated in complex ways, usually including feedback links that propagate influences in complicated chains, that make reaching conclusions by simple structural analysis an utterly impossible task. Kosko's work was the first serious attempt to introduce dynamics into Axelrod's 1970's pioneering work on Cognitive Maps (CM) [3]. With cognitive mapping Axelrod introduced a graphic way to express real-world qualitative dynamic social systems from the viewpoint of social decision makers. CMs consisted on ordered graphs representing concepts (the entities that are relevant for the system in question) and the relations between those concepts. For several years CM analysis was simply structural and consisted in methods to extract information based on the way the concepts were interconnected [19]. These methods allowed essentially identifying what were the key concepts in the modeled systems. No dynamical analysis regarding the evolution of the systems through time was possible even though CM were meant to represent dynamic systems.

Work supported by national funds through Fundação para a Ciência e a Tecnologia (FCT) under reference UID/CEC/50021/2019, grant SFRH/BSAB/136312/2018 and project LISBOA-01-0145-FEDER-031474.

R. B. Kearfott et al. (Eds.): IFSA 2019/NAFIPS 2019, AISC 1000, pp. 201–213, 2019.
https://doi.org/10.1007/978-3-030-21920-8_19

Since the introduction of Fuzzy Cognitive Maps (FCM) by Kosko [15–17], hundreds of works have applied FCM or FCM variants to the most diverse fields. However, FCM have limitations in what concerns the modeling and simulation of the dynamics of CM [8, 9, 12]. Several FCM extensions or variations have been proposed throughout the years to deal with those limitations. Some of them stay very true to the basic mechanisms of FCM, while others present a rather different approach. Usually each approach has its own designation (e.g.: DCN – Dynamical Cognitive Networks [20], E-FCM – Extended Fuzzy Cognitive Maps [14], RB-FCM – Rule Based Fuzzy Cognitive Maps [12], etc.); in this work we use the term Dynamic Cognitive Maps (DCM) to roughly cover such approaches.

One of the things that most DCM have in common is the fact that their cognitive configuration is usually static, i.e., once the cognitive map is modeled, neither its concept structure, nor the relations that define the system behavior, change through time. This fact hinders DCM capabilities to model complex real-world systems temporal behavior, especially when the time span to be modeled and simulated is very large. Acampora and Loio [1] explicitly addressed the need for mechanisms that would allow modeling such behavior with the introduction of Timed Automata FCM (TA-FCM) in 2011. However, the notion of cognitive map Meta-states, introduced in 2006 [5, 6], can implicitly model such mechanisms on a qualitative and simpler way. This paper shows how those principles can be applied to implement Evolving Dynamic Cognitive Maps (Ev-DCM).

2   Dynamic Cognitive Maps

2.1   Fuzzy Cognitive Maps

Fuzzy Cognitive Maps (FCM), as introduced by Kosko [16, 17], are meant to be a combination of Neural Networks and Fuzzy Logic that allow us to predict the change of the concepts represented in Causal Maps. The graphical illustration of FCM is a signed directed graph with feedback, consisting of n nodes and weighted interconnections. Nodes of the graph stand for the concepts that are used to describe the behavior of the system and they are connected by signed and weighted arcs representing the causal relationships that exist among concepts. FCM representation can usually be reduced to a $n \times n$ matrix where each position contains a weight representing the causal relation between the concept indicated in the respective row and the concept indicated in the respective column. FCM have been extensively studied and applied. Extensive details about FCM representation, inference and the associated semantics can be found in [8].

The dynamics of a FCM, which basically consist in its evolution in time, are modeled on an iterative way: time is discrete and the current value of each concept is computed based on the values of its inputs in the previous iteration. The update of the values of each concept for the current iteration must occur only after all concepts have been calculated. As the FCM evolves through time, the map might reach equilibrium, converging to a single state or a finite cycle of states. It is important to notice that "time" should be considered essential when modeling a FCM, since the rate of change on a social system (or in fact in most real world systems) cannot be infinite; i.e., when simulating a FCM, one cannot assume that the value of a concept can change from its minimum to its maximum value on a single iteration unless this iteration represents a large enough amount of time.

When seen as a whole, FCM allows the answer to what-if questions in causal maps: what happens to a system if some of its concepts change, or if new concepts are introduced or removed.

2.2 Fuzzy Cognitive Maps Extensions and Variations

E-FCM [14] were the first attempt to address FCM obvious limitations. They explicitly address timing, the monotonic/symmetric causality and a few other, although not all, FCM issues [7, 8]. In E-FCM, the weights are replaced by a non-linear function that also includes time delays. E-FCM are a vastly superior alternative to FCM even if their complexity necessarily hinders their application.

Dynamical Cognitive Networks (DCN) [20], are an interesting proposal that attempted to overcome several FCM issues but that unfortunately were not consistently marketed or developed, and therefore their real capabilities as a FCM alternative to model social systems were never shown.

Fuzzy Cognitive Networks (FCN) [18] are one of the few exceptions to the DCM static weights norm: FCN update their weights and reach new equilibrium points based on the continuous interaction with the system they describe. The updating method takes into account the input and the output from the real system and combines them in a vector that is used by an updating algorithm to obtain the new weight matrix. However, one should note that temporal issues are not in the base of the development of the proposed mechanism.

Timed Automata FCM (TA-FCM) [1] use a double-layered temporal granularity to improve the time representation in qualitative system dynamics. The low-level layer is based on FCMs timing mechanisms such as the base time (B-time) [11], and the high-level layer (T-time), that acts as a supervisor, provides FCMs with two timing mechanisms, the *cognitive era* and the *cognitive configuration* that allow the representation of a generic system as an entity that crosses a sequence of time periods (cognitive eras). Each cognitive era represents the longest interval time in which the system does not change its cognitive configuration. TA-FCM were the first FCM to formally address the need of change in cognitive configuration. TA-FCM are based on timed automata theory [2] and represent complex systems by means of a collection of states and transitions depending upon temporal events. TA-FCM are based on a solid mathematical foundation that provide system verification techniques able to be used early in the system design cycle in order to detect logical bugs before the system is implemented. On the other hand, they largely increase the complexity of using FCMs, denying one of their main advantages.

A rather different alternative to FCM are Rule Based Fuzzy Cognitive Maps (RB-FCM). Due to their specificity and relevance to the theme of the paper, they will be described separately in the following section.

2.3   Rule Based Fuzzy Cognitive Maps

RB-FCM allow a representation of the dynamics of complex real-world qualitative systems with feedback, and the simulation of events and their influence in the system [12]. They can be represented as fuzzy directed graphs with feedback, and are composed of fuzzy nodes (Concepts), and fuzzy links (Relations). RB-FCM are true cognitive maps since are not limited to the representation of causal relations. Concepts are fuzzy variables described by linguistic terms, and Relations are defined with fuzzy rule bases.

RB-FCM are essentially iterative fuzzy rule based systems where fuzzy mechanisms to deal with feedback were added, timing mechanisms were introduced, new ways to deal with uncertainty propagation were defined, and several kinds of concepts (Levels, Variations) and relations (Causal, Inference, Alternatives, Probabilistic, Opposition, Conjunction, etc.) were proposed in order to cope with the complexity and diversity of the dynamic qualitative systems being modeled [4, 9–12]. Among new contributions brought by RB-FCM, there is a new fuzzy operation, the Fuzzy Carry Accumulation, which is essential to model the mechanisms of qualitative causal relations (FCR – Fuzzy Causal Relations) while maintaining the simplicity and versatility of FCM [8]. In order to better model the dynamics of the modeled systems, two main classes of Concepts were proposed: Levels, that represent the absolute values of system entities (e.g., LInflation is Good); and Variations, that represent the change in value of a system entity in a given amount of time (e.g., VInflation increased very much).

3   Expressing Time in Dynamic Cognitive Maps

Time is probably the most essential factor when modeling a dynamic system. However, most DCM approaches seem to ignore this fact [8]. In order to maintain consistency in the process of modeling the dynamics of a qualitative system, it is necessary to develop and introduce timing control mechanisms. Several DCM mechanisms that allow the representation of time flow, delays, and the inhibition of certain relations when they have no influence on a given instant have been proposed in [11] and implemented in RB-FCM:

- Implicit Time: Time in DCM must be implicit in every relation. Therefore, the responsibility of maintaining temporal coherence in the process of modeling a system relies heavily in the modeler. He or she must not only be responsible to find the nature and characteristics of the relation but must also ensure that the magnitude of the relation is adequate to the time interval that it represents.
- Base time (*B-Time*): Each system iteration must represent the flow of a given amount of time. *B-Time* represents the "resolution" of the model, i.e., the highest level of temporal detail that a simulation can provide in the modeled system. *B-Time* must always be implicit while defining each relation, especially in causal relations. The choice of *B-Time* is highly dependent on the real-world system being modeled. It could be one hour, one day, two days, one week, one year, one century or any other time period. It depends on the desired or advisable level of detail, complexity and intended long-term analysis of the system. Shorter *B-Times* usually need more

detailed and complex rule bases and imply a much more careful approach to the precision and validity of the rules. Longer *B-Times* should provide more valid long-term simulations, but short-term detail, precision and validity will possibly be sacrificed.

- Other possible time intervals: Since some relations only make sense at intervals larger than *B-Time*, they can be associated with a time interval that indicates at which iterations they should resolve.
- Delays: Applied when the effect of a relation is not immediate is only experienced after some iterations (e.g. the effect of Oil price on fuel price).

4   Finite State Machines

A Finite State Machine (FSM) [21, 22], often simply called state machine, is a mathematical abstraction that can be used to design digital logic or computer programs. It is a behavior model composed of a finite number of states (usually graphically represented by a circle), transitions between those states (graphically represented by arrows), a set of inputs used to define transition conditions, and actions (outputs) that can be either associated to a state or to a transition. It is equivalent to a flow graph in which one can inspect the way logic runs when certain conditions are met. The operation of a FSM begins from one of the states, and goes through transitions that depend on the set of inputs on the present state. A FSM can reach a final state or might simply evolve on a finite cycle (or cycles) of states. In the present approach, one uses a particular type of FSM commonly referred as transducers, that includes Moore and Mealy machines [22].

Despite the graphic similarities, FSM and DCM are entirely different mechanisms: in FSM only one state (circle) is active in each instant, while in DCM all concepts (circles) are active simultaneously.

5   Evolving Dynamic Cognitive Maps

5.1   The Need for Ev-DCM

The need for Evolving Dynamic Cognitive Maps arises from several implementation issues, all connected with time.

As mentioned in Sect. 2, temporal issues are one of the most important aspects when properly modelling the dynamics of a real-world DCM. In complex cognitive maps, the time modelling related problems usually begin when a decision regarding the duration of each iteration of the simulation (*B-Time*) has to be made: since it is hard to see the effects of policies on social sceneries in short time, system simulations often must cover a large time span (from several months to a few years); however, it is often necessary to model some actions and decisions in much shorter time spans in the exact same system (sometimes even hourly). For example, on a DCM of a system where one attempts to model how purse-seine fishing vessel skippers react to government fishing bans and policies [13, 23], opting for *B-Time* = 1 day would prevent the modelling of the skipper behaviour during the daily fishing trips which is exactly when the skipper takes the decisions that affect the fishing ecosystem; on the other hand, opting to model the system on a hourly basis would imply simulation runs consisting of a few thousand to tens of thousands of iterations. The problem associated with the latter option, is that the high number of feedback links would mean that results would be highly unreliable (in the same sense that it is unreliable to make long term weather predictions). The solution to this problem uses a temporal hierarchical approach, similar in concept (but not in form) to [1]: in the present example, a "fishing" day would consist of several DCM, each being simulated on an hourly basis, and the long-term evolution would be simulated on a daily basis.

Besides the temporal-scale problem, it is natural to expect the concepts and relations composing a DCM (the cognitive configuration) to change through time: people change their minds, their opinions, the way they act as they grow older; society evolves; drastic socio-economic-political changes in the world make humans and society adapt to new realities and start taking them as normal, etc. For example, the price of 30 USD per oil barrel was considered normal in the year 2000; In the same epoch, the current price of over 100 USD would be deemed extremely high, unacceptable and even catastrophic for the economy; However, in a time span of less than 10 years there were sudden increases in oil price (e.g. 60 USD/barrel in 2002; 130 USD/barrel in 2008), resulting from a conjunction of world events. Such events led politics, economists and society in general to adapt and find those prices perfectly normal and acceptable. After a period of gradual oscillations, the oil price suddenly crashed (2014) to year 2K levels, and another period of adjustment began. A cognitive map including oil barrel price as a concept would have to somehow evolve through time in order to incorporate these changes – e.g., the linguistics terms and membership functions of a fuzzy concept *Oil_price* would have to evolve (*Medium* would become associated with "around 100 USD" instead of "around 30 USD").

In some systems the evolution can be so drastic after some key events that the resulting CM hardly bears any resemblance with the initial system. As an example, one can use the previously mentioned purse-seine fishing vessel skipper CM: the cognitive map that models skipper behaviour while he is at port deciding if it is worth to leave to a fishing trip, is totally different from the map modelling its behaviour while he is managing the fishing operations.

In brief, one can find three major reasons for the need of Evolving DCM (Ev-DCM):

1 – The need to model different temporal scales in the same CM;
2 – The natural evolution of the cognitive configuration through time;
3 – Drastic changes.

5.2 Ev-DCM Definition

A Mealy FSM transducer-based mechanism was developed to support the definition and implementation of Ev-DCM. Each state of this FSM based mechanism is called *meta-state*.

Definition 1: An Ev-DCM consists of several meta-states organized as a deterministic FSM transducer where each meta-state is a DCM. An Ev-DCM is a septuple

$$(\Sigma, \Gamma, MS, ms_0, \delta, \omega, b_{time}),\tag{1}$$

where:

- Σ is the input alphabet. It defines the inputs for the transition conditions (events).
- Γ is the output alphabet. It defines the possible outputs for the EV-DCM and consists of all possible states of all the DCM contained in the EV-DCM.
- *MS* is a finite and non-empty set of meta-states. Each meta-state is a DCM.
- ms_0 is the initial meta-state and an element of *MS*, i.e., $ms_0 \in MS$.
- δ is a possibly time dependent closed state transition function, $\delta : MS \times \Sigma \rightarrow MS$. It defines the conditions for the transitions and the next meta-state according to the set of events.
- ω is the output function, and it is a function of the current meta-state and the input alphabet, i.e., $\omega : MS \times \Sigma \rightarrow \Gamma$. It corresponds to the simulation of the active DCM.
- b_{time} is the current iteration and is based on B-Time as defined in Sect. 3.
- Only one meta-state can be active in each b_{time} instant.

Definition 2: The DCM associated with each meta-state is simulated only and only if the corresponding meta-state is active. Such DCM is referred to as "current DCM".

An Ev-DCM is therefore composed of a finite number of states (here called meta-states), possible transitions, a set of inputs used to define transition conditions (events), and actions (outputs). Meta-states *(MS)*, outputs *(Γ)* and output generation *(ω)* are addressed in Sect. 5.3. Transitions *(δ)* and events *(Σ)* are addressed in Sect. 5.4.

5.3   Meta-States and Outputs

A meta-state is a DCM. In the case of a FCM, a meta-state is simply defined by two matrixes, the $n \times 1$ state matrix containing the values of each of the n concepts in the current iteration, and an $n \times n$ matrix containing the weights defining the relations. More complex DCM, like for example RB-FCM, need more complex structures (including rule bases and inference mechanisms).

The use of the term *meta-state* results from the fact that each DCM is itself defined by its state, which is the set of the values of its concepts on a given iteration (given by the current iteration $n \times 1$ matrix on a FCM). Therefore, state refers to the present values of the current DCM, and meta-state refers to which DCM is active.

The possible states of all DCM that compose the Ev-DCM correspond to the set of outputs *(Γ)* of the Ev-DCM. The output function ω is the simulation of the current DCM until the occurrence of an event that generates a transition to a new meta-state.

Although *MS* is finite for operational reasons, the set of all possible *MS* is infinite, since there is no limit to the number of different DCMs.

The FSM is used to represent the evolution of an Ev-DCM. Since each meta-state can be represented by totally distinct DCMs, this mechanism allows a huge versatility in the type of changes that can be modeled through meta-state transitions, from simple alterations in one relation (the equivalent of changing the weight of a relation on FCMs), to a completely different FCM, where some or even all concepts and relations might be different. Although other works have proposed more complex mechanisms to change FCM relations during simulation [1, 18], the Meta-state approach capability to totally alter all components of a DCM is novel, and at the same time, incredibly simple.

Figure 1 shows an example of the FSM used to model the active daily life of a purse-seine fishing vessel skipper [23]. *MS* is the set of five meta-states (each has a corresponding DCM) that are used to represent the stages composing the daily activity (see further details in Sect. 6).

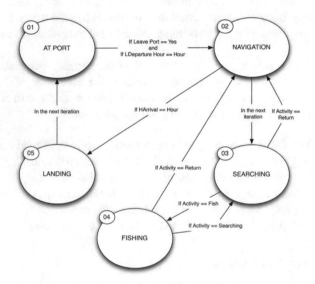

Fig. 1. System Meta-states: Each meta-state contains a DCM that is simulated when its meta-state is active. Only one Meta state is active on a given instant. Transitions between Meta states occur when the active DCM reaches certain conditions.

The following operations on DCM components have been defined and implemented in Ev-DCM (namely in RB-FCM) in order to allow for a large variety of *MS*:

- Relations – Add relation; Remove relation; Modify the relation between one or more concepts. In FCM these operations consist of adding, removing or changing the weight of a relation. Adding a relation consists in replacing a '0' in the FCM matrix by a value in the interval [−1,1]); removing a relation consists in replacing one of the non-zero elements of the matrix by '0'; modifying a relation consist in changing one of matrix non-zero elements. In more complex DCM, the changes can as complex as the DCM allows it. For example, in RB-FCMs, in addition to adding or removing relations, one can change the rules in the fuzzy rule base or even change the type of relation, e.g., evolve from a causal relation to an inference relation [9];

- Concepts – Add concept; Remove concept; Modify concept.
  Concepts can be added, removed or modified. Introduction and deletion usually imply changes in the associated relations. In FCM only additions and removals are possible. The addition of a concept results in changing the FCM $n \times n$ matrix into a $(n+1) \times (n+1)$ matrix. The removal consists in reducing its size to $(n-1) \times (n-1)$. In RB-FCM, one can also alter the membership functions and linguistic terms used to define the concept, as exemplified in the Oil price per barrel example (see Sect. 5.1).

5.3.1   Meta-State Activity Phase

As in any FSM, only one meta-state, referred to as *current meta-state*, can be active on a given Ev-DCM on a given instant. However, it is convenient to allow a concept from a given meta-state to get its inputs from concepts in other meta-states in order to avoid unnecessary concept replication. In such cases, the value the concept gets is the one assumed by the antecedent concept in the last iteration its meta-state was active. An example can be seen in Fig. 2 between concepts *Landings* and *Revenue*, where revenue from the fishing trip is calculated on meta-state *FISHING* based partially on the value of the amount of fish captured in meta-state *LANDING*.

5.4   Events and Transition Function

The inputs Σ that are used to define the transitions of the Ev-DCM are obtained from the present values of the Ev-DCM concepts. E.g., on a FCM, Σ is obtained from the $n \times 1$ state matrix.

Transitions (δ), i.e., the change of the active status from the current meta-state to a new meta-state, can occur depending on two types of events:

- Unconditional (or timed) transitions $\delta t : MS \times b_{time} \rightarrow MS$ – transition to the following meta-state occurs after a certain previously defined number of iterations triggered by b_{time} (corresponding to a given time interval).
- Conditional transitions, $\delta c : MS \times \Sigma \rightarrow MS$ – transition to the following meta-state occurs when one or more of the following events occur:
  - a concept reaches a given value;
  - a logical condition involving the values of several concepts becomes true;
  - a mathematical function combining the values of one or more concepts reaches a given value.

The set of input events is highly flexible and obviously dependent on the used DCMs. In Ev-FCM, a conditional transition δc corresponds to checking or operating the values in the state matrix, and given a true result, change the active status to a new FCM according to a transitions table. Changing the active status is equivalent to replace the current weight and state matrix to the new weight and state matrixes characterizing the new FCM.In generic or more complex Ev-DCM, transitions can be implemented through a set of rules that indicate the possible next meta-states according to the occurring event.

Concerning the effect of a transition, there is no limit to the amount of changes that can occur on a DCM when switching to a new meta-state. In some Evolving DCMs, smooth transitions are expected – small changes in the relations, small changes in the membership functions defining the concepts, etc.; in others, the transition of a meta-state to the following one might represent a change to a totally different cognitive map. For example, each of the five meta-states represented in Fig. 1 use five very different cognitive maps that model five very different stages in the skipper daily activity (Fig. 2, see Sect. 6).

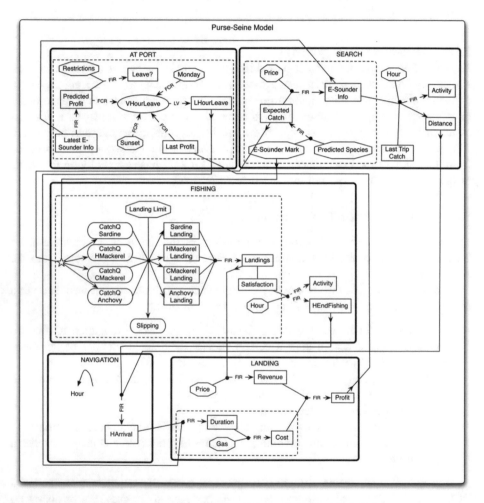

Fig. 2. Purse-seine Fishing Skipper Behavior Ev-DCM.

6 Application Example

As an example of an application of the presented approach to model, we use a Purse Seine Fishing Skipper Behavior Ev-DCM. This CM was originally proposed in [13], was used in a real world case study presented in [23], and has been used as an example throughout this paper on how to model Evolving DCM. The goal of this DCM is to obtain a qualitative dynamic model based on real world data taken from a real-world qualitative system: the day to day behavior of purse seine fishing fleet skippers. Purse seine fishing is fishing using a large fishing net that hangs in the water due to weights along the bottom edge and floats along the top. The purse-seine is a preferred technique for capturing fish species which school, or aggregate, close to the surface, such as sardines, mackerel, anchovies, herring, etc. Purse-seine fishing on the Portuguese coast is done on small ships with a cargo capacity of less than 20000 kg. On purse-seine fishing, skippers look for fish using an echo-sounder while the ship is on the move. After fish is located, a purse net is dropped and used to capture the fish. The net dimension is around 1000 m long by 120 m tall. A fishing operation takes around 2 h to complete. Skippers act independently from other skippers in the fleet, although cooperation (information exchange regarding fisheries) is possible and common. Several fishing operations can be made in one day. A fishing day starts after 5 PM, and usually ends early in the morning, although end time can vary a lot due to several factors.

Figure 1 shows the 5 different meta-states and the transition conditions that simulate one fishing day. Each fishing day starts with the system in the *At_Port* state, where decisions whether to start a fishing trip and when to leave port must be made. Transition to the next Meta-state happens at a given hour (unconditional transition), if concept *LeavePort* is true. After leaving port, the skipper heads on a given direction (*Navigation*) and starts looking for fish using the echo sounder (*Search*). While searching, and according to the echo sounder results, a decision must be made on whether to start a fishing operation, continue searching or return to port. If one continues searching, the Meta-state remains the same; otherwise one changes to one of the following Meta-states. During the fishing operation (*Fishing Op*), fish is captured using a purse-net. At the end of a fishing operation, the skipper must decide on keep searching (change to the meta-state *Search*) or return to port (change to the meta-state *Navigation* with concept Activity stating to return). At the end of the fishing day, the ship arrives to port *(Landing)* where the catch is unloaded and sold at the auction market, and the skipper makes a balance of the fishing day.

Figure 2 shows a simplified representation of the whole Ev-DCM. All types of transitions and structural modifications described in Sect. 4 are present in the map. It is possible to observe the diversity of each meta-state, and how such diversity would be impossible to model using other previously proposed DCM approaches.

7   Conclusion

Despite their simplicity, the proposed meta-state based mechanisms are a rather effective and tested proposal to model and simulate temporal evolving DCM. Even though meta-states were specifically developed with RB-FCM in mind, they are based in very well-known techniques (finite state machines, logical statements and simple math equations) that can easily be adapted and implemented by most dynamic cognitive map developers/modelers without the need for a steep learning curve and without overtly complicating the simpler DCM approaches (like FCM).

When compared to other solutions like the TA-FCM, the proposed Ev-DCM clearly lack the strong mathematical background and system verification techniques capability, which are obviously strong assets. On the other hand, it gains in features (virtually all components of a FCM can be evolved through time), versatility (can be applied to all DCM) and simplicity of use, making it a very strong option.

References

1. Acampora, G., Loia, V.: On the temporal granularity in fuzzy cognitive maps. IEEE Trans. Fuzzy Syst. **9**(6), 1040–1057 (2011)
2. Alur, R.: A theory of timed automata. Theor. Comput. Sci. **126**, 183–235 (1994)
3. Axelrod, R.: The Structure of Decision: Cognitive Maps of Political Elites. Princeton University Press, Princeton (1976)
4. Carvalho, J.P., Tomé, J.A.B.: Rule based fuzzy cognitive maps – fuzzy causal relations. In: Mohammadian, M. (ed.) Computational Intelligence for Modelling, Control and Automation: Evolutionary Computation & Fuzzy Logic for Intelligent Control, Knowledge Acquisition & Information Retrieval. IOS Press, Amsterdam (1999)
5. Carvalho, J.P., Carola, M., Tome, J.A.: Forest fire modelling using rule-based fuzzy cognitive maps and Voronoi based cellular automata. In: Proceedings of the 25th International Conference of the North American Fuzzy Information Processing Society, NAFIPS 2006, Montreal, Canada (2006)
6. Carvalho, J.P., Carola, M., Tomé, J.A.: Using rule based fuzzy cognitive maps to model dynamic cell behaviour in Voronoi based cellular automata. In: Proceedings of the WCC I2006 – 2006 IEEE World Congress on Computational Intelligence, pp. 1503–1510 (2006)
7. Carvalho, J.P.: On the semantics and the use of fuzzy cognitive maps in social sciences. In: Proceedings of the WCCI 2010 – 2010 IEEE World Congress on Computational Intelligence, Barcelona, pp. 2456–2461 (2010)
8. Carvalho, J.P.: On the semantics and the use of fuzzy cognitive maps and dynamic cognitive maps in social sciences. Fuzzy Sets Syst. **214**, 6–19 (2013)
9. Carvalho, J.P., Tomé, J.A.: Fuzzy mechanisms for qualitative causal relations. In: Seising, R. (ed.) Views on Fuzzy Sets and Systems from Different Perspectives. Philosophy and Logic, Criticisms and Applications. Studies in Fuzziness and Soft Computing. Springer, Berlin (2009). Chapter 19
10. Carvalho, J.P., Tomé, J.A.: Rule based fuzzy cognitive maps in socio-economic systems. In: Proceedings of the IFSA-EUSFLAT 2009 - International Fuzzy systems Association World Congress, European Society for Fuzzy Logic and Technology International Conference, pp. 1821–1826 (2009)
11. Carvalho, J.P., Tomé, J.A.: Rule based fuzzy cognitive maps - expressing time in qualitative system dynamics. In: Proceedings of the 2001 FUZZ-IEEE Conference, Melbourne, Australia (2001)
12. Carvalho, J.P., Tomé, J.A.: Rule based fuzzy cognitive maps – qualitative systems dynamics. In: Proceedings of the 19th International Conference of the North American Fuzzy Information Processing Society, NAFIPS 2000, Atlanta, pp. 407–411 (2000)
13. Carvalho, J.P., Wise, L., Murta, A., Mesquita, M.: Issues on dynamic cognitive map modelling of purse-seine fishing skippers behavior. In: Proceedings of the WCCI 2008 – 2008 IEEE World Congress on Computational Intelligence, Hong-Kong, pp. 1503–1510 (2008)
14. Hagiwara, M.: Extended fuzzy cognitive maps. In: Proceedings of IEEE International Conference on Fuzzy Systems, pp. 795–801 (1992)

15. Kosko, B.: Fuzzy cognitive maps. Int. J. Man-Mach. Stud **24**(1), 65–75 (1986)
16. Kosko, B.: Fuzzy Thinking. Hyperion, Santa Clara (1993)
17. Kosko, B.: Neural Networks and Fuzzy Systems: A Dynamical Systems Approach to Machine Intelligence. Prentice-Hall International Editions, Upper Saddle River (1992)
18. Kottas, T.L., Boutalis, Y.S., Christodoulou, M.A.: Fuzzy cognitive network: a general framework. Intell. Decis. Technol **1**, 183–196 (2007)
19. Laukkanen, M.: Conducting causal mapping research: opportunities and challenges. In: Eden, C., Spender, J.-C. (eds.) Managerial and Organisational Cognition. Sage, Thousand Oaks (1998)
20. Miao, Y., Liu, Z., Siew, C., Miao, C.: Dynamical cognitive network - an extension of fuzzy cognitive map. IEEE Trans. Fuzzy Syst. **9**(5), 760–770 (2001)
21. Minsky, M.: Computation: Finite and Infinite Machines, 1st edn. Prentice-Hall, Upper Saddle River (1967)
22. Sipser, M.: Introduction to the Theory of Computation, Second Edition, International Edition, Thomson Course Technology (2006)
23. Wise, L., Murta, A., Carvalho, J.P., Mesquita, M.: Qualitative modelling of fishermen's behaviour in a pelagic fishery. Ecol. Model. **228**, 112–122 (2012)

Fuzzy Set Similarity Between Fuzzy Words

Valerie Cross[(✉)] and Valeria Mokrenko

Computer Science and Software Engineering, Miami University,
Oxford, OH 45056, USA
{crossv,mokrenvi}@miamioh.edu

Abstract. Fuzzy set similarity measures determine the similarity between two fuzzy sets. Semantic similarity measures determine the similarity between concepts within an ontology. Research in determining sentence similarity has used semantic similarity between fuzzy words that have been structured in an ontology based on data provided by human experts to create fuzzy sets for these fuzzy words. The research uses four different kinds of fuzzy set similarity measures, three that are standard ones for type-1 fuzzy sets and another one based on the distance between defuzzified and normalized COGs for type-2 fuzzy sets and examines both the Pearson and Spearman correlations among these different fuzzy set similarity measures on fuzzy words from four of the six categories established in past sentence similarity research. The results show that all standard fuzzy set measures are highly correlated though some categories of fuzzy words result in higher correlations than others. The standard fuzzy set similarity measures have lower correlation with the one developed for similarity between type-2 fuzzy sets using their defuzzified and normalized COGs.

Keywords: Fuzzy set similarity · Sentence similarity · Semantic similarity · Computing with words (CWW) · Geometric distance-based measures · Partial matching measures · Jaccard fuzzy set similarity measure

1 Introduction

Natural language communication between humans and machines is a primary objective inartificial intelligence (AI) research. One aspect that is challenging this research is addressing the vagueness or imprecision inherent in natural language. A method for translating vague or imprecise words is needed for machines to be able to understand human communications. One approach is computing with words (CWW) [1]. CWW uses fuzzy set theory to quantify these words by creating fuzzy set representations for them. These fuzzy sets specify a membership function and the word is referred to as a fuzzy word. CWW techniques can then be used on these fuzzy words to improve communication and understanding between humans and machines. CWW provides a framework by which fuzzy words can be quantified, scaled against each other and then become machine representable.

When trying to understand communication between two entities, a major issue is determining the similarity between the words that are being used by each entity. Overall understanding is based on somewhat agreeing on the meanings of the words being used in the communication. For short text, researchers have developed sentence

© Springer Nature Switzerland AG 2019
R. B. Kearfott et al. (Eds.): IFSA 2019/NAFIPS 2019, AISC 1000, pp. 214–223, 2019.
https://doi.org/10.1007/978-3-030-21920-8_20

similarity measures that incorporate word pair similarity along with other overall sentence similarity measures. An early approach to measure word pair similarity was latent semantic analysis [2]. Statistical analysis is based on the occurrences of the word pairs in the blocks within a large corpus. LSA creates semantic vectors and calculates similarity between these vectors.

With the rise of the Semantic Web, ontologies [3] became a key knowledge representation means. Within ontologies, semantic similarity measures are used to determine the similarity between word pairs [4]. STASIS [5] examined the use of semantic similarity measures within the context of the WordNet ontology to measure text similarity. For short pieces of text, STASIS uses the semantic similarity measure in [6] between each word pair, one word from each text, to create a semantic vector. This semantic measure is calculated as

$$S(w1, w2) = e^{-\alpha l} * \frac{e^{\beta h} - e^{-\beta h}}{e^{\beta h} + e^{-\beta h}}$$

where l represents the path length between the two words in the ontology and h represents the depth of their common subsumer. The STASIS sentence similarity also incorporates corpus statistics within the semantic vector. STASIS, however, did not address vagueness or imprecision in words, i.e., it did not handle fuzzy words.

The researchers' next efforts focused on this challenge and produced FAST (Fuzzy Algorithm for Similarity Test) [7] to measure text similarity that contains fuzzy words. To quantify fuzzy words, it was necessary to develop a data set based on human judgments of similarity. These fuzzy words are organized into six different categories: *age, size/distance, frequency, goodness, membership level* and *temperature*. Using a type-1 fuzzy set representation, a fuzzy word is scaled and an ontology is created from these scaled values, one ontology for each of the six categories. Once an ontology exists, then the semantic similarity measure, found in [6] and also used for STASIS, is used between fuzzy words that are in the same category. For FAST, the parameters α and β were set to 0.2 and 0.6, respectively and were determined empirically. In their research FAST showed an improvement over LSA and STASIS in a comparative experimental study which assessed how well FAST sentence similarity judgments correlated with those of human subjects' sentence similarity judgments.

FUSE [8] (FUzzy Similarity mEasure) took basically the same approach as FAST except type-2 interval fuzzy sets are used since the researchers proposed that type-1 fuzzy sets could not capture the uncertainty of humans [9]. The size of the vocabulary of fuzzy words for FUSE was increased by 57% over that of FAST. To provide more distance between nodes in the ontology and to handle this increased vocabulary, the ontology built from sets became more complex. Instead of only five nodes in an ontology, the ontology uses eleven nodes, and each node has a smaller range of the normalized and scaled COG values representing the fuzzy words. In both FAST and FUSE human evaluators were used to create the fuzzy set representations for the fuzzy words. In their experimental comparison using three different data sets each consisting of sentence pairs, FUSE performed better than STASIS and FAST as measured by correlation to human judgments of sentence similarity.

In this paper, the primary focus is to analyze and compare the results of several different standard fuzzy set similarity measure between type-1 fuzzy sets. These measures are used on type-1 fuzzy sets created from the data collected from the human evaluators for fuzzy words. This approach is simpler than those used in FAST and FUSE, which use semantic similarity within an ontology. Because FUSE uses type-2 fuzzy sets, the study also uses a COG distance based similarity for type 2 interval fuzzy sets on the *size-distance* category. Since the COGs for fuzzy words in the other three categories were not available from the FUSE researchers, only this category could be studied and compared with fuzzy set similarities for type-1 fuzzy sets for this same category.

In a previously submitted paper, the correlation between these fuzzy set similarity measures and the semantic similarity measures produced by FAST and FUSE are reported for 20 different fuzzy word pairs. The study in this paper selects four categories from the six and produces fuzzy set similarity measures between all pairs of fuzzy words in each of the categories. The similarity results are analyzed with respect to both Pearson and Spearman correlations among the fuzzy set similarity measures. Other comparisons are made within and across the similarities for the different categories to determine any patterns in the fuzzy set similarity results.

The paper organization is as follows; Sect. 2 explains how the data was collected and how this data is interpreted and turned into a simple triangular membership function. Section 3 reviews the existing fuzzy set similarity measures used in this study. Section 4 describes the experimental design and the analysis of the results that compare the correlations among the different fuzzy set similarity measures. Finally, Sect. 5 presents the conclusions and future work.

2 Representing Fuzzy Words

In both FAST and FUSE, human evaluators were used to create the fuzzy sets representing the words in each of the six categories. The process used in FAST has several steps. First, the six categories were established as previously specified. Next, human subjects populated each of the six categories with fuzzy words. Then, subjects had to quantify each fuzzy word by specifying a single value, a point in the 0 to 10 scale where the membership function for that fuzzy word would be highest. For each fuzzy word, a mean and a standard deviation were calculated over all the subjects' ratings.

To evaluate the fuzzy set similarity measures, a membership function is created for each fuzzy word. Data from the type-1 fuzzy sets, specifically the defuzzified value or mean and the standard deviation, for each fuzzy word was acquired from the FAST researchers A pseudo triangular fuzzy set was created where the membership degree at the mean value is 1.0. A normal probability density distribution with ±3 standard deviations away from the mean were used for the endpoints of the triangular fuzzy set since 99.7% of the data is within three standard deviations of the mean. Figure 1 shows the membership function for *centre* with a mean of 4.93 and a standard deviation of 0.5.

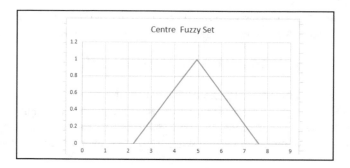

Fig 1. Centre fuzzy set

3 Fuzzy Set Similarity Measures Between Fuzzy Words

The following brief discussion on the three fuzzy set similarity measures is taken from [9]. A detailed examination and thorough review of a variety of fuzzy set similarity measures is in [10].

3.1 Zadeh's Sup-Min

Zadeh's consistency index, also referred to as the sup-min or partial matching index falls into the set-theoretic category of fuzzy similarity measures. It roughly estimates the similarity between two fuzzy sets by finding at what domain values they intersect and determines their similarity by taking the highest membership degree among their intersection points. Given two fuzzy sets A and A', similarity between the two is determined as

$$S_{Zadeh}(A, A') = \sup_{u \in U} T(A'(u), A(u)) \tag{1}$$

where T can be any t-norm, but usually the minimum is used for the t-norm. It is referred to as a partial matching index since it only provides an estimated similarity value between the two fuzzy sets based on a single point shared by the two fuzzy sets. If the two fuzzy sets do not overlap, their similarity is zero. This characteristic of the similarity measure might not be suitable for certain applications, for example, one where a similarity is needed even where there is no overlap between the fuzzy sets.

3.2 Jaccard

The fuzzy Jaccard similarity measure [11] is defined as a fuzzy extension of the Jaccard index between two crisp sets by replacing set cardinality with fuzzy set cardinality. This fuzzy set similarity measure is also in the set theoretic category but provides a more comprehensive view of similarity between the two fuzzy sets. This is because all elements in both fuzzy sets are taken into account not just the one intersection point as in sup-min. Given two fuzzy sets A and A′, similarity between the two is determined as

$$S_{Jaccard}(A, A') = |A \cap A'|/|A \cup A'| \tag{2}$$

The similarity is measured by the proportion of the area of the intersection of the two fuzzy sets to the area of the union of the two fuzzy sets. Although this similarity measure takes a more comprehensive look at the fuzzy sets than sup-min, it has the same characteristic of producing a zero when there is no overlap between the two fuzzy sets.

3.3   Geometric Fuzzy Similarity Based on Dissemblance Index

Set theoretic fuzzy set similarity measures do not consider the distance of the fuzzy set A′ from A. With the geometric fuzzy similarity measure [12], the distance between the two sets is the basis for determining their similarity. This distance is based on the dissemblance index that measures the distance between two real intervals. If V = [v_1, v_2] and W = [w_1, w_2], then

$$DI(V, W) = (|v_1 - w_1| + |v_2 - w_2|)/[2(\beta_2 - \beta_1)] \tag{3}$$

where [β_1, β_2] is an interval that contains both V and W. The factor $2(\beta_2 - \beta_1)$ is necessary to produce a normalized degree of dissemblance such that $0 \leq DI(V, W) \leq 1$. The dissemblance index consists of two components, the left and right sides of each interval and may be generalized to fuzzy intervals.

A pair of boundary functions Land Rand parameters (r_1, r_2, λ, ρ) define a fuzzy interval N. The core of N, the values for which $\mu_N(r) = 1.0$ is the interval [r_1, r_2]. Parameters λ and ρ are used to define the left L and the right R boundary functions and the support of N, the values for which $\mu_N(r) \geq 0$, which is [$r_1 - \lambda$, $r_2 + \rho$]. The L function and the R function define the membership functions for elements in the intervals [$r_1 - \lambda$, r_1] and [r_2, $r_2 + \rho$], respectively. If L is positively sloping and linear and R is negatively sloping and linear then the interval N is a trapezoidal fuzzy membership function. Calculating the fuzzy dissemblance index between A and A′ is done as an integration over α in the range 0 to 1 as

$$fDI(A'(u), A(u)) = \left[\int |L_{A'}(\alpha) - L_A(\alpha)| + |R_{A'}(\alpha) - R_A(\alpha)|d\alpha\right]/[2(\beta_2 - \beta_1)] \tag{4}$$

where [β_1, β_2] is an interval that contains both A′ and A. fDI calculates a dissimilarity measure between the two fuzzy intervals based on a normalized distance. It can be converted into a similarity measure between the fuzzy intervals as

$$S_{GeoSim}(A, A') = 1 - fDI(A(u), A'(u)) \tag{5}$$

With this similarity measure, even though A and A′ may not overlap, a nonzero similarity value is produced since the distance between the two sets is used.

3.4   Similarity Using Type-2 COG Distances

For the type-2 interval fuzzy sets used in FUSE, a single value defuzzified COG was produced by adapting Mendel's footprint of uncertainty (FOU) method [13]. For each word in the *size-distance* category, the COG was determined using the lower FOU and upper FOU. The COGs were then scaled into the range [−1, +1]. To see how well a measure based solely on the distance between these scaled COG values worked, the following simple similarity measure between is also used in this study:

$$S_{Type2-Dist}(A, A') = 1 - abs(COG_{Scaled}(A) - COG_{Scaled}(A'))/2 \qquad (6)$$

The distance between the two defuzzified and normalized centers of gravity is normalized by the size of the scaled interval [−1, +1]. Calculating this similarity measure between pairs of fuzzy words in the *size-distance* category provides a means of determining how well it correlates with the other type-1 fuzzy set similarity measures for this category.

4   Experimental Results and Analysis on Word Categories

In [9] 20 pairs of fuzzy words were used that were taken from sentence pairs in part of the study done to investigate the performance of FAST [7]. The results in [9] showed that the three fuzzy set similarity measures for these 20 pairs had very high correlation with the FUSE's semantic similarity measure used within its 11 node category ontologies. These fuzzy set similarity measures had even higher correlation with FUSE than FAST which used the same semantic similarity measure as FUSE but only had 5 node category ontologies.

Our current study compares three fuzzy set similarity measures on type-1 fuzzy sets using four of the six categories to determine how well these measures correlate with each other on measuring the similarity of fuzzy words. The selected categories are *level of membership, frequency, age,* and *size-distance* each with 21, 25, 30, and 50 fuzzy words, respectively. The similarity between all pairs of words within each of the categories is computed so that each category has a large number of fuzzy word pair similarities for which the results are analyzed. For example, several of the fuzzy words in the *frequency* category at the low end of frequency are *never, barely, scarcely,* and *rarely.* In the medium frequency group, examples are *occasionally, periodically, normally* and *regularly.* In the high end of frequency, examples are *repeatedly, consistently, constantly,* and *always.* To better investigate the results, both the Pearson correlation and the Spearman correlation are calculated. The following table shows the Pearson correlation and the Spearman rank correlation of these three similarity measures for each of the three fuzzy word categories where G-Z is for GeoSim and Zadeh correlation, G-J is for GeoSim and Jaccard correlation and Z-J is for the Zadeh and Jaccard correlations.

Table 1. Pearson and Spearman Correlations between Similarities for each Category

Pearson	Correlation				Spearman	Correlation			
	Age	Level	Frequency	Size-Dist		Age	Level	Frequency	Size-Dist
G-Z	0.96491	0.9769	0.9882719	0.98406	G-Z	0.96906	0.97767	0.993258	0.98948704
G-J	0.91493	0.95914	0.9608052	0.97945	G-J	0.97869	0.99566	0.998996	0.99925188
Z-J	0.83399	0.88936	0.9249872	0.93251	Z-J	0.93009	0.95996	0.999886	0.98574759

Since the Spearman correlation is computed on ranks of the similarity for the fuzzy word pairs, it shows monotonic relationships while the Pearson correlation is computed on true values and, thus, depicts linear relationships. Calculating both can provide more information on the relationships among the fuzzy set similarity measures. The results in Table 1 indicate that all three fuzzy set similarity measures are highly correlated.

For each of the nine cases, three type-1 fuzzy similarity measures over the selected categories, the Spearman correlations are higher than the Pearson correlations, though for the G-Z combination, the difference between the two correlations is very slight, at most around 0.005 for the *age, frequency* and *size-distance* categories and only 0.00077 for the *level* category. This closeness between the two different correlations over all the categories indicates that GeoSim and Zadeh measures, do not distinguish a difference in the type of relationship between these two similarity measures over the fuzzy word pairs. The differences between the two correlations for the other two pairs of similarity measures is higher. For the G-J correlations, the range is 0.0198 for *size-distance* to 0.06376 for *age*. For the Z-J correlations, the range difference is even higher from 0.05324 for *size-distance* to 0.0961 for *age*.

For the Pearson correlations, the G-Z correlations are highest across all the categories when compared to the other fuzzy set similarity pairs' correlations. The Z-J Pearson correlations are the lowest over all the categories. It is interesting to note that GeoSim (G) and Zadeh(Z) fuzzy set similarity measures are very different. GeoSim is a distance based similarity measure between the membership functions while the Zadeh measure is a simple one point agreement where the two fuzzy sets intersect at the highest membership degree. Yet these two have the highest correlation over all categories. The Z-J pair having the lowest Pearson correlations over all the categories appear significant since these values are quite lower than for G-Z and G-J similarity pair correlations. Both Zadeh and Jaccard measures are based on an intersection of the two fuzzy sets where the Zadeh measure looks at only one point in the intersection and the Jaccard measure uses the area of the intersection in its calculation. The GeoSim measure, even though it is distance-based correlates better with both the Zadeh and Jaccard measures which are both based on fuzzy set intersection.

Looking across the various categories of Pearson correlations, the *age* category has the lowest correlations followed by *level* for the three fuzzy set similarity pairs. The highest correlations are for the *size-distance* category for two of the three fuzzy set similarity pairs. For the G-Z pair, however, *size-distance* is about 0.0042 lower than that of the *frequency* category.

For the Spearman correlations, G-J are the highest for three of the four categories. Only for the *frequency* category is Z-J very slightly higher. Interestingly, the Z-J has the lowest correlations for the other three categories, especially lower for the *age* and *level* categories. The *age* category has the lowest Spearman correlations over the four categories. The *frequency* category has higher correlations over all similarity pairs than those for two of the three other word categories. For the G-J pair, however, *frequency* correlation is around 0.0003 lower than the *size-distance* category.

To summarize, across fuzzy-set similarity pairs, the GeoSim measure in general has the highest correlations for both Pearson and Spearman with Zadeh and Jaccard measures respectively except for *frequency* category where Z-J has a slightly higher Spearman correlation. The Z-J similarity pair, however, has the lowest Pearson correlations for all categories and the lowest Spearman correlation for three of the four categories, i.e., the *frequency* category just previously mentioned where it was ever so slightly the highest.

Across categories, for both Pearson and Spearman, the correlations are lowest for *age* followed by *level*. The highest are *size-distance* for Pearson except for *frequency* for the G-Z pair and for Spearman, the highest is *frequency* except for *size-distance* for the G-J pair.

The fourth fuzzy set similarity measure that uses the distances between the COGs for type 2 interval fuzzy sets was only used with the *size-distance* category. The defuzzified and normalized COGs for the fuzzy words in the other three categories were not provided by the FUSE researchers. Below are the correlations of the three fuzzy set similarity measures used on the type 1 fuzzy sets with the type-2 COG distance based fuzzy set similarity (T) on type-2 interval fuzzy sets for this category (Table 2).

Table 2. Pearson and Spearman Correlation for Type2Dist measure and Other Three (only for the *size-distance* category)

Pearson	Correlation size-dist	Spearman	Correlation size-dist
G-T	0.7893243	G-T	0.8072024
Z-T	0.7990748	Z-T	0.8109709
J-T	0.7509547	J-T	0.8066398

All correlations in this table are much lower than those in Table 1. The Spearman correlations are higher than those of Pearson correlations as seen in Table 1 but with a smaller range of difference from 0.012 for Z-T to 0.057 for J-T than those found for correlations among the three standard fuzzy set measures. The Zadeh measure correlates the highest with the type-2 COG distance based measure for both Pearson and Spearman. The Jaccard measure has the lowest correlations but only slightly lower for the Spearman correlations for the type-2 COG distance based measure. The lower correlations in general with the type-2 COG distance based are expected. The similarity measurement is occurring on two different kinds of fuzzy sets, type-1 for the three standard fuzzy set similarity measures and type-2 interval fuzzy sets for the COG distance based measure. The consistent part is that Spearman correlations are higher as seen in Table 1 and both GeoSim and Zadeh correlate better with the COG distance-based measure than the Jaccard measure.

5 Conclusions

This study examines the correlations among three standard fuzzy set similarity measures using the type-1 fuzzy sets created for the fuzzy words established for four of the six different categories and used in both the FAST and FUSE research. Both Pearson and Spearman correlations are provided for pairwise comparison of these three measure. The results show that these fuzzy set similarity measures are highly correlated for this task. The Pearson correlations are above 0.91 for most pairs of these measures except for the Zadeh and Jaccard pair. Measure for two categories, *age* and *level*, however, are still above 0.83. The Spearman correlations are higher than the Pearson ones and are above 0.93 for all pairs of these measures with the GeoSim and Jaccard pair having the highest except for the *frequency* category. In general, the geometric distance based GeoSim correlates better with the two intersection-based measures Zadeh and Jaccard than these two correlate with each other.

Both the Pearson and Spearman correlations are in general highest for the *frequency* category followed by the *size-distance* category and lowest for *age* followed by *level*. This suggests that further research and study needs to be done to determine the differences in the spread and distribution of fuzzy words in these lower correlation categories. Possibly the difference in the number of words in each category has an effect since *level* and *age* have 21 and 30 words, respectively where *size-distance* 50 words.

For the *size-distance* category, both Pearson and Spearman correlations of the three standard fuzzy set similarity measures with the type-2 distance based similarity measures are low when compared to those correlations among the three standard ones. They range from 0.751 to 0.811, and the differences between the two different kinds of correlation are much smaller than for those between the three standard measures. It is believed that the difference in similarity between type-1 fuzzy sets and similarity between the type-2 interval fuzzy sets just using a COG distance is a major factor in the low correlations of the standard fuzzy set measures with the type-2 COG distance measure.

Further work is planned to experiment with replacing the semantic similarity measures in FAST and FUSE with these fuzzy set similarity measures and determine how well they perform in the task of measuring sentence similarity. FAST and FUSE have established sentence pairs with human judgments of their sentence similarity. By performing experiments with the test sets of sentence pairs, a comparison can be made between the performance of fuzzy set similarity measures and the semantic similarity measure used in both FAST and FUSE in measuring sentence similarity. These fuzzy set similarity measures are much easier to calculate than the process of creating ontologies for the fuzzy word categories in order to use a semantic similarity measure. If one or more of these fuzzy set similarity measures have the same or better performance than the semantic similarity measure, the step of creating ontologies for the various categories of fuzzy words is not needed to measure sentence similarity.

Acknowledgment. This continued research is based on the data provided for the previous research in [9] by researchers Keeley Crockett and Naeemeh Adel. We want to thank them for their assistance in making this research possible.

References

1. Zadeh, L. (ed.): Computing with Words in Information/Intelligent Systems. Springer, Berlin (1999)
2. Landauer, T., Foltz, P., Laham, D.: An introduction to latent semantic analysis. Discourse Processes 25(3), 259–284 (1998)
3. Gruber, T.R.: A translation approach to portable ontology specifications. Knowl. Acquis. 5 (2), 199–220 (1993)
4. Budanitsky, A., Hirst, G.: Evaluating WordNet-based measures of lexical semantic relatedness. Comput. Linguist. 32, 13–47 (2006)
5. Li, Y., Mclean, D., Bandar, Z., O'Shea, J., Crockett, K.: Sentence similarity based on semantic nets and corpus statistics. IEEE Trans. Knowl. Data Eng. 18(8), 1138–1150 (2006)
6. Li, Y., Bandar, Z., McLean, D.: An approach for measuring semantic similarity between words using multiple information sources. IEEE Trans. Knowl. Data Eng. 15(4), 871–882 (2003)
7. Chandran, D., Crockett, K.A., McLean, D., Bandar, Z.: FAST: a fuzzy semantic sentence similarity measure. In: International Conference on Fuzzy Systems, FUZZ-IEEE (2013)
8. Adel, N., Crockett, K.A., Crispin, A., Chandran, D., Carvalho, J.P.: FUSE (Fuzzy Similarity Measure) - a measure for determining fuzzy short text similarity using Interval Type-2 fuzzy sets. In: International Conference on Fuzzy Systems, FUZZ-IEEE, pp. 1–8 (2018)
9. Cross, V., Mokrenko, V., Krockett, K., Adel, N.: Ontological and fuzzy set similarity between perception-based words. Submitted to Fuzz-IEEE (2019)
10. Cross, V.: An analysis of fuzzy set aggregators and compatibility measures, 264 p. Ph.D. Dissertation, Computer Science and Engineering, Wright State University, Dayton, OH, March 1993
11. Jaccard, P.: The distribution of the flora in the alpine zone. New Phytol. 11, 37–50 (1912)
12. Cross, V., Sudkamp, T.: Geometric compatibility modification. Fuzzy Sets Syst. 84(3), 283–299 (1996)
13. Hao, M., Mendel, J.M.: Encoding words into normal interval type-2 fuzzy sets: HM approach. IEEE Trans. Fuzzy Syst. 24(4), 865–879 (2016)

Simulating the Behaviour of Choquet-Like (pre) Aggregation Functions for Image Resizing in the Pooling Layer of Deep Learning Networks

Camila Dias[1]([⊠]), Jessica Bueno[1], Eduardo Borges[1], Giancarlo Lucca[1], Helida Santos[1], Graçaliz Dimuro[1], Humberto Bustince[2], Paulo Drews Jr.[1], Silvia Botelho[1], and Eduardo Palmeira[3]

[1] Centro de Ciências Computacionais, Universidade Federal do Rio Grande, Rio Grande, Brazil
cmdias@outlook.com.br, jessica_bsaldivia@hotmail.com, gracaliz@gmail.com
[2] Dpt. Estad., Informatica y Matematicas, ISC, Universidad Publica de Navarra, Pamplona, Spain
bustince@unavarra.es
[3] Dpt. Ciências Exatas e Tecnológicas, Universidade Estadual de Santa Cruz, Ilhéus, Brazil
espalmeira@uesc.br

Abstract. The data volume expansion has generated the need to develop efficient knowledge extraction techniques. Most problems that are processed by these techniques have complex information to be identified and use different machine learning methods, such as Convolutional and Deep Learning Network. These networks may use a variety of aggregation functions to resize images in the pooling layer. This paper presents a study of the application of aggregation functions based on the generalizations of the Choquet integral, namely, the novel Choquet-like (pre) aggregation functions, in image dimensional reduction, simulating the pooling layer of a Deep Learning Networks. This paper is the natural evolution of the initial study where only the standard Choquet integral was applied. We compare the behaviour of such functions with the usual ones used in the literature, namely, the maximum and the arithmetic mean. A quantitative evaluation is done over an image dataset by using different image quality measures to compare the results.

1 Introduction

In recent years, mainly due to the web revolution, the volume of data available for analysis has grown substantially. This huge data volume has generated the need to develop more efficient knowledge extraction techniques [7].

The techniques applied to solve these problems, such as genetic algorithms and Convolutional Neural Networks (CNNs), use several functions inside different layers that are employed in their architectures. Deep Learning Network

© Springer Nature Switzerland AG 2019
R. B. Kearfott et al. (Eds.): IFSA 2019/NAFIPS 2019, AISC 1000, pp. 224–236, 2019.
https://doi.org/10.1007/978-3-030-21920-8_21

(DLN) has gained greater attention, standing out as a new area of machine learning research [10]. Deep learning can be broadly defined as: *"a class of machine learning techniques that analyze many layers of nonlinear data processing for extraction and change of supervised or unsupervised aspects and for recognition and classification of paradigms"* [9].

In the context of DLNs, the pooling layer is responsible to reduce the spatial size of the representation, in order to reduce the amount of parameters and computation in the network, and it uses aggregation functions such as the *maximum and arithmetic mean* to perform the data dimensional reduction [31].

However, recent papers in the literature have discussed the Choquet integral [8] as an aggregation function that is able to consider the relationship among the data to be aggregated [26]. Then, in order to improve the aggregation of meaningful information without degrading its discriminative power in image processing within the pooling layer in DLNs, we propose to replace the common functions in the literature by some generalizations of the Choquet integral [8], that is, Choquet-like (pre) aggregation functions [23–26], following the initial work in [11] regarding the standard Choquet integral. Such generalizations have presented excellent performance in fuzzy rule based classification systems [19].

The general purpose is to analyze if the generalizations of the Choquet integral may outperform the *maximum function* in the context of image restoration using DLNs, inside the pooling layer. This is a long term objective.

As a short term objective, it is desirable to analyse whenever the generalizations of the Choquet integral may perform well in image resizing outside the network. That is, we are going to simulate the pooling layer, developing an experimental study where some generalizations of the Choquet integral are used to replace the maximum and the mean in the pooling layer. A quantitative evaluation is done over an image dataset (IIIT 5K-Word dataset [27]) by using different image quality measures to compare the results with the usual pooling functions.

This article is organized as follows. In Sect. 2 we present the generalizations of the Choquet integral and Sect. 3 discusses the image processing in the context of image resizing. In Sect. 4, the process of resizing images is detailed. In Sect. 5, the analysis and experimental results are discussed. Section 6 is the Conclusion.

2 Generalizations of the Choquet Integral

Aggregation is a process of combining different numeric values returning a single value. The operator that performs this task is called an aggregation function [16].

Definition 21. *A function* $A : [0,1]^n \to [0,1]$ *is said to be an* n*-ary* aggregation function *whenever the following conditions hold:* **(A1)** *Boundary Conditions:* $A(0,\ldots,0) = 0$ *and* $A(1,\ldots,1) = 1$; **(A2)** *Monotonicity:* A *is non-decreasing in each argument:* $A(x_1,\ldots,x_n) \leq A(y_1,\ldots,y_n)$ *whenever* $x_i \leq y_i$, *for all* $i \in \{1,\ldots,n\}$.

Sometimes a function need not to be increasing in all domain to perform well in applications:

Definition 22 [6]. *Let $\vec{r} = (r_1, \ldots, r_n)$ be a real n-dimensional vector, $\vec{r} \neq \vec{0}$. A function $FA : [0,1]^n \to [0,1]$ is directionally increasing with respect to $\vec{r}$ ($\vec{r}$-increasing, for short) if for all $(x_1, \ldots, x_n) \in [0,1]^n$ and $c > 0$ such that $(x_1 + cr_1, \ldots, x_n + cr_n) \in [0,1]^n$ it holds that $F(x_1 + cr_1, \ldots, x_n + cr_n) \geq F(x_1, \ldots, x_n)$. Similarly, one defines an $\vec{r}$-decreasing function.*

Definition 23 [26]. *Let $\vec{r} = (r_1, \ldots, r_n)$ be a real n-dimensional vector, $\vec{r} \neq \vec{0}$. A function $PA : [0,1]^n \to [0,1]$ is said to be an n-ary $\vec{r}$-pre-aggregation function if: **(PA1)** F is $\vec{r}$-increasing; **(PA2)** F satisfies the boundary conditions: **(A2)** (i) and (ii).*

Definition 24 [22]. *An aggregation function $T : [0,1]^2 \to [0,1]$ is a t-norm if the following conditions hold, for all $x, y, z \in [0,1]$: **(T1)** Commutativity: $T(x,y) = T(y,x)$; **(T2)** Associativity: $T(x, T(y,z)) = T(T(x,y), z)$; **(T3)** Boundary condition: $T(x,1) = x$.*

Definition 25 [1]. *An aggregation function $C : [0,1]^2 \to [0,1]$ is a copula if it satisfies the following conditions, for all $x, x', y, y' \in [0,1]$ with $x \leq x'$ and $y \leq y'$: **(C1)** $C(x,y) + C(x',y') \geq C(x,y') + C(x',y)$; **(C2)** $C(x,0) = C(0,x) = 0$; **(C3)** $C(x,1) = C(1,x) = x$.*

Definition 26. *A function $F : [0,1]^n \to [0,1]$ is called* idempotent *if for any $x \in [0,1]$, $F(x,x,\ldots,x) = x$. F is said to be* averaging, *if $\min \boldsymbol{x} \leq F(\boldsymbol{x}) \leq \max \boldsymbol{x}$, for all $\boldsymbol{x} \in [0,1]^n$.*

In the context of aggregation functions, idempotency and averaging behavior are equivalent concepts. However, this is not true for pre-aggregation functions.

Some other properties may be required for aggregation functions $F : [0,1]^2 \to [0,1]$: **(LAE)** F is said to be left 0-absorbent: $\forall y \in [0,1] : F(0,y) = 0$; **(RNE)** Right Neutral Element: $\forall x \in [0,1] : F(x,1) = x$.

The most common aggregation functions used in applications such as classification and image processing are the maximum [21,26], the minimum t-norm [12,17], the product t-norm [26], the arithmetic mean [32] and copulas in general [24].

In DLNs aggregation functions are used in the pooling layer. The data received at the pooling layer [14] of a DLN are sub-scaled from small regions to produce a map of smaller features as input to the next level of the network. Currently, the most used pooling aggregation function are the arithmetic mean and the maximum [15]. The max pooling usually presents better results when compared to the mean pooling.

Definition 27. *Let $N = \{1, \ldots, n\}$ and 2^N be the set of subsets of N. The function $\mathfrak{m} : 2^N \to [0,1]$ is a* fuzzy measure *if, for the whole set $A, B \subseteq N$, the following conditions hold: **(m1)** Boundary Conditions: $\mathfrak{m}(\emptyset) = 0$ and $\mathfrak{m}(N) = 1$; **(m2)** Monotonicity: $\mathfrak{m}(A) \leq \mathfrak{m}(B)$ whenever $A \subseteq B$.*

The fuzzy measure adopted in this work is the power measure [3], defined by

$$\mathfrak{m}_P(A) = \left(\frac{|A|}{n} \right)^q, \text{ where } q > 0. \tag{1}$$

The value of q can be defined by an expert or learned using some genetic algorithm, which is able to determine the best value for it according to the problem being treated.

Definition 28. *Let* $\mathfrak{m} : 2^N \rightarrow [0,1]$ *be a fuzzy measure. The* discrete Choquet integral *of* $\boldsymbol{x} \in [0,1]^n$ *with respect to a fuzzy measure* $\mathfrak{m}$ *is a function* $\mathfrak{C}_\mathfrak{m}$: $[0,1]^n \rightarrow [0,1]$, *defined by:*

$$\mathfrak{C}_\mathfrak{m}(\boldsymbol{x}) - \sum_{i=1}^{n} \left(x_{(i)} - x_{(i-1)} \right) \cdot \mathfrak{m}(A_{(i)}), \tag{2}$$

where $\vec{x} = (x_{(1)}, \ldots, x_{(n)})$ *is a non-decreasing permutation of the input* $\boldsymbol{x}$, *that is,* $0 \leq x_{(1)} \leq \ldots \leq x_{(n)}$, $x_{(0)} = 0$, *and* $A_{(i)} = \{(i), \ldots, (n)\}$ *is a subset of the indices of* $n - i + 1$ *largest components of* $\vec{x}$.

The Choquet integral (2) satisfies the conditions of Definition 21, then it is an aggregation function. The Choquet integral can be written in the expanded form:

$$\mathfrak{C}_\mathfrak{m}(\mathbf{x}) = \sum_{i=1}^{n} \left(x_{(i)} \cdot \mathfrak{m}(A_{(i)}) - x_{(i-1)} \cdot \mathfrak{m}(A_{(i)}) \right). \tag{3}$$

The Choquet integral can be generalized by copulas C, generating a family of aggregation functions called CC-integrals [24]. This generalization is constructed by changing the operator of the product of the classic Choquet integral in its expanded form (3) by a copula C.

Definition 29 [24]. *Take a fuzzy measure* $\mathfrak{m} : 2^N \rightarrow [0,1]$ *and a copula* C : $[0,1]^2 \rightarrow [0,1]$. *The Choquet integral based on the* copula C *and with respect to the fuzzy measure* $\mathfrak{m}$ *is the function* $C_\mathfrak{m}^C : [0,1]^n \rightarrow [0,1]$, *called CC-integral, defined by*

$$C_\mathfrak{m}^C(\boldsymbol{x}) = \sum_{i=1}^{n} C\left(x_{(i)}, \mathfrak{m}(A_{(i)}) \right) - C\left(x_{(i-1)}, \mathfrak{m}(A_{(i)}) \right). \tag{4}$$

Applying the minimum (which is a t-norm and a copula) in Eq. (4), one obtains the $C_{\min}$-integral [12]:

$$C_\mathfrak{m}^{\min}(\mathbf{x}) = \sum_{i=1}^{n} \left(\min\{x_{(i)}, \mathfrak{m}(A_{(i)})\} - \min\{x_{(i-1)}, \mathfrak{m}(A_{(i)})\} \right). \tag{5}$$

Definition 210. *Let* $\mathfrak{m} : 2^N \to [0,1]$ *be a fuzzy measure and* $M : [0,1]^2 \to [0,1]$ *be a function such that* $M(x,0) = 0$ *for all* $x \in [0,1]$. *In this way, the* classic Choquet integral *extends for the function given by:*

$$\mathfrak{C}_{\mathfrak{m}}^{M}(x) = \sum_{i=1}^{n} M\left(x_{(i)} - x_{(i-1)}, \mathfrak{m}(A_{(i)})\right). \tag{6}$$

Theorem 21 [26]. *Let* $M : [0,1]^2 \to [0,1]$ *be a function such that for all* $x, y \in [0,1]$ *satisfies* $M(x,y) \le x$, $M(x,1) = x$, $M(0,y) = 0$ *and* M *is* $(0,1)$-*non-decreasing. So, for some fuzzy measure* $\mathfrak{m}$, *classic Choquet integral is an idempotent and averaging pre-aggregation function.*

As an example, consider the Hamacher product, which is a t-norm and also a copula $T_{Ham} : [0,1]^2 \to [0,1]$ defined as follows:

$$T_{Ham}(x,y) = \begin{cases} 0 & \text{if } x = y = 0 \\ \dfrac{xy}{x+y-xy} & \text{otherwise} \end{cases} \tag{7}$$

Then, the t-norm T_{Ham}, considering the Definition 210, can be applied to Theorem 21, where T_{Ham} satisfies such conditions and we obtain the following idempotent and averaging pre-aggregation function:

$$C_{\mathfrak{m}}^{T_{Ham}}(\vec{x}) = \sum_{i=1}^{n} \begin{cases} 0, \text{ if } x_{(i)} = x_{(i-1)} \text{ and } \mathfrak{m}(A_{(i)}) = 0 \\ \dfrac{\left(x_{(i)} - x_{(i-1)}\right) \cdot \mathfrak{m}(A_{(i)})}{x_{(i)} - x_{(i-1)} + \mathfrak{m}(A_{(i)}) - \left(x_{(i)} - x_{(i-1)}\right) \cdot \mathfrak{m}(A_{(i)})}, \text{ otherwise} \end{cases} \tag{8}$$

In the following, we present the method for constructing a family of pre-aggregation functions defined by generalizing the discrete Choquet Integral using left 0-absorbent functions $\mathfrak{F} : [0,1]^2 \to [0,1]$, obtaining the so-called $\mathfrak{C}_{\mathfrak{F}}$-integrals.

Definition 211. *Let* $F : [0,1]^2 \to [0,1]$ *be a bivariate function and* $\mathfrak{m} : 2^N \to [0,1]$ *be a fuzzy measure. The* Choquet-like integral based on F with respect to $\mathfrak{m}$, *called* C_F-integral *is the function* $C_{\mathfrak{m}}^{F} : [0,1]^n \to [0,1]$, *defined, for all* $x \in [0,1]^n$, *by*

$$\mathcal{C}_{\mathfrak{m}}^{F}(\vec{x}) = \min\left\{1, \sum_{i=1}^{n} F\left(x_{(i)} - x_{(i-1)}, \mathfrak{m}\left(A_{(i)}\right)\right)\right\}. \tag{9}$$

Theorem 22 [25]. *For any fuzzy measure* $\mathfrak{m} : 2^N \to [0,1]$ *and left 0-absorbent* **(RNE)***-function* $F : [0,1]^2 \to [0,1]$, $\mathfrak{C}_{\mathfrak{m}}^{F}$ *is a* $\vec{1}$-*pre-aggregation function.*

For example, using the left 0-absorbent function

$$F_{NA}(x,y) = \begin{cases} x, & \text{if } x \le y \\ \min\left\{\dfrac{x}{2}, y\right\}, & \text{otherwise} \end{cases} \tag{10}$$

we obtain the following C_F-integral

$$\mathcal{C}_{\mathrm{m}}^{FNA}(\vec{x}) = \min\left\{1, \sum_{i=1}^{n}\left\{\begin{array}{ll} x_{(i)} - x_{(i-1)}, & \text{if } x_{(i)} - x_{(i-1)} \leq \mathrm{m}(A_{(i)}) \\ \min\left\{\dfrac{x_{(i)} - x_{(i-1)}}{2}, \mathrm{m}(A_{(i)})\right\}, & \text{otherwise} \end{array}\right.\right\}$$

(11)

3 Image Processing in the Context of Image Resizing

The state-of-the-art of digital image processing techniques presents significant opportunities for treatment and analysis. There are several applications made especially for the study of terrestrial resources [28], industrial and environmental areas [20], for example.

Image reduction technique is commonly used to speed up processing or to reduce the storage or transmission cost [4,30]. In the context of DLNs, it is applied in the pooling layer, using the maximum or the arithmetic mean to aggregate pixel values of an image. The maximum draws the most important features and captures the strongest activation, disregarding all other values in the pooling area. The arithmetic mean takes into account all activations in a pooling area with equal contributions [32].

In this paper, alternative (pre) aggregation functions are proposed considering some generalizations of the Choquet integral [8]. As the standard Choquet integral, such generalizations are defined based on a fuzzy measure, which represents the degree of relationship between the elements to be aggregated [5]. In this way, the Choquet integral's significance occurs due to its model to consider the importance of each attribute to be aggregated, as well as their interactions.

The input dataset of this paper is composed by images that simulate the pooling layer of a CNN only with the filter application, that is, without applying the whole network architecture.

Related to image processing, the values to be aggregated are the pixels between 0 and 255 displayed in a window, instead of an array or vector of numerical values. These values were normalized between 0 and 1 to use the Choquet integral and its generalizations (3), (5), (8), (11). Thus, in this sense, it is understood that the more information of the relationship among the pixels to be aggregated, the better is the resulting image. The classic Choquet integral, considering image processing, can capture information in predominantly darker regions of an image, since it requires ordering by the difference of the value of the pixels in the neighborhood [11]. In addition, to facilitate the analysis of the results, the input images are converted to grayscale. In order to evaluate the resulting image quality, different measures [13] are applied.

The resizing is then used to return the same size of the input image. To perform a resizing, the nearest function in Matlab software is applied. The technique is used after aggregation with the purpose of comparing the input image with the resulting image, since it is only possible to apply image quality measures once the input image has the same size as the output image. We observe that this resizing method

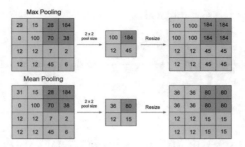

Fig. 1. The process steps of maximum and mean aggregation and resizing of the output image.

is not the best one that may be found in the literature, such as image magnification using interval information [2]. However, since the resizing is not the focus of this work, we decided to use it due to its simplicity (Fig. 1).

4   Problem Formulation and Methodology

We propose to use the Choquet integral and its generalizations using several parameters (stride, window size and fuzzy power measure exponent value (q) demonstrated in Eq. 1) chosen by experts.

After performing the experiments with the different aggregation functions, image quality measurements are applied. To perform that, seven quality measures [13] were adopted such as: Average Difference (AD), Structural Content (SC↓), Normalized Cross-Correlation (NK↑), Maximum Difference (MD↓), Normalized Absolute Error (NAE↓), Mean Squared Error (MSE↓) and Peak Signal to Noise Ratio (PSNR↑). These measures are intended to evaluate how much the output image matches the input image. The arrow ↑ denotes the higher the value the higher the quality, i.e. how similar the images are. In contrast, ↓ means the lower the value, the better the quality of the output image.

Consider an image $M \times N$, where M is the number of rows and N the number of columns. Besides, $P(i,j)$ represents a pixel of the original image and $\widehat{P}(i,j)$ represents resized modified image pixel.

These measures are defined below:

$$AD = \frac{1}{MN} \sum_{i=1}^{M} \sum_{j=1}^{N} \left(P(i,j) - \widehat{P}(i,j) \right), \tag{12}$$

$$SC = \frac{\sum_{i=1}^{M} \sum_{j=1}^{N} (P(i,j))^2}{\sum_{i=1}^{M} \sum_{j=1}^{N} \left(\widehat{P}(i,j) \right)^2}, \tag{13}$$

$$NK = \frac{\sum_{i=1}^{M} \sum_{j=1}^{N} \left(P(i,j) \times \widehat{P}(i,j) \right)}{\sum_{i=1}^{M} \sum_{j=1}^{N} (P(i,j))^2}, \tag{14}$$

$$MD = Max|P(i,j) - \widehat{P}(i,j)|, \text{ for } i \in \{1, 2, ..., M\} \text{ and } j \in \{1, 2, ..., N\}. \quad (15)$$

$$NAE = \frac{\sum_{i=1}^{M} \sum_{j=1}^{N} |O\left((P(i,j)) - O\left(\widehat{P}(i,j)\right)\right)|}{\sum_{i=1}^{M} \sum_{j=1}^{N} |O\left(P(i,j)\right)|}, \quad (16)$$

$$MSE = \frac{1}{MN} \sum_{i=1}^{M} \sum_{j=1}^{N} \left(P(i,j) - \widehat{P}(i,j)\right)^2, \quad (17)$$

$$PSNR = 10 \log_{10} \frac{(2^n - 1)^2}{\sqrt{MSE}}, \quad (18)$$

When the quality measures are computed, hypothesis tests are applied for a statistical analysis of the results, where the purpose of the statistical analysis is to verify if there is statistical significance between the treatments applied in the data [29]. For this, two hypotheses are presented:

- Null Hypothesis: H_0, if this hypothesis is rejected, the alternative hypothesis is accepted;
- Alternative Hypothesis: H_1.

In this paper, non-parametric tests were applied, since there is no warranty of data normality and homogeneity. The non-parametric Friedman test [18] is applied to point statistical differences between a group of results, that is, between the aggregation functions used.

Let μ be the median of the image quality measure of each aggregator. The Friedman hypothesis test is formulated as follows:

- Null Hypothesis: $\mu_1 = \mu_2 = ... = \mu_6$
- Alternative Hypothesis: there is difference between some of the medians compared.

In order to perform the hypothesis tests, the level of significance was defined as 5%, that is, $\alpha = 0.05$ means it has maximum 5% of chance to reject the null hypothesis if it is true. The measure adopted to accept or reject the hypothesis is the $p - valor$, where if $p - valor < \alpha$ the null hypothesis is rejected, in case $p - valor > \alpha$ the null hypothesis is accepted. It is emphasized that all the experiments were executed in Matlab software.

5   Results and Discussion

In order to compare the maximum, arithmetic mean, classic Choquet integral and its generalizations: $\mathfrak{C}_m^{FNA}$, $\mathfrak{C}_m^{min}$ and $\mathfrak{C}_m^{THam}$, we used 12 images from the IIIT 5K-Word dataset [27].

For each of these 12 images, experiments were performed considering the parameters: window size (2×2, 3×3 and 4×4) and stride (2 and 3). We observe that using different window sizes and strides, it allows to capture interaction characteristics between multiple windows. In addition, the parameter q of the

Table 1. Average results obtained from the image quality measures among the 12 images and all the 6 variants (window size: 2×2, 3×3 and 4×4; and stride: 2 and 3). The letters shown as exponent mean the equality or statistical difference of the results. The results with letters in common are not statistically different. In the case of the NK and SC measurements the maximum function did not obtain different statistics of the best results.

Quality measure	$\mathfrak{C}_m$	$\mathfrak{C}_m^{THam}$	$\mathfrak{C}_m^{min}$	$\mathfrak{C}_m^{FNA}$	Max	Mean
AD (q=0.1)	-28.19	-28.25	$\mathbf{-18.54}^b$	-25.87	-28.17	$\mathbf{1.49}^a$
↓ MD (q=0.1)	150.69	$\mathbf{149.35}^b$	188.47	162.94	$\mathbf{134.31}^a$	182.67
↓ MSE (q=0.7)	$\mathbf{2390}^a$	3421	2581	3865	4104	$\mathbf{1494}^a$
↓ NAE (q=0.7)	$\mathbf{0.31}^b$	0.35	0.34	0.37	0.35	$\mathbf{0.20}^a$
↑ NK (q=0.1)	$\mathbf{1.14}^a$	$\mathbf{1.13}^a$	1.06	1.12	1.12^a	0.91
↑ PSNR (q=0.7)	$\mathbf{14.95}^b$	13.71	14.53	13.13	13.23	$\mathbf{17.22}^a$
↓ SC (q=0.1)	$\mathbf{0.70}^a$	$\mathbf{0.70}^a$	0.83	0.72	0.72^a	1.10

fuzzy measure was varied, chosen by a specialist, applying the values 0.1, 0.3, 0.5 and 0.7.

We have applied seven quality measures: AD (the result should be closer to zero), SC ↓, NK ↑, MD ↓, NAE ↓, MSE ↓ and PSNR ↑.

A statistical analysis was performed on the means (Table 1) to verify if there is statistically difference between the results.

As it can be seen in Table 1 the best results for AD, MD and MSE were mean, maximum and mean respectively. The second best results for the same image quality measures were $\mathfrak{C}_m$, $\mathfrak{C}_m^{min}$, $\mathfrak{C}_m^{THam}$ and $\mathfrak{C}_m^{FNA}$. Statistically they did not tie. In the case of MSE results the classic Choquet integral exceeded the maximum function, which is the most used in DLNs applications. In measurements of image quality NAE and NK the best results were arithmetic mean and $\mathfrak{C}_m$ and secondly the best results were the $\mathfrak{C}_m$ and the generalization that consider the $\mathfrak{C}_m^{THam}$. In the case of the NAE measurement it takes the same comparison of the MSE measurement.

It can be observed that for the MD measure most of the results of the maximum arithmetic function will be better than the other aggregation and preaggregation functions. As well the AD measure, the mean will always stand out.

There was no statistically significant difference among the $\mathfrak{C}_m$, $\mathfrak{C}_m^{THam}$ and the Maximum functions. Finally, in the last two functions used, PSNR and SC, the best results were arithmetic mean and in the case of SC image quality measure $\mathfrak{C}_m$ and $\mathfrak{C}_m^{THam}$ obtained the lowest results, standing out from the others. In the case of the PSNR measure there is a statistical difference and in SC's case the $\mathfrak{C}_m$ and $\mathfrak{C}_m^{THam}$ functions are equal statistically, jointly with the maximum function.

The functions derived from the Choquet integral presented the best visual details in the experimental results (the Choquet integral function in Fig. 2 and the $\mathfrak{C}_m^{min}$ function in Fig. 3). It can be observed in the resulting images that the standard Choquet integral and its considered generalizations are more representative to identify the object (in this case the words described in the image), leaving the image less dirty, blurry or serrated.

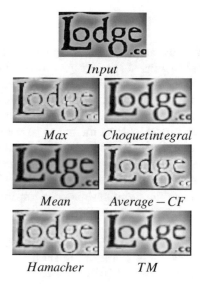

Fig. 2. Results of max, mean, Choquet integral and its generalizations obtained through the experiments applied to one of the images of the IIIT 5K-Word data set called 138_6 (after resizing process). The parameters used were: window size $= 4 \times 4$, stride $= 2$ and power fuzzy measure exponent $= 0.7$

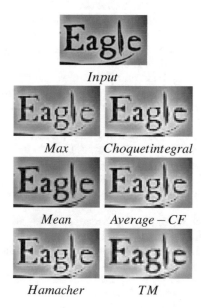

Fig. 3. Results of max, mean, Choquet integral and its generalizations obtained through the experiments applied to one of the images of the IIIT 5K-Word data set called 127_3 (after resizing process). The parameters used were: window size $= 2 \times 2$, stride $= 2$ and power fuzzy measure exponent $= 0.1$

6   Conclusion

This paper presented a study of the implementation of (pre) aggregation functions based on the generalizations of the Choquet integral for image resizing, simulating their behaviour in the DLN pooling layer. We compared the behaviour of such functions with the usual ones used in the literature, namely, the maximum and the arithmetic mean. A quantitative evaluation was done over an image dataset by using different image quality measures to compare the results.

Future work is immediately concerned with the use of such functions directly in the pooling layer of a DLN, with an additional feature that is to use the net to learn the fuzzy measure directly from the data and check whether the high diversity of the trained classifiers can estimate or improve an ensemble formed by the set.

Acknowledgments. Supported by CAPES/Brasil, CNPq/Brazil (proc. 305882/2016-3), FAPERGS (TO 17/2551-0000872-3) and the Spanish Ministry of Science and Technology (under project TIN2016-77356-P (AEI/FEDER, UE)).

References

1. Alsina, C., Frank, M.J., Schweizer, B.: Associative Functions: Triangular Normsand Copulas. World Scientific, Singapore (2006)
2. Jurio, A., Pagola, M., Mesiar, R., Beliakov, G., Bustince, H.: Image magnification using interval information. IEEE Trans. Image Process. **20**(11), 3112–3123 (2011)
3. Barrenechea, E., Bustince, H., Fernandez, J., Paternain, D., Sanz, J.A.: Using the Choquet integral in the fuzzy reasoning method of fuzzy rule-based classification systems. Axioms **2**(2), 208–223 (2013)
4. Beliakov, G., Bustince, H., Paternain, D.: Image reduction using means on discrete product lattices. IEEE Trans. Image Process. **21**(3), 1070–1083 (2012)
5. Beliakov, G., Sola, H.B., Sanchez, T.C.: A Practical Guide to Averaging Functions. Springer, Cham (2016)
6. Bustince, H., Fernandez, J., Kolesárová, A., Mesiar, R.: Directional monotonicity of fusion functions. Eur. J. Oper. Res. **244**(1), 300–308 (2015)
7. Chen, H., Chiang, R., Storey, V.: Business intelligence and analytics: from big data to big impact. MIS Q. **36**(4), 1165–1188 (2012)
8. Choquet, G.: Theory of capacities. Annales de l'Institut Fourier **5**, 131–295 (1953–1954)
9. Jaswal, D., Vishvanathan, S., Kp, S.: Image classification using convolutional neural networks. Int. J. Sci. Eng. Res. **5**, 1661–1668 (2014)
10. Deng, L., Yu, D., et al.: Deep learning: methods and applications. Found. Trends® Signal Process. **7**(3–4), 197–387 (2014)
11. Dias, C.A., Bueno, J.C.S., Borges, E.N., Botelho, S.S.C., Dimuro, G.P., Lucca, G., Fernandéz, J., Bustince, H., Drews Jr., P.L.J.: Using the Choquet integral in the pooling layer in deep learning networks. In: Fuzzy Information Processing, pp. 144–154. Springer, Cham (2018). (Best Paper Awards - 2nd place at NAFIPS 2018)
12. Dimuro, G.P., Lucca, G., Sanz, J.A., Bustince, H., Bedregal, B.: CMin-Integral: a Choquet-like aggregation function based on the minimum t-norm for applications to fuzzy rule-based classification systems. In: Aggregation Functions in Theory and in Practice, pp. 83–95. Springer, Cham (2018)

13. Eskicioglu, A.M., Fisher, P.S.: Image quality measures and their performance. IEEE Trans. Commun. **43**(12), 2959–2965 (1995)
14. Fukushima, K.: Neocognitron-a self-organizing neural network model for a mechanism of pattern recognition unaffected by shift in position. Biol. Cybern. **15**, 106–115 (1981)
15. Goodfellow, I., Bengio, Y., Courville, A.: Deep Learning. MIT Press, Cambridge (2016)
16. Grabisch, M., Marichal, J., Mesiar, R., Pap, E.: Aggregation Functions. Cambridge University Press, Cambridge (2009)
17. Hamel, P., Lemieux, S., Bengio, Y., Eck, D.: Temporal pooling and multiscale learning for automatic annotation and ranking of music audio, pp. 729–734 (2011)
18. Hodges, J.L., Lehmann, E.L.: Rank methods for combination of independent experiments in analysis of variance. Ann. Math. Stat. **33**(2), 482–497 (1962)
19. Ishibuchi, H., Nakashima, T., Nii, M.: Pattern classification with linguistic rules. In: Fuzzy Sets and Their Extensions: Representation, Aggregation and Models, vol. 220, pp. 1077–1095 (2008)
20. Jia, Y., Shelhamer, E., Donahue, J., Karayev, S., Long, J., Girshick, R., Guadarrama, S., Darrell, T.: Caffe: Convolutional architecture for fast feature embedding. In: Proceedings of the 22nd International Conference on Multimedia, pp. 675–678 (2014)
21. Kim, Y.: Convolutional neural networks for sentence classification. In: Proceedings of the Conference on Empirical Methods in Natural Language Processing, Doha, Qatar, pp. 1746–1751 (2014)
22. Klement, E.P., Mesiar, R., Pap, E.: Triangular Norms. Kluwer Academic Publisher, Dordrecht (2000)
23. Lucca, G., Dimuro, G.P., Fernandez, J., Bustince, H., Bedregal, B., Sanz, J.A.: Improving the performance of fuzzy rule-based classification systems based on a nonaveraging generalization of CC-integrals named $c_{F_1 F_2}$-integrals. IEEE Trans. Fuzzy Syst. **27**(1), 124–134 (2019)
24. Lucca, G., Sanz, J.A., Dimuro, G.P., Bedregal, B., Asiain, M.J., Elkano, M., Bustince, H.: CC-integrals: Choquet-like copula-based aggregation functions and its application in fuzzy rule-based classification systems. Knowl.-Based Syst. **119**, 32–43 (2017)
25. Lucca, G., Sanz, J.A., Dimuro, G.P., Bedregal, B., Bustince, H., Mesiar, R.: C_F-integrals: a new family of pre-aggregation functions with application to fuzzy rule-based classification systems. Inf. Sci. **435**, 94–110 (2018)
26. Lucca, G., Sanz, J.A., Dimuro, G.P., Bedregal, B., Mesiar, R., Bustince, A.K.R.H.: Preaggregation functions: construction and an application. IEEE Trans. Fuzzy Syst. **24**(2), 260–272 (2016)
27. Mishra, A., Alahari, K., Jawahar, C.: Scene text recognition using higher order language priors. In: 23rd British Machine Vision Conference (2012)
28. Mizoguchi, T., Ishii, A., Nakamura, H., Inoue, T., Takamatsu, H.: Lidar-based individual tree species classification using convolutional neural network. In: Videometrics, Range Imaging, and Applications XIV, vol. 10332, p. 103320O. International Society for Optics and Photonics (2017)
29. Morettin, P.A., Bussab, W.O.: Estatística Básica. Editora Saraiva (2017)
30. Paternain, D., Fernández, J., Bustince, H., Mesiar, R., Beliakov, G.: Construction of image reduction operators using averaging aggregation functions. Fuzzy Sets Syst. **261**, 87–111 (2015)

31. Scherer, D., Muller, A., Behnke, S.: Evaluation of pooling operations in convolutional architectures for object recognition. In: 20th International Conference on Artificial Neural Networks, Thessaloniki, Greece, pp. 92–101 (2010)
32. Yu, D., Wang, H., Chen, P., Wei, Z.: Mixed pooling for convolutional neural networks. In: International Conference on Rough Sets and Knowledge Technology, pp. 364–375 (2014)

Integrating Antonyms in Fuzzy Inferential Systems via Anti-membership

Scott Dick[1](✉) ⬤ and Peter Sussner[2]

[1] University of Alberta, Edmonton, AB T6G 1H9, Canada
dick@ece.ualberta.ca
[2] University of Campinas, Campinas, SP 13083-859, Brazil
sussner@ime.unicamp.br

Abstract. Numerous authors have proposed extending fuzzy inferential systems to include the antonyms in fuzzy rules. To date, however, those efforts require significant changes to the nature of a linguistic variable, directly implying substantial additional computation. We propose a new mechanism for incorporating antonyms into fuzzy rules, based on allowing negative-valued memberships along with two new union and intersection operations developed by Dick et al. We prove that these operations form a total ordering over $[-1,1]$, and then show how they integrate antonyms into fuzzy rules seamlessly and require little additional computation.

Keywords: Fuzzy inferential systems · Linguistic variables · Antonyms · Lattice theory

1 Introduction

Antonyms, negations, and opposites in general appear to be important elements of human reasoning and cognition. Negativity bias refers to the greater influence negative information has on human perception compared to positive information. This has been repeatedly demonstrated in psychological studies. Functional MRI studies show that brain activity is significantly higher in specific locations when processing negative information vs. positive [13, 17]. Decades of research furthermore show that many concepts are easily understood through examples of opposing pairs, considered as opposite ends of a "spectrum," e.g. the Semantic Differential scale in psychology [5, 12]. A small number of machine learning researchers have explored the idea of negative rules, e.g. [1, 3]. They suggest that positive assertions (if A then B) are fairly inefficient at expressing prohibitions or the absence of something, whereas negative assertions are highly efficient at them. Negative rules might simply attenuate positive ones [3], or might fire independently [1]. Intuitionistic fuzzy sets draw memberships from the unit square $[0,1] \times [0,1]$, with the axes representing membership and non-membership degrees, respectively [2]. Bipolar fuzzy sets are similar, but represent non-membership as an element of $[-1,0]$ [18].

Unsurprisingly, antonyms have also been of significant interest to the fuzzy systems community for a long time [4, 5, 7, 14–16]. Several approaches to representing

© Springer Nature Switzerland AG 2019
R. B. Kearfott et al. (Eds.): IFSA 2019/NAFIPS 2019, AISC 1000, pp. 237–245, 2019.
https://doi.org/10.1007/978-3-030-21920-8_22

antonyms in fuzzy systems have been proposed. For example, de Soto et al. [5] propose that antonym should be an "inner" operation

$$\mu_{ant-P}(x) = \mu(A(x)) \tag{1}$$

where P is a linguistic value, $ant\ P$ is the antonym of P (another linguistic value) and A (x) is a function that maps the support of P into the support of $ant\ P$. They define $ant\ P$ as being the "polar opposite" of P in a bipolar linguistic variable; i.e. the value appearing in the same ordinal position as P if the ordering of terms is reversed. This basic idea was expanded into a "computing with antonyms" approach in [15], in which the inner operation $A(x)$ from [5] is a reflection about the midpoint of the universe of discourse (assumed to be a finite interval of the real line). The ideas of computing with antonyms were applied to robot navigation in [8]. These robots map their environment as a grid of cells, marked as either empty or occupied. These are treated as imprecise linguistic values, and furthermore antonyms. When a cell is marked as both, this contradiction is detected and resolved by further sensor evidence.

Tizoosh's *opposite fuzzy sets* and *opposition-based computing* [14] also start by defining antonymy as a function over the support of P. Opposition in this case is a generalization of reflection about a midpoint; the metric for finding that midpoint is allowed to be an arbitrary (possibly nonlinear) function. The support of $ant\ P$ is the set of points opposing the support of P; the memberships for opposing points are equal.

Antonyms are also a part of Novak's "fuzzy logic in the broader sense," which sees to formalize a theory of human reasoning [11]. He again treats antonyms as a "polar opposite" a term formalized by reference to the class of *evaluating syntagms*, linguistic terms that identify a position in an ordered scale. The extrema of a set of evaluating syntagms, along with the neutral (midpoint) element form a *basic linguistic trichotomy*. (Note that in linguistics, the antonym is usually considered to induce three values: a predicate, it's antonym, and the *don't-care* middle ground; negations, however, follow the law of excluded middle and induce only two values, the predicate and its negation [15]) Linguistic terms are associated with *intensions*, which relate the term to a fuzzy set on the universe of discourse.

Despite the above, there has been little work on integrating antonyms into practical Fuzzy inferential systems. One reason might be the reliance on reflecting about the midpoint of an interval to find $ant\ P$; while this is acceptable in a uniform, symmetric partition of a universe of discourse, many practical LVs are anything but. The process of tuning Membership Functions (MFs) to maximize performance often skews those MFs and distorts their supports. Evaluating the ant operator (as called for in [15]) using a reflection might thusly result in a fuzzy set associated with no linguistic value in the LV.

Our solution is instead to allowing membership values in the range $[-1,1]$; those in the range $[-1,0]$ were dubbed *anti-memberships* in [6]. We can then reflect the memberships in a linguistic variable about 0, producing fuzzy sets having pure anti-memberships for the antonym of each term; these reflected fuzzy sets are labeled as antonyms of the original terms, only in reverse sort order. We then use a symmeterized version of the $min()$ and $max()$ operators (introduced in [6]) to execute a fuzzy infer-ence. However, anti-memberships are not from the familiar chain $[-1,1]$ ordered by the

"<" of real numbers; instead, they are generally ordered by the absolute value of the membership. We show that this is an alternative total ordering of $[-1,1]$; and that using them we can construct a type-1 fuzzy inference engine with minimal additional overhead that integrates antonyms into fuzzy linguistic rules. As we will see, our extensions are furthermore fully compatible with existing type-1 fuzzy rulebases.

The remainder of this paper is organized as follows. In Sect. 2, we show that the *absmin()* and *absmax()* functions totally order the set $[-1,1]$. In Sect. 3 we describe the extensions our approach will require for a type-1 Fuzzy inferential system (FIS). In Sect. 4 we provide examples of how such an FIS would work, both for existing rulebases and for rules that include antonyms. We close with a summary and discussion of future work in Sect. 5.

2 A Total Ordering of Memberships ∪ Anti-memberships

In this section, we prove that the functions *absmin()* and *absmax()* form a distributive lattice over $[-1,1]$; that this lattice is bounded, with infimum 0 and supremum +1; and that the lattice is a total ordering of $[-1,1]$.

Definition 1 ([6]): The function absmax is given by:

$$absmax(x, y) = \begin{cases} x & if\ |x| \geq |y| \\ y & if\ |x| < |y| \\ |x| & if\ |x| = |y| \wedge x \neq y \end{cases} \tag{2}$$

Definition 2 ([6]): The function absmin is given by:

$$absmin(x, y) = \begin{cases} x & if\ |x| \leq |y| \\ y & if\ |x| > |y| \\ |y| & if\ |x| = |y| \wedge x \neq y \end{cases} \tag{3}$$

The ordering of $[-1,1]$ above is based on the absolute value of the arguments. In the case of equal absolute values, we define the sign of the arguments as a secondary ordering, with negative values less than positive ones. By definition, the operators are idempotent. In [6], we proved that the two operators are associative and distributive; commutativity is obvious from the definitions. We now examine their properties on the interval $[-1,1]$.

Theorem 1: The operations *absmax()* and *absmin()* form the join and meet, respectively, of a distributive lattice over $[-1,1]$.

Proof
The algebraic structure $([-1,1], absmax(), absmin())$ is a lattice if the join and meet are commutative, associative, idempotent and absorptive. The lattice is distributive if the join and meet are distributive over one another. The first three properties, and distributivity, have been established, and so we simply prove absorption. For brevity, we

will only discuss absorption of *absmin*() by *absmax*(); the dual proof is similar. Let a, $b \in [-1,1]$. By definition,

$$a \vee (a \wedge b) = a \tag{4}$$

$$absmax(a, absmin(a, b)) = a \tag{5}$$

By distributivity from [6],

$$absmin(absmax(a, a), absmax(a, b)) \tag{6}$$

If *absmax(a,b)* = *b*, then the expression reduces to *absmin(a,b)*, which is necessarily *a*. Else, we have *absmin(a,a)* = *a*. Thus, absorption is proven, and ([−1,1], *absmax*(), *absmin*()) is thus a distributive lattice.

Lemma 1: The lattice ([−1,1], *absmax*(), *absmin*()) is bounded, with infimum 0 and supremum +1.

Proof
Let $a \in [-1,1]$. The following are obvious from Definitions (1–2):

$$absmax(1, a) = 1 \tag{7}$$

$$absmax(0, a) = a \tag{8}$$

$$absmin(1, a) = a \tag{9}$$

$$absmin(0, a) = 0 \tag{10}$$

By Eqs. (7–8), +1 is the supremum of the lattice. Likewise, by Eqs. (9–10) 0 is the infimum of the lattice.

Theorem 2: The lattice ([−1,1], *absmax*(), *absmin*()) is a total ordering of [−1,1].

Proof
By Theorem (1), ([−1,1], *absmax*(), *absmin*()) is a lattice, and thus a partial order over [−1,1]. To prove that it is a total order, we need only prove the connex property: for any $a,b \in [-1,1]$, either $a \leq b$ or $b \leq a$. In terms of the join and meet for our lattice, this means:

$$absmax(a, b) = a \oplus absmax(a, b) = b \tag{11}$$

$$absmin(a, b) = a \oplus absmin(a, b) = b \tag{12}$$

with $\oplus$ the exclusive-OR relation. Both (11) and (12) are trivially true for all a, b in our lattice, and so ([−1,1], *absmax*(), *absmin*()) is a total ordering of [−1,1].

3 Constructing Antonyms Using Anti-memberships

Our approach requires extensions to the semantic rule of a linguistic variable, and a corresponding extension of the inference engine and defuzzifier. We discuss each of these below.

3.1 Anti-memberships in Linguistic Variables

A linguistic variable is a 5-tuple (X, U, T, S, M) where X is the name of a variable, U is the universal set of values for X, T is the set of linguistic terms for X, S is a syntactic rule for generating elements of T (usually a context-free grammar), and M is a semantic rule that associates each term in T with a fuzzy subset of U. The grammar of S usually specifies a set of *atomic* terms, as well as hedges that modify those atomic terms. The semantic rule then associates each atomic term with a fuzzy set, and each hedge with a function $h: [0,1] \rightarrow [0,1]$ that will modify the membership function of the fuzzy set[9].

Plainly, existing approaches to antonyms in LVs create challenges. At the semantic level, they require that the support of a fuzzy set be changed to its "polar opposite," which is usually calculated via reflection about the midpoint of the universe of discourse. At the syntactic level, *ant P* is usually an element of T that does not share the same atomic term as P. This necessarily means that a rewrite rule for antonyms must include more than just the modifier *ant* on the LHS; context is required, and thus evaluating *ant P* (as demanded in[15]) requires S to be a phrase-structure grammar. This further assumes that a linguistic term with a matching support exists; for optimized rulebases, there is an excellent chance that this is not so.

Our proposal differs in both respects. At the semantic level, we directly assign *ant P* to a fuzzy set having pure anti-membership (see Fig. 1). This is a simple table lookup (optionally followed by computing hedges in the usual manner). This also gives the fuzzy system designer full control over what fuzzy sets are chosen as the antonyms of each other. We suggest that the antonym fuzzy sets be chosen as reflections of the "positive" fuzzy sets about $\mu = 0$, as in Fig. 1. The antonym sets are then labeled in reverse sort order compared to the positive ones, matching the usual semantics of fuzzy antonyms but not requiring a re-computation of the supports. This also allows the semantic operation of evaluating an antonym to simply be multiplying the anti-membership function by −1. This also has the effect of guaranteeing that, when the antonym is evaluated, a compatible linguistic term and fuzzy set exist. However, the syntactic rule S will still need to be extended to a phrase structure grammar.

It is worth noting that, while [15] asserts that an antonym must be evaluated, this is only one point of view. An alternative viewpoint would be that such an evaluation is beside the point; the evaluation process changes a negative expression to a positive one, *which is often not desirable*. Humans, as noted, frequently use negative information in building mental models or decision-making. The semantic rule of Fig. 1 easily supports either approach.

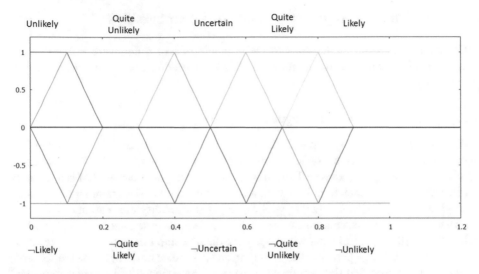

Fig. 1. A linguistic variable with five trapezoidal membership functions for "Likelihood," which is conceived of as a probability (adapted from [10]). Note that the partition of the universe of discourse ([0,1]) is non-uniform and non-symmetric. Antonyms are marked with the symbol ¬. Anti-memberships for the antonyms of the linguistic terms are the reflections for the fuzzy sets about μ = 0.

3.2   Anti-memberships in Fuzzy Inferential Systems

The proposed changes to LVs in Sect. 3.1 require only a simple alteration to the commonly-used *min-max* inference engine: substituting *absmin*() and *absmax*() for the *min*() and *max*() operations, respectively. Consider fuzzy IF-THEN rules of the form

$$IF \ x_1 \ is \ A_{1i} \ and \ \dots \ and \ x_n \ is \ A_{nj} \ THEN \ y \ is \ B_k \tag{13}$$

where $x_1, \dots, x_n$ are observations drawn from n orthogonal universes of discourse $U_1, \dots, U_n$, A_{pi} is the i-th fuzzy subset of U_p, $y \in V$ is the system output drawn from V, and B_k is the k-th fuzzy subset of V. We compute the firing strength of this rule ω as [9]:

$$\omega = absmin\left(A_{1i}(x_1), \dots, A_{nj}(x_n)\right) \tag{14}$$

Assuming this is the r-th rule in a fuzzy rulebase, we compute the inference result as a fuzzy set $B'(y)$ using the compositional rule of inference[9]:

$$B'(y) = \underset{r}{absmax} \ absmin(\omega_r, B_r(y)) \tag{15}$$

Defuzzification also needs to account for anti-memberships. For example, we can modify the centroid defuzzifier to use the absolute value of x, as follows [9]:

$$\frac{\int_a^b x \cdot |\mu(x)| dx}{\int_a^b |\mu(x)| dx} \qquad (16)$$

Plainly, when anti-memberships are not present, Eqs. (14–16) reduce to the standard max-min fuzzy inferencing for type-1 fuzzy sets.

4   Example

Let's consider an example of using antonyms in approximate reasoning from [10]. A group of scientists was asked to rate the likelihood (treated as being a probability) of genetically modified (GM) foods being dangerous to humans. A five-level scale for likelihood was provided (*unlikely, quite unlikely, uncertain, quite likely, likely*), with each term being associated with a trapezoidal fuzzy set; the five terms formed a uniform, symmetric partition of the universe of discourse [0,1].

In our example (see Fig. 1), we have altered the linguistic variable to reflect an asymmetry humans might perceive between high and low probabilities. The fuzzy set for "Unlikely" had its support reduced to just [0,0.2], while that for "Likely" is [0.7,1]. Similarly, we expanded the support for "Quite unlikely" to [0,0.5], while that for "Quite likely" is [0.5,0.9]. We are attempting to capture the notion that people perceive technically improbably things as more relevant to them (hence a narrowing of "Unlikely"), while being more accepting (fatalistic?) about technically probable events actually happening. While this is certainly not a rigorous argument, the key point is that these concepts are not necessarily symmetric with each other in a numerical sense.

Lawry [10] found that the scientists on average judged GM foods to be *unlikely* ($\mu = 1$) or *quite unlikely* ($\mu - 0.3$) to be dangrous. He then asked what this implied for the statement "GM foods are safe." This was accomplished by taking the antonym of the two linguistic terms, at the same memberships. Let us do so as well, and then defuzzify the result.

As the reader can verify, Unlikely/1.0 + Quite unlikely/0.3 defuzzifies to a value of $x = 0.03$ in Fig. 1. When we then take the antonyms (symbolized by $\neg$), we have $\neg$Unlikely/−1.0 + $\neg$Quite unlikely/−0.3. Following the methods from Sect. 3.2, these defuzzify to $x = 0.87$. If we evaluated the antonyms before taking the semantic rule, we would have Likely/1.0 + Quite likely/0.3. This again defuzzifies to $x = 0.87$. From this example, we see that anti-memberships, applied according to our approach, are able to model antonyms even for evaluating syntagms with unequal and asymmetric supports between their terms.

5   Summary and Future Work

In this paper, we proposed a new method for representing antonyms in a linguistic variable. Anti-memberships (in the range $[-1,0]$) are proposed to complement the usual membership functions. We propose symmetrized versions of the $max()$ and $min()$ functions, and show that these create a total ordering of $[-1,1]$ different from the usual chain ordered by the real-valued "<". We discuss modification to the syntactic and semantic rules of a linguistic variable, and to the inference engine and defuzzifier of a fuzzy inferential system, required to incorporate antonyms into fuzzy rules. Finally, an example demonstrates our proposed approach.

In future work, we intend to explore the utility of antonyms in modeling and control of a variety of phenomena. One promising direction would be to integrate our antonym-based approach with neuro-fuzzy or genetic-fuzzy systems, and determine if, and to what degree, including antonyms improves their accuracy. Another direction would be to employ antonyms in minimizing fuzzy logical predicates (fuzzy switching functions, fuzzy rules, etc.)

References

1. Antonie, M.-L., Zaiane, O.R.: An associative classifier based on positive and negative rules. In: ACM Data Mining and Knowledge Discovery (2004)
2. Atanassov, K.T.: Intuitionistic fuzzy sets. Fuzzy Sets Syst. **20**, 87–96 (1986)
3. Branson, J.S., Lilly, J.H.: Incorporation, characterization, and conversion of negative rules into fuzzy inference systems. IEEE Trans. Fuzzy Syst. **9**, 253–268 (2001)
4. De Soto, A.R.: On automorphisms, synonyms, antonyms in fuzzy set theory. In: ITHURS (1996)
5. de Soto, A.R., Trillas, E.: On antonym and negate in fuzzy logic. Int. J. Intell. Syst. **14**, 295–303 (1999)
6. Dick, S., Yager, R., Yazdanbakhsh, O.: On pythagorean and complex fuzzy set operations. IEEE Trans. Fuzzy Syst. **24**, 1009–1021 (2016)
7. Garcia-Honrado, I., Trillas, E.: An essay on the linguistic roots of fuzzy sets. Inf. Sci. **181**, 4061–4074 (2011)
8. Guadarrama, S., Ruiz-Mayor, A.: Approximate robotic mapping from sonar data by modeling perceptions with antonyms. Inf. Sci. **180**, 4164–4188 (2010)
9. Klir, G.J., Yuan, B.: Fuzzy Sets and Fuzzy Logic: Theory and Applications. Prentice Hall PTR, Upper Saddle River (1995)
10. Lawry, J.: A methodology for computing with words. Int. J. Approximate Reasoning **28**, 51–89 (2001)
11. Novak, V.: Antonyms and linguistic qualifiers in fuzzy logic. Fuzzy Sets Syst. **124**, 335–351 (2001)
12. Osgood, C.E., Suci, G.J., Tannenbaum, P.H.: The Measurement of Meaning. University of Illinois Press, Chicago (1957)
13. Smith, N.K., Larsen, J.T., Chartrand, T.L., Cacioppo, J.T.: Being bad isn't always good: affective context moderates the attention bias toward negative information. J. Pers. Soc. Psychol. **90**, 210–220 (2006)
14. Tizhoosh, H.: Opposite fuzzy sets with applications in image processing. In: IFSA-EUSFLAT, Lisbon, Portugal (2009)

15. Trillas, E., Moraga, C., Guadarrama, S., Cubillo, S., Castineira, E.: Computing with antonyms. In: Nikravesh, M., et al. (eds.) Forging New Frontiers: Fuzzy Pioneers I, pp. 133–153. Springer, Berlin (2007)
16. Trillas, E., Riera, T.: Towards a representation of synonyms and antonyms by fuzzy sets. BUSEFAL **5**, 42–68 (1980)
17. Vaish, A., Grossmann, T., Woodward, A.: Not all emotions are created equal: the negativity bias in social-emotional development. Psych. Bull. **134**, 383–403 (2008)
18. Zhang, W.-R.: (Yin) (Yang) bipolar fuzzy sets. In: IEEE International Conference on Fuzzy Systems, Anchorage, AK, USA (1998)

Some Notes on the Addition of Interactive Fuzzy Numbers

Estevão Esmi$^{(\boxtimes)}$, Laécio Carvalho de Barros, and Vinícius Francisco Wasques

Institute of Mathematics, Statistics and Scientific Computing,
University of Campinas, Campinas, Brazil
{eelaureano,laeciocb}@ime.unicamp.br, vwasques@outlook.com

Abstract. This paper investigates some fundamental questions involving additions of interactive fuzzy numbers. The notion of interactivity between two fuzzy numbers, say A and B, is described by a joint possibility distribution J. One can define a fuzzy number $A+_J B$ (or $A-_J B$), called J-interactive sum (or difference) of A and B, in terms of the sup-J extension principle of the addition (or difference) operator of the real numbers. In this article we address the following three questions: (1) Given fuzzy numbers B and C, is there a fuzzy number X and a joint possibility distribution J of X and B such that $X +_J B = C$? (2) Given fuzzy numbers A, B, and C, is there a joint possibility distribution J of A and B such that $A +_J B = C$? (3) Given a joint possibility distribution J of fuzzy numbers A and B, is there a joint possibility distribution N of $(A +_J B)$ and B such that $(A +_J B) -_N B = A$? It is worth noting that these questions are trivially answered in the case where the fuzzy numbers A, B and C are real numbers, since the fuzzy arithmetic $+_J$ and $-_N$ are extension of the classical arithmetic for real numbers.

1 Introduction

A fuzzy (sub)set A of a non-empty set Ω is characterized by a function $\mu_A : \Omega \rightarrow [0,1]$ called of membership function of A, where $\mu_A(x)$ represents the membership degree of x in A. For notation convenience, we alternatively use the symbol $A(x)$ instead of $\mu_A(x)$. We say that a fuzzy set A is contained in a fuzzy set B or, that B is greater or equal than A, denoted by $A \subseteq B$, if $A(x) \leq B(x)$ for all $x \in \Omega$. The α-cut of A is defined as $[A]_\alpha = \{x \in \Omega; | A(x) \geq \alpha\}$ for every $\alpha \in (0, 1]$. If Ω is topological space such as $\mathbb{R}$, then we define $[A]_0$ as the closure of $supp(A) = \{x \in \Omega : A(x) > 0\}$, that is, $\overline{supp(A)}$ [1].

A fuzzy set A of $\mathbb{R}$ is said to be a fuzzy number if $[A]_\alpha$ is a non-empty, bounded and closed interval of $\mathbb{R}$, say $[A]_\alpha = [a_\alpha^-, a_\alpha^+]$, for every $\alpha \in [0, 1]$ [1]. We denote the class of fuzzy numbers by the symbol $\mathbb{R}_{\mathscr{F}}$. Typical examples of fuzzy numbers include trapezoidal, triangular, and Gaussian fuzzy numbers. A trapezoidal fuzzy number $(a; b; c; d)$, for $a \leq b \leq c \leq d$, is given by

© Springer Nature Switzerland AG 2019
R. B. Kearfott et al. (Eds.): IFSA 2019/NAFIPS 2019, AISC 1000, pp. 246–257, 2019.
https://doi.org/10.1007/978-3-030-21920-8_23

$$(a; b; c; d)(x) = \begin{cases} 1, & \text{if } x \in [b, c] \\ \frac{x-a}{b-a}, & \text{if } x \in [a, b) \\ \frac{d-x}{d-c}, & \text{if } x \in (c, d] \\ 0, & \text{otherwise} \end{cases}.$$

The set of real numbers and, more general, the set of non-empty, bounded, and closed intervals of $\mathbb{R}$ are contained in the class of trapezoidal numbers since $(a; a; b; b) \equiv [a, b]$ for all $a \leq b$. If $b = c$ then we speak of triangular fuzzy number that is simply denoted by the symbol $(a; b; d)$. Note that every real number a can uniquely be identified as the triangular fuzzy number $(a; a; a)$. The Gaussian fuzzy number $(m, \sigma, c)_G$, with $\sigma, c > 0$, is given by

$$(m, \sigma, c)_G(x) = \begin{cases} e^{\frac{-(m-x)^2}{\sigma^2}}, & \text{if } x \in [m - c, m + c] \\ 0, & \text{otherwise} \end{cases}.$$

A fuzzy set J of $\mathbb{R}^2$ is said to be a *joint possibility distribution* of the fuzzy numbers A and B if [3,4,6]

$$A(z) = \bigvee_{y \in \mathbb{R}} J(z, y) \quad \text{and} \quad B(z) = \bigvee_{x \in \mathbb{R}} J(x, z), \quad \forall z \in \mathbb{R}. \tag{1}$$

Here, the symbols $\bigvee$ and $\bigwedge$ denote the infimum and supremum operators, respectively. The fuzzy numbers A and B are also called marginal distributions of J. Furthermore, the fuzzy numbers A and B are called *non-interactive* if

$$J(x, y) = A(x) \wedge B(y) \tag{2}$$

for all $(x, y) \in \mathbb{R}^2$. Otherwise, A and B are called J-*interactive* or simply interactive.

Let J be a joint possibility distribution of A and B. The sup-J extension of a function $f : \mathbb{R}^2 \to \mathbb{R}$ at (A, B) is the fuzzy set $f_J(A, B)$ of $\mathbb{R}$ such that

$$f_J(A, B)(z) = \bigvee_{(x,y) \in f^{-1}(z)} J(x, y), \forall z \in \mathbb{R}, \tag{3}$$

where $f^{-1}(z) = \{(x, y) \in \mathbb{R}^2 \mid z = f(x, y)\}$ [3,6]. By definition, $0 = \bigvee \emptyset$. If f is a continuous function, then an application of Nguyen's theorem [7] ensures that $[f_J(A, B)]_\alpha = f([J]_\alpha)$ for every $\alpha \in [0, 1]$. One can define the J-interactive sum of A and B as the fuzzy set

$$A +_J B := s_J(A, B),$$

where $s(x, y) = x + y$ for all $x, y \in \mathbb{R}$ [3,4]. Similarly, the J-interactive difference of A and B is given by the fuzzy set

$$A -_J B := d_J(A, B),$$

where $d(x, y) = x - y$ for all $x, y \in \mathbb{R}$ [3,4]. Moreover, for $\lambda \in \mathbb{R}$ and $A \in \mathbb{R}_{\mathscr{F}}$, we defined λA the fuzzy number such that $[\lambda A]_\alpha = \{\lambda x \mid x \in [A]_\alpha\} = \lambda [A]_\alpha$ for all $\alpha \in [0, 1]$.

Theorem 1. *Let J be a joint possibility distribution of fuzzy numbers A and B and let $\bar{J}$ and $J^\star$ be the fuzzy sets of $\mathbb{R}^2$ given by*

$$\bar{J}(x,y) = J(y,x) \tag{4}$$

and

$$J^\star(x,y) = J(x,-y) \tag{5}$$

for all $(x,y) \in \mathbb{R}^2$. The fuzzy set $\bar{J}$ is a joint possibility distribution of B and A and $A +_J B = B +_{\bar{J}} A$. Moreover, the fuzzy set $J^\star$ is a joint possibility distribution of A and $(-B)$ and $A -_J B = A +_{J^\star} (-B)$.

Proof. For all $z \in \mathbb{R}$, we have that

$$(A +_J B)(z) = \bigvee_{z=x+y} J(x,y) = \bigvee_{z=x+y} \bar{J}(y,x) = (B +_{\bar{J}} A)(z)$$

and

$$(A -_J B)(z) = \bigvee_{z=x-y} J(x,y) = \bigvee_{z=x+(-y)} J^\star(x,-y) = (A +_{J^\star} (-B))(z).$$

The standard addition of A and B arises when A and B are non-interactive, that is, J is given as in Eq. (2). In this case, we simply use the symbol $A + B$ instead of $A +_J B$ and every α-cut of $A + B$ are given by [1]:

$$[A + B]_\alpha = [a_\alpha^- + b_\alpha^-, a_\alpha^+ + b_\alpha^+]. \tag{6}$$

Let $C, B \in \mathbb{R}_{\mathscr{F}}$, if there exists $A \in \mathbb{R}_{\mathscr{F}}$ such that $A + B = C$ then A is called of the Hukuhara difference (or, for short, H-difference) of C and B and it is also denoted by $C -_H B$ [8]. Recall that H-difference is not defined for all pair of fuzzy numbers since for every $B \in \mathbb{R}_{\mathscr{F}}$ the mapping $F_B : \mathbb{R}_{\mathscr{F}} \to \mathbb{R}_{\mathscr{F}}$ given by $F_B(A) = A + B$ for all $A \in \mathbb{R}_{\mathscr{F}}$ is not surjective. For example, the H-difference between $(0; 1; 2; 3)$ and $(0; 1; 3)$ is not defined because there exists no $A \in \mathbb{R}_{\mathscr{F}}$ such that $(0; 1; 3) + A = (0; 1; 2; 3)$. The standard addition and H-difference satisfy the following interesting property:

$$(A + B) -_H B = A \tag{7}$$

for all $A, B \in \mathbb{R}_{\mathscr{F}}$. Moreover, by definition, if $C -_H B$ exists, then $(C -_H B) + B = C$.

The generalized Hukuhara (of gH-difference, for short) extends the Hukuhara difference [9]. The gH-difference of the fuzzy numbers C and B, when it exists, is denoted by $C -_{gH} B$, and its α-cuts are given by

$$[C -_{gH} B]_\alpha = \left[(c_\alpha^- - b_\alpha^-) \wedge (c_\alpha^+ - b_\alpha^+), (c_\alpha^- - b_\alpha^-) \vee (c_\alpha^+ - b_\alpha^+) \right]$$

for all $\alpha \in [0,1]$. The H-difference of $C = (0; 1; 2)$ and $B = (0; 1; 3)$ does not exists, however, $(0; 1; 2) -_{gH} (0; 1; 3) = (-1; 0; 0)$ but $(-1; 0; 0) + (0; 1; 3) \neq$

$(0; 1; 2)$, that is, $(C -_{gH} B) + B \neq C$. Moreover, the gH-difference is not defined for all pair of fuzzy numbers. For instance, the gH-difference of $(0; 1; 2; 3)$ and $(0; 1; 3)$ does not exists.

In contrast, the generalized difference (or g-difference) between two arbitrary fuzzy numbers is always defined and extends the gH-difference [2]. For $C, B \in \mathbb{R}_{\mathscr{F}}$, the g-difference of C and B is a fuzzy number $C -_g B$ whose α-cuts, for $\alpha \in [0, 1]$, are given by:

$$[C -_g B]_\alpha = \left[\bigwedge_{\beta \in [\alpha, 1]} (c_\beta^- - b_\beta^-) \wedge (c_\beta^+ - b_\beta^+), \bigvee_{\beta \in [\alpha, 1]} (c_\beta^- - b_\beta^-) \vee (c_\beta^+ - b_\beta^+) \right].$$

For example, we have that $(0; 1; 2; 3) -_g (0; 1; 3) = [0, 1]$ and $((0; 1; 2; 3) -_g (0; 1; 3)) + (0; 1; 3) = (0; 1; 3; 4) \neq (0; 1; 2; 3)$. This last example shows that the equality

$$(C -_g B) + B = C \tag{8}$$

may not hold true.

Note that the Hukuhara difference of C and B corresponds to the unique solution of the following equation:

$$X + B = C \tag{9}$$

where $X \in \mathbb{R}_{\mathscr{F}}$. However, this equation may not has a solution.

In the next sections we investigate two variations of Equation (9). The first one consists of relaxation of the hypothesis of X and B be non-interactive. More precisely, we investigate solutions for the equation

$$X +_J B = C \tag{10}$$

where $X \subset \mathbb{R}_{\mathscr{F}}$ and J is a joint possibility distribution of X and B. In Sect. 2, we provide the maximal solution for Eq. (10).

Note that in Eq. (9) we have that X and B are marginal distributions of the joint possibility distribution $J_\wedge(x, y) = X(x) \wedge B(y)$ for all $(x, y) \in \mathbb{R}^2$. This mean that the free variable in Eq. (9) is the marginal distribution X. In Sect. 3, we consider X as a fixed marginal distribution, say $X = A$, and the joint possibility distribution J as free variable. More precisely, for $A, B, C \in \mathbb{R}_{\mathscr{F}}$, we consider the following equation

$$A +_J B = C \tag{11}$$

where J is a joint possibility distribution of A and B. Section 3 investigates when a fuzzy number C can be written as a J-interactive sum of A and B.

Let J be a joint possibility distribution of two fuzzy numbers A and B. Motivated by Eq. (7), Sect. 4 studies the existence of a joint possibility distribution N of $(A +_J B)$ and B that satisfies the following equality:

$$(A +_J B) -_N B = A. \tag{12}$$

2 Given Fuzzy Numbers B and C, Is There a Fuzzy Number X and a Joint Possibility Distribution J of X and B Such that $X +_J B = C$?

In this section we study Eq. (10). In other words, for given $B, C \in \mathbb{R}_{\mathscr{F}}$, we are interested to know if there exists a fuzzy number X and a joint possibility distribution J of X and B such that the fuzzy number C corresponds to the J-interactive sum of X and B, that is, $X +_J B = C$.

Theorem 2. *Let B, C be fuzzy numbers. The fuzzy set S of $\mathbb{R}^2$ whose membership function is defined by*

$$S(x, y) = B(y) \wedge C(x + y), \quad \forall (x, y) \in \mathbb{R}^2. \tag{13}$$

is a join possibility distribution with marginals distributions $X \in \mathbb{R}_{\mathscr{F}}$ and B such that $X +_S B = C$ and $[X]_\alpha = [c_\alpha^- - b_\alpha^+, c_\alpha^+ - b_\alpha^-]$ for all $\alpha \in [0, 1]$. Moreover, X and S are the maximal solution of (10), i.e., if there are a fuzzy number $\tilde{X}$ and a joint possibility distribution $\tilde{J}$ of $\tilde{X}$ and B where $\tilde{X} +_{\tilde{j}} B = C$, then $\tilde{J} \subseteq S$ and $\tilde{X} \subseteq X$.

Proof. Let X be the fuzzy set whose membership function is given by

$$X(x) = \bigvee_{y \in \mathbb{R}} B(y) \wedge C(x + y), \quad \forall x \in \mathbb{R}.$$

Let $0 < \alpha \le 1$. If $y \in [B]_\alpha$ and $z \in [C]_\alpha$ then $S(z - y, y) \ge \alpha$ which implies that $z - y \in [X]_\alpha$. Thus, we have that $[c_\alpha^- - b_\alpha^+, c_\alpha^+ - b_\alpha^-] = \{z - y \mid z \in [C]_\alpha, y \in [B]_\alpha\} \subseteq [X]_\alpha$. Now, if $x \in [X]_\alpha$ then there exists (y_n) such that $\lim_{n \to \infty} B(y_n) \wedge C(x + y_n) \ge \alpha$. For n sufficient large we have $B(y_n) \ge B(y_n) \wedge C(x + y_n) > \frac{\alpha}{2}$. This implies that (y_n) is bounded and, therefore, there exists a convergent subsequence $(y_{n_k}) \in [B]_{\frac{\alpha}{2}}$, say $\lim_{k \to \infty} y_{n_k} = \hat{y} \in [B]_{\frac{\alpha}{2}}$. If $\beta := B(\hat{y}) < \alpha$ then for $\beta < \rho < \alpha$ and for k sufficient large we have that $y_{n_k} \in [B]_\rho$ because $B(y_{n_k}) \ge B(y_{n_k}) \wedge C(x + y_{n_k}) \ge \rho$. Since $[B]_\rho$ is a bounded closed interval, we have $\hat{y} \in [B]_\rho$ which produces the following contradiction: $B(\hat{y}) < \rho \le B(\hat{y})$. Therefore, we must have $B(\hat{y}) \ge \alpha$. Similarly, we can show that $x + \hat{y} \in [C]_\alpha$. These last observations imply that $x \in [c_\alpha^- - y_\alpha^+, c_\alpha^+ - y_\alpha^-]$. Hence, we conclude that $[X]_\alpha = [c_\alpha^- - b_\alpha^+, c_\alpha^+ - b_\alpha^-]$ for all $\alpha \in (0, 1]$. Finally, we obtain that

$$[X]_0 = \overline{supp(X)} = \overline{\bigcup_{\alpha \in (0,1]} [X]_\alpha} = \left[\bigwedge_{\alpha \in (0,1]} c_\alpha^- - b_\alpha^+, \bigvee_{\alpha \in (0,1]} c_\alpha^+ - b_\alpha^- \right] = \left[c_0^- - b_0^+, c_0^+ - b_0^- \right].$$

Let $\bar{y} \in [B]_1$ and $\bar{z} \in [C]_1$. For every $y, z \in \mathbb{R}$, we have

$$B(y) \ge \bigvee_{x \in \mathbb{R}} S(x, y) \ge S(\bar{z} - y, y) = C(\bar{z}) \wedge B(y) = B(y)$$

and

$$C(z) \ge \bigvee_{z = x + y} S(x, y) \ge S(z - \bar{y}, \bar{y}) = C(z) \wedge B(\bar{y}) = C(z).$$

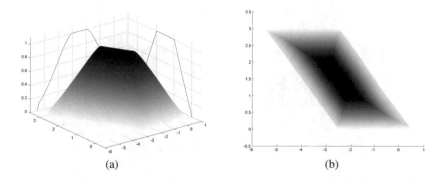

Fig. 1. (a) The joint possibility distribution S given as in Eq. (13), with $B = (0; 1; 2; 3)$ and $C = (-1; 1; 1.5)_G$, and its marginal distributions X and B. (b) The top-view of the joint possibility distribution S.

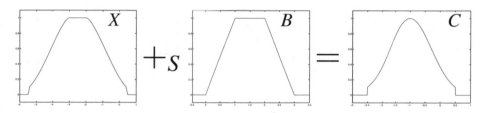

Fig. 2. Visual representations of S-interactive sum of X of B, where $B = (0; 1; 2; 3)$, $C = (-1; 1; 1.5)_G$, and X and S are given as in Fig. 1.

Let $\tilde{J}$ be a joint possibility distribution of $\tilde{X} \in \mathbb{R}_{\mathscr{F}}$ and B such that $\tilde{X} +_{\tilde{J}} B = C$. For $x, y \in \mathbb{R}$ and $z = x + y$, we have $B(y) = \bigvee_{w \in \mathbb{R}} \tilde{J}(w, y) \geq \tilde{J}(x, y)$ and $C(z) = \bigvee_{u+v=z} \tilde{J}(u, v) \geq \tilde{J}(x, y)$. Thus, we conclude that $S(x, y) = B(y) \wedge C(x+y) \geq \tilde{J}(x, y)$ for all $(x, y) \in \mathbb{R}^2$, which implies that $X(x) = \bigvee_{y \in \mathbb{R}} S(x, y) \geq \bigvee_{y \in \mathbb{R}} \tilde{J}(x, y) = \tilde{X}(x)$ for all $x \in \mathbb{R}$.

Example 1. Consider $B = (0; 1; 2; 3)$ and $C = (-1; 1; 1.5)_G$. Figure 1 exhibits the greatest fuzzy number X and joint possibility distribution S of X and B given as in Theorem 2. Figure 2 presents a visual representation of the $X +_S B = C$.

An interest example arises when $C = B$, that is, when we consider the equation

$$X +_J B = B. \tag{14}$$

Note that this equation has the trivial solution $0 + B = B$ and Theorem 2 provides its the greatest solution.

Example 2. Consider the Gaussian fuzzy numbers $B = C = (1; 1; 1)_G$. Figure 3 exhibits the greatest fuzzy number X and joint possibility distribution S of X and B given as in Theorem 2. A visual representation of the $X +_S B = B$ is illustrated in Fig. 4.

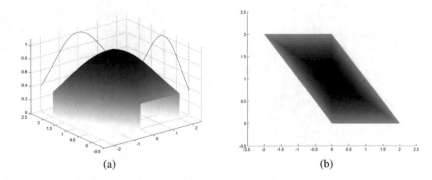

(a)   (b)

Fig. 3. (a) The joint possibility distribution S given as in Eq. (13), with $B = C = (1; 1; 1)_G$, and its marginal distributions X and B. (b) The top-view of the joint possibility distribution S.

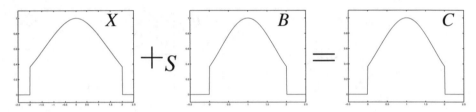

Fig. 4. Visual representations of S-interactive sum of X of B, where $B = C = (1; 1; 1)_G$, and X and S are given as in Fig. 3.

As we can observe in Example 2, Eq. (10) may have more than one solution. The next corollary is an immediate consequence of Theorems 1 and 2.

Corollary 1. *Let $A, B, C \in \mathbb{R}_{\mathscr{F}}$ and let R, T and V be fuzzy sets of $\mathbb{R}^2$ given by*

$$R(x, y) = A(x) \wedge C(x+y), \quad T(x, y) = B(y) \wedge C(x-y), \quad V(x, y) = A(x) \wedge C(x-y)$$
(15)

for all $(x, y) \in \mathbb{R}^2$. We have that

1. *R is a join possibility distribution with marginals distributions A and $X \in \mathbb{R}_{\mathscr{F}}$ such that $A +_R X = C$;*
2. *T is a join possibility distribution with marginals distributions B and $X \in \mathbb{R}_{\mathscr{F}}$ such that $X -_T B = C$;*
3. *V is a join possibility distribution with marginals distributions A and $X \in \mathbb{R}_{\mathscr{F}}$ such that $A -_V X = C$.*

Proof. From Theorem 2, we have that $X +_S A = C$ with $S(y, x) = A(x) \wedge C(y + x)$, $\forall (y, x) \in \mathbb{R}^2$. Theorem 1 implies that $X +_S A = C = A +_{\bar{S}} X$ and $\bar{S}(x, y) = S(y, x) = R(x, y)$ for all $(x, y) \in \mathbb{R}^2$. Theorem 2 ensures that $X +_S (-B) = C$ with $S(x, y) = B(-y) \wedge C(x+y)$, $\forall (x, y) \in \mathbb{R}^2$. Theorem 1 implies that $X +_S (-B) = C = X -_{S^\star} B$ and $S^\star(x, y) = S(x, -y) = B(y) \wedge C(x-y) = T(x, y)$ for all $(x, y) \in \mathbb{R}^2$. Similarly, one can prove the item 3.

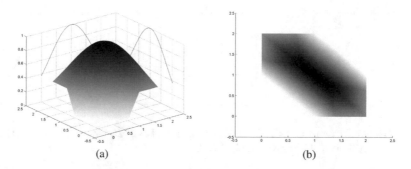

(a)   (b)

Fig. 5. (a) The joint possibility distribution M given as in Eq. (16), with $A = B = (1; 1; 1)_G$ and $C = (1; 2; 3)$. The fuzzy numbers A and B are the marginal distributions of M. (b) The top-view of the joint possibility distribution M.

3 Given Fuzzy Numbers A, B, and C, Is There a Joint Possibility Distribution J of A and B Such that $A +_J B = C$?

This section studies Eq. (11). In particular, we establish necessary and sufficient conditions for a fuzzy number C be written as a J-interactive sum of two given fuzzy numbers A and B. For example, there exists no joint possibility distribution J of the real numbers 1 and 2 such that $1 +_J 2 = (2; 3; 3.5)$. However, it is not obvious if there exists or not a joint possibility distribution J of $(1; 1; 1)_G$ and $(1; 1; 1)_G$ such that $(1; 1; 1)_G +_J (1; 1; 1)_G = (1; 2; 3)$.

Theorem 3. *[5] Let A, B, and C be fuzzy numbers. If there exists a joint possibility distribution J of A and B such that $A +_J B = C$, then the fuzzy set M of $\mathbb{R}^2$ given by*

$$M(x, y) = A(x) \wedge B(y) \wedge C(x + y) \tag{16}$$

is a joint possibility distribution of A and B such that $J \subseteq M$ and $A +_M B = C$.

The next corollary is an immediate consequence of Theorem 3.

Corollary 2. *[5] Let A, B, and C be fuzzy numbers. There exists a joint possibility distribution J of A and B such that $A +_J B = C$ if, and only if, the fuzzy set M given as in Eq. (16) is a joint possibility distribution of A and B and $A +_M B = C$.*

Example 3. Let $A = B = (1; 1; 1)_G$ and $C = (1; 2; 3)$. The joint possibility distribution M given by Eq. (16) and its marginal distributions are depicted in Fig. 5. One can observe in Fig. 6 that A and B are the marginal distributions of M. Moreover, the M-interactive sum of A and B produces the fuzzy number C.

Example 4. Let $A = (-1; -0.5; 0; 1)$, $B = (-1; -0.5; 0.5; 1)$ and $C = (-1.5; 0; 0.5; 1)$. The joint possibility distribution M given by Eq. (16) and its

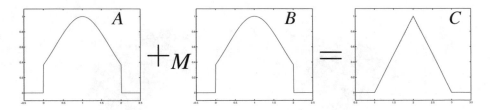

Fig. 6. Visual representations of M-interactive sum of A of B, where $A = B = (1;1;1)_G$ and $C = (1;2;3)$, and M is given as in Fig. 5.

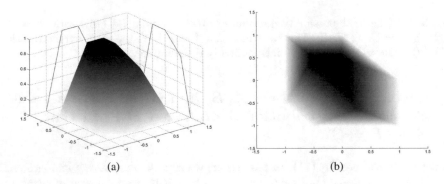

(a)   (b)

Fig. 7. (a) The fuzzy set M given as in Eq. (16), with $A = (-1; -0.5; 0; 1)$, $B = (-1; -0.5; 0.5; 1)$ and $C = (-1.5; 0; 0.5; 1)$. The fuzzy number B is not a marginal distribution of M. (b) The top-view of the joint possibility distribution M.

marginal distributions are depicted in Fig. 7. One can observe in Fig. 7 that B is not a marginal distribution of M. From Corollary 2, there exists no joint possibility distribution J of A and B such that $(-1; -0.5; 0; 1) +_J (-1; -0.5; 0.5; 1) = (-1.5; 0; 0.5; 1)$.

4   Given a Joint Possibility Distribution J of Fuzzy Numbers A and B, Is There a Joint Possibility Distribution N of $(A +_J B)$ and B Such that $(A +_J B) -_N B = A$?

In this section, we focus on Eq. (12). Let C be the J-interactive sum of fuzzy numbers A and B. A natural question that arises here is if there exists a joint possibility distribution N of C and B such that $A = C -_N B$. Theorem 4 provides an affirmative answer to this question, establishing a formula for the maximal solution.

Theorem 4. [5] Let J be a joint possibility distribution of two fuzzy numbers A and B such that $C := A +_J B \in \mathbb{R}_\mathscr{F}$. The fuzzy set N of $\mathbb{R}^2$ whose membership function is given by

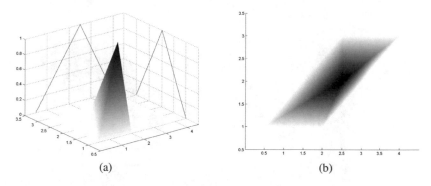

(a)   (b)

Fig. 8. (a) The joint possibility distribution N given as in Eq. (17), with $A = (-0.5; 0.5; 1)$, $B = (1; 2; 3)$, and $C = A + B = (0.5; 2.5; 4)$. Note that C and B are the marginal distributions of N. (b) The top-view of the joint possibility distribution N.

$$N(z, y) = A(z - y) \wedge B(y) \wedge C(z), \quad \forall (z, y) \in \mathbb{R}^2, \tag{17}$$

is a joint possibility distribution of C and B and $A = C -_N B$. Moreover, if there exists a joint possibility distribution $\tilde{J}$ of C and B such that $A = C -_{\tilde{J}} B$, then $\tilde{J} \subseteq N$.

Recall that the standard addition of two fuzzy numbers A and B coincides with the non-interactive addition (*i.e.* via the joint possibility distribution J given as in Eq. (2)). By Theorem 4, we obtain $C -_H B = A = C -_N B$ where $C = A + B$ and N is given by Eq. (17). The next example illustrates this last observation.

Example 5. Let $A = (-0.5; 0.5; 1)$, $B = (1; 2; 3)$, and $C = A + B = (0.5; 2.5; 4)$. The joint possibility distribution N given by Eq. (17) and its marginal distributions C and B are depicted in Fig. 8. As expected from Theorem 4, the equality $C -_N B = A$ holds true.

Example 6 considers the J-interactive sum of two fuzzy numbers A and B where J is given by the Cartesian product of A and B based on the drastic t-norm (t_D) [4].

Example 6. Let $A = (1; 2; 3)$ and $B = (-1; 1; 1.5)_G$ be J-interactive, where $J(x, y) = x \, t_D \, y$ for all $(x, y) \in \mathbb{R}^2$, and let $C = A +_J B$. The joint possibility distribution N given by Eq. (17) and its marginal distributions C and B are depicted in Fig. 9. From Theorem 4, we have $C -_N B = A$.

Next corollary is a consequence of Theorem 4 and Corollary 1.

Corollary 3. *[5] Let $A, B, C \in \mathbb{R}_{\mathscr{F}}$. We have that*

$$A +_M B = C \Leftrightarrow A = C -_N B. \tag{18}$$

where M and N are the fuzzy sets of $\mathbb{R}^2$ given respectively by (16) and (17).

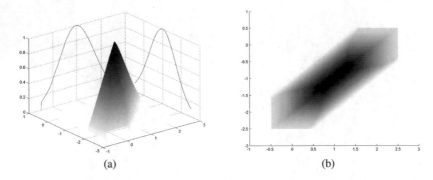

(a)   (b)

Fig. 9. (a) The joint possibility distribution N given as in Eq. (17), with $A = (1; 2; 3)$, $B = (-1; 1; 1.5)_G$ and $C = A +_J B$. Note that C and B are the marginal distributions of N. (b) The top-view of the joint possibility distribution N.

5   Concluding Remarks

In Sects. 2–4, we investigate fundamental questions regarding additions and subtractions of (interactive) fuzzy numbers such as the ones raised in Abstract. Some of these questions are not completely satisfactory answered using only the well-established standard addition and the g-/gH-/H-difference. For example, for every $B, C \in \mathbb{R}_{\mathscr{F}}$ satisfying $(b_0^+ - b_0^-) > (c_0^+ - c_0^-)$, we have there exists no $X \in \mathbb{R}_{\mathscr{F}}$ such that $X + B = C$. Thus, for every $B \in \mathbb{R}_{\mathscr{F}} \setminus \mathbb{R}$ we have that set $\{X + B \mid X \in \mathbb{R}_{\mathscr{F}}\}$ is a proper subset of $\mathbb{R}_{\mathscr{F}}$ which may restrict the use of this addition in some applications. In contrast, when one uses the concept of interactivity, some interesting questions that arise naturally can be answered in the affirmative. Section 2 states that for every $B, C \in \mathbb{R}_{\mathscr{F}}$ there exist a fuzzy number X and a joint possibility distribution S of X and B such that $X +_S B = C$.

The relaxation of the requirement of non-interactivity between the operands implies that every equation of the form $X +_J B = C$ has at least one solution for given $B, C \in \mathbb{R}_{\mathscr{F}}$. It is worth noting that such a relaxation does not increasing the numbers of free variables in this equation since X is a marginal distribution of J. Thus, we are still having a free variable that is the joint possibility distribution J. Theorem 2 presents the maximal solution for the equation $X +_J B = C$. However, one can impose a fixed form for the marginal distribution X of J, say $X = A \in \mathbb{R}_{\mathscr{F}}$. Section 2 establishes necessary and sufficient conditions for the equation $A +_J B = C$ has a solution. In this case, in contrast to standard addition, Example 3 shows that a triangular fuzzy number can be written as an interactive addition of two Gaussian fuzzy numbers. Finally, in Sect. 4, we state that if a fuzzy number C is given by means of a J-interactive sum of the fuzzy numbers A and B, then the fuzzy number A corresponds to the N-interactive difference of C and B (see Theorem 4). Thus, we have $(A +_J B) -_N B = A$ for every joint possibility distribution J of A and B.

Acknowledgements. This work was partially supported by FAPESP under grant no. 2016/26040-7 and CNPq under grants no. 306546/2017-5 and 142414/2017-4.

References

1. Barros, L.C., Bassanezi, R.C., Lodwick, W.A.: A First Course in Fuzzy Logic, Fuzzy Dynamical Systems, and Biomathematics - Theory and Applications. Studies in Fuzziness and Soft Computing, vol. 347. Springer, Berlin (2017)
2. Bede, B., Stefanini, L.: Generalized differentiability of fuzzy-valued functions. Fuzzy Sets Syst. **230**, 119–141 (2013)
3. Carlsson, C., Fullér, R., Majlender, P.: Additions of completely correlated fuzzy numbers. In: Proceedings of the 2004 IEEE International Conference on Fuzzy Systems, vol. 1, pp. 535–539 (2004)
4. Dubois, D., Prade, H.: Additions of interactive fuzzy numbers. IEEE Trans. Autom. Control **26**(4), 926–936 (1981)
5. Esmi, E., Barros, L.C.: Arithmetic operations for interactive fuzzy numbers (2019). Submitted for publication
6. Fullér, R., Majlender, P.: On interactive fuzzy numbers. Fuzzy Sets Syst. **143**(3), 355–369 (2004)
7. Nguyen, H.T.: A note on the extension principle for fuzzy sets. J. Math. Anal. Appl. **64**(2), 369–380 (1978)
8. Puri, M., Ralescu, D.: Differentials of fuzzy functions. J. Math. Anal. Appl. **91**, 552–558 (1983)
9. Stefanini, L., Bede, B.: Generalized Hukuhara differentiability of interval-valued functions and interval differential equations. Nonlinear Anal.: Theory Methods Appl. **71**(3–4), 1311–1328 (2009)

Fuzzy Rule-Based Classification with Hypersphere Information Granules

Chen Fu$^{(\boxtimes)}$ and Wei Lu

School of Control Science and Engineering, Dalian University of Technology,
Dalian City, People's Republic of China
fuchen@mail.dlut.edu.cn, luwei@dlut.edu.cn

Abstract. Fuzzy rule-based classification has been studied by a number of classification architectures. In this study, hypersphere information granules are used to form initial fuzzy classification model in an intuitive and interpretative way. The principle of justifiable granularity offers a certain way to optimizing information granules while facing the coverage and specificity criteria. By engaging a synergy of the principle of justifiable granularity and migrating prototypes, the refined classification model is constructed for better classification performance. A series of experiments concerning synthetic datasets and comparative studies are also implemented to exhibit the feasibility and effectiveness of the proposed classification method.

1 Introduction

Fuzzy classifications [7,10,16] are commonly encountered in the constructs of machine learning because of their capabilities of flexibly building models in many applications. The concept of fuzzy classification and ensuing fuzzy classifiers are proposed by Zadeh and Kalaba [2]. In particular, fuzzy classifiers coming in rule-based form whose transparent architectures augmented by numerous gradient-based and evolutionary techniques have become a central point in developing classification. Classification accuracy is the key indicator of classifier performance. In the continuous improvement of classification accuracy, the fuzzy classification can be easily observed through various applications such as medical systems [15], intrusion detection [20] and earthquake predictions [1] and financial applications [18] among others. In spite of these existing numerous applications, there are still some issues besides enhancing classification accuracy with respect to figuring out the architecture of formed fuzzy classifier and simplifying classification rules formation process. In order to seek for a universal coverage on these issues of the eventually induced classification model, based on the concept of information granules, this approach works by adjusting the number of the formed hypersphere information granules, optimizing them with the principle of justifiable granularity and migrating the prototypes.

   The proposed method includes two main stages. At the first stage, an initial fuzzy classification model is established by engaging a synergy of FCM

© Springer Nature Switzerland AG 2019
R. B. Kearfott et al. (Eds.): IFSA 2019/NAFIPS 2019, AISC 1000, pp. 258–269, 2019.
https://doi.org/10.1007/978-3-030-21920-8_24

algorithm [3] and the information granules [8,24]. The numerical input data is partitioned into several regions by invoking FCM. Fuzzy hypersphere information granules can be formed based on the partitions with prototypes. The second stage focuses on the refinement of the initial model. Here the positions of prototypes formed at the first stage are migrated to further improve the classification performance in some supervised way, in which two significant objectives should be considered. One is that the formed hypersphere information granules can well represent the corresponding data. The other is that the refined model has better classification accuracy on the entire data-base. Especially for the first issue, the principle of justifiable granularity [12–14] is considered, which offers an option to form information granules based on both coverage and specificity criterion. So far, the refined fuzzy granular classification model is finally constructed. Compared to some well-known classifiers, several key facets of the resulting classification model are identified:

- Clustering helps to generate hypersphere information granules intelligibly and transparently, which is used to form the initial classification model
- The number of the generated hypersphere information granules has important impacts on the performance of the constructed classification model.
- The synergy of the principle of justifiable granularity and prototypes migration strategy helps enhancing the performance of the classifier.

The paper is structured as follows: Sect. 2 introduces a way of developing and optimizing hypersphere information granules based on the principle of justifiable granularity. The framework of the two-stage fuzzy rule-based classification model is presented in Sect. 3. In Sect. 4, the experiments on three synthetic data sets and comparison with other classifiers are presented. Section 5 provides some helpful conclusions.

2 Designing Information Granules with the Principle of Justifiable Granularity

In this section, we first introduce the fundamental of building information granules, viz. the principle of justifiable granularity. Next, based on the requirements of the subsequent classification model, we add the parameter modification of the principle of justifiable granularity and establish the optimized objective function in the process of forming information granules.

2.1 Information Granules and the Principle of Justifiable Granularity

Information granules [10,24] are fundamental entities used in all pursuits of Granular Computing [8], which play a pivotal role in human cognitive and decision-making activities. They are instrumental in forming abstract entities in which complex phenomena could be described, modeled, and interpreted.

Fuzzy clustering are commonly used to generate information granules. In a nutshell, large collection of numeric data can be transformed into a small number of information granules through clustering, where the generated information granules own a concise and abstract description of original data.

The principle of justifiable granularity offers an alternative to clustering methods in realizing a significant way to construct information granules on a basis of available experimental evidence. The constructed information granule becomes an abstraction of the available experimental evidence. The underlying rationale behind the principle is to deliver a concise and abstract description of the data such that (i) the produced granule is *justified* on basis of the available experimental data, and (ii) the granule owns a well-defined *semantics* meaning that it can be easily distinguished from the others. These two are expressed by the criterion of coverage and the criterion of specificity. Coverage states how much data samples are embraced in the formed information granule. Specificity stresses the semantics of the information granule. The definition of coverage and specificity depends on the formal nature of information granule to be formed. The information granules obtained after clustering are in the form of hypersphere, denoted as Ω^ρ (see Fig. 1 $\rho \in [0,1]$ stands for the radius of the hypersphere), which are formed on a basis of a collection of normalized data $\mathbf{x}_1, \mathbf{x}_2, \cdots, \mathbf{x}_N \in [0,1]^m$. The coverage is measured in terms of the amount of data within Ω^ρ which involves some distance function, namely

$$Cov(\Omega^\rho) = card\{\mathbf{x}_k | \|\mathbf{x}_k - \mathbf{v}\| \le \rho\}, \tag{1}$$

$card\{\cdot\}$ denotes the cardinality of Ω^ρ, $\mathbf{x}_k$ and $\mathbf{v}$ are the data and the representative positioned in the space of $[0,1]^m$. The specificity of Ω^ρ, says $Sp(\Omega^\rho)$, is set as a decreasing function of the size of the hypersphere information granule. In limit cases, if Ω^ρ is composed of only one element, $Sp(\Omega^\rho)$ reaches the highest value for 1. If Ω^ρ is the entire space, $Sp(\Omega^\rho)$ returns zero. The implementation of specificity comes as

$$Sp(\Omega^\rho) = 1 - \rho, \tag{2}$$

where the radius ρ is regarded as the information granularity of Ω^ρ.

The key task is to build an information granule so that both the coverage and specificity can achieve the highest values. Apparently, the two criteria are in conflict: increasing the specificity will reduce the coverage and vice versa. A viable solution is to take the product of them two and optimize the information granule while the product attains its maximal value. Therefore, we have

$$arg \max_{\rho, \mathbf{v}}\{Cov(\Omega^\rho)Sp(\Omega^\rho)\}. \tag{3}$$

2.2 Granulation Based on the Principle of Justifiable Granularity with Auxiliary Information

With the above explanation, it is obvious that both the clustering and the principle of justifiable granularity can be used to produce information granules. However, it is worth noticing that the difference between them. First, clustering

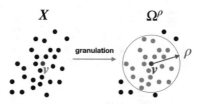

Fig. 1. The information granule Ω^ρ for two-dimensional data $\mathbf{X}$ expresses in form of hypersphere with prototype $\mathbf{v}$ and radius ρ

produces at least two information granules (clusters) and the principle of justifiable granularity leads to only a single one. Second, and also the most important is that the principle of justifiable granularity can be regarded as the follow-up optimization step of the clustering result.

Since the information granules involved in the proposed classification model are generated by clustering in an unsupervised way and used as antecedents for subsequent optimization with the principle of justifiable granularity. The coverage $Cov(\Omega^\rho)$ should be modified (discounted) by the diversity of the data included in Ω^ρ where this diversity is quantified in the form of the entropy function $entr(\Omega^\rho)$

$$Cov'(\Omega^\rho) = Cov(\Omega^\rho)(1 - entr(\Omega^\rho)), \tag{4}$$

where $entr(\Omega^\rho) = -\sum p(\mathbf{x}_k)log(p(\mathbf{x}_k))$. In case of the classification problem, Eq. (4) penalizes the amount of the data partitioned into Ω^ρ and not being homogeneous in terms of class membership $p(\mathbf{x}_k)$. The higher the entropy, the lower the coverage $Cov(\Omega^\rho)$. It's easy to observe that $p(\mathbf{x}_k)$ represents the membership degree(weight) of individual data versus the prototype $\mathbf{v}$ of information granule Ω^ρ). Different positions of prototype $\mathbf{v}$ result in different weights $p(\mathbf{x}_k)$ of individual data. Accordingly, we modify the optimization in Eq. (3) as

$$\arg\max_{\rho,\mathbf{v}}\{Cov(\Omega^\rho)Sp(\Omega^\rho)(1 - entr(\Omega^\rho))\}. \tag{5}$$

For a collection of numeric data, we can obtain some partitions and prototypes through invoking clustering algorithm. Then the hypersphere information granules can be produced and optimized on the principle of justifiable granularity in Eq. (5).

3   Fuzzy Granular Classification Model Based on the Principle of Justifiable Granularity

In this section, we elaborate on a two-stage fuzzy rule-based classification model based on the principle of justifiable granularity. The blueprint of the entire modeling process is presented in Fig. 2. The modeling process is first to establish an initial fuzzy classification model by engaging a synergy of FCM

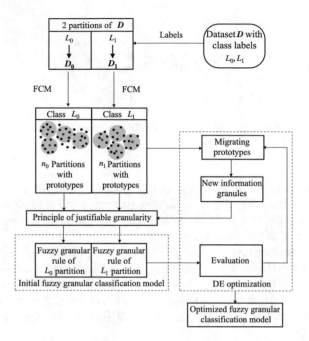

Fig. 2. The process of the proposed fuzzy rule-based classification model

algorithm and hypersphere information granules, and then to refine the initial model by migrating the position of prototypes based on the principle of justifiable granularity. In the construction of the model, a normalized dataset $\mathbf{D} = \{\mathbf{x}_k = (x_{k1}, x_{k2}, \cdots, x_{km}) | k = 1, 2, \cdots, N\}$ with two known labels L_0, L_1 is considered, where $\mathbf{x}_k \in [0, 1]^m$.

3.1 The Design of the Initial Classifier

To make this study easy to follow, we first briefly recall the generic topology of two-class fuzzy rule-based classifier. A two-class fuzzy rule-based classifier is a mapping from the m-dimensional feature space $\{\mathbf{x}_k | k = 1, 2, \cdots, N, \mathbf{x}_k \in [0, 1]^m\}$ (m is set as 2 below for ease of expression) to a classification label space $\{L_0, L_1\}^2$. The "If-then" rules [9,17] are expressed as:

$$\text{If } \mathbf{x}_k \text{ is } \Omega_i^\rho \text{ then } \mathbf{x}_k \text{ is } L_i, i = 0, 1, \tag{6}$$

where Ω_i^ρ is a cluster of dataset $\mathbf{D}$ (expressed as a hypersphere information granule) with label L_i. Next we introduce the process of constructing the initial fuzzy granular classification model based on the hypersphere information granules.

To form the initial classification model, we first split the data D into some partitions in light of the class labels (L_0, L_1) by applying FCM algorithm and transfer the partitions into hypersphere information granules based on the principle of justifiable granularity, see Fig. 3. Specially, each class of data in dataset

$\mathbf{D}$ is clustered separately by directly invoking FCM where the number of clusters is a multiplicity of the number of classes (says, $c = 2, 4, 8, 12$). Apparently, c determines the number of produced information granules. Once the clustering is completed, $2c$ prototypes $\mathbf{v}_1^{L_0}, \mathbf{v}_2^{L_0}, \cdots, \mathbf{v}_c^{L_0}$ and $\mathbf{v}_1^{L_1}, \mathbf{v}_2^{L_1}, \cdots, \mathbf{v}_c^{L_1}$ are returned. Therefore, for the kth element in L_0 partition, its membership degree versus the ith prototype $\mathbf{v}_i^{L_0}$ is determined by the formula

$$u_{ki} = \frac{1}{\sum_{s=1}^{c} \left(\frac{||\mathbf{x}_k - \mathbf{v}_i^{L_0}||}{||\mathbf{x}_k - \mathbf{v}_s^{L_0}||} \right)^{\frac{2}{d-1}}}. \tag{7}$$

In Eq. (7), d is the fuzzification coefficient, whose value is commonly set as 2. Next, we take the case of $c = 2$ as an example to introduce how to establish the hypersphere information granules. Since there are four ($c \times 2$) produced clusters $\mathbf{D}_1^{L_0}, \mathbf{D}_2^{L_0}, \mathbf{D}_3^{L_1}, \mathbf{D}_4^{L_1}$, we use the cluster centers as prototypes and form four corresponding information granules with certain hypersphere radius base on the principle of justifiable granules, says, $\Omega_{\rho_1}^{\mathbf{v}_1^{L_0}}$, $\Omega_{\rho_2}^{\mathbf{v}_2^{L_0}}$, $\Omega_{\rho_3}^{\mathbf{v}_3^{L_1}}$, $\Omega_{\rho_4}^{\mathbf{v}_4^{L_1}}$. Among them, $\Omega_{\rho_1}^{\mathbf{v}_1^{L_0}}$ stands for a hypersphere information granule with the prototype $\mathbf{v}_1^{L_0}$ whose radius is ρ_1, the same goes for other ones, see Fig. 3. Note that, these hypersphere cannot intersect. Therefore, in order to make the information granules better represent the corresponding partitions, the values of the radius should be as higher as possible under the condition that the sum of the radius of two adjacent hypersphere is less than the Euclidean distance of the corresponding prototypes. The produced information granule $\Omega_{\rho_1}^{\mathbf{v}_1^{L_0}}$ depicts the main characteristic of elements in $\mathbf{D}_1^{L_0}$. In the sequel, note that the partition $\mathbf{D}_1^{L_0}$ is closely related with the class label L_0. We directly articulate the information granule produced in L_0 class label such that a fuzzy granular classification rule describing the relation between the partition $\mathbf{D}_1^{L_0}$ and the class label L_0 is formed under the level ρ_1 information granularity, viz.,

$$\text{If } \mathbf{x}_k \text{ is } \Omega_{\rho_1}^{\mathbf{v}_1^{L_0}} \text{ then } \mathbf{x}_k \text{ is } L_0, \tag{8}$$

where $\mathbf{x}_k$ is $\Omega_{\rho_1}^{\mathbf{v}_1^{L_0}}$ means that the element $\mathbf{x}_k$ falls within the border of the hypersphere $\Omega_{\rho_1}^{\mathbf{v}_1^{L_0}}$, that is $d(\mathbf{x}_k, \mathbf{v}_1^{L_0}) \leq \rho_1$. $d(\mathbf{x}_k, \mathbf{v}_1^{L_0})$ means the Euclidean distance of $\mathbf{x}_k$ from $\mathbf{v}_1^{L_0}$ (the same as below). Further, for other partitions of the dataset $\mathbf{D}$, the corresponding fuzzy granular classification rules expressed in the form of Eq. (8) can be also obtained according to the above-mentioned process. At this point, from Fig. 3, it's also worth noting that there are elements that do not fall within any hypersphere information granules, their labels can be obtained by calculating the Euclidean distance of them from all produced prototypes, says, $d(\mathbf{x}_k, \mathbf{v}_1^{L_0}), d(\mathbf{x}_k, \mathbf{v}_2^{L_0}), d(\mathbf{x}_k, \mathbf{v}_3^{L_1}), d(\mathbf{x}_k, \mathbf{v}_4^{L_1})$ and compare $d(\mathbf{x}_k, \mathbf{v}_1^{L_0}) + d(\mathbf{x}_k, \mathbf{v}_2^{L_0})$ with $d(\mathbf{x}_k, \mathbf{v}_3^{L_1}) + d(\mathbf{x}_k, \mathbf{v}_4^{L_1})$. Apparently, $\mathbf{x}_k$ belongs to the nearest prototypes' class. So far the initial fuzzy granular classification model consisting of "If-then" rules can be constructed, that is,

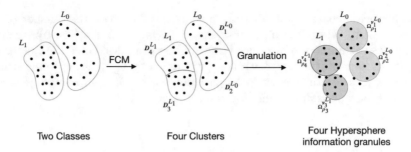

Fig. 3. An example of the process of clustering on basis of two-dimension data $(c = 2)$

$$\text{If } \mathbf{x}_k \text{ is } \Omega_{\rho_j}^{\mathbf{v}_j^{L_i}} \text{ then } \mathbf{x}_k \text{ is } L_i, \ i = 0, 1$$

$$\text{If } \sum_{j=2i+1}^{2i+2} d(\mathbf{x}_k, \mathbf{v}_j^{L_i}) \text{ is the minimum then } \mathbf{x}_k \text{ is } L_i, \ i = 0, 1. \tag{9}$$

In order to quantify the performance of the formed initial classification model, here the classification accuracy, which is one of the most common evaluation index encountered in the classification problem, is considered. It is defined as follows:

$$Q_{Acc} = \sum_{k=1}^{N_s} \frac{f(\mathbf{x}_k)}{N_s} \times 100 \text{ with } f(\mathbf{x}_t) = \begin{cases} 1, & if \ \hat{L}_i = L_i \\ 0, & if \ \hat{L}_i \neq L_i \end{cases}, \tag{10}$$

where N_s is the number of the test samples. L_i and $\hat{L}_i$ are the true class label and the predicted class label of the test sample $\mathbf{x}_k$ $(k = 1, 2, \cdots, N_s)$, respectively.

3.2   The Refine of the Initial Classifier by Migrating Prototypes

In the construction of the initial fuzzy granular classification model, the resulted hypersphere information granules maybe not well represent the corresponding partitions due to Eq. (5), which results in the degeneration of performance of the initial fuzzy rule-based classification model. Thus in this stage, its main task is to refine the performance of the initial model. It is worthy to note that the partition of dataset $\mathbf{D}$ is closely related with the position of prototypes. Following this, the initial fuzzy granular classification model can be further refined by continuously migrating the four prototypes $\mathbf{v}_1^{L_0}, \mathbf{v}_2^{L_0}, \mathbf{v}_3^{L_1}, \mathbf{v}_4^{L_1}$ obtained from the first stage in a supervised way. Apparently, the refinement of the initial classification model is essentially an optimization problem. Let $\varepsilon_{i1}, \varepsilon_{i2}$ become the variate levels of the ith $(i = 1, 2, 3, 4)$ prototype on individual dimensionality (attribute). Based on this migration scheme of prototypes and the objective function for principle of justifiable granularity (see Eq. (5) in Sect. 2), the above-mentioned optimization problem is formally expressed in the form

$$\arg \max_{\Delta \mathbf{v}_1^{L_0}, \Delta \mathbf{v}_2^{L_0}, \Delta \mathbf{v}_3^{L_1}, \Delta \mathbf{v}_4^{L_1}} \{Q_{Acc} + \frac{1}{4} \sum_{j=1}^{4} Cov(\Omega_{\rho_j}^{\mathbf{v}_j^{L_i}}) Sp(\Omega_{\rho_j}^{\mathbf{v}_j^{L_i}})(1 - entr(\Omega_{\rho_j}^{\mathbf{v}_j^{L_i}}))\},$$

(11)

subject to $\mathbf{v}_j^{L_i} + \Delta \mathbf{v}_j^{L_i} \in [0,1]^2$ ($j = 1,2,3,4$) and $d(\mathbf{v}_1^{L_0}, \mathbf{v}_2^{L_0}) \leq (\rho_1 + \rho_2)$, $d(\mathbf{v}_3^{L_1}, \mathbf{v}_4^{L_1}) \leq (\rho_3 + \rho_4)$.

The objective function of this optimization problem contains two meanings: one is the formed hypersphere information granules attains best representation of data under the guidance of the principle of justifiable granularity. The other is that the classification accuracy reaches a better level. To solve the optimization problem Eq. (11), the differential evolution (DE) [23], a population-based optimization technology becomes a sound option. After solving the optimization problem, the corresponding refined fuzzy granular classification model is completely obtained, viz.,

$$\text{If } \mathbf{x}_k \text{ is } \Omega_{\rho_j, opt}^{\mathbf{v}_j^{L_i}} \text{ then } \mathbf{x}_k \text{ is } L_i, \ i = 0,1$$

$$\text{If } \sum_{j=2i+1}^{2i+2} d(\mathbf{x}_k, \mathbf{v}_{j, opt}^{L_i}) \text{is the minimum then } \mathbf{x}_k \text{ is } L_i, \ i = 0,1,$$

(12)

where $\Omega_{\rho_j, opt}^{\mathbf{v}_j^{L_i}}$ is the optimized hypersphere fuzzy information granule of the jth partition in class L_i formed by migrating prototypes.

4 Experimental Studies

In this section, a series of experimental studies are presented to illustrate the construction and performance of the fuzzy classifier. The experiments involved three synthetic datasets which are characteristic in machine learning research. Experimental data sets are normalized and randomly divided into ten partitions with *10-folder cross-validation model*, which randomly combining seven of them (70%) as training set and the remaining one (30%) as test set. The number of clusters of FCM is selected to be a multiplicity of the number of classes (says $c = 2,4,8,12$). The parameter m is set as 2. Thus we can carry out the classification experiments on all data sets and determine the average produced by the ten partitions. The values of the parameters of the DE algorithm used to optimize Eq. (11) are reported in Table 1. Here the symbols Np, Gm, CR and F express the number of populations, the number of generations, the crossover constant and the mutation scaling factor, respectively. These parameters are obtained by a trial-and-error method. We also experimented with some other combinations of these parameters but no visible improvements of the values of objective function Eq. (11) were reported.

Table 1. Parameters used in the experiments.

Parameter	Np	Gm	CR	F
Value	60	100	0.8	0.2

4.1 Data Sets

Three synthetic datasets are involved in the experiments to fully validate the performance of the proposed classification model. The synthetic datasets are labeled in terms of their geometric appearance and each data set is composed of 400 patterns, see Fig. 4, samples belonging to two classes are marked in different colors. In order to better reveal the ability of the principle of justifiable granularity to filter useful information in data, 10% standard deviation of Gaussian noise are added to the data in Circles and Moons datasets. These datasets are very commonly used in machine learning research to validate the performance of various classifiers. To make it easier and more intuitive to compare the performance of the proposed method, comparison experiments involved with other seven classifiers are performed.

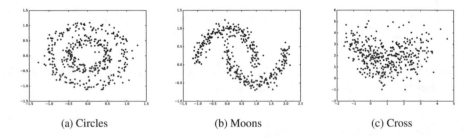

(a) Circles   (b) Moons   (c) Cross

Fig. 4. Synthetic data sets

4.2 Experiment Results and Comparison with Well-Known Classifiers

We experimented on the above three synthetic datasets in light of the proposed method. The corresponding results are reported in Tables 2 and 3.

In Table 2, the symbols Acc_{Tr} and Acc_{Tst} stand for the classification accuracy for the training set and the testing set, respectively. We can see clearly that parameter c (determine the number of formed hypersphere information granules) plays a decisive role in classification performance, which can be adjusted as an important parameter for classifying datasets with different distributions. In Table 3, we report the best average value of our method and compare them with seven well-known classifiers, namely k-nearest neighbor classifier (kNN)[6], Linear support vector machine (L-SVM)[19], SVM with radial basis function (R-SVM)[21], decision trees (DT)[5], random forest (RF)[4], neural network (Nr-N)[22] and quadratic discriminant analysis (QDA)[11]. We compare our average

Table 2. Experimental results for synthetic datasets under different numbers of clusters c.

c	Circles Acc_{Tra}	Circles Acc_{Tst}	Moons Acc_{Tra}	Moons Acc_{Tst}	Cross Acc_{Tra}	Cross Acc_{Tst}
2	52.37 ± 3.22	48.71 ± 4.63	68.39 ± 4.34	69.11 ± 5.85	64.22 ± 4.85	65.21 ± 4.34
4	65.55 ± 4.11	62.46 ± 5.67	84.23 ± 2.11	84.19 ± 2.66	72.17 ± 3.85	74.69 ± 6.28
8	86.34 ± 1.88	85.78 ± 2.50	94.45 ± 1.64	93.63 ± 2.03	73.84 ± 2.15	75.66 ± 1.76
12	98.90 ± 1.26	99.32 ± 2.33	99.39 ± 0.92	98.48 ± 1.80	84.81 ± 2.65	86.72 ± 2.15
20	97.96 ± 1.58	96.08 ± 2.75	98.82 ± 1.21	97.42 ± 2.70	78.42 ± 3.15	77.55 ± 4.37

Table 3. Experimental results comparison.

	kNN	L-SVM	R-SVM	DT	RF	Nr-N	QDA	Ours
Circles	99.38	81.87	100	93.75	96.88	84.83	84.38	99.32
Moons	99.38	49.38	99.38	96.88	96.88	98.75	96.25	98.48
Cross	83.13	86.88	86.88	85.00	86.25	86.88	86.88	86.72

results with the historical best values of the well-known classifiers, our approach exhibits competitive performance. Specially, for datasets whose regions do not intersect but cannot be linearly separable (Circles and Moons), kNN and R-SVM achieve satisfactory results due to the characteristics of their models. Our proposed method keep satisfactory performance in nonlinear separable datasets while performing better than KNN in intersecting dataset (Cross). With regard to R-SVM, we achieve the similar classification performance which the model based on information granules is more interpretable and simple.

As mentioned above, the number of the hypersphere information granules has important impacts on classification accuracy while forming classification model. Further, the principle of justifiable granularity offers a certain way to optimizing the hypersphere information granules and results in the production of well-representative information granules. The principle addresses the coverage and specificity well and helps to resist noise effects. Besides, the migration of prototypes helps to refine the model along with the principle of justifiable granularity. Therefore through adjusting the class parameters c of the initial clustering we can form the best performance model.

5 Conclusions

The paper proposed the fuzzy rule-based classification model with hypersphere information granules. The construction of the model consists of two phases: (i) the constructions of an initial model by the synergy of clustering and forming information granules, (ii) the refinement of the initial model by the synergy of the principle of justifiable granularity and migrating prototypes. A series of experiments are implemented on several synthetic datasets and the results show the proposed method can perform competitive classification accuracy. Some helpful conclusions are summarized as follow: (i) the hypersphere information granules helps constructing initial model intuitively and interpretatively and their numbers can be adjusted due to different datasets; (ii) the principle of justifiable granularity plays a vital role in forming classification models and (iii) the prototypes migration facilitate to enhance the performance of the constructed model.

Acknowledgements. This research was supported by the Natural Science Foundation of China under Grant No. 61876029.

References

1. Asencio-Cortés, G., Martínez-Álvarez, F., Morales-Esteban, A., Reyes, J.: A sensitivity study of seismicity indicators in supervised learning to improve earthquake prediction. Knowl.-Based Syst. **101**, 15–30 (2016)
2. Bellman, R., Kalaba, R., Zadeh, L.: Abstraction and pattern classification. J. Math. Anal. Appl. **13**(1), 1–7 (1966)
3. Bezdek, J.C.: Pattern Recognition with Fuzzy Objective Function Algorithms. Springer, Heidelberg (2013)
4. Bonissone, P., Cadenas, J.M., Garrido, M.C., Díaz-Valladares, R.A.: A fuzzy random forest. Int. J. Approximate Reasoning **51**(7), 729–747 (2010)
5. Breiman, L.: Classification and Regression Trees. Routledge, Abingdon (2017)
6. Cover, T.M., Hart, P.E.: Nearest neighbor pattern classification. IEEE Trans. Inf. Theory **13**(1), 21–27 (1967)
7. Duda, R.O., Hart, P.E., Stork, D.G.: Pattern Classification. Wiley, Hoboken (2012)
8. Fujita, H., Gaeta, A., Loia, V., Orciuoli, F.: Resilience analysis of critical infrastructures: a cognitive approach based on granular computing. IEEE Trans. Cybern. **99**, 1–14 (2018)
9. Ishibuchi, H., Nozaki, K., Tanaka, H.: Distributed representation of fuzzy rules and its application to pattern classification. Fuzzy Sets Syst. **52**(1), 21–32 (1992)
10. Ishibuchi, H., Nakashima, T., Nii, M.: Classification and Modeling with Linguistic Information Granules: Advanced Approach to Linguistic Data Mining. Springer, Heidelberg (2006)
11. Mika, S., Ratsch, G., Weston, J., Scholkopf, B., Mullers, K.R.: Fisher discriminant analysis with kernels. In: Neural Networks for Signal Processing IX: Proceedings of the 1999 IEEE Signal Processing Society Workshop (cat. no. 98th8468), August, pp. 41–48. IEEE (1999)
12. Pedrycz, W., Amato, A., Di Lecce, V., Piuri, V.: Fuzzy clustering with partial supervision in organization and classification of digital images. IEEE Trans. Fuzzy Syst. **16**(4), 1008–1026 (2008)
13. Pedrycz, W., Al-Hmouz, R., Morfeq, A., Balamash, A.: The design of free structure granular mappings: the use of the principle of justifiable granularity. IEEE Trans. Cybern. **43**(6), 2105–2113 (2013)
14. Pedrycz, W.: Granular computing for data analytics: a manifesto of human-centric computing. IEEE/CAA J. Automatica Sinica **5**(6), 1025–1034 (2018)
15. Pota, M., Esposito, M., De Pietro, G.: Designing rule-based fuzzy systems for classification in medicine. Knowl.-Based Syst. **124**, 105–132 (2017)
16. Roubos, J.A., Setnes, M., Abonyi, J.: Learning fuzzy classification rules from labeled data. Inf. Sci. **150**(1–2), 77–93 (2003)

17. Sanz, J.A., Fernández, A., Bustince, H., Herrera, F.: Improving the performance of fuzzy rule-based classification systems with interval-valued fuzzy sets and genetic amplitude tuning. Inf. Sci. **180**(19), 3674–3685 (2010)
18. Sanz, J.A., Bernardo, D., Herrera, F., Bustince, H., Hagras, H.: A compact evolutionary interval-valued fuzzy rule-based classification system for the modeling and prediction of real-world financial applications with imbalanced data. IEEE Trans. Fuzzy Syst. **23**(4), 973–990 (2015)
19. Scholkopf, B., Smola, A.J.: Learning with Kernels: Support Vector Machines, Regularization, Optimization, and Beyond. MIT Press, Cambridge (2002)
20. Shalaginov, A., Franke, K.: Big data analytics by automated generation of fuzzy rules for Network Forensics Readiness. Appl. Soft Comput. **52**, 359–375 (2017)
21. Shawe-Taylor, J., Cristianini, N.: An Introduction to Support Vector Machines and Other Kernel-based Learning Methods, vol. 204. Cambridge University Press, Cambridge (2000)
22. Specht, D.F.: Probabilistic neural networks. Neural Netw. **3**(1), 109–118 (1990)
23. Xu, X., Li, Y.: Comparison between particle swarm optimization, differential evolution and multi-parents crossover. In: 2007 Proceedings of International Conference on Computational Intelligence and Security (CIS 2007), December, pp. 124–127. IEEE (2007)
24. Zadeh, L.A.: Fuzzy sets and information granularity. Adv. Fuzzy Set Theory Appl. **11**, 3–18 (1979)

Multiple Modeling and Fuzzy Switching Control of Fixed-Wing VTOL Tilt-Rotor UAV

Yunus Govdeli[1], Anh Tuan Tran[2], and Erdal Kayacan[3(✉)]

[1] Nanyang Technological University,
50 Nanyang Avenue, Singapore 639798, Singapore
yunus002@e.ntu.edu.sg
[2] Department of Mechanical Systems Engineering,
Nagoya University, Nagoya 464-8603, Japan
tuan.tran@mae.nagoya-u.ac.jp
[3] Department of Engineering, Aarhus University, 8200 Aarhus, Denmark
erdal@eng.au.dk

Abstract. This paper presents a detailed aerodynamic modeling technique along with a fuzzy switching multi-model guidance and control strategy for a custom blended wing-body tilt-rotor unmanned aerial vehicle (UAV). The tilt-rotor configuration affects the overall aerodynamic characteristics significantly and hence, a comprehensive mathematical model is required. Thus, after a configuration selection and design process, a 6DOF mathematical model including a thrust model, propeller effects on the wings and free-stream aerodynamics is developed in this work. Due to a higher tendency of inherently unstable motion with increasing rotor angles and rapidly changing aerodynamic characteristics, multiple models are generated to cover the whole tilting regime. For the control of tilting motion, gain-scheduled flight controllers are preferred. While switching between the models, a fuzzy logic based algorithm is employed. The proposed technique is able to successfully control the aircraft for a full flight envelope from hover to landing. It is also evident from the results that, in the same tilting duration, the fuzzy switching algorithm helps the UAV to show a superior tracking performance when compared to the linear switching technique at a fixed tilting rate.

1 Introduction

Over the past several decades, unmanned aerial vehicles (UAVs) have gained significant attention by proving their usefulness as tools for various practical applications such as surveillance [1], search and rescue [2], 3D mapping [3], structural inspection [4], and many others. Rotary wing UAVs are vertical take-off and landing (VTOL) vehicles where the lift generation and maneuvering solely rely on the thrust generated by propellers [5]. They do not require any runway to be airborne and are practical in challenging areas [6]. On the other hand, fixed-wing UAVs provide a much longer flight endurance resulting in a larger monitoring

© Springer Nature Switzerland AG 2019
R. B. Kearfott et al. (Eds.): IFSA 2019/NAFIPS 2019, AISC 1000, pp. 270–284, 2019.
https://doi.org/10.1007/978-3-030-21920-8_25

area [7]. Therefore, rotary wing systems are preferred mostly for small-scale applications whereas the fixed-wing UAVs are employed for longer missions. In addition to the above-mentioned configurations, the fixed-wing VTOL UAVs combine the capabilities of both of the configurations with sufficient cruise flight endurance and VTOL features [8,9]. In this paper, a blended wing-body configuration with VTOL capability is selected and designed as seen in Fig. 1a. The four front rotors are able to tilt 90° while the two back rotors remain in the same position. It also houses two symmetrically positioned aileron and elevator control surfaces for maneuvering. To facilitate the control of the UAV, the differential thrust between the rotors is also utilized. Its approximate mission path is also presented in Fig. 1b.

Our UAV is designed to be stable for the steady level flight condition. For a conventional fixed-wing UAV, the aerodynamic behavior is quite linear as the angle of attack does not reach very adverse levels [10]. However, with increasing rotor tilting angles, the thrust and the propeller wake behind the rotors act onto the wings at a continuously increasing angle. This results in a behavioral change in the aerodynamics of the wing surfaces [11,12]. Hence, for a particular portion of the wings, assumptions including low-speed aerodynamics cannot be made. Therefore, the aerodynamic models for the wake region, wake-free region, and the thrust are separately designed for our tilt-rotor UAV in this paper.

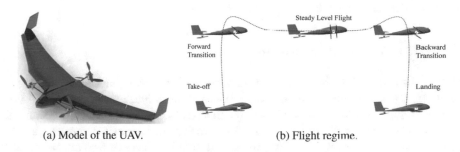

(a) Model of the UAV.   (b) Flight regime.

Fig. 1. (a) Tilt-rotor blended wing-body VTOL UAV model and (b) the flight regime. The UAV takes off vertically, transits to the steady level flight, then transits back to the hover mode, and lands.

In general, a UAV control scheme is in a cascaded form, which consists of two levels, namely, guidance and attitude control. Here, the generation of references for the attitude dynamics varies with respect to the selected control approach. Dynamic inversion or the obtainment of inverse dynamics between the inner and outer loops is a common strategy [13], yet, it is computationally demanding to generate the inverse models, which also include significant assumptions [14]. On the other hand, there are simpler path-following algorithms, i.e., carrot-chasing algorithms, nonlinear guidance laws, line-of-sight based algorithms, which ensure that the UAV remains on a predefined trajectory. In this paper, the carrot-chasing algorithm is selected to generate circular paths [15].

The dynamics of a fixed-wing VTOL UAV changes significantly from hover to steady level flight due to the aerodynamic effects. Therefore, multiple linear models are generated to cover the whole tilting regime and then trim conditions are calculated around these linearization points. There are ten models representing the cases for 10° intervals of tilting. Conventional PID controllers are designed at each linearized model. For the tilting angles between two consecutive trim points, controller gains are calculated via linear interpolations. The tilt-rotor dynamics is considerably different around each of these trim points as introduced above. The stable behavior of the UAV changes depending on the aerodynamic characteristics, and the wake region of the flow behind the propellers. Hence, tilting the rotors at a fixed tilting rate may not result in an optimal solution. With being proven as an efficient technique for nonlinear systems [16], the idea in this paper is to employ a fuzzy switching algorithm between the models so that the transition becomes smoother and less oscillatory. Keeping the tilting rate slower when necessary, and keeping it fast when possible, the tilting can be completed in the same total duration with the fixed tilting rate case, but with smaller deviations from the calculated trim states.

The paper is organized as follows. The mathematical model of the UAV is derived including the thrust and propeller downstream effects in Sect. 2. In Sect. 3, the trim analysis and proposed controller structures are presented. Section 4 showcases the simulation studies and their discussions. Finally, conclusions are drawn in Sect. 5.

2   Mathematical Modeling

This section includes the derivation of a nonlinear mathematical model of the UAV. In our approach, we focus on an in-depth aerodynamic model of the flow over the wings. Unlike other applications, in this paper, the thrust generation and the propeller downstream are modeled separately for a more accurate aerodynamic representation. Specifically, the aerodynamic effects of free airstream and propeller-induced airstream are modeled using the aerodynamic data obtained from CFD in ANSYS Fluent and DATCOM analyses [17]. These datasets are used to generate aerodynamic lookup-tables for lift, drag, and moment coefficients along each axis. Furthermore, the thrust generation is modeled using the propeller disk theory and embedded into the aerodynamic model. The assumptions made for the derivations are highlighted in each respective section. The general equations of motion for the UAV are obtained using Newton's second law by assuming that the XZ plane of the UAV body axis system is the symmetry plane, the mass of the UAV remains constant, the UAV is a rigid body, and the Earth is the inertial reference. Two coordinate systems are used to describe the states of the UAV. The fixed inertial frame points to North-East-Down (NED) of the Earth. Secondly, in Fig. 2, the body frame $(O_B; X_B; Y_B; Z_B)$ is located at the center of gravity (CG) of the UAV with the X_B-axis pointing to the front and the Y_B-axis to the right side of the system. Hence, the equations of motion are obtained as

$$
\begin{aligned}
m(\dot{u} + qw - rv) &= F_{GX} + F_{AX} + F_{TX}, & \dot{p}I_{xx} + qr(I_{zz} - I_{yy}) - (\dot{r} + pq)I_{xz} &= L_A + L_T, \\
m(\dot{v} + ru - pw) &= F_{GY} + F_{AY} + F_{TY}, & \dot{q}I_{yy} - pr(I_{zz} - I_{xx}) + (p^2 - r^2)I_{xz} &= M_A + M_T, \\
m(\dot{w} + pv - qu) &= F_{GZ} + F_{AZ} + F_{TZ}, & \dot{r}I_{zz} + pq(I_{yy} - I_{xx}) + (qr - \dot{p})I_{xz} &= N_A + N_T,
\end{aligned}
\tag{1}
$$

where F_* denote forces, L_*, M_*, N_* stand for moments. The subscripts A and T stand for aerodynamic and thrust components, respectively, while gravitational forces are shown with F_{G_*} [18]. On the left-hand side, the mass is indicated as m, while I_* denote the moments of inertia. The translational speed and rotational rate along X_B, Y_B and Z_B are shown with u, v, w and p, q, r, respectively. Finally, the gravitational force components, body angular velocities and the translational equations can be represented as follows

$$
\begin{bmatrix} F_{GX} \\ F_{GY} \\ F_{GZ} \end{bmatrix} = R \begin{bmatrix} 0 \\ 0 \\ mg \end{bmatrix}, \quad
\begin{bmatrix} p \\ q \\ r \end{bmatrix} = R \begin{bmatrix} \dot{\phi} \\ \dot{\theta} \\ \dot{\psi} \end{bmatrix}, \quad
\begin{bmatrix} \dot{x} \\ \dot{y} \\ \dot{z} \end{bmatrix} = R_E \begin{bmatrix} u \\ v \\ w \end{bmatrix},
\tag{2}
$$

where R and R_E are Euler rotation and transformation matrices. Here, ϕ, θ, ψ stand for roll, pitch and yaw angles, respectively, while x, y, and z denote the displacements in the fixed inertial frame [18]. In the upcoming subsections, the forces and moments generated by the propellers and the aerodynamic effects are calculated.

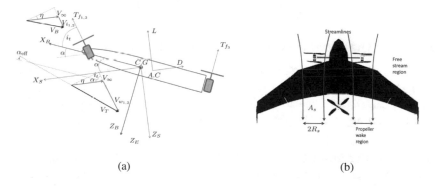

(a)   (b)

Fig. 2. Earth, body, stability axes (a). Free stream and propeller wake regions on the wing (b).

2.1 Forces and Moments Generated by the Thrust of Propellers

In this subsection, the thrust generation model is discussed in detail. As highlighted in Fig. 2, the propeller upstream velocity V_B along the body axis is a resultant vector of the free-stream velocity V_∞ and the propeller induction effect

V_i. In a similar fashion, the total downstream velocity V_T is obtained using the output velocity V_w and V_∞. Thus, the effective angle of attack α_{eff} is the angle between the total flight velocity vector and the body axis of the system. Here, i_t is the tilting angle of the front rotors and α is the aerodynamic angle of attack. Considering the upstream and downstream speeds of the propellers, we can write the conservation of momentum and energy equations in the direction normal to the propeller disk [19,20], and then the thrust T_f generated by the front propellers can be determined as

$$T_f = 2\rho A V_i \sqrt{(V_\infty \sin(i_t + \alpha))^2 + (V_\infty \cos(i_t + \alpha) + V_i)^2}, \tag{3}$$

where the induced velocity V_i can appropriately be calculated as in [19,20]. Accordingly, the velocity of the flow behind the propeller, V_T, and its angle of attack to a proportion of the wing, α_{eff}, can be obtained as

$$V_T = \sqrt{\left(V_w \xi \sin\left(i_t + \alpha\right)\right)^2 + \left(V_w \xi \cos\left(i_t + \alpha\right) + V_\infty \cos \beta\right)^2}, \tag{4}$$

$$\alpha_{\text{eff}} = \arctan \frac{V_\infty \sin \alpha - V_w \sin i_t}{V_\infty \cos \alpha + V_w \cos i_t}, \quad V_w = 2V_i, \quad \xi = \frac{\pi}{2} - i_t - \alpha. \tag{5}$$

After the completion of the derivation of forces and moments caused by the wake region, the induction caused by the rotating propeller can be acquired as below

$$F_{TX} = \frac{(T_{f_1} + T_{f_2}) \cos(i_t)}{m}, \qquad L_T = (T_{f_1} - T_{f_2}) \sin(i_t) l_1,$$

$$F_{TY} = 0, \qquad M_T = (T_{f_1} + T_{f_2}) \sin(i_t) l_2 - T_{f_3} l_3, \tag{6}$$

$$F_{TZ} = \frac{-(T_{f_1} + T_{f_2}) \sin(i_t) - T_{f_3}}{m}, \quad N_T = (T_{f_1} - T_{f_2}) \cos(i_t) l_1.$$

2.2   Aerodynamic Forces and Moments on the Wing

Multiple aerodynamic models are generated and integrated over the wing to obtain the total aerodynamic forces and moments. The main components of these models are the free-stream forces/moments and the ones generated on the propeller wake regions. This approach is very similar to the one in [11].

2.2.1   Aerodynamic Forces and Moments in the Free-Stream Region

The first component of the total aerodynamic model is the free-stream model. This model is generated using the analysis results from DATCOM. The aerodynamic coefficients here show the characteristics of a conventional fixed-wing model. Hence, considering the obtained aerodynamic data, (7) and (8) can be written as follows

$$\bar{q}^{\text{fs}} = \frac{1}{2}\rho V_\infty^2, \qquad C_L^{\text{fs}} = C_{L_0}^{\text{fs}} + C_{L_{\delta_e}}^{\text{fs}}(\delta_e) + \frac{\bar{c}^{\text{fs}}}{2V_\infty}\left(C_{L_{\dot{\alpha}}}^{\text{fs}}\dot{\alpha} + C_{L_q}^{\text{fs}}q\right),$$

$$C_D^{\text{fs}} = C_{D_\alpha}^{\text{fs}}(\alpha) + C_{D_{\delta_e}}^{\text{fs}}(\delta_e), \quad C_M^{\text{fs}} = C_{m_\alpha}^{\text{fs}}\alpha + C_{m_{\delta_e}}^{\text{fs}}\delta_e + \frac{\bar{c}^{\text{fs}}}{2V_\infty}\left(C_{m_{\dot{\alpha}}}^{\text{fs}}\dot{\alpha} + C_{m_q}^{\text{fs}}q\right),$$

$$\tag{7}$$

$$C_l^{\text{fs}} = C_{l_\beta}^{\text{fs}}\beta + C_{l_{\delta_a}}^{\text{fs}}\delta_a + C_{l_{\delta_r}}^{\text{fs}}\delta_r + \frac{b^{\text{fs}}}{2V_\infty}\left(C_{l_p}^{\text{fs}}p + C_{l_r}^{\text{fs}}r\right),$$

$$C_{Y_w} = C_{Y_\beta}^{\text{fs}}\beta + C_{Y_{\delta_a}}^{\text{fs}}\delta_a + C_{Y_{\delta_r}}^{\text{fs}}\delta_r + \frac{b^{\text{fs}}}{2V_\infty}\left(C_{Y_p}^{\text{fs}}p + C_{Y_r}^{\text{fs}}r\right), \qquad (8)$$

$$C_n^{\text{fs}} = C_{n_\beta}^{\text{fs}}\beta + C_{n_{\delta_a}}^{\text{fs}}\delta_a + C_{n_{\delta_r}}^{\text{fs}}\delta_r + \frac{b^{\text{fs}}}{2V_\infty}\left(C_{n_p}^{\text{fs}}p + C_{n_r}^{\text{fs}}r\right),$$

where the superscript 'fs' stands for free stream. Since these forces will be evaluated along the body axes, it is convenient to obtain the corresponding force coefficients as

$$C_X^{\text{fs}} = -C_D^{\text{fs}}\cos(\alpha) + C_L^{\text{fs}}\sin(\alpha), \quad C_Z^{\text{fs}} = -C_D^{\text{fs}}\sin(\alpha) - C_L^{\text{fs}}\cos(\alpha). \qquad (9)$$

Finally, the forces and moments generated on the entire wing by the free-stream airflow can be calculated as

$$X_B^{\text{fs}} = C_X^{\text{fs}}\bar{q}^{\text{fs}}S^{\text{fs}}, \qquad Y_B^{\text{fs}} = C_Y^{\text{fs}}\bar{q}^{\text{fs}}S^{\text{fs}}, \qquad Z_B^{\text{fs}} = C_Z^{\text{fs}}\bar{q}^{\text{fs}}S^{\text{fs}},$$
$$L_B^{\text{fs}} = C_l^{\text{fs}}\bar{q}^{\text{fs}}S^{\text{fs}}b^{\text{fs}}, \quad M_B^{\text{fs}} = C_m^{\text{fs}}\bar{q}^{\text{fs}}S^{\text{fs}}\bar{c}^{\text{fs}}, \quad N_B^{\text{fs}} = C_n^{\text{fs}}\bar{q}^{\text{fs}}S^{\text{fs}}b^{\text{fs}}. \qquad (10)$$

2.2.2 Aerodynamic Forces and Moments in the Propeller Wake Region

The aerodynamic forces, X_B^{wake}, Y_B^{wake}, Z_B^{wake} and moments, L_B^{wake}, m_B^{wake}, N_B^{wake}, in the propeller wake region can be calculated similarly to Sect. 2.2.1. However, in this case, the velocity of the flow after propellers V_T and the effective angle of attack α_{eff} should be considered. Moreover, since α_{eff} can be large due to the tilting motions of rotors, the aerodynamic coefficient data collected from DATCOM, which is only valid in the low angle of attack region, cannot be used in this case. Instead, a number of computational fluid dynamics analyses have been done using ANSYS Fluent software to estimate the aerodynamic coefficients of the wing part in the propeller wake region. Finally, the total aerodynamic forces including all possible propeller induction and downstream effects can be summarized as

$$F_{Ax} = X_B^{\text{fs}} + X_B^{\text{wake}}, \quad F_{Ay} = Y_B^{\text{fs}} + Y_B^{\text{wake}}, \quad F_{Az} = Z_B^{\text{fs}} + Z_B^{\text{wake}},$$
$$L_A = L_B^{\text{fs}} + L_B^{\text{wake}}, \quad M_A = M_B^{\text{fs}} + M_B^{\text{wake}}, \quad N_A = N_B^{\text{fs}} + N_B^{\text{wake}}. \qquad (11)$$

3 Tilt-Rotor Flight Control

This section describes the methodology for the control of the tilt-rotor UAV. Considering the fact that the wings remain at low angles of attack, the UAV shows the characteristics of a fixed-wing system. However, with increasing rotor angles, it has a tendency to act more like a multi-rotor system. Due to the complex aerodynamic characteristics of the proposed configuration, the aircraft is inherently unstable for high tilt angles, e.g., 70°, 80°, and 90°. As such, the controller structure consists of a level of stability augmentation systems (SASs), control augmentation systems (CASs), and guidance laws as shown in Fig. 3. Control inputs are control surface inputs and the differential thrust generated by the rotors [21,22]. To control the pitch motion, elevator deflection and the differential thrust between the front and back rotors are used. In the same fashion, lateral and directional control is carried out using the aileron input and the thrust difference between the right-hand side and left-hand side rotors. There is no separate rudder unit on this system so the rolling and yawing motion are controlled together. The significant aircraft design parameters are listed in Table 1.

After modeling the 6DOF dynamics of the UAV, multiple trim points covering the whole tilting regime are generated. In total, there are ten linear models representing the trim states with $10°$ intervals. Conventional PID controllers are designed for each linearized model. SAS gains are calculated to stabilize the system at a particular linear model using the cost function in [22]. CAS and the guidance loop gains are calculated to control the attitude angles and the position of the UAV, respectively, using the second cost function in [22]. Following the gain calculations, the simulations are carried out with the nonlinear model of the system. To find out the gains in-between the trim points, the controller gains are linearly interpolated. For this purpose, the method in [21] and [22] are used. Then, to check the robustness of the controllers, the controller gains are tested for the neighboring linear models as well. The tilt schedule is generated using a fixed tilt rate by inputing the tilt angle to the simulator from $90°$ to $0°$.

For the guidance of the UAV, the longitudinal control is evaluated separately than the lateral-directional control. As explained in the Introduction, there are numerous approaches developed for the guidance of a UAV. Most of these algorithms function in XY plane for a simpler calculation of a reference attitude. In this paper, we focus on a carrot-chasing algorithm. UAVs mostly track combinations of straight lines and circular paths. The carrot-cashing algorithm is a method developed to follow such paths. However, by its nature, it is a 2D track-

Table 1. System configuration and design parameters.

Specification	
Airfoil type	MH61
Take-off weight m	3.8 kg
Wingspan b	1.61 m
Mean aerodynamic chord $\bar{c}$	0.272 m
Distance between front propellers	0.6 m
Distance between front and rear propellers	0.6 m
Moment of inertia I_{xx}	0.3794 kgm^2
Moment of inertia I_{yy}	0.1002 kgm^2
Moment of inertia I_{zz}	0.4755 kgm^2
Moment of inertia I_{xz}	-0.0064 kgm^2

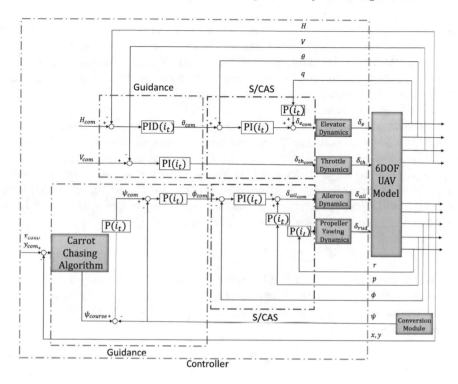

Fig. 3. Switching and cascaded controller structure of the tilt-rotor UAV. The switching algorithm adjusts the controller gains and rotor tilting rate di_t/dt with respect to the instantaneous tilt angle i_t. The outer loops represent the guidance control whereas the inner loops are used for the attitude control. The actuators and throttle dynamics are also taken into account when calculating the control input to the 6DOF UAV model.

ing algorithm and it assumes that the aircraft speed and attitude controllers are fully functional. Since we already have a functional set of altitude, speed, and attitude controllers for our UAV, we can obtain a complete 3D guidance and control for the system. The basic idea in the carrot-chasing algorithm is to update the location of a virtual target point with respect to the current position of the UAV. Accordingly, a reference course angle is generated such that the UAV follows any pre-defined path like a "rabbit" chasing a "carrot" [15].

3.1   Fuzzy Switching Logic Between Models

As explained thoroughly in the aerodynamic modeling sections, a portion of the wing is under the effect of the propeller wake and thrust force, which causes a high angle of attack flow on it. Consequently, the aerodynamic characteristic of the UAV for the low tilting angles are significantly different from the high tilting angles. Therefore, the trim states obtained over the course of tilting show distinctive features. It is also worth noting that the controller gains are calculated via linear interpolations between each trim point. This results in the fact that the switches between two consecutive models take different durations until the UAV is stabilized. Therefore, a nonlinear tilting function accounting for this effect is necessary. The idea in this paper is to generate this nonlinear function using a fuzzy logic-based switching algorithm and to reduce the deviations from the calculated trim conditions.

The switching algorithm is designed as a function of the tilting angle. As reflected in Fig. 4c, each trim point has a corresponding trim speed. Therefore, three Gaussian membership functions (MFs) representing the partial membership of the tilting angle i_t to a set of low μ_1, average μ_2 and high μ_3 angles are defined as shown in Fig. 4a. Here, the centers and the width of the MFs are $[c_1, c_2, c_3] = [0, 45, 90]$ and $[\sigma_1, \sigma_2, \sigma_3] = [30, 20, 30]$, respectively. Using a Sugeno type fuzzification, the ith fuzzy rule can be written as [23]

$$IF\ i_t\ is\ \tilde{A}_i;\ \ THEN\ di_t/dt = k_i, i = 1; \ldots, I, \tag{12}$$

where I is the number of MFs. The tilting rate di_t/dt is the output variable, k_i is the consequent part of the fuzzy system, and $\tilde{A}_i$ is a type-1 fuzzy set. Then, the corresponding tilting rate is calculated by (13). The resulting distribution is given in Fig. 4b.

$$\frac{di_t}{dt} = \frac{\sum_{i=1}^{I} \mu_i k_i}{\sum_{i=1}^{I} \mu_i}. \tag{13}$$

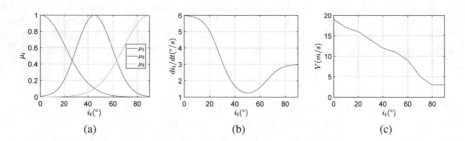

(a)   (b)   (c)

Fig. 4. Selected membership functions for the tilt angle (a) and its corresponding tilting output (b). The MFs are used to generate the tilting rate di_t/dt with respect to the instantaneous tilting angle i_t. The MF selection is carried out considering the relationship between the tilt angle and speed interpolations (c).

4   Simulation Results and Discussions

In this section, the simulation results of the fixed and fuzzy switching rates are presented. For the tilting motion, the switching duration between the models is very critical. The UAV is able to balance the loads at each model defined in $10°$ intervals. Each model is represented by a model number as in Fig. 5a. The highest model number, 9, represents the near hover condition whereas the model number 0 corresponds to the steady level flight condition. Furthermore, one should also note that it takes an average of 30 s to balance the UAV at each model as shown in Fig. 5b. The reason of balancing the loads at each trim point is to show the stability of the UAV at each trim point. By this manner, it is proven that the UAV would be able to maneuver at each trim point and we would have control at every tilting angle. During the tilting period, the altitude reference of the UAV is kept the same at 10 m as highlighted in Fig. 5d. The corresponding reference pitch angle θ_{ref} is also shown in Fig. 5e. From these graphs, it is noticed that the oscillations at the altitude and pitch angle depend strictly on the magnitude of tilting rate. When the tilting rate is kept low throughout the transition, then the magnitudes of the oscillations can be reduced significantly. However, this results in a much longer duration for tilting of the rotors. Therefore, in order to keep the tilting duration at a practical level, the fixed tilting rate di_t/dt is selected as $3°/s$. Hence, by decreasing the tilting rate where the oscillations are adverse, we can obtain a better performance. For this purpose, the fuzzy algorithm described in the previous section is employed. All Fig. 5 results compare the performances of the fixed and fuzzy tilting rates. As shown in Figs. 5d, e and f, for the switches between 6th and 4th models, $3°/s$ tilting rate is too high and causes an oscillatory response. This is the region where the aerodynamic forces start getting more

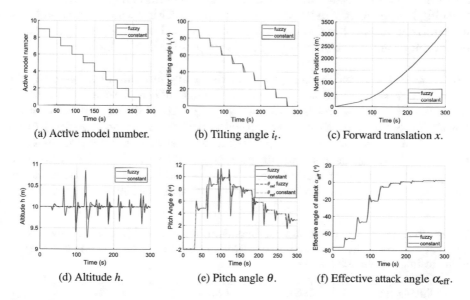

(a) Active model number.   (b) Tilting angle i_t.   (c) Forward translation x.

(d) Altitude h.   (e) Pitch angle θ.   (f) Effective attack angle α_{eff}.

Fig. 5. Comparison of the cases of the fixed tilt rate $di_t/dt = 3°$ and the fuzzy logic based tilting for an intermittent tilting case. (a) highlights the initial tilting position of the UAV whereas (b) shows the change in the tilt angle i_t. The forward translation (c), altitude (d), and the desired pitch angle tracking (e) are also compared for the fuzzy and constant tilting cases. The corresponding efficient angle of attack changes are reflected in (f).

prominent and the UAV transients from the multi-rotor dominant dynamics
to a more fixed-wing dominant dynamics. Therefore, it is very sensitive to the
tilting rate in this range, and a slower tilting is preferred as dictated by the fuzzy
switching algorithm. The further tilting angles are not susceptible to the tilting
rate as much as in this region. Thus, the tilting rate is increased even higher
than $3°/s$ for further tilting angles of which results are presented in Table 2. It is
evident that the maximum error in altitude and pitch angle tracking is decreased
by 45.2% and 5.5%, respectively. Besides, the mean absolute error (MAE) and
the root mean squared error (RMSE) are reduced by 24.4% and 31.3% for the
altitude and 20.3% and 23.7% for the pitch angle, respectively.

Continuous Tilting: After checking the behavior of the fuzzy switching algorithm
in intermittent tilting, the continuous tilting case is investigated. As shown in
Fig. 6a, in this case, the time spent between each model is varying for the fuzzy
switching algorithm. The resulting rotor tilting behavior is shown in Fig. 6b,
where both fuzzy and constant tilting strategies take the same total duration
but in different regimes. The distance covered during the transition is shown in
Fig. 6c, where fuzzy algorithm takes 41 m less than the constant tilting rate case.
The differences of oscillations in the altitude can be seen in Fig. 6d and the pitch
angle tracking is highlighted in Fig. 6e. Although the fuzzy switching algorithm
outperforms the constant switching algorithm, the performance of the pitch angle
tracking is significantly reduced in the continuous tilting when compared to the
intermittent tilting. Finally, the change of α_{eff} is given in Fig. 6f. The control
performance in the continuous tilting case is presented in Table 3.

3D Trajectories: After the completion of the tilting motion, a complete circular
trajectory tracking result is presented for the sake of completeness of the 6DOF
maneuver. Figures 7a and b showcase the intermittent tilting and the continuous tilting cases, respectively. The initial oscillations are caused by tilting of the
rotors. The last group of oscillations is due to the start of the loiter flight. Since
the rolling and yawing motions are controlled only with the differential thrust
and aileron inputs, there are naturally some oscillations. As the carrot-chasing
algorithm does not have any control on the altitude, the reference lines are also

Table 2. Comparison of results for maximum error (MAX), mean absolute error
(MAE), and root mean squared error (RMSE) of h and θ for the whole transition
flight in intermittent tilting case.

Tilting type	MAX		MAE		RMSE	
	h (m)	θ (°)	h (m)	θ (°)	h (m)	θ (°)
Fixed rate switching	0.8511	5.6510	0.0726	0.3797	0.1590	0.9277
Fuzzy rate switching	0.4663	5.3970	0.0549	0.3027	0.1092	0.7080
Difference (%)	45.2	4.5	24.4	20.3	31.3	23.7

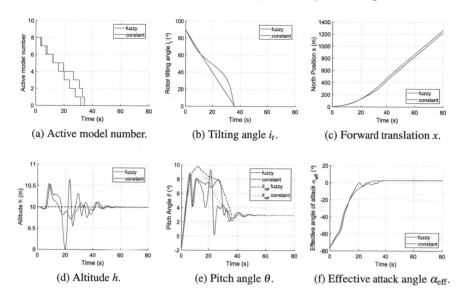

(a) Active model number.   (b) Tilting angle i_t.   (c) Forward translation x.

(d) Altitude h.   (e) Pitch angle θ.   (f) Effective attack angle α_{eff}.

Fig. 6. Comparison of the cases of the fixed tilt rate $di_t/dt = 2.5°$ and the fuzzy logic based tilting for a continuous tilting case. (a) highlights the initial tilting position of the UAV whereas (b) shows the change in the tilt angle i_t. The forward translation (c), altitude (d) and the desired pitch angle tracking (e) are also compared for the fuzzy and constant tilting cases. The corresponding efficient angle of attack changes are reflected in (f).

Table 3. Comparison of results for maximum error (MAX), mean absolute error (MAE), and root mean squared error (RMSE) of h and θ for the whole transition flight in continuous tilting case.

Tilting type	MAX		MAE		RMSE	
	h (m)	θ (°)	h (m)	θ (°)	h (m)	θ (°)
Fixed rate switching	0.9940	5.7360	0.0955	0.6016	0.2069	1.2934
Fuzzy rate switching	0.5409	4.8832	0.0653	0.5271	0.1212	1.0256
Difference (%)	45.6	14.9	31.6	12.4	41.4	20.7

oscillating at the beginning of the circular motion. However, with oscillations in altitude dying out due to the longitudinal flight controller, the carrot-chasing algorithm starts generating horizontal references and UAV tracks down the reference successfully.

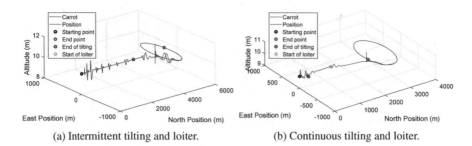

(a) Intermittent tilting and loiter.   (b) Continuous tilting and loiter.

Fig. 7. Complete tilting followed by a loiter flight. The UAV starts its motion at the hover condition. In (a) the rotors are tilted 10° every 30 s intermittently whereas in (b), they are tilted continuously in the first 36 s. The initial picks in the reference lines highlighted in red are caused by the coupled rolling-yawing motion when starting the loiter flight.

5   Conclusions

In this study, a multi-model flight control of a tilt-rotor blended wing-body UAV with a fuzzy switching algorithm is presented. First, a 6DOF nonlinear model of the UAV is generated. Then, conventional PID controllers are designed at ten different trim states which represent the flight conditions at ten different tilt angles of the rotors. For the angles in-between the trim models, linear interpolations are carried out to calculate the controller gains. As the controller inputs, the differential thrust and control surface inputs are considered simultaneously. The rotors are first tilted at 3° fixed tilting rate and then with a varying tilting rate defined by the fuzzy algorithm. The results indicate that, for the intermittent tilting case, the fuzzy algorithm helps to reduce the RMSE of the oscillations by 31.3% and 41.4% for the altitude and 23.7% and 20.7% for the pitch angle tracking, respectively. The main benefit of the fuzzy algorithm is the fact that the tilting rate is reduced where the oscillations are significant. In the end, a case study of a full transition from hover to forward and then to a loiter flight is presented. It is evident from the results that the UAV is capable of maneuvering on the given trajectory successfully. As a future study, the linear interpolations between the models will also be fuzzified for a better trajectory tracking performance.

Acknowledgements. This research is supported by the National Research Foundation, Prime Ministers Office, Singapore under its Medium-Sized Centre funding scheme.

References

1. Tomic, T., Schmid, K., Lutz, P., Domel, A., Kassecker, M., Mair, E., Grixa, I.L., Ruess, F., Suppa, M., Burschka, D.: Toward a fully autonomous UAV: research platform for indoor and outdoor urban search and rescue. IEEE Robot. Autom. Mag. **19**(3), 46–56 (2012)
2. Faessler, M., Fontana, F., Forster, C., Mueggler, E., Pizzoli, M., Scaramuzza, D.: Autonomous, vision-based flight and live dense 3D mapping with a quadrotor micro aerial vehicle. J. Field Robot. **33**(4), 431–450 (2016). https://onlinelibrary.wiley.com/doi/abs/10.1002/rob.21581
3. Nex, F., Remondino, F.: UAV for 3D mapping applications: a review. Appl. Geom. **6**(1), 1–15 (2014). https://doi.org/10.1007/s12518-013-0120-x
4. Bircher, A., Alexis, K., Burri, M., Oettershagen, P., Omari, S., Mantel, T., Siegwart, R.: Structural inspection path planning via iterative viewpoint resampling with application to aerial robotics. In: Proceedings of 2015 IEEE International Conference on Robotics and Automation (ICRA), May 2015, pp. 6423–6430 (2015)
5. Austin, R.: Unmanned Aircraft Systems: UAVS Design, Development and Deployment, vol. 54. Wiley, Hoboken (2011)
6. Govdeli, Y., Wong, Z.W., Kayacan, E.: Additive manufacturing of unmanned aerial vehicles: current status, recent advances, and future perspectives. In: Proceedings of the 2nd International Conference on Progress in Additive Manufacturing (Pro-AM 2016), pp. 39–48 (2016)
7. Beard, R.W., Kingston, D., Quigley, M., Snyder, D., Christiansen, R., Johnson, W., McLain, T., Goodrich, M.: Autonomous vehicle technologies for small fixed-wing UAVs. J. Aerosp. Comput. Inf. Commun. **2**(1), 92–108 (2005)
8. Low, J.M., Govdeli, Y., Ravindrababu, S., Kayacan, E.: On the comparison of diamond honeycomb and 3D-Kagome structures for 3D printed UAVs. In: Proceedings of the 2nd International Conference on Progress in Additive Manufacturing (Pro-AM 2018), pp. 341–346 (2018)
9. Lyu, X., Gu, H., Wang, Y., Li, Z., Shen, S., Zhang, F.: Design and implementation of a quadrotor tail-sitter VTOL UAV. In: Proceedings of 2017 IEEE International Conference on Robotics and Automation (ICRA), May 2017, pp. 3924–3930 (2017)
10. Nelson, R.C.: Flight Stability and Automatic Control, vol. 2. WCB/McGraw Hill, New York (1998)
11. Yuksek, B., Vuruskan, A., Ozdemir, U., Yukselen, M., Inalhan, G.: Transition flight modeling of a fixed-wing VTOL UAV. J. Intell. Robot. Syst. **84**(1–4), 83–105 (2016)
12. Ravindrababu, S., Govdeli, Y., Wong, Z.W., Kayacan, E.: Evaluation of the influence of build and print orientations of unmanned aerial vehicle parts fabricated using fused deposition modeling process. J. Manuf. Process. **34**, 659–666 (2018)
13. Singh, S., Padhi, R.: Automatic path planning and control design for autonomous landing of UAVs using dynamic inversion. In: Proceedings of 2009 American Control Conference, June 2009, pp. 2409–2414 (2009)
14. Hervas, J.R., Reyhanoglu, M., Tang, H., Kayacan, E.: Nonlinear control of fixed-wing UAVs in presence of stochastic winds. Commun. Nonlinear Sci. Numer. Simul. **33**, 57–69 (2016)
15. Sujit, P., Saripalli, S., Sousa, J.B.: Unmanned aerial vehicle path following: a survey and analysis of algorithms for fixed-wing unmanned aerial vehicless. IEEE Control Syst. **34**(1), 42–59 (2014)
16. Sehab, R.: Fuzzy PID supervision for a nonlinear, system: design and implementation. In: Proceedings of NAFIPS 2007 - 2007 Annual Meeting of the North American Fuzzy Information Processing Society, June 2007, pp. 36–41 (2007)

17. Williams, J.E., Vukelich, S.R.: The USAF stability and control digital datcom: users manual. Techncial report, vol. I. McDonnell Douglas Astronautics, Co St Louis, MO (1979)
18. Napolitano, M.R.: Aircraft Dynamics: From Modeling to Simulation. Wiley, Hoboken (2012)
19. Johnson, W.: Rotorcraft Aeromechanics. Cambridge Aerospace Series. Cambridge University Press, Cambridge (2013)
20. Nguyen, H.D., Yu, L., Mori, K.: Aerodynamic characteristics of quadrotor helicopter. In: Proceedings of AIAA Flight Testing Conference, AIAA AVIATION Forum, pp. 31–41 (2017)
21. Sato, M., Muraoka, K.: Flight controller design and demonstration of quad-tilt-wing unmanned aerial vehicle. J. Guid. Control Dyn. **38**(6), 1071–1082 (2014)
22. Tran, A.T., Sakamoto, N., Sato, M., Muraoka, K.: Control augmentation system design for quad-tilt-wing unmanned aerial vehicle via robust output regulation method. IEEE Trans. Aerosp. Electron. Syst. **53**(1), 357–369 (2017)
23. Kayacan, E., Khanesar, M.A.: Chapter 7 - Sliding mode control theory-based parameter adaptation rules for fuzzy neural networks. In: Kayacan, E., Khanesar, M.A. (eds.) Fuzzy Neural Networks for Real Time Control Applications, pp. 85–131. Butterworth-Heinemann, Oxford (2016). http://www.sciencedirect.com/science/article/pii/B9780128026878000074

Design of Optimal Fuzzy Controllers for Autonomous Mobile Robots Using the Grey Wolf Algorithm

Eufronio Hernandez, Oscar Castillo$^{(\boxtimes)}$, and Jose Soria

Computer Science in the Graduate Division, Tijuana Institute of Technology,
Tijuana, BC, Mexico
ocastillo@tectijuana.mx

Abstract. Through the advance of technology, every day new methods or computational techniques emerge that allow us to solve problems in different areas, such as medicine, engineering, even in any industrial process. Optimization is of vital importance in this industry, the main objective being to find the best possible solution to the problem. In this work we propose to use the Grey Wolf Optimizer (GWO), which is a metaheuristic inspired by the hunting behavior and leadership hierarchy of grey wolves, in addition to analyzing and explaining the proposed methodology for the optimization of fuzzy controllers for mobile autonomous robots.

Keywords: Grey Wolf Optimizer (GWO) ·
Autonomous mobile robots · Fuzzy controllers · Optimization ·
Fuzzy system · Bio-inspired algorithm

1 Introduction

Bio-inspired computing is based on using analogies with natural or social systems to solve complex problems. The bio inspired algorithms simulate the behavior of natural systems for the design of non-deterministic heuristic methods of search, learning and behavior [34].

The optimization is characterized by finding the best solution for any problem computationally speaking whether maximization or minimization, for its realization we need bio inspired algorithms that simulate in a natural way some task that seeks to obtain a solution. In addition to having certain stages, such as inputs (variables), development or process (fitness) and an output (cost) [13].

There are several bio inspired algorithms that help us solve these problems, such as: flower pollination algorithm [35], bat algorithm [36], firefly algorithm [32], cuckoo search [33], bacterial foraging [1], artificial bee colony [14], ant colony optimization [28], particle swarm [15], genetic algorithms [10], these algorithms are the most used in this area.

© Springer Nature Switzerland AG 2019
R. B. Kearfott et al. (Eds.): IFSA 2019/NAFIPS 2019, AISC 1000, pp. 285–295, 2019.
https://doi.org/10.1007/978-3-030-21920-8_26

This research is being carried out in order to present a different way of performing fuzzy controller's optimization using the gray wolf algorithm, thus improving its performance.

This paper has been ordered as follows: in Sect. 2 the literature review, in Sect. 3 the proposed method is shown, in Sect. 3 a methodology details is presented, in Sect. 4 fuzzy logic controllers, in Sect. 5 the results and discussions short summary, in Sect. 6 the conclusions obtained after obtaining the results made with fuzzy system.

2   Literature Review

In this section the basic concepts are shown to obtain an overview of the proposed method:

2.1   Grey Wolf Optimizer

Seyedali Mirjalili proposed the GWO in 2014, basically is responsible for simulating the natural behavior of a pack of wolves to hunt their prey, in other words, mimics the leadership hierarchy and hunting mechanism of the gray wolf, you can say that within this herd there are different kinds of wolves like: alpha, beta, delta and omegas as shown in Fig. 1 [22].

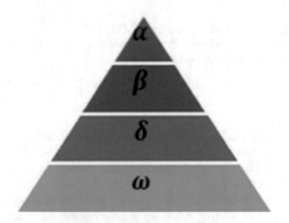

Fig. 1. Pyramid of hierarchy.

The hunting phases of GWO are listed as follows [23]:

Tracking, chasing, and approaching the prey. Pursuing, encircling, and harassing the prey until it stops moving. Attack towards the prey.

In this section some concepts are detailed to obtain a better panorama about the proposed method:

Optimization of fuzzy controllers for autonomous mobile robots using the Grey Wolf Optimizer.

2.2   Encircling Prey

Wolves surround the prey during the hunt. To mathematically represent this behavior, the following equations are presented (1) and (2):

$$D = |CX_p(t) - X(t)| \tag{1}$$

$$X(t+1) = X_p(t) - AD \tag{2}$$

t: indicates the current iteration, A and C: are coefficients, : is the position vector of the prey, $\boldsymbol{X}$: indicates the position vector of a grey wolf. The coefficients A and C, are calculated as follows, by using Eqs. (3) and (4) and the variable D represents the distance:

$$A = 2ar_1 - a \tag{3}$$

$$C = 2r_2 \tag{4}$$

a: is a number that linearly decrease from 2 to 0 over the course of iterations, are random numbers in $[0, 1]$.

2.3   Hunting Prey

Gray wolves naturally have the ability to recognize the location of the prey and surround them; the alpha wolf is sometimes responsible for guiding the hunt, sometimes involving the beta and delta wolves. To mathematically simulate this, we consider that alpha (best solution) beta and delta have a better knowledge about the possible location of the dam. As maybe seen in Eqs. (5), (6) and (7).

So, we keep the first three best solutions obtained so far and force the other search agents (omegas) to update their positions according to the position of the best search agents.

In essence, the following formulas are proposed.

$$D_\alpha = |C_1 X_\alpha - X| \qquad D_\beta = |C_2 X_\beta - X| \qquad D_\delta = |C_3 X_\delta - X| \tag{5}$$

$$X_1 = X_\alpha - A_1(D_\alpha) \qquad X_2 = X_\beta - A_2(D_\beta) \qquad X_3 = X_\delta - A_3(D_\delta) \tag{6}$$

$$X(t+1) = \tfrac{x_1+x_2+x_3}{3} \tag{7}$$

2.4   Attacking Prey

Gray wolves attack the prey when it stops moving. To mathematically model this characteristic, the value of the parameter "a" is decreased within the algorithm. When the values of "A" are within the range $[-1, 1]$ (Fig. 2).

a) If $|A| < 1$, then the prey is attacked (Exploitation)
b) If $|A| > 1$, then the prey is searched (Exploration)

Pseudo code of the GWO:

Initialize the gray wolf population $X_i(i = 1, 2, ..., n)$

Initialize a, A and C

Calculate the fitness of each search agent

X_α = The best search agent

X_β = The second best search agent

X_δ = The third best search agent

While ($t <$ Maximum number of iterations)

    For, each search agent

    Update the position of the current search agent using equation (7)

    End For

Update a, A, and C

Calculate the fitness of all search agents

Update X_α, X_β, X_δ

$t = t + 1$

End While

Return X_α

Fig. 2. Pseudocode GWO

2.5   Fuzzy Logic

Fuzzy logic was proposed by LA Zadeh in 1965, from the University of California, Berkeley, based on a work on the theory of fuzzy sets, allows the representation of human knowledge, to treat inaccurate information by combining rules or linguistic values, based on the theory of fuzzy sets, such as: "very low", "low", " medium", "high", "very high" [37–39].

Traditional logic only accepts the use of two values, completely true or completely false, for example the proposition "tomorrow will be warm" must be true or false. However, the information that people use contains a certain degree of uncertainty. In fuzzy logic there is the possibility of having a degree of truth within a range of 0 to 1, for example in the proposition "the president of Mexico is old", this can have a degree of accuracy of 0.7, then the fuzzy logic try to measure that degree and help the computer understand that information. It is worth mentioning that fuzzy systems transform these into mathematical models, with the aim of facilitating the work of the designer and the computer, resulting in more real representations.

2.6 Autonomous Robot Movil

They are systems capable of interacting with a high degree of autonomy within the limits of their environment, being able to move in any environment according to their programming. This leads to its use in areas such as: space exploration, wastewater treatment, and tasks that seem tedious or heavy for humans [7, 16].

Robots in the industry have been stealthily evolving from stationary machines to refined mobile platforms for the performance of numerous automation jobs. That is why robotics has achieved its greatest success in the world of industrial manufacturing. Currently the robot arms, or manipulators, are configured in a specific position on the assembly line, in order to perform jobs that require high speed and precision, such as moving heavy products, accommodating some type of material in order, etc.

Autonomous mobile robots are able to navigate difficult environments to make a path. One of the advantages is that they require little data such as external input to maneuver, which highlights an important capacity. The construction areas are inherently rough. Some innovations are the integration of intelligence systems, such as neural networks, bio inspired algorithms, some vision tools; the latter is equipped with multiple sensors helping them to perceive a dynamic environment in real time. Autonomous mobile robots have great industrial potential so it is very feasible to implement them in various areas.

Proposed Method
In particular, the work will focus on obtaining the best values for the fuzzy system. In Fig. 3 the proposed model is presented [26].

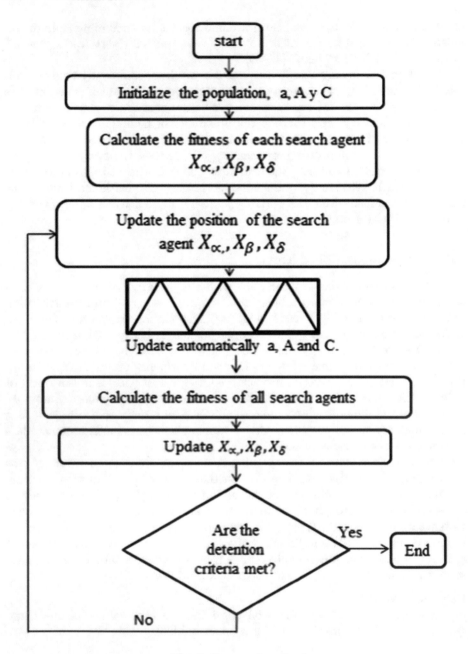

Fig. 3. Proposed method

3   Methodology

A system is created to optimize fuzzy controllers for mobile autonomous robots using the gray wolf optimizer. Through the combination of 2 methods, the use of a bio-inspired algorithm and a fuzzy system will allow to reach the desired trajectory for the robots.

As a first step we will obtain the parameters of the membership functions to optimize, and then we execute the GWO algorithm with the fuzzy system parameters for optimization, so we will obtain the optimized fuzzy controller and finally we will see a simulated fuzzy controller, marking the desired path, as shown in Fig. 4.

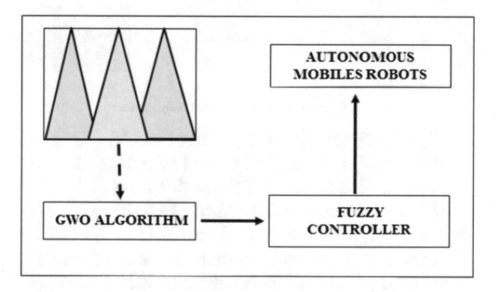

Fig. 4. Methodology method

4 Fuzzy Logic Controller

These rules of the fuzzy system describe a relation between the linguistic variables based on the membership functions, taking into account the inputs and outputs [2,6,8,27].

A fuzzy Mamdani type system is designed containing two linear error (ev) and angular error (ew) inputs, which comprise a range of [−1 1], which contain three membership functions, two trapezoidal at the ends and one triangular in the middle. So it has two outputs Torque 1 and Torque 2, [−1 1]. They contain three triangular membership functions, as shown in Fig. 5, [5,7,29,30].

This fuzzy controller has 9 if-then rules, which are shown Fig. 6, [3,24,31].

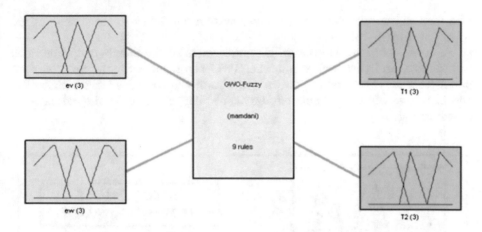

Fig. 5. Autonomous mobile robot fuzzy system

1. If (ev is N) and (ew is N) then (T1 is N) (T2 is N)
2. If (ev is N) and (ew is Z) then (T1 is N) (T2 is Z)
3. If (ev is N) and (ew is P) then (T1 is N) (T2 is P)
4. If (ev is Z) and (ew is N) then (T1 is Z) (T2 is N)
5. If (ev is Z) and (ew is Z) then (T1 is Z) (T2 is Z)
6. If (ev is Z) and (ew is P) then (T1 is Z) (T2 is P)
7. If (ev is P) and (ew is N) then (T1 is P) (T2 is N)
8. If (ev is P) and (ew is Z) then (T1 is P) (T2 is Z)
9. If (ev is P) and (ew is P) then (T1 is P) (T2 is P)

Fig. 6. Rules robot fuzzy system

5 Results and Discussion

Different experiments were carried out with the GWO, and the following parameters were used:

(a) Search agent number: 50
(b) Iterations: 5000

The mean square error was used as the aptitude function, as see Eq. (8), with which it is intended to minimize the error of the desired trajectory.

$$MSE = \frac{1}{n} \sum_{i=1}^{n} \left(\widehat{y}_i - y_i \right)^2 \tag{8}$$

Next, Table 1 is presented, where it is observed that 15 experiments were performed, of which the best result that it gave us was number 15, with 5000 iterations, with a total of 50 search agents, giving as well as error 0.000044249828665182.

Table 1. GWO parameters for each experiment and results of Error.

No.	Iterations	Agents	Error
1	500	55	0.000741061
2	400	50	7.29585E−05
3	1000	60	0.003882946
4	600	45	0.064997856
5	450	65	0.018010558
6	500	45	0.000626393
7	800	40	0.021905084
8	550	56	0.019565891
9	4000	60	3.19E−04
10	3000	70	0.0021119
11	2000	75	0.001740619
12	1000	65	0.008165245
13	800	55	0.000436007
14	700	60	0.000417135
15	5000	50	0.42498E−05

6   Conclusions and Future Work

In this work, until now the algorithm of the wolves was analyzed, need to deepen and perform more tests. The tests were carried out with the fuzzy system in the mobile autonomous robot plant; the optimization has given good results. To carry out the optimization of the fuzzy systems, the GWO was used, and with it 15 different experiments were carried out, changing the value of the parameters, so that can say, partially, that good results have been obtained. As future work, will try to perform some tests for simulation, so can make comparisons with other meta-heuristics, also perform the statistical test. In addition, we could use type-2 fuzzy logic as in [9,11,12,17,25]. We can also consider applying the proposed method in different areas of application like in [4,18–21].

Acknowledgment. It is widely appreciated to the Consejo Nacional de Ciencia y Tecnología and Tecnológico Nacional de Mexico/Tijuana Institute of Technology for the time, resource, space and facilities provided for the development of this work.

References

1. Das, S., Biswas, A., Dasgupta, S., Abraham, A.: Bacterial foraging optimization algorithm: theoretical foundations, analysis, and applications. Found. Comput. Intell. **3**, 23–55 (2009)
2. Soria: Comparative study of bio-inspired algorithms applied to the optimization of type-1 and type-2 fuzzy controllers for an autonomous mobile (2012)
3. Abdalla, T.Y., Abdulkareem, A.: A PSO optimized fuzzy control scheme for mobile robot path tracking. Int. J. Comput. Appl. **76**(2), 11–17 (2013)
4. Aguilar, L., Melin, P., Castillo, O.: Intelligent control of a stepping motor drive using a hybrid neuro-fuzzy ANFIS approach. Appl. Soft Comput. **3**(3), 209–219 (2003)
5. Astudillo, L., Melin, P., Castillo, O.: Chemical optimization algorithm for fuzzy controller design (2014)
6. Carvajal, O., Castillo, O., Soria, J.: Optimization of membership function parameters for fuzzy controllers of an autonomous mobile robot using the flower pollination algorithm. J. Autom. Mob. Robot. Intell. Syst **12**(1), 44–49 (2018)
7. Castillo, O., Melin, P., Montiel, O., Sepulveda, R., Pedrycz, W.: Theoretical Advances and Applications of Fuzzy Logic and Soft Computing. Springer, Tijuana (2007)
8. Castillo, O., Neyoy, H., Soria, J., García, M., Valdez, F.: Dynamic fuzzy logic parameter tuning for ACO and its application in the fuzzy logic control of an autonomous mobile robot. Int. J. Adv. Robot. Syst. (2013). https://doi.org/10.5772/54883
9. Cázarez-Castro, N.R., Aguilar, L.T., Castillo, O.: Designing type-1 and type-2 fuzzy logic controllers via fuzzy Lyapunov synthesis for nonsmooth mechanical systems. Eng. Appl. AI **25**(5), 971–979 (2012)
10. Goldberg, D.: Genetic Algorithms in Search, Optimization and Machine Learning. Addison-Wesley, Boston (1987)
11. González, C.I., Melin, P., Castro, J.R., Castillo, O., Mendoza, O.: Optimization of interval type-2 fuzzy systems for image edge detection. Appl. Soft Comput **47**, 631–643 (2016)
12. González, C.I., Melin, P., Castro, J.R., Mendoza, O., Castillo, O.: An improved sobel edge detection method based on generalized type-2 fuzzy logic. Soft Comput. **20**(2), 773–784 (2016)
13. Gupta, P., Cambini, R., Appadoo, S.S.: Recent advances in optimization theory and applications (2018)
14. Karaboga, D., Basturk, B.: A powerful and efficient algorithm for numerical function optimization: artificial bee colony (ABC) algorithm. J. Glob. Optim. **39**, 459–471 (2007)
15. Kennedy, J., Eberhart, R.: Particle swarm optimization, pp. 1942–1948 (1995)
16. Martinez, R., Castillo, O., Aguilar, L.T.: Optimization of interval type-2 fuzzy logic controllers for a perturbed autonomous wheeled mobile robot using genetic algorithms. Inf. Sci **179**(13), 2158–2174 (2009)
17. Melin, O.P.: Intelligent systems with interval type-2 fuzzy logic. Int. J. Innov. Comput. Inf. Control **4**(4), 771–783 (2008)
18. Melin, P., Amezcua, J., Valdez, F., Castillo, O.: A new neural network model based on the LVQ algorithm for multi-class classification of arrhythmias. Inf. Sci **279**, 483–497 (2014)

19. Melin, P., Castillo, O.: Modelling, Simulation and Control of Non-linear Dynamical Systems: An Intelligent Approach Using Soft Computing and Fractal Theory. CRC Press, Boca Raton (2001)
20. Melin, P., Castillo, O.: Adaptive intelligent control of aircraft systems with a hybrid approach combining neural networks, fuzzy logic and fractal theory. Appl. Soft Comput. 3(4), 353–362 (2003)
21. Melin, P., Sánchez, D., Castillo, O.: Genetic optimization of modular neural networks with fuzzy response integration for human recognition. Inf. Sci 197, 1–19 (2012)
22. Mirjalili, S., Mirjalili, S.M., Lewis, A.: Grey wolf optimizer. Adv. Eng. Softw. 69, 46–61 (2014)
23. Muro, C., Escobedo, R., Spector, L., Coppinger, R.: Wolf-pack (Canis lupus) hunting strategies emerge from simple rules in computational simulations. Behav. Process. 88, 192–197 (2011)
24. Olivas, F., Valdez, F., Castillo, O., Gonzales, C., Martinez, G., Melin, P.: Ant colony optimization with dynamic parameter adaptation based on interval type-2 fuzzy logic systems. Appl. Soft Comput. 53, 74–87 (2016)
25. Ramírez, C.L., Castillo, O., Melin, P., Díaz, A.R.: Simulation of the bird age-structured population growth based on an interval type-2 fuzzy cellular structure. Inf. Sci 181(3), 519–535 (2011)
26. Rodriguez, L., Castillo, O., Soria, J.: A Study of parameters of the grey wolf optimizer algorithm for dynamic adaptation with fuzzy logic. Springer, Tijuana (2017)
27. Sanchez, M., Castillo, O., Castro, J.: Generalized type-2 fuzzy systems for controlling a mobile robot and a performance comparison with interval type-2 and type-1 fuzzy systems. Expert Syst. Appl. 42, 5904–5914 (2015)
28. Socha, M.D.K.: An introduction to ant colony optimization. In: Handbook of Metaheuristics, vol. 26. IRIDIA, Brussels (2006). ISSN 1781-3794
29. Sombra, A., Valdez, F., Melin, P., Castillo, O.: A new gravitational search algorithm using fuzzy logic to parameter adaptation. Cancun, México (2013)
30. Soto, C., Valdez, F., Castillo, O.: A review of dynamic parameter adaptation methods for the firefly algorithm. In: Nature-Inspired Design of Hybrid Intelligent Systems (2007)
31. Valdez, F., Melin, P., Castillo, O.: Evolutionary method combining particle swarm optimization and genetic algorithms using fuzzy logic for decision making (2009)
32. Yang, X.S.: Firefly Algorithm, Lévy Flights and Global Optimization BT - Research and Development in Intelligent Systems, vol. XXVI (2010)
33. Yang, X.S., Deb, S.: Cuckoo Search via Lévy Flights (2009)
34. Yang, X.S., Karamanoglu, M.: Swarm intelligence and bio-inspired computation: an overview. In: Swarm Intelligence and Bio-Inspired Computation, pp. 3–23 (2013)
35. Yang, X.S., Karamanoglu, M., He, X.: Flower pollination algorithm: a novel approach for multiobjective optimization. Eng. Optim. 46(9), 1222–1237 (2014)
36. Yang, X.S.: Bat: algorithm: literature review and applications. Int. J. Bio-Inspired Comput. 5(3), 141–149 (2013)
37. Zadeh, L.: Fuzzy sets. Inf. Control 8, 338–353 (1965)
38. Zadeh, L.: The concept of a linguistic variable and its application to approximate reasoning—I. Inform. Sci 8, 199–249 (1975)
39. Zadeh, L.: Fuzzy logic. IEEE Comput. Mag. 1, 83–93 (1988)

Comparative Study of Fuzzy Controller Optimization with Dynamic Parameter Adjustment Based on Type 1 and Type 2 Fuzzy Logic

Marylu L. Lagunes$^{(\boxtimes)}$, Oscar Castillo, Fevrier Valdez, and Jose Soria

Tijuana Institute of Technology, Tijuana, BC, Mexico
marylu.lara@tectijuana.edu.mx, ocastillo@tectijuana.mx

Abstract. This paper presents a comparison of fuzzy controller optimization results using dynamic parameter adjustment Type 1 (T1) and Interval Type 2 (T2) fuzzy logic to the Firefly Algorithm (FA). The FA is used for optimizations parameters of the membership functions in the fuzzy controllers. The dynamic adjustment is applied to the randomness parameter of the search space, which represents the exploration of the method, avoiding stagnation or premature convergence. The FA generates the values that the parameters of the membership functions take for optimization use in the fuzzy systems for control. The control plants have one or more input variables that are processed and result in one or more output variables, it would be very difficult to model the human reasoning in equations to achieve a machine acquires the knowledge acquired by humans. For that reason the fuzzy logic that generates that insertion is used as if it were human reasoning.

1 Introduction

The main contribution focuses on the performance of the FA [1,2] to generate an efficient parameter data vector for the minimization of error in optimization problems, using T1 and T2 fuzzy logic for the dynamic adjustment of the alpha parameter (exploration). The generated values, optimize the parameters of the membership functions of fuzzy controllers. To perform the dynamic adjustment a simple fuzzy model of one input representing the iterations and one output that results in the values of the randomness parameter was developed. The parameters of the optimized membership functions are values of fuzzy controllers, used in plants to perform processes in general industrial, such as controlling temperatures, water flow, etc. In order to transmit human reasoning in this type of processes, fuzzy logic is used as a source of uncertainty for making decisions and evaluating the input and output variables of the plant. There are related works on optimization using metaheuristics such as [3–9]. This paper is structured as follows: Sect. 2 describes T1 and T2 fuzzy logic, and presents the Firefly Algorithm (FA), Sect. 3 describes the fuzzy dynamic adjustment model, Sect. 4 shows

© Springer Nature Switzerland AG 2019
R. B. Kearfott et al. (Eds.): IFSA 2019/NAFIPS 2019, AISC 1000, pp. 296–305, 2019.
https://doi.org/10.1007/978-3-030-21920-8_27

the fuzzy controllers to be optimized, Sect. 5 details the experimentation, Sect. 6 shows the results obtained from the fuzzy controller comparison and finally, Sect. 7 describes the conclusions.

2   Fuzzy Logic

Fuzzy logic Zadeh in [10,11] was established with the principle of incompatibility between precision and complexity, when the analysis and the way of solving problems by humans, could not be interpreted easily, so that a machine would do the same. As some processes are not compatible with the complexity of human reasoning.

The T2 fuzzy logic is represented by an uncertainty trace (FOU) that can be visualized as the interval between these two T1 membership functions. This type fuzzy logic was described by Zadeh [12] and later on its use and popularity with Mendel and Liang [13,14]. It is used to model uncertainty and imprecision [15,16].

2.1   Firefly Algorithm

The FA has basic 3 rules [17,18].

a. Fireflies are of the same genus, so any firefly could be attracted to any other firefly.
b. The attractiveness depends on the brightness and it is minimized depending on the distance between the fireflies, the firefly that shines the least will move towards the one that shines the most, if there is not one that shines more, they will move at random.
c. The glow of a firefly is given by its function of adequacy.

3   Model of Fuzzy Dynamic Parameter Adjustment

Dynamic parameter adjustment is applied to the FA to help improve the performance in the search space, with a good exploration the method has a good chance of finding the optimal objective function. For this reason a fuzzy system with a range [0 1] was designed, with iteration as input, and alpha as output, with low, medium and high as linguistic variables respectively, in each of the triangular membership functions. Fig. 1 shows the input and Fig. 2 the output of the T1 fuzzy system [19].

To compare the behavior of the method, we designed the same model with T2 fuzzy logic and observe which contributes more, in improving the behavior of firefly algorithm in the optimization. Figures 3 and 4 show the T2 fuzzy system.

$$Iteration = \frac{Current\ Iteration}{Maximum\ of\ Iterations}, \qquad (1)$$

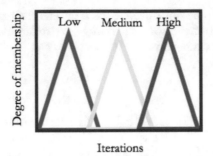

Fig. 1. Input 1, T1 system fuzzy logic

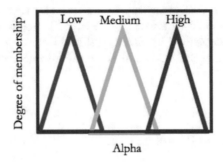

Fig. 2. Output 1, T1 system fuzzy logic

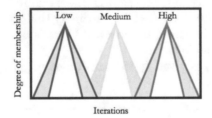

Fig. 3. Input 1, T2 system fuzzy logic

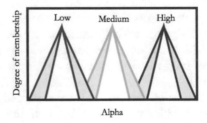

Fig. 4. Output 1, T2 system fuzzy logic

4   Fuzzy Controllers

4.1   Water Level Control in a Tank

With this model we control a water tank to be more specific the water level. Input one is the water level error and the water level change rate as input two, and each of these inputs uses 3 membership functions. It has an output that represents the speed with which the control valve closes or opens, and uses 5 membership functions [20]. In Table 1 the rules of the fuzzy inference system are observed (Fig. 5).

Table 1. Water level control in the tank rules

Number	Rules
1	If level is okay, then valve is noChange
2	If level is low, then valve is openFast
3	If level is high, then valve is closeFast
4	If level is okay, and rate is positive, then valve is closeSlow
5	If level is okay, and rate is negative, then valve is openSlow

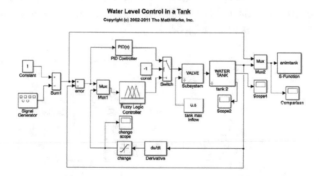

Fig. 5. Water level control in a tank plant

4.2   Temperature Control in a Shower

This fuzzy inference system has the temperature error as input one and the flow error as input two, and each input uses 3 membership functions. The speed of closes or openes the cold and hot water valves are the two outputs respectively, using 5 membership functions each, [20]. In the Table 2 the rules of the fuzzy inference system are observed.

Table 2. Temperature control in a shower rules

Number	Rules
1	If temp is cold, and flow is soft, then cold is open_Slow, hot is open_Fast
2	If temp is cold, and flow is good, then cold is close_Slow, hot is open_Slow
3	If temp is cold, and flow is hard, then cold is close_Fast, hot is close_Slow
4	If temp is good, and flow is soft, then cold is open_Slow, hot is open_Slow
5	If temp is good, and flow is good, then cold is steady, hot is steady
6	If temp is good, and flow is hard, then cold is close_Slow, hot is close_Slow
7	If temp is hot, and flow is soft, then cold is open_Fast, hot is open_Slow
8	If temp is hot, and flow is good, then cold is open_Slow, hot is close_Slow
9	If temp is hot, and flow is hard, then cold is close_Slow, hot is close_Fast

5   Experimentation

We experimented with the controller of the Water level control in a tank and the Temperature control in a shower, performing 5 experiments varying the population, iteration and dynamically the alpha parameter. Each of these 5 experiments were executed 30 times, with the same parameters, to make a comparison of the behavior and results of the optimization using T1 and T2. As can be noticed in Table 3, the parameters used in the FA for the controller with T1 and T2 fuzzy logic are the same parameters (Fig. 6).

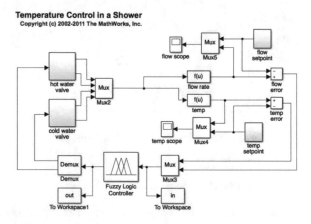

Fig. 6. Temperature control in a shower plant

Table 3. Firefly algorithm parameters

Experiment	Firefly	Iterations	Beta	Gamma	Alpha
1	15	2000	1	0.1	D
2	20	1500	1	0.1	D
3	25	1200	1	0.1	D
4	30	750	1	0.1	D
5	35	600	1	0.1	D

6   Simulations Results

The results obtained in the optimization of the parameters of the membership functions, the fuzzy controllers of the Water level control in a tank and Temperature control in a shower, using dynamic adjustment of the alpha parameter with T1 fuzzy logic, doing 5 experiments of 30 executions each, are present the following in Table 4.

Table 4. Type 1 fuzzy logic results

Experiment	Water level control in a Tank		Temperature control in a Shower	
	T1 MSE (Best)	Average	T1 MSE (Best)	Average
1	0.0885	0.7178	0.0671	0.5463
2	0.0664	0.6413	0.0299	0.9274
3	0.0785	0.8104	0.0139	0.5085
4	0.0664	0.5758	0.0739	0.6930
5	0.0364	0.9274	0.0139	0.5085

The experimentation was done with T1 fuzzy logic, but to observe its behavior and comparison was also done with T2 fuzzy logic, for which the same parameters were used as with T1. Table 5 shows the results obtained with T2.

Table 5. Type 2 fuzzy logic results

Experiment	Water level control in a Tank		Temperature control in a Shower	
	T2 MSE (Best)	Average	T2 MSE (Best)	Average
1	0.0012	0.1921	0.0041	0.1292
2	0.0049	0.1410	0.0039	0.0353
3	0.0283	0.1638	0.0278	0.1960
4	0.0133	0.1592	0.0014	0.0634
5	0.0035	0.1399	0.0010	0.1907

6.1   Statistical Test

With the results obtained, the statistical test was performed, for each of the optimized controllers, to compare the values that resulted from using T1 and T2 fuzzy logic for the dynamic adjustment of the alpha parameter in the FA method, as shows in the following Tables 6 and 7.

$$z = \frac{(\bar{x}_1 - \bar{x}_2) - (\bar{\mu}_1 - \bar{\mu}_2)}{\sigma_{\bar{x}_1 - \bar{x}_2}} , \qquad (2)$$

Table 6. Parameters statistical test water level control in a tank

Parameter	Value
Level of significance	95
H_a	$\mu_1 < \mu_2$
H_0	$\mu_1 \geq \mu_2$
Difference	−0.436
z (Observed value)	−5.929
z (Critical value)	−1.645
p-value (one-tailed)	<0.0001
alpha	0.05

Table 7. Statistical test results for water tank

Water level control in a Tank				
Dynamic adjustment	Number of samples	Mean	Standard deviation	Alpha
T1	30	0.576	0.363	0.05
T2	30	0.140	0.175	0.05

The Tables 8 and 9 show the parameters and results obtained in the statistical test with temperature control in a shower.

Table 8. Parameters statistical test temperature control in a shower control

Parameter	Value
Level of significance	95
H_a	$\mu_1 < \mu_2$
H_0	$\mu_1 \geq \mu_2$
Difference	−0.436
z (Observed value)	−7.356
z (Critical value)	−1.645
p-value (one-tailed)	<0.0001
alpha	0.05

Table 9. Statistical test results obtained

Temperature control in a shower				
Dynamic adjustment	Number of samples	Mean	Standard deviation	Alpha
T1	30	0.509	0.350	0.05
T2	30	0.035	0.040	0.05

7   Conclusions

In this paper we proposed fuzzy dynamic adjustment of the alpha parameter for, developing a fuzzy controller of one input and one output, with the objective of optimizing the fuzzy controller membership functions, these are optimized by a data vector generated and tested by the firefly algorithm. As it was observed previously, the results in the comparison of type 1 and type 2 fuzzy logic are very similar, therefore it is intended to add noise to the plants to observe their behavior in the optimization. However, the results were favorable with the error that was generated before being optimized. As future work, will need to perform more experiments with variations in values of some parameters and we could use general type-2 fuzzy logic [26–29] or use other applications as in [21–25].

References

1. Yang, X.S., He, X.: Firefly algorithm: recent advances and applications. Int. J. Swarm Intell. **1**, 321–354 (1995)
2. Yang, X.S.: Nature-Inspired Metaheuristic Algorithms. Luniver Press, London (2010)
3. Amador-Angulo, L., Castillo, O.: Comparative analysis of designing differents types of membership functions using bee colony optimization in the stabilization of fuzzy controllers. In: Fuzzy Logic in Intelligent System Design, vol. 648, p. 131. Springer, Heidelberg (2017)
4. Lagunes, M.L., Castillo, O., Soria, J.: Methodology for the optimization of a fuzzy controller using a bio-inspired algorithm. In: Nature Inspired Design of Hybrid Intelligent Systems, vol. 667, pp. 551–571. Springer, Heidelberg (2017)
5. Bernal, E., Castillo, O., Soria, J.: Imperialist competitive algorithm with dynamic parameter adaptation applied to the optimization of mathematical functions. In: Nature Inspired Design of Hybrid Intelligent Systems, vol. 667, pp. 329–341. Springer, Heidelberg (2017)
6. Rodríguez, L., Castillo, O., Soria, J.: IA study of parameters of the grey wolf optimizer algorithm for dynamic adaptation with fuzzy logic. In: Nature Inspired Design of Hybrid Intelligent Systems, vol. 667, pp. 371–390. Springer, Heidelberg (2017)
7. Peraza, C., Valdez, F., Castillo, O.: Improved method based on type-2 fuzzy logic for the adaptive harmony search algorithm. In: Fuzzy Logic Augmentation of Neural and Optimization Algorithms, vol. 749, pp. 29–37. Springer, Heidelberg (2018)

8. Lagunes, M.L., Castillo, O., Valdez, F., Soria, J., Melin, P.: Parameter optimization for membership functions of type-2 fuzzy controllers for autonomous mobile robots using the firefly algorithm. In: Fuzzy Information Processing, vol. 813, pp. 569–679. NAFIPS (2018)

9. Lagunes, M.L., Castillo, O., Soria, J.: Optimization of membership function parameters for fuzzy controllers of an autonomous mobile robot using the firefly algorithm. In: Fuzzy Logic Augmentation of Neural and Optimization Algorithms, vol. 749, pp. 199–206. Springer, Heidelberg (2018)

10. Zadeh, L.A.: Fuzzy sets. Inf. Control. **8**(3), 338–353 (1965)

11. Zadeh, L.A.: Fuzzy logic. Computer (Long. Beach. Calif.) **21**(3), 83–93 (1988)

12. Zadeh, L.A.: The concept of a linguistic variable and its application to approximate reasoning-III. Inf. Sci. (NY) **9**(1), 43–80 (1975)

13. Karnik, N.N., Mendel, J.M., Liang, Q.: Type-2 fuzzy logic systems. IEEE Trans. Fuzzy Syst. **7**(6), 643–658 (1999)

14. Liang, Q., Mendel, J.M.: Interval type-2 fuzzy logic systems: theory and design. IEEE Trans. Fuzzy Syst. **8**(5), 535–550 (2000)

15. Castillo, O., Melin, P., Kacprzyk, J., Pedrycz, W.: Type-2 fuzzy logic: theory and applications. In: 2007 IEEE International Conference on Granular Computing (GRC 2007), p. 145 (2007)

16. Perez, J., Valdez, F., Castillo, O.: Modification of the bat algorithm using type-2 fuzzy logic for dynamical parameter adaptation. In: Nature Inspired Design of Hybrid Intelligent Systems, vol. 667, pp. 343–355. Springer, Heidelberg (2017)

17. Soto, C., Valdez, F., Castillo, O.: A review of dynamic parameter adaptation methods for the firefly algorithm. In: Nature Inspired Design of Hybrid Intelligent Systems, vol. 667, pp. 285–295. Springer, Heidelberg (2017)

18. Solano-Aragón, C., Castillo, O.: Optimization of benchmark mathematical functions using the firefly algorithm. In: Recent Advances on Hybrid Approaches for Designing Intelligent Systems, vol. 547, pp. 177–189. Springer, Heidelberg (2014)

19. Ochoa, P., Castillo, O., Soria, J.: Fuzzy differential evolution method with dynamic parameter adaptation using type-2 fuzzy logic. In: 8th International Conference on Intelligent Systems, pp. 113–118. IEEE (2016)

20. Water Level Control in a Tank - MATLAB: Simulink Example - MathWorks America Latina. https://la.mathworks.com/help/fuzzy/examples/water-level-control-in-a-tank.html. Accessed 04 July 2018

21. Leal Ramírez, C., Castillo, O., Melin, P., Rodríguez Díaz, A.: Simulation of the bird age-structured population growth based on an interval type-2 fuzzy cellular structure. Inf. Sci. **181**(3), 519–535 (2011)

22. Cázarez-Castro, N.R., Aguilar, L.T., Castillo, O.: Designing type-1 and type-2 fuzzy logic controllers via fuzzy Lyapunov synthesis for nonsmooth mechanical systems. Eng. Appl. of AI. **25**(5), 971–979 (2012)

23. Castillo, O., Melin, P.: Intelligent systems with interval type-2 fuzzy logic. Int. J. Innov. Comput. Inf. Control 4(4), 771–783 (2008)

24. Mendez, M., Castillo, O.: Interval type-2 TSK fuzzy logic systems using hybrid learning algorithm, information and control. In: The 14th IEEE International Conference on Fuzzy Systems, FUZZ 2005, pp. 230–235 (2005)

25. Melin, P., Castillo, O.: Intelligent control of complex electrochemical systems with a neuro-fuzzy-genetic approach. IEEE Trans. Industr. Electron. **48**(5), 951–955 (2001)

26. Melin, P., González, C.I., Castro, J.R., Mendoza, O., Castillo, O.: Edge-detection method for image processing based on generalized type-2 fuzzy logic. IEEE Trans. Fuzzy Syst. **22**(6), 1515–1525 (2014)

27. González, C.I., Melin, P., Castro, J.R., Castillo, O., Mendoza, O.: Optimization of interval type-2 fuzzy systems for image edge detection. Appl. Soft Comput. **47**, 613–643 (2016)
28. González, C.I., Melin, P., Castro, J.R., Mendoza, O., Castillo, O.: An improved sobel edge detection method based on generalized type-2 fuzzy logic. Soft Comput. **20**(2), 773–784 (2016)
29. Ontiveros, E., Melin, P., Castillo, O.: High order α-planes integration: a new approach to computational cost reduction of general type-2 fuzzy systems. Eng. Appl. AI **74**, 186–197 (2018)

An Interpretation of the Fuzzy Measure Associated with Choquet Calculus for a HIV Transference Model

Beatriz Laiate[1]([⊠]), Rosana M. Jafelice[2], Estevão Esmi[1], and Laécio C. Barros[1]

[1] Department of Applied Mathematics, State University of Campinas,
Campinas, São Paulo, Brazil
`beatrizlaiate@gmail.com`, `eelaureano@gmail.com`, `laeciocb@ime.unicamp.br`
[2] Federal University of Uberlândia, Uberlândia, Minas Gerais, Brazil
`rmotta@ufu.br`

Abstract. This paper presents two dynamics that describe the transference between the classes of symptomatic and asymptomatic of an HIV-seropositive population. These models can be formulated using Choquet calculus where the underlying fuzzy measure is intrisically connected with the phenomena, and that is the main focus of this article. In particular, the fuzzy measures associated with Choquet calculus are obtained from the transference rates in both cases: with or without antiretroviral treatment.

1 Introduction

Choquet calculus, as proposed in [15,16], is defined in terms of fuzzy measures. This paper shows that the corresponding fuzzy measure can be connected with some parameters of the model of the phenomenon under consideration. Here, we focus on the study of the dynamics of an HIV-seropositive population with and without treatment.

HIV (human immunodeficient virus) belongs to retrovirus family and is responsible for causing AIDS (Acquired Immune Deficiency Syndrome) in human body. HIV acts primarily infecting CD4+ T lymphocytes, a class of immune cells. Once the virus reaches the bloodstream, it interacts with CD4+ receptors and breaks down the protein envelope that surrounds viral RNA (Ribonucleic Acid). These interactions initiate a fusion of the viral and cellular membrane. The process of virus reproduction involves several stages, which can be disrupted or minimized by specific types of antiretroviral drugs. For this, Herz et al. [7] proposed a model for intracellular dynamics involving CD4+ T cells and viral load in patients undergoing antiretroviral therapy. This model considers a combination of two types of inhibitors as well as the intracellular and phamacological delays. Intracellular delay is defined by the mode of operation of the combined treatment of Antirretrovirals (ARVs), which acts by blocking the intermediary processes between the contact of HIV with the lymphocytes and the release of free viruses into the bloodstream. The pharmacological delay is defined as the

R. B. Kearfott et al. (Eds.): IFSA 2019/NAFIPS 2019, AISC 1000, pp. 306–317, 2019.
https://doi.org/10.1007/978-3-030-21920-8_28

time interval required for the absorption, distribution and penetration of the drug into the target cells of the virus.

The process of viral reproduction in the human body in the absence of treatment causes, over time, a drastic decrease in the concentration of CD4+ T cells, caused mainly by apoptosis (cell destruction process due to HIV infection). As described in Sect. 3, the HIV-positive individuals are classified in asymptomatic or symptomatic, depending on the clinical stage in which they are. According to [17], treatment with ARVs allows the same life expectancy to the HIV-seropositive patient as an HIV-negative person of the same age. Usually, symptomatic individuals that adhered between 95% and 100% of treatment right after their first symptoms may become asymptomatic from three to six months. The clinical progression of treatment depends on individual factors, however, it is important to quantify the return rate of individuals under treatment to asymptomatic stage [11].

This paper studies the dynamics of transference and return between the symptomatic and asymptomatic phases considering the absence and presence of ARV treatment, respectively. We interpret the transference model of Anderson [1] using Choquet calculus [15] whereby the integral of a function is defined with respect to a fuzzy measure. We proposed a model of return to the asymptomatic phase for a HIV-seropositive population under treatment. We consider the effectiveness of the treatment as a measure of recovery of the minimum concentrations of CD4+ cells in the peripheral blood. In this case, we define the rate of return as a function of the pharmacological delay, which is assigned to the time interval necessary for the effect of the drug to occur. In both cases, asymptomatic population variation is described as an equation with Choquet derivative [12, 15], in which the fuzzy measure assumes biological sense, namely, measure whose derivative is interpreted as transfer rate and return.

2   Choquet Calculus

In order to define a fuzzy measure, we first need to introduce the concept of σ-algebra.

Definition 1 [3]. Let Ω be a non-empty set. A family $\mathscr{A}$ of subsets of Ω is called *σ-algebra* if it satisfies the following axioms:

(σ_1) The empty set $\emptyset \in \mathscr{A}$;
(σ_2) If an event $A \in \mathscr{A}$, then its complement $A' \in \mathscr{A}$;
(σ_3) If the events $A_1, A_2, \ldots \in \mathscr{A}$, then $\bigcup_{i=1}^{\infty} \in \mathscr{A}$.

A fuzzy measure is defined as follows.

Definition 2 [3,13]. Let $\mathscr{A}$ be σ-algebra on $\Omega \neq \emptyset$. A set-valued function $\mu : \mathscr{A} \to [0,1]$ is called a *fuzzy measure* if

1. $\mu(\emptyset) = 0$ and $\mu(\Omega) = 1$;
2. $\mu(A) \leq \mu(B)$ whenever $A \subseteq B$.

The triple $(\Omega, \mathscr{A}, \mu)$ is called *fuzzy measure space*.

Definition 3 [13]. Let $(\Omega, \mathscr{A}, \mu)$ a fuzzy measure space such that $\mu(\Omega) < \infty$. The *Choquet integral* of a measurable function g with respect to μ is

$$(C) \int_\Omega f d\mu = \int_0^\infty \mu(g > t) dt + \int_{-\infty}^0 [\mu(g \geq t) - \mu(\Omega)]\, dt, \qquad (1)$$

and for $A \in \mathscr{A}$,

$$(C) \int_A g d\mu = \int_0^\infty [\mu((g > t) \cap A) dt] + \int_{-\infty}^0 [\mu((g \geq t) \cap A) - \mu(A)]\, dt. \qquad (2)$$

Here, we focus on the case where $\Omega = \mathbb{R}$ and $A = [0,t]$ which the σ-algebra $\mathscr{A}$ given by the power set of $\mathbb{R}$, i. e., $\mathscr{A} = 2^\mathbb{R}$. Note that the Choquet integral of $g : \mathbb{R}^+ \to \mathbb{R}^+$ with respect to a fuzzy measure μ on $[0,t]$ is given by

$$(C) \int_{[0,t]} g d\mu = \int_0^\infty \mu(\{\tau | g(\tau) \geq r\} \cap [0,t]) dr, \qquad (3)$$

since the integral of the negative part is zero in Eq. (2).

In this article, we focus on the family of fuzzy measures established in [15], [16]. Specifically, we consider the following family of fuzzy measures:

Definition 4 [15]. Let $m : \mathbb{R}^+ \longrightarrow \mathbb{R}^+$ an increasing and differentiable function with the initial condition $m(0) = 0$ and ρ a Lebesgue measure [8]. A fuzzy measure μ_m as a *distorted Lebesgue measure* is defined by

$$\mu_m(\cdot) = m(\rho(\cdot)), \qquad (4)$$

where $\mu_m([a,b]) = m(\rho([a,b])) = m(b - a)$ whenever $[a,b] \subseteq \mathbb{R}$ and $b \geq a$.

If we assume that the function $h(\tau) = \mu([0,\tau])$ is differentiable with respect to τ, then we have $h'(\tau) = \frac{\partial}{\partial \tau} \mu([0,\tau]) \doteq \mu'([0,\tau])$ whenever $0 \leq \tau \leq t$ [15]. The next theorem establishes conditions for the Choquet integral of a continuous function to coincide with its Riemann integral.

Theorem 1 [12]. *Let $g : \mathbb{R} \to \mathbb{R}$ be non increasing; then the Choquet integral of g with respect to a fuzzy measure μ on $[0,t]$ is given by*

$$(C) \int_{[0,t]} g d\mu = \int_0^t \mu'([0,\tau]) g(\tau) d\tau. \qquad (5)$$

In particular, if $\mu = \mu_m$, then

$$(C) \int_{[0,t]} g d\mu_m = \int_0^t m'(\tau) g(\tau) d\tau. \tag{6}$$

Proof. [12].

Definition 5 [12,15]. Given a continuous $f : \mathbb{R} \to \mathbb{R}$, consider the following Choquet integral equation with respect to μ_m:

$$f(t) = f(0) + (C) \int_{[0,t]} g d\mu_m. \tag{7}$$

The Choquet derivative of f with respect to a fuzzy measure $\boldsymbol{\mu_m}$ is defined by the solution g of Eq. (7) and it is represented as

$$\frac{df}{d\mu_m}(t) = g(t). \tag{8}$$

If such a g exists, f is said to be *differentiable with respect to μ_m or μ_m-differentiable.*

3   HIV Transference Dynamics

Two clinical phases are established between the virus infection and the manifestation of the symptoms. The body manifests a brief phase of acute infection upon contact with HIV, characterized by an abrupt decay in CD4+ concentration levels, as well as a considerable increase in viral concentration levels. The response of the immune system is given by the increase of the concentration of HIV antibodies, whose biological role neutralizes the action of infected T lymphocytes. This period lasts around 4 weeks and is followed by a long period characterized by the continued loss of CD4+ T cells, named asymptomatic clinical phase. In this second period, there is a gradual decrease in CD4+ levels, during which the viral concentration in the blood is controlled by the immune system itself. Below a concentration threshold of CD4+ (given by 200 cells/μL), the immunodeficiency symptoms begin to manifest, initiating the clinical (symptomatic) phase. This phase occurs after an average time of 10 years, as depicted in Fig. 1 [14].

Although there are many clinical categories for diagnosis, when CD4+ T cell counts 200 cells/μL or below, the patient is diagnosed as having AIDS [5]. In healthy patients, blood levels of CD4+ T cells count around 1000 cells/μL and are quickly replenished in the body when some decrease in their concentration is perceived. Therefore, the concentration of CD4+ T cells remains the major indicator for AIDS diagnosis, replacing the appearance of the symptoms for this [4].

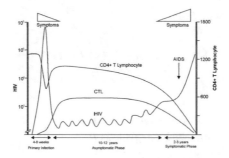

Fig. 1. Time course of natural history of HIV infection [10, 14].

Some studies have shown that CD4+ cell counts are the best indicator for the therapeutic response. Data from collaborative studies, when examining patients who started antiretroviral therapy, correlated the progression in CD4+ cell count with the likelihood of developing AIDS or dying in the course of treatment [4]. There is an inverse correlation between the number of peripheral blood CD4+ cells and the risk of diseases associated with HIV infection, specifically type 1.

In this paper we consider the effectiveness of ART at the instant t as the percentage of individuals who reached the minimum CD4+ count of ($c \geq 350$ cells/μL) after t weeks under treatment.

3.1 Symptomatic and Asymptomatic Classes

In this section two dynamics are proposed for a HIV-seropositive population without vital dynamics, i.e., not subject to births or deaths in the considered period. Thus, the mathematical models presented aims to account for relatively short periods. The class models assume classification of individuals in the two clinical categories, asymptomatic and symptomatic, according to the clinical progression of HIV infection. The membership of individuals in each class is exclusively based on CD4+ T lymphocytes count in peripheral blood. We denote by $x = x(t)$ the proportion of asymptomatic individuals and by $y = y(t)$ the proportion of symptomatic individuals at time t, so that $x(t) + y(t) = 1$, $t \geq 0$. The transference rate to the symptomatic stage of the HIV-seropositive population without treatment is represented by λ. However, for the HIV-seropositive population under treatment, the return rate ω represents immunosuppression, followed by the immune recovery. Figure 2 depicts both dynamics as class models.

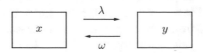

Fig. 2. Class diagram: λ is the transference rate for symptomatic clinical stage and ω is the return rate to asymptomatic clinical stage.

The transference dynamics suggests that, under these conditions, $\frac{1}{\lambda}$ is the mean time that HIV-seropositive individuals without treatment remain asymptomatic. According to Fig. 1, $\frac{1}{\lambda} \approx 10$ years, corresponding to the period of the asymptomatic clinical stage. On the other hand, return dynamics suggests that $\frac{1}{\omega}$ is the mean time that HIV-seropositive individuals remain symptomatic when undergoing antiretroviral therapy. Next, we present the mathematical models that represent the dynamics of populations x and y.

3.1.1   Transference Dynamics Without Treatment

Consider a HIV-seropositive population with no-antiretroviral therapy ($\omega = 0$ in Fig. 2). Anderson [1] introduced that the conversion rate λ of the infection to the manifestation of AIDS in the organism, as a function of time, in the natural history of HIV, would be the transference from the asymptomatic stage to the symptomatic. The model can be written as

$$\begin{cases} \dfrac{dx(t)}{dt} = -\lambda(t)x(t) \\[4mm] \dfrac{dy(t)}{dt} = \lambda(t)x(t), \end{cases} \tag{9}$$

Most HIV-seropositive individuals are asymptomatic or mildly asymptomatic [2]. This is represented by the initial conditions when $x(0) = 1 - \varepsilon$ and $y(0) = \varepsilon$, where ε is a sufficiently small proportion of the HIV-seropositive population. From (9), $\frac{dx(t)}{dt} + \frac{dy(t)}{dt} = 0$ and, therefore, $\frac{d(x(t)+y(t))}{dt} = 0$, i.e., the condition $x(t) + y(t) = 1$ is satisfied. Therefore, $x(t) = 1 - y(t)$ so that the dynamics of the symptomatic population are fully described by

$$\frac{dy(t)}{dt} - \lambda(t)(1 - y(t)), \quad y(0) = \varepsilon, \tag{10}$$

whose solution can be written as

$$y(t) = y(0) + \int_0^t \lambda(s)(1 - y(s))d\tau, \tag{11}$$

i.e.,

$$y(t) = \varepsilon + \int_0^t \lambda(s)(1 - y(s))d\tau. \tag{12}$$

Note that

$$\frac{d}{dt}(1 - y(t)) = -y'(t) = -\lambda(t)x(t) < 0, \forall t \geq 0 \tag{13}$$

and, therefore, according to (6)

$$\int_0^t \lambda(s)(1 - y(s))d\tau = (C)\int_{[0,t]} (1 - y) \, d\mu_m, \tag{14}$$

where $m'(\tau) = \lambda(\tau)$ whenever $\tau \in [0, t]$. By Theorem 1, Eq. (14) is equivalent to equation

$$y(t) = \varepsilon + (C) \int_{[0,t]} (1 - y) \, d\mu_m. \tag{15}$$

According to Definition 3, the Initial Value Problem (IVP) (10) can be written as a IVP with Choquet derivative

$$\frac{dy}{d\mu_m}(t) = 1 - y(t), \quad y(0) = \varepsilon. \tag{16}$$

This model presents the intrinsic measure $m = m(t)$ where $m'(t) = \lambda(t)$, $t \geq 0$, so that $m(t) = \int_0^t \lambda(s) ds$, that is, $m = m(t)$ is the cumulative of the transference rate of the disease up to date t.

Note that $m(0) = 0$ and that the higher the rate λ, the greater the values of m. According to collaborative studies [4] it is possible to infer low CD4+ count population to present higher values for the transfer rate λ, and consequently, high value for m. Likewise, $\frac{1}{m'}$ indicates the mean time of the HIV-seropositive individuals being in the clinical asymptomatic stage. Therefore, we can afirm the greater fuzzy measure μ_m the more severe the process of illness of the HIV-seropositive population.

3.1.2   Return Dynamics with Treatment

Considering the cycle of reproduction of HIV in the human organism, it is observed that in previous studies the effect of treatment by ARVs is represented by an intracellular delay [7]. This delay is given by the time interval between the time of infection or the T lymphocytes and the production of free viruses released into the bloodstream. In [9] was considered that such delay depends on the characteristics of each patient, including factors such as frequency and adherence to medical treatment.

In this subsection we present a return model of the HIV-seropositive population from the symptomatic to the asymptomatic stage. The return rate, in this case, is associated with the percentage of individuals that reached the immunological reconstitution due to the treatment. As in previous studies [7], we consider that the drug will have an important effect on the body after a period of time $\tau > 0$, from which is expected an improvement in the clinical progression of the infection. This interval of time is called pharmacological delay.

First we consider that population dynamics in the initial pharmacological range do not have transfer or return rates [7], since $[0, \tau]$ is the interval of time required for the drug to have an initial effect on the body. We define an initial function [6] $\varphi : [0, \tau] \longrightarrow [0, 1]$ such that $x = x(t)$ coincides with φ in $[0, \tau]$ and we can write

$$x(t) = \varphi(t), \ y(t) = \phi(t), \ 0 \leq t \leq \tau \tag{17}$$

where $\phi(t)$ and $\varphi(t)$ are constants for $0 \leq t \leq \tau$.

Also, we are assuming that for $t \geq \tau$, the derivatives $\frac{dx}{dt}$ and $\frac{dy}{dt}$ at time t are determined by the symptomatic population that started treatment on the date $t - \tau$. Considering the pharmacological delay, we propose the following system of differential equations:

$$\begin{cases} \dfrac{dx(t)}{dt} = \omega(t - \tau)y(t - \tau) \\[2mm] \dfrac{dy(t)}{dt} = -\omega(t - \tau)y(t - \tau), \end{cases} \tag{18}$$

where $\omega = \omega(t - \tau)$ is the return rate from the asymptomatic clinical stage to the asymptomatic clinical stage at time t. Equation (18) describes the dynamics of the HIV-seropositive population when $t \geq \tau$, since we are assuming that the individuals started the treatment at time $t = 0$. Note that by adding the equations member to member, we have $\frac{dx(t)}{dt} + \frac{dy(t)}{dt} = 0$, so that $x(t) + y(t) = 1$ for $t \geq \tau$. Whereas $\phi(t), \varphi(t)$ are constant functions so that $\varphi(t) = x(\tau)$ and $\phi(t) = 1 - x(\tau)$ for $0 \leq t \leq \tau$, the dynamics of $x(t)$ is completely described by

$$\begin{cases} \dfrac{dx(t)}{dt} = \omega(t - \tau)(1 - x(t - \tau)), & t \geq \tau \\[2mm] x(t) = x(\tau), & 0 \leq t < \tau. \end{cases} \tag{19}$$

From (19), it is found for $t \geq \tau$ that

$$\frac{dx(t)}{dt} = \omega(t - \tau)(1 - x(t - \tau)), \ t \geq \tau. \tag{20}$$

Integrating both members of Eq. (20), it is found that

$$x(t) = x(\tau) + \int_\tau^t \omega(s - \tau)(1 - x(s - \tau))ds. \tag{21}$$

Equation (21) is equivalent to the following integral equation

$$x(t) = x(\tau) + (C) \int_{[0,t]} 1 - x(s - \tau)\, d\mu_m, \tag{22}$$

which, by the identity $(C) \int_{[0,t]} g(s - \tau)d\mu_m = \int_\tau^t m'(s - \tau)g(s - \tau)ds$ can be rewritten as

$$x(t) = x(0) + (C) \int_{[0,t]} 1 - x(s - \tau)\, d\mu_m, \tag{23}$$

where the generator $m = m(t)$ of the fuzzy measure μ_m is such that $m'(s - \tau) = \omega(s - \tau)$, $s \geq \tau$. Thus, Eq. (23) can be rewritten in the form of differential equation with Choquet derivative,

$$\frac{dx}{d\mu_m}(t) = 1 - x(t - \tau), \ x(0) = x(\tau), \tag{24}$$

where $x(\tau)$ is known.

In this case, since $\frac{1}{m'}$ is the mean time that individuals under treatment remain symptomatic and $m'(s - \tau) = \omega(s - \tau)$, $\omega(s)$ individuals leave the symptomatic stage per unit of time. Thus one can say that $m(t - \tau) = \int_{\tau}^{t} \omega(s - \tau)ds$ is the number of individuals who initiated treatment at time τ and are recovered at time t.

Therefore, we can say that the effectiveness of the treatment is intrinsically linked to the fuzzy measure used in Choquet calculus. Note that the Choquet derivative, for the case of Eq. (24), holds for $v \geq 0$. This implies that the return dynamics starts to exist from the time $t = \tau$, when, by hypothesis, the effect of the drug becomes effective. For $0 \leq v \leq \tau$, it is expected that x and y of the population in question remain constant, that is, there are no changes in the clinical stages. Therefore, the initial history for the Eq. (24) with Choquet derivative is given by $x(t) = x(\tau)$ for $0 \leq t \leq \tau$.

4   Analysis of the Solution for the Class Model

Next, we analyze the solutions of (16) and (24), interpreting the measure m and, as a consequence, the fuzzy measure adopted in each specific problem.

4.1   Transference Dynamics Without Treatment

Consider the differential equation given by IVP (10)

$$\frac{dy(t)}{dt} = \lambda(t)(1 - y(t)). \tag{25}$$

Applying the variable separation approach, we can find the solution of the IVP (16), from which we can write

$$y(t) = 1 - \frac{(1 - \varepsilon)}{e^{m(t)}} = 1 - (1 - \varepsilon)e^{-m(t)}, \, t > 0, \tag{26}$$

where $m'(t) = \lambda(t)$. Therefore, $x(t) = (1 - \varepsilon)e^{-m(t)}$, $t > 0$.

Note that since $m(t) > 0$ when $t > 0$, the population of symptomatic individuals only grows over time. At the same time, if $m(t) \to \infty$, then $y(t) \to 1$ and, consequently, $x(t) \to 0$. This solution suggests that HIV-seropositive individuals without treatment necessarily migrate to the symptomatic stage after a sufficiently long period of time, as suggested in Fig. 1.

4.2   Return Dynamics with Treatment

Differently from the case without treatment, where the solution of the model is given by (26), the case with treatment is a delay IVP and it has no explicit solution for the asymptomatic population $x(t)$. Then, through Taylor expansion, we present an approximate solution of (19) for the population $x(t)$. From Eq. (20), we have that the asymptomatic HIV-seropositive population $x = x(t)$ under treatment can be described by

$$x'(t) = \omega(t - \tau)(1 - x(t - \tau)). \tag{27}$$

Taking the first order approximation from Taylor expansion, Eq. (27) results in

$$x'(t) = \frac{\omega(t - \tau)(1 - x(t))}{1 - \omega(t - \tau)\tau}, \, t \geq \tau. \tag{28}$$

Then, we obtain the IVP

$$\begin{cases} x'(t) = \frac{\omega(t-\tau)(1-x(t))}{1-\omega(t-\tau)\tau}, \, t \geq \tau \\ x(0) = x(\tau) \end{cases} \tag{29}$$

whose solution, given by

$$x(t) = 1 + (x(\tau) - 1)e^{-\int_\tau^t \frac{\omega(s-\tau)}{1-\omega(s-\tau)\tau} ds} \tag{30}$$

approximates the solution of (27).

Analyzing the solution of (27) from the approximated solution (30), we consider the cases in which $\omega\tau > 1$ or $\omega\tau < 1$.

- If $\omega\tau > 1 \Leftrightarrow 1 - \omega\tau < 0$, then from (30) we see that $x(t)$ must be smaller than $x(\tau)$ whenever $t \geq \tau$. Therefore, the treatment does not show efficiency since the asymptomatic population decreases with time. Indeed, $\omega\tau > 1$ is equivalent to $\tau > \frac{1}{\omega}$, since $\omega > 0$ always. For the return model this would mean that the time for the drug to take effect in the organism would be longer or equal to the mean time that the individuals remain symptomatic after the start of the treatment. As the return to asymptomatic stage occurs due to the effect of the drug, this premise could not be true in general.
- If $\omega\tau < 1 \Leftrightarrow 1 - \omega\tau > 0$, then from (30), we have that $x(t)$ grows apart from $x(\tau)$. In fact, $\omega\tau < 1$ is equivalent to $\tau < \frac{1}{\omega} = \frac{1}{m'}$, where $m'(t) = \lambda(t)$ whenever $t \geq \tau$, as established in (24). This means that the higher the growth rate of $m = m(t)$, less is the time required for the drug to be absorbed by the organism. The intrinsic fuzzy measure to the model of return consequently determines the pharmacological delay of the treatment. Thus, the approximate solution by Eq. (30) qualitatively has the same behavior as the solution of Eq. (24), since $x(t) \to 1$ when $t \to \infty$. Therefore, the treatment shows efficiency and suggests that individuals become asymptomatic after a long time period of treatment.

5   Final Comments

It has been studied in this work two transference models of a HIV-seropositive population for symptomatic and asymptomatic clinical stage from Choquet Calculus, one without treatment and other with treatment.

For the case without treatment, we write the variation of the symptomatic population through Choquet derivative, in which the derivative of the distorted function of the fuzzy measure represents the transfer rate. In the modeling with treatment, the derivative of the distortion function of the fuzzy measure represents the rate of return to the asymptomatic stage. For the first case, Choquet integral of the symptomatic population is defined in the interval $[0, t]$ since it is expected that the transfer to the symptomatic clinical stage can begin when $t > 0$. However, the anti-viral drug treatment suggests the return to asymptomatic stage is possible only after the time of onset of the drug effect, that is, when $t \geq \tau$. Therefore, it seems pertinente that Choquet integral of the asymptomatic population in this case is defined in the interval $[0, t - \tau]$.

Differential equations via Choquet, reveal that the used fuzzy measure has characteristics intrinsic to the modeled phenomena. In the case of transference dynamics without treatment, the used fuzzy measure determines the severity of HIV-seropositive individuals become ill, since the higher the growth rate of $m = m(t)$, the shorter the mean time individuals remain asymptomatic. For the return dynamics with antiretroviral therapy, the fuzzy measure is related to the pharmacological delay of the drug, so that the higher the growth rate, the less time needed for the absorption of ARV, since $\frac{1}{m'}$ acts as the upper bound for this delay. In addition, the growth rate of m is as higher as shorter the mean time that individuals under treatment remain symptomatic.

Acknowledgements. The authors would like to thank Prof. Dr. Francisco Hideo Aoki (HC-Unicamp-Brazil). This work was partially supported by CAPES under grant no. 1696945, by CNPq under grant no. 306546/2017-5, and by Fapesp under grant no. 2016/26040-7.

References

1. Anderson, R.M., Medley, G.F., May, R.M., Johnson, A.M.: A preliminary study of the transmission dynamics of the human immunodeficiency virus (HIV), the causative agent of AIDS. Math. Med. Biol. **3**, 229–263 (1986)
2. Avert: HIV and AIDS in Brazil. Global information and education on HIV and AIDS. https://www.avert.org/professionals/hiv-around-world/latin-america/brazil. Cited 15 Jan 2019
3. Barros, L.C., Bassanezi, R.C., Lodwick, W.A.: A first course in fuzzy logic. In: Fuzzy Dynamical Systems, and Biomathematics. Springer, Heidelberg (2017)
4. Battegay, M., Nüesch, R., Hirschel, B., Kaufmann, G.R.: Immunological recovery and antiretroviral therapy in HIV-1 infection. Lancet Inf. Dis. **6**(5), 280–287 (2006)
5. Centers for Disease Control and Prevention. About HIV/AIDS. https://www.cdc.gov/hiv/basics/whatishiv.html. Cited 27 Jan 2019
6. Hale, J.K., Lunel, S.M.V.: Introduction to Functional Differential Equations. Springer, Heidelberg (2013)
7. Herz, A.V.M., Bonhoeffer, S., Anderson, R.M., May, R.M., Nowak, M.A.: Viral dynamics *in vivo*: limitations on estimates of intracellular delay and virus decay. Proc. Natl. Acad. Sci. **93**, 7247–7251 (1996)
8. Honig, C.S.: A integral de Labesgue e suas aplicacões. IMPA (1977)

9. Jafelice, R.M., Barros, L.C., Bassanezi, R.C.: Study of the dynamics of HIV under treatment considering fuzzy delay. Comp. Appl. Math. **33**(1), 45–61 (2014). https://doi.org/10.1007/s40314-013-0042-6
10. Jafelice, R.M., Barros, L.C., Bassanezi, R.C., Gomide, F.: Methodology to determine the evolution of asymptomatic HIV population using fuzzy set theory. Int. J. Uncertainty Fuzziness Knowl. Based Syst. **13**(1), 39–58 (2005). https://doi.org/10.1142/S0218488505003308
11. Jafelice, R.M.: Modelagem Fuzzy para Dinâmica de Transferência de Soropositivos para HIV em Doenca Plenamente Manifesta. Thesis, Faculdade de Engenharia Elétrica e de Computacão, Unicamp (2003)
12. Laiate, B.: Cálculo de Choquet com Aplicacões em Dinâmica de Populacões. Dissertation, Instituto de Matemática, Estatística e Computacão Científica, Unicamp (2017)
13. Nguyen, H.T., Walker, E.A.: A First Course in Fuzzy Logic. Chapman and Hall/CRC, New York (2005)
14. Perelson, A., Nelson, P.: Mathematical analysis of HIV-1 dynamics in vivo. SIAM Rev. **41**, 3–44 (1999). https://doi.org/10.1137/S0036144598335107
15. Sugeno, M.: A way to choquet calculus. IEEE Trans. Fuzzy Syst. **23**, 1439–1457 (2015). https://doi.org/10.1109/TFUZZ.2014.2362148
16. Sugeno, M.: A note on derivatives of functions with respect to fuzzy measures. Fuzzy Sets Syst. **222**, 1–17 (2013). https://doi.org/10.1016/j.fss.2012.11.003
17. The Joint United Nations Programme on HIV/AIDS (UNAIDS). http://www.unaids.org. Cited 27 Dec 2018

A Ranking Method of Hexagonal Fuzzy Numbers Based on Their Possibilistic Mean Values

Worrawate Leela-apiradee[1]([✉]) and Phantipa Thipwiwatpotjana[2]

[1] Department of Mathematics and Statistics, Faculty of Science and Technology, Thammasat University, Pathum Thani 12121, Thailand
worrawateleela@gmail.com
[2] Department of Mathematics and Computer Science, Faculty of Science, Chulalongkorn University, Bangkok 10330, Thailand
phantipa.t@chula.ac.th

Abstract. A hexagonal fuzzy number (HFN) with its membership function as a nonlinear function, which is a generalization of triangular fuzzy numbers, trapezoidal fuzzy numbers, linear pentagonal fuzzy numbers and linear hexagonal fuzzy numbers, is defined in this paper. Cardinality of HFN is applied to achieve an algorithm for classifying types of HFNs. In addition, we present a ranking method for those fuzzy numbers based on their possibilistic mean values. Therefore, an explicit formula of the possibilistic mean value of HFN is proposed.

Keywords: Hexagonal fuzzy number · Possibilistic mean value · Ranking

1 Introduction

This paper starts with the concept of fuzzy number, which is a special case of a normal, convex fuzzy set of the real line.

Definition 1 (See Zadeh [23]). A **fuzzy set** $\tilde{A}$ on a universal set X is defined by
$$\tilde{A} = \{(x, \mu_{\tilde{A}}(x)) : x \in X\},$$
where $\mu_{\tilde{A}} : X \to [0, 1]$ is called the **membership function** of $\tilde{A}$.

Definition 2 (See Hanss [13]). A fuzzy set $\tilde{A}$ on the set of real numbers $\mathbb{R}$ is called a **fuzzy number** if its membership function $\mu_{\tilde{A}}$ satisfies the following properties:

(i) $\mu_{\tilde{A}}$ is piecewise continuous in its domain.
(ii) $\tilde{A}$ is a normal fuzzy set, i.e., there exists $x_0 \in \mathbb{R}$ such that $\mu_{\tilde{A}}(x_0) = 1$.

© Springer Nature Switzerland AG 2019
R. B. Kearfott et al. (Eds.): IFSA 2019/NAFIPS 2019, AISC 1000, pp. 318–329, 2019.
https://doi.org/10.1007/978-3-030-21920-8_29

(*iii*) $\tilde{A}$ is a convex fuzzy set, i.e., for any $x_1, x_2 \in \mathbb{R}$,

$$\mu_{\tilde{A}}(\lambda x_1 + (1-\lambda)x_2) \geq \min\{\mu_{\tilde{A}}(x_1), \mu_{\tilde{A}}(x_2)\} \text{ for all } 0 \leq \lambda \leq 1.$$

Two types of fuzzy number, namely, triangular fuzzy number with three parameters and trapezoidal fuzzy number with four parameters are widely applied to solve real-world problems in many fields of science and technology. However, if the problems are concerned with more than four parameters, pentagonal fuzzy number with five parameters established in [5,15] and linear hexagonal fuzzy number (LHFN) with six parameters established in [4,11] play the important rule in dealing with these problems. The LHFN was developed in fuzzy matrix, fuzzy linear system, fuzzy linear programming problem, fuzzy transportation problem and fuzzy inventory model, which can be reviewed in the following literature.

Dinagar and Narayanan [7] defined determinant of hexagonal fuzzy matrices containing LHFN in their entries and verified some relevant properties. Recently, orthogonal hexagonal fuzzy matrices have been defined in [21].

Fuzzy linear system whose coefficient matrix and right-hand side vector contain LHFN in their components was studied in [8,22]. To solve the system, matrix inversion method was used in [8], while singular value decomposition method was used in [22].

Sahaya and Revathy [17] introduced fuzzy linear programming problem based on LHFNs. They solved a basic feasible solution and the optimal solution by using simplex method. A ranking procedure of LHFNs was developed in [16] to achieve the optimal solution of fuzzy linear programming problem with multi-objective functions.

Thamaraiselvi and Santhi [20] solved fuzzy transportation problem in which the values of transportation costs, demands and supplies are represented as LHFNs. They obtained a fuzzy basic feasible solution and the optimal solution by using Vogel's approximation method and fuzzy zero point method, respectively. With the same problem, [3,18] solved a fuzzy basic feasible solution and the optimal solution by using the best candidate method and centroid ranking technique, respectively. Elumalai et al. [9] developed robust ranking method, which is an improvement of Vogel's approximation method, and applied zero suffix method to solve the optimal solution of the problem.

Based on LHFNs, fuzzy inventory model was analyzed in [6]. The development of the model with multi items based on LHFNs was found by Dhanam and Kalaiselvi in [4].

Moreover, the LHFN was mentioned in many articles on fuzzy games and fuzzy project network. For instance, the solution of fuzzy games with pure strategies solved by using minimax principle was discussed in [14]. Selvakumari and Sowmiya [19] presented a method for finding critical path in fuzzy project network using Pascal's triangle graded mean integration when the duration time of each activity is represented as LHFN.

In this paper, we propose a new definition of a HFN. A classification of HFN together with its algorithm are presented in Sect. 2. We introduce in Sect. 3

possibilistic mean value of HFNs and derive it as an explicit formula in order to apply the formula in a ranking method. At the end of this section, a numerical example is illustrated. The conclusion of this article is addressed in the last section.

2   Hexagonal Fuzzy Number

The hexagonal fuzzy number is a fuzzy number representing a situation when it can be split into five subcategories and concerned with six parameters $(a_1, a_2, a_3, a_4, a_5, a_6)$. For example, Dinagar and Narayanan proposed in [8] a study of human beings in terms of energy and enthusiasm for the human lifespan in the range of ages $[a_1, a_6]$, which is classified into different five stages as follows:

- Young (the range of ages $[a_1, a_2)$ years),
- Early Adulthood
  (the range of ages $[a_2, a_3)$ years),
- Middle Adulthood
  (the range of ages $[a_3, a_4)$ years),
- Later Adulthood
  (the range of ages $[a_4, a_5)$ years),
- Old (the range of ages $[a_5, a_6)$ years).

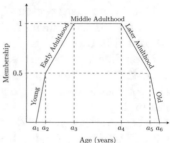

The membership function of their study is depicted as a linear hexagon as above.

In real-world problems with five subcategories, their membership functions may not be illustrated as the linear hexagons. To solve the problems with non-linear hexagons, we need to introduce, in Definition 3, the concept of hexagonal fuzzy number whose membership function is more general than LHFN's membership function presented in [4,11]. Furthermore, we classify HFNs, in Definition 4, into three characteristics based on the shape of their cardinality. This classification is adapted from Bodjanova's idea in [1].

Definition 3. Given a_1, a_2, a_3, a_4, a_5 and a_6 be any real numbers such that $a_1 < a_2 < a_3 \leq a_4 < a_5 < a_6$. Let n be a positive integer. Then, $\tilde{a} = (a_1, a_2, a_3, a_4, a_5, a_6; r)_n$ with $0 < r < 1$ is called a **hexagonal fuzzy number (HFN)** if its membership function $\mu_{\tilde{a}}$ is defined as (1), which can be depicted as Fig. 1.

$$\mu_{\tilde{a}}(x) = \begin{cases} 0, & \text{if } x < a_1; \\ r\left(\frac{x-a_1}{a_2-a_1}\right)^n, & \text{if } a_1 \leq x \leq a_2; \\ r + (1-r)\left(\frac{x-a_2}{a_3-a_2}\right)^n, & \text{if } a_2 \leq x \leq a_3; \\ 1, & \text{if } a_3 \leq x \leq a_4; \\ r + (1-r)\left(\frac{a_5-x}{a_5-a_4}\right)^n, & \text{if } a_4 \leq x \leq a_5; \\ r\left(\frac{a_6-x}{a_6-a_5}\right)^n, & \text{if } a_5 \leq x \leq a_6; \\ 0, & \text{if } x > a_6. \end{cases} \tag{1}$$

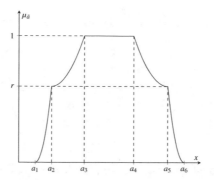

Fig. 1. A hexagonal fuzzy number $\tilde{a} = (a_1, a_2, a_3, a_4, a_5, a_6; r)_n$

For the case $n = 1$, the HFN $\tilde{a} = (a_1, a_2, a_3, a_4, a_5, a_6; r)_1$ is represented as a linear hexagonal fuzzy number (LHFN).

Necessary conditions for being triangular fuzzy number, trapezoidal fuzzy number and linear pentagonal fuzzy number of LHFN are provided in the following remark.

Remark 1. Let $\tilde{a} = (a_1, a_2, a_3, a_4, a_5, a_6; r)_1$ be a LHFN.

1. If $a_3 = a_4$ and $\frac{a_2 - a_1}{a_3 - a_1} = r = \frac{a_6 - a_5}{a_6 - a_4}$, then the membership function $\mu_{\tilde{a}}$ can be written as follows:

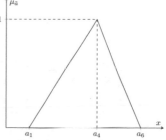

$$\mu_a(x) = \begin{cases} 0, & \text{if } x < a_1; \\ \frac{x - a_1}{a_4 - a_1}, & \text{if } a_1 \leq x < a_4; \\ \frac{a_6 - x}{a_6 - a_4}, & \text{if } a_4 \leq x \leq a_6; \\ 0, & \text{if } x > a_6, \end{cases}$$

that is, $\tilde{a}$ turns into a triangular fuzzy number $\tilde{a} = (a_1, a_4, a_6)$.

2. If $\frac{a_2 - a_1}{a_3 - a_1} = r = \frac{a_6 - a_5}{a_6 - a_4}$, then the membership function $\mu_{\tilde{a}}$ can be written as follows:

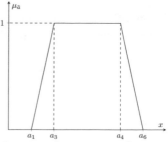

$$\mu_{\tilde{a}}(x) = \begin{cases} 0, & \text{if } x < a_1; \\ \frac{x - a_1}{a_3 - a_1}, & \text{if } a_1 \leq x \leq a_3; \\ 1, & \text{if } a_3 \leq x \leq a_4; \\ \frac{a_6 - x}{a_6 - a_4}, & \text{if } a_4 \leq x \leq a_6; \\ 0, & \text{if } x > a_6, \end{cases}$$

that is, $\tilde{a}$ turns into a trapezoidal fuzzy number $\tilde{a} = (a_1, a_3, a_4, a_6)$.

3. If $a_3 = a_4$, then the membership function $\mu_{\tilde{a}}$ can be written as follows:

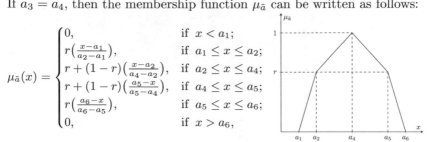

$$\mu_{\tilde{a}}(x) = \begin{cases} 0, & \text{if } x < a_1; \\ r\left(\frac{x-a_1}{a_2-a_1}\right), & \text{if } a_1 \leq x \leq a_2; \\ r + (1-r)\left(\frac{x-a_2}{a_4-a_2}\right), & \text{if } a_2 \leq x \leq a_4; \\ r + (1-r)\left(\frac{a_5-x}{a_5-a_4}\right), & \text{if } a_4 \leq x \leq a_5; \\ r\left(\frac{a_6-x}{a_6-a_5}\right), & \text{if } a_5 \leq x \leq a_6; \\ 0, & \text{if } x > a_6, \end{cases}$$

that is, $\tilde{a}$ turns into a linear pentagonal fuzzy number $\tilde{a} = (a_1, a_2, a_4, a_5, a_6; r)$.

The **cardinality** of $\tilde{a} = (a_1, a_2, a_4, a_5, a_6; r)_n$ with its membership function $\mu_{\tilde{a}}$, denoted by card $\tilde{a}$, can be written as

$$\begin{aligned} \text{card}\,\tilde{a} &= \int_{a_1}^{a_6} \mu_{\tilde{a}}(x)\,dx \\ &= \int_{a_1}^{a_3} \mu_{\tilde{a}}(x)\,dx + \int_{a_3}^{a_4} 1\,dx + \int_{a_4}^{a_6} \mu_{\tilde{a}}(x)\,dx \\ &\qquad\qquad (\text{since } \mu_{\tilde{a}}(x) = 1 \text{ for all } x \in [a_3, a_4]) \\ &= A_1(\tilde{a}) + (a_4 - a_3) + A_2(\tilde{a}), \end{aligned} \tag{2}$$

where

$$A_1(\tilde{a}) = \int_{a_1}^{a_3} \mu_{\tilde{a}}(x)\,dx \text{ and } A_2(\tilde{a}) = \int_{a_4}^{a_6} \mu_{\tilde{a}}(x)\,dx. \tag{3}$$

Definition 4. Let $\tilde{a} = (a_1, a_2, a_3, a_4, a_5, a_6; r)_n$ be a HFN with its membership function $\mu_{\tilde{a}}$. The classification of $\tilde{a}$ is presented as follows:

1. $\tilde{a}$ is a **HFN with heavy left tail** if $A_1(\tilde{a}) > 0.5\,\text{card}\,\tilde{a}$,
2. $\tilde{a}$ is a **HFN with heavy right tail** if $A_2(\tilde{a}) > 0.5\,\text{card}\,\tilde{a}$,
3. $\tilde{a}$ is a **HFN with light tails** if $\max\{A_1(\tilde{a}), A_2(\tilde{a})\} \leq 0.5\,\text{card}\,\tilde{a}$, which is further distinguished into two specific types:
   - If $A_1(\tilde{a}) = A_2(\tilde{a})$, then $\tilde{a}$ has **equally light tails**.
   - If $A_1(\tilde{a}) \neq A_2(\tilde{a})$, then $\tilde{a}$ has **unequally light tails**.

The terms card $\tilde{a}$, $A_1(\tilde{a})$ and $A_2(\tilde{a})$ are computed by (2) and (3).

To be able to classify the HFNs according to Definition 4 easily, we need to explicitly calculate the values of integral $A_1(\tilde{a})$ and $A_2(\tilde{a})$ by substituting the membership function $\mu_{\tilde{a}}$ according to (1). Then

$$
\begin{aligned}
A_1(\tilde{a}) &= \int_{a_1}^{a_2} \mu_{\tilde{a}}(x)\,dx + \int_{a_2}^{a_3} \mu_{\tilde{a}}(x)\,dx \\
&= \int_{a_1}^{a_2} r\left(\frac{x-a_1}{a_2-a_1}\right)^n dx + \int_{a_2}^{a_3} r + (1-r)\left(\frac{x-a_2}{a_3-a_2}\right)^n dx \\
&= \frac{r}{(a_2-a_1)^n}\int_{a_1}^{a_2}(x-a_1)^n\,dx + r(a_3-a_2) + \frac{1-r}{(a_3-a_2)^n}\int_{a_2}^{a_3}(x-a_2)^n\,dx \\
&= \frac{r}{n+1}(a_2-a_1) + r(a_3-a_2) + \frac{1-r}{n+1}(a_3-a_2) \\
&= \frac{r}{n+1}(a_2-a_1) + \frac{nr+1}{n+1}(a_3-a_2),
\end{aligned}
$$

that is,

$$
A_1(\tilde{a}) = \frac{1}{n+1}\big[r(a_2-a_1) + (nr+1)(a_3-a_2)\big]. \tag{4}
$$

In a similar fashion as the above derivation, we obtain the area $A_2(\tilde{a})$ as

$$
A_2(\tilde{a}) = \frac{1}{n+1}\big[(nr+1)(a_5-a_4) + r(a_6-a_5)\big]. \tag{5}
$$

Using Definition 4 with the formulas of card $\tilde{a}$, $A_1(\tilde{a})$ and $A_2(\tilde{a})$ as shown in (2), (4) and (5), respectively, we accomplish a method for classifying HFNs in the algorithm below.

Algorithm 1. A classification of HFNs.

Step 1: Input a HFN $\tilde{a} = (a_1, a_2, a_3, a_4, a_5, a_6; r)_n$.

Step 2: Calculate the values $A_1(\tilde{a})$, $A_2(\tilde{a})$ and card $\tilde{a}$ as the following formulas:

$$
A_1(\tilde{a}) = \frac{1}{n+1}\big[r(a_2-a_1) + (nr+1)(a_3-a_2)\big],
$$

$$
A_2(\tilde{a}) = \frac{1}{n+1}\big[(nr+1)(a_5-a_4) + r(a_6-a_5)\big],
$$

$$
\text{card}\,\tilde{a} = A_1(\tilde{a}) + (a_4-a_3) + A_2(\tilde{a}).
$$

Step 3: Classify the HFNs by using Definition 4.

- If $A_1(\tilde{a}) > 0.5\,\text{card}\,\tilde{a}$, stop and display "$\tilde{a}$ is a HFN with heavy left tail".
- If $A_2(\tilde{a}) > 0.5\,\text{card}\,\tilde{a}$, stop and display "$\tilde{a}$ is a HFN with heavy right tail".
- Otherwise, i.e., $\max\{A_1(\tilde{a}), A_2(\tilde{a})\} \le 0.5\,\text{card}\,\tilde{a}$,
  - If $A_1(\tilde{a}) = A_2(\tilde{a})$, stop and display "$\tilde{a}$ has equally light tails".
  - Otherwise, stop and display "$\tilde{a}$ has unequally light tails".

---

3   A Ranking Method of HFNs

As we reviewed in the first section, the ranking of fuzzy numbers is an important procedure in dealing with, for example, fuzzy linear programming problem and fuzzy transportation problem. In this section, we focus on a ranking method of our proposed HFNs based on their possibilistic mean values. Let us introduce the possibilistic mean value in the following definition.

Definition 5 (See Fullér and Majlender [10]). Let $\tilde{a}$ be a fuzzy number with its α-cut set $\tilde{a}^\alpha = [a_L^\alpha, a_U^\alpha]$. A function $f : [0, 1] \to \mathbb{R}$ is said to be a **weighting function** if f is non-negative, monotone increasing and satisfies the following normalization condition

$$\int_0^1 f(\alpha)\, d\alpha = 1.$$

The f-**weighted possibilistic mean value** of $\tilde{a}$ is defined by

$$m_f(\tilde{a}) = \int_0^1 f(\alpha) \cdot \frac{a_L^\alpha + a_U^\alpha}{2}\, d\alpha.$$

Note that if the weighting function f is given by $f(\alpha) = 2\alpha$ for any $\alpha \in [0, 1]$, then the value $m_f(\tilde{a})$ becomes

$$m(\tilde{a}) = \int_0^1 \alpha(a_L^\alpha + a_U^\alpha)\, d\alpha, \tag{6}$$

which is known as **possibilistic mean value** of $\tilde{a}$ mentioned in [2].

To evaluate the possibilistic mean value of a HFN $\tilde{a}$ according to (6), we need to know how the α-cut set of $\tilde{a}$ is. This proposition is proved in order to find a closed form of that set.

Proposition 1. *Let $\tilde{a} = (a_1, a_2, a_3, a_4, a_5, a_6; r)_n$ be a HFN. The α-cut set of $\tilde{a}$ is represented as*

$$\tilde{a}^\alpha = \begin{cases} \left[a_1 + \left(\frac{\alpha}{r}\right)^{1/n}(a_2 - a_1), a_6 - \left(\frac{\alpha}{r}\right)^{1/n}(a_6 - a_5)\right], & \text{if } \alpha \in [0, r]; \\ \left[a_3 - \left(1 - \left(\frac{\alpha - r}{1 - r}\right)^{1/n}\right)(a_3 - a_2), a_4 + \left(1 - \left(\frac{\alpha - r}{1 - r}\right)^{1/n}\right)(a_5 - a_4)\right], & \text{if } \alpha \in [r, 1]. \end{cases}$$

Proof. Let $\alpha \in [0, 1]$. Then, the membership function $\mu_{\tilde{a}}$ shown in (1) implies

$$\tilde{a}^\alpha = \{x \in \mathbb{R} : \mu_{\tilde{a}}(x) \geq \alpha\} = X_1 \cup X_2 \cup X_3 \cup X_4 \cup X_5,$$

where

$$\begin{aligned} X_1 &= \{x \in [a_1, a_2] : r\left(\tfrac{x - a_1}{a_2 - a_1}\right)^n \geq \alpha\}, \\ X_2 &= \{x \in [a_2, a_3] : r + (1 - r)\left(\tfrac{x - a_2}{a_3 - a_2}\right)^n \geq \alpha\}, \\ X_3 &= \{x \in [a_3, a_4] : \alpha \leq 1\} = [a_3, a_4], \\ X_4 &= \{x \in [a_4, a_5] : r + (1 - r)\left(\tfrac{a_5 - x}{a_5 - a_4}\right)^n \geq \alpha\} \text{ and} \\ X_5 &= \{x \in [a_5, a_6] : r\left(\tfrac{a_6 - x}{a_6 - a_5}\right)^n \geq \alpha\}. \end{aligned}$$

Moreover, the set X_1, X_2, X_4 and X_5 can be simplified as follows:

$$X_1 = \{x \in [a_1, a_2] : x \geq a_1 + \left(\tfrac{\alpha}{r}\right)^{1/n}(a_2 - a_1)\}$$
$$= \{x \in \mathbb{R} : a_2 \geq x \geq a_1 + \left(\tfrac{\alpha}{r}\right)^{1/n}(a_2 - a_1) \geq a_1\}$$
$$= \{x \in \mathbb{R} : x \geq a_1 + \left(\tfrac{\alpha}{r}\right)^{1/n}(a_2 - a_1) \text{ where } \alpha \in [0, r]\},$$
$$X_2 = \{x \in [a_2, a_3] : x \geq a_2 + \left(\tfrac{\alpha-r}{1-r}\right)^{1/n}(a_3 - a_2)\}$$
$$= \{x \in \mathbb{R} : a_3 \geq x \geq a_2 + \left(\tfrac{\alpha-r}{1-r}\right)^{1/n}(a_3 - a_2) \geq a_2\}$$
$$= \{x \in \mathbb{R} : x \geq a_2 + \left(\tfrac{\alpha-r}{1-r}\right)^{1/n}(a_3 - a_2) \text{ where } \alpha \in [r, 1]\},$$
$$X_4 = \{x \in [a_4, a_5] : x \leq a_5 - \left(\tfrac{\alpha-r}{1-r}\right)^{1/n}(a_5 - a_4)\}$$
$$= \{x \in \mathbb{R} : a_4 \leq x \leq a_5 - \left(\tfrac{\alpha-r}{1-r}\right)^{1/n}(a_5 - a_4) \leq a_5\}$$
$$= \{x \in \mathbb{R} : x \leq a_5 - \left(\tfrac{\alpha-r}{1-r}\right)^{1/n}(a_5 - a_4) \text{ where } \alpha \in [r, 1]\},$$
$$X_5 = \{x \in [a_5, a_6] : x \leq a_6 - \left(\tfrac{\alpha}{r}\right)^{1/n}(a_6 - a_5)\}$$
$$= \{x \in \mathbb{R} : a_5 \leq x \leq a_6 - \left(\tfrac{\alpha}{r}\right)^{1/n}(a_6 - a_5) \leq a_6\}$$
$$= \{x \in \mathbb{R} : x \leq a_6 - \left(\tfrac{\alpha}{r}\right)^{1/n}(a_6 - a_5) \text{ where } \alpha \in [0, r]\}.$$

Therefore, the values of α can be distinguished in two cases below.

Case I: If $\alpha \in [0, r]$, then $a_1 + \left(\tfrac{\alpha}{r}\right)^{1/n}(a_2 - a_1) \leq x \leq a_6 - \left(\tfrac{\alpha}{r}\right)^{1/n}(a_6 - a_5)$, that is,

$$x \in \left[a_1 + \left(\tfrac{\alpha}{r}\right)^{1/n}(a_2 - a_1), a_6 - \left(\tfrac{\alpha}{r}\right)^{1/n}(a_6 - a_5)\right]. \tag{7}$$

Case II: If $\alpha \in [r, 1]$, then

$$x \geq a_2 + \left(\tfrac{\alpha-r}{1-r}\right)^{1/n}(a_3 - a_2)$$
$$= a_3 - (a_3 - a_2) + \left(\tfrac{\alpha-r}{1-r}\right)^{1/n}(a_3 - a_2)$$
$$= a_3 - \left(1 - \left(\tfrac{\alpha-r}{1-r}\right)^{1/n}\right)(a_3 - a_2)$$

and

$$x \leq a_5 - \left(\tfrac{\alpha-r}{1-r}\right)^{1/n}(a_5 - a_4)$$
$$= a_4 + (a_5 - a_4) - \left(\tfrac{\alpha-r}{1-r}\right)^{1/n}(a_5 - a_4)$$
$$= a_4 + \left(1 - \left(\tfrac{\alpha-r}{1-r}\right)^{1/n}\right)(a_5 - a_4),$$

that is,

$$x \in \left[a_3 - \left(1 - \left(\tfrac{\alpha-r}{1-r}\right)^{1/n}\right)(a_3 - a_2), a_4 + \left(1 - \left(\tfrac{\alpha-r}{1-r}\right)^{1/n}\right)(a_5 - a_4)\right]. \tag{8}$$

The expressions (7) and (8) complete the proof of the proposition.

Using the α-cut set in Proposition 1, the possibilistic mean value of a HFN $\tilde{a}$ can be derived as follows:

$$
\begin{aligned}
m(\tilde{a}) &= \int_0^r \alpha(a_L^\alpha + a_U^\alpha)\, d\alpha + \int_r^1 \alpha(a_L^\alpha + a_U^\alpha)\, d\alpha \\
&= \int_0^r \alpha\left[a_1 + \left(\tfrac{\alpha}{r}\right)^{1/n}(a_2 - a_1) + a_6 - \left(\tfrac{\alpha}{r}\right)^{1/n}(a_6 - a_5)\right] d\alpha \\
&\quad + \int_r^1 \alpha\left[a_3 - \left(1 - \left(\tfrac{\alpha-r}{1-r}\right)^{1/n}\right)(a_3 - a_2) + a_4 + \left(1 - \left(\tfrac{\alpha-r}{1-r}\right)^{1/n}\right)(a_5 - a_4)\right] d\alpha \\
&= \int_0^r \alpha(a_1 + a_6)\, d\alpha + (a_2 - a_1 - a_6 + a_5)\int_0^r \alpha\left(\tfrac{\alpha}{r}\right)^{1/n} d\alpha \\
&\quad + \int_r^1 \alpha(a_3 + a_4)\, d\alpha + (a_5 - a_4 - a_3 + a_2)\int_r^1 \alpha\left[1 - \left(\tfrac{\alpha-r}{1-r}\right)^{1/n}\right] d\alpha \\
&= \frac{r^2}{2}(a_1 + a_6) + \frac{nr^2}{2n+1}(a_2 - a_1 - a_6 + a_5) + \frac{1-r^2}{2}(a_3 + a_4) \\
&\quad + (a_5 - a_4 - a_3 + a_2)\int_r^1 \left[\alpha - \alpha\left(\tfrac{\alpha-r}{1-r}\right)^{1/n}\right] d\alpha,
\end{aligned}
$$

where

$$
\begin{aligned}
\int_r^1 \left[\alpha - \alpha\left(\tfrac{\alpha-r}{1-r}\right)^{1/n}\right] d\alpha &= \frac{1-r^2}{2} - \frac{1}{(1-r)^{1/n}}\int_r^1 \alpha(\alpha - r)^{1/n}\, d\alpha \\
&= \frac{1-r^2}{2} - \frac{1}{(1-r)^{1/n}}\int_0^{1-r} (u+r)u^{1/n}\, du \\
&\qquad\qquad\qquad\qquad\qquad \text{(by letting } u = \alpha - r) \\
&= \frac{1-r^2}{2} - \frac{1}{(1-r)^{1/n}}\left[\frac{n(1-r)^{\frac{1}{n}+2}}{2n+1} + \frac{nr(1-r)^{\frac{1}{n}+1}}{n+1}\right] \\
&= \frac{1-r^2}{2} - \frac{n(1-r)^2}{2n+1} - \frac{nr(1-r)}{n+1}.
\end{aligned}
$$

Therefore, an explicit formula of $m(\tilde{a})$ is represented as

$$
\begin{aligned}
m(\tilde{a}) = {}&\frac{r^2}{2}(a_1 + a_6) + \frac{nr^2}{2n+1}(a_2 - a_1 - a_6 + a_5) + \frac{1-r^2}{2}(a_2 + a_5) \\
&- \left[\frac{n(1-r)^2}{2n+1} + \frac{nr(1-r)}{n+1}\right](a_5 - a_4 - a_3 + a_2).
\end{aligned} \tag{9}
$$

Based on the possibilistic mean values of HFNs, we achieve the following method for ranking HFNs, which is developed from Goetschel and Voxman [12].

Definition 6. Let $\tilde{a}$ and $\tilde{b}$ be any two HFNs. The ranking of $\tilde{a}$ and $\tilde{b}$ is defined by their possibilistic mean values as follows:

(i) $\tilde{a}$ is **greater than** $\tilde{b}$, denoted by $\tilde{a} \succ \tilde{b}$, if $m(\tilde{a}) > m(\tilde{b})$,

(ii) $\tilde{a}$ is **less than** $\tilde{b}$, denoted by $\tilde{a} \prec \tilde{b}$, if $m(\tilde{a}) < m(\tilde{b})$,

(iii) $\tilde{a}$ is **equal to** $\tilde{b}$, denoted by $\tilde{a} \approx \tilde{b}$, if $m(\tilde{a}) = m(\tilde{b})$,

where the values $m(\tilde{a})$ and $m(\tilde{b})$ related to $\tilde{a}$ and $\tilde{b}$ are computed by (9).

Moreover, we demonstrate the following example to show how HFNs are classified and ranked according to Algorithm 1 and Definition 6, respectively.

Example 1. Let $\tilde{a} = (1, 3, 4, 5, 8, 10; 0.4)_1$, $\tilde{b} = (1, 2, 5, 7, 8, 9; 0.72)_3$, $\tilde{c} = (1, 3, 4, 4, 5, 7; 0.5)_2$ and $\tilde{d} = (2, 5, 8.5, 9, 11, 12; 0.2)_1$ be given four HFNs, which are illustrated in Fig. 2. Let us first provide their classification according to Algorithm 1. By calculating $A_1(\tilde{a})$, $A_2(\tilde{a})$ and card $\tilde{a}$, we obtain

$$A_2(\tilde{a}) = 2.5 > 2.3 = (0.5)(4.6) = 0.5\,\mathrm{card}\,\tilde{a},$$

which means that $\tilde{a}$ is a HFN with heavy right tail. The rest $\tilde{b}, \tilde{c}$ and $\tilde{d}$ can be done as follows. Since

$$\max\{A_1(\tilde{b}), A_2(\tilde{b})\} = \max\{2.55, 0.97\} = 2.55 \leq 2.76 = (0.5)(5.52) = 0.5\,\mathrm{card}\,\tilde{b}$$

and $A_1(\tilde{b}) = 2.55 \neq 0.97 = A_2(\tilde{b})$, then $\tilde{b}$ is a HFN with unequally light tails. Since

$$\max\{A_1(\tilde{c}), A_2(\tilde{c})\} = \max\{1, 1\} = 1 - (0.5)(2) = 0.5\,\mathrm{card}\,\tilde{c}$$

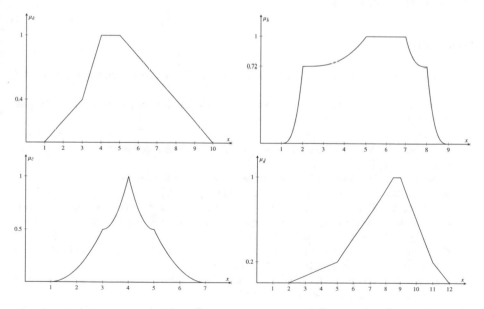

Fig. 2. The HFNs $\tilde{a}$ (above left), $\tilde{b}$ (above right), $\tilde{c}$ (below left) and $\tilde{d}$ (below right).

and $A_1(\tilde{c}) = 1 = A_2(\tilde{c})$, then $\tilde{c}$ is a HFN with equally light tails. Furthermore, we have

$$A_1(\tilde{d}) = 2.4 > 2.1 = (0.5)(4.2) = 0.5\,\mathrm{card}\,\tilde{d},$$

that is, $\tilde{d}$ is a HFN with heavy left tail.

Let us next rank the HFNs $\tilde{a}, \tilde{b}, \tilde{c}$ and $\tilde{d}$ according to our proposed ranking method in Definition 6. By using (9), we obtain the possibilistic mean values of $\tilde{a}, \tilde{b}, \tilde{c}$ and $\tilde{d}$ as

$$m(\tilde{a}) = 5.02, \quad m(\tilde{b}) = 5.3696, \quad m(\tilde{c}) = 4 \text{ and } m(\tilde{d}) = 8.4267,$$

i.e., $m(\tilde{c}) < m(\tilde{a}) < m(\tilde{b}) < m(\tilde{d})$. Therefore, the given HFNs can be ranked as $\tilde{c} \prec \tilde{a} \prec \tilde{b} \prec \tilde{d}$.

4 Conclusion

This article works on a new concept of HFN including its classification and its ranking method. The procedure for classifying was developed in Algorithm 1 from the concept of its cardinality. We obtained an explicit formula of possibilistic mean value of HFN as shown in (9) and applied it in our ranking method as shown in Definition 6.

Acknowledgements. The authors would like to thank to the referees for valuable comments and suggestions.

References

1. Bodjanova, S.: Median value and median interval of a fuzzy number. Inf. Sci. **172**, 73–89 (2005)
2. Carlsson, C., Fullér, R.: On possibilistic mean value and variance of fuzzy numbers. Fuzzy Sets Syst. **122**, 315–326 (2001)
3. Christi, A., Malini, D.: Solving transportation problems with hexagonal fuzzy numbers using best candidates method and different ranking techniques. Int. J. Eng. Res. Appl. **6**, 76–81 (2016)
4. Dhanam, K., Kalaiselvi, T.: Multi-item inventory model using hexagonal fuzzy number. Int. J. Math. Soft Comput. **6**, 93–105 (2016)
5. Dhanam, K., Parimaladevi, M.: Cost analysis on a probabilistic multiitem inventory model with ranking of circumcenter of the centroid of pentagonal fuzzy number. Int. J. Pure Appl. Math. **113**, 1–9 (2017)
6. Dinagar, D.S., Kannan, J.R.: On fuzzy inventory model with allowable shortage. Int. J. Pure Appl. Math. **99**, 65–76 (2015)
7. Dinagar, D.S., Narayanan, U.H.: On determinant of hexagonal fuzzy number matrices. Int. J. Math. Appl. **4**, 357–363 (2016)
8. Dinagar, D.S., Narayanan, U.H.: On inverse of hexagonal fuzzy number matrices. Int. J. Pure Appl. Math. **115**, 147–158 (2017)

9. Elumalai, P., Prabu, K., Santhoshkumar, S.: Fuzzy transportation problem using hexagonal fuzzy numbers by robust ranking method. Emperor Int. J. Financ. Manag. Res. UGC Jr. **45308**, 52–58 (2017)
10. Fullér, R., Majlender, P.: On weighted possibilistic mean and variance of fuzzy numbers. Fuzzy Sets Syst. **136**, 363–374 (2003)
11. Ghadle, K.P., Pathade, P.A.: Solving transportation problem with generalized hexagonal and generalized octagonal fuzzy numbers by ranking method. Glob. J. Pure Appl. Math. **13**, 6367–6376 (2017)
12. Goetschel, R., Voxman, W.: Elementary fuzzy calculus. Fuzzy Sets Syst. **18**, 31–43 (1986)
13. Hanss, M.: Applied Fuzzy Arithmetic. Springer, Heidelberg (2005)
14. Hussain, R.J., Priya, A.: Solving fuzzy game problem using hexagonal fuzzy number. J. Comput. **1**, 53–59 (2016)
15. Mondal, S.P., Mandal, M.: Pentagonal fuzzy number, its properties and application in fuzzy equation. Future Comput. Inform. J. **2**, 110–117 (2017)
16. Rajarajeswari, P., Sahaya, S.A.: Ranking of hexagonal fuzzy numbers for solving multi objective fuzzy linear programming problem. Int. J. Comput. Appl. **84**, 14–19 (2013)
17. Sahaya, S.A., Revathy, M.: A new ranking on hexagonal fuzzy numbers. Int. J. Fuzzy Log. Syst. **6**, 1–8 (2016)
18. Sahaya, S.A., Vijayalakshmi, K.R.: Fuzzy transportation problem using symmetric hexagonal fuzzy numbers. IOSR J. Math. **12**, 76–81 (2016)
19. Selvakumari, K., Sowmiya, G.: Fuzzy network problems using job sequencing technique in hexagonal fuzzy numbers. Int. J. Adv. Eng. Res. Dev. **4**, 116–121 (2017)
20. Thamaraiselvi, A., Santhi, R.: Optimal solution of fuzzy transportation problem using hexagonal fuzzy numbers. Int. J. Sci. Eng. Res. **6**, 40–45 (2015)
21. Vasanthi, L., Elakkiya, E.S.: Orthogonal fuzzy matrix using hexagonal fuzzy numbers. Int. J. Res. Appl. Sci. Eng. Technol. **6**, 2349–2352 (2018)
22. Venkatesan, A., Virginraj, A.: Algorithm for fully fuzzy linear system with hexegonal fuzzy number matrix by singular value decomposition. Int. J. Comput. Algorithms **3**, 218–222 (2014)
23. Zadeh, L.A.: Fuzzy sets. Inf. Control **8**, 338–353 (1965)

A Neglected Theorem - Numerical Analysis via Variable Interval Computing

Tsau Young Lin[✉]

San Jose State University, San Jose, CA 95192, USA
ty.lin@sjsu.edu, prof.tylin@gmail.com

Abstract. A neglected theorem in numerical computing is proved. In science labs, the computing of a scientific formula $y_0 = E(x_0)$ is often *approximated by computing* an 'adequately close' finite decimal f, namely, $E(f)$. In essence, numerical computing is approximate computing: A finite decimal, say $f = 0.333$, can be rounded from any reals in $[0.3325, 0.333)$, called rounding interval; so a computation of f is an approximate computing for any reals in this rounding interval. A real number, based on Cantor's infinite decimal representation, can be regarded as a convergent sequence of rounding intervals. The computing of such sequences is variable interval computing – hope to be a useful approach to numerical analysis.

1 Introduction

A 'standard' routine that computes $y = E(x)$ at x_0 in a science lab will be examined closely.

Example 1. *A Faulty Folklore*

1. *Suppose we are "not", using a ruler to measure the diameter of a test tube. By reading the ruler's markings, we may observe that:*
   a. *the best approximation of this diameter is, say $f = 0.333$.*
   b. *the 'theoretical' or 'unknown true' value x_0 is located in, say $[0.3325, 0.3335) = \rho(0.333) = \rho(f)$, called the rounding interval of f (i.e., $x_0 = \tau(f)$)*
2. *We will express this observation by saying*

   "$f = 0.333$ is 'adequately close to' x_0" (i.e., $\tau(f) \in \rho(f)$.)

3. *In the science lab, $E(f) = E(0.333)$ is often computed by computers*

   E is a computable function

   Let the computed value be $E(0.333) = 0.666$.
4. *Then the "standard claim" would be that the "unknown true value" $E(x_0)$ is located in the rounding interval of $E(0.333)$. i.e., $E(x_0) \in [0.6655, 0.6665) = \rho(0.666) = \rho(E(0.333)) = \rho(E(f))$. In plain words,*

   "$E(f) = 0.666$ is 'adequately close to' $E(x_0) = \tau(E(f))$."

© Springer Nature Switzerland AG 2019
R. B. Kearfott et al. (Eds.): IFSA 2019/NAFIPS 2019, AISC 1000, pp. 330–339, 2019.
https://doi.org/10.1007/978-3-030-21920-8_30

Our goal of explicitly listing this 'standard' routine is to pointing out a few missing points. In his book, Ridgway Scott gives a bullet and explains ([8] Chap. 1, p. 1):

- "the effects of finite-precision arithmetic (a.k.a. round-off error).

The author explains: "The third topic is still not well understood at the most basic level, in the sense that there is not a well-established mathematical model for finite-precision arithmetic."

First, we would like to observe a few key points:

1. an explicitly given finite decimal f,
2. an unknown true value $x_0 = \tau(f)$,
3. a rounding interval ρf,
4. a relation, $\tau(f) \in \rho(f)$,
5. a scientific formula $y = E(x)$,
6. an unjustified principle, replacing the computation of $y = E(x)$ at x_0 by $E(f)$, and
7. an unjustified claim: $E(x_0) \in \rho(E(f))$

The primary results are:(1) Item 6 may not be true, in general and (2) propose to replace it by Theorem 1. Implicitly, we are trying to recommand that the variable interval computing is a candidate of the missing mathematical model.

1.1 Analysis

Heuristic Approximation 'Theorem:' (HAT)

$E(f)$ is 'adequately close to' $E(x_0)$ whenever f is 'adequately close to' x_0. Or formally

$$E(\tau(f)) \in \rho(E(f)) \text{ whenever } \tau(f) \in \rho(f). \tag{1}$$

Each numerical computing will be ended with a regular rounding, at that point, we more or less wish that HAT is true. In this note, we will show that it is true with modification; see Theorem 1.

Let us analyze a simple case. Suppose the scientific formula E given above is a simple addition, namely, $E(x) = x + x$. Let us trace how the unknown true value $x_0 = \tau(f)$ is moved in this simple addition. In Example 1, the approximate value is: $f = 0.333$.

$$E(x_0) = x_0 + x_0 = \tau(0.333) + \tau(0.333) \doteq 0.333 + 0.333 = 0.666$$

Let us use the calculus of complexes [11] to trace the sum of $\tau(f)$'s.

$$
\begin{aligned}
E(\tau(0.333)) &= \tau(0.333) + \tau(0.333) \in \rho(0.333) \oplus \rho(0.333) \\
&= [0.3325, 0.3335) \oplus [0.3325, 0.3335) \\
&= \{x_1 + x_2 \mid x_1 \in X_1 \text{ and } x_2 \in X_2\} \\
&= [0.6650, 0.6670) \tag{2}
\end{aligned}
$$

In other words, $E(\tau(0.333)) \in [0.6650, 0.6670) \not\subseteq [0.6655, 0.6665) = \rho(0.666)$
This concludes:

$$E(\tau(f)) = \tau(0.333) + \tau(0.333) \text{ may not belong to } \rho(0.666) = \rho(E(f)) \quad (3)$$

The probability distribution of the 'unknown true value x_0' is normally unknown. Without it, the probability that "$E(x_0) \in \rho(E(f))$" is 0.5. The computation given above indicates that the 'theoretical' or

the 'the unknown true value' $E(x_0)$ is located in $[0.6650, 0.6670)$;

the length of this interval is 0.0020. On the other hand, the regular rounding assumes that $E(x_0)$ is located in $[0.6655, 0.6665$; the length of this interval is 0.0010. So the probability is the ratio 0.5.; this is the case when E is a single sum. For two sums, it will be 0.25, etc.

The probability will decrease towards zero,

when the number of the summands increases. An explicit counter example is presented here.

Example 2. *A Counter Example for Heuristic Approximation 'Theorem'*

A rule quoted from some website: "When adding or subtracting, we go by the number of decimal places (i.e., the number of digits on the right side of the decimal point) rather than by the number of significant digits. Identify the quantity having the smallest number of decimal places, and use this number to set the number of decimal places in the answer."

Suppose $E(x) = x + \ldots x$ (adding 2000 times.) For illustration, let us assume hypothetically we do know

1. 'the unknown true value' $x_0 = \tau(0.333) = 0.33344444\ldots$. Then,
2. $E(f) = 666.000$ (based on the quoted rule.)
3. $y_0 = E(x_0) = 0.33344444\ldots + \ldots = 0.6668888\ldots$
   (2000 additions, where $\tau(0.333)$ denote the 'the unknown true value' x_0.

$$E(x_0)) \notin \rho(E(x_0) = \rho(666.000) = 665.9995 \quad 666.005).$$

Observe that $\rho(E(f))$ does not include '$E(\tau(f))$.'

1.2   Main Theorems

So the Heuristic Approximation 'Theorem' is basically invalid. Theorem 1 is a very weak form of a valid theorem, along the same direction. Based on the concept of regular rounding, we create the Cantor Neighborhood System (CMS). A Cantor Neighborhood System at p is a local based of the topological neighborhood system at p ([4], Chap. 1, Ex. B).

To state the theorems, we also need few terms. A scientific formula $y = E(x)$ is said to be "*nice at* x_0," if E is continuous at x_0 and $E(x_0)$ is a computable number. A number ε is said to be the the maximal acceptable radius of errors if ε is the acceptable maximal distance between the all approximate points and the exact desirable point.

Theorem 1. *Approximation Theorem.*

Suppose $E(x)$ is 'nice' at x_0 and $\varepsilon = 1/10^p$ (p is an integer) is the maximal acceptable radius of errors. Then, there exists a Cantor neighborhood $\rho(f_n, x_0)$ with diameter $\delta = 1/10^n$ such that

q_p *is 'adequately close' to$E(x_0)$ whenever f_n is 'adequately close' to x_0.*

Or formally,

$$E(x_0) \in \rho(q_p, E(x_0)) \text{ whenever } x_0 \in \rho(f_n, x_0), \tag{4}$$

where $q_p = k_1 k_2 \ldots k_m.h_1 h_2 \ldots h_p$ is the result of rounding $E(f_n)$ to p decimal places.

Remark 1: The condition, E is continuous at x_0, gives us the Cantor neighborhood $\rho(f_n) = [c, d]$; however, that is not enough. We also need an explicit finite decimal that is near $E(x_0)$; such a requirement is guaranteed by the condition, '$E(x_0)$ is a computable number.'

Remark 2: This theorem is a theoretical one; it does tell us that a valid f_n does exist. In typical applications, we need stronger 'nice' at x_0.

Corollary 1. *Suppose, in addition, $E(x)$ is continuous on a finite closed interval $C = [c_1, c_2]$, where $x_0 \in (c_1, c_2)$. Then, with the given diameter $\varepsilon = 1/10^p$, there exists a small closed Cantor neighborhood $\overline{\rho(f_n, x_0)} = [c, d](\subseteq C)$ with diameter $\delta = 1/10^n$ such that*

q_p *is 'adequately close' to $E(x_0)$ whenever f_n is 'adequately close' to x_0.*

Or formally,

$$E(x_0) \in \rho(q_p, E(x_0)) \text{ whenever } x_0 \in \rho(f_n, x_0) = [c, d], \tag{5}$$

where $q_p = k_1 k_2 \ldots k_m.h_1 h_2 \ldots h_p$ is the result of rounding $E(f_n)$ to p decimal places.

Proof: Observe that we do not impose the second condition of 'nice' explicitly. So we need to show that $E(x_0)$ is a computable number; by Weistrass approximation theorem, for any $\varepsilon_1(< \varepsilon)$ that we choose, there is a polynomial $p(x)$ such that $|E(x) - p(x)| < \varepsilon_1 \ \forall x \in [c, d]$. Next, we will choose a finite decimal $g \in [c, d]$ such that $|p(x_0) - p(g)| < \varepsilon - \varepsilon_1$. Observe that we have $|E(x_0) - p(g)| \leq |E(x_0) - p(x_0)| + |p(x_0) - p(g)| < \varepsilon_1 + (\varepsilon - \varepsilon_1) = \varepsilon$. So $p(g)$ is the desirable finite decimal (the rational umber r in the proof of Theorem 1). The result of rounding $p(g)$ to p decimal places is the desirable q_p. QED.

Remark 3: Since $p(x)$ is an explicitly known polynomial on $[c, d]$, through $p(g)$ we can check if f_n is a valid candidate. This corollary is useful in practices.

Proof of Theorem 1

Observe that given such an ε implies: a unique open Cantor neighborhood open-$\rho(q_p, E(x_0)) = (a, b)$ with diameter ε is also given. By the continuity of E at x_0 (first condition of 'nice'), there exists a *closed* Cantor neighborhood closed-$\rho(f_n, x_0) = [c, d]$ such that ([4], Chap. 3, Theorem 1, Item (e), p. 86))

$$E(x) \in (a, b) \text{ whenever } x \in [c, d]. \tag{6}$$

Next, we note that $E(x_0)$ is a computable number (second condition of 'nice',) then there exists a computable function which, given any positive rational error bound ε_1, produces a rational number r such that $|r - E(x_0)| \leq \varepsilon_1$, where $\varepsilon_1 = \min(\text{dist}(E(x_0), a), \text{dist}(E(x_0), b))$, where $\text{dist}(E(x_0), a)$ is the distance between the two points $E(x_0)$ and a, and a similar story for $\text{dist}(E(x_0), b)$. Such an r, by the simple geometry of the three points, $E(x_0)$, a, and b, is lying in the interval (a, b). Now let us take q_p to be the result of rounding r to p decimal places; this q_p does meet the condition:

$$E(x_0) \in \rho(q_p, E(x_0)) \text{ whenever } x_0 \in \rho(f_n, x_0), \tag{7}$$

QED

Theorem 2

$$(\mathscr{R}_C, \oplus, \odot, >) \cong (\mathscr{R}, +, \times, >)$$

as complete ordered fields.

Roughly this theorem implies that using variable interval computing, we can recapture the real number system.

2   Decimals and Cantor Neighborhood Systems (CNS)

The object of this section is to review two basic concepts: Cantor's infinite decimal representation of a real number, and the topology of real numbers using such representations.

- Any real number x can be represented by an infinite decimal number of the following form

$$x \equiv k_1 k_2 \ldots k_m . h_1 h_2 \ldots h_n \ldots, \tag{8}$$

where each k and each h are one of the digits $0, 1, 2, \ldots 9$. $k_1 k_2 \ldots k_m$ is the integer part and $h_1 h_2 \ldots h_n$ is the decimal part. All sequences are infinite (no n exists such that $h_n, h_{(n-1)} \ldots$ are all zeros). A finite decimal 2.50 is replaced 2.4999 . . ..

By an infinite decimal, we mean it represents a convergent sequence of finite decimals. Based on such a sequence, we will construct a local base, called Cantor

Neighborhood System, of the topological neighborhood system (TNS) at x_0; see [4], p. 50 and Chap. I, Ex. B, respectively. Most of the following are from [7] and [6]; We hope our summarizations are clearer and better.

Here, we will illustrate the idea by examples.

Example 3. *Examples of Cantor Neighborhood Systems*

Ex1. Suppose $x = 0.666666\ldots$ is the Cantor representation of a real number. Then we may consider the following two concepts.
   a. the Cantor Sequence; each finite decimal has been rounded

$$S(0.666\ldots) = \{0.7, 0.67, 0.667, 0.6667\ldots\}$$

that consists of finite decimals that are obtained by rounding off the long tail of x.
   b. the Cantor Neighborhood System (CNS) of x that is a local base

$$\mathscr{N}_\rho(0.666\ldots) = \{\rho(0.7),\ \rho(0.67), \rho(0.667), \rho(0.6667)\ldots\},$$

that consists of the rounding intervals of all the finite decimals in the Cantor Sequence.

Ex2. Here is another example, $x = 0.3333\ldots$.
   a. its Cantor Sequence; each finite decimal has been rounded

$$S(0.333\ldots) = \{0.33, 0.33, 0.333, 0.3333\ldots\}$$

   b. its Cantor Neighborhood System of x is

$$\mathscr{N}_\rho(0.333\ldots) = \{\rho(0.3),\ \rho(0.33), \rho(0.333), \rho(0.3333)\ldots\}.$$

They are Cantor neighborhood systems of $1/3$ and $2/3$.

With these examples, we will set up the formal definitions for Cantor Sequence and Cantor Neighborhood System at x.

Definition 1. *Suppose a real number x has the following infinite decimal representation:*

$$x \equiv k_1 k_2 \ldots k_m.h_1 h_2 \ldots h_n \ldots,$$

1. *Then its Cantor Sequence is as follows; the overline indicates that each finite decimal has been rounded.*

$$S(x) = \{\overline{k_1 \ldots k_m.h_1},\ \overline{k_1 k_2 \ldots k_m.h_1, h_2},\ \ldots\}$$

2. *So its Cantor Neighborhood System of x is a sequence of rounding intervals. Again the overline indicates that each finite decimal has been rounded*

$$\mathscr{N}_\rho(x) = \{\rho(\overline{k_1 \ldots k_m.h_1}), \rho(\overline{k_1 \ldots k_m.h_1 h_2})\ldots\}, \tag{9}$$

3. *The Cantor Neighborhood System for the real numbers* $\mathscr{R}$:

$$\mathscr{N}_\rho(\mathscr{R}) = \{\mathscr{N}_\rho(x) \mid \quad \forall x \in \mathscr{R}\}$$

Proposition 1. *It is important to observe that the decimal representation of any "unknown true value" x_0 is infinite, so it cannot be the end points of any rounding intervals. More precisely x_0 is in the interior of every Cantor neighborhood of x_0. ow, we have clearly defined the CNS of each point, that may be regarded as a variable interval represents the point.*

3   Algebraic and Topological Structure of CNS

The goal of this section is to review the algebraic and topological structures of the real number system $\mathscr{R}$ in terms of Cantor neighborhood systems.

A Cantor neighborhood or a Cantor neighborhood system is called an open, a (left-closed and right-open), and a closed Cantor neighborhood or Cantor neighborhood system (CNS) if each interval is an open, a (left-closed and right-open), and a closed respectively; the adjective in the middle one that is embraced by parenthesis will be regarded as default, it will be skipped. All three kinds of CNS are local bases of the topological neighborhood system of x. A Cantor neighborhood is an interval. A Cantor neighborhood system (CNS) of x is a sequence of (decreasing) intervals, in notation (from Eq. (9))

$$\mathscr{N}_\rho(x) = \{u \mid u = \rho(\overline{k_1 \ldots k_m . h_1}), \rho(\overline{k_1 \ldots k_m . h_1 h_2}) \ldots\}$$

In set theoretical notation, we use a variable u to refer to an arbitrary member. Such an u can be regarded as a variable interval or variable length interval. So by 'variable interval computing', we mean the computing of such sequences $\mathscr{N}_\rho(x)$; observe that the variable u represents an arbitrary element (a rounding interval) and etc. Observe that the lengths of these intervals (called rounding intervals) vary through $1/10$, $1/10^2$, $\ldots\ldots$; the sequence $\mathscr{N}_\rho(x)$ is convergent to a single real number x. So we may studies the rounding intervals with length less than a given ε etc.; "The so-called dependency problem is a major obstacle to the application of interval arithmetic." (Wikipedia) Our approach might reduce it dramatically; we hope.

To see this, in this section, we want to examine the process of the 'limits' (or idealization [9]) of these intervals.

First, we need to decouple the two operators + and × of the real number $\mathscr{R}$.

Definition 2. *Algebraic system with two operators $(E, \circ_1, \circ_2)$ is called a bi-operator algebra.*

The real number $\mathscr{R}$, the power set $2^{\mathscr{R}}$ and the power set of power set $2^{2^{\mathscr{R}}}$ are all bi-operator algebras [5]. Following this algebraic line, we define $\mathscr{R}$ algebraically ([1] p. 98).

Theorem 3. *The real number system* $(\mathscr{R}, +, \times)$ *is a complete ordered field.*

We will outline the key steps in introducing complete ordered field structure into CNS (Theorem 2) and TNS (Theorem 4); these are summarized from [7] and [6].

$$TNS : \mathscr{R}_T = \{\bar{p} \mid \bar{p} = \text{TNS}(p), p \in R\}. \tag{10}$$

$$CNS : \mathscr{R}_C = \{\tilde{p} \mid \tilde{p} = \text{CNS}(p), p \in R\}. \tag{11}$$

First, we introduce external algebraic operations in $T1$ and $C1$.

T1. $\forall \, \bar{p}, \, \bar{q} \in \mathscr{R}_T$,

$$\bar{p} \oplus' \bar{q} \equiv \{N(p) + N(q) \mid N(p) \in \bar{p}, N(q) \in \bar{q}\}$$

$$\bar{p} \odot' \bar{q} \equiv \{N(p) \cdot N(q) \mid N(p) \in \bar{p}, N(q) \in \bar{q}\},$$

where

$$N(p) + N(q) = \{r_1 + r_2 | r_1 \in N(p), r_2 \in N(q)\},$$

$$N(p) \cdot N(q) = \{r_1 \cdot r_2 | r_1 \in N(p), r_2 \in N(q)\},$$

C1. $\forall \, \tilde{p}, \, \tilde{q} \in \mathscr{R}_C$,

$$\bar{p} \oplus' \bar{q} \equiv \{N_\rho(p) + N_\rho(q) \mid N_\rho(p) \in \tilde{p}, N_\rho(q) \in \tilde{q}\}$$

$$\bar{p} \odot' \bar{q} \equiv \{N_\rho(p) \cdot N_\rho(q) \mid N_\rho(p) \in \tilde{p}, N_\rho(q) \in \tilde{q}\},$$

where

$$N_\rho(p) + N_\rho(q) = \{r_1 + r_2 | r_1 \in N_\rho(p), r_2 \in N_\rho(q)\},$$

$$N_\rho(p) \cdot N_\rho(q) = \{r_1 \cdot r_2 | r_1 \in N_\rho(p), r_2 \in N_\rho(q)\}$$

Second, we observe 4 inclusions in T_2, T_3, C_2, C_3; they are induced by the continuities of the addition and multiplication.

T2. $\forall \, \bar{p}, \, \bar{q} \in \mathscr{R}_T$,

$$\bar{p} \oplus' \bar{q} \subseteq \overline{p + q};$$

$$\bar{p} \odot' \bar{q} \subseteq \overline{p \cdot q}.$$

C2. $\forall \, \tilde{p}, \, \tilde{q} \in \mathscr{R}_C$,

$$\tilde{p} \oplus' \tilde{q} \subseteq \widetilde{p + q};$$

$$\tilde{p} \odot' \tilde{q} \subseteq \widetilde{p \cdot q}.$$

T3. New algebraic structure defined by: $\forall \, \bar{p}, \, \bar{q} \in \mathscr{R}_T$,

$$\bar{p} \oplus \bar{q} \equiv \overline{p + q} : \quad \bar{p} \odot \bar{q} \equiv \overline{p \cdot q}$$

Third, we introduce two abstract operations,$\oplus$ and $\odot$ in T_3 and C_3. They are abstract operations between $\text{TNS}(\bar{p})$ and $\text{TNS}(\bar{q})$ and between $\text{CNS}(\tilde{p})$ and $\text{CNS}(\tilde{q})$

T3. a topological structure defined by the inequality,

$$\bar{p} < \bar{q} \Leftrightarrow p < q,$$

T3. Based on the inequality, topological structures, such as TNS and CNS have been defined previously.

C3. New algebraic structure defined by: $\forall\ \tilde{p},\ \tilde{q} \in \mathscr{R}_C$,

$$\tilde{p} \oplus \tilde{q} \equiv \widetilde{p+q} : \quad \tilde{p} \odot \tilde{q} \equiv \widetilde{p \cdot q}$$

C3. a topological structure defined by the inequality,

$$\tilde{p} < \tilde{q} \Leftrightarrow p < q,$$

C3. Based on the inequality, topological structures, such as TNS and CNS have been defined previously.

Here are main theorems

Theorem 4.

$$(\mathscr{R}_T,\ \oplus,\ \odot, >) \cong (\mathscr{R},\ +,\ \times, >)$$

as complete ordered fields.

Proof: From the remarks around Definition 2, the isomorphism $\cong$ (as a map $\longrightarrow$) is one-to-one homomorphism with respect to each operator. So $\cong$ is isomorphism as bi-operator algebras. The identities (of the axioms of well ordered fields) should be transformed from right-hand-side to left-hand-side. So the bi-operator isomorphism becomes isomporphism of complete ordered fields. QED.

Proof of Theorem 2. The same proof is valid for this theorem.

4   Conclusions

Recall that Theorem 2 is:

$$(\mathscr{R}_C,\ \oplus,\ \odot, >) \cong (\mathscr{R},\ +,\ \times, >)$$

as complete ordered fields.

Theorem 2 reflects the structure of CNS; it is the limit case of interval computing. In other words, the theorem is about the intervals of all lengths $\forall\ \varepsilon$. By choosing a particular set of ε's, this isomorphism is reduced to a near-isomorphism; that is where the problem of numerical analysis sits. Theorem 1 and Corollary 1 are formulated in appropriate level of ε. Our approach, dealing with near-isomorphisms, may be a right way toward a model for finite-precision arithmetic [8].

References

1. Birkhof, G., MacLane, S.: A Survey of Modern Algebra. MacMillan Publishers, London (1970)
2. Hopcroft, J., Ullman, J.: Introduction to Automata Theory, Languages and Computation (1979)
3. Kearfott, R.B.: Interval Computations: Introduction, Uses, and Resources. https://interval.louisiana.edu/preprints/survey.pdf
4. Kelley, J.: General Topology, Van Nostrand (Chap. 3, Theorem 1, Item e) (1955)
5. Kuroki, N.: On power semigroups. Proc. Japan Acad. **47**, 449 (1971)
6. Lin, T.-Y.: A mathematical theory of fuzzy numbers - granular computing approach. In: RSFDGrC 2013, pp. 208–215 (2013)
7. Lin, T.-Y.: A paradox in rounding errors approximate computing for big data. In: SMC 2015, pp. 2567–2573 (2015)
8. Scott, L.R.: Numerical Analysis. Princeton University Press, Princeton (2011)
9. Smale, S.: Some remarks on the foundations of numerical analysis. SIAM Rev. **32**(2), 211–220 (1990). Society for Industrial and Applied Mathematics
10. Wilder, R.: The Foundation of Mathematics, p. 82. Wiley, Hoboken (1952)
11. Zassenhaus, H.: Group Theory. Chelsea Publishing Co. (1958). (Dover 1999)

Information Transmission
and Nonspecificity in Feature Selection

Pasi Luukka[(⊠)] and Christoph Lohrmann

School of Business and Management, Lappeenranta University of Technology,
Lappeenranta, Finland
{pasi.luukka,christoph.lohrmann}@lut.fi

Abstract. In this paper we propose a novel feature selection method
which is based on fuzzy measures. More specifically, we apply a similarity measure to form similarity matrices from the data and apply nonspecificity on similarity degrees in order to conduct feature selection. To
measure how relevant a particular feature is, we apply an information
transmission measure. We exemplify our method on a simple artificial
case to demonstrate its ability to select informative features. Moreover,
we test our method on two real world data sets, the chronic kidney disease and the diabetic retinopathy Debrecen dataset. The nonspecificity-
based feature selection method leads for both datasets to improvements
in the mean classification performance. In comparison with the popular
ReliefF algorithm and the Fisher Score, the new method reaches competitive results and also accomplishes the highest mean accuracy for both
datasets.

1 Introduction

The era of big data has brought data's dimensionality to be of growing interest. On one hand, more features can improve classification performance, but, on
the other hand, the important question arises which of the features are possibly
irrelevant or redundant from classification point of view. While the availability
of data is nowadays less of a concern, at the same time, relevance of features
has become the prevalent question. To address these challenges, many different dimensionality reduction techniques have been proposed by researchers [4].
In general, one can divide dimensionality reduction techniques into two categories: feature selection and feature extraction. Basic feature extraction techniques include e.g. Principal Component Analysis (PCA) [7] and Linear Discriminant Analysis (LDA), [5], [16]. In this paper, the focus is on feature selection.
Feature selection can be conducted in three different types: wrapper, embedded
and filter methods [3]. Wrapper and embedded methods interact with classifiers
to obtain the feature subset with the best classification performance, whereas
filter methods do not interact with classifiers. In general, better classification
accuracies can be achieved with wrapper or embedded methods, whereas filter
methods have lower computational cost. In this paper we concentrate on the
filter approach.

© Springer Nature Switzerland AG 2019
R. B. Kearfott et al. (Eds.): IFSA 2019/NAFIPS 2019, AISC 1000, pp. 340–350, 2019.
https://doi.org/10.1007/978-3-030-21920-8_31

Information theory is widely deployed in filter methods. Here, especially mutual information is frequently applied [20]. In particular, methods using Shannon's information entropy [18] are common, but also fuzzy entropy based methods [14], [12] are applied. In this paper, we approach this subject from nonspecificity and information transmission point of view [9], which, to our knowledge, has not been done before. Moreover, in our proposed feature selection method we connect similarity measures and ideal vectors of classes with nonspecificity for the first time.

2   Mathematical Background

Next we shortly go through the concepts of nonspecificity, information transmission and similarity, so that the nonspecificity−based feature selection method is clearer to follow.

2.1   Nonspecificity and Information Transmission

Nonspecificity is a type of uncertainty which is connected to the sizes (cardinalities) of relevant sets of alternatives. It was first conceived in terms of classical set theory and was defined using the function

$$U(A) = c \cdot \log_b(Card(A)) \tag{1}$$

where $Card(A)$ denotes the cardinality of a finite nonempty set A, where b and c are positive constants. When $b = 2$ and $c = 1$ we obtain $U(A) = \log_2(Card(A))$ and uncertainty is measured in bits. Function U is called a Hartley function [6]. When the Hartley function U is applied to nonempty subsets of a given finite universal set X it has the form $U : P(X) - \{\emptyset\} \to \mathbb{R}^+$, where $P(X)$ denotes the power set of the universal set and $0 \leq U(A) \leq \log_2(Card(A))$. Large sets result in less specific predictions than their smaller counterparts and singletons are fully specific ($U(A) = 0$ if set A consists of a single element).

This function can easily be extended to the fuzzy set theory [21] context where cardinalities are now obtained by using membership degree $A(x)$ of element x belonging to fuzzy set A as $Card(A) = \sum_{x \in X} A(x)$. A common choice for the logarithm function in such a case is the natural logarithm.

If set A is reduced to set B then the amount of uncertainty-based information transmission between these sets can be obtained by $T(A, B) = U(A) - U(B)$. The difference here is viewed as the amount of reduced uncertainty. This information transmission can also be written as $T(A, B) = \log(\frac{Card(A)}{Card(B)})$. If $Card(B) = 1$ we obtain $T(A, B) = U(A)$ which can be viewed as the amount of information needed to characterize one element of set A.

2.2   Similarity

In data handling, many procedures begin with a mutual comparison of the data samples. In the simple case, if the data are point-shaped and compared by pairs, the comparison consists of obtaining a relation $\mu_R : \mathbb{R} \times \mathbb{R} \rightarrow [0,1]$. Usually, at first a mapping of the elements into the unit interval is needed and actual similarity measures conduct a mapping $s : [0,1] \times [0,1] \rightarrow [0,1]$. There exist several different similarity measures [2]. Here, we apply the similarity introduced in [22] (see also [13]). Similarity of two elements can be defined as

$$s(x,y) = 1 - |x - y| \tag{2}$$

where $x, y \in [0,1]$. This can be extended by following the generalized Łukasiewicz structure as [15]

$$s(x,y) = (1 - |x^p - y^p|)^{1/p} \tag{3}$$

where parameter $p \in \mathbb{R}^+$ can be used to adjust the strength of the similarity for the data at hand.

3   Nonspecificity-Based Feature Selection Method

In the following a step-by-step description of the novel nonspecificity algorithm and a simple example on an artificial dataset is presented.

3.1   Step-by-Step Procedure of the Algorithm

Next, we go into detail about how to use similarity, nonspecificity and information transmission in the context of feature selection by introducing a short algorithm of the process. N denotes the number of classes, n denotes the overall number of samples, n_i denotes the number of samples in class i and D denotes the number of features.

1: Normalize the data into unit interval. $X^D \rightarrow [0,1]^D$
2: Obtain ideal vectors v_i for each class i from the training set data.

$$v_{i,d} = \left(\frac{1}{n_i} \sum_{x \in X_i} x_d \right), \quad d = 1, \cdots, D, \quad i = 1, \cdots, N \tag{4}$$

where x_d denotes the value of the feature d of samples x and we restrict the mean calculation to only those samples that belong to class i.
3: Obtain similarities $S(x_{j,d}, v_{i,d})$ with $x_j \in X_i$ between data sample vectors x_j belonging to class i and the ideal vector v_i. Where for the dth feature

$$S(x_{j,d}, v_{i,d}) = (1 - | (x_{j,d})^p - (v_{i,d})^p |)^{\frac{1}{p}} \tag{5}$$

$d = 1, \ldots, D, \quad i = 1, \ldots, N, \quad j = 1, \ldots, n$
4: Obtain nonspecificity w.r.t. class i for each feature d by using similarities obtained in step 3 belonging to class i.

$$u_{1,i,d} = \ln \left(\sum_{x_j \in X_i} S(x_{j,d}, v_{i,d}) \right) \tag{6}$$

where subscript 1 denotes class specific nonspecificities.

5: Obtain similarities $S(x_{j,d}, v_{i,d})$ with $x_j \in X$. As opposed to step 3, now we compute the similarities for all observations x_j with the ideal vector v_i independent from the class labels. Where for the dth feature

$$S(x_{j,d}, v_{i,d}) = (1 - |(x_{j,d})^p - (v_{i,d})^p|)^{\frac{1}{p}} \tag{7}$$

6: Obtain nonspecificity w.r.t. class i for each feature d by using similarities obtained in step 5.

$$u_{2,i,d} = \ln \left(\sum_{x_j \in X} S(x_{j,d}, v_{i,d}) \right) \tag{8}$$

where X denotes the entire data set and where subscript 2 denotes the nonspecificity that is not class specific.
7: Obtain information transmission value T_d for each feature d using the two nonspecificities computed in step 4 and 6.

$$T_d = \sum_{i=1}^{N} (u_{1,i,d}/n_i - u_{2,i,d}/n) \tag{9}$$

8: Reduce the number of features from the data by selecting $k < D$ features having k highest information transmission values T.

3.2   Detailed Example of the Procedure

Next, we present a simple example where one of the features is irrelevant and demonstrate how the procedure to find this variable works. Table 1 consists of a data set with four sample vectors with three features (f_1, f_2, f_3) and class label information in the fourth column. The first two samples belong to class 1 and the last two samples to class 2. The third feature is irrelevant for the classification task.

Table 1. Simple data set

f_1	f_2	f_3	C
1.1	1.5	1	1
0.9	2.5	3	1
2.0	3	0.5	2
2.1	5	3.5	2

Subsequently, we present the feature selection technique in the step by step manner for our example data.

1: **Normalize the data into unit interval.** $X^D \to [0, 1]^D$. This is now done in Table 2.

Table 2. Normalized data set

f_1	f_2	f_3	C
0.1667	0	0.1667	1
0	0.2857	0.8333	1
0.9167	0.4286	0	2
1	1	1	2

2: **Obtain ideal vectors v_i for each class i from the training data set.**
Here, we used the arithmetic mean. Mean vectors are calculated for each class as:

$$v_1 = [0.0833, 0.1429, 0.5]$$

$$v_2 = [0.9583, 0.7143, 0.5]$$

3: **Obtain similarities $S(x_{j,d}, v_{i,d})$ between the ideal vectors and the data sample vectors belonging to class i.** This is now done in Table 3.

Table 3. Similarities where similarity is obtained between the ideal vectors v_i and samples within classes

Similarity	f_1	f_2	f_3
$S(x_{1,d}, v_{1,d})$	0.9167	0.8571	0.6667
$S(x_{2,d}, v_{1,d})$	0.9167	0.8571	0.6667
$S(x_{3,d}, v_{2,d})$	0.9583	0.7143	0.5
$S(x_{4,d}, v_{2,d})$	0.9583	0.7143	0.5

4: **Obtain nonspecificity w.r.t. class i for each feature d by using similarities obtained in step 3 belonging to class i.** These are now in Table 4.

Table 4. Nonspecificities u_1

Nonspecificity	f_1	f_2	f_3
$u_{1,1,d}$	0.6061	0.5390	0.2877
$u_{1,2,d}$	0.6506	0.3567	0

Where e.g. for feature f_1 the nonspecificity is $u_{1,1,1} = ln(0.9167 + 0.9167) = 0.6061$.

5: **Obtain similarities $S(x_{j,d}, v_{i,d})$ between the ideal vectors v_i and all data sample vectors x_j.** Results are displayed in Table 5.

Table 5. Similarities where similarity is obtained between the ideal vectors v_i and all sample vectors x_j

Similarity	f_1	f_2	f_3
$S(x_{1,d}, v_{1,d})$	0.9167	0.8571	0.6667
$S(x_{2,d}, v_{1,d})$	0.9167	0.8571	0.6667
$S(x_{3,d}, v_{1,d})$	0.1667	0.7143	0.5
$S(x_{4,d}, v_{1,d})$	0.0833	0.1429	0.5
$S(x_{1,d}, v_{2,d})$	0.2083	0.2857	0.6677
$S(x_{2,d}, v_{2,d})$	0.0417	0.5714	0.6677
$S(x_{3,d}, v_{2,d})$	0.9583	0.7143	0.5
$S(x_{4,d}, v_{2,d})$	0.9583	0.7143	0.5

6: **Obtain nonspecificity w.r.t. class i for each feature d by using similarities obtained in step 5.** These are now in Table 6.

Table 6. Nonspecificities u_2

Nonspecificity	f_1	f_2	f_3
$u_{2,1,d}$	0.7340	0.9445	0.8473
$u_{2,2,d}$	0.7732	0.8267	0.8473

7: **Obtain information transmission value T_d for each feature d using the two nonspecificities computed in step 4 and 6.** For information transmission after normalizing nonspecificities, we obtain

$$T = [0.2515, 0.0052, -0.2798]$$

Here e.g. $T_1 = 0.6061/2 - 0.734/4 + 0.6506/2 - 0.7732/4 = 0.2515$.

8: **Reduce the number of features from the data by selecting $k < D$ features having k highest information transmission values T.** Now choosing the two most important features results in selecting the first two features and discarding the third feature, which was the irrelevant feature.

4   Experimental Results and Analysis

In this section, the two real-world datasets and the classifier for the comparison of the filter methods are presented. Moreover, the training procedure with details on the datasplit and parameter selection is discussed.

4.1 Real-World Data Sets

The real-world data sets used in the research are the chronic kidney disease data set and the diabetic retinopathy Debrecen dataset, which are both freely available from the UCI Repository of Machine Learning Database [11]. The chronic kidney disease data set [19] is a dataset consisting of 24 variables and a binary class label that indicates whether a patient has chronic kidney disease or not. After removing observations for which at least one feature value was missing, the dataset contained 156 observations. The second data set, the diabetic retinopathy Debrecen data set [1], encompasses 19 features and a binary class label that identifies samples that show signs of diabetic retinopathy and those who do not show any signs of it. The data set contains 1151 observations without any missing values.

4.2 Similarity Classifier

In our experimental results, we are using a similarity classifier [15] to classify the real world data sets. Next, we shortly describe the main principles of this procedure.

The goal of the similarity based classifiers is to classify (a set of) samples into classes as accurately as possible, premised on the degree of similarity of each sample to each particular class. Suppose a data matrix is to be classified into N different classes. The initial step is to find ideal vectors for each class, which are supposed to represent a class well. For class i, such a vector is denoted as $\mathbf{v_i} = (v_{i,1}, v_{i,2}, ..., v_{i,D})$. We observe that there are several ways of determining these ideal vectors, $\mathbf{v_i}$, for example one can use the generalized mean

$$v_{i,d} = \left(\frac{1}{n_i} \sum_{\mathbf{x} \in X_i} x_d^m \right)^{\frac{1}{m}}, \quad \forall d = 1, \cdots, D, \quad i = 1, \cdots, N \quad (10)$$

where the parameter m (comes from the generalized mean) is fixed $\forall i, d$. Moreover, n_i denotes the number of samples in class i. To determine into which class any arbitrary sample $x_j \in X$ belongs, it is compared to the ideal vectors of all classes. The similarity between a sample $\mathbf{x}$ and an ideal vector of a given class $\mathbf{v_i}$ is given by

$$S\langle \mathbf{x_j}, \mathbf{v_i} \rangle = \left(\frac{1}{D} \sum_{d=1}^{D} \left(1 - |x_{j,d}^p - v_{i,d}^p| \right)^m \right)^{\frac{1}{m}} \quad (11)$$

The sample $x_j \in X$ is assigned to the class for which it has the highest similarity value, e.g., in accordance with:

$$Class(\mathbf{x_j}) = \arg \max_{i=1,\dots,N} S\langle \mathbf{x_j}, \mathbf{v_i} \rangle \quad (12)$$

4.3   Training Procedure

The proposed nonspecificity-based feature selection algorithm is tested on the two real-world data sets and compared to the ReliefF algorithm [10,17], a popular filter method that is an extension of the relief algorithm [8], as well as the Fisher Score, a similarity-based filter method [4]. All filter approaches were applied on the entire dataset to obtain a feature ranking. In order to compare the filter methods, for each filter method the lowest ranking features were removed on a step-by-step basis and the classification accuracy was calculated with a similarity classifier. For the similarity classifier, standard parameters (p = 1 and m = 1) and a data split into 70% training data and 30% test data were used. For each step, a random subset was generated 1000 times and the mean accuracy was determined as the average accuracy on the random test data subsets. All calculations were implemented with MATLAB^{TM} - software.

4.4   Results

The results on the Kidney disease data set demonstrate that the nonspecificity-based feature selection can outperform the ReliefF (with 10 nearest hits and misses) as well as the Fisher Score. The result is displayed in Fig. 1.

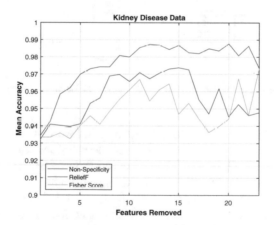

Fig. 1. Mean classification accuracies on the Kidney disease data set.

Initially, ReliefF is characterized by mean accuracy values that are close to those of the new feature selection method. However, it experiences a sudden decline for the removal of the 17th feature, where the difference in mean accuracy deteriorates from 0.99% points to 2.68% points. The difference to the novel nonspecificity−based method remains at a high level. The largest difference between the nonspecificity−based algorithm and ReliefF occurs after the removal of the 20th feature, where the new algorithm is with 98.77% mean accuracy 4.23% points more accurate. Overall, this is also the removal step where the highest mean accuracy 98.77% with the nonspecificity—based feature selection is observed (20 out of 24 features removed). In contrast to that, ReliefF never reaches an accuracy higher than 97.38% (15 removed features) and the Fisher Score never exceeds a mean accuracy of 97.32% (23 removed features). It is also noteworthy, that the nonspecificity—based feature selection algorithm ends up with the same best feature as the Fisher Score, whereas ReliefF's highest ranking choice leads to a clearly lower mean accuracy with the similarity classifier.

The mean accuracies on the second data set, the diabetic retinopathy data, highlight, that all three methods show comparable classification performances on this data set. The results are presented in Fig. 2.

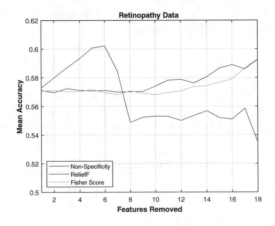

Fig. 2. Mean classification accuracies on the diabetic retinopathy data set.

The nonspecificity-based feature selection can improve the initial mean accuracy by removing up to 6 features, whereas the classification accuracy with the feature subset with ReliefF and Fisher Score tends to be almost constant until that point. Between 8 to 18 removed features, the Fisher Score and ReliefF lead with on average around 2.4% and 2.9% points to a better feature subset than the new approach. Overall, the nonspecificity-based feature selection leads to the highest mean accuracy of 60.21% (6 removed features). In contrast to that, the highest performance with the Fisher Score is 59.25% (18 removed features) and with ReliefF 59.29% (18 removed features). Overall, the results of all approaches are within few percentage points.

5   Conclusions

In this paper, we have presented a novel feature selection technique based on nonspecificity and information transmission. In this method, we also combine the use of ideal vectors as class representatives, and similarity matrices calculated from the data and these ideal vectors. Nonspecificity is computed from these similarity matrices and the information transmission value is computed from normalized nonspecificity values.

Our method is validated with two real world data sets. The performance on the two real-world data sets demonstrated that the nonspecificity-based feature selection can identify irrelevant features and can improve the classification performance for both data sets. In comparison to ReliefF and the Fisher Score, the novel algorithm leads in both applications to the highest mean classification accuracy.

For future research, we plan to combine nonspecificity with strive which is measuring possible conflicting information. Besides this we also plan to use fuzzy sets in membership computations instead of similarity.

References

1. Antal, B., Hajdu, A.: Diabetic retinopathy debrecen data set (2014). https://archive.ics.uci.edu/ml/datasets/Diabetic+Retinopathy+Debrecen+Data+Set
2. Bandemer, H., Näther, W.: Fuzzy Data Analysis. Kluwer Academic Publishing, Norwell (1992)
3. Blum, A., Langley, P.: Selection of relevant features and examples in machine learning. Artif. Intell. **97**, 245–271 (1997)
4. Duda, R.O., Hart, P.E., Stork, D.G.: Pattern Classification. Wiley, Hoboken (2012)
5. Fisher, R.A.: The use of multiple measurements in taxonomic problems. Ann. Eugenics **7**(2), 179–188 (1936)
6. Hartley, R.V.L.: Transmission of information. Bell Syst. Tech. J. **8**(3), 535–563 (1928)
7. Jolliffe, I.: Principal Component Analysis. Springer, Heidelberg (1986)
8. Kira, K., Rendell, L.A.: A practical approach to feature selection. In: Proceedings of the Ninth International Workshop on Machine Learning (1992). https://doi.org/10.1016/S0031-3203(01)00046-2
9. Klir, G.J., Yuan, B.: Fuzzy Sets and Fuzzy Logic, Theory and Applications. Prentice Hall, Upper Saddle River (1995)
10. Kononenko, I., Simec, E., Robnik-Sikonja, M.: Overcoming the myopia of inductive learning Algorithms with RELIEFF. Appl. Intell. **7**, 39–55 (1997)
11. Lichman, M.: UCI Machine Learning Repository (2013). Accessed 5 Nov 2018. http://archive.ics.uci.edu/ml
12. Lohrmann, C., Luukka, P., Jablonska-Sabuka, M., Kauranne, T.: A combination of fuzzy similarity measures and fuzzy entropy measures for supervised feature selection. Expert Syst. Appl. **110**, 216–236 (2018)
13. Łukasiewicz, J.: Selected Work. Cambridge University Press, Cambridge (1970)
14. Luukka, P.: Feature selection using fuzzy entropy measures with similarity classifiers. Expert Syst. Appl. **38**, 4600–4607 (2011)
15. Luukka, P., Saastamoinen, K., Könönen, V.: A classifier based on the maximal fuzzy similarity in the generalized Łukasiewicz-structure. In: Proceedings of 10th IEEE International Conference on Fuzzy Systems (2001)
16. McLachlan, G.J.: Discriminant Analysis and Statistical Pattern Recognition. Wiley Interscience (2004)

17. Robnik-Sikonja, M., Kononenko, I.: Theoretical and empirical analysis of ReliefF and RReliefF. Mach. Learn. **53**(1–2), 23–69 (2003). https://doi.org/10.1023/A: 1025667309714
18. Shannon, C.E.: A mathematical theory of communication. Bell Syst. Tech. J. **27**(3), 379–423 (1948)
19. Soundarapandian, P., Rubini, L.: Chronic Kidney Disease Data Set (2015)
20. Vergara, J.R., Estevez, P.A.: A review of feature selection methods based on mutual information. Neural Comput. Appl. **24**, 175–186 (2014)
21. Zadeh, L.: Fuzzy sets. Inf. Control **8**(3), 338–353 (1965)
22. Zadeh, L.: Similarity relations and fuzzy orderings. Inf. Sci. **3**(1), 177–200 (1971)

Effect of Maximizing Recall and Agglomeration of Feedback on Accuracy

Ross MacDonald[1], Nikita Neveditsin[1], Pawan Lingras[1(✉)], and Trent Hillard[2]

[1] Department of Mathematics and Computing Science, Saint Mary's University,
Halifax, Nova Scotia, Canada
pawan@cs.smu.ca
[2] Green Power Labs Inc., Halifax, Nova Scotia, Canada

Abstract. This paper studies multiple aspects of modeling user preference in a heterogeneous environment, where different individuals describe their level of comfort with the temperatures in different rooms in a building. The study shows that sampling based on fuzzy clustering provides the best approach for addressing the imbalance resulting from limited feedback points. Another issue addressed in the paper is a comparison of models based on agglomerated dataset for the entire building versus datasets for individual rooms. Ideally, personalized models for individual rooms should provide the best models. However, the number of feedback points for individual rooms is much smaller resulting in even larger imbalance in data. In many diagnostic situations in engineering and health sciences, recalling the critical decisions is more important than the prediction accuracy. The paper studies the quality of modeling by maximizing recall versus maximizing the AUC of models.

1   Introduction

One of the major issues in developing supervised learning to model user preferences is the amount of feedback. The problem is further exacerbated in a temporal dataset, where the complete dataset consists of every point over a given time period. For example, in a commercial building, users may give feedback about the room temperatures as normal, too hot, too cold from time to time. The time points where there is no feedback are assumed to be normal conditions. The dataset for a room will have a strong bias towards normal. In the dataset used in this study, the number of feedback points are as small as 0.06% of the dataset. Any supervised learning model that predicts all time points to be normal will have 99.94% accuracy. There are three different approaches to reduce the imbalance: upsampling, downsampling, and agglomeration of the datasets. In upsampling, we replicate the feedback points with "too hot" and "too cold" until they represent a reasonable proportion in the dataset. In downsampling, we pick a subset of feedback points with "normal" designation so that they represent a reasonable proportion in the dataset. In a previous study [1], we proposed the

R. B. Kearfott et al. (Eds.): IFSA 2019/NAFIPS 2019, AISC 1000, pp. 351–361, 2019.
https://doi.org/10.1007/978-3-030-21920-8_32

use of conventional and fuzzy clustering techniques for downsampling. The time points with "normal" designation were clustered based on similarity in weather conditions. Representatives of these clusters were retained in the dataset. The results were compared with a number of sampling techniques including upsampling and random downsampling. The downsampling using fuzzy clustering was found to provide the best results. In that previous study [1], we combined the feedback points from different rooms into a single agglomerated dataset for the building. The agglomeration does reduce the imbalance in the dataset. However, different rooms in the building are impacted differently by the weather and the occupants of these rooms do not have the same temperature preferences. This means that the agglomerated dataset is heterogeneous both in terms of external weather conditions and user preferences.

In this study we use the experience from previous study [1] to test the effect of agglomeration on both the accuracy and recall of the feedback. The results provide an indication of how easy it is to transfer a machine learning model from one room/person to another. The paper further studies the importance of different metrics in such an engineering problem. The two metrics that are commonly used for evaluating quality of a machine learning model are accuracy and recall. Accuracy measures the proportional number of times the model predict "normal" or "exceptional" conditions. In our case, "exceptional" conditions are feedback that the room is "too hot" or "too cold". In many diagnostic problems in engineering and medicine, ability to identify (or recall) all the "exceptional" conditions is more important than the accuracy. For example, while diagnosing cancer, physicians would rather predict a benign case as malignant than misdiagnose a malignant condition as benign. That is, a physician would prefer to have a 100% recall of malignant conditions. Similarly in our study, we want to recall as many (preferably all) "exceptional" conditions while maintaining as high an accuracy as possible. This study shows how to maximize recall and studies the effect of different level of recall on the accuracy.

2   Review of Literature

2.1   Sampling

In order to avoid class imbalance side-effects, upsampling of the data points of each minority class can be performed in order to compensate for the class imbalance issues [2]. By randomly adding additional data points that are repeats of the minority classes until all classes are equally represented, it will force the models to perform more insightful fitting. Upsampling will be accomplished through the use of the preProcess method in the Caret library in R [3]. Caret accomplishes upsampling by randomly sampling the minority classes with replacement, until the number of samples of the minority classes are approximately equivalent to the majority class. Upsampling alone has the potential of introducing inaccuracies in the model, and thus all analysis will be done with repeated cross validation in order to ensure a consistent model is produced.

Similarly, downsampling is also an option to address a class imbalance issue. In the case of downsampling, random samples of the majority classes are discarded until the number of data points in each class is approximately equal to the number of samples in the minority class [2]. Downsampling can also exhibit inaccuracies in the model, especially when the minority class sample size is extremely low. Therefore, all downsampling analysis will also be performed with repeated cross validation in order to ensure consistent model performance.

3   Sampling Using Clustering

Ideally, we want to use all the observations in our analysis. However, sometimes it is computationally infeasible to include all the observations. Many times, it is even more difficult to collect all the observations. In our case, the problem is slightly different. Machine learning models are normally designed to maximize the accuracy or precision of its predictions, Since 99.94% of our observations are classified as "Comfortable", predicting everything as Comfortable will lead to an enviable precision of 99.94%. In practical application, we do not want to miss prediction of any feedback that is either "Too cold" or "Too warm". That is, we want to maximize the recall of the classes "Too cold" and "Too warm". In order to achieve this, we want to reduce the number of 218,499 observations from the "Comfortable" class down to about 130 using downsampling. Alternatively, we can increase the number of "Too cold" and "Too warm" observations to 218,499. Since our goal is to minimize the number of feedback received from the user, downsampling may make more sense. Random downsampling may not necessarily cover the entire range of observations from the "Comfortable" class. In this section, we propose partitioning the observation space for the "Comfortable" class into clusters. A representative from each cluster will be used as an observation for the "Comfortable" class.

We approach this clustering based downsampling using crisp clustering using k-means by creating 60 clusters. k-means clustering is one of the most popular statistical clustering techniques [4,5]. The objective of the algorithm is to assign n objects to k clusters. We will then use the centroid of these clusters as an observation that is classified as "Comfortable".

Conventional clustering assigns various objects to precisely one cluster. A fuzzy generalization of the clustering uses a fuzzy membership function to describe the degree of membership (ranging from 0 to 1) of an object to a given cluster. There is a stipulation that the sum of the fuzzy memberships of an object to all the clusters must be equal to 1. The algorithm was first proposed by Dunn [6]. Subsequently, a modification was proposed by Bezdek [7]. Fuzzy c-means clustering tends to accommodate outliers a little better than the crisp clustering. Therefore, we will use fuzzy c-means to again create 80 clusters. We will pick the observation with the highest fuzzy membership for each cluster as an observation that is classified as "Comfortable". Additionally, we will employ Unsupervised Fuzzy Competitive Learning (UFCL) alongside c-means as a comparison solution. UFCL implements a variation on the FCM algorithm used by c-means fuzzy clustering. The minimization formula is executed only on the ith term of the set instead of on each index within the set. The simplification of the minimization formula from FCM to UFCL thus creates a form of unsupervised stochastic approximation that closely resembles FCM but is capable of providing better results in some circumstances [8].

3.1 Evaluation Metrics

The most comprehensive performance measurement of a classification model is through the analysis of the Receiver Operator Characteristic Curves (ROC) [9]. The calculations leading to the creation of an ROC curve for a particular model takes into consideration several aspects of the performance of the model. These considerations are as follows:

- The recall (Sensitivity) of the Model(1) – The number of correct predictions of an event over the total number of occurrences of the event. This measurement is also known as the true positive rate of predictions of the model [9].
- The Specificity(2) – The number of correct predictions of the absence of an event over the total number of absences of the event. This measurement is also known as the true-negative rate of the model [9].

$$Sensitivity = Recall = \frac{TruePositivePredictions}{TotalRealPositiveEvents} \tag{1}$$

$$Specificity = \frac{TrueNegativePredictions}{TotalRealNegativeEvents} \tag{2}$$

By plotting the sensitivity versus the specificity of a model, a ROC curve is created. The primary goal in this analysis is to maximize the performance of the model by maximizing both the sensitivity as well as the specificity. In doing so, the Area Under the Curve (AUC) must be maximized. The maximum possible value for an AUC is 1, and thus model performance will be based on the model which produces the highest AUC value [9].

The secondary performance objective in this research will be to maximize the sensitivity of the model. In doing so, this maximizes the number of correct predictions of the model overall, which is ultimately the purpose of the research.

4 Study Data

The data consists of observations of internal and external conditions recorded every fifteen minutes. The data is collected from eighteen rooms on three floors. There are three types of feedback:

- Comfortable
- Too Cold
- Too Warm

Total number of observations were 218,629. We received 133 feedback: (Comfortable, 3), (Too cold, 79), (Too warm, 51). We assume "Comfortable" as the feedback when users do not provide any feedback. That means 99.94% of the observations corresponded to the Comfortable class. Table 1 shows the distribution of feedback for each room.

The independent variables included:

- Date
- Time
- weather variables
- building characteristics
- room state variables

The solar radiation is a sum of all the surfaces that are externally facing, including external walls, windows, and roof. This sum takes into account the geographical location of a room in the building.

Table 1. Feedback summary by room

Room	Comfortable	Too Cold	Too Warm
118	0	6	0
119	0	5	0
120	0	1	4
206	0	2	0
207	0	2	0
226	0	3	0
228	0	0	22
232	0	0	1
234	0	3	0
235	0	11	5
300	1	11	0
301	0	0	15
306	1	11	1
320	1	2	0
326	0	1	0
328	0	0	2
332	0	20	0
334	0	1	0

5   Predicting Comfort Level Feedback for the Agglomerated Dataset

In order to test the quality of sampling with the proposed approach, Ross et al. [1] tried a number of machine learning techniques and sampling techniques to predict the feedback for a given set of observations.

Figure 1 provides a comparison of AUC performance for different sampling techniques.

In the figure, "rf" stands for Random Forest, "sl" means support vector machines with linear kernel, "nn" corresponds to neural networks. For sampling, ufcl (UFCL) and cmeans (Fuzzy c-means) are two competing fuzzy clustering approaches.

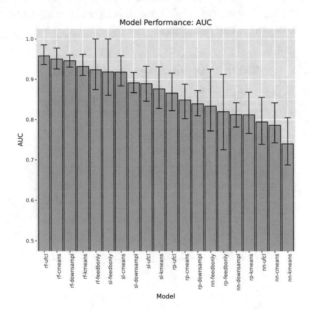

Fig. 1. AUC performance comparison across each analysis

Here we can see both of the fuzzy clustering methods of downsampling, c-means as well as UFCL, render the best performing models for maximizing the AUC measurement. The UFCL fuzzy clustering model performs best with an AUC of 95.8%.

Likewise, Fig. 2 provides a comparison of recall performance across each analysis.

Here we can see both of the fuzzy clustering methods of downsampling, c-means as well as UFCL, render the best performing models for maximizing the recall measurement. The c-means fuzzy clustering model switches spots this time with the UFCL model and demonstrates the best recall of 86.5%.

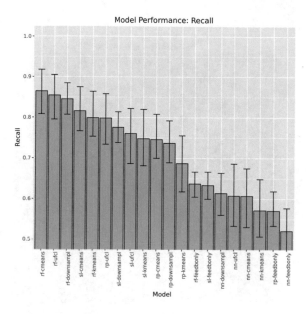

Fig. 2. Recall performance comparison across each analysis

Based on conclusions from Ross et al. [1] we will use fuzzy clustering and random forest for the remaining analysis in this paper.

6   Comparison of Recall and AUC for Individual Datasets Versus Agglomerated Dataset

Different rooms in a building are affected differently by weather and hence require different temperature adjustments. Moreover, occupants of these rooms do not necessarily have the same expectations for room temperatures. Ideally, it will be better to develop a separate model for each room. However, the number of feedback points are further reduced in such cases. We first look at the AUC and recall for two types of models: universal model for the building and individual models for each room.

Optimizing first for recall, we test for the recall performance of each room model against each other room. Figure 3 shows the best results for this basic analysis.

We observe that recall performance is not ideal when treating each room independently with models built from each other room.

Similarly, optimizing for AUC, we test for the recall performance of each room model against each other room. Figure 4 shows the best results for this basic analysis.

We, again, observe that recall performance is not ideal when treating each room independently with models built from each other room.

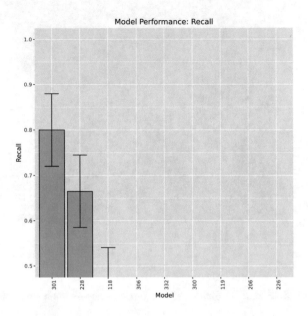

Fig. 3. Recall performance comparison by individual rooms optimizing for recall

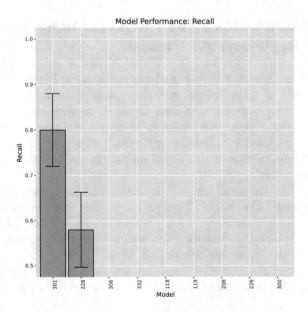

Fig. 4. Recall performance comparison by individual rooms optimizing for AUC

Next, Table 2 shows overall AUC and recall for the entire building and individual rooms for both the models.

Table 2. Comparison of models based on agglomerated and individual datasets - maximizing AUC

Room	Rm. AUC	Aggl. AUC	Rm. Recall	Aggl. Recall
118	0.9998	0.9395	1.00	1.0000
119	0.9995	0.9633	1.00	1.0000
120	1.0000	0.6648	1.00	0.0000
206	1.0000	0.8648	1.00	1.0000
226	1.0000	0.8472	1.00	1.0000
228	0.9965	0.9074	1.00	0.9333
232	1.0000	0.8649	1.00	1.0000
235	0.9946	0.7990	1.00	0.8125
300	0.9998	0.9555	1.00	1.0000
301	0.9962	0.9658	1.00	1.0000
306	0.9949	0.8833	1.00	1.0000
328	1.0000	0.9896	1.00	1.0000
332	0.9992	0.8264	1.00	1.0000
building	0.9918	0.9918	0.98	0.9800

We observe in Table 2 that by taking the agglomerated results of each model performance across rooms and using the best candidates in each case, we can build a composite model that performs quite well in most situations.

Most machine learning models measure the quality of predictions using the accuracy based on a measure such as AUC. However, in many diagnostic situations, one may be more interested in recalling all the exceptional points at the expense of accuracy. For example, in predicting a mission critical fault in an engineering system, it is important to catch all the faulty situations and investigate further. This may mean that we will end up identifying normal situation as a faulty from time to time. The prediction of user feedback is one such situation. In our case, the exceptional points are when the feedback is "Too warm" or "Too cold". The training process can be modified to maximize recall as opposed to maximizing AUC.

Thus, Table 3 shows the results for maximizing recall using the same agglomerating approach above.

We see in Table 3 that by agglomerating the results across each of the rooms and building a composite model of the best performing models in each case, we can achieve our goals in maximizing recall and producing a reliable model for properly predicting the important feedback points of "Too warm" and "Too cold".

Table 3. Comparison of models based on agglomerated and individual datasets - maximizing recall

Room	Rm. AUC	Aggl. AUC	Rm. Recall	Aggl. Recall
118	0.9824	0.9335	1.00	1.0000
119	0.9982	0.9592	0.75	1.0000
120	N/A	0.6745	N/A	0.0000
206	1.0000	0.8346	1.00	1.0000
226	1.0000	0.8444	1.00	1.0000
228	0.9903	0.9023	1.00	0.9333
232	N/A	0.8526	N/A	1.0000
235	N/A	0.7895	N/A	0.8125
300	0.9989	0.9564	1.00	1.0000
301	0.9979	0.9645	1.00	1.0000
306	0.9932	0.8820	1.00	1.0000
328	N/A	0.9806	N/A	1.0000
332	0.9990	0.8241	1.00	1.0000
building	0.9905	0.9905	0.98	0.9800

Comparing the results in Tables 2 and 3, we see that models built through a target of maximizing AUC also produce models that maximize recall and thus perform slightly better than maximizing only for recall alone. This is the best of both worlds and means that we choose the AUC maximized models for the composite model going forward.

7   Conclusion

In this paper, we describe a problem of addressing imbalance of data in a building energy management system, where users provide feedback on their comfort level. The users do not provide any feedback for 99.94% of the observations. A machine learning model will simply classify all the observations as "Comfortable" and achieve 99.94% precision. Our goal is to maximize recall of all the feedback points. In order to cover all the range of observations, the study proposed the use of crisp and fuzzy clustering. The number of clusters were approximately the same as the number of feedback points. Centroid of a crisp cluster was used as an observation classified as "Comfortable". An observation with highest fuzzy membership for a fuzzy cluster was classified as "Comfortable". The results were compared with no sampling, upsampling, random sampling.

Using the results from [1], Fig. 2 shows that the best downsampling method for our application was c-means. Figure 2 also demonstrates that the Random Forest algorithm provides us with the highest recall performance. C-means and Random Forest were then used for the remainder of this investigation.

We note that if AUC or recall are optimized on individual rooms, the resulting recall performance is not ideal, as depicted in Figs. 3 and 4. This can be attributed to the fact that each room is affected by variables in different ways: weather, occupants, structure, HVAC system, etc. These findings tell us that we must treat each room separately and build a composite model using the best performing model from each room.

Table 2 demonstrates the optimization possible when looking for the best AUC performance. However, given the application of this research, recall is much more important than overall performance that AUC provides. Thus, the analysis was run again for maximizing recall and Table 3 shows the surprising results, that recall alone is not as high as the previous model fitting.

Finally, we can conclude from these findings that a c-means downsampled data set fit with a Random Forest algorithm can be used to train a model for each room to optimize not only the recall, but also the AUC, for a building comfort feedback system. These results can now be used to feed into a system that can automatically adjust the building temperature pre-emptively such that comfort feedback is no longer required, thus optimizing building comfort for the occupants.

References

1. MacDonald, R., Neveditsin, N., Lingras, P., Qin, Z., Hillard, T.: Sampling using fuzzy and crisp clustering to improve recall of building comfort feedback. In: International Conference on Fuzzy Systems, vol. 1, no. 1, pp. 1–6 (2019)
2. Burez, J., Van den Poel, D.: Handling class imbalance in customer churn prediction. Expert Syst. Appl. **36**(3), 4626–4636 (2009). http://dx.doi.org/10.1016/j.eswa.2008.05.027
3. Kuhn, M.: The caret package (2018). https://cran.r-project.org/package=caret
4. Hartigan, J.A., Wong, M.A.: Algorithm AS136: a k-means clustering algorithm. Appl. Stat. **28**, 100–108 (1979)
5. MacQueen, J.: Some methods for classification and analysis of multivariate observations. In: Proceedings of Fifth Berkeley Symposium on Mathematical Statistics and Probability, vol. 1, pp. 281–297 (1967)
6. Dunn, J.C.: A fuzzy relative of the ISODATA process and its use in detecting compact well-separated clusters. Cybernetics **3**(3), 32–57 (1973). http://dx.doi.org/10.1080/01969727308546046
7. Bezdek, J.C.: Pattern Recognition with Fuzzy Objective Function Algorithms. Kluwer Academic Publishers, Norwell (1981)
8. Pal, N.R., Bezdek, J.C., Hathaway, R.J.: Sequential competitive learning and the fuzzy c-means clustering algorithms. Neural Netw. **9**(5), 787–796 (1996)
9. Kuhn, M.: Applied Predictive Modeling. Springer, New York (2013)

Generalized Pre-aggregations

Luis Magdalena[1]([✉]) [iD], Daniel Gómez[2] [iD], Javier Montero[3] [iD],
Susana Cubillo[1] [iD], and Carmen Torres[1] [iD]

[1] ETSI Informáticos, Universidad Politécnica de Madrid,
Boadilla del Monte, Madrid, Spain
{luis.magdalena,susana.cubillo}@upm.es, ctorres@fi.upm.es
[2] Faculty of Statistics, Complutense University of Madrid, Madrid, Spain
dagomez@estad.ucm.es
[3] Faculty of Mathematics, Complutense University of Madrid, Madrid, Spain
monty@mat.ucm.es

Abstract. In this paper we propose an extension of the concept of pre-aggregation at two diferent levels. On the one hand, we extend the definition of pre-aggregation to a more general framework where the space $[0, 1]$ is replaced by a totally ordered set $\mathcal{T}$ with maximum and minimum value. On the other hand, since $\mathcal{T}$ could be even discrete, we generalize the concept of monotonicity for functions going from $\mathcal{T}^n$ into $\mathcal{T}$. In order to do so we introduce the concept of conditioned monotonicity based on the chains in $\mathcal{T}^n$ and generalizing that of directional monotonicity.

Keywords: Aggregation functions · Pre-aggregation functions ·
Directional monotonicity · Conditioned monotonicity

1   Introduction

Aggregation functions [4] play a central role in information fusion and decision making. Most of the effort in the area has been concentrated on aggregating variables that take values on $[0, 1]$ interval, and do so by using functions that are monotonic. But some recent extensions are exploring the possibility of relaxing both conditions. On the one hand, generalized aggregation functions consider a complete lattice instead of an interval [7], or it is even possible to work on a bounded poset when working with type-2 fuzzy sets [8]. On the other hand, monotonicity can also be relaxed (or extended) to define new properties as weak monotonicity [10], or directional monotonicity [2], which allows monotonicity along (some) fixed ray. The concept of directional monotonicity is then the basis for the definition of pre-aggregation functions [5].

In this paper we will explore those two ideas (extending the universe of the variable and relaxing monotonicity) to generalize the concept of pre-aggregation at both levels. On the one hand, we will extend the definition of pre-aggregation by replacing the interval $[0, 1]$ by a totally ordered set $\mathcal{T}$ with maximum and minimum value. This extension allows us to deal with a more general aggregation

© Springer Nature Switzerland AG 2019
R. B. Kearfott et al. (Eds.): IFSA 2019/NAFIPS 2019, AISC 1000, pp. 362–370, 2019.
https://doi.org/10.1007/978-3-030-21920-8_33

of objects. On the other hand, since $\mathcal{T}$ could be even discrete, we generalize the concept of monotonicity for functions going from $\mathcal{T}^n$ into $\mathcal{T}$. In order to do so we reinterpret the concept of monotonicity in terms of chains or linear orders in $\mathcal{T}^n$, by defining the idea of conditioned monotonicity. The new general concept of conditioned monotonicity becomes a generalization of the concept of directional monotonicity even in the case that the space was the interval $[0, 1]$, since the definition based on chains allows us to define other classes of spatial growing as shapes-growing.

The rest of the paper is structured as follows. Next section describes some preliminary concepts related to orders and aggregations. Section 3 introduces the concept of conditioned monotonicity. The idea is then adapted in Sect. 4 to some special cases related to particular properties of the considered set of chains, as the case with prototype based partitions. Section 5 uses the concept of conditioned monotony to define the concept of generalized pre-aggregation functions. Finally, some conclusions are presented.

2 Preliminaries

2.1 Partial Order Sets and Linear Orders

In this subsection we will introduce some definitions and notations related to total and partial orders that will be used in this work.

Definition 1. A **strict partial order** [6,9] $(X, <)$ is a binary relation over the set X satisfying the following properties:

1. **Anti-reflexive** For all $x \in X$ $x \not< x$.
2. **Transitive** $x < y$ and $y < z$ implies $x < z$.

Given a subset $L \subseteq X$, we will say that L is a **chain or a strict linear order** of $(X, <)$ if and only if for any pair of elements $x, y \in L$, with $x \neq y$ we have that $x < y$ or $y < x$. When either $x < y$ or $y < x$ we say that x and y are comparable elements.

We will denote by $\mathcal{P}_<(X)$ the set of all possible chains in $(X, <)$.

In a similar way it is possible to define a (non strict) partial order.

Definition 2. A (non strict) **partial order** [9] $(X, <=)$ is a binary relation over the set X satisfying the following properties:

1. **Reflexive** $x <= x$.
2. **Transitive** $x <= y$ and $y <= z$ implies $x <= z$.
3. **Antisymmetric** $x <= y$ and $y <= x$, implies $x = y$.

Given a subset $L \subseteq X$, we will say that L is a **chain or a linear order** of $(X, <=)$ if and only if for any pair of elements $x, y \in L$, we have that $x <= y$ or $y <= x$. As in the case of strict orders, we will refer to those pairs of elements were $x <= y$ or $y <= x$ as comparable elements.

In a partial order there could be pairs of elements that are not comparable, but every two elements are comparable in a linear order.

We will denote by $\mathcal{P}_{<=}(X)$ the set of all possible chains in $(X, <=)$.

2.2   Aggregation Operators

An aggregation operator in $[0,1]$ is usually defined as a function

$$Ag_n : [0,1]^n \to [0,1]$$

where it is usually assumed the following properties:

1. Ag_n is monotone, non decreasing.
2. $Ag_n(0,\dots,0) = 0$.
3. $Ag_n(1,\dots,1) = 1$.

There exist other proposals of aggregation processes in which the three previous conditions imposed to the function Ag_n are relaxed. One of the recent studies focuses on the relaxing the monotonicity of the function. In [10], it was proposed the notion of weak monotonicity to extend classical monotonicity property. In this case, monotonicity is required only along the direction of the first quadrant diagonal. This concept of weak monotonicity was further extended in [2] by introducing the notion of directional monotonicity. By the interchange of classical and directional monotonicity in the conditions of classical aggregation functions we have a new class of aggregation operators: the pre-aggregation functions. In the following two definitions we formalized both ideas.

Definition 3. Let f be a function from $[0,1]^n$ to $[0,1]$. And let r be an element of $\mathbb{R}^n$. Then we will say that f is **r-growing** if for all $x \in [0,1]^n$ and for all $\lambda \geq 0$, the following holds:

$$\text{If}\quad x + \lambda r \in [0,1]^n \quad \text{then}\quad f(x + \lambda r) \geq f(x).$$

Once the r-growing concept is defined, the pre-aggregation is presented as follows.

Definition 4. [5] A mapping $Ag_n : [0,1]^n \to [0,1]$ is an n-dimensional **pre-aggregation** function if it satisfies:

1. There exists a real vector $r \in [0,1]^n$ with $r \neq 0$ such that Ag_n is r-growing.
2. $Ag_n(0,\dots,0) = 0$ and $Ag_n(1,\dots,1) = 1$.

Moreover, n-ary fusion functions have been proposed by some authors as any mapping $Ag_n : [0,1]^n \to [0,1]$ (see, e.g., [2]).

Some of these previous definitions has been extended to a more general class of situations replacing the lattice $[0,1]$ into a more general scenario. In the following definition (see for example [1,7]), we present a general definition of classical aggregation operators.

Definition 5. Let $\mathcal{T}$ be a complete lattice with a maximum and minimum element ($1_{\mathcal{T}}$ and $0_{\mathcal{T}}$ respectively) and let $\mathcal{T}^n$ be the natural lattice of n elements of type $\mathcal{T}$ (i.e. $\mathcal{T}^n = \underbrace{\mathcal{T} \times \dots \times \mathcal{T}}_{n\ times}$). A **generalized aggregation function** is a mapping $Ag : \mathcal{T}^n \to \mathcal{T}$ such that it satisfies:

1. $Ag(\underbrace{0_T, 0_T ..., 0_T}_{n \ times}) = 0_T$ and $Ag(\underbrace{1_T, 1_T, .., 1_T}_{n \ times}) = 1_T$.

2. Ag is monotonic with respect to the lattice's order.

Remark 1. Obviously, this previous definition of aggregation operators permits to deal with more general objects, as, for example, the digital images. An image is a matrix of $n \times m$ pixels in $\{0, 1, 2, ..., 255\}$. Or subsets of pixels in which its position is also considered.

It is clear that any classical aggregation operator is a pre-aggregation operator as well as a generalized aggregation operator. Pre-aggregation functions has been very useful in different aggregation processes (see for example [3]) when the aggregation is done for values over the complete lattice $[0, 1]$. The way in which this notion could be extended to deal with more general objects (we think) merits to be explored.

To do so, the key question to be analyzed is how to extend the concept of generalized aggregation function to a more general class in which the monotonicity was relaxed.

3   The Concept of Conditioned Monotonicity

Our main idea now is to extend the concept of pre-aggregation to a more general universe than $[0, 1]^n$, as it was previously done with aggregations [1].

The definition of pre-aggregations relies on the concept of r-growing functions that at the end drives to that of directional monotonicity. Consequently we first need to adapt this idea of directional monotonicity to a universe where *directions* may not exist. To do so we will first transform the concept of *directional monotonicity* into the more general one of *conditioned monotonicity*.

We will now work with the totally ordered set (linearly ordered set) $(T, <)$ with maximum (1_T) and minimum (0_T). $(T, <)$ as well as $(T^n, <^n)$ are complete lattices, being the later a poset. According to Definition 5, any monotonic function $F : T^n \to T$, with $Ag(\underbrace{0_T, 0_T ..., 0_T}_{n \ times}) = 0_T$ and $Ag(\underbrace{1_T, 1_T, .., 1_T}_{n \ times}) = 1_T$ will be a generalized aggregation. Let us now relax the restriction of monotonicity to achieve a more flexible definition.

Definition 6. Let $(T, <)$ be a totally ordered set with a maximum and a minimum element (1_T and 0_T respectively), let $(T^n, <^n)$ be the natural lattice of n elements of type T, let $\mathcal{P}_<(T^n)$ be the set of all totally ordered subsets (chains) of T^n, and let C by a subset of $\mathcal{P}_<(T^n)$. Consequently $C = \{C_i : i \in I\}$ where C_i are chains of T^n.

A mapping $F : T^n \to T$ is said to be a C-**conditioned growing function** if it satisfies that $\forall a, b \in T^n$, if $a < b$ and $\exists C_i \in C$ such that $a, b \in C_i$, then $F(a) \leq F(b)$.

A mapping $F : T^n \to T$ is said to be a C-**conditioned decreasing function** if it satisfies that $\forall a, b \in T^n$, if $a < b$ and $\exists C_i \in C$ such that $a, b \in C_i$, then $F(a) \geq F(b)$.

In other words, a function will be said to be a C-conditioned growing function if $F : C_i \to T$ is a growing function, for every C_i in C.

It is important to notice that being T a totally ordered set, the image of F is totally ordered, i.e., $\forall x, y \in T^n$ either $F(x) \geq F(y)$ or $F(x) \leq F(y)$ (both of them when $F(x) = F(y)$). Similarly, any two elements in C_i are comparable (chains are linearly ordered sets).

Definition 7. Let $(T, <)$ be a totally ordered set with a maximum and a minimum element (1_T and 0_T respectively), let $(T^n, <^n)$ be the natural lattice of n elements of type T, let $\mathcal{P}_<(T^n)$ be the set of all totally ordered subsets (chains) of T^n, and let C by a subset of $\mathcal{P}_<(T^n)$.

A mapping $F : T^n \to T$ is said to be a C-**conditioned monotonic function** if it is either a C-conditioned growing function or a C-conditioned decreasing function.

Theorem 1. *Let $(T, <)$ be a totally ordered set with a maximum and a minimum element (1_T and 0_T respectively), let $(T^n, <^n)$ be the natural lattice of n elements of type T, let $\mathcal{P}_<(T^n)$ be the set of all totally ordered subsets (chains) of T^n.*

If $F : T^n \to T$ is a $\mathcal{P}$-conditioned monotonic function, i.e., it is a C-conditioned monotonic function with $C = \mathcal{P}_<(T^n)$, then F is a monotonic function.

Proof. $F : T^n \to T$ will be a $\mathcal{P}$-conditioned monotonic function if and only if it is either a $\mathcal{P}$-conditioned growing function or a $\mathcal{P}$-conditioned decreasing function.

Let first assume that $F : T^n \to T$ is a $\mathcal{P}$-conditioned growing function. As $\mathcal{P}_<(T^n)$ is the set of all totally ordered subsets (chains) of T^n, then $F : C_i \to T$ will be a growing function, for every chain (C_i) in T^n. On the other hand, for every pair of elements $a, b \in T^n$, such that $a < b$, there will be, at least, a chain $C_i \in \mathcal{P}_<(T^n)$ such that $a, b \in C_i$. And being $F : C_i \to T$ a growing function, then $F(a) \leq F(b)$. Consequently, for every a, b in T^n, if $a < b$ then $F(a) \leq F(b)$, and the function is monotonic.

Otherwise, if $F : T^n \to T$ is not a $\mathcal{P}$-conditioned growing function, then it will be a $\mathcal{P}$-conditioned decreasing function. In that case for every pair of elements $a, b \in T^n$ such that $a < b$, there will be, at least, a chain $C_i \in \mathcal{P}_<(T^n)$ such that $a, b \in C_i$. And being $F : C_i \to T$ a decreasing function, then $F(a) \geq F(b)$. Consequently, for every a, b in T^n, if $a < b$ then $F(a) \geq F(b)$, and the function is also monotonic.

In summary, the concept of conditioned monotonicity allows a broad scope of analysis depending on the relation between C and $\mathcal{P}_<(T^n)$. The more elements are added to C, the closer will be the conditioned monotonicity to monotonicity. As a matter of fact, according to Theorem 1, when $C = \mathcal{P}_<(T^n)$, the C-conditioned monotonicity will become monotonicity. On the other hand, if $C = \emptyset$, then the C-conditioned monotonicity eliminates any class of monotonicity as is done in the fusion functions.

4 Conditioned Monotonicity on Prototype-Based Partitions

Once defined the concept of conditioned monotonicity, and established its upper bound in monotonicity when $C = \mathcal{P}_<(T^n)$, it could be of interest to consider the situation with some particular subsets (C) of $\mathcal{P}_<(T^n)$.

The first idea could be to consider a C such that the union of all elements in C (being each one a totally ordered subset of T^n) was T^n. In this case, C will be a family of nonempty subsets of T^n whose union is T^n, i.e., a cover of T^n. As a result, every element x in T^n will be an element of, at least, one chain (C_i) in C. This seems to be a good starting point for the concept of generalized pre-aggregation, since it ensures that every element in T^n could be a matter of analysis. But, at the same time, there could be situations where due to some specific characteristics of the problem, not every element of T^n is of interest. So, working on C covering T^n could be interesting but is not mandatory.

If C is a cover of T^n, and F is C-conditioned growing, we can assure that for any $x \in T^n$ there at least a linear order $C_x \in C$, with $x \in C_x$. Nevertheless, it could be happen that this linear order be non unique. So, it could be also interesting to restrict a little bit more the previous concept by adding the condition that the subsets in C were pairwise disjoint. In this case, the cover of T^n becomes a partition of T^n.

If C is a partition of T^n, and F is C-conditioned growing, we can assure that for any $x \in T^n$ there exist a unique linear order $C_x \in C$, with $x \in C_x$. Let us note that the elements of C_x, are the only elements of T^n that are comparable with x in the function F. In the r–directional monotonicity defined in [2], the role of C_x is taken by the straight line with direction r that contains the element x.

Now the question is how to obtain these partitions in a natural way. In this sense, we can add some additional restrictions, e.g., that the partition was not any partition but a partition whose components were generated on the basis of a certain property or condition. We will call this kind of partition a prototype-based partition, and will refer to the building property or condition as the prototype of the partition.

An example of such a prototype based partition on $[0, 1]^n$ will be that generated by a prototype being a straight line with a certain slope. This situation corresponds to the so called directional monotonicity, being the direction that of the line (prototype).

But it can be easily extended to other kind of patterns or conditions as a parabola or any other curve or variety, with the only restriction that the prototype should generate chains (linearly ordered sets).

5 Generalized Pre-aggregation Functions

Definition 8. Let $(T, <)$ be a totally ordered set with a maximum and a minimum element (1_T and 0_T respectively), let $(T^n, <^n)$ be the natural lattice of n

elements of type T, let $\mathcal{P}_<(T^n)$ be the set of all totally ordered subsets (chains) of T^n, and let C by a subset of $\mathcal{P}_<(T^n)$.

A **generalized pre-aggregation function over** C is a mapping $Ag : T^n \to T$ such that it satisfies:

1. $Ag(\underbrace{0_T, 0_T ..., 0_T}_{n \; times}) = 0_T$ and $Ag(\underbrace{1_T, 1_T, .., 1_T}_{n \; times}) = 1_T$.
2. Ag is a C-conditioned monotonic function.

Proposition 1. *Let $(T, <)$ be a totally ordered set with a maximum and a minimum element (1_T and 0_T respectively), let $(T^n, <^n)$ be the natural lattice of n elements of type T, and let $\mathcal{P}_<(T^n)$ be the set of all totally ordered subsets (chains) of T^n.*

A generalized pre-aggregation function over $\mathcal{P}_<(T^n)$ is a generalized aggregation function.

Proof. Direct from Theorem 1.

Definition 9. A mapping $Ag : T^n \to T$ is said to be a **covering generalized pre-aggregation function** over C when it is a generalized pre-aggregation function over C, and C covers T^n.

Definition 10. A mapping $Ag : T^n \to T$ is said to be a **partitioning generalized pre-aggregation function** over C when it is a generalized pre-aggregation function over C, and C is a partition of T^n.

Proposition 2. *Any pre-aggregation function in the sense of Definition 4 [5] is a partitioning generalized pre-aggregation function.*

Proof. If $Ag : [0,1]^n \to [0,1]$ is a pre-aggregation function, there exists a real vector $r \in [0,1]^n$ with $r \neq 0$ such that Ag is r-growing.

Let us now define a relation R in $[0,1]^n \times [0,1]^n$ as: xRy when $\exists \lambda \in \mathbb{R}$ such that $y - x = \lambda r$, i.e., when $x \neq y$, the line defined by the two points is parallel to (has the same direction that) vector r.

The relation R is an equivalence relation:

- $\forall x \in [0,1]^n, xRx.\ x - x = 0 = 0r.$
- $\forall x, y \in [0,1]^n, xRy \implies yRx.\ y - x = \lambda r \implies x - y = -\lambda r.$
- $\forall x, y, z \in [0,1]^n, xRy$ and $yRz \implies xRz.\ y - x = \lambda_1 r$ and $z - y = \lambda_2 r \implies z - x = (\lambda_1 + \lambda_2)r.$

Consequently, R induces a partition over $[0,1]^n$, where each class of the partition is made up of the points in a line parallel to vector r.

So we have a partition of $[0,1]^n$ and we need to show now that every set in the partition (every class) is a chain, i.e., a totally ordered set. This last fact trivially holds since the vector r is non negative vector ($r_i \geq 0$) by definition and thus preserve the natural order of $[0,1]^n$ avoiding incomparabilities.

Proposition 3. *Any partitioning generalized pre-aggregation function is a covering generalized pre-aggregation function.*

Proof. By definition, a partition of a set should be such that its union is equal to the set, i.e., covers the set. So, a partition of $\mathcal{T}^n$ is also a cover of $\mathcal{T}^n$, and consequently a partitioning generalized pre-aggregation function is a covering generalized pre-aggregation function.

6   Conclusions

In this paper we propose an extension of the concept of pre-aggregation at two different levels. On the one hand, we extend the definition of pre-aggregation to a more general framework where the space [0, 1] is replaced by a totally ordered set $\mathcal{T}$ with maximum and minimum value. This extension allows us to deal with a more general aggregation of objects rather than values into the [0, 1] interval. On the other hand, since $\mathcal{T}$ could be even discrete, we generalize the concept of monotonicity for functions going from $\mathcal{T}^n$ into $\mathcal{T}$. In order to do so we introduce a definition based on chains or linear orders. The new general concept of growing (and monotonicity) is a generalization of the concept of r-growing functions even in the case that the space was the interval [0, 1], since the definition based on chains allows us to define other classes of spatial growing as shapes-growing. This spatial idea of shape-growing is formalized with the concept of prototype that can be viewed as an equivalence relation between objects that partition the set $\mathcal{T}^n$ into subspaces in such a way that growing (monotonicity) of the function is satisfied at least through these subspaces, as happens with the straight lines in the usual concept of pre-aggregation.

Acknowledgment. This research has been partially supported by the Government of Spain (grant TIN2015-66471-P), the Government of Madrid (grant S2013/ICCE-2845), Complutense University (UCM Research Group 910149 and grant PR26/16-21B-3) and Universidad Politécnica de Madrid.

References

1. Bustince, H., Herrera, F., Montero, J. (eds.): Fuzzy Sets and Their Extensions: Representation, Aggregation and Models. Intelligent Systems from Decision Making to Data Mining, Web Intelligence and Computer Vision. Studies in Fuzziness and Soft Computing, vol. 220. Springer, Heidelberg (2008)
2. Bustince, H., Fernandez, J., Kolesárová, A., Mesiar, R.: Directional monotonicity of fusion functions. Eur. J. Oper. Res. **244**, 300–308 (2015)
3. Bustince, H., Barrenechea, E., Sesma-Sara, M., Lafuente, J., Dimuro, G.P., Mesiar, R., Kolesárová, A.: Ordered directionally monotone functions: justification and application. IEEE Trans. Fuzzy Syst. **26**(4), 2237–2250 (2018)
4. Calvo, T., Kolesárová, A., Komorníková, M., Mesiar, R.: Aggregation operators: properties, classes and construction methods. Stud. Fuzziness Soft Comput. **97**, 3–104 (2002)
5. Lucca, G., Sanz, J.A., Dimuro, G.P., Bedregal, B., Mesiar, R., Kolesárová, A., Bustince, H.: Preaggregation functions: construction and an application. IEEE Trans. Fuzzy Syst. **24**(2), 260–272 (2016)

6. González-Pachón, J., Gómez, D., Montero, J., Yáñez, J.: Soft dimension theory. Fuzzy Sets Syst. **137**(1), 137–149 (2003)
7. Montero, J., González-del-Campo, R., Garmendia, L., Gómez, D., Rodríguez, J.T.: Computable aggregations. Inf. Sci. **460**, 439–449 (2018)
8. Torres-Blanc, C., Cubillo, S., Hernández, P.: Aggregation operators on type-2 fuzzy sets. Fuzzy Sets Syst. **324**, 74–90 (2018)
9. Trotter, W.T.: Combinatorics and Partially Ordered Sets: Dimension Theory, vol. 59. Johns Hopkins University Press, Baltimore (1992)
10. Wilkin, T., Beliakov, G.: Weakly monotonic averaging functions. Int. J. Intell. Syst. **30**, 144–169 (2015)

STRESSDIAG: A Fuzzy Expert System for Diagnosis of Stress Types Including Positive and Negative Rules

Mai Thi Nu[1], Nguyen Hoang Phuong[2(✉)], and Hoang Tien Dung[1]

[1] Center for Application of Health IT, Ministry of Health, Ba Đình, Vietnam
mainuitmoh@gmail.com, dunggm@gmail.com
[2] Thang Long University, Hanoi, Vietnam
nhphuong2008@gmail.com

Abstract. A fuzzy rule based expert system STRESSDIAG is presented for diagnosis of stress types including positive and negative rules. After designing and building a suitable inference engine for this system, to create effective knowledge base consisting of more than 700 positive rules for confirmation of conclusion and of more than 100 negative rules for exclusion of the same conclusion. How the rule base is constructed, managed and used are focused on for diagnosis of diagnosis of stress types such as light stress, middle stress, serious stress and serious stress with mental disorder. The inference engine shows how to combine positive and negative rules. The first evaluation of STRESSDIAG is presented by the medical expert's group in the field of mental diseases in Vietnam and confirmed that STRESSDIAG diagnoses with a high accuracy.

Keywords: Fuzzy expert systems · Combination of positive and negative rules · Diagnosis of stress types

1 Introduction

The stress disorders is one of the widely spread diseases in the world. This disease gets about 3–5% of population in the world. In Vietnam, the number of patients of mental disorders gets about 3,8% of population of Vietnam. One problem facing medical doctors is how to recognize the types of stress because diagnosing well the stress types medicine doctors can understand the level of a disease and based on this level of disease, the correct treatment should be applied. The stress type diagnosis mistakes can be very harmful. If a medicine doctor misses eight syndrome diagnosis or diagnoses with not accuracy result, then no correct treatment is applied, and the patient's health worsens. On the other hand, if a traditional medicine doctor mistakenly diagnoses eight syndrome diagnosis, the the patient is going through a useless a disease treatment that means treatment methods such as herbal plants are applied not correctly and meanwhile his disease worsens his health. It is therefore important to diagnose stress types diagnosis correctly [15–17].

STRESSDIAG is a computer program which inputs the patient's symptoms, and uses its inference engine and the knowledge base provided by medicine doctors to determine whether the patient has any type of stress. In developing this system, we faced the following problem: symptoms such as weight losing, decreasing appetite,

© Springer Nature Switzerland AG 2019
R. B. Kearfott et al. (Eds.): IFSA 2019/NAFIPS 2019, AISC 1000, pp. 371–381, 2019.
https://doi.org/10.1007/978-3-030-21920-8_34

serious stress in morning etc., are not precisely defined, while a computer requires the input data to be exact and digitized. Therefore, we must model the uncertainty of symptoms so that the computer will be able to handle the medical input data. For such modeling, STRESSDIAG uses fuzzy logic and approximate reasoning methods, two mathematical tools specialized in handling and processing real Vietnamese medicine's fuzziness. The paper is organized as follows: Sect. 2 presents a structure and knowledge base of the system. Section 3 gives an example of the system's diagnosis process and of the way the knowledge base is used in the process. The evaluation of the system's ability by medical expert's group is given in Sect. 4, and the system's application scope in Sect. 5. Conclusions and future plans are discussed in Sect. 6.

2   Structure of the System

STRESSDIAG was developed by using Visual C#.NET programming language and Data base Microsoft SQL Server 2012 and Webserver Microsoft Information Server. From user's viewpoint, STRESSDIAG is equipped with the nice and friendly graphic interface. The most important components of the system are: knowledge base, reasoning engine, knowledge acquisition and an explanation. Let us describe these subsystems.

2.1   Knowledge Base

The knowledge base of STRESSDIAG contains rules provided by the traditional medicine doctors in National Hospital of mental diseases in Thuong Tin Province, Vietnam. These rules come in three different forms:

Positive form: $E_k \rightarrow D_j(\mu^c_{R_{SD}}(E_k, D_j))$

Negative form: $E_k \rightarrow \neg D_j(\mu^e_{R_{SD}}(E_k, D_j))$

Where E_k is an elementary conjunction of symptoms S_i in form of $E_k = S_1 \&, \ldots, S_m$, for each i, i = 1,…,m.

D_j is a stress type of mental diseases.

Assume that $\mu^c_{R_{SD}}(E_k, D_j) = 0$ or $\mu^e_{R_{SD}}(E_k, D_j) = 0$, where $\mu^c_{R_{SD}}(E_k, D_j)$, $\mu^e_{R_{SD}}(E_k, D_j)$ are two different fuzzy weights of fuzzy rules in [0, 1]. It is impossible that E_k both confirms and excludes D_j. More precisely:

(i) $\mu^c_{R_{SD}}(E_k, D_j) = 0$ means the elementary conjunction E_k of symptoms S_i definitely not confirms the conclusion of D_j

(ii) $\mu^c_{R_{SD}}(E_k, D_j) = 1$ means the elementary conjunction E_k of symptoms S_i definitely confirms the conclusion of D_j

(iii) $0 < \mu^c_{R_{SD}}(E_k, D_j) < 1$ means the elementary conjunction E_k of symptoms S_i confirms the conclusion of D_j with some fuzzy degree.

It is similar for the case of $\mu^e_{R_{SD}}(E_k, D_j)$.

Intermediary form: $E_k \rightarrow S_h(\mu^c_{R_{SS}}(E_i, S_h))$

(i) $\mu^c_{R_{SS}}(E_k, S_h) = 0$ means the elementary conjunction E_k of symptoms S_i definitely not confirms the symptom S_h

(ii) $\mu^c_{R_{SS}}(E_k, S_h) = 1$ means the elementary conjunction E_k of symptoms S_i definitely confirms the symptom S_h

(iii) $0 < \mu_{R_{SS}}^c(E_k, S_h) < 1$ means the elementary conjunction E_k of symptoms S_i confirms the symptom S_h with some fuzzy degree.

In rules of all three types, the premise E_k is a set of patient's symptoms. Rules in positive form are used to make a positive diagnosis, i.e. a diagnosis confirming eight rules diagnosis. Rules in negative form are used to make a negative diagnosis, i.e. a diagnosis excluding eight rules diagnosis. Rules of the third type formalize the way doctors reason: "If a patient is suffering from these symptoms, so he should be suffering from this other symptom because this symptom is usually observed together with the first group of symptom".

At present, STRESSDIAG's knowledge base contain more than 800 rules; it contains the basic rule base for diagnosis of 268 positive rules and 70 negative rules of serious stress type with mental disorder and of 229 positive rules and 10 negative rules of serious stress type, of 123 positive rules and 10 negative rules of light stress type, of 144 positive rules and 20 negative rules of light stress type. The number of rules is increasing very fast due to continuing knowledge acquisition.

2.2 Inference Engine

In general, the system accepts fuzzy descriptions of the patient's symptoms $S_i (i = 1,\ldots m)$ where $\mu_{R_{PS}}(P_q, S_i)$ is a fuzzy degree and it takes the value in $[0,1]$.

- $\mu_{R_{PS}}(P_q, S_i) = 1$ means symptom S_i surely present for patient P_q.
- $\mu_{R_{PS}}(P_q, S_i) = 0$ means symptom S_i surely absent for patient P_q.

$0 < \mu_{R_{PS}}(P_q, S_i) < 1$ means symptom S_i present for patient P_q with some degree.

We use the above-mentioned rules to determine the possibility of stress types diagnosis. This is done in in three steps.

On the first step, we use all positive rules and compute the degree $\mu_{R_{PD}}^c(P_q, D_j)$ of confirming D_j by using as formula:

$$\mu_{R_{PD}}^c(P_q, D_j) = Max_{E_k'} Min[\mu_{R_{PS}}(P_q, E_k'); \mu_{R_{SD}}^c(E_k', D_j)],$$

where E_i' varies over all elementary conjunctions of symptoms for which $\mu_{R_{SD}}^c(E_k', D_j)$ is positive, $\mu_{R_{PS}}(P_q, E_k')$ is a degree to which a patient P_q satisfies the premise of the rule E_k', and $\mu_{R_{SD}}^c(E_k', D_j)$ is the degree with which the conclusion of this rule confirms D_j.

On the second step, we determine the degree $\mu_{R_{PD}}^e(P_q, D_j)$ of excluding D_j by using as formula:

$$\mu_{R_{PD}}^e(P_q, D_j) = Max_{E_k'} Min[\mu_{R_{PS}}(P_q, E_k'); \mu_{R_{SD}}^e(E_k', D_j)].$$

On the third step, we combine these two degree into a single degree:

$$\mu_{R_{PD}}^{tot}(P_q, D_j) = \mu_{R_{PD}}^c(P_i, D_j) \ominus \mu_{R_{PD}}^e(P_i, D_j)$$

in $[-1, 1]$.

The operation $\ominus$ is a group operation defined by

$$x \ominus y = x \oplus -y$$

Operation $\oplus$ is an ordered Abelian group operation on
$[-1, 1]$. We can use an operation from the medical expert system MYCIN [19], in which the MYCIN group operation $\oplus$ on $[-1, 1]$ is defined as follows:

$$x \oplus y = x + y + x * y \quad \text{for } x, y \geq 0$$
$$x \oplus y = x + y - x * y \quad \text{for } x, y \leq 0$$

$$x \oplus y = \frac{x + y}{1 - \min(|x|, |y|)} \quad \text{for all other} \quad x, y.$$

Finally, the consultation results are the following:
The total degree $\mu_{R_{PD}}^{tot}(P_q, D_j) = 1$ means Absolutely Confirmation of conclusion of D_j.

1. The total degree $\mu_{R_{PD}}^{tot}(P_q, D_j)$ such that $0.6 \leq \mu_{R_{PD}}^{tot}(P_q, D_j) < 1$ means Almost Confirmation of conclusion of D_j.
2. The total degree $\mu_{R_{PD}}^{tot}(P_q, D_j)$ such that
   $\varepsilon \leq \mu_{R_{PD}}^{tot}(P_q, D_j) < 0.6$ means Possible Confirmation of conclusion of D_j.
3. The total degree $\mu_{R_{PD}}^{tot}(P_q, D_j)$ such that
   $-\varepsilon \leq \mu_{R_{PD}}^{tot}(P_q, D_j) < \varepsilon$ means "unknown" about Confirmation of conclusion of D_j.
4. The total degree $\mu_{R_{PD}}^{tot}(P_q, D_j)$ such that $-0.6 \leq \mu_{R_{PD}}^{tot}(P_q, D_j) < -\varepsilon$ means Possible Exclusion of conclusion of D_j.
5. The total degree $\mu_{R_{PD}}^{tot}(P_q, D_j)$ such that $-1 < \mu_{R_{PD}}^{tot}(P_q, D_j) < -0.6$ means Almost Exclusion of conclusion of D_j.
6. The total degree $\mu_{R_{PD}}^{tot}(P_q, D_j) = -1$ means Absolutely Exclusion of conclusion of D_j.
   Where ε is a heuristic value. Let recall that D_j consists of four types of stress such as such as light stress, middle stress, serious stress and serious stress with mental disorder.

2.3   Knowledge Acquisition

In our system, we used two sources of rules:

2.3.1   Rules from Experts
Most rules are formed by traditional medicine doctors from National Hospital of mental diseases No. 1. To form these rules, we listed all the symptoms seen in stress types diagnosis patients (there are about more 40 such symptoms. We sorted these symptoms by the frequency of their occurrence in eight rules diagnosis patients. Then, we formed all possible combinations of the most frequent symptoms and some combinations

involving less frequent symptoms, and asked the medicine doctors to estimate the degree with which this combination of symptoms confirms or excludes stress types diagnosis.

2.3.2   Statistical Approach

In this approach, for each combination of symptoms, instead of asking a medical doctor, we look into the database of already diagnosed patients, find all the patients who had these symptoms, and estimate the possibility degree as, e.g. the proportion of those who had stress types. This approach is efficient and fast, but to get statistically justified estimates, we must have a large database of patient's records with correct diagnosis, and the existing database is sometimes not large enough.

2.3.3   Verifying the Rule Base

Traditional medical doctors are not perfect, and, as a result, their rules may not exactly correct. Similarly, statistical rules are gathered from a limited data, and some of them may therefore be wrong. It is desirable to maintain the correctness of the rule base and to avoid conflicts between the rules.

The accuracy of expert rules depend on the medicine doctor's skills. At present, we simply combine all medical doctors' rules together; in the future, we are planning to test the medical doctors' diagnostic abilities and to "weight" rules proposed by different medicine doctors based on their different diagnostic abilities.

At present, the rule base is maintained by a group consisting of leading medical doctors in mental disorders. These doctors evaluate all the rules, including the statically rules, and eliminate rules which they believe to be false. After that, the system is applied to different patients, and the results are shown to the medical doctors' group. If the medical doctors' group sees a wrong diagnosis, it proposes a way to correct the rules.

To avoid the conflict between the rules, every time a new rule is being added to the rule base, STRESSDIAG checks whether this new rule is in conflict with any of existing rules; if there is a conflict, the doctors' group decides which of the two conflicting rules to keep.

2.4   Explanation

To make it easier to understand the reasoning process and the diagnostic result, STRESSDIAG must be able to explain how and why it comes to a certain conclusion about possibility of being stress types of a patient. During the diagnosis, the reasoning engine browses the rule base and marks all the matched rules. When the diagnosis is completed, the explanation is formed by collecting all the matched rules and every step of reasoning using each matched rule. In this explanation, STRESSDIAG shows its final conclusion, all sets of patient's symptoms which were used in reasoning, and the rules which matched with each set.

As a result, the users can see the intermediate diagnostic conclusions from all three steps of the diagnosing process, and the way the rules affect the final conclusion.

3   An Example of the Performance of the System

The domain experts use the knowledge acquisition module to enter the positive rules and negative rules for stress types diagnosis.

Figure 1 presents some functionalities of the system such as knowledge acquisition, knowledge base containing the positive rules and negative rules and a functionality of Diagnosis.

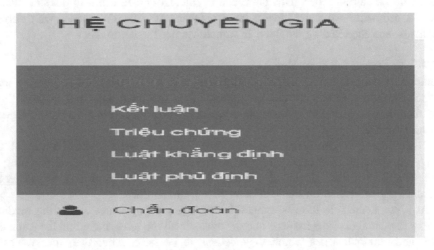

Fig. 1. Functionalities of the system

Figure 2 shows the functionality of acquisition of stress types into the systems. They are code F00: No stress, F13: the stress types as light stress, F14: middle stress, F15: serious stress and F16: serious stress with mental disorder.

kết luận

STT ▼	Mã ▼	Tên
0	F00	Không trầm cảm
1	F13	Trầm cảm nhẹ
2	F14	Trầm cảm vừa
3	F15	Trầm cảm nặng
4	F16	Trầm cảm nặng có loạn thần

Fig. 2. Knowledge acquisition of stress types

Figure 3 shows the functionality of acquisition of symptoms of stress types into the systems. They are No. 1: Increasing complexion, No. 2: Losing all interest and pleasant, No. 3: increasing energy, No. 4: increasing attention and No. 5: Increasing self-respectful and self-confident, No. 6: Feeling guilt, no worthy.

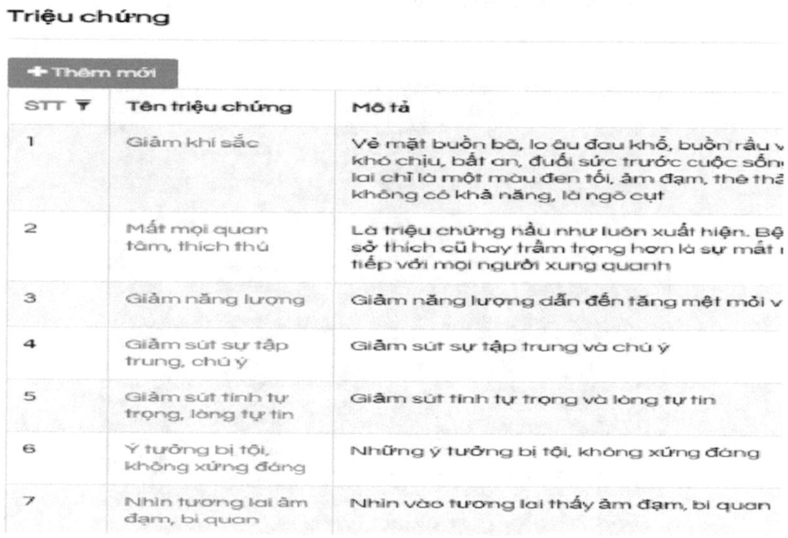

Triệu chứng

STT ▼	Tên triệu chứng	Mô tả
1	Giảm khí sắc	Vẻ mặt buồn bã, lo âu đau khổ, buồn rầu v khó chịu, bất an, đuối sức trước cuộc sốn lại chỉ là một màu đen tối, âm đạm, thể th không có khả năng, là ngõ cụt
2	Mất mọi quan tâm, thích thú	Là triệu chứng hầu như luôn xuất hiện. Bệ sở thích cũ hay trầm trọng hơn là sự mất tiếp với mọi người xung quanh
3	Giảm năng lượng	Giảm năng lượng dẫn đến tăng mệt mỏi v
4	Giảm sút sự tập trung, chú ý	Giảm sút sự tập trung và chú ý
5	Giảm sút tinh tự trọng, lòng tự tin	Giảm sút tinh tự trọng và lòng tự tin
6	Ý tưởng bị tội, không xứng đáng	Những ý tưởng bị tội, không xứng đáng
7	Nhìn tương lai âm đạm, bi quan	Nhìn vào tương lai thấy âm đạm, bi quan

Fig. 3. Knowledge acquisition of symptoms

Luật khẳng định

Kết luận *   Trầm cảm nhẹ

STT   1000   Mã số   1000

Triệu chứng *   Giảm khí sắc   Mất mọi quan tâm, thích thú   Giảm năng lượng
Giảm sút sự tập trung, chú ý   Giảm sút tinh tự trọng, lòng tự tin
Ý tưởng bị tội, không xứng đáng   Nhìn tương lai âm đạm, bi quan

Nếu *   Giảm khí sắc; Mất mọi quan tâm, thích thú; Giảm năng lượng; Giảm sút sự tập trung, chú ý; Giảm sút tinh tự trọng, lòng tự tin; Ý tưởng bị tội, không xứng đáng; Nhìn tương lai âm đạm, bi quan

✔ Hiệu lực

Lưu   ← Quay lại

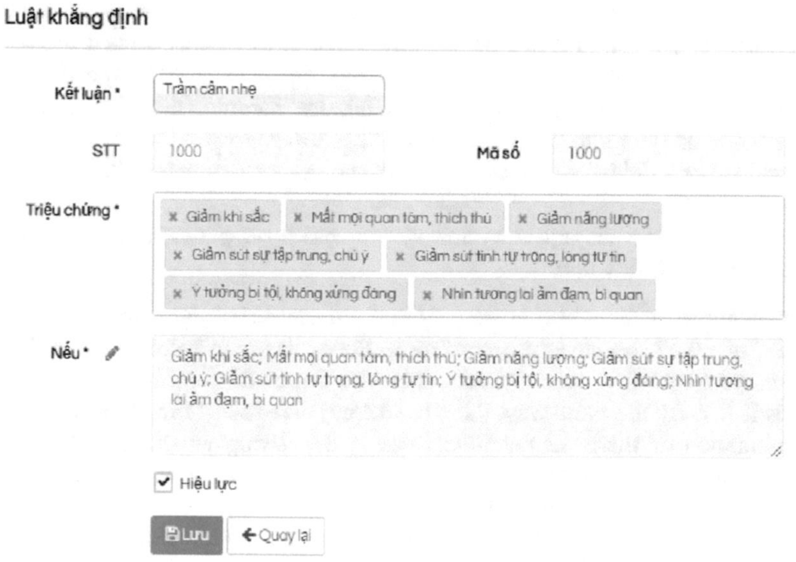

Fig. 4. Knowledge acquisition of positive rules

Figure 4 represents some kind positive rules stored in knowledge base. For example, IF Increasing complexion, Losing all interest and pleasant, increasing energy, increasing attention, Increasing self-respectful and self-confident, Feeling guilt, no worthy, Feeling gray future THEN Light Stress with degree 1.

Let us look for an example of making a diagnosis as follow:

Assume that a patient has the following symptom as described in Fig. 5.

Fig. 5. Input of patient's symptoms

The value of symptom "Increasing complexion" takes a degree 0.7, the value of symptom "Losing all interest and pleasant" takes a degree 0.8, the value of symptom "increasing energy" takes a degree 0.7, the value of symptom "increasing attention" takes a degree 0.6 and the value of symptom "Increasing self-respectful and self-confident" takes a degree 0.5, the value of symptom "Sleeping disorder" takes a degree 0.8; the value of symptom "Eating disorder" takes a degree 0.8; the value of symptom "Having idea of suicide" takes a degree 0.8; the value of symptom "Having imagine" takes a degree 0.8.

The result of the diagnosis by the system based on patient's input symptoms is calculated as the followings:

1. For diagnosis of light stress, we gets that a diagnosis of light stress with degree of belief of −0.67, because of according to the explanation mechanism of the system there are 52 positive rules "fired", then the degree of combination of 52 positive rules is 0.7. By the same way, there are 12 negative rules "fired", then the degree of combination of these 12 negative rules is 0.9. Using an ordered Abelian group operation described above, the total degree of combination of positive and negative rules for conclusion of light stress is −0.67.

2. For diagnosis of middle stress, we gets that a diagnosis of middle Stress with degree of belief of −0.63, because of according to the explanation mechanism of the system there are 40 positive rules "fired", then the degree of combination of 40

positive rules is 0.6. By the same way, there are 4 negative rules "fired", then the degree of combination of these 4 negative rules is 0.85. Using an ordered Abelian group operation described above, the total degree of combination of positive and negative rules for conclusion of middle stress is −0.63.

3. For diagnosis of serious stress, we gets that a diagnosis of serious stress with degree of belief of −0.33, because of according to the explanation mechanism of the system there are 62 positive rules "fired", then the degree of combination of 62 positive rules is 0.7. By the same way, there are 12 negative rules "fired", then the degree of combination of these 12 negative rules is 0.8. Using an ordered Abelian group operation described above, the total degree of combination of positive and negative rules for conclusion of serious stress is −0.33.

4. For diagnosis of serious stress with mental disorder, we gets that a diagnosis of serious stress with mental disorder with degree of belief of 0.7, because of according to the explanation mechanism of the system there are only 44 positive rules "fired", then the degree of combination of 44 positive rules is 0.7. No

The final conclusion of diagnosis is the patient get "serious stress with mental disorder" with degree 0,7 means "almost confirmation" of conclusion of "serious stress with mental disorder".

4   Evaluation

To test STRESSDIAG, we applied it to several hundred patients and compared the system's diagnoses with the diagnoses of leading medicine doctor's group.

We also used hundred of archived patient's records from the National Hospital of mental diseases as the medical input data for STRESSDIAG, and then compared the system's conclusions with the recorded diagnoses. In the vast majority of cases, STRESSDIAG's diagnosis was the same as recorded by the traditional medicine doctor's group. There were differences in diagnosis only in the few cases in which patients had rare symptoms which we originally did not take into consideration in our design of the system. This drawback of our system was corrected by updating the list of symptoms, the list of set of symptoms, by adding new rules, by slightly adjusting the possibility values in several existing rules. After this correction, the system worked well on all recorded patient data.

Our general conclusion is that STRESSDIAG diagnoses with a high accuracy. Its only drawback seems to be that it requires more than 800 rules; the results would be more intuitively clear and probably computationally faster if we could somehow compress these rules into fewer more general rules.

5   STRESSDIAG's Application Scope

Once the rule base of STRESSDIAG is completed, it would be able to diagnose as well as an experienced medicine doctor. We expect it to diagnose even better than an experienced medicine doctor, because our system contains knowledge provided not by a single medicine doctor, but by many leading experienced doctors, and it also uses a large body of reliable records of previously correctly diagnosed patients.

STRESSDIAG can be used as a consultation tool for stress types diagnosis not only in Vietnam, but anywhere, in the world, especially, in the oriental countries such as China, Korea, Japan, India etc.

In Vietnam, medicine expert in mental diseases in country side is decreasing due to the modern medicine is developed years by years. It is a reason, there is a lack of experienced medicine doctors. Therefore STRESSDIAG is a very useful consultation tool for stress types diagnosis, especially, in the provinces and country side in Vietnam.

One application of STRESSDIAG is in education. Students and inexperienced doctors can use STRESSDIAG as a good teacher available anywhere at any time.

6 Conclusions

This paper presents an overview of STRESSDIAG project. In this paper, we described the key facts about stress types diagnosis, and the reason why a medical expert system would be good solution to solve the problem of diagnosing eight syndromes of the stress types diagnosis. We described the general structure of the system, the system's knowledge base, and the diagnosis process which uses this knowledge base. The advantage of this work is that we included the negative rules into the system because the negative rules can excludes the diagnosis of the stress types therefore the result of diagnosis is more accuracy. On the other side, STRESSDIAG is more closing to thinking of medical doctors in the field of mental disorders.

In order to improve STRESSDIAG, we are currently working in two research directions. First, we keep maintaining and updating STRESSDIAG's rule base. Second, we are trying to improve the form of positive and negative rules by including the importance of symptoms and reasoning engine so that it can diagnoses with higher accuracy and speed.

References

1. Phuong, N.H.: Towards Intelligent Systems for Integrated Western and Eastern Medicine. The Gioi Publishers, Hanoi (1997)
2. Shortliffe, E.H.: Computer Based Medical Consultation: MYCIN. Am. Elsevier, New York (1976)
3. Adlassnig, K.-P.: CADIAG-2: computer – assisted medical diagnosis using fuzzy subsets. In: Gupta, M.M., Sanchez, E. (eds.) Approximate Reasoning in Decision Analysis, pp. 219–247. North-Holland Publishing Company, Amsterdam (1982)
4. Daniel, M., Hajek, P., Phuong, N.H.: CADIAG-2 and MYCIN-like systems. Int. J. Artif. Intell. Med. 9, 241–259 (1997)
5. Kandel, A.: Fuzzy Expert Systems. CRC Press, Boca Raton (2000)
6. Shortliffe, E., Buchanan, B., Feigenbaum, E.: Knowledge engineering for medical decision making: a review of computer-based clinical decision aids. In: Proceedings of IEEE, vol. 69, p. 1207 (1997)

7. Giaratano, J., Riley, G.: Expert Systems: Principles and Programming. PWS Publishing Company (1994)
8. Miller, R.A., Pople, H.E., Myers, J.D.: INTERNIST-1, an experimental computer-based diagnostic consultant for general internal medicine. New Engl. J. Med. **307**(8), 468–476 (1982)
9. Zadeh, L.A.: Fuzzy sets. Inf. Control **8**, 338–353 (1965)
10. Zadeh, L.A.: The role of fuzzy logic in the management of uncertainty in expert systems. Fuzzy Sets Syst. **11**, 199 (1983)
11. Phuong, N.H.: Fuzzy set theory and medical expert systems. survey and model. In: Proceedings of SOFSEM 1995. Theory and Practice in Informatics. LNCS, vol. 1012, pp. 431–436. Springer, Heidelberg (1995)
12. Phuong, N.H., Kreinovich, V.: Fuzzy logic and its applications in medicine. Int. J. Med. Inform. **62**, 165–173 (2001)
13. Hajek, P., Havranek, T., Jirousek, R.: Uncertain Information Processing in Expert Systems. CRC Press, Boca Raton (1992)
14. Nu, M.T., Phuong, N.H., Hirota, K.: Modeling a fuzzy rule based expert system combining positive and negative knowledge for medical consultations using the importance of symptoms. In: Proceedings of IFSA-SCIS 2017, 27–30 June 2017, Otsu, Japan (2017)
15. https://www.who.int/mental_health/management/depression/en/
16. ICD - 10, Medical Publisher, Vietnam (2010). (in Vietnamese and in English)
17. http://giadinh.net.vn/y-te/30-nguoi-viet-nam-bi-roi-loan-tam-than2018.htm

Functors Among Categories of *L*-fuzzy Partitions, *L*-fuzzy Pretopological Spaces and *L*-fuzzy Closure Spaces

Jiří Močkoř[(⊠)] and Irina Perfilieva

IRAFM, University of Ostrava, 30. dubna 22, 701 03 Ostrava 1, Czech Republic
{Jiri.Mockor,irina.perfilieva}@osu.cz

Abstract. In the present work we consider several categories with a weaker structure than that of an *L*-valued topology, namely the categories of Čech *L*-fuzzy interior spaces, Čech *L*-fuzzy closure spaces, categories of *L*-fuzzy pretopological spaces and *L*-fuzzy co-pretopological spaces, the category of reflexive *L*-fuzzy relations and, finally, the category of spaces with *L*-fuzzy partitions. We connect all these categories and some of their subcategories using commutative diagrams of functors.

1 Introduction

In the recent literature on *L*-valued fuzzy sets we observed a paradigm shift from the notion of a fuzzy set to that of a fuzzy partition. This opens the door for various spaces where a fuzzy partition is connected with particular structures as fuzzy approximation spaces, fuzzy rough sets, fuzzy topologies, etc. All these structures are characterized using various languages, so that is not easy to understand their interrelationship. In [10], we have attempted to find a common characterization of say, fuzzy structured spaces, using the so called operator point of view. We were motivated by the idea that all these spaces are based on the abstract notion of "closeness", and that they are introduced using axiomatically expressed properties of this notion.

In [4], we started from the well known Kuratowski closure and interior operators and considered their relational representation. The latter helped us to establish the connection between Kuratowski operators and lattice-based F-transforms. Similarly, in [10], we showed that weaker closure and interior operators, named after Čech, can be also represented using F-transforms. These facts mean that the lattice-based F-transforms can be used in parallel with closure and interior operators as their canonical representation.

In this paper, we raise our research on the level of categories and propose to connect all the above considered fuzzy structured spaces by functors among the corresponding categories. In detail, we considered the categories of sets with: Čech closure and interior operators, *L*-fuzzy pretopology and *L*-fuzzy co-pretopology, reflexive *L*-fuzzy relations and finally, *L*-fuzzy partitions. For all these categories, we find the corresponding functors and connect them using

© Springer Nature Switzerland AG 2019
R. B. Kearfott et al. (Eds.): IFSA 2019/NAFIPS 2019, AISC 1000, pp. 382–393, 2019.
https://doi.org/10.1007/978-3-030-21920-8_35

commutative diagrams, we show that in all these cases, the "starting" category is the one with L-fuzzy partitions.

We hope that the results of these research will be used in applications to image and data analysis, time series, decision making, etc., where the notion of closeness is significantly relaxed due to intensive research in the direction of dimensionality reduction and non-local operations [5,6].

2 Preliminaries

We will be using the similar denotation to that introduced in [10]. For the sake of comprehension, we repeat some useful notions and refer to [1,3] for additional details regarding residuated lattices.

Definition 1. A *residuated lattice L is an algebra $(L, \wedge, \vee, \otimes, \rightarrow, 0, 1)$ such that*

(i) $(L, \wedge, \vee, 0, 1)$ *is a bounded lattice with the least element 0 and the greatest element 1;*
(ii) $(L, \otimes, 1)$ *is a commutative monoid, and*
(iii) $\forall a, b, c \in L, \ a \otimes b \le c \iff a \le b \rightarrow c.$

A residuated lattice $(L, \wedge, \vee, \otimes, \rightarrow, 0, 1)$ is *complete* if it is complete as a lattice.

The following is the derived unary operations of *negation* $\neg$:

$$\neg a = a \rightarrow 0,$$

A residuated lattice L is called an *integral, commutative Girard-monoid* [7], if its negation is involutive, i.e.,

$$\neg\neg a = a.$$

Throughout this paper, a complete residuated lattice $L = (L, \wedge, \vee, \otimes, \rightarrow, 0, 1)$ will be fixed. Therefore in many places, the explicit reference to L will be omitted. We refer to [3,7] for some useful properties of L that will be used in the following text without explicit citation.

Let X be a nonempty set and L^X a set of all L-fuzzy sets (=L-valued functions) of X. For all $\alpha \in L$, $\underline{\alpha}(x) = \alpha$ is a constant L-fuzzy set on X. For all $u \in L^X$, the $core(u)$ is a set of all elements $x \in X$, such that $u(x) = 1$. An L-fuzzy set $u \in L^X$ is called *normal*, if $core(u) \ne \emptyset$. An L-fuzzy set $\chi^X_{\{y\}} \in L^X$ is a *singleton*, if it has the following form

$$\chi^X_{\{y\}}(x) = \begin{cases} 1, & \text{if } x = y, \\ 0, & \text{otherwise.} \end{cases}$$

Assume that $f : X \rightarrow Y$ is a map. Then, according to Zadeh's extension principle f can be extended to the operators $f_Z^{\rightarrow} : L^X \rightarrow L^Y$ and $f_Z^{\leftarrow} : L^Y \rightarrow L^X$ such that for $u \in L^X, v \in L^Y, y \in Y,$

$$f_Z^{\rightarrow}(u)(y) = \bigvee_{x, f(x) = y} u(x), \quad f_Z^{\leftarrow}(v) = v \circ f.$$

Definition 2. *The map* $i : L^X \to L^X$ *is called a Čech (L-fuzzy) interior operator,if for every* $\underline{\alpha}, u, v \in L^X$, *it fulfills*

1. $i(\underline{\alpha}) = \underline{\alpha}$,
2. $i(u) \leq u$,
3. $i(u \wedge v) = i(u) \wedge i(v)$.

We say that a Čech interior operator $i : L^X \to L^X$ is a *strong Čech-Alexandroff interior operator*, if

$$i(\underline{\alpha} \to u) = \underline{\alpha} \to i(u) \quad \text{and} \quad i(\bigwedge_{j \in J} u_j) = \bigwedge_{j \in J} i(u_j).$$

We remark that the notion of *Čech (L-fuzzy) closure operator* can be introduced using the immanent duality of L.

Definition 3. *Let* X *be a nonempty set. An L-fuzzy relation Ron* X *is a fuzzy subset of* $X \times X$,*i.e.,* $R \in L^{X \times X}$.*An L-fuzzy relation R is called reflexive, if*

$$\forall \, x \in X, \quad R(x,x) = 1.$$

We remind the notion of an L-fuzzy pretopological space and L-fuzzy co-pretopological space as it has been introduced in [11].

Definition 4. *An L-fuzzy pretopology on* X *is a set of functions* $\tau = \{p_x \in L^{L^X} : x \in X\}$,*such that for all* $u,v \in L^X, \alpha \in L$ *and* $x \in X$,

1. $p_x(\underline{\alpha}) = \alpha$,
2. $p_x(u) \leq u(x)$,
3. $p_x(u \wedge v) = p_x(u) \wedge p_x(v)$.

We say that an L-fuzzy pretopological space (X, τ) is a *strong Čech-Alexandroff L-fuzzy pretopological space*, if

$$p_x(\underline{\alpha} \to u) = \alpha \to p_x(u) \quad \text{and} \quad p_x(\bigwedge_{j \in J} u_j) = \bigwedge_{j \in J} p_x(u_j).$$

Definition 5. *An L-fuzzy co-pretopology on* X *is a set of functions* $\eta = \{p^x \in L^{L^X} : x \in X\}$, *such that for all* $u,v \in L^X, \alpha \in L$ *and* $x \in X$,

1. $p^x(\underline{\alpha}) = \alpha$,
2. $p^x(u) \geq u(x)$,
3. $p^x(u \wedge v) = p^x(u) \wedge p^x(v)$.

We say that an L-fuzzy co-pretopological space (X, τ) is a   *strong Čech-Alexandroff L-fuzzy co-pretopological space*, if

$$p^x(\underline{\alpha} \otimes u) = \alpha \otimes p^x(u) \quad \text{and} \quad p^x(\bigvee_{j \in J} u_j) = \bigvee_{j \in J} p^x(u_j).$$

Finally, we recall the notion of an L-fuzzy partition (see [2,7]).

Definition 6. *A set $\mathcal{A}$ of normal fuzzy sets $\{A_\xi : \xi \in \Lambda\}$ in X is an L -fuzzy partition of X, if*

1. *the corresponding set of ordinary subsets $\{core(A_\xi) : \xi \in \Lambda\}$ is a partition of X, and*
2. *$core(A_{\xi_1}) = core(A_{\xi_2})$ implies $A_{\xi_1} = A_{\xi_2}$.*

Instead of the index set Λ from $\mathscr{A}$ we use $|\mathscr{A}|$.

3   Basic Categories

As it can be expected, by defining the appropriate morphisms between the structures listed in the introductory section we will get categories. In this article, we will deal primarily with the following categories

1. The category **CInt** of sets with Čech L-fuzzy interior operators,
2. The category **CClo** of sets with Čech L-fuzzy closure operators
3. The category **FPrTop** of L-fuzzy pretopological spaces,
4. The category **FcoPreTop** of L-fuzzy co-pretopological space,
5. The category **FRel** of sets with reflexive L-fuzzy relations
6. The category **SFP** of spaces with L-fuzzy partitions

and also some of their subcategories. The objects and morphisms of these categories are defined by the following definition.

Definition 7. *In what follows by X, Y we denote sets from the standard category* **Set**.

1. *The category* **CInt** *is defined by*
   a. *Objects are pairs (X, i), where $i : L^X \to L^X$ is a Čech L-fuzzy interior operator,*
   b. *$f : (X, i) \to (Y, j)$ is a morphism, if*

$$\forall v \in L^Y, x \in X \quad j(v)(f(x)) \le i(f_Z^\leftarrow(v))(x).$$

2. *The category* **CClo** *is defined by*
   a. *Objects are pairs (X, c), where $c : L^X \to L^X$ is a Čech L-fuzzy closure operator,*
   b. *$f : (X, c) \to (Y, d)$ is a morphism, if $f : X \to Y$ in* **Set***, and*

$$\forall u \in L^X, \quad f_Z^\rightarrow(c(u)) \le d(f_Z^\rightarrow(u)).$$

3. *The category* **FPreTop** *is defined by*
   a. *Objects are L-fuzzy pretopological spaces (X, τ),*
   b. *$f : (X, \tau) \to (Y, \sigma)$ is a morphism, where $\tau = \{p_x \in L^{L^X} : x \in X\}$, $\sigma = \{q_y \in L^{L^Y} : y \in Y\}$, if $f : X \to Y$ in* **Set***, and*

$$\forall v \in L^Y, \quad q_{f(x)}(v) \le p_x(f_Z^\leftarrow(v)).$$

4. *The category* **FcoPreTop** *is defined by*
   a. *Objects are L-fuzzy co-pretopological spaces* (X, τ),
   b. $f : (X, \tau) \to (Y, \sigma)$ *is a morphism, where* $\tau = \{p^x \in L^{L^X} : x \in X\}$, $\sigma = \{q^y \in L^{L^Y} : y \in Y\}$, *if* $f : X \to Y$ *in* **Set**, *and*

$$\forall u \in L^X, \quad q^{f(x)}(f_Z^{\rightarrow}(u)) \geq p^x(u).$$

5. *The category* **FRel** *is defined by*
   a. *Objects are pairs* (X, R), *where* R *is a reflexive L-fuzzy relation on* X,
   b. $f : (X, R) \to (Y, S)$ *is a morphism, if* $f : X \to Y$ *is a map in* **Set** *and*

$$\forall x, x' \in X \quad R(x, x') \leq S(f(x), f(x')).$$

6. *The category* **SFP** *is defined by*
   a. *Objects are sets with a L-fuzzy partition* $(X, \mathscr{A})$,
   b. $(f, \sigma) : (X, \mathscr{A}) \to (Y, \mathscr{B})$ *is a morphism if* $f : X \to Y$ *and* $\sigma : |\mathscr{A}| \to |\mathscr{B}|$ *are maps and for each* $\alpha \in |\mathscr{A}|$,

$$f_Z^{\rightarrow}(A_\alpha) \leq B_{\sigma(\alpha)}.$$

We consider the following full subcategories of the above categories:

1. The subcategory **sACCo** of **CClo** with strong Čech-Alexandroff L-fuzzy closure operators as objects,
2. The subcategory **sACInt** of **CInt** with strong Čech-Alexandroff L-fuzzy interior operators as objects,
3. The subcategory **sAFPreTop** of **FPreTop** with strong Čech-Alexandroff L-fuzzy pretopological spaces as objects,
4. The subcategory **sAFcoPreTop** of **FcoPreTop** with strong Čech-Alexandroff L-fuzzy co-pretopological spaces,

In the paper [8] another definition of a category of sets with reflexive L-valued fuzzy relations was introduces. Recall that for a L-fuzzy relation R in a set X, the upper $\overline{R}_X$ and lower $\underline{R}_X$ L-approximation operators $L^X \to L^X$ are defined by

$$\forall u \in L^X, \quad \overline{R}_X(u)(x) = \bigvee_{t \in X} R(x, t) \otimes u(t) \quad \text{and} \quad \underline{R}_X(u)(x) = \bigwedge_{t \in X} R(x, t) \to u(t).$$

Definition 8. *([8]) The category* **L-APP** *is defined by*

1. *Objects are pairs* (X, R), *where* X *is a set and* R *is a reflexive L-valued fuzzy relation.*
2. $f : (X, R) \to (Y, S)$ *is a morphism if* $f : X \to Y$ *is a map and for all* $v \in L^Y$, *hold*

$$f_Z^{\leftarrow}(\underline{S}_Y(v)) \leq \underline{R}_X(f_Z^{\leftarrow}(v)) \quad \text{and} \quad f_Z^{\leftarrow}(\overline{S}_Y(v)) \geq \overline{R}_X(f_Z^{\leftarrow}(v)).$$

Although the definition of the category **L-APP** is more complicated, it is, in fact, similar to the category **FRel**.

Proposition 1. *Let L be a complete Girard monoid. Then the category* **L-APP** *is isomorphic to the category* **FRel**.

Proof. Since the objects of both categories are identical, we need to prove that isomorphisms of both categories are also identical. Let $f : (X, R) \to (Y, S)$ be a morphism in the category **FRel**. Then for arbitrary $v \in L^Y, x \in X$ we have

$$f_Z^{\leftarrow}(\underline{S}_Y(v))(x) \leq \bigvee_{t \in X} S(f(x), f(t)) \to v(f(t)) \leq \bigvee_{t \in X} R(x, t) \to v(f(t)) =$$

$$\underline{R}_X(f_Z^{\leftarrow}(v))(x),$$

$$f_Z^{\leftarrow}(\overline{S}_Y(v))(x) = \bigvee_{z \in Y} S(f(x), z) \otimes v(z) \geq$$

$$\bigvee_{t \in X} S(f(x), f(t)) \otimes v(f(t)) \geq \bigvee_{t \in X} R(x, t) \otimes v(f(t)) = \overline{R}_X(f_Z^{\leftarrow}(v))(x).$$

On the other hand, let f be a morphism in **L-APP**. Then for arbitrary $x, x' \in X$ we have

$$\neg S(f(x), f(x')) = \bigwedge_{z \in Y} S(f(x), z) \to \neg\chi^Y_{\{f(x')\}}(z) =$$

$$\underline{S}_Y(\neg\chi^Y_{\{f(x')\}})(f(x)) = f_Z^{\leftarrow}(\underline{S}_Y(\neg\chi_{\{f(x')\}}))(x) \leq$$

$$\underline{R}_X(f_Z^{\leftarrow}(\neg\chi_{\{f(x')\}}))(x) = \bigwedge_{t \in X} R(x, t) \to \neg\chi_{\{f(x')\}}(f(x)) = \neg R(x, x').$$

Since L is a Girard monoid, $R(x, x') \leq S(f(x), f(x'))$ holds and f is a morphism in the category **FRel**.

$\square$

In the following theorem we present the first type of relationships among our categories.

Theorem 1. *The following diagram of functors commutes:*

$$\textbf{SFP} \xrightarrow{W} \textbf{sAFcoPreTop} \xrightarrow{F'} \textbf{sACClo} \xrightarrow{1_{\textbf{sACClo}}} \textbf{sACClo}$$

(with vertical arrows: $\downarrow$ on the left, $\downarrow$ in the middle, $\uparrow H$ on the right)

$$\textbf{FcoPreTop} \xrightarrow{F} \textbf{CClo} \xrightarrow{M} \textbf{FRel},$$

where F, F' are isomorphisms of categories, F' is a restriction of F and $\hookrightarrow$ are embedding functors.

Proof. (1) We define the functor $F : \textbf{FcoPrTop} \to \textbf{CClo}$ such that for a morphism $f : (X, \tau) \to (Y, \sigma)$ in **FcoPrTop**, where $\tau = \{p^x \in L^{L^X} : x \in X\}$, $\sigma = \{q^y \in L^{L^Y} : y \in Y\}$, we put

$$F(X, \tau) = (X, c) \quad \text{and} \quad F(Y, \sigma) = (Y, d),$$

$$\forall u \in L^X, v \in L^Y, \quad c(u)(x) = p^x(u) \text{ and } d(v)(y) = q^y(v),$$

$$F(f) = f.$$

According to [10], $(X, c) \in \mathbf{CClo}$. We show that $f : (X, c) \to (Y, d)$ is a morphism in **CClo**. In fact, for $u \in L^X$ we obtain

$$f_{\overrightarrow{Z}}(c(u))(y) = \bigvee_{x \in X, f(x) = y} c(u)(x) = \bigvee_{x, f(x) = y} p^x(u) \le q^{f(x)}(f_{\overrightarrow{Z}}(u)) = d(f_{\overrightarrow{Z}}(u))(y).$$

Hence, F is a functor. We define the inverse functor $F^{-1} : \mathbf{CClo} \to \mathbf{FcoPrTop}$ such that for a morphism $g : (X, c) \to (Y, d)$ in **CClo**, we put

$$F^{-1}(X, c) = (X, \tau) \quad \text{and} \quad F^{-1}(Y, d) = (Y, \sigma),$$
$$\tau = \{t^x \in L^{L^X} : x \in X\} \quad \text{and} \quad \sigma = \{s^y \in L^{L^Y} : y \in Y\},$$
$$u \in L^X, v \in L^Y \quad \text{and} \quad t^x(u) = c(u)(x), \quad s^y(v) = d(v)(y),$$
$$F^{-1}(g) = g.$$

Then it can be proven similarly that $f : (X, \tau) \to (Y, \sigma)$ is a morphism in **FcoPrTop** and it is clear that F and F^{-1} are inverse functors.

(2) Let F' be the restriction of F to the subcategory **sA FcoPreTop**. It is clear that F' is also an isomorphism functor.

(3) Let $f : (X, R) \to (Y, S)$ be a morphism in **FRel** and let the functor $H : \mathbf{FRel} \to \mathbf{sACClo}$ be defined by

$$H(X, R) = (X, c), \quad \text{and} \quad F(Y, S) = (Y, d),$$
$$\forall u \in L^X, \quad c(u)(x) = \bigvee_{z \in X} R(x, z) \otimes u(z)$$
$$H(f) = f.$$

According to results from [8, 10] and many others, $(X, c) \in \mathbf{sACClo}$. We show that f is a morphism in **sACClo**. For $y \in Y$ we have

$$f^{\overrightarrow{}}(c(u))(y) = \bigvee_{z, f(z) = y} c(u)(y) = \bigvee_{z, f(z) = y} \bigvee_{t \in X} R(z, t) \otimes u(t) \le$$
$$\bigvee_{z, f(z) = y} \bigvee_{t \in X} S(y, f(t)) \otimes u(t) = \bigvee_{t \in X} S(y, f(t)) \otimes u(t) \le$$
$$\bigvee_{s \in Y} S(y, s) \otimes \bigvee_{t, t \in f^{-1}(s)} u(t) = \bigvee_{s \in Y} S(y, s) \otimes f_{\overrightarrow{Z}}(u)(s) = d(f_{\overrightarrow{Z}}(u))(y).$$

Hence f is a morphism in **sACClo** and H is a functor.

(4) Let $f : (X, c) \to (Y, d)$ be a morphism in **sACClo** and let us define the functor $M : \mathbf{sACClo} \to \mathbf{FRel}$ by

$$M(X, c) = (X, R),$$
$$\forall x, y \in X, \quad R(x, y) = c(\chi^X_{\{y\}})(x),$$
$$M(f) = f.$$

Clearly, R is a reflexive L-relation. We prove that $f : (X, R) \to (Y, S)$ is a morphism in **FRel**, where $(Y, S) = H(Y, d)$. In fact, we have

$$R(x, y) = c(\chi^X_{\{y\}})(x) \leq \bigvee_{z \in X, f(z) = f(x)} c(\chi^X_{\{y\}})(z) = f_Z^{\to}(c(\chi^X_{\{y\}}))(f(x)) \leq$$

$$d(f_Z^{\to}(\chi^X_{\{y\}}))(f(x)) = d(\chi^Y_{\{f(x)\}})(f(x)) = S(f(x), f(y)).$$

Therefore, H is a functor.

(5) We prove that for arbitrary $(X, d) \in$ **sACClo**, $MH(X, d) = (X, d)$. In fact let $u \in L^X$ and let $MH(X, d) = (X, \overline{d})$. From the definition of H, M it follows that

$$\overline{d}(u)(x) = \bigvee_{z \in X} u(z) \otimes d(\chi^X_{\{z\}})(x).$$

According to [9], we can write

$$u = \bigvee_{z \in X} \underline{u(z)} \otimes \chi^X_{\{z\}}.$$

where $\underline{\alpha}$ is a constant function from L^X with the value α. Since d is a strong Čech-Alexandroff operator, we obtain

$$d(u)(x) = d(\bigvee_{z \in X} \underline{u(z)} \otimes \chi^X_{\{z\}})(x) = \bigvee_{z \in X} d(\underline{u(z)})(x) \otimes d(\chi_z)(x) =$$

$$\bigvee_{z \in X} u(z) \otimes d(\chi_{\{z\}})(x) = \overline{d}(u)(x).$$

(6) Let $(f, \sigma) : (X, \mathcal{A}) \to (Y, \mathcal{B})$ be a morphism in **SFP**. Then we set

$$W(X, \mathcal{A}) = (X, \{p^x \subset L^{L^X} : x \in X\}), \quad W(f, \sigma) = f$$

where

$$\forall u \in L^X, \quad p^x(u) = F^\uparrow_{X, \mathcal{A}}(u)(w_X(x)),$$

where $w_X : X \to |\mathcal{A}|$ is the function such that $w_X(x)$ is the unique element of $|\mathcal{A}|$ such that $x \in core(A_{w_X(x)})$. It is easy to prove that $W(X, \mathcal{A})$ is a strong Čech-Alexandroff L-fuzzy co-pretopological space. We need only to prove that f is a morphism in the category **sAFcoPreTop**. Since (f, σ) is a morphism, the following diagram commutes:

$$\begin{array}{ccc} X & \xrightarrow{w_X} & |\mathcal{A}| \\ {\scriptstyle f}\big\downarrow & & \big\downarrow{\scriptstyle \sigma} \\ Y & \xrightarrow{w_Y} & |\mathcal{B}|. \end{array}$$

In fact, for $x \in X$ we have

$$1 = A_{w_X(x)}(x) \leq f_Z^{\to}(A_{w_X(x)})(f(x)) \leq B_{\sigma(w_X(x))}(f(x)),$$

and it follows $w_Y(f(x)) = \sigma(w_X(x))$. Let $W(Y, \mathscr{B}) = (Y, \{q^y \in L^{L^Y} : y \in Y\})$. Then for arbitrary $x \in X$ and $u \in L^X$ we have

$$q^{f(x)}(f_Z^{\rightarrow}(u)) = F_{Y,\mathscr{B}}^{\uparrow}(f_Z^{\rightarrow}(u))(w_Y(f(x))) = \bigvee_{y \in Y} f_Z^{\rightarrow}(u)(y) \otimes B_{w_Y(f(x))}(y) \geq$$

$$\bigvee_{z \in X} u(z) \otimes B_{\sigma w_X(x)}(f(x)) \geq \bigvee_{z \in X} u(z) \otimes A_{w_X(x)}(z) = F_{X,\mathscr{A}}^{\uparrow}(u)(w_X(x)) = p^x(u).$$

Hence f is a morphism and W is a functor.

  □

Theorem 2. *Let L be a complete Girard monoid. Then the following diagram of functors commutes:*

$$\textbf{SFP} \xrightarrow{V} \textbf{sAFPreTop} \xrightarrow{G'} \textbf{sACInt} \xrightarrow{1_{\textbf{sACInt}}} \textbf{sACInt}$$

$$\textbf{FPreTop} \xrightarrow{G} \textbf{CInt} \xrightarrow{K} \textbf{FRel}.$$

where G, G' are isomorphisms of categories, G' is a restriction of G and $\hookrightarrow$ are embedding functors.

Proof. (1) We define the functor $G : \textbf{FPreTop} \to \textbf{CInt}$ such that for a morphism $f : (X, \tau) \to (Y, \sigma)$ in **FPreTop**, where $\tau = \{p_x \in L^{L^X} : x \in X\}$, $\sigma = \{q_y \in L^{L^Y} : y \in Y\}$, we put

$$G(X, \tau) = (X, i) \quad \text{and} \quad G(Y, \sigma) = (Y, j),$$
$$\forall u \in L^X, x \in X, \quad i(u)(x) = p_x(u),$$
$$G(f) = f.$$

According the results from [10], $(X, i) \in \textbf{CInt}$. Then $f : (X, i) \to (Y, j)$ is a morphism in **CInt**, since

$$\forall v \in L^Y, \quad j(v)(f(x)) = q_{f(x)}(v) \leq p_x(f_Z^{\leftarrow}(v)) = p_x(f^{\leftarrow}) = i(f_Z^{\leftarrow}(v))(x).$$

For any morphism $f : (X, i) \to (Y, j)$ in **CInt**, the inverse functor $G^{-1} : \textbf{CInt} \to \textbf{FPreToP}$ is defined by

$$G(X, i) = (X, \tau), \quad \text{and} \quad \tau = \{t_x \in L^{L^X} : x \in X\},$$
$$\forall u \in L^X, \quad t_x(u) = i(u)(x),$$
$$G(f) = g.$$

It is clear that G^{-1} is a functor and it is inverse to G.

  (2) Let G' be the restriction of G to the subcategory **sAFPreTop**. It is clear that G' is also an isomorphism functor.

(3) Let $f : (X, R) \to (Y, S)$ is a morphism in **FRel**. The functor K is defined by

$$K(X, R) = (X, i) \quad \text{and} \quad K(f) = f,$$
$$\forall u \in L^X, x \in X, \quad i(u)(x) = \bigwedge_{t \in X} R(x, t) \to u(t).$$

Then according to [10], $(X, i) \in \mathbf{sACInt}$. We show that $f : (X, i) \to (Y, j)$ is a morphism in **sACInt**. In fact, for arbitrary $v \in L^Y, x \in X$, we have

$$j(v)(f(x)) = \bigwedge_{z \in Y} S(f(x), z) \to v(z) \leq \bigwedge_{t \in X} S(f(x), f(t)) \to v(f(t)) \leq$$
$$\bigwedge_{t \in X} R(x, t) \to v(f(t)) = \bigwedge_{t \in X} R(x, t) \to f_Z^\leftarrow(v)(x) = i(f_Z^\leftarrow(v))(x).$$

Hence, K is a functor.

(4) Let $f : (X, i) \to (Y, j)$ be a morphism in **CInt**. Then the functor $N : \mathbf{CInt} \to \mathbf{FRel}$ is defined by

$$N(X, i) = (X, R) \quad \text{and} \quad N(f) = f,$$
$$\forall x, t \in X, \quad R(x, t) = \neg i(\neg \chi_{\{t\}}^X)(x),$$

where $\neg$ is the negation operator in the Girard monoid. It is clear that R is reflexive. In fact, if $R(x, x) = \neg(i(\neg \chi_{\{x\}}^X))(t) = \alpha$, then we have $\neg \alpha = i(\neg \chi_{\{x\}}^X)(x) \leq \neg \chi_{\{x\}}^X(x) = 0_L$ and it follows that $\alpha = 1_L$. We prove that f is a morphism in **FRel**. Let $N(Y, j) = (Y, S)$. Then we need to prove $R(x, t) \leq S(f(x), f(t))$,. i.e.,

$$\neg(i(\neg \chi_{\{t\}}^X))(x) \leq \neg(j(\neg \chi_{\{f(x)\}}^Y))(f(x)). \tag{1}$$

From $\chi_{\{t\}}^X(x) \leq \chi_{\{f(t)\}}^Y(f(x))$ we obtain

$$\neg \chi_{\{t\}}^X(x) \geq \neg \chi_{\{f(t)\}}^Y(f(x)) = f_Z^\leftarrow(\neg \chi_{\{f(t)\}}^Y)(x).$$

Since f is a morphism in **CInt**, we have $j(v)(f(x)) \leq i(f_Z^\leftarrow(v))(x)$. Therefore, for $v := \neg \chi_{\{f(t)\}}^Y$ we obtain

$$j(\neg \chi_{\{f(t)\}}^Y)(f(x)) \leq i(f_Z^\leftarrow(\neg \chi_{\{f(t)\}}^Y))(x) \leq i(\neg \chi_{\{t\}}^X)(x)$$

and the inequality (1) holds.

(5) Let $(X, i) \in \mathbf{sACInt}$. We show that $KN(X, i) = (X, i)$. In fact, let $KN(X, i) = (X, j)$, where

$$\forall u \in L^X, \quad j(u)(x) = \bigwedge_{t \in X} \neg i(\neg \chi_{\{t\}}^X)(x) \to u(t).$$

The following identity can be proven simply for arbitrary $u \in L^X$:

$$u = \bigwedge_{t \in X} \overline{\neg u(t) \to \neg \chi_{\{t\}}}.$$

Since i is a strong Čech-Alexandroff interior operator, we have

$$i(u)(x) = i\left(\bigwedge_{t\in X} \overline{\neg u(t) \to \neg\chi^X_{\{t\}}}\right)(x) = \bigwedge_{t\in X} i\left(\overline{\neg u(t) \to \neg\chi^X_{\{t\}}}\right)(x) =$$

$$\bigwedge_{t\in X} \neg u(t) \to i(\neg\chi^X_{\{t\}})(x) = \bigwedge_{t\in X} \neg(i(\neg\chi^X_{\{t\}})(x)) \to u(t) =$$

$$\bigwedge_{t\in X} R(x,t) \to u(t) = j(u)(x).$$

(6) Let $(f,\sigma) : (X,\mathscr{A}) \to (Y,\mathscr{B})$ be a morphism in the category **SFP**. We define the functor $V : \mathbf{SFP} \to \mathbf{sAFPreTop}$ by

$$V(X,\mathscr{A}) = (X, \{p_x \in L^{L^X} : x \in X\}) \quad \text{and} \quad V(f,\sigma) = f,$$

$$\forall u \in L^X, \quad p_x(u) = F^{\downarrow}_{X,\mathscr{A}}(u)(w_X(x)) = \bigwedge_{z\in X} A_{w_X(x)}(z) \to u(z).$$

It can be proven easily that $V(X,\mathscr{A})$ is a strong Čech-Alexandroff pretopological space and we prove only that f is a morphism in **sAFPreTop**. In fact, for arbitrary $x \in X, v \in L^Y$ we have

$$q_{f(x)}(v) = \bigwedge_{t\in Y} B_{w_Y(f(t))}(t) \to v(t) \leq \bigwedge_{z\in X} B_{w_Y(f(x))}(f(z)) \to v(f(t)) =$$

$$\bigwedge_{z\in X} B_{\sigma w_X(x)}(f(z)) \to v(f(t)) \leq \bigwedge_{z\in X} f^{\to}_Z(A_{w_X(x)})(f(z)) \to v(f(z)) \leq$$

$$\bigwedge_{z\in X} A_{w_X(x)}(z) \to v(f(z)) = p_x(f^{\leftarrow}_Z(v)).$$

Hence, V is a functor. Therefore, the diagram of functors commutes.

□

Acknowledgements. This research was partially supported by the project 18-06915S provided by the Grant Agency of the Czech Republic.

References

1. Höhle, U.: Fuzzy sets and sheaves. Part I, basic concepts. Fuzzy Sets Syst. **158**, 1143–1174 (2007)
2. Močkoř, J.: Spaces with fuzzy partitions and fuzzy transform. Soft Comput. **13**(21), 3479–3492 (2017)
3. Novák, V., Perfilijeva, I., Močkoř, J.: Mathematical Principles of Fuzzy Logic. Kluwer Academic Publishers, Boston (1999)
4. Perfilieva, I., Singh, A.P., Tiwari, S.P.: On the relationship among F-transform, fuzzy rough set and fuzzy topology. Soft Comput. **21**, 3513–3523 (2017)

5. Perfilieva, I., Hurtik, P.: The F-transform plus PCA dimensionality reduction with application to pattern recognition in large databases. In: Proceedings of 2018 IEEE Symposium Series on Computational Intelligence (SSCI 2018), pp. 1020–1026 (2018)
6. Perfilieva, I.: F-transform-based optimization for image restoration (inpainting). In: Proceedings of 2018 IEEE International Conference on Fuzzy Systems, FUZZ-IEEE 2018, 8–13 July 2018, Rio de Janeiro, Brazil (2018)
7. Perfilieva, I.: Fuzzy transforms: theory and applications. Fuzzy Sets Syst. **157**, 993–1023 (2006)
8. Ramadan, A.A., Li, L.: Categories of lattice-valued closure (interior) operators and Alexandroff L-fuzzy topologies. Iranian J. Fuzzy Syst. (2018). https://doi.org/10.22111/IJFS.2018.4239
9. Rodabaugh, S.E.: Powerset operator foundation for poslat fuzzy SST theories and topologies. In: Höhle, U., Rodabaugh, S.E. (eds.) Mathematics of Fuzzy Sets: Logic, Topology and Measure Theory, The Handbook of Fuzzy Sets Series, vol. 3, pp. 91–116. Kluwer Academic Publishers, Boston (1999)
10. Singh, A.P., Tiwari, S.P., Perfilieva, I.: F-transform, L-fuzzy partitions and l-fuzzy pretopological spaces: an operator oriented view (Submitted)
11. Zhang, D.: Fuzzy pretopological spaces, an extensional topological extension of FTS. Chin. Ann. Math. **3**, 309–316 (1999)

On Modeling of Generalized Syllogisms with Intermediate Quantifiers

Petra Murinová$^{(\boxtimes)}$

Institute for Research and Applications of Fuzzy Modeling, NSC IT4Innovations,
University of Ostrava, Ostrava, Czech Republic
petra.murinova@osu.cz

Abstract. In our previous papers, we introduced a general principle in fuzzy natural logic in which the class of intermediate quantifiers can be introduced, and proved all of 105 generalized syllogisms. We also proposed generalized Peterson's square of opposition with generalized definitions of contrary, contradictory, sub-contrary and subalterns. This approach is devoted to designing generalized Peterson's rules which will be used for verification of the validity of generalized syllogisms with intermediate quantifiers based on its position inside in generalized Peterson's square of opposition.

1 Introduction

The main goal of this paper is to analyze generalized syllogisms with intermediate quantifiers using generalized Peterson's rules which Peterson defined in [21] as a generalization of Aristotle's rules. In our previous work [7], we syntactically proved 105 generalized Peterson's syllogisms based on higher order fuzzy logic, namely, Łukasiewicz fuzzy type theory (Ł-FTT) which is based on Łukasiewicz algebra. Recall that all syntactical proofs were constructed using axioms of Ł-FTT and deduction rules. In [9], we introduced a structure of generalized syllogisms consisting of all 105 generalized syllogisms which stand among Aristotles's ones.

In connection with these results, also the extended syllogistic reasoning by adding new quantifiers was proposed (see, e.g., [4,18,22,25]). The extensions were proposed in several directions. The generalization of the syllogistic reasoning in the four classical figures by replacing classical quantifiers by the generalized (fuzzy) ones was introduced in [21] and later generalized in [7].

The main objective of this paper is to propose an algorithm for an analysis of validity of generalized syllogisms which will be constructed based on basic Peterson's rules: two *involving distribution*, two *involving quality* and two derived rules which characterize *quantity*.

The idea of this paper is using proposing algorithm to analyze the validity and the invalidity of generalized syllogisms with "basic" intermediate quantifiers *"All", "Almost all", "Most", "Many", "Some"*. The assumption of four figures

© Springer Nature Switzerland AG 2019
R. B. Kearfott et al. (Eds.): IFSA 2019/NAFIPS 2019, AISC 1000, pp. 394–405, 2019.
https://doi.org/10.1007/978-3-030-21920-8_36

with five basic quantifiers leads to 4000 possible generalized syllogisms. Verification of the validity of 105 generalized syllogisms can be obtained by several methods:

- Applying of Venn's diagram
- Using graded Peterson's square of opposition
- To find formal (syntactical) proofs of corresponding syllogisms
- Using Aristotle's and Peterson's rules

The verification of the validity of 105 (24 Aristotle's and 81 intermediate) syllogisms with intermediate quantifiers using Venn's diagram was introduced in [21] in Sect. 3. Peterson interpreted main Aristotle's syllogisms and selected species of intermediate syllogisms.

The next method for the verification of the validity of syllogisms is based on "Peterson's square" of opposition. The first version of the square of opposition generalizing Aristotle's one with "Few", "Many" and "Most" was introduced by Peterson in [20] in 1979. In [23], Thompson extended Peterson's approach by the intermediate quantifier "Most" which was used [21] in the sense "More than half" unlike Peterson used the "Most" as synonymous with the "Almost all". A model of the Peterson's square in mathematical fuzzy logic was introduced in [8]. The authors showed that this square can be interpreted using the formalism of higher-order fuzzy logic (the fuzzy type theory). The verification of all valid syllogisms, introduced in [24], is based on the relations of contrary, contradictory, sub-contrary and subalterns among the intermediate quantifiers which form Peterson's square.

As we mentioned above, the next possibility to verify validity of all the 105 generalized syllogisms with intermediate quantifiers, is to find mathematical (syntactical) proof of every syllogism in many-valued logic. In [13], the first formalization with mathematical definitions of basic intermediate quantifiers was proposed. We also proposed a definition of the validity of syllogisms which was formulated syntactically. It means that all generalized syllogisms which were proved in [7] hold in *every* model.

Another method of how to verify the validity of syllogisms will be discussed in this paper. We very shortly introduce an algorithm that generates all forms of possible syllogisms. Then using generalized Peterson's rules we will verify the feasibility of all inferred forms of generalized syllogisms.

2   Preliminaries: Fuzzy Natural Logic and Intermediate Quantifiers

In this section, we will briefly recall a few main concepts of the fuzzy natural logic (FNL)[1] and definitions of generalized intermediate quantifiers which form generalized Peterson'square of opposition. Fuzzy natural logic is a group of three main mathematical theories containing a model of the semantics of some parts of

---

[1] It follows the concept of natural logic which was proposed by Lakoff in [5].

natural language. The fuzzy natural logic founds applications in interpretation and linguistic characterization of times series (see [14]).

There are main theories (until now) which form FNL:

- Theory of evaluative linguistic expressions (T^{Ev}).
- Theory of fuzzy/linguistic IF-THEN rules and logical inference from them.
- Theory of fuzzy generalized quantifiers (T^{IQ}), generalized Aristotle's syllogisms and generalized Peterson's square of opposition.

The reader can find details in several papers [8, 11, 12].

In this paper, we will be interested in the theory of fuzzy generalized quantifiers with their generalized syllogisms and generalized Peterson's square. Our model of intermediate quantifiers is based on the concept of evaluative linguistic expressions. A motivation, fundamental assumptions and the formalization of their theory is in detail presented in [12].

2.1   Theory of Evaluative Linguistic Expressions

Evaluative linguistic expressions are expressions of natural language such as *small, medium, big, very short, more or less deep, quite roughly strong, extremely high*, etc. The formal theory of the semantics of expressions of natural language is a special theory of the higher order fuzzy logic[2] which was proposed in [12] based on Łukasiewicz MV-algebra (see [3, 15]) with delta operation.

We consider a set *EvExpr* of evaluative linguistic expressions. Very important group of *EvExpr* are *simple evaluative expressions*, which are considered in the following simple form:

$$\langle \text{linguistic hedge}\rangle\langle \text{TE-adjective}\rangle \tag{1}$$

where ⟨TE-adjective⟩ is an evaluative adjective (e.g., *good, interesting*), a gradable adjective (e.g., *small, warm*), and possibly also other specific kind of adjective. The ⟨linguistic hedge⟩ (it will be denoted by ν) is an intensifying adverb that makes the meaning of the evaluative expression either more or less specific. Typical examples are *extremely, significantly, very* (narrowing hedge), *more or less, roughly, quite roughly, very roughly* (widening hedge) or *rather* (specifying hedge).

In this paper, we will assume the special group of evaluative expressions which are used in definitions of intermediate quantifiers below. We will work with expressions *small (Sm)* and *big (Bi)* which will be modified by hedges *extremely (Ex)* and *very (Ve)*.

For this paper is not important to define all properties, axioms and other definitions. For the detail we recommend to see full paper ([12]).

---

[2] This fuzzy logic is called *fuzzy type theory* and was proposed in [11].

2.2   Intermediate Quantifiers

Intermediate quantifiers are expressions of natural language (*a large part of, most, almost all, many, etc.*) which belong among generalized quantifiers whose theory was studied in [19, 21]. The basic idea of intermediate quantifiers is: *intermediate quantifiers are classical general or existential quantifiers, but the universe of quantification is modified, and the modification can be imprecise.* Its semantical properties with their syllogisms were analyzed in [21].

A mathematical model of intermediate quantifiers proposed below[3] consists of the assumption that the quantifiers are just *classical* quantifiers (universal or existential), but the universe of quantification is modified by corresponding evaluative linguistic expression.

Definition 1. An intermediate quantifier of type $\langle 1, 1 \rangle$ interpreting the sentence

"$\langle$Quantifier$\rangle$ B's are A".

is one of the following formulas:

$$(Q_{Ev}^{\forall} x)(B, A) \equiv (\exists z)[(\forall x)((B|z)\, x \Rightarrow Ax) \wedge Ev((\mu B)(B|z))], \qquad (2)$$

$$(Q_{Ev}^{\exists} x)(B, A) \equiv (\exists z)[(\exists x)((B|z)x \wedge Ax) \wedge Ev((\mu B)(B|z))]. \qquad (3)$$

By $(B|z)$[4] we denote only those fuzzy sets that are "cuts" of the universe B, i.e., fuzzy sets $z \subseteq B$ such that for each element x, the membership degree $z(x) = B(x)$.

By putting of a concrete evaluative linguistic expressions we obtain five *positive* and five *negative* intermediate quantifiers. For example, for the definition of the quantifier "Almost all" we use the expression *extremely big*. Then the quantifier **P** says that we consider the greatest fuzzy set $(B|z)$ being subset of B such that all its elements have the property A and $(B|z)$ is "extremely big" (in the sense of the measure μ). The meaning of the other intermediate quantifiers is similar. On the other hand to define the quantifier "Many" means to use the expression *not small*.

2.3   Categorical Syllogism

The categorical syllogism is a special kind of logical argument in which the conclusion is inferred from two premises: the major premise (first) and minor premise (second). The syllogisms will be written as triples of formulas $\langle P_1, P_2, C \rangle$. The intermediate syllogism is obtained from any traditional syllogism when replacing one or more of its formulas by formulas containing intermediate quantifiers.

Theory of syllogistic reasoning deals with quantifiers of type $\langle 1, 1 \rangle$ (will be denoted by Q_1, Q_2, Q_3) divided into four figures. Let P, S, M be formulas such that:

---

[3] Recall that the quantifiers defined below are quantifiers of type $\langle 1, 1 \rangle$.

[4] The precise mathematical formula which represents commented pro perty is defined in [10] by formula (13).

- S represents *subject*,
- P denotes *predicate*,
- M is *middle formula*.

We assume the following structure of four figures:

Figure I	**Figure II**	**Figure III**	**Figure IV**
Q_1 M is P	Q_1 P is M	Q_1 M is P	Q_1 P is M
Q_2 S is M	Q_2 S is M	Q_2 M is S	Q_2 M is S
Q_3 S is P	Q_3 S is P	Q_3 S is P	Q_3 S is P

where the first line in each figure is the *major premise* P_1, the second line is the *minor premise* P_2 and the third line is the conclusion C.

Typical example of classical Aristotle's syllogism is as follows:

$$P_1 : \text{All men are mortal.}$$
$$\textbf{AAA}\text{-I (Barbara)} \, P_2 : \text{All Greeks are men.}$$
$$C: \text{All Greeks are mortal.}$$

3   Peterson's Rules for Aristotle's Syllogisms

As we mentioned above there are a few ways how to verify the validity of syllogisms. Traditionally, it means applying six Aristotle's rules, the validity of classical syllogisms could be determined. In [21], Peterson proposed four "basic" rules, two involving *distribution* and two involving *quality*. He showed, that two additional rules of *quantity* were derivable from four basic ones.

3.1   Peterson's (Classical) Rules

The main objective of this subsection is to introduce Peterson's six rules and later explain its meaning with concrete examples. To apply Peterson's rules is not necessary to know precise mathematical definitions of quantifiers. Important is to know a position inside Aristotle's square which has been analyzed using propositional logic [6,17] as well as in classical first-order logic [1,2,16,21]. It means, we need to know if the quantifier is *affirmative or negative*, and if it is *universal or particular*.

At first, we start with two distributive rules as follow:

1. *Rules of Distribution*

   (R1)   The middle formula must be distributed at least once.
   (R2)   No formula is distributed in the conclusion unless it is distributed in one premise.

To use these rules, we have to know what distribution is. In Aristotle's syllogisms, the traditional notion of distribution is binary. If the quantifier **O** occurred in a categorical which said something about every member of the formula's extension, then the formula was said to be *distributed*. Otherwise, the formula said to be *undistributed*. So, there are only *two choices*-distributed or undistributed. Hence, subjects of universals and predicates of negatives were distributed, while all other positions of formula were undistributed. It means that the subject formulas of universal formulas (**A** and **E**) and the predicate formulas of negative formulas (**E** and **O**) are *distributed*. The other formulas are not distributed.

	Quantifier	Quantity	Quality	Distributed
All S are P	**A**	universal	affirmative	S
No S are P	**E**	universal	negative	S and P
Some S are P	**I**	particular	affirmative	none
Some S are not P	**O**	particular	negative	P

Example 1 (Rule (R1)). The rule (R1) explains a formal mistake in which a syllogism contains a middle formula that is not distribute in either premise. There is an example of invalid syllogism:

$$P_1 : \text{All sharks are fish.}$$
AAA-III $\underline{P_2 : \text{All salmon are fish.}}$
$$C : \text{All salmon are sharks.}$$

Example 2 (Rule (R2)). The rule (R2) explains a formal mistake in which a syllogism contains a middle formula that is not distribute in either premise. There is an example of invalid syllogism:

$$P_1 : \text{All horses are animals.}$$
AOO-I $\underline{P_2 : \text{Some dogs are not horses.}}$
$$C : \text{Some dogs are not animals.}$$

2. Rules of Quality

(R3)   At least one premise is affirmative.
(R4)   The conclusion is negative if and only if one premise is.

The rule (R3) says that no conclusion draw from two negative premises. The rule (R4) says that negative premise requires a negative conclusion, and a negative conclusion requires a negative premise.

Example 3 (Rule (R3)). There is an example of negative syllogism:

$$P_1 : \text{No birds are mammals.}$$
EOI-I $\underline{P_2 : \text{Some dogs are not birds.}}$
$$C : \text{Some dogs are birds.}$$

Example 4 (Rule (R4)). Similarly, the rule (R4) says that any syllogism having exactly one negative formula is invalid. There is an example of invalid syllogism:

$$P_1 : \text{All crows are birds.}$$
AOI-I $P_2 : \text{Some wolves are not crows.}$
$$\overline{ C : \text{Some wolves are birds.}}$$

Thus, a negative conclusion cannot follow from positive premises.

3. *Rules of Quantity*

(R5) At least one premise is universal.
(R6) If either premise is particular, the conclusion is particular.

To infer a valid syllogism from two particular premises is not possible. Similarly, the conclusion of a syllogism which consists one particular premise cannot be universal. Concrete invalid forms of syllogisms will be introduced below.

4 Generalized Peterson's Rules and Inferred Forms

This section focus on the main task of this paper, which is to introduce and apply generalized Peterson's rules for the verification of the validity of generalized syllogisms with intermediate quantifiers. Recall that generalized intermediate quantifiers form generalized (**5**)-square of opposition which was constructed as a generalization of Aristotle's one in [8] and gives us all necessary information about quantifiers[5]. All important properties of intermediate quantifiers will be summarized below.

Similarly as in classical approach, which was introduced in Subsect. 3.1, positions of all quantifiers play an important role for using quality as well as distribution's rules which characterize a relationship between subjects (S) and predicates (P). These two mentioned formulas are contained in the conclusion of every syllogism, while they never occur together in the premises, and it means, their relationship must be determined by the middle formula which is guaranteed by distribution's rules.

4.1 Generalized Peterson's Rules for Intermediate Syllogisms

To work with the validity of generalized syllogisms with intermediate quantifiers, we need to reformulate classical rules which were defined in the previous section. How it will be seen below, the main rule, which will be reformulated, is the rule of distribution, while rules of quality remain unchanged. As we mentioned above rules of quantity were inferred from "basic" ones, it means that they could be derived form reformulated rules of distribution and quality.

1. *Rules of Distribution*

(R1) In a valid syllogism, the sum of distribution indices (DIs) for the middle formula must exceed 5.
(R2) No formula may be more nearly distributed in the conclusion than it is in the premises (i.e., no formula may bear a higher DIs in the conclusion than it bears in the premises).

---

[5] We refer to our detailed papers where **5**-square could be found.

An assumption of five basic intermediate quantifiers forming general **5**-square of opposition leads to the maximum DI of 5. Note that the numerical indexes (DI1 through DI5) explain the relationships between quantifiers. It means, for example, that formulas with DI5 are more distributed than those formulas with DI4. Since 5 is the maximum DI than this value of distribution index will be assigned to the traditional formulas. It will be the case that the subjects of all universal quantifiers (**A** and **E**) and predicates of all negative intermediate quantifiers **E, O, B, D** and **G** have a DI5. Let us assign DI1 to the subjects of particular and the predicates of affirmative which were in the classical approach called undistributed. Form the monotonicity of positive quantifiers (it is inside of the generalized square fulfilled by subalterns property) the predicates of **P, T** and **K** are all assigned DI of 1. Finally, one more by monotonicity, the predicates of the predominant, majority, common affirmative quantifiers (**P, T, K**) have DI of 4, 3, 2, respectively.

2. *Rules of Quality*

> (R3)   At least one premise is affirmative.
>
> (R4)   The conclusion is negative if and only one premise is.

3. *Rules of Quantity*

> (R3)   At least one premise must have a quantity of majority (T or D) or higher.
>
> (R4)   If any premise is non-universal, then the conclusion must have a quantity that is less than or equal to that premise.

Table 1. Properties of the quantifiers used in the algorithm

	Quantifier	Quantity	Quality	DI Subject	DI predicate
All S are P	**A**	Universal	Affirmative	DI5	DI1
No S are P	**E**	Universal	Negative	DI5	DI5
Almost all S are P	**P**	Predominant	Affirmative	DI4	DI1
Almost all S are not P	**B**	Predominant	Negative	DI4	DI5
Most S are P	**T**	Majority	Affirmative	DI3	DI1
Most S are not P	**D**	Majority	Negative	DI3	DI5
Many S are P	**K**	Common	Affirmative	DI2	DI1
Many S are not P	**G**	Common	Negative	DI2	DI5
Some S are P	**I**	Particular	Affirmative	DI1	DI1
Some S are not P	**O**	Particular	Negative	DI1	DI5

4.2   Inferred Forms of Generalized Syllogisms

The last subsection is devoted to a presentation of some examples of inferred forms (see Table 1) of generalized syllogisms. Below, we introduce selected forms of generalized syllogisms which fulfilled or violate rules (*R*1)–(*R*4).

Table 2. Inferred forms of generalized syllogism, **x** represents any quantifier

R1	R2	R3	R4	R1, R2, R3, R4
Exx TTTT	**EEx** TTTT	**Axx** TTTT	**ExE** TTTT	**AAA** TFFF
DAx FTTT	**DxI** TTTT	**DAx** TTTT	**DxB** TTTT	**ADG** FTFF
TPx FTFF	**TxI** TTTT	**Txx** TTTT	**TPA** TTTT	**TPI** FFTF
. . .	. . .	. . .	. . .	. . .

4.3   Sketch of the Algorithm

We continue with the short presentation of algorithm and the explanation generalized Peterson's rules. At first, we explain distribution rules (R1 and R2). By $(DIP)_{P_1}$ we denote the distribution index of the predicate in the major premise, by $(DIP)_{P_2}$ we understand the distribution index of the predicate in the minor premise and bz $(DIP)_C$ we denote the distribution index of the predicate in the conclusion. Analogously we can introduce symbols for the middle formula and the subject.

From the $\langle P_1, P_2, C, Fig \rangle$ we can obtain the structure of distribution indexes for the corresponding figure Fig as follows:

$$P_1 : \ (DIP)_{P_1} \ (DIM)_{P_1}$$
$$P_2 : \ (DIS)_{P_2} \ (DIM)_{P_2}$$
$$\overline{C : \ (DIS)_C \ \ (DIP)_C}$$

The rule R1 can be now verified by summing up the DI's for the middle formulas.

$$(DIM)_{P_1} + (DIM)_{P_2} > 5 \tag{4}$$

The rule R2 can be now verified by comparing of DI's between premise and conclusion.

$$(DIS)_{P_2} \geq (DIS)_C \quad \text{and} \quad (DIP)_{P_1} \geq (DIP)_C \tag{5}$$

The rules R3 and R4 can be easily seen from the Table 1 by quality property.

4.4   Examples of Intermediate Syllogisms

The main objective of this subsection is to explain the meaning of generalized rules with concrete examples of generalized syllogisms. At first, we start with an explanation of the example of the valid syllogism with two universal affirmative premises of Figure-I:

AAx-I:

P_1 : All people who suffer form high blood pressure take medicine.

$\underbrace{\qquad\qquad}_{middle\ formula}$ $\underbrace{\quad}_{predicate}$

P_2 : All people in stress have high blood pressure.

$\underbrace{}_{subject}$ $\underbrace{\qquad}_{middle\ formula}$

C : *All, Almost all, Most, Many, Some* people who are in stress take medicine.

$\underbrace{}_{subject}$ $\underbrace{}_{predicate}$

At first, let us notice positions of the subject, the predicate and the middle formula. Let us go to discuss all the rules as follow:

- (R1): Recall that to fulfill the first rule means to be interested in the middle formula. In this case, the conclusion is not important. From Table 1 it follows that (DIs) for the middle formula is 6 and rule (R1) is fulfilled. There is a deeper explanation of why 6. The middle formula in the first premise is in the position of the subject and it has (DIs) 5 if universal (**A**) is assumed. Similarly, (DIs) of the middle formula in the second premise is 1, because it is in the position of the predicate.
- (R2): Now we will be interested in (DIs) index of the subject and the predicate in both premises as well as in the conclusion. From Table 2 it follows that:
  - subject: (DIs) of the subject in minor premise is 5 and in the conclusion is 1.
  - predicate:(DIs) of the predicate in major premise is 5 and in the conclusion is 1.

  Let us realize that if the rule (R2) will be fulfilled with the universal conclusion then it will be fulfilled with every conclusion because (DIs) of the conclusion of other positive quantifiers will be smaller than (DIs) of premises.
- (R3) and (R4) are trivially fulfilled.

We continue with the example of non-trivial syllogism (both premises consist of intermediate quantifiers) of Figure-III. We will see that the presented syllogism will be valid with the particular conclusion only. The quaternion of $\{T, F\}$ represents the feasibility and violation of the individual rules for the corresponding figures (Table 3).

PPx-III:

P_1 : Almost all shares of companies grow with growing economy.

$\underbrace{\qquad}_{middle formula}$ $\underbrace{\quad}_{predicte}$

P_2 : Almost all shares of companies grow while companies are on World Stock Markets.

$\underbrace{\qquad}_{middle formula}$ $\underbrace{\quad}_{subject}$

C : Q_3 companies being on World Stock Markets have growing economy.

$\underbrace{}_{subject}$ $\underbrace{}_{predicate}$

- (R1): This rule is trivially fulfilled for all conclusions, because (DIs) of the middle formula is equal to 8.

- (R2):
  - subject: (DIs) of the subject in minor premise is 1.
  - predicate:(DIs) of the predicate in major premise is 1.

  For the valid syllogism, we have to find quantifier Q_3 which has (DIs) at most 1 for the subject as well as for the predicate. It is fulfilled for "Some".
- (R3) and (R4) are trivially fulfilled.

We can conclude that the syllogism with the quantifier "Almost all" in both premises is valid with the quantifier "Some" in the conclusion only.

Table 3. Forms of generalized syllogism with intermediate quantifiers

	R1	R2	R3	R4	R1, R2, R3, R4
PPA	FFTF	FFFF	TTTT	TTTT	FFFF
PPE	FFTF	FFFF	TTTT	FFFF	FFFF
PPP	FFTF	TTFF	TTTT	TTTT	FFFF
PPB	FFTF	FFFF	TTTT	FFFF	FFFF
PPT	FFTF	TTFF	TTTT	TTTT	FFFF
PPD	FFTF	FFFF	TTTT	FFFF	FFFF
PPK	FFTF	TTFF	TTTT	TTTT	FFFF
PPG	FFTF	FFFF	TTTT	FFFF	FFFF
PPI	FFTF	TTTT	TTTT	TTTT	FFTF
PPO	FFTF	FFFF	TTTT	FFFF	FFFF

5   Conclusion

In this paper, we focused on the verification of generalized intermediate quantifiers applying of generalized Peterson's rules. We introduced the sketch of the algorithm which inferred all of 105 valid forms of generalized syllogisms. The main objective for the future is to extend this approach by new quantifiers "A few", "Several" and "A little" and to apply proposing algorithm with the theory of syllogistic reasoning for an interpretation of natural data.

Acknowledgements. The paper has been supported by the project "LQ1602 IT4Innovations excellence in science".

References

1. Afshar, M., Dartnell, C., Luzeaux, D., Sallantin, J., Tognetti, Y.: Aristotle's square revisited to frame discovery science. J. Comput. **2**, 54–66 (2007)
2. Brown, M.: Generalized quantifiers and the square of opposition. Notre Dame J. Formal Logic **25**, 303–322 (1984)

3. Cignoli, R.L.O., D'Ottaviano, I.M.L., Mundici, D.: Algebraic Foundations of Many-valued Reasoning. Kluwer, Dordrecht (2000)
4. Dubois, D., Prade, H.: On fuzzy syllogisms. Comput. Intell. **4**, 171–179 (1988)
5. Lakoff, G.: Linguistics and natural logic. Synthese **22**, 151–271 (1970)
6. Prade, H., Miclet, L.: Analogical proportions and square of oppositions. In: Laurent, A., et al. (ed.) Proceedings of 15th International Conference on Information Processing and Management of Uncertainty in Knowledge-Based Systems, 15–19 July, Montpellier, CCIS, vol. 443, pp. 324–334 (2014)
7. Murinová, P., Novák, V.: A formal theory of generalized intermediate syllogisms. Fuzzy Sets Syst. **186**, 47–80 (2013)
8. Murinová, P., Novák, V.: Analysis of generalized square of opposition with intermediate quantifiers. Fuzzy Sets Syst. **242**, 89–113 (2014)
9. Murinová, P., Novák, V.: The structure of generalized intermediate syllogisms. Fuzzy Sets Syst. **247**, 18–37 (2014)
10. Murinová, P., Novák, V.: On the model of "many" in fuzzy natural logic and its position in the graded square of opposition. Fuzzy Sets Syst
11. Novák, V.: On fuzzy type theory. Fuzzy Sets Syst. **149**, 235–273 (2005)
12. Novák, V.: A comprehensive theory of trichotomous evaluative linguistic expressions. Fuzzy Sets Syst. **159**(22), 2939–2969 (2008)
13. Novák, V.: A formal theory of intermediate quantifiers. Fuzzy Sets Syst. **159**(10), 1229–1246 (2008)
14. Novák, V.: Linguistic characterization of time series. Fuzzy Sets Syst. **285**, 52–72 (2015)
15. Novák, V., Perfilieva, I., Močkoř, J.: Mathematical Principles of Fuzzy Logic. Kluwer, Boston (1999)
16. Parsons, T.: Things that are right with the traditional square of opposition. Log. Univers. **2**, 3–11 (2008)
17. Pellissier, R.: "Setting" n-opposition. Log. Univers. **2**, 235–263 (2008)
18. Pereira-Fariña, M., Díaz-Hermida, F., Bugarín, A.: On the analysis of set-based fuzzy quantified reasoning using classical syllogistics. Fuzzy Sets Syst. **214**, 83–94 (2013)
19. Peters, S., Westerståhl, D.: Quantifiers in Language and Logic. Claredon Press, Oxford (2006)
20. Peterson, P.L.: On the logic of "few", "many" and "most". Notre Dame J. Formal Logic **20**(155–179), 155–179 (1979)
21. Peterson, P.: Intermediate Quantifiers. Logic, linguistics, and Aristotelian semantics. Ashgate, Aldershot (2000)
22. Schwartz, D.G.: Dynamic reasoning with qualified syllogisms. Artif. Intell. **93**, 103–167 (1997)
23. Thompson, B.E.: Syllogisms using "few", "many" and "most". Notre Dame J. Formal Logic **23**, 75–84 (1982)
24. Turunen, E.: An algebraic study of peterson's intermediate syllogisms. Soft Comput. (2014). https://doi.org/10.1007/s00500-013-1216-2
25. Zadeh, L.A.: Syllogistic reasoning in fuzzy logic and its applications to usuality and reasoning with dispositions. IEEE Trans. Syst. Man Cybern. **15**, 754–765 (1985)

An Approach to Pulse Symbols Based Fuzzy Reasoning in Diagnosis of Traditional Vietnamese Medicine Including the Importance of Symptoms

Nguyen Hoang Phuong[1(✉)], Anh Nguyen[2], and Truong Hong Thuy[3]

[1] Thang Long University, Hanoi, Vietnam
nhphuong2008@gmail.com
[2] Auburn University, Auburn, AL, USA
anhnguyen@auburn.edu
[3] Thai Nguyen University of Medicine and Pharmacy, Thai Nguyen, Vietnam
truongthihongthuy@tump.edu.vn

Abstract. We propose a fuzzy model of reasoning for a pulses Based Disease Symptoms Diagnosis of Traditional Vietnamese Medicine including the importance of symptoms. Symptoms are represented as pulse symbols and disease symptoms which are regarded as fuzzy sets. The value of a membership degree in the fuzzy set of a pulse symptom or a disease symptom is given in [0, 1]. Diagnostic process is based on pulse symbols by triple Cun-Quan-Chi of traditional Vietnamese medicine. Based on pulse symbols, the diagnostic process provides the physician with disease symptoms with some types absolutely confirmation, almost confirmation, possible confirmation and not confirmation. An example of pulse based disease symptom diagnosis including the importance of symptoms by Left hand in Cun 1st position is given. The result in the example shows that the obtained diagnosis is more reasonable and it closes to thinking of the traditional medicine practitioners.

Keywords: Fuzzy systems · Pulses based diagnosis · Traditional Vietnamese · Medicine

1 Introduction

The traditional Vietnamese medicine is considered as a complement method for western medicine. The traditional Vietnamese medicine is based on philosophy of Yin and Yang, the five elements, energy, blood and body fluid. "When Yin wins a victory, Yang will be diseased; when Yang wins a victory, Yin will be diseased". To treat a disease, we must make a balance between Yang and Yin [4, 5]. The nature of the diagnosis of the traditional Oriental medicine is normally based on experiences of traditional practitioners and the most information of the traditional Oriental medicine are fuzzy in nature such as pale face, red eyes etc. Therefore fuzzy logic can be used to model and develop diagnosis of the traditional Oriental medicine. The concepts of Fuzzy sets were proposed by Zadeh in 1965 [2] and the role of fuzzy logic in the

© Springer Nature Switzerland AG 2019
R. B. Kearfott et al. (Eds.): IFSA 2019/NAFIPS 2019, AISC 1000, pp. 406–416, 2019.
https://doi.org/10.1007/978-3-030-21920-8_37

management of uncertainty in expert systems was discussed in [3]. In 1980, a Fuzzy Logical Model of Computer-Assisted Medical Diagnosis proposed in [12]. Phuong et al. have shown that fuzzy logic is a useful mathematical tool for medical applications in [1]. We review some works applying the fuzzy logic in the field of modern, traditional Oriental which combines the Western and traditional Oriental medicine in diagnosis and treatment. In the work [9], foundations for traditional Oriental Medicine was proposed. On the other side, some methods of interval computation and neural networks are applied in diagnosis and treatment of Oriental traditional medicine as in [18] - Interval-Based Expert Systems and their use for Traditional Oriental Medicine. In general, in medical diagnosis, based on symptoms appearing in the patient, the conclusion of the disease will be given. Each symptom contributes for the conclusion of disease with its importance degree. For example, for diagnosis of Tuberculosis, the symptom coughing with blood is more importance than the symptom chest pain. In [10], we presented a reasoning model for the pulse based diagnosis. The aim of this paper, we propose an approach to pulse based reasoning in diagnosis including the importance of symptoms. The paper is organized as follows:

In Sect. 2, some concepts of pulse based diagnosis are reviewed. In Sect. 3, we propose methods of diagnosis based on pulse symbols by triple Cun - Guan - Chi including the importance of symptoms. In Sect. 4, examples of pulse based diagnosis using the proposed model are given. Finally, we discuss conclusions and further works.

2   Recall Some Notions of Pulse Diagnosis

2.1   Location and Types of Pulses

Let's recall some notions of pulse based diagnosis of traditional Vietnamese medicine. Along with inspecting (conducting a general observation of the patient), auscultation and olfaction (i.c., listening and smelling), and questioning (obtaining information about a patient's medical history and symptoms), pulse diagnosis is considered an essential part of the practice of traditional Oriental medicine [7, 11]. It has been practiced in China and Japan and Vietnam for centuries, and while it is difficult to master and considered some what subjective by physicians in the West, it remains an important diagnostic tool by both Traditional Vietnamese Medicine practitioners and patients.

Let's bring back some of his concepts of Pulse Based Diagnosis (PBD) from [6]. Pulses reflect a movement of vital energy Qi and blood. It means pulses are places where the blood is always concentrated and the blood always circulates with actuation of vital energy.

Therefore, pulse-based diagnosis is to understand the transformation of vital energy and blood in meridians, understand physiology, pathology of Zang Fu organs in the body.

Based on the above concepts, PBD is to consider a prosperousness - declination of Yang and Yin in the body, a deficiency - excess of vital energy Qi in order to diagnose exactly syndromes, selecting suitable meridians and acupuncture points and applying a suitable treatment.

PBD involves the use of three fingers to press the radical artery of the wrist, with the throbbing segment of the radical artery divided into three sections, namely, the distal section, middle section and proximal section. To take the pulse is to determine the pulse rate, force, wave and so on.

Location of the Pulse: The Guan (Second) Position is found opposite the styloid process of the radius, the Cun Position is found between the Guan Position and the wrist and the Chi position is found at a point equal the distance between Guan and Cun. Bellow are common pulse locations and related meridians in Table 1:

Table 1 The common pulse locations and related meridians

	Left wrist	Right wrist
Cun - 1st position	HT/SI	LU/LI
Guan - 2nd position	LV/GB	SP/ST
Chi - 3rd position	KD/UB	PC/TH

Where HT stands for Heart, SI stands for Small Intestine, LV stands for Liver, GB stands for Gallbladder, KD stands for Kidney, UB stands for Bladder. LU stands for Lung, LI stands for Large Intestine, SP stands for Spleen, ST stands for Stomach, PC stands for Pericardium and TH stand for Triple Burner.

Pulse Types

The followings are main 29 disease pulses which are in frequent use in clinical practice today [10]:

1. Superficial pulse (Phu)	16. Hollow pulse (Khau)
2. Deep pulse (Tram)	17. Tympanic pulse (Cach)
3. Slow pulse (Tri)	18. Firm pulse (Dong)
4. Rapid pulse (Sac)	19. Soft pulse (Nhu)
5. Slippery pulse (Hoat)	20. Weak pulse (Nhuoc)
6. Irregular pulse (Sap)	21. Scattered pulse (Tan)
7. Deficiency pulse (Hu)	22. Small pulse (Te)
8. Substantive pulse (Thuc)	23. Hidden pulse (Phuc)
9. Long pulse (Truong)	24. Throbbing pulse (Dong)
10. Short pulse (Doan)	25. Abrupt pulse (Suc)
11. Faint pulse (Vi)	26. Irregularly Intermittent pulse (Ket)
12. Full pulse (Hong)	27. Regularly Intermittent pulse (Doi)
13. Tense pulse (Khan)	28. Large pulse (Dai)
14. Slow pulse (Hoan)	29. Swift pulse (Tat)
15. Stringy pulse (Huyen)	

2.2   Disease Symptoms Diagnosis Based on Pulse Symbols in Triple Cun Guan - Chi

2.2.1   Pulse Diagnosis by Left Hand in Cun 1st Position

2.2.1.1 IF a patient has the following Pulse symbols: *Substantive pulse* (Thuc), *Stringy pulse* (Huyen), *Superficial pulse (Phu)* THEN one can get one of the Disease symptoms: Heat wind, Blood Heart, primary Allergy, arm pain, face paralytics, voice loss.
2.2.1.2 IF a patient has the following Pulse symbols: *Rapid pulse (Sac)*, *Slippery pulse* (Hoat) THEN one can get one of the Disease symptoms: Heat symptoms, heat Pericardium and heart, head pain, heart rhythm disorder, worry, voice loss.
2.2.1.3 IF a patient has the following Pulse symbols: *Deep pulse (Tram)*, *Tense pulse* (Khan) THEN one can get one of the Disease symptoms: Cold syndrome, interior syndrome, flutter.
2.2.1.4 IF a patient has the following Pulse symbols: *Deep pulse (**Tram**)*, Faint pulse (**Vi**), Weak pulse (**Nhuoc**)

THEN one can get one of the Disease symptoms: Deficiency syndromes, Insomnia, Speak in one's sleep.

2.2.2   Pulse Diagnosis by Left Hand in Guan 2nd Position

2.2.2.1 IF a patient has the following Pulse symbols: *Superficial pulse (**Phu**)*, *Substantive pulse (**Thuc**)*, *Stringy pulse* (Huyen) THEN one can get one of the Disease symptoms: Red eyes, loss of eyesight, headache.
2.2.2.2 IF a patient has the following Pulse symbols: *Rapid pulse (Sac)*, *Tense pulse* (Khan) THEN one can get one of the Disease symptoms: sinew hard, outer nerve, blood deficiency, Bad blood circulation, Pain of side of man's chest.
2.2.2.3. IF a patient has the following Pulse symbols: *Deep pulse (Tram)*, *Stringy pulse (Huyen)*, *Rapid pulse (Sac)* THEN one can get one of the Disease symptoms: stagnation, paralytics.
2.2.2.4 IF a patient has the following Pulse symbols: Faint pulse (Vi), Weak pulse (Nhuoc), Scattered pulse (Tan) THEN one can get one of the Disease symptoms: deficiency liver, loss of eyesight, blood deficiency.

One can see more about Pulse diagnosis by Left hand in Chi 3rd position, Pulse diagnosis by Right hand in Cun 1st position, Pulse diagnosis by Right hand in Guan 2nd position and Pulse diagnosis by Left hand in Chi 3rd position in [13].

3   Proposed Method of Diagnosis Based on Pulse Symbols by Triple Cun-Guan - Chi Including the Importance of Symptoms

Based on the concepts of pulse diagnosis based on pulse symbols in triple Cun - Guan - Chi by experience of Prof. Nguyen Tai Thu described above, we define some definitions bellows:

Definition 1: A fuzzy patient data for patient P_q for all pulses $PUS_i (i = 1, \ldots m)$ and diseases symptom $D_j (j = 1, \ldots, n)$ are a fuzzy degrees $\mu_{R_{PS}}(P_q, PUS_i)$. It takes the value in $[0, 1]$.

- $\mu_{R_{PS}}(P_q, PUS_i) = 1$ means symptom PUS_i surely present for patient P_q.
- $\mu_{R_{PS}}(P_q, PUS_i) = 0$ means symptom PUS_i surely absent for patient P_q.
- $0 < \mu_{R_{PS}}(P_q, PUS_i). < 1$ means symptom PUS_i present for patient P_q with some degree.

Definition 2: Let us have an elementary conjunctions $E - PUS_h$ of pulses PUS_i in form of

$$E - PUS_h = PUS_1 \&, \ldots, PUS_m$$

If for each i, i = 1, ..., m then we define the value of an elementary conjunction $E - PUS_h$ of pulses PUS_i by

$$\mu_{R_{PS}}(P_q, E - PUS_h) = \min_{PUS_i \in E - PUS_h}$$
$$(\mu_{R_{PS}}(P_q, PUS_1), \ldots, \mu_{R_{PS}}(P_q, PUS_m))$$

On the other side, to simplify the practical situation, we assume that conjunction of propositions is not included negated propositions.

Definition 3: The knowledge base including the degrees of importance of symptoms is defined as the following:

A rule base for diagnosis of disease symptoms given by $\mu_{R_{S-DIAG}}^c(E - PUS_h, D_j)$ consists of rules:

$$E - PUS_h \rightarrow DIAG_j(\mu_{R_{S-DIAG}}^c(E - PUS_h, D_j), \mu_{R_{SimpD}}^c(Simp_i, D_j))$$

More precisely:

(i) $\mu_{R_{S-DIAG}}^c(E - PUS_h, D_j) = 0$ means the elementary conjunction $E - PUS_h$ of pulses PUS_i excludes the diagnosis of possibility of disease symptom D_j.
(ii) $\mu_{R_{S-DIAG}}^c(E - PUS_h, D_j) = 1$ means the elementary conjunction $E - PUS_h$ of pulses confirms the diagnosis of possibility of disease symptom D_j.
(iii) $0 < \mu_{R_{S-DIAG}}^c(E - PUS_h, D_j) < 1$ means the elementary conjunction $E - PUS_h$ of pulses PUS_i confirms the diagnosis of possibility of disease symptom D_j with some fuzzy degree.

For the degree of importance of symptoms $\mu_{R_{SimpD}}^c(Simp_i, D_j)$ which indicates the importance of i-th symptom for confirmation of diagnosis D_j and they take the fuzzy value in $[0, 1]$.

- $\mu^c_{R_{SimpD}}(Simp_i, D_j) = 0$ means that the pulse PUS_i is not important vis-a-vis that disease D_j.
- $\mu^c_{R_{SimpD}}(Simp_i, D_j) = 1$ means that the pulse PUS_i is absolute important vis-a-vis that disease D_j.
- $0 < \mu^c_{R_{SimpD}}(Simp_i, D_j) < 1$ indicates the important degree of the pulse PUS_i vis-a-vis that disease D_j with some degree in [0, 1].

In practice, there are some methods to determine the weights $\mu^c_{R_{SimpD}}(Simp_i, D_j)$ such as (a) it is based on experience of medical experts, (b) applying a statistics method, c) Using a genetic algorithm to find the weight that minimize the error between the system's outputs and the samples in the case there is a good set of training samples [10].

Definition 4: Given a patient data, the degree for confirmation of the diagnosis of disease symptom D_j by patient P_q from observed pulses PUS_i is:

$$\mu^c_{R_{P-DIAG}}(P_i, D_j) =$$
$$Max_{E'_q} Min[\min[\mu_{R_{PS}}(P_q, E - PUS'_h), \mu^c_{R_{SimpD}}(Simp'_i, D_j)];$$
$$\mu^c_{R_{S-DIAG}}(E - PUS'_h, D_j)]$$

$E - PUS'_h$ and $Simp'_q$ varies over all elementary conjunctions of symptoms.
One can see the representation of the degree $\mu^c_{R_{P-DIAG}}(P_i, D_j)$ in [0, 1] in Graph 1.

$$0 \underline{\hspace{3cm}} 0.5 \underline{\hspace{3cm}} 1$$

$$\mu^c_{R_{P-DIAG}}(P_i, D_j)$$

Graph 1: Representation of $\mu^c_{R_{P-DIAG}}(P_i, D_j)$

Finally, the consultation results are the following:

1. The degree $\mu^c_{R_{P-DIAG}}(P_i, D_j) = 1$ means Absolutely Confirmation of the disease symptom D_j.
2. The degree $\mu^c_{R_{P-DIAG}}(P_i, D_j)$ such that $0.6 \leq \mu^c_{R_{P-DIAG}}(P_i, D_j) < 1$ means Almost Confirmation of the disease symptom D_j.
3. The degree $\mu^c_{R_{P-DIAG}}(P_i, D_j)$ such that $\varepsilon \leq \mu^c_{R_{P-DIAG}}(P_i, D_j) < 0.6$ means Possible Confirmation of the disease symptom D_j.
4. The degree $\mu^c_{R_{P-DIAG}}(P_i, D_j)$ such that $0 \leq \mu^c_{R_{P-DIAG}}(P_i, D_j) < \varepsilon$ means "unknown" of the disease symptom D_j.
5. The degree $\mu^c_{R_{P-DIAG}}(P_i, D_j) = 0$ means Absolutely Exclusion of the disease symptom D_j.

ε is the value which closes to 0 and can take $\varepsilon = 0.02$

4 Examples of Pulse Based Diagnosis Using the Importance of Symptoms

Pulse Diagnosis by Left Hand in Cun 1st Position

Given the rule base for disease symptom diagnosis based on pulses:

Rule 1: IF *Substantive pulse* (Thuc), *Stringy pulse* (Huyen), *Superficial pulse* (Phu) THEN possibility of heat wind symptom (weight = 1)

Rule 2: IF *Stringy pulse* (Huyen) THEN possibility of heat wind symptom (weight = 0.3)

Rule 3: IF *Substantive pulse* (Thuc) THEN possibility of heat wind symptom (weight = 0.4)

Rule 4: IF *Superficial pulse* (Phu) THEN possibility of heat wind symptom (weight = 0.2)

Rule 5: IF *Substantive pulse* (Thuc), *Stringy pulse* (Huyen) THEN possibility of blood heart symptom (weight = 0.7)

Rule 6: IF *Substantive pulse* (Thuc), *Superficial pulse* (Phu) THEN possibility of primary Allergy symptom (weight = 0.6)

In Case 1: All degrees of importance of symptoms are the same, i.e. they take the degree value which is 1.

Given Pulse symptoms observed by patient: $\mu_{R_{PS}}(P_q, \text{Substantive pulse}) = 0.8$

$$\mu_{R_{PS}}(P_q, \text{Superficial pulse}) = 0.6$$

All degrees of importance of symptoms are 1.

- $\mu^c_{R_{SimpD}}(\text{Substantive pulse, heat wind symptom}) = 1.$
- $\mu^c_{R_{SimpD}}(\text{Stringy pulse, heat wind symptom}) = 1.$
- $\mu^c_{R_{SimpD}}(\text{Superficial pulse, heat wind symptom}) = 1.$
- $\mu^c_{R_{SimpD}}(\text{Substantive pulse, primary Allergy}) = 1.$
- $\mu^c_{R_{SimpD}}(\text{Superficial pulse, primary Allergy}) = 1.$

Inference process:

1. According to definition 4, the Rule 3, Rule 4 and Rule 6 fired, then we calculate:
   The weight of the Rule 3:

- $\mu^c_{R_{P-DIAG}}(P_i, \text{heat wind}) = \min(\min[1, 0.8]; 0.4) = 0.4.$
   The weight of the Rule 4:
- $\mu^c_{R_{P-DIAG}}(P_i, \text{heat wind}) = \min(\min[1, 0.6]; 0.2) = 0.2.$
   The weight of heat wind symptom is calculated by the following:
- $\mu^c_{R_{P-DIAG}}(P_i, \text{heat wind}) = \max(0.4; 0.2) = 0.4.$
   The weight of the Rule 6:
- $\mu^c_{R_{P-DIAG}}(P_i, \text{Allergy}) = \min(\min[1, 0.8]; \min[1, 0.6]; 0.6) = 0.6.$

Conclusion: The weight of primary Allergy symptom is 0.6.

It means in the case that all degrees of importance of symptoms are the same and they take the degree value which is 1 and if the patient got the substantive pulse with degree of 0.8 and Superficial pulse with degree of 0.6 then the patient P_q has the weight of heat wind symptom 0.4 a possibility of Almost Confirmation of disease symptom of primary allergy symptom with their degree of belief of 0.6.

In Case 2: All degrees of importance of symptoms are not the same, i.e. they take the different degree values in [0, 1].

Given Pulse symptoms observed by patient: $\mu_{R_{PS}}(P_q, \text{Substantive pulse}) = 0.8$ $\mu_{R_{PS}}(P_q, \text{Superficial pulse}) = 0.6$

All degrees of importance of symptoms are 1.

- $\mu^c_{R_{SimpD}}(\text{Substantive pulse, heat wind symptom}) = 0.3.$
- $\mu^c_{R_{SimpD}}(\text{Stringy pulse, heat wind symptom}) = 0.5.$
- $\mu^c_{R_{SimpD}}(\text{Superficial pulse, heat wind symptom}) = 0.4.$
- $\mu^c_{R_{SimpD}}(\text{Substantive pulse, primary Allergy}) = 0.7.$
- $\mu^c_{R_{SimpD}}(\text{Superficial pulse, primary Allergy}) = 0.5.$

Inference process:

1. According to definition 4, the Rule 3, Rule 4 and Rule 6 fired, then we calculate: The weight of the Rule 3:

- $\mu^c_{R_{P-DIAG}}(P_i, \text{heat wind}) = \min(\min[0.3, 0.8]; 0.4) = 0.3.$
  The weight of the Rule 4:
- $\mu^c_{R_{P-DIAG}}(P_i, \text{heat wind}) = \min(\min[0.4, 0.6]; 0.2) = 0.2.$
  The weight of heat wind symptom is calculated by the following:
- $\mu^c_{R_{P-DIAG}}(P_i, \text{heat wind}) = \max(0.3; 0.2) = 0.3.$
  The weight of the Rule 6:
- $\mu^c_{R_{P-DIAG}}(P_i, \text{Allergy}) = \min(\min[0.7, 0.8]; \min[0.5, 0.6]; 0.6) = 0.6.$

Conclusion: The weight of primary Allergy symptom is 0.6.

It means in the case that all degrees of importance of symptoms are the same and they take the different degree values and if the patient got the substantive pulse with degree of 0.8 and Superficial pulse with degree of 0.6 then the patient P_q has the weight of heat wind symptom **0.3** a possibility of Almost Confirmation of disease symptom of primary allergy symptom with their degree of belief of **0.6**.

Discussions: In case 1 of the example above, given Pulse symptoms observed by patient as the followings: $\mu_{R_{PS}}(P_q, \text{Substantive pulse}) = 0.8 \ \mu_{R_{PS}}(P_q, \text{Superficial pulse}) = 0.6$.

And All degrees of importance of symptoms are the same, i.e. they take the degree value which is 1, then the patient P_q has the weight of heat wind symptom 0.4 a possibility of Almost Confirmation of disease symptom of primary allergy symptom with their degree of belief of 0.6.

In case 2, we included the importance of symptom in the rule base i.e. we have the degree of importance of symptom *Substantive pulse* $\mu^c_{R_{SimpD}}$ (Substantive pulse, heat wind symptom) = 0.3 means the importance of symptom *Substantive pulse* for "heat wind symptom" with possible confirmation and the degree of importance of symptom *Substantive pulse* $\mu^c_{R_{SimpD}}$ (Substantive pulse, primary Allergy) = 0.7 means the importance of symptom *Substantive pulse* for "primary Allergy symptom" with almost confirmation.

Similarly, we have the degree of importance of symptom *Superficial pulse* $\mu^c_{R_{SimpD}}$ (Superficial pulse, heat wind symptom) = 0.4 means the importance of symptom *Superficial pulse* for "heat wind symptom" with possible confirmation then the patient P_q has the weight of heat wind symptom 0.3 a possibility of Almost Confirmation of disease symptom of primary allergy symptom with their degree of belief of 0.

It is clear that if we include the importance of symptoms in the model of reasoning for a pulses Based Disease Symptoms Diagnosis of Traditional Vietnamese Medicine, the conclusions of disease symptoms are more distinguish between different conclusions of disease symptoms. On the other side, the model including the importance of symptoms is more close to thinking of the traditional medicine doctors in pulses Based Disease Symptoms Diagnosis.

Based on the above method in [10], the traditional medicine experts have developed 576 rules for diagnosis of 18 disease symptoms by Left hand in Cun 1st position, 112 rules for diagnosis of 20 disease symptoms by Left hand in Guan 2nd position, 73 rules for diagnosis of 15 disease symptoms by Left hand in Chi 3rd position, 35 rules for diagnosis of 7 disease symptoms by Right hand in Cun 1st position, 24 rules for diagnosis of 8 disease symptoms by Right hand in Guan 2nd position and 242 rules for diagnosis of 8 disease symptoms by Left hand in Chi 3rd position [18].

At present, we implement this proposed model of reasoning for pulse based diagnosis in the computer by C++ programming language and we evaluate the results by medical doctors of Traditional Vietnamese medicine in Vietnam.

Example of symptom input into the program: A patient has Substantive pulse (weight: 0.5), Stringy pulse (weight: 0.7),. Superficial pulse (weight: 0.8) at Cun left hand and Substantive pulse (weight: 0.5), Superficial pulse (weight: 0.7), Rapid pulse (weight: 0.8) at Cun right hand (Fig. 1).

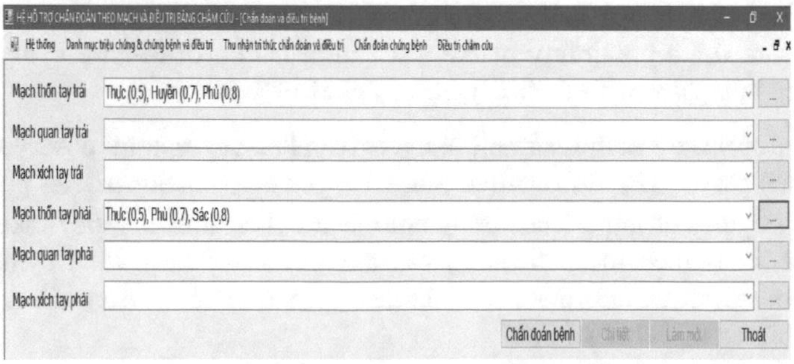

Fig. 1 Interface to input pulse symptoms

After clicking on the Diagnose button in the program, the system will compare the rules in the knowledge base and the program will calculate the weights of the diagnoses that are satisfied with the premise of the law and make a diagnosis of the highest weight disease syndrome. Users can view specific diagnostic processes by clicking on the Details button in the program. The first results shows that the diagnosis is suitable of the diagnosis results of medical doctors.

5 Conclusions

We have proposed a fuzzy model of reasoning including the importance of symptoms for a pulses Based Disease Symptoms Diagnosis of traditional Vietnamese medicine. Some examples provided to show how the model is worked. The results of example with including the importance of symptoms show that the conclusion of diagnosis of disease symptoms is more reasonable and more accuracy. In order to improve the program, we are currently working in two research directions. First, we keep maintaining and updating system's rule base. Second, we are trying to improve the system by including negative rules.

References

1. Phuong, N.H., Kreinovich, V.: Fuzzy logic and its applications in medicine. Int. J. Med. Inf. **62**, 165–173 (2001)
2. Zadeh, L.A.: Fuzzy sets. Inf. Control **8**, 338–353 (1965)
3. Zadeh, L.A.: The role of fuzzy logic in the management of uncertainty in expert systems. Fuzzy Sets Syst. **11**, 199 (1983)
4. Van Ky, T.: Handbook of Diagnosis and Treatment: Internal Traditional Medicine. DaNang Publisher (2015). (in Vietnamese)
5. Lu, H.C.: Chinese Natural Cures: Traditional Methods for Remedies and Preventions. Black Dog & Leventhal Publishers, Inc. (1986)
6. Nguyen, T.T.: Applying theory of Oriental Medicine in Treatment research by Acupuncture. Vietnam Acupuncture Institute (1990). (in Vietnamese)
7. Phuong, N.H.: Design of a fuzzy system for diagnosis and treatment of integrated western and eastern medicine. Int. J. Gen Syst **30**(2), 219–239 (2001)
8. Phuong, N.H., Starks, S.A., Kreinovich, V.: Interval-based expert systems and their use for traditional oriental medicine. In: Phuong, N.H., Ohsato, A. (eds.) Proceedings of Vietnam-Japan Bilateral Symposium on Fuzzy Systems and Applications, VJFUZZY 1998, Halong Bay, Vietnam, 30 September–2 October 1998, pp. 697–703 (1998)
9. Phuong, N.H., Starks, S.A., Kreinovivh, V.: Towards foundations for traditional oriental medicine. In: Phuong, N.H., Ohsato, A. (eds.) Proceedings of Vietnam-Japan Bilateral Symposium on Fuzzy Systems and Applications, VJFUZZY 1998, Halong Bay, Vietnam, 30 September–2 October 1998, pp. 704–708 (1998)

10. Phuong, N.H., Nguyen, A., Thuy, T.T.H.: A fuzzy model of diagnosis and treatment by acupuncture based on pulse symbols by triple Cun - Quan - Chi of traditional oriental medicine. In: Proceedings of 2018 IEEE International Conference on Fuzzy Systems (FUZZ-IEEE), 8–13 July 2018, Rio de Janeiro, Brasil, pp. 647–654 (2018)
11. Shu, J.-J., Sun, Y.: Developing classification indices for Chinese pulse diagnosis. Complement. Ther. Med. **15**(3), 190–198 (2007)
12. Adlassnig, K.-P.: A fuzzy logical model of computer-assisted medical diagnosis. Methods Inf. Med. **19**, 141–148 (1980)

Fuzzy Transform in Time Series Decomposition

Linh Nguyen$^{(\boxtimes)}$ and Vilém Novák

Institute for Research and Applications of Fuzzy Modelling, NSC IT4Innovations,
University of Ostrava, 30. dubna 22, 701 03 Ostrava 1, Czech Republic
{Linh.Nguyen,Vilem.Novak}@osu.cz

Abstract. In this paper, we provide a method for applying the fuzzy transform of higher degree to time series decomposition. We assume that a time series can be decomposed into a trend-cycle, a seasonal component and an irregular fluctuation, we devote theoretical justifications for decomposing it into an additive model. Several examples are consider to demonstrate our methodology.

1 Introduction

The fuzzy transform (F-transform) was introduced by Perfilieva in [17] with the purpose to include fuzzy models into the approximation theory. It was then generalized by means of polynomials to the so-called fuzzy transform of higher degree (F^m-transform, $m \in \mathbb{N}$) in [18]. The F^m-transform consists of two phases: *direct* and *inverse*. The direct F^m-transform maps a function from the L^2 space to a family of polynomials up to degree m, called the direct F^m-transform *components*. They are determined with respect to basic functions of a fuzzy partition. Note that the F^m-transform components are the best approximations of the original function with respect to weights determined by basic functions. The inverse F^m-transform is defined as the weighted average of the F^m-transform components. Recently, a new representation of the direct F^m-transform components, based on monomials, was introduced in [7,9]. This makes the fuzzy transform of higher degree simple to use, and therefore motivates for its applications, especially, to time series analysis [6,8,11].

Time series decomposition is one of the major tasks in time series analysis. It is a significant approach to model a time series with respect to its patterns (e.g., trend, seasonality, etc) that are understandable. The decomposition provides valuable knowledge for practical purposes, for instance, to forecast a time series or to mine its essential information [12,13,15,16,19]. Under the assumption that a time series can be characterized by a trend-cycle, a seasonal component and an irregular fluctuation, a decomposition technique provides a methodology to model it according to a sum or a product of these components. The main issue is to estimate the trend-cycle and the seasonal component. There are two main approaches for the time series decomposition consisting of model-based approach (e.g., XII-ARIMA model, state space model, etc.) and nonparametric approach

© Springer Nature Switzerland AG 2019
R. B. Kearfott et al. (Eds.): IFSA 2019/NAFIPS 2019, AISC 1000, pp. 417–428, 2019.
https://doi.org/10.1007/978-3-030-21920-8_38

(e.g., Seasonal Trend decomposition based on Loess (SLT), Singular Spectrum Analysis (SSA), etc.). For more details about these methods, we refer to [1,3,4, 19].

The F-transform has been applied to several issues in time series analysis (see [12–14, 16] and elsewhere). In these investigations, the F-transform was mainly used for the trend-cycle (nonparametric) estimation. Recently, together with the mentioned theoretical developments, the F^m-transform has been used as a more general technique to the estimation of the trend-cycle [6]. The theoretical background of this is based on the assumption that the trend-cycle slightly changes its course. This assumption narrows the class of time series that the F^m-transform can be applied for estimating of the trend-cycle. To make it around, in this paper we consider another assumption based on the smoothness of the latter. Additionally, let us note that a successful estimation of the trend-cycle requires a significant suppression or better elimination of high frequencies (the seasonal component) and the irregular fluctuation. In the use of the F^m-transform, these requirements were theoretically investigated in [6,8,14]. An interesting question raises whether the fuzzy transform of higher degree can be, at the same time, applied to estimation of the seasonal component. The answer is positive. In this paper, we will present a procedure for estimation of the seasonal component using fuzzy transform technique. The technique is a combination of the decomposing of seasonal data (the data obtained after the subtraction of the estimated trend-cycle) into sub-series based on its period and the applying to them the fuzzy transform. The idea is based on the technique proposed in [19]. This paper is a continuation of the research in [14] about the application of the F-transform to trend-cycle estimation, and moreover, contributes mathematical backgrounds to the recently published paper [11] about forecasting of time series using new fuzzy techniques.

The paper is organized as follows. Section 2 is devoted to the basic concepts: the F^m-transform and the time series models. The main contribution is discussed in Sect. 3. The fourth section is devoted to the illustration of our proposal, and the last section is left for conclusions.

2   Preliminaries

By $\mathbb{N}$, $\mathbb{Z}$ and $\mathbb{R}$ we denote the set of natural numbers ($\mathbb{N} = \{0, 1, \ldots\}$), integers and real numbers, respectively.

2.1   Higher Degree Fuzzy Transform

The fundamental concept of the theory of fuzzy transform is that of *fuzzy partition*. In this paper, we restrict our analysis to a very simple type of fuzzy partitions, determined with respect to a generating function.

Definition 1. A real-valued function $K : \mathbb{R} \rightarrow [0, 1]$ is said to be a *generating function* if it is continuous, even, non-increasing in $[0, \infty]$ and vanishing outside of $(-1, 1)$, i.e., $K(x) > 0$ iff $x \in (-1, 1)$.

Two generating functions, frequently used in applications of the fuzzy transform, are in the following example.

Example 1. The functions K^{tr}, $K^{rc} : \mathbb{R} \to [0,1]$ defined by

$$K^{tr}(t) = \max(1 - |t|, 0) \tag{1}$$

$$K^{rc}(t) = \begin{cases} \frac{1}{2}(1 + \cos(\pi t)), & -1 \le t \le 1; \\ 0, & \text{otherwise,} \end{cases} \tag{2}$$

for any $t \in \mathbb{R}$, are called the *triangle* and *raised cosine* generating functions, respectively.

Definition 2. Let K be a generating function, h and r be positive real constants, and let $t_0 \in \mathbb{R}$. The set $\mathbf{A} = \{A_z \mid z \in \mathbb{Z}\}$ where

$$A_z(t) = K\left(\frac{t - t_0 - zr}{h}\right)$$

is called *a simple uniform fuzzy partition of the real line determined by the quadruplet (K, h, r, t_0)* if to any $t \in \mathbb{R}$, there exists $z \in \mathbb{Z}$ such that $A_z(t) > 0$. The parameters h, r and t_0 are called the *bandwidth, shift* and the *central node,* respectively. Let $t_z = t_0 + zr$, $z \in \mathbb{Z}$. We call t_z and A_z the *z-th node* and the *z-th basic function.*

Since the central node t_0 has little influence in applications of the fuzzy transform to time series analysis, we will consider only a simple uniform fuzzy partitions with $t_0 = 0$. A fuzzy partition, from now on, is determined by the triplet (K, h, r) of a generating function, a bandwidth and a shift, respectively.

Let $L^2_{loc}(\mathbb{R})$ be a set of all functions that are square integrable on any closed sub-interval of the real line. Below, we recall the representation of the direct and inverse phases of the higher degree fuzzy transform [7,9].

Definition 3. Let $f \in L^2_{loc}(\mathbb{R})$, $m \in \mathbb{N}$, and $\mathbf{A}$ be a simple uniform fuzzy partition of the real line determined by the triplet (K, h, r). The *direct fuzzy transform of degree m (the direct F^m-transform) of f with respect to $\mathbf{A}$* is the set

$$\mathbf{F}^m[f|\mathbf{A}] = \{F^m_z[f] \mid z \in \mathbb{Z}\}$$

where, for any $z \in \mathbb{Z}$,

$$F^m_z[f](t) = C_{z,0} + C_{z,1}(t - t_z) + \ldots + C_{z,m}(t - t_z)^m, \quad t \in [t_z - h, t_z + h] \tag{3}$$

determined by

$$(C_{z,0}, C_{z,1}, \ldots, C_{z,m})^T = (\mathscr{H}_m)^{-1} \cdot (\mathscr{L}_m)^{-1} \cdot \mathscr{Y}_{m,z}$$

with $\mathscr{H}_m = \text{diag}(1, h^1, \ldots, h^m)$, $\mathscr{L}_m = (Z_{ij})$ is an $(m+1) \times (m+1)$ invertible matrix defined by

$$Z_{ij} = \int_{-1}^{1} \tau^{i+j-2} K(\tau) d\tau, \quad i, j = 1, \ldots, m+1,$$

and $\mathcal{Y}_{m,z} = (Y_{z,1}, \ldots, Y_{z,m+1})^T$ defined by

$$Y_{z,\ell} = \int_{-1}^{1} f(h\tau + t_z) \cdot \tau^{\ell-1} K(\tau) d\tau, \quad \ell = 1, \ldots, m+1.$$

The polynomial $F_z^m[f]$ is called the *z-th component of the direct F^m-transform* $\mathrm{F}^m[f|\mathbf{A}]$.

Definition 4. Let $f \in L^2_{loc}(\mathbb{R})$ and $\mathbf{A} = \{A_z \mid z \in \mathbb{Z}\}$ be a simple uniform fuzzy partition of the real line. Let $\mathrm{F}^m[f|\mathbf{A}]$ be the direct F^m-transform of f with respect to $\mathbf{A}$. The *inverse fuzzy transform of degree m (inverse F^m-transform)* of f with respect to $\mathrm{F}^m[f|\mathbf{A}]$ is defined by

$$\hat{f}_{\mathbf{A}}^m(t) = \frac{\sum_{z\in\mathbb{Z}} F_z^m[f](t) \cdot A_z(t)}{\sum_{z\in\mathbb{Z}} A_z(t)}, \quad t \in \mathbb{R}. \tag{4}$$

Besides the notation $\hat{f}_{\mathbf{A}}^m$, we also write $\hat{f}_h^m$ to denote the inverse F^m-transform of a function f if the bandwidth of the used fuzzy partition is the main parameter.

2.2   Time Series Modeling

Let $(\Omega, \mathscr{F}, P)$ be a probability space. By a time series, we understand a complex-valued stochastic process

$$X : \mathbb{T} \times \Omega \to \mathbb{C}$$

where $\mathbb{T} \subseteq \mathbb{R}$. Quite often, we consider $\mathbb{T} = \{0, \ldots, N\} \subset \mathbb{N}$ being a finite set of numbers interpreted as time moments (cf. [2,5]). We also often consider only real part of $\mathbb{C}$ and speak about *real-valued stochastic process*. Our fundamental assumption is that the time series can be decomposed into four components:

$$X(t,\omega) = Tr(t) + C(t) + S(t) + R(t,\omega), \quad t \in \mathbb{T}, \omega \in \Omega \tag{5}$$

where Tr is a *trend*, C is a *cyclic* component, S is a *seasonal* component and R is a random *noise*. Note that Tr, C, S are ordinary functions not having stochastic character. The model (5) is often simplified by considering a *trend-cycle* component TC which means that it includes two components from (5):

$$TC(t) = Tr(t) + C(t).$$

The seasonality S characterizes influences of seasonal factors (e.g., the quarter of the year, the month, or day of the week) to a time series. These effects are repeated and of a fixed and known period. Consequently, this component is usually modeled by a (sum of) periodic function $S(t)$ with a known period. Finally, the irregular component reflects irregular properties exhibiting in a time series.

The model (5) is called *additive decomposition model*, is usually applied to time series that the magnitudes of their seasonal fluctuations do not clearly vary during the time domain. The other way is called *multiplicative decomposition model* at which $X(t, \omega)$ is modeled as a product of its components as follows:

$$X(t, \omega) = TC(t) \cdot S(t) \cdot R(t, \omega).$$

This model is suitable when the magnitudes of the seasonality factors vary according to time moments. However, it can be readily transformed into an additive one by applying the logarithmic transformation. If we fix $\omega \in \Omega$, we obtain one *realization* of the time series (5) which is an ordinary function $X : \mathbb{T} \to \mathbb{R}$. Hence, in practice we can consider a time series to be a set of values

$$\{X(t) \mid t \in \mathbb{T}\}.$$

3   Decomposition Using the Higher Degree Fuzzy Transform

3.1   Estimation of the Trend-Cycle

Let X be a time series (5). In this section, we provide a method enabling us to estimate components of X using the fuzzy transform of higher degree.

A successful estimation of the trend-cycle TC requires a significant suppression or even elimination of the seasonal component S and the irregular fluctuation R. There were several papers about this issue, e.g., [6,8,10,14] which provide theoretical justification showing that the seasonal and irregular components can be successfully suppressed by applying the fuzzy transform. More precisely, for any $m \geq 0$, we obtain that

$$\lim_{h \to \infty} \hat{S}_h^m(t) = 0, \quad \text{uniformly on } \mathbb{R}, \tag{6}$$

and

$$\lim_{h \to \infty} \hat{R}_h^m(t) = 0 \quad \text{for any } t \in \mathbb{T} \tag{7}$$

with high probability. In [6], we proved that if the modulus of continuity $\omega(TC, \delta)$ is small for a large positive number δ (i.e., the trend-cycle $TC(t)$ does not change abruptly its course), it can be successfully estimated by using higher-degree fuzzy transform.

Theorem 1. *Let X be a time series (5), $\hat{X}_h^m$, $\hat{R}_h^m$ and $\hat{R}_h^m$, $m \geq 0$, be the inverse F^m-transforms of X, S and R, respectively, with respect to a simple uniform fuzzy partition $\mathbf{A}$ determined by a triplet (K, h, r). Then, for any $t \in \mathbb{R}$, it holds that*

$$\left| \hat{X}_h^m(t) - TC(t) \right| \leq C_m \cdot \omega(TC, 2h) + \left| \hat{S}_h^m(t) \right| + \left| \hat{R}_h^m(t) \right|, \tag{8}$$

where C_m is a positive constant depending on m.

The following theorem shows that the higher-degree fuzzy transform is a suitable tool for estimation of the trend cycle, provided that it is a smooth function.

Theorem 2. *Let X be a time series (5), $\hat{X}_h^m$, $\hat{R}_h^m$ and $\hat{R}_h^m$, $m \geq 0$, be the inverse F^m-transforms of X, S and R, respectively, with respect to a simple uniform fuzzy partition $\mathbf{A}$ determined by a triplet (K, h, r). Suppose that the trend-cycle TC is $(m+1)$-times differentiable on $\mathbb{R}$ so that $\left| TC^{(m+1)}(t) \right| \leq \zeta$, $\zeta > 0$, $t \in [t_z - h, t_z + h]$ and $z \in \mathbb{Z}$, where t_z is the z-th node of $\mathbf{A}$. Then, for any $t \in \mathbb{R}$, it holds that*

$$\left| \hat{X}_h^m(t) - TC(t) \right| \leq \frac{D_m \zeta \cdot h^{m+1}}{(m+1)!} + \left| \hat{S}_h^m(t) \right| + \left| \hat{R}_h^m(t) \right|, \tag{9}$$

where D_m is a positive constant depending on m.

Sketch of the Proof: Applying the Taylor's Theorem to the trend-cycle function TC and using the fact that the F^m-transform preserves polynomials of a degree up to m, we obtain that

$$\left| TC(t) - \widehat{TC}_h^m(t) \right| \leq \frac{D_m \zeta \cdot h^{m+1}}{(m+1)!}, \quad \text{for any } t \in \mathbb{R}, \tag{10}$$

where $\widehat{TC}_h^m$ is the inverse F^m-transform of TC with respect to $\mathbf{A}$. From the additive model of X, one can prove that

$$\left| \hat{X}_h^m(t) - TC(t) \right| \leq \left| \widehat{TC}_h^m(t) - TC(t) \right| + \left| \hat{S}_h^m(t) \right| + \left| \hat{R}_h^m(t) \right|, \quad \text{for any } t \in \mathbb{R}.$$

From this, using (10), we obtain the desired result.

From (6) and (7), one can see that $\left| \hat{S}_h^m(t) \right|$ and $\left| \hat{R}_h^m(t) \right|$ are decreasing with the increase of the bandwidth h. At the same time, if the assumptions of Theorem 1 and Theorem 2 are fulfilled then conclude that the trend-cycle can be well estimated by the inverse F^m-transform of X.

3.2   Estimation of the Seasonal Component

In this subsection, we assume that $\mathbb{T} = \{0, 1, \ldots, N\}$. Since the trend-cycle can be estimated using the higher-degree fuzzy transform, we can replace X by the time series

$$Y(t) = X(t) - \hat{X}_h^m(t), \quad t \in \mathbb{T}.$$

The time series Y can be decomposed into the seasonal component S and the irregular fluctuation R

$$Y(t) = S(t) + R(t), \quad t \in \mathbb{T}. \tag{11}$$

Below, we propose a method how the seasonal component S can be estimated. For simplicity, we assume that S consists of one periodic function only.

Assume that p is the periodicity of S. At first, we determine p time series Y_j, $j = 0, \ldots, p - 1$ as follows:

$$Y_j(t) = Y(j + pt), \quad t = 0, 1, \ldots, N_j$$

where $N_j = \lfloor \frac{N-j}{p} \rfloor$[1] and call them *sub-series* of Y. It is easy to see that the j-th sub-series Y_j can be decomposed into the additive form

$$Y_j(t) = S_j(t) + R_j(t), \quad t = 0, 1, \ldots, N_j$$

where $S_j(t) = S(j + pt)$ and $R_j(t) = R(j + pt)$. Since the periodicity of the seasonal component S is p, the term S_j is either a constant, or changes its course only slightly. Indeed, in practice the amplitude of the seasonality usually changes according to time moments, i.e., influences of the seasonal factors are changeable. However, such fluctuations are slight. As a result, S_j can be considered as a trend-cycle of the j-th sub-series Y_j. Based on the results of the previous subsection, it can be well estimated by applying the fuzzy transform (F^0-transform[2]) to the sub-series Y_j, i.e., S_j is estimated by $\widehat{Y_j}$ for $j = 0, \ldots, p - 1$. Since the defined relation of S and S_j, $j = 0, \ldots, p - 1$, i.e.,

$$S_j(t) = S(j + pt), \quad t = 0, 1, \ldots, N_j,$$

S can be reconstructed from $\{ S_j \mid j = 0, \ldots, p - 1 \}$ by the following formula:

$$S(t) - S_j(kp), \quad t \in \mathbb{T},$$

where $j = \mathrm{mod}(t, p)$ and $k = \frac{t-j}{p}$. Consequently, the seasonal component S can be estimated on the basis of $\{ \widehat{Y_j} \mid j = 0, \ldots, p - 1 \}$. Namely, S is estimated by $\tilde{S}$ where

$$\tilde{S}(t) = \widehat{Y_j}(kp), \quad t \in \mathbb{T},$$

where $j = \mathrm{mod}(t, p)$ and $k = \frac{t-j}{p}$.

4   Illustrations

This section is devoted to illustrative examples of the theoretical justifications considered in Sect. 3. Namely, we apply the higher-degree fuzzy transform to estimation of the trend-cycle and the seasonal component (the irregular fluctuation is the remainder of a time series after extraction of the estimated trend-cycle and seasonal component). Both artificial and real time series are chosen for this issue. To show how good is the proposed method, we compare it with the well-known STL method (Seasonal-Trend decomposition procedure based on LOESS).

For the sake of simplicity, we restrict our computation to the use of triangular simple uniform fuzzy partitions whose the shift is a half of the corresponding bandwidth, i.e., $r = h/2$.

---

[1] The largest integer less than or equal to $\frac{N-j}{p}$.

[2] By smoothness of S_j, we do not need to use higher-degree fuzzy transform.

Example 2. Let the time series $X(t)$ be generated by the following formula

$$X(t) = 0.008(t - 30)^2 + 50\sin\left(\frac{\pi}{6}t\right) - 30\sin\left(\frac{\pi}{3}t\right) + R(t)$$

on the set of integers $\mathbb{T} = \{0, 1, \ldots 239\}$, where $R(t)$ is a realization of the autoregressive stochastic process ξ determined by

$$\xi(t) = 0.5\xi(t - 1) + 0.3\xi(t - 2) - 0.7\xi(t - 3) + 0.2\xi(t - 4) + \varepsilon(t)$$

where $\varepsilon(t) \sim WN(0, 16)$. Its trend-cycle, seasonal component and irregular fluctuation are modeled by $0.008(t - 30)^2$, $50\sin\left(\frac{\pi}{6}t\right) - 30\sin\left(\frac{\pi}{3}t\right)$ and $R(t)$, respectively. It is easy to see that the period of the seasonality is $p = 12$. Figure 1 displays decomposition results where the trend-cycle is estimated by the inverse F^2-transform of X with respect to the fuzzy partition determined with the bandwidth $h = 20$, and the seasonal component is estimated by the technique in Subsect. 3.2 with the period $p = 12$ and the F^0-transform computed with the fuzzy partition determined with the bandwidth $h = 12$. A detailed comparison of the proposed technique with the STL method by mean of RMSE is provided in Table 1.

Table 1. The estimation errors

Method\Components	Trend-cycle	Seasonal component	Irregular component
Fuzzy transform	5.0596	7.1928	8.9101
STL	5.7248	7.2278	9.2991

From the previous example, we can see that a time series can be successfully decomposed into an additive decomposition model by using the fuzzy transform technique. The method can be definitely comparable with the well-known STL method and, even in this example, it defeats the STL method.

Below, we consider another example where we apply the proposed technique to a real time series.

Example 3. We choose the quarterly time series

- Quarterly gross fixed capital expenditure—private non-dwelling construction, Australia: millions of dollars, 1984/85 prices. (Q3 1960 – Q2 1988).

This time series has a clear trend-cycle and seasonal component where the period of the seasonality is $p = 4$. Nevertheless, these components can not be exactly known. We use the F^3-transform with respect to the fuzzy partition determined with the bandwidth $h = 6$ for estimating the trend-cycle. At the same time, we apply the procedure introduced in Subsect. 3.2 to the estimation of the latter component. In this issue, we use the F^0-transform with respect to the fuzzy partition determined with the bandwidth $h = 4$. The obtained results are depicted

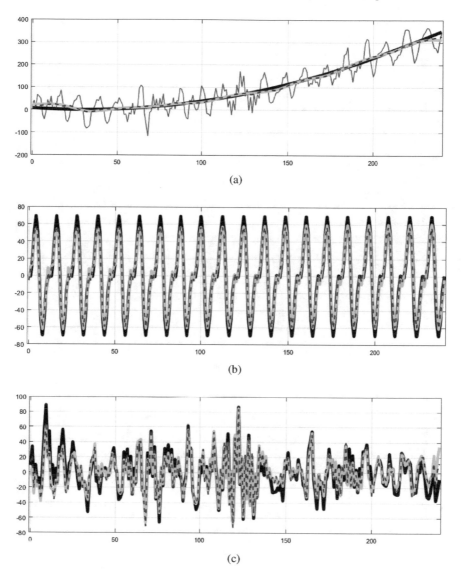

Fig. 1. Estimations of the trend-cycle (a), seasonal component (b), and the remainder (c) using the fuzzy transform (thick-gray lines) and using the STL method (dashed lines). The given components are the thick-dark lines.

in Fig. 2 together with the results from the STL method. In this case, we can not evaluate the estimation error because of the unknown of the exact components. Therefore, we leave to the reader the task of assessing the quality of the used methods.

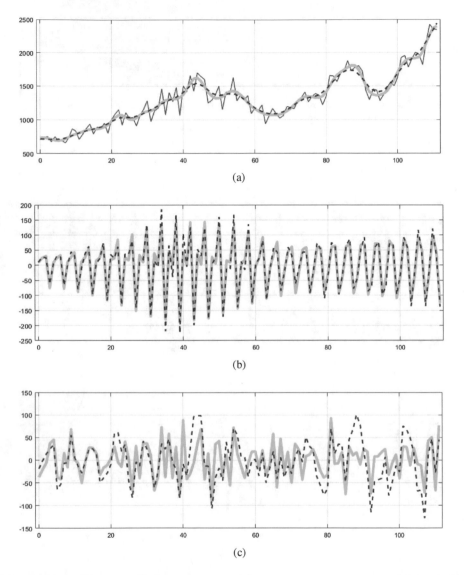

Fig. 2. The estimations of the trend-cycle (a), the seasonal component (b), and the remainder (c) using fuzzy transform method (thick-gray lines) and using STL method (dashed lines).

From the previous two examples, one can see that the proposed method can be in all respects compared with the STL method and, as in the former example, it seems to be better than the STL one. Note, that even better results could be attained by the fuzzy transform technique after a deeper analysis of the optimal setting of the bandwidth. This task, however, has been left to our future research. It is important to emphasize that the computational complexity of the fuzzy transform is much lower than that of the LOESS used in the STL method (see [7]). Additionally, this paper provides theoretical background to the novel forecasting method introduced in [11] where its efficiency was confirmed.

5   Conclusion

In this paper, we provided a methodology for time series decomposition using the higher-degree fuzzy transform. Namely, under the assumptions that a time series can be additively decomposed into the trend-cycle, seasonal component and irregular fluctuation, we have shown that these components can be well estimated using the higher degree fuzzy transform. The proposed method was demonstrated both on artificial as well as real time series. The experimental results were compared with the results obtained using the well-known STL method. The results are fully comparable (the difference between the results of both methods are almost invisible) but the F-transform has much lower computational complexity. Hence, we are convinced that the fuzzy transform is a promising technique for time series analysis.

Acknowledgements. This work is supported by the project GA ČR No. 18-13951S.

References

1. Alexandrov, T., Bianconcini, S., Dagum, E.B., Maass, P., McElroy, T.S.: A review of some modern approaches to the problem of trend extraction. Econom. Rev. **31**(6), 593–624 (2012)
2. Anděl, J.: Statistical Analysis of Time Series. SNTL, Praha (1976). (in Czech)
3. Dagum, E.B.: The X-II-ARIMA seasonal adjustment method. Catalogue No. 12-564E (1980)
4. Godolphin, E., Triantafyllopoulos, K.: Decomposition of time series models in state space form. Comput. Stat. Data Anal. **50**, 2232–2246 (2006)
5. Hamilton, J.D.: Time Series Analysis. Princeton University Press, Princeton (1994)
6. Holčapek, M., Nguyen, L.: Trend-cycle estimation using fuzzy transform of higher degree. Iran. J. Fuzzy Syst. **15**(7), 23–54 (2018)
7. Holčapek, M., Nguyen, L., Tichý, T.: Polynomial alias higher degree fuzzy transform of complex-valued functions. Fuzzy Sets Syst. **342**, 1–31 (2017)
8. Holčapek, M., Nguyen, L.: Suppression of high frequencies in time series using fuzzy transform of higher degree. In: 16th International Conference on Information Processing and Management of Uncertainty in Knowledge-Based Systems, IPMU 2016, Eindhoven, The Netherlands, pp. 705–716 (2016)
9. Nguyen, L., Holčapek, M., Novák, V.: Multivariate fuzzy transform of complex-valued functions determined by monomial basis. Soft Comput. **21**(13), 3641–3658 (2017)
10. Nguyen, L., Holčapek, M.: Higher degree fuzzy transform: application to stationary processes and noise reduction. In: Advances in Fuzzy Logic and Technology 2017, EUSFLAT 2017, pp. 1–12. Springer, Warsaw (2017)
11. Nguyen, L., Novák, V.: Forecasting seasonal time series based on fuzzy techniques. Fuzzy Sets Syst. (2018). https://doi.org/10.1016/j.fss.2018.09.010

12. Novák, V.: Linguistic characterization of time series. Fuzzy Sets Syst. **285**, 52–72 (2016)
13. Novák, V.: Mining information from time series in the form of sentences of natural language. Int. J. Approximate Reasoning **78**, 192–209 (2016)
14. Novák, V., Perfilieva, I., Holčapek, M., Kreinovich, V.: Filtering out high frequencies in time series using f-transform. Inf. Sci. **274**, 192–209 (2014)
15. Novák, V., Perfilieva, I., Romanov, A., Yarushkina, N.: Time series grouping and trend forecast using f^1-transform and fuzzy natural logic. In: Marco se Moraes, R., Kerre, E.E., dos Santos Machado, L., Lu, J. (eds.) Decision Making and Soft Computing, pp. 143–148. World Scientific (2014)
16. Novák, V., Štěpnička, M., Dvořák, A., Perfilieva, I., Pavliska, V., Vavříčková, L.: Analysis of seasonal time series using fuzzy approach. Int. J. Gen. Syst. **39**, 305–328 (2010)
17. Perfilieva, I.: Fuzzy transforms: theory and applications. Fuzzy Sets Syst. **157**, 993–1023 (2006)
18. Perfilieva, I., Daňková, M., Bede, B.: Towards a higher degree f-transform. Fuzzy Sets Syst. **180**, 3–19 (2011)
19. Theodosiou, M.: Forecasting monthly and quarterly time series using STL decomposition. Int. J. Forecast. **27**, 1178–1195 (2011)

A Formal Model of the Intermediate Quantifiers "A Few", "Several" and "A Little"

Vilém Novák and Petra Murinová[✉]

Institute for Research and Applications of Fuzzy Modeling, NSC IT4Innovations,
University of Ostrava, 30. dubna 22, 701 03 Ostrava 1, Czech Republic
{Vilem.Novak,Petra.Murinova}@osu.cz

Abstract. In this paper, we extend the model of intermediate quantifiers by three new ones, namely "a few, a little" and "several". We proved some of the fundamental properties of these quantifiers and relations to the other ones. We also demonstrate that they naturally fall in the generalized square of opposition.

1 Introduction

In our previous papers (see, e.g., [4,7] and others), we developed a theory of intermediate quantifiers and focused on several selected ones, namely *all, almost all, most, many* and *some*. The quantifiers *all* and *some* are classical. Nontrivial intermediate ones are *almost all, most* and *many*. All of them characterize big part of the universe of discourse. There are, however, also intermediate quantifiers whose meaning corresponds to the opposite, i.e., small part of the universe. Typical examples of them are *several, (a) few* and *(a) little*. The first two quantifiers can be joined with a countable noun while the latter with an uncountable one. There is a difference between *a few* and *few* (*a little* and *little*) in wider context when the former is a simple quantifier characterizing a certain number or amount corresponding to *some*, while the latter carry also a negative evaluation expressing non-satisfaction with a given amount. Because we do not have (so far) formal means to express such evaluation, we will consider only the first cases.

According to Cambridge or Merriam Webster dictionaries, the meaning of *a few or several* corresponds to more than 1 or 2 and is *less than many*. Recall that quantifiers characterize certain sizes of (fuzzy) sets. In our theory, we model size using the concept of measure. Because, countability or uncountability is hidden in it, there is no formal difference between *a few* and *a little*. By Longman thesaurus, we understand *several* more than *a few* to not a large number. In general, *a few* could mean two or three while *several* three or four. But this heavily depends on the number of elements of the universe; cf. "a few of well dressed women people at a party attended by seven people" with "several books are detective stories (on a shelf with 9 books)".

© Springer Nature Switzerland AG 2019
R. B. Kearfott et al. (Eds.): IFSA 2019/NAFIPS 2019, AISC 1000, pp. 429–441, 2019.
https://doi.org/10.1007/978-3-030-21920-8_39

It is interesting that *some* is more specific: it always means at least one, but maybe all which cannot hold with "few, several, or many". We will touch this relation at the end of this paper when presenting the generalized square of opposition. Recall that the first version of the square of opposition generalizing Aristotle's one with "Few", "Many" and "Most" was introduced by Peterson in [10] in 1979. A model of the Peterson's square in mathematical fuzzy logic was introduced in [4]. The quantifier "few" was from the point of the theory of syllogistic reasoning analyzed in [11].

2   Preliminaries

2.1   Fuzzy Type Theory

The theory of intermediate quantifiers has been developed in higher-order fuzzy logic (fuzzy type theory; Ł-FTT) with Łukasiewicz MV-algebra of truth values. The basic syntactical objects of Ł-FTT are classical, namely the concepts of *type* and *formula* (see [1]). The atomic types are ε (elements) and o (truth values). General types are defined as follows: if α, β are types then $(\beta\alpha)$ is a type. We denote types by Greek letters and the set of all types by *Types*. The set of all formula of a type α is denoted by *Types*$_\alpha$.

The *language* of Ł-FTT denoted by J, consists of variables $x_\alpha, \ldots$, special constants $c_\alpha, \ldots$ ($\alpha \in$ *Types*), the symbol λ, and brackets. We will consider the following concrete special constants: $\mathbf{E}_{(o\alpha)\alpha}$ (fuzzy equality) for every $\alpha \in$ *Types*, $\mathbf{C}_{(oo)o}$ (conjunction), $\mathbf{D}_{(oo)}$ (delta operation on truth values) and the description operator F $\iota_{\varepsilon(o\varepsilon)}$.

Formulas are formed of variables, constants (each of specific type), and the symbol λ. Thus, each formula A is assigned a type and we write it as A_α. A set of formulas of type α is denoted by *Form*$_\alpha$. The set of all formulas is *Form* $= \bigcup_{\alpha \in Types}$ *Form*$_\alpha$[1]. If $B \in$ *Form*$_{\beta\alpha}$ and $A \in$ *Form*$_\alpha$ then $(BA) \in$ *Form*$_\beta$. Similarly, if $A \in$ *Form*$_\beta$ and $x_\alpha \in J$, $\alpha \in$ *Types*, is a variable then $(\lambda x_\alpha A) \in$ *Form*$_{\beta\alpha}$.

The truth values form a linearly ordered MV$_\Delta$-algebra (see [2,9]). A special case is the standard Łukasiewicz MV$_\Delta$-algebra $\mathscr{L} = \langle [0,1], \vee, \wedge, \otimes, \rightarrow, 0, 1, \Delta \rangle$ where $\vee$ is minimum, $\wedge$ is maximum, $\otimes$ is Łukasiewicz conjunction, $\rightarrow$ is Łukasiewicz disjunction and $\Delta(a) = 1$ for $a = 1$, otherwise is equal to zero. By $\oplus$ we denote Łukasiewicz disjunction.

Interpretation of formulas is the following. If $\mathscr{M}$ is a model then $\mathscr{M}(A_o) \in M_o$ is a truth value, $\mathscr{M}(A_\varepsilon) \in M_\varepsilon$ is some element and $\mathscr{M}(A_{\beta\alpha}) : M_\alpha \longrightarrow M_\beta$ is a function. For example, $\mathscr{M}(A_{o\alpha}) : M_\alpha \longrightarrow M_o$ is a fuzzy set and $\mathscr{M}(A_{(o\alpha)\alpha}) : M_\alpha \times M_\alpha \longrightarrow M_o$ a fuzzy relation.

---

[1] To improve readability of formulas, we quite often write the type only once in the beginning of the formula and then omit it. Alternatively, we write $A \in$ *Form*$_\alpha$ to emphasize that A is a formula of type α and do not repeat its type again.

Theorem 1 ([5,6]). *Let T be a consistent theory, $A \in Form_o$ and $\mathbf{u}_{1,\alpha}, \dots, \mathbf{u}_{n,\alpha}$, $\alpha \in Types$ be new special constants that do not belong to the language $J(T)$. If $T \vdash (\exists x_{1,\alpha}) \cdots (\exists x_{n,\alpha}) \boldsymbol{\Delta} A$ then $T \cup \{A_{x_{1,\alpha},\dots,x_{n,\alpha}}[\mathbf{u}_{1,\alpha}, \dots, \mathbf{u}_{n,\alpha}]\}$ is a conservative extension of T.*

2.2   Evaluative Linguistic Expressions

The theory of intermediate quantifiers is based on the theory of evaluative linguistic expressions that are expressions of natural language such as *small, medium, big, very short, more or less deep, quite roughly strong, extremely high,* etc. There is a formal theory of them presented in detail in [6] and less formally including formulas for the direct computation in [8].

The semantics of evaluative linguistic expressions is formulated in a special formal theory T^{Ev} of L-FTT . Its language J^{Ev} has the following special symbols:

(i) The constants $\top, \bot \in Form_o$ for truth and falsity and $\dagger \in Form_o$ for the middle truth value.

(ii) A special constant $\sim \in Form_{(oo)o}$ for an additional fuzzy equality on the set of truth values L.

(iii) A set of special constants $\boldsymbol{\nu}, \dots \in Form_{oo}$ for linguistic hedges and a set of triples of additional constants $\mathbf{a}_{\boldsymbol{\nu}}, \mathbf{b}_{\boldsymbol{\nu}}, \mathbf{c}_{\boldsymbol{\nu}}, \dots \in Form_o$ where each triple is associated with one hedge $\boldsymbol{\nu}$. The J^{Ev} is supposed to contain the following special constants: $\{Ex, Si, Ve, ML, Ro, QR, VR\}$ that represent the linguistic hedges (*extremely, significantly, very, roughly, more or less, rather, quite roughly, very roughly*, respectively).

The evaluative expressions are construed by special formulas $Sm \in Form_{oo(oo)}$ (*small*), $Me \in Form_{oo(oo)}$ (*medium*), $Bi \in Form_{oo(oo)}$ (*big*), and $Ze \in Form_{oo(oo)}$ (*zero*) that can be extended by several selected linguistic hedges. Recall that a *hedge*, i.e., usually adverb such as "very, significantly, about, roughly", etc. is in general construed by a formula $\boldsymbol{\nu} \in Form_{oo}$ with specific properties. To classify that a given formula is a hedge, we introduced a formula $Hedge \in Form_{o(oo)}$. Then $T^{Ev} \vdash Hedge\, \boldsymbol{\nu}$ means that $\boldsymbol{\nu}$ is a hedge. We assume that the following is provable: $T^{Ev} \vdash Hedge\, \boldsymbol{\nu}$ for all $\boldsymbol{\nu} \in \{Ex, Si, Ve, ML, Ro, QR, VR\}$.

The evaluative linguistic expression is represented in the theory T^{Ev} by one of the following formulas: $Sm\,\boldsymbol{\nu}, Me\,\boldsymbol{\nu}, Bi\,\boldsymbol{\nu}, Ze\,\boldsymbol{\nu} \in Form_{oo}$ where $\boldsymbol{\nu}$ is a hedge. For example, $Sm\,Ve$ is a formula construing the evaluative expression "very small". We will also consider an *empty hedge* $\bar{\boldsymbol{\nu}}$ that is always present in front of *small, medium* and *big* if no other hedge is given. The connective $\boldsymbol{\Delta}_{oo}$ represents the expression "utmost". If, for explanation, we do not need to know the concrete evaluative expression then we will use the metavariable Ev. Evaluative expressions characterize certain imprecisely determined positions on a bounded linearly ordered scale. As there is an infinite number of scales, we must first specify the *context*, in which we characterize them (cf. [6]).

This is formally represented by a function $w : E \longrightarrow M$. Less formally, we define the context as a triple of numbers $v_L, v_S, v_R \in M$ such that $v_L < v_S < v_R$

(the ordering on M is induced by w). Then $x \in w$ iff $x \in [v_L, v_S] \cup [v_S, v_R]$. For simplification, we usually omit in the theory of intermediate quantifiers the context. More details can be found in the above cited literature.

Let $\boldsymbol{\nu}_{1,oo}, \boldsymbol{\nu}_{2,oo}$ be two hedges, i.e., $T^{\mathrm{Ev}} \vdash \mathit{Hedge}\ \boldsymbol{\nu}_{1,oo}$ and $T^{\mathrm{Ev}} \vdash \mathit{Hedge}\ \boldsymbol{\nu}_{2,oo}$. We define a relation of partial ordering of hedges by

$$\ll := \lambda p_{oo} \lambda q_{oo} \cdot (\forall z_o)(pz \Rightarrow qz). \tag{1}$$

Lemma 1. *The following ordering the specific hedges can be proved.*

$$T^{\mathrm{Ev}} \vdash \boldsymbol{\Delta} \ll Ex \ll Si \ll Ve \ll \bar{\boldsymbol{\nu}} \ll ML \ll Ro \ll QR \ll VR. \tag{2}$$

Proof. The properties $T^{\mathrm{Ev}} \vdash Ex \ll Si, \dots, T^{\mathrm{Ev}} \vdash QR \ll VR$ are provable on the basis of the definition of the corresponding hedges (cf. [5,6]). To prove that $T^{\mathrm{Ev}} \vdash \boldsymbol{\Delta} \ll Ex$, let $\mathcal{M} \models T^{\mathrm{Ev}}$ be a model. Then $\mathcal{M}_p(\boldsymbol{\Delta} x_o) = 0$ whenever $\mathcal{M}_p(x_o) < 1$. Hence, $\mathcal{M}_p((\forall x_o)(\boldsymbol{\Delta} x_o \Rightarrow Ex\, x_o)) = 1$ which implies $T^{\mathrm{Ev}} \vdash (\forall x_o)(\boldsymbol{\Delta} x_o \Rightarrow Ex\, x_o)$ by completeness. $\square$

Theorem 2. *Let* $\boldsymbol{\nu}_{1,oo}, \boldsymbol{\nu}_{2,oo}$ *be hedges such that* $T^{\mathrm{Ev}} \vdash \boldsymbol{\nu}_{1,oo} \ll \boldsymbol{\nu}_{2,oo}$. *Then*

$$Sm\, \boldsymbol{\nu}_1 \subseteq Sm\, \boldsymbol{\nu}_2, \qquad Me\, \boldsymbol{\nu}_1 \subseteq Me\, \boldsymbol{\nu}_2, \qquad Bi\, \boldsymbol{\nu}_1 \subseteq Bi\, \boldsymbol{\nu}_2. \tag{3}$$

3   The Theory of Intermediate Quantifiers

Recall that by intermediate quantifiers, we understand expressions of natural language such as "many, most, almost all, a few, several", etc. Their meaning lays between the meaning of the classical quantifiers "for all" ($\forall$) and "exists" ($\exists$). They are modeled by selected formulas of a special formal theory T^{Ev} of Ł-FTT . These formulas express quantification over the universe represented by a fuzzy set whose size is characterized by a measure. This is specified in the following definition.

Definition 1. Let $R \in \mathit{Form}_{o(o\alpha)(o\alpha)}$ be a formula where $\alpha \in \mathit{Types}$ is an arbitrary type.

(i) A formula $\mu \in \mathit{Form}_{o(o\alpha)(o\alpha)}$ defined by

$$\mu_{o(o\alpha)(o\alpha)} \equiv \lambda z_{o\alpha}\, \lambda x_{o\alpha}\, (R z_{o\alpha}) x_{o\alpha} \tag{4}$$

represents a *measure on fuzzy sets* in the universe of type $\alpha \in \mathit{Types}$ if it has the following properties:
  (M1) $\boldsymbol{\Delta}(x_{o\alpha} \subseteq z_{o\alpha})\,\&\,\boldsymbol{\Delta}(y_{o\alpha} \subseteq z_{o\alpha})\,\&\,\boldsymbol{\Delta}(x_{o\alpha} \subseteq y_{o\alpha}) \Rightarrow ((\mu z_{o\alpha}) x_{o\alpha} \Rightarrow (\mu z_{o\alpha}) y_{o\alpha})$,
  (M2) $\boldsymbol{\Delta}(x_{o\alpha} \subseteq z_{o\alpha}) \Rightarrow ((\mu z_{o\alpha})(z_{o\alpha} \setminus x_{o\alpha}) \equiv \neg(\mu z_{o\alpha}) x_{o\alpha})$,
  (M3) $\boldsymbol{\Delta}(x_{o\alpha} \subseteq y_{o\alpha})\,\&\,\boldsymbol{\Delta}(x_{o\alpha} \subseteq z_{o\alpha})\,\&\,\boldsymbol{\Delta}(y_{o\alpha} \subseteq z_{o\alpha}) \Rightarrow ((\mu z_{o\alpha}) x_{o\alpha} \Rightarrow (\mu y_{o\alpha}) x_{o\alpha})$
  where $x_{o\alpha}, y_{o\alpha}, z_{o\alpha}$ are variables representing fuzzy sets.

(ii) The following formula characterizes *measurable fuzzy sets* of a given type α:

$$\mathbf{M}_{o(o\alpha)} \equiv \lambda z_{o\alpha} \cdot \Delta\neg(z_{o\alpha} \equiv \emptyset_{o\alpha}) \,\&\, \Delta(\mu z_{o\alpha}) z_{o\alpha} \,\&$$
$$(\forall x_{o\alpha})(\forall y_{o\alpha})\Delta((\text{M1})\,\&\,(\text{M3}))\,\&\,(\forall x_{o\alpha})\Delta(\text{M2}) \quad (5)$$

where (M1)–(M3) are the axioms from (i).

On the basis of this definition we can prove the following lemma (cf. [5]).

Lemma 2. *Let* $\vdash \mathbf{M}(B_{o\alpha})$ *and* $\vdash C_{o\alpha} \equiv \emptyset$. *Then* $\vdash (\mu B)C \equiv \bot$.

For the definition of the intermediate quantifier, we need a special operation "cut of a fuzzy set" for the given fuzzy sets $y, z \in Form_{o\alpha}$[2]:

$$y|z \equiv \lambda x_\alpha \cdot zx \,\&\, \Delta(\Upsilon(zx) \Rightarrow (yx \equiv zx)). \quad (6)$$

The meaning of this formula becomes clear form the following lemma.

Lemma 3. *Let* $y, z \in Form_{o\alpha}$ *be variables,* $\mathcal{M}$ *be a model and p an assignment such that* $B = \mathcal{M}_p(y) \subseteq_\sim M_\alpha$, $Z = \mathcal{M}_p(z) \subseteq_\sim M_\alpha$. *Then for any* $m \in M_\alpha$

$$\mathcal{M}_p(y|z)(m) = (B|Z)(m) = \begin{cases} B(m), & \text{if } B(m) = Z(m), \\ 0, & \text{if } Z(m) \neq B(m). \end{cases}$$

One can see that the operation $B|Z$ "cuts" B by taking only those $m \in M_\alpha$ from the fuzzy set B whose membership $B(m)$ is equal to $Z(m)$, otherwise $(B|Z)(m) = 0$. If there is no such element then $B|Z = \emptyset$. We can thus take various fuzzy sets Z to "pick up proper elements" from B.

Lemma 4. *Let* $B, Y, Z \subseteq_\sim M$. *Then the following holds true:*

(a) $B|Z \subseteq B$ *and* $B|Z = Z|B$.

Proof. (a) Let $m \in M$. Then $(B|Z)(m) = (Z|B)(m) = B(m) = Z(m)$, or $(B|Z)(m) = (Z|B)(m) = 0$.

Definition 2. Let $\mathscr{S} \subseteq Types$ be a selected set of types and $P = \{R \in Form_{o(o\alpha)(o\alpha)} \mid \alpha \in \mathscr{S}\}$ be a set of new constants. Let T be a consistent extension of the theory T^{Ev} in the language $J(T) \supseteq J^{\text{Ev}} \cup P$. We say that the theory T contains *intermediate quantifiers* w.r.t. the set of types $\mathscr{S}$ if for all $\alpha \in \mathscr{S}$ the following is provable:

(i)

$$T \vdash (\exists z_{o\alpha})\mathbf{M}_{o(o\alpha)} z_{o\alpha}. \quad (7)$$

---

[2] The special (derived) formula $\Upsilon_{oo}A_o$ says that A_o in every model has a non-zero truth value and $\hat{\Upsilon}_{oo}A_o$ that A_o has a general truth value (i.e., neither false 0, nor true 1).

(ii)
$$T \vdash (\forall z_{o\alpha})(\exists x_{o\alpha})(\mathbf{M}_{o(o\alpha)} z_{o\alpha} \Rightarrow (\boldsymbol{\Delta}(x_{o\alpha} \subseteq z_{o\alpha}) \,\&\, \hat{\Upsilon}((\mu z_{o\alpha}) x_{o\alpha})). \quad (8)$$

We will denote the theory due to Definition 2 by T^{IQ} fix a selected set of types $\mathscr{S}$.

Definition 3. Let $Ev \in Form_{oo}$ be a formula representing some evaluative linguistic expression, $z \in Form_{o\alpha}$, $x \in Form_{\alpha}$ be variables and $A, B \in Form_{o\alpha}$ be formulas such that $T^{\mathrm{IQ}} \vdash \mathbf{M}_{o(o\alpha)} B$, $\alpha \in \mathscr{S}$.

(i) An intermediate quantifier of type $\langle 1, 1 \rangle$ is one of the following formulas:

$$(Q_{Ev}^{\forall} x)(B, A) \equiv (\exists z)[(\forall x)((B|z)\, x \Rightarrow Ax) \wedge Ev((\mu B)(B|z))], \quad (9)$$

$$(Q_{Ev}^{\exists} x)(B, A) \equiv (\exists z)[(\exists x)((B|z)x \wedge Ax) \wedge Ev((\mu B)(B|z))]. \quad (10)$$

Either of the quantifiers (9) or (10) construes the sentence

$$\langle \text{Quantifier} \rangle \ B\text{'s are } A.$$

(ii) An intermediate quantifier of type $\langle 1, 1 \rangle$ with presupposition is one of the following formulas:

$$({}^{*}Q_{Ev}^{\forall} x)(B, A) \equiv (\exists z)[((\exists x)(B|z)x \,\&\, (\forall x)((B|z)\, x \Rightarrow Ax)) \wedge Ev((\mu B)(B|z))], \quad (11)$$

$$({}^{*}Q_{Ev}^{\exists} x)(B, A) \equiv (\exists z)[(\neg(\exists x)(B|z)x \,\nabla\,((\exists x)((B|z)x) \wedge Ax) \wedge Ev((\mu B)(B|z)))]. \quad (12)$$

These quantifiers construe the same sentence above but, moreover, assume that $B|z$ is non-empty.

The formula $B_{o\alpha}$ in (i)–(iii) represents a *universe of quantification*.

If we replace the metavariable Ev in (9)–(12) by a formula representing a specific evaluative linguistic expression we obtain definition of the concrete intermediate quantifier. Namely, we will consider the following quantifiers:

(A) "All B's are A": $(Q_{Bi\,\boldsymbol{\Delta}}^{\forall} x)(B, A)$ **(E)** "No B's is A": $(Q_{Bi\,\boldsymbol{\Delta}}^{\forall} x)(B, \neg A)$
(P) "Almost all B's are A": $(Q_{Bi\,Ex}^{\forall} x)(B, A)$
    (B) "Almost all B's are not A": $(Q_{Bi\,Ex}^{\forall} x)(B, \neg A)$
(T) "Most B's are A": $(Q_{Bi\,Ve}^{\forall} x)(B, A)$ **(D)** "Most B's are not A": $(Q_{Bi\,Ve}^{\forall} x)(B, \neg A)$
(K) "Many B's are A": $(Q_{\neg\,Sm}^{\forall} x)(B, A)$**(G)** "Many B's are not A": $(Q_{\neg\,Sm}^{\forall} x)(B, \neg A)$
(I) "Some B's are A": $(Q_{Bi\,\boldsymbol{\Delta}}^{\exists} x)(B, A)$ **(O)** "Some B's are not A": $(Q_{Bi\,\boldsymbol{\Delta}}^{\exists} x)(B, \neg A)$

(for the details see [3,4]). By [5, Theorem 6(a)], **(A)** reduces to the classical formula $(\forall x)(Bx \Rightarrow Ax)$ (and similarly **(E)**).

Lemma 5. *Let $\mathscr{M} \models T^{IQ}$ be a model, p an assignment and $B, z \in Form_{o\alpha}$ be formulas. Let $Ev \in \{Bi\Delta, BiSi, BiVe, \neg(Sm\,\bar{\nu})\}$. If $\mathscr{M}_p(Ev((\mu B)(B|z))) > 0$ then there exists an element $m \in M_\alpha$, such that*

$$\mathscr{M}_p((\exists x)(B|z)x) \geq \mathscr{M}_p(B|z)(m) > 0 \tag{13}$$

holds in any model $\mathscr{M} \models T^{IQ}$ and for every assignment $p \in$ Asg.

Proof. Let $\mathscr{M} \models T^{IQ}$ be a model. If $\mathscr{M}_p(Ev((\mu B)(B|z))) > 0$ then $\mathscr{M}_p((\mu B)(B|z)) > 0$ because none of the considered evaluative expressions allows zero value of this measure. Using Lemma 2 we conclude that the fuzzy set $\mathscr{M}_p(B|z)$ is non empty. ∎

Lemma 6. *Let $\mathscr{M} \models T^{IQ}$ be a model, p an assignment and $A, B, \in Form_{o\alpha}$ be formulas and $Ev \in \{Bi\Delta, BiSi, BiVe, \neg(Sm\,\bar{\nu})\}$. Then the following is true in every model $\mathscr{M} \models T^{IQ}$:*

- $\mathscr{M}((^*Q^\forall_{Ev}\,x)(B, A)) = \mathscr{M}((Q^\forall_{Ev}\,x)(B, A))$
- $\mathscr{M}((^*Q^\forall_{Ev}\,x)(B, \neg A)) = \mathscr{M}((Q^\forall_{Ev}\,x)(B, \neg A))$

Proof. Using [4, Lemma 5] by completeness theorem $\mathscr{M}_p((^*Q^\forall_{Ev}\,x)(B, A)) \leq \mathscr{M}_p((Q^\forall_{Ev}\,x)(B, A))$ holds in every model $\mathscr{M} \models T^{IQ}$ and for every assignment p.

Let there be a model $\mathscr{M} \models T^{IQ}$ and an assignment p such that $B = \mathscr{M}_p(B) \subseteq_{\sim} M_{o\alpha}$, $Z = \mathscr{M}_p(z) \subseteq_{\sim} M_{o\alpha}$ Let $\mathscr{M}_p((Q^\forall_{Ev}\,x)(B, A)) = a > 0$, otherwise the opposite implication is trivially fulfilled (by $T^{IQ} \vdash \bot \Rightarrow A$). Then from (9) it follows that $\mathscr{M}_p(Ev((\mu B)(B|z))) > 0$. From Lemma 5 there is an element $m \in M_\alpha$ such that

$$\mathscr{M}_p((\exists x)(B|z)x) \geq (B|Z)(m) > 0.$$

We conclude that $\mathscr{M}((Q^\forall_{Ev}\,x)(B, A)) \leq \mathscr{M}((^*Q^\forall_{Ev}\,x)(B, A))$. ∎

Theorem 3. *Let $A_{o\alpha}, B_{o\alpha}$ be formulas. Then the following is provable in the theory T^{IQ} for the above defined quantifiers:*

$$T^{IQ} \vdash (\mathbf{A}) \Rightarrow (\mathbf{P}), T^{IQ} \vdash (\mathbf{P}) \Rightarrow (\mathbf{T}), T^{IQ} \vdash (\mathbf{T}) \Rightarrow (\mathbf{K})$$

Proof. The implications in this theorem follow from the definition (9) using (3) and the properties of L-FTT . ∎

At the end of this section, we recall from [4] definitions of basic relations among formulas which define generalized square of opposition.

Definition 4. *Let T be a consistent theory of L-FTT and $P_1, P_2 \in Form_o$ be closed formulas of type o [3].*

---

[3] Recall that **&** is interpreted by Lukasiewicz conjunction $\otimes$ and ∇ is interpreted by Lukasiewicz disjunction $\oplus$.

(i) We say that P_2 is *subaltern of P_1 in the theory T* (P_1 is *superaltern of P_2 in the theory T*) if $T \vdash P_1 \Rightarrow P_2$.
(ii) They are *contraries in the theory T* if $T \vdash \neg(P_1 \mathbin{\&} P_2)$.
(iii) They are *sub-contraries in the theory T* if $T \vdash (P_1 \nabla P_2)$.
(iv) They are *contradictories in the theory T* if both $T \vdash \neg(\Delta P_1 \mathbin{\&} \Delta P_2)$ as well as $T \vdash \Delta P_1 \nabla \Delta P_2$.

4   "Several, A Few, A Little" in the Square of Opposition

We will define the quantifiers "A few, Several, A little" in the similar way as above.

Definition 5. Let T^{IQ} be a theory containing intermediate quantifiers w.r.t. a set of types $\mathscr{S}$, $z \in Form_{o\alpha}$, $x \in Form_{\alpha}$ and $A, B \in Form_{o\alpha}$ be the same as in Definition 3. The following are new quantifiers:

(**F**) "A few (A little) B's are A": $(^*Q^{\forall}_{Sm\ Si}\, x)(B, A)$.
(**V**) "A few (A little) B's are not A": $(^*Q^{\forall}_{Sm\ Si}\, x)(B, \neg A)$.
(**S**) "Several B's are A": $(^*Q^{\forall}_{Sm\ Ve}\, x)(B, A)$.
(**Z**) "Several B's are not A": $(^*Q^{\forall}_{Sm\ Ve}\, x)(B, \neg A)$.

There is no formal difference between "A few" and "A little". The actual difference manifests itself only when a specific model $\mathscr{M}$ is considered. Namely, if the support of the fuzzy set $\mathscr{M}(B_{o\alpha})$ is countable then (i) construes "A few". If it is uncountable then it construes "A little".

Let us remark the question whether presupposition in both quantifiers should be considered. Indeed, the empirical observation reveals necessity to exclude empty fuzzy set $\mathscr{M}(B_{o\alpha}|z_{o\alpha})$. For example:

We saw several does grazing in the meadow

cannot mean that there was no doe in the meadow. Alternatively, we can introduce a special evaluative expression *"positive ⟨hedge⟩small"* that is formally defined by

$$^{+}Sm\,\boldsymbol{\nu} \equiv \lambda t_o \cdot (Sm\,\boldsymbol{\nu})t_o \mathbin{\&} \neg(Ze\,\bar{\boldsymbol{\nu}})t_o. \tag{14}$$

Clearly, $^{+}Sm\,\boldsymbol{\nu} \in Form_{oo}$.

In a similar way as in Lemma 6, we can prove:

Lemma 7. *Let $Ev \in \{Sm\ Si, Sm\ Ve\}$. Then $T^{IQ} \vdash (^*Q^{\forall}_{Ev}\, x)(B, A) \equiv Q^{\forall}_{^{+}Ev}\, x)(B, A)$.*

In the sequel, we will consider the quantifiers due to Definition 5 defined using the evaluative expression (14).

4.1   Sub-altern of "A Few" and "Several"

Let us introduce the following property:

$$\mathbf{SQ}_{o(o\alpha)} \equiv \lambda z_{o\alpha} \cdot (\forall x_{o\alpha})(\exists y_\alpha)\mathbf{\Delta}[(z|y) \subseteq (z|x) \, \&$$
$$(\neg \, Sm \, \bar{\boldsymbol{\nu}}((\mu z)(z|x)) \Rightarrow \, ^+ Sm \, Si((\mu z)(z|y)))] \quad (15)$$

This property says that inside a given fuzzy set z there is a significantly small non-empty fuzzy set $z|y$ (the size of fuzzy sets is measured using the measure μ). Such a property is quite natural but it is not implied by the axioms of measure. On the other hand, we cannot accept (15) as an axiom holding for all fuzzy sets $z_{o\alpha}$. The reason is that if we consider fuzzy sets over a universe with small number of elements (e.g., 3–5) then we can determine (fuzzy) sets having sufficiently large measure but we can at the same time hardly find (fuzzy) sets with significantly small measure. For example, if we consider a set with 4 elements then a set with 2 elements has the standard measure equal to 0.5 and a singleton 0.25. The former can represent "not small" but the latter can hardly be evaluated as "significantly small".

Theorem 4. *Let $B_{o\alpha}, A_{o\alpha}$ be formulas. Then the following is provable in T^{IQ}:*

(a) $T^{IQ} \vdash (\mathbf{S}) \Rightarrow (\mathbf{I})$,
(b) $T^{IQ} \vdash (\mathbf{F}) \Rightarrow (\mathbf{S})$.
(c) Let $T^{IQ} \vdash \mathbf{SQ}_{o(o\alpha)} B_{o\alpha}$. Then $T^{IQ} \vdash Q^\forall_{\neg \, Sm \, \bar{\boldsymbol{\nu}}} x)(B, A) \Rightarrow Q^\forall_{+ \, Sm \, Si} x)(B, A)$
(i.e., $T^{IQ} \vdash (\mathbf{K}) \Rightarrow (\mathbf{F})$).

Proof. (a)

(L.1) $\vdash ((B|z)x \Rightarrow Ax) \Rightarrow ((B|z)x \wedge Ax)$ $\qquad$ (provable in Ł-FTT)
(L.2) $T^{IQ} \vdash (B|z)x \Rightarrow Bx$ $\qquad$ (consequence of Lemma 4(a))
(L.3) $T^{IQ} \vdash (B|z)x \wedge Ax \Rightarrow Bx \wedge Ax$ $\qquad$ (L.2, properties of Ł-FTT)
(L.4) $T^{IQ} \vdash ((B|z)x \Rightarrow Ax) \Rightarrow ((B|z) \Rightarrow (Bx \wedge Ax))$
$\qquad\qquad\qquad\qquad\qquad\qquad$ (L.1, L.3, properties of Ł-FTT)
(L.5) $T^{IQ} \vdash (\forall x)((B|z)x \Rightarrow Ax) \Rightarrow ((\exists x)(B|z) \Rightarrow (\exists x)(Bx \wedge Ax))$
$\qquad\qquad\qquad\qquad\qquad\qquad$ (L.4, properties of quantifiers)
(L.6) $T^{IQ} \vdash ((\forall x)((B|z)x \Rightarrow Ax) \& (\exists x)(B|z)) \wedge (Sm \, Ve)((\mu B)(B|z))$
$\qquad\qquad\qquad \Rightarrow (\exists x)(Bx \wedge Ax))$ $\qquad$ (L.5, properties of quantifiers)
(L.7) $T^{IQ} \vdash (\exists z)[((\forall x)((B|z)x \Rightarrow Ax) \& (\exists x)(B|z)) \wedge (Sm \, Ve)((\mu B)(B|z))]$
$\qquad\qquad\qquad \Rightarrow (\exists x)(Bx \wedge Ax))$ $\qquad$ (L.6, properties of quantifiers)
(L.8) $T^{IQ} \vdash (\exists z)[(\forall x)((B|z)x \Rightarrow Ax) \wedge (Sm \, Ve)((\mu B)(B|z))] \Rightarrow (\exists x)(Bx \wedge Ax))$
$\qquad\qquad\qquad\qquad\qquad\qquad$ (L.7, Lemma 7)

(b) This implication follows from (9) and (3).
(c) After rewriting the proposition, we have

$$(\exists z')[(\forall x)((B|z')\,x \Rightarrow Ax) \; \wedge \neg \, Sm \, \bar{\boldsymbol{\nu}}((\mu B)(B|z'))] \Rightarrow$$
$$(\exists z)[(\forall x)((B|z)\,x \Rightarrow Ax) \wedge Sm \, Si((\mu B)(B|z))]. \quad (16)$$

By the assumption and Theorem 1, we will extend the theory T^{IQ} by a new constant $\mathbf{u}_{o\alpha}$ and consider an arbitrary $z_{o\alpha}$. Then

$$T^{IQ} \vdash (B|\mathbf{u} \subseteq B|z)\,\&[\neg\, Sm\,\bar{\boldsymbol{\nu}}((\mu B)(B|z)) \Rightarrow\,^+ Sm\, Si((\mu B)(B|\mathbf{u}))]. \quad (17)$$

From (17) we derive $[(\forall x)((B|z)\,x \Rightarrow Ax)] \Rightarrow (\forall x)((B|\mathbf{u})\,x \Rightarrow Ax)$. From this and the second part of (17) we obtain

$$[(\forall x)((B|z)\,x \Rightarrow Ax)\, \wedge\neg\, Sm\,\bar{\boldsymbol{\nu}}((\mu B)(B|z))] \Rightarrow$$
$$[(\forall x)((B|\mathbf{u})\,x \Rightarrow Ax)\,\&^+ Sm\, Si((\mu B)(B|\mathbf{u}))]$$

by the properties of Ł-FTT . Formula (16) is obtained from the latter using axiom of existence, rule of generalization and the quantifier properties.   □

4.2   Sub-contrary of "A Few" and "Several"

It has been justified in [5] that we should exclude certain fuzzy sets from their role as universes of quantification. Therefore, we will introduce the following property.

$$\mathbf{Reg}_{o(o\alpha)} \equiv \lambda z_{o\alpha} \cdot (\exists x_\alpha)\varDelta zx\,\&$$
$$(\exists v_{o\alpha})[(\forall x_\alpha)(\Upsilon(z|v)x \Rightarrow \Upsilon((z|v)x)^2)\,\&((\mu z)(z|v) > \dagger) \quad (18)$$

If $T^{IQ} \vdash \mathbf{Reg}\, B_{o\alpha}$ then we will call formula $B_{o\alpha}$ *regular*. By $\mathscr{S}^B$ we denote a set of types α that do not contain the type o.

The relation of subcontrary means that we can find elements of the universe that have a given property A as well as those having $\neg A$. We argue that this happens with the quantifier *some* and in special cases also with *many*. The situation with *a few* and *several* is similar. As we evaluate sizes of (fuzzy) sets using measure, we cannot be sure that in a big fuzzy set we find a sufficiently small one. For example in a set of 5 elements, 2-3 mean already "many" but it has no sense to speak about "several". Therefore, we have to consider the property (15).

Lemma 8. *Let* $B, A, z, z' \in Form_{o\alpha}$, $\alpha \in \mathscr{S}^B$ *and assume that* $T^{IQ} \vdash \mathbf{Reg}\, B \wedge$ $\mathbf{SQ}\, B$. *Let us consider the following extension of the theory* T^{IQ}:

$$\bar{T} = T^{IQ} \cup \{(\exists z)(\exists z')\varDelta[(\neg(Sm\,\bar{\boldsymbol{\nu}})((\mu B)(B|z)))\,\&(\neg(Sm\,\bar{\boldsymbol{\nu}})((\mu B)(B|z')))\,\&$$
$$\neg(\exists x)((B|z)x\,\&\neg Ax)\,\&\neg(\exists x)((B|z')x\,\&\,Ax)]\}. \quad (19)$$

Then the quantifiers

(a) $[(Q^{\forall}_{(Sm\,Si)}\,x)(B, A), (Q^{\forall}_{(Sm\,Si)}\,x)(B, \neg A)]$, *(i.e.* [$\mathbf{F}, \mathbf{V}$]*)*
(b) $[(Q^{\forall}_{(Sm\,Ve)}\,x)(B, A), (Q^{\forall}_{(Sm\,Ve)}\,x)(B, \neg A)]$, *(i.e.* [$\mathbf{S}, \mathbf{Z}$]*)*

are sub-contraries in $\bar{T}$.

Proof. It is proved in [5] that $\bar{T}$ is consistent.

(a) Let $\mathbf{u}_{o\alpha}, \mathbf{u}'_{o\alpha}, \mathbf{v}_{o\alpha}, \mathbf{v}'_{o\alpha} \notin J(\bar{T})$ be new constants and T' be the following conservative extension of $\bar{T}$ due to Theorem 1:

$$T' = \bar{T} \cup \{(\neg(Sm\,\bar{\nu})((\mu B)(B|\mathbf{u}))\,\&\,(\neg(Sm\,\bar{\nu})((\mu B)(B|\mathbf{u}'))\,\&$$
$$\neg(\exists x)((B|\mathbf{u})x\,\&\,\neg Ax)\,\&\,\neg(\exists x)((B|\mathbf{u}')x\,\&\,Ax),$$
$$(B|\mathbf{v}) \subseteq (B|\mathbf{u})\,\&\,(\neg Sm\,\bar{\nu}((\mu B)(B|\mathbf{u})) \Rightarrow {}^+Sm\,Si((\mu B)(B|\mathbf{v}))),$$
$$(B|\mathbf{v}') \subseteq (B|\mathbf{u}')\,\&\,(\neg Sm\,\bar{\nu}((\mu B)(B|\mathbf{u}')) \Rightarrow {}^+Sm\,Si((\mu B)(B|\mathbf{v}')))\}.$$

From this we obtain

$$T' \vdash (\forall x)((B|\mathbf{v})\,x \Rightarrow Ax) \wedge {}^+Sm\,Si((\mu B)(B|\mathbf{v})), \qquad (20)$$
$$T' \vdash (\forall x)((B|\mathbf{v}')\,x \Rightarrow Ax) \wedge {}^+Sm\,Si((\mu B)(B|\mathbf{v}')) \qquad (21)$$

using the properties of quantifiers and rules of FTT. From (20) and (21) using theorem on variants, $\exists$-substitution axiom and the obvious implication $\vdash A_o\,\&\,B_o \Rightarrow A_o \,\nabla\, B_o$ we obtain

$$\bar{T} \vdash (\exists z)[(\forall x)((B|\mathbf{v})\,x \Rightarrow Ax) \wedge {}^+Sm\,Si((\mu B)(B|\mathbf{v}))]\,\nabla$$
$$(\exists z')[(\forall x)((B|\mathbf{v}')\,x \Rightarrow \neg Ax) \wedge {}^+Sm\,Si((\mu B)(B|\mathbf{v}'))]. \qquad (22)$$

(b) can be proved similarly as (a) using Theorem 4(b).

Theorem 5. *Let $B, A, z, z' \in Form_{o\alpha}$. Then there exists a model $\mathcal{M} \models T^{IQ}$ in which the quantifiers*

(a) $[(Q^{\forall}_{(Sm\,Si)}\,x)(B,A), (Q^{\forall}_{(Sm\,Si)}\,x)(B,\neg A)]$, (i.e. $[\mathbf{F}, \mathbf{V}]$)
(b) $[(Q^{\forall}_{(Sm\,Ve)}\,x)(B,A), (Q^{\forall}_{(Sm\,Ve)}\,x)(B,\neg A)]$, (i.e. $[\mathbf{S}, \mathbf{Z}]$)

are sub-contraries.

Proof. Let $\bar{T}$ be a theory from Lemma 8. As it is consistent, it has a model $\mathcal{M} \models \bar{T}$ which, at the same is a model of T^{IQ}.

The straight lines in **5**-square below mark contradictories, the dashed lines contraries and the dotted lines sub-contraries. The arrows indicate the subaltern (Table 1).

Table 1. Generalized square of opposition with "A few", "Several" and "A little"

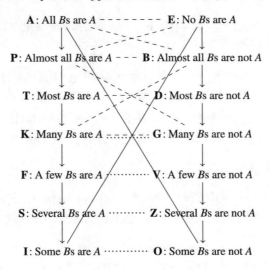

5   Conclusion

In this paper, we extend the formal model of intermediate quantifiers by three new ones, namely "a few", "a little" and "several". We proved some of their properties and demonstrated that they naturally fall in the extended square of opposition.

Acknowledgements. The work was supported from ERDF/ESF by the project "Centre for the development of Artificial Intelligence Methods for the Automotive Industry of the region" No. CZ.02.1.01/0.0/0.0/17-049/0008414.

References

1. Andrews, P.: An Introduction to Mathematical Logic and Type Theory: To Truth Through Proof. Kluwer, Dordrecht (2002)
2. Cignoli, R.L.O., D'Ottaviano, I.M.L., Mundici, D.: Algebraic Foundations of Many-Valued Reasoning. Kluwer, Dordrecht (2000)
3. Murinová, P., Novák, V.: A formal theory of generalized intermediate syllogisms. Fuzzy Sets Syst. **186**, 47–80 (2013)
4. Murinová, P., Novák, V.: Analysis of generalized square of opposition with intermediate quantifiers. Fuzzy Sets Syst. **242**, 89–113 (2014)
5. Murinová, P., Novák, V.: The theory of intermediate quantifiers in fuzzy natural logic revisited and the model of "Many". Fuzzy Sets Syst. (Submitted)
6. Novák, V.: A comprehensive theory of trichotomous evaluative linguistic expressions. Fuzzy Sets Syst. **159**(22), 2939–2969 (2008)
7. Novák, V.: A formal theory of intermediate quantifiers. Fuzzy Sets Syst. **159**(10), 1229–1246 (2008)

8. Novák, V., Perfilieva, I., Dvořák, A.: Insight into Fuzzy Modeling. Wile, Hoboken (2016)
9. Novák, V., Perfilieva, I., Močkoř, J.: Mathematical Principles of Fuzzy Logic. Kluwer, Boston (1999)
10. Peterson, P.L.: On the logic of "few", "many" and "most". Notre Dame J. Form. Log. **20**, 155–179 (1979)
11. Thompson, B.E.: Syllogisms using "few", "many" and "most". Notre Dame J. Form. Log. **23**, 75–84 (1982)

Robust Evolving Granular Feedback Linearization

Lucas Oliveira[1,2]([✉]), Anderson Bento[2], Valter Leite[2], and Fernando Gomide[1]

[1] Department of Computer Engineering and Automation,
School of Electrical and Computer Engineering,
University of Campinas, Campinas, SP, Brazil
gomide@dca.fee.unicamp.br
[2] Department of Mechatronics Engineering,
Federal Center for Technological Education of Minas Gerais, Divinópolis, MG, Brazil
lqsoliveira@cefetmg.br, bentoavb@gmail.com, valter@ieee.org

Abstract. This paper develops an adaptive feedback linearization approach to control nonlinear systems under model mismatch conditions. The approach uses the participatory learning modeling algorithm to estimate the nonlinearities from data streams online, and the certainty equivalence principle to compute the control signal. Simulation experiments with the classic surge tank level control benchmark show that evolving robust granular feedback linearization outperforms exact feedback linearization.

Keywords: Robust control · Evolving systems · Feedback linearization

1 Introduction

Nowadays, automation and control are major components of most industrial processes and systems such as energy, aircraft, robots, communications, and transportation. The majority of industrial machines are nonlinear uncertain systems, reason why design and synthesis of controllers is a challenge [5]. Exact feedback linearization (EFL) is a powerful nonlinear method used to control a specific class of nonlinear systems [10,11,20]. However, exact feedback controllers lack robustness when they face model mismatches and uncertain systems [7,8].

Many papers have addressed strategies to add robustness in closed-loop feedback linearized systems. For instance, [23] develops an indirect adaptive fuzzy rule-based approach to estimate and to compute online the control input to track a reference signal. Alternatively, [17] suggests an indirect adaptive fuzzy control algorithm that is robust against reconstruction errors for single-input-single-output nonlinear dynamical systems with unknown nonlinearities. In [18] a biomimicry of social bacterial foraging approach is used to develop an indirect adaptive controller. Recently, a scheme developed upon the notion of model reference adaptive control and evolving fuzzy participatory learning algorithm (ePL) [13] was developed in [16].

© Springer Nature Switzerland AG 2019
R. B. Kearfott et al. (Eds.): IFSA 2019/NAFIPS 2019, AISC 1000, pp. 442–452, 2019.
https://doi.org/10.1007/978-3-030-21920-8_40

This paper introduces a novel approach, namely robust evolving granular feedback linearization (ReGFL), using the notion of indirect adaptive control, similarly as in [17] and [23]. Differently, from the previous approaches, ReGFL employs the ePL modeling algorithm to estimate online the values of the functions that appear in the control law. The ReGFL also relies on the certainty equivalence principle [3]. The ReGFL control algorithm is used in the control loop of a surge tank, a classic benchmark adopted by many authors [2,16,19,21] to evaluate the performance of feedback linearization. The ReGFL is compared against the EFL. The results are quantified using the integral absolute error (IAE), and the integral of time-weighted absolute error (ITAE), indexes adopted in process control engineering to measure the performance of control loops. The results show that ReGFL outperforms EFL.

The remainder of the paper is organized as follows. Sections 2 and 3 overview the concepts of evolving participatory learning and exact feedback linearization. Section 4 formulates the problem addressed in this paper and develops a feedback linearization solution using the ePL algorithm. Section 5 summarizes the results produced by ReGFL and EFL during the surge tank level control. Finally, Sect. 6 summarizes the contributions of the paper and suggests issues to be addressed in the future.

2 Granular Evolving Participatory Learning Modeling

Evolving systems are a class of adaptive structures with learning and self-organization abilities [1,12]. They update the structural components and respective parameters on demand using stream data as input. The idea is to self-adapt the system to improve its performance and update knowledge from information contained in the stream data [15].

The evolving participatory learning algorithm was introduced in [13] within an adaptive systems modeling framework. Models developed by ePL are a collection of functional fuzzy rules of the form:

$$\mathcal{R}_i : \textbf{IF } z(k) \text{ is } \text{A}^i \textbf{ THEN } y_i(k) = f_i(z(k))$$
$$i = 1, \cdots, c(k)$$

where $\mathcal{R}_i$ is the i-th fuzzy rule, $c(k)$ is the number of fuzzy rules at the k-th step, $z(k)$ is the input, $y_i(k)$ is the output of the i-th rule, A^i is the membership function of the antecedent, and $f_i(z(k))$ is a function of the input $z(k)$. The ePL clusters data stream and simultaneously estimates the parameters of a local model assigned to each cluster [13,16]. The main user specified parameters ePL are the learning rate $\alpha \in [0,1]$, the arousal update rate $\beta \in [0,1]$, the arousal index threshold value $\tau \in [0,1]$, and the compatibility threshold value $\lambda \in [0,1]$. Cluster centers are the modal values of Gaussian membership functions of the fuzzy rule antecedents:

$$\mu_i(k) = e^{-\frac{\|z(k) - v_i(k)\|^2}{\delta_r}}, \tag{1}$$

where $v_i(k) \in [0,1]^p$ is the i-th cluster center at the k-th step, p is the dimension of the input space, $\| \cdot \|$ is the Euclidean norm, and δ_r is a positive value that settles the zone of influence (spread) of the i-th local model. After initialization, the participatory learning algorithm verifies at each processing step whether a new cluster must be created, an old cluster should be revised to account for the new data, or if redundant clusters must be deleted [13]. The cluster structure is updated using the compatibility measure $\rho_i(k) \in [0,1]$, and the arousal index $a_i(k) \in [0,1]$. This paper uses the following expressions

$$\rho_i(k) = 1 - \frac{\|z(k) - v_i(k)\|}{\sqrt{p}} \tag{2}$$

$$a_i(k+1) = a_i(k) + \beta(1 - \rho_i(k) - a_i(k)) \tag{3}$$

If the values of the arousal indexes are such that

$$\underset{j\,=\,1,\,\cdots,\,c(k)}{\text{argmin}} \ \{a_j(k+1)\} > \tau$$

then a new cluster is created. Otherwise, the cluster center most compatible with the current input data is updated using:

$$v_s(k+1) = v_s(k) + \alpha \left(\rho_s(k)\right)^{1-a_s} (z(k) - v_s(k))$$
$$s = \underset{j\,=\,1,\,\cdots,\,c(k)}{\text{argmax}} \ \{\rho_j(k)\} \tag{4}$$

The parameters of the function of the rule consequent can be updated using the recursive least square (RLS). The ePL clustering verifies whether redundant clusters have been created checking the compatibility measures $\rho_{ij}(k)$ between current clusters centers i and j:

$$\rho_{ij}(k) = 1 - \frac{\|v_i(k) - v_j(k)\|}{\sqrt{p}} \tag{5}$$

and if it is greater than or equal to the compatibility threshold value λ, namely:

$$\rho_{ij}(k) \geq \lambda \tag{6}$$

then the cluster center $v_j(k)$ is declared redundant and removed. Otherwise, the current cluster structure remains as is. The output of the model at step k is the weighted average of the local models:

$$y(k) = \frac{\sum_{i=1}^{c(k)} \mu_i(k) y_i(k)}{\sum_{i=1}^{c(k)} \mu_i(k)} \tag{7}$$

3   Exact Feedback Linearization

Consider the single-input-single-output nonlinear system:

$$\dot{x}_1 = x_2$$
$$\dot{x}_2 = x_3$$
$$\cdots$$
$$\dot{x}_n = f(\mathbf{x}) + g(\mathbf{x})u$$
$$y = x_1 \tag{8}$$

or, in an equivalent form:

$$x^{(n)} = f(\mathbf{x}) + g(\mathbf{x})u$$
$$y^{(n)} = x_n \tag{9}$$

with $\mathbf{x} = \begin{bmatrix} x_1 \ x_2 \cdots x_n \end{bmatrix}^T \in D \subseteq \mathcal{R}^n$ the state vector, u and y the input and output of the system, $x^{(n)}$ and $y^{(n)}$ the n-th derivative of the state and of the output. $f(\mathbf{x})$ and $g(\mathbf{x}) \in D \subseteq \mathcal{R}^n$ are nonlinear functions of the state. We assume that $f(\mathbf{x})$ and $g(\mathbf{x})$ are smooth vector fields on $\mathcal{R}^n$, that is, they are infinitely differentiable functions [20, pp. 385]. The system (8) is controllable if $g(\mathbf{x}) \neq 0 \quad \forall \mathbf{x} \in D \subseteq \mathcal{R}^n$ in the control input [10]:

$$u = \frac{1}{g(\mathbf{x})}[v - f(\mathbf{x})] \tag{10}$$

where v is a linear control law. Notice that plugging (10) in (9) we cancel the nonlinearities and get the linear system [11]:

$$x^{(n)} = v. \tag{11}$$

Closed-loop stability is guaranteed if we choose $v = -K\mathbf{x}$ with K Hurwitz [4,9], that is, the state feedback control law produces a closed-loop system with all its roots strictly in the left-half complex plane [6], which means exponentially stable dynamics, namely $\mathbf{x}(t) \to 0$ as $t \to \infty$ [22].

4   Robust Evolving Granular Feedback Linearization

This section suggests a new robust evolving granular feedback linearization controller using the certainty equivalence principle [22]. Because in practice functions $f(\mathbf{x})$ and $g(\mathbf{x})$ of (8) are rarely known [20], the idea is to use estimates $\hat{f}(\mathbf{x}, \mathbf{e})$ and $\hat{g}(\mathbf{x}, \mathbf{e})$ of f and g and apply the following control law:

$$u = \frac{1}{\hat{g}(\mathbf{x}, \mathbf{e})}[v - \hat{f}(\mathbf{x}, \mathbf{e})] \tag{12}$$

From the system model (8) and control input (12) we get:

$$x^{(n)} = v + [f(\mathbf{x}) - \hat{f}(\mathbf{x}, \mathbf{e})] + [g(\mathbf{x}) - \hat{g}(\mathbf{x}, \mathbf{e})]u. \tag{13}$$

To develop a state feedback control law such that the output y asymptotically tracks a smooth reference signal $r(t)$, we choose the linear control law as follows:

$$v = r^{(n)} - \mathbf{Ke} \tag{14}$$

where $r^{(n)}$ is the n-th derivative of $r(t)$, $\mathbf{e} = \mathbf{r} - \mathbf{x} = \begin{bmatrix} e_1 & \dot{e}_1 & \cdots & e^{(n-1)} \end{bmatrix}^T$, and $\mathbf{r} = \begin{bmatrix} r & \cdots & r^{(n-1)} \end{bmatrix}^T$ is the vector of the reference values for each state x_i, and $\mathbf{K}$ is Hurwitz. Therefore, because $e = r(t) - y$, $e^{(n)} = r^{(n)} - y^{(n)}$, from (13) we obtain the following expression for the error dynamics:

$$e^{(n)} = \mathbf{Ke} - [f(\mathbf{x}) - \hat{f}(\mathbf{x}, \mathbf{e})] - [g(\mathbf{x}) - \hat{g}(\mathbf{x}, \mathbf{e})]u \tag{15}$$

Let functions estimates be such that $\hat{f}(\mathbf{x}, \mathbf{e}) = f(\mathbf{x}) + w_f$ and $\hat{g}(\mathbf{x}, \mathbf{e}) = g(\mathbf{x}) + w_g$, where w_f and w_g are the estimation errors of the ePL algorithm. Thus, the error dynamics can be expressed as:

$$e^{(n)} = \mathbf{Ke} + w \tag{16}$$

where $w = w_f + w_g u$. Equation (16) can be rewritten as:

$$\dot{\mathbf{e}} = A_c \mathbf{e} + B_c w \tag{17}$$

where

$$A_c = \begin{bmatrix} 0 & 1 & 0 & 0 & \cdots & 0 \\ 0 & 0 & 1 & 0 & \cdots & 0 \\ \vdots & \vdots & \vdots & \vdots & \ddots & \vdots \\ 0 & 0 & 0 & 0 & \cdots & 1 \\ k_n & k_{n-1} & \cdots\cdots\cdots & k_1 \end{bmatrix} \text{ and } B_c = \begin{bmatrix} 0 \\ 0 \\ \vdots \\ 0 \\ 1 \end{bmatrix}.$$

The new control loop topology suggested herein is shown in Fig. 1. Notice that the plant state $\mathbf{x}$, the error vector $\mathbf{e}$, and the linear control signal v are inputs to the ePL algorithm which produces estimates $\hat{f}(\mathbf{x}, \mathbf{e})$ and $\hat{g}(\mathbf{x}, \mathbf{e})$ to compute the control signal u.

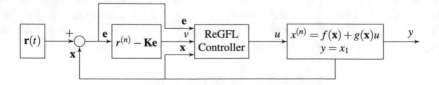

Fig. 1. Robust evolving granular feedback linearization ReGFL.

More specifically, the control signal u is produced by the ePL algorithm using functional fuzzy rules of the form:

IF $z(k)$ is A^i **THEN** $\hat{f}_i(\mathbf{x}, \mathbf{e}, k) = \gamma_f^i(k)\,\xi(k)$ **AND** $\hat{g}_i(\mathbf{x}, \mathbf{e}, k) = \gamma_g^i(k)\xi(k)$

where $\gamma_f^i(k)$ and $\gamma_g^i(k)$ are vectors of parameters, $z(k) = \begin{bmatrix} \mathbf{x}(k)\, e_1(k) \end{bmatrix}^T$, and $\xi(k) = \begin{bmatrix} \mathbf{x}^T(k)\, \mathbf{e}^T(k)\, 1 \end{bmatrix}^T$. The fuzzy rules use affine functions as local approximators of f and g. As mentioned in Sect. 2, the RLS can be used to estimate the parameters $\gamma_f^i(k)$ and $\gamma_g^i(k)$, and compute values $\hat{f}_i(\mathbf{x}, \mathbf{e}, k)$ and $\hat{g}_i(\mathbf{x}, \mathbf{e}, k)$ for $\hat{f}(\mathbf{x}, \mathbf{e})$ and $\hat{g}(\mathbf{x}, \mathbf{e})$ at each step k. In particular, we use the RLS algorithm with forgetting factor [14]:

$$
\begin{aligned}
\Upsilon_f^i(k) &= \frac{\Phi_f^i(k-1)\xi(k)}{\xi^T(k)\Phi_f^i(k-1)\xi(k) + \zeta} \\
\gamma_f^i(k) &= \gamma_f^i(k-1) + \Upsilon_f^i(k)\left[\left(K_p e_1(k) + K i_f \Sigma_{j=0}^k e_1(k)\right) - \xi^T(k)\gamma_f^i(k-1)\right] \\
\Phi_f^i(k) &= \frac{1}{\zeta}\left[\Phi_f^i(k-1) - \frac{\Phi_f^i(k-1)\xi(k)\xi^T(k)\Phi_f^i(k-1)}{\xi^T(k)\Phi_f^i(k-1)\xi(k) + \zeta}\right]
\end{aligned}
\tag{18}
$$

and

$$
\begin{aligned}
\Upsilon_g^i(k) &= \frac{\Phi_g^i(k-1)\xi(k)}{\xi^T(k)\Phi_g^i(k-1)\xi(k) + \zeta} \\
\gamma_g^i(k) &= \gamma_g^i(k-1) + \Upsilon_g^i(k)\left[\left(K i_g \Sigma_{j=0}^k e_1(k)\right) - \xi^T(k)\gamma_g^i(k-1)\right] \\
\Phi_g^i(k) &= \frac{1}{\zeta}\left[\Phi_g^i(k-1) - \frac{\Phi_g^i(k-1)\xi(k)\xi^T(k)\Phi_g^i(k-1)}{\xi^T(k)\Phi_g^i(k-1)\xi(k) + \zeta}\right]
\end{aligned}
\tag{19}
$$

where $\Upsilon_q^i(k)$ is the Kalman gain of the i-th rule at the k th step ($q = f$ or g), $\Phi_q^i(k)$ is the covariance matrix, and ζ the forgetting factor.

At each step k, the values of the functions $\hat{f}(\mathbf{x}, \mathbf{e})$ and $\hat{g}(\mathbf{x}, \mathbf{e})$ are found using (7), and next the of the control signal computed using (12). Algorithm 1 details the robust evolving granular feedback linearization (ReGFL) control algorithm.

5   Simulation Results

The surge tank depicted in Fig. 2 is a classic benchmark used to evaluate and compare feedback linearization approaches [2, 16, 19, 21]. In particular, we will use the instance suggested in [22]. The tank dynamics is:

$$
\dot{h} = \frac{-c\sqrt{2gh}}{A(h)} + \frac{1}{A(h)}u
\tag{20}
$$

where h is the level (m), g is the gravity constant (m/s^2), c is the cross-sectional area of the output pipe (m^2), u is the input (m^3/s), and $A(h)$ is the cross-sectional area of the tank at h (m^2), given by $A(h) = ah + b$, where a and b

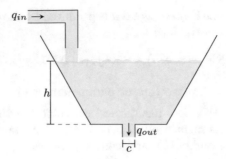

Fig. 2. Surge tank.

Algorithm 1. ReGFL control algorithm.

1: Input: $v(k), \mathbf{e}(k), \mathbf{x}(k), \quad k = 1, \cdots$
2: Output: $u(k)$
3: Choose initial values for $\alpha, \beta, \tau, \lambda \in [0,1]$
4: Choose $K_p, Ki_f, Ki_g, \delta_r$ and ζ
5: Set the initial clusters to $c(1) \geq 2, \quad v_i(k) \leftarrow \left[\mathbf{x}(k)\ e_1(k)\right]^T$ and $\quad i = 1, \cdots, c(1)$
6: Set the initial arousal index $a_i(k) \leftarrow 0, \quad i = 1, \cdots, c(1)$
7: **while** $k \leq \infty$ **do**
8: $\quad \left[v(k)\ \mathbf{e}(k)\ \mathbf{x}(k)\right] \leftarrow$ Read new data
9: $\quad z(k) \leftarrow \left[\mathbf{x}(k)\ e_1(k)\right]^T$
10: $\quad \xi(k) \leftarrow \left[\mathbf{x}^T(k)\ \mathbf{e}^T(k)\ 1\right]^T$
11: $\quad$ **for** $i = 1$ to $c(k)$ **do**
12: $\quad\quad$ Compute the compatibility index $\rho_i(k)$ using (2)
13: $\quad\quad$ Compute the arousal index $a_i(k+1)$ using (3)
14: $\quad$ **end for**
15: $\quad$ **if** $\underset{j=1,\cdots,c(k)}{\text{argmin}} \{a_j(k)\} > \tau$ **then**
16: $\quad\quad v^{c(k)+1} \leftarrow z(k)$
17: $\quad\quad c(k) \leftarrow c(k) + 1$
18: $\quad$ **else**
19: $\quad\quad$ Update the most compatible cluster $v_s(k)$ using (4)
20: $\quad\quad$ Update parameter vectors $\gamma_f^i(k)$ and $\gamma_g^i(k)$ using (18) and (19)
21: $\quad$ **end if**
22: $\quad$ **for** $i = 1$ to $c(k) - 1$ **and** $j = i + 1$ to $c(k)$ **do**
23: $\quad\quad$ Compute the compatibility index $\rho_{ij}(k)$ using (5)
24: $\quad\quad$ **if** $\rho_{ij}(k) \geq \lambda$ **then**
25: $\quad\quad\quad$ Delete the cluster center $v_j(k)$ and rule $\mathcal{R}_j$
26: $\quad\quad\quad c(k) \leftarrow c(k) - 1$
27: $\quad\quad$ **end if**
28: $\quad$ **end for**
29: $\quad$ **for** $i = 1$ to $c(k)$ **do**
30: $\quad\quad$ Compute firing degree $\mu_i(k)$ using (1)
31: $\quad$ **end for**
32: $\quad$ Compute estimates $\hat{f}(\mathbf{x}, \mathbf{e}, k)$ and $\hat{g}(\mathbf{x}, \mathbf{e}, k)$ using (7)
33: $\quad$ Compute control signal $u(k)$ using (12)
34: **end while**

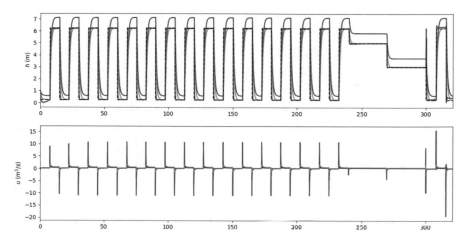

Fig. 3. Feedback linearization in surge tank level control: Square waveform.

are known constants. Assuming Euler approximation, the discrete model of the tank is:

$$h(k+1) = h(k) + T\left[-\frac{c\sqrt{2gh(k)}}{A(h(k))} + \frac{1}{A(h(k))}u(k)\right] \qquad (21)$$

where T is the sampling time. Tank simulation uses (21) and a state feedback controller $v = -Ke$ with gain set at $K = 1.25$ to stabilize the closed-loop system. The constants are $a = 0.01$, $b = 0.2$ and $c = 0.05$, and the sampling time $T = 0.1$ s. Notice that the sampling time is short enough to approximate reasonably well the continuous dynamics of the tank [19]. The ePL parameters are set as $\alpha = 0.005$, $\beta = 0.000125$, $\tau = 0.0075$, $\lambda = 0.85$, $\delta_r = 0.25$ and $\zeta = 0.98$. The values of these parameters were chosen as in [15, Cap. 4]. Parameter update formulas use $K_p = 0.55$, $Ki_f = 0.01$ and $Ki_g = 0.04$. Moreover, the actuator saturates at ± 50 m^3/s, that is, the control inputs are constrained as follows [22]:

$$u_s(k) = \begin{cases} 50 & \text{if } u_s(k) > 50 \\ u_s(k) & \text{if } -50 \leq u_s(k) \leq 50 \\ -50 & \text{if } u_s(k) < -50 \end{cases} \qquad (22)$$

To evaluate the performance of the ReGFL, we consider three scenarios characterized three reference signals, a square, a sawtooth, and a triangular waveform, respectively, as in [2].

Simulation results for each scenario are depicted in Figs. 3, 4, and 5. The following convention is adopted: control performance with nominal tank parameter with EFL (green line); control performance with the value of the tank parameter c 50% smaller than the nominal value with EFL (red line) and with ReGFL (blue line). The reference is depicted in black dashed line. The EFL uses the same state feedback gain as ReGFL.

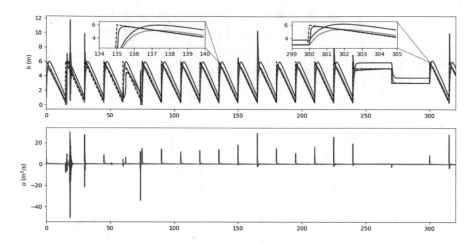

Fig. 4. Feedback linearization in surge tank level control: Sawtooth waveform.

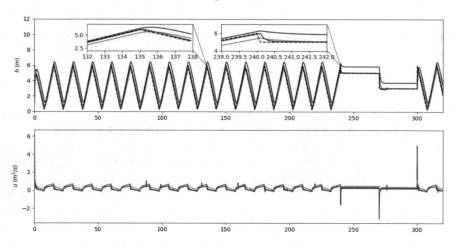

Fig. 5. Feedback linearization in surge tank level control: Triangular waveform.

Looking at Figs. 3, 4 and 5 we can visually verify that when a precise tank model is used, the EFL behaves precisely as expected (green line). However, when there is modeling mismatch (due to 50% variation in the value of c), one clearly observes the online adaptivity and superior performance of ReGFL. Alternatively, performance evaluation of the closed-loop system behavior can be quantified using the integral absolute error (IAE) and integral of time-weighted absolute error (ITAE), as usual in process control. Table 1 summarizes the results and shows that the ReGFL outperforms EFL. In Table 1, the lower the index value, the better the performance.

Table 1. Controllers performance.

Waveform	Method	$IAE(\%)$	$ITAE(\%)$	Rules
Square	EFL Nominal	164.4	22500	-
	EFL Mismatch	326.6	50850	-
	ReGFL Mismatch	48.0	6720	2
Sawtooth	EFL Nominal	132.6	18190	-
	EFL Mismatch	323.3	49780	-
	ReGFL Mismatch	72.1	7090	3
Triangular	EFL Nominal	142.3	19300	-
	EFL Mismatch	334.5	37890	-
	ReGFL Mismatch	46.3	6315	3

6   Conclusion

This paper developed a robust granular feedback linearization adaptive control approach based on evolving participatory learning and the certainty equivalence principle. The evolving granular feedback controller was evaluated using level control of the classic surge tank benchmark. Its performance was compared against the exact feedback linearization. Simulation results have shown that the evolving robust granular feedback linearization outperforms exact feedback linearization. Future research should compare the controller developed herein with alternative adaptive controllers, pursue convergence analysis of the algorithm, and derive estimates of performance bounds under load disturbances and parameter drifts.

Acknowledgments. The authors acknowledge the Brazilian National Council for Scientific and Technological Development (CNPq) for grant 305906/2014-3, and the Federal Center for Technological Education of Minas Gerais (CEFET-MG) for their support.

References

1. Angelov, P.: Autonomous Learning Systems: From Data Streams to Knowledge in Real-time, 1st edn. Wiley, Hoboken (2013)
2. Banerjee, S., Chakrabarty, A., Maity, S., Chatterjee, A.: Feedback linearizing indirect adaptive fuzzy control with foraging based online plant model estimation. Appl. Soft Comput. **11**(4), 3441–3450 (2011)
3. Van de Water, H., Willems, J.: The certainty equivalence property in stochastic control theory. IEEE Trans. Autom. Control **26**(5), 1080–1087 (1981)
4. DeJesus, E.X., Kaufman, C.: Routh-Hurwitz criterion in the examination of eigenvalues of a system of nonlinear ordinary differential equations. Phys. Rev. A **35**, 5288–5290 (1987)
5. Dinh, T.Q., Marco, J., Yoon, J.I., Ahn, K.K.: Robust predictive tracking control for a class of nonlinear systems. Mechatronics **52**, 135–149 (2018)

6. Dorf, R.C., Bishop, R.H.: Modern Control Systems, 9th edn. Prentice-Hall Inc., Upper Saddle River (2000)
7. Esfandiari, F., Khalil, H.K.: Output feedback stabilization of fully linearizable systems. Int. J. Control **56**(5), 1007–1037 (1992)
8. Freidovich, L.B., Khalil, H.K.: Performance recovery of feedback-linearization-based designs. IEEE Trans. Autom. Control **53**(10), 2324–2334 (2008)
9. Ho, M.T., Datta, A., Bhattacharyya, S.P.: An elementary derivation of the Routh-Hurwitz criterion. IEEE Trans. Autom. Control **43**(3), 405–409 (1998)
10. Isidori, A.: Nonlinear Control Systems, 3rd edn. Springer, London (1995)
11. Khalil, H.: Nonlinear Systems, 3rd edn. Prentice Hall, Upper Saddle River (2002)
12. Leite, D., Palhares, R., Campos, V., Gomide, F.: Evolving granular fuzzy model-based control of nonlinear dynamic systems. IEEE Trans. Fuzzy Syst. **23**(4), 923–938 (2015)
13. Lima, E., Hell, M., Ballini, R., Gomide, F.: Evolving Fuzzy Modeling Using Participatory Learning, pp. 67–86. Wiley, Hoboken (2010)
14. Ljung, L.: System Identification: Theory for the User, 2nd edn. Prentice-Hall Inc., Upper Saddle River (1999)
15. Lughofer, E.: Evolving Fuzzy Systems, 1st edn. Springer, Heidelberg (2011)
16. Oliveira, L., Leite, V., Silva, J., Gomide, F.: Granular evolving fuzzy robust feedback linearization. In: Evolving and Adaptive Intelligent Systems, Ljubljana, June 2017
17. Park, J., Seo, S., Park, G.: Robust adaptive fuzzy controller for nonlinear system using estimation of bounds for approximation errors. Fuzzy Sets Syst. **133**(1), 19–36 (2003)
18. Passino, K.: Biomimicry of bacterial foraging for distributed optimization and control. IEEE Control Syst. Mag. **22**(3), 52–67 (2002)
19. Passino, K., Yurkovich, S.: Fuzzy Control, 1st edn. Addison-Wesley, Boston (1997)
20. Sastry, S.: Nonlinear Systems - Analysis, Stability and Control, 1st edn. Springer, Heidelberg (1999)
21. Silva, J., Oliveira, L., Gomide, F., Leite, V.: Avaliação experimental da linearização por realimentação granular evolutiva. In: Proceedings Fifth Brazilian Conference on Fuzzy Systems, Fortaleza, CE, Brazil, June 2018
22. Slotine, J., Li, W.: Applied Nonlinear Control, 1st edn. Prentice Hall, Upper Saddle River (1991)
23. Wang, L.: Stable adaptive fuzzy controllers with application to inverted pendulum tracking. IEEE Trans. Syst. Man Cybern. Part B **26**(5), 677–691 (1996)

Comparative Analysis of Type-1 Fuzzy Inference Systems with Different Sugeno Polynomial Orders Applied to Diagnosis Problems

Emanuel Ontiveros-Robles[✉], Patricia Melin, and Oscar Castillo

Tijuana Institute of Technology,
Calzada Tecnologico s/n, Tomas Aquino, 22414 Tijuana, B.C., Mexico
emanuel.ontiveros18@tectijuana.edu.mx, {pmelin,ocastillo}@tectijuana.mx

Abstract. Fuzzy Logic has been implemented successfully for different kind of problems. One of the interesting problems that had been solved with Fuzzy Logic is the classification problem, however, there exist an opportunity to improve this system to be competitive in the realm of classifications problems with respect another kind of methods for example Artificial Neural Networks. The present paper is focused in a specific application of classification problems, the diagnosis systems, this problem consists in training an intelligent system to learn the relationship between symptoms and diagnosis. This kind of problems are usually based in powerful non-linear methods for example Modular Neural-Networks or complex hybrids models, however, in this paper are applied the Type-1 Takagi Sugeno Fuzzy Systems (TSK) but analyzing the improvement of their performance by increasing the order of the Sugeno polynomial, the objective is to evaluate if is possible to improve the performance of the TSK systems applied in diagnosis problems. The conventional Takagi-Sugeno Fuzzy Systems are based in the aggregation of first-order polynomial but it is interesting to observe the effect of increase the order of this polynomial, the TSK Fuzzy Diagnosis Systems are evaluated by their accuracy obtained in ten benchmark dataset of the UCI Dataset Repository, for different kind of diseases and different difficult levels.

1 Introduction

Nowadays, the Fuzzy Logic representation is well accepted in different kind of applications, for example, control applications [1–7], image processing [8,9], industrial applications [17,18], and others. An interesting field of computational system is the diagnosis system, that is a classification problem, in the current paper it is evaluated the accuracy of Fuzzy Systems in diagnosis problems, for example the problems presented in [10–12], it is interesting to observe that the Fuzzy Systems are not commonly used for this kind of problems, usually, the diagnosis systems are solved helped with powerful algorithms or methods for example, Modular Neural Networks or hybrid methods, but, in this paper, we

© Springer Nature Switzerland AG 2019
R. B. Kearfott et al. (Eds.): IFSA 2019/NAFIPS 2019, AISC 1000, pp. 453–465, 2019.
https://doi.org/10.1007/978-3-030-21920-8_41

explore the performance of the Takagi-Sugeno Fuzzy Systems for this problems. The Takagi-Sugeno Fuzzy Systems are frequently used for modelling non-lineal problems, and offer versatility to be combined with different kind of methods or systems. For example, they are the used in the Adaptive Neuro Fuzzy Inference Systems (ANFIS) that is a hybrid method of Fuzzy Systems with Artificial Neural Networks. The organization of the present paper is explained as follows: Sect. 2 a brief introduction to Takagi-Sugeno Fuzzy Systems, that are the fundament of the paper, Sect. 3 talks about the Takagi-Sugeno Fuzzy Diagnosis Systems, that are the proposed approach to diagnosis based in fuzzy logic, Sect. 4 present the experimental results realized with ten datasets of UCI Dataset repository and finally Sect. 5 presents the conclusion of the work.

2 Takagi-Sugeno Fuzzy Systems

Fuzzy Logic is a logic model that expands the conventional logic [16], that handle two membership values, and introduces the concept of membership degree, this degree is within [0, 1]. This is expressed in Eq. (1).

$$A = \{(x, \mu_A(c)) | x \exists X\},\tag{1}$$

where $\mu_A(c)$ is the membership degree of x to the set A. This evolution of conventional logic allows to model the human knowledge in special systems called Fuzzy Inference Systems through rules that are usually called knowledge base, these rules are related with special logical operators called T-Norm and S-Norm and relates the membership degree associated with the input signals features (big, small, close, far) with output signals features (go forward, go back). There is a different kind of Fuzzy Inference Systems, for example, Mamdani FIS, Tsukamoto FIS and Takagi-Sugeno FIS. The present paper is focused in the Takagi-Sugeno FISs, this kind of FIS had been implemented successfully in several problems because their output is polynomial. The TSK Fuzzy rules are expressed as follows (Eq. (2)).

$$R^I : if\ x_1\ is\ A_1^i\ and\ x_2\ is\ A_2^i\ ...\ x_n\ is\ A_n^i\ then$$
$$f^i = f_i(x_1, x_2...x_n; \widetilde{a^l}) = f_i(\overrightarrow{x}; \overrightarrow{a^i}) = a_0^i + a_1^i x_1 + ... + a_n^i x_n,\tag{2}$$

The output of the fuzzy system is computed by the following Eq. (3)

$$\frac{\sum_{i=1}^c R^i f^i}{\sum_{i=1}^c R^i},\tag{3}$$

Although the conventional form of TSK output denoted for (3) is a first-order polynomial, this approach can be expanded to a high-order polynomial output, that are proposed in [17] and their expression is described in (4).

$$f_i(\overrightarrow{x}; \overrightarrow{a}^i) = a_0^i + a_1^i x_1 + ... + a_n^i x_n + a_{n+1}^i x_1^2 + ... + a_{2n}^i x_n^2 + ...a_{(m-1)n)}^i x_1^m + ... + a_{mn}^i x_n^m,\tag{4}$$

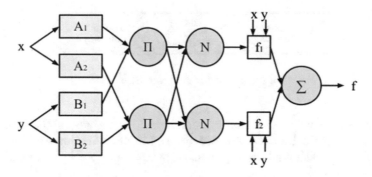

Fig. 1. ANFIS architecture

That can be simplified as follows (5)

$$f_i(\overrightarrow{x};\overrightarrow{a}^i) = a_0^i + \sum_{k=1}^{m}\sum_{s=1}^{n} a_{((k-1)n+s)}^i x_s^k,$$ (5)

3   Takagi-Sugeno Fuzzy Diagnosis System

For the present paper, it is proposed to use the architecture of the Neuro Fuzzy Inference System proposed for Jang in [18], the architecture consist on a neural network with five layers and it is illustrated in Fig. 1.

Layer 0: Inputs

$$X = [x_{p,1}, ..., x_{p,n}],$$ (6)

Layer 1: Fuzzification

$$\mu_{F_i^l}(x_{p,i}) = e^{\frac{(x_{p,i}+m_{l,i})^2}{\sigma_{l,i}^2}},$$ (7)

Layer 2: Compute firing strengths (T-Norm as Product) Remembering the rules are defined by the number of input membership functions and are the combination of these.

$$\alpha_{p,r} = \prod_i \mu_{F_i^{lR}},$$ (8)

Layer 3: Normalization

$$\phi_R = \frac{\alpha_{p,R}}{\sum_{i=1}^{m}\alpha_{i,R}},$$ (9)

Layer 4: Sugeno Polynomial

$$f_i(\overrightarrow{x};\overrightarrow{a}^i) = \phi_0^i + \sum_{k=1}^{m}\sum_{s=1}^{n} \phi_{((k-1)n+s)}^i x_s^k,$$ (10)

Fig. 2. Steps for generate TSK FDS

It is proposed to be used Gaussian membership functions in the inputs. This kind of membership function are widely used in classification approaches with fuzzy logic, the mathematical representation can be appreciated in Eq. (11).

$$gaussmf(x, \sigma, c) = e^{\frac{-(x-c)^2}{2(\sigma^2)}},$$ (11)

The process to generate the TSK FDS is divided in three steps and these are presented in Fig. 2. This process consists in obtaining the parameters of the proposed TSK FDS, the first step is obtaining the centers, this centers can be obtained by the implementation of clustering algorithms, for example, k-nearest neighbors, Fuzzy C-Means or Subtractive method, the centers of the Gaussian membership functions are the center of the clusters obtained in this step, in this case we proposed to use FCM. The second step is obtaining the standard deviations of the Gaussian membership functions, these parameters can be obtained with statistical methods for example, in this case, we obtain the standard deviations helped in a non-lineal regression. Finally, the third step is obtaining the TSK Coefficients, these values can be obtained with optimization methods or mathematical methods, in this case we proposed to use least-square error.

4   Experimental Results

This Section introduces the experiments realized in order to evaluate the impact of the Sugeno polynomial order in the accuracy of the fuzzy diagnosis system. The experiments in evaluates the accuracy obtained for the FDS with different orders of Sugeno polynomial, and reporting the average of 30 experiments, on the other hand, there are also documented the performance in the training data in order to observe if there exist any relationship for training data and the accuracy for test data, it is important to note that the 70 The experiments are used as benchmark problems ten datasets that can be founded in the UCI Dataset Repository, these datasets correspond to different kind of diseases and are widely used for test different classification methods. The benchmark datasets that are proposed to be used are listed in Table 1.

Figures 3, 4, 5, 6, 7, 8, 9, 10, 11 and 12 illustrate the obtained results, the results are measured in accuracy and the graphs represent the accuracy obtained for different order of Sugeno polynomial.

For the WBCD dataset can be observed the overfitting effect on the system, the increasing of the Sugeno polynomial improve the accuracy for the training data however, the performance of the testing data is in decreasing. For

Table 1. Benchmark datasets

Dataset	Attributes	Instances
Breast Cancer Wisconsin (Original) Data Set	10	699
Breast Cancer Wisconsin (Diagnostic) Data Set	32	569
Haberman's Survival Data	3	306
Mammographic Mass Data Set	6	906
Pima Indians Diabetes Data Set	9	768
Fertility Data Set	10	100
Immunotherapy Data Set	8	90
Cryotherapy Data Set	7	90
Breast Cancer Coimbra	10	116
Heart dataset	13	270

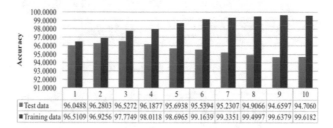

Fig. 3. WBCD dataset- Results

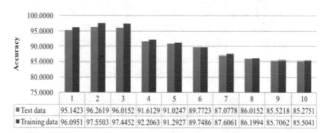

Fig. 4. WDCB dataset- Results

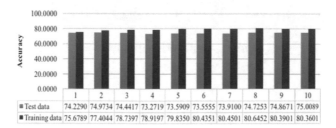

Fig. 5. Haberman dataset- Results

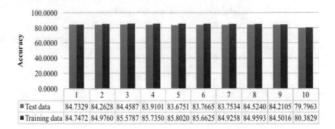

Fig. 6. Mammograms dataset- Results

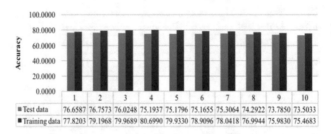

Fig. 7. Diabetes dataset- Results

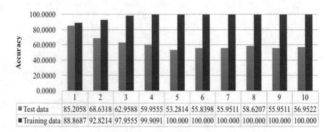

Fig. 8. Fertility dataset- Results

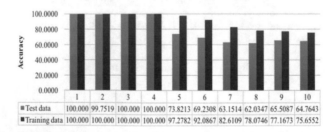

Fig. 9. Immunotherapy- Results

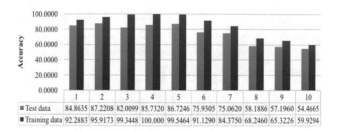

	1	2	3	4	5	6	7	8	9	10
▪ Test data	84.8635	87.2208	82.0099	85.7320	86.7246	75.9305	75.0620	58.1886	57.1960	54.4665
▪ Training data	92.2883	95.9173	99.3448	100.000	99.5464	91.1290	84.3750	68.2460	65.3226	59.9294

Fig. 10. Cryotherapy Data Set- Results

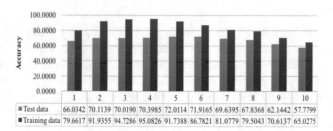

	1	2	3	4	5	6	7	8	9	10
▪ Test data	66.0342	70.1139	70.0190	70.3985	72.0114	71.9165	69.6395	67.8368	62.1442	57.7799
▪ Training data	79.6617	91.9355	94.7286	95.0826	91.7388	86.7821	81.0779	79.5043	70.6137	65.0275

Fig. 11. Breast Cancer Coimbra- Results

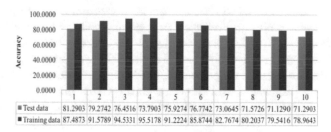

	1	2	3	4	5	6	7	8	9	10
▪ Test data	81.2903	79.2742	76.4516	73.7903	75.9274	76.7742	73.0645	71.5726	71.1290	71.2903
▪ Training data	87.4873	91.5789	94.5331	95.5178	91.2224	85.8744	82.7674	80.2037	79.5416	78.9643

Fig. 12. Heart dataset- Results

the WDCB dataset can be observed the that the performance of training data and testing data are similar however, the increasing of the Sugeno polynomial decrease the performance obtained. The performance obtained for Haberman dataset is very similar for the different Sugeno polynomial order, however can be observed an improvement with the increasing of the polynomial order. The performance obtained for the Mammograms dataset is very similar for the different Sugeno polynomial order, and the best result can be obtained for the first order. The Diabetes dataset have a very similar performance for the different Sugeno polynomial order, however the performance tends to decrease with the increasing of the polynomial order. In the Fertility dataset can be appreciated the effect of the overfitting, the test data performance decrease with the improvement of the training data performance, the best result can be found with the first order of Sugeno polynomial. For the Immunotherapy dataset the performance is perfect by low orders of Sugeno polynomial but decrease with the increasing

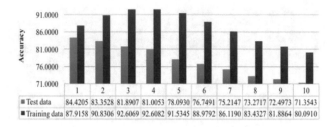

	1	2	3	4	5	6	7	8	9	10
Test data	84.4205	83.3528	81.8907	81.0053	78.0930	76.7491	75.2147	73.2717	72.4973	71.3543
Training data	87.9158	90.8306	92.6069	92.6082	91.5345	88.9792	86.1190	83.4327	81.8864	80.0910

Fig. 13. Results summary

of the polynomial order. In this case, for the Cryotherapy Data Set the performance is better with medium orders of Sugeno polynomial. Similar than the last dataset, for the Breast Cancer Coimbra the performance is better with medium orders of Sugeno polynomial. Finally, for the Heart dataset the better performance can be obtained with the first order of Sugeno Polynomial. As you can observe, the performance does not change significantly for the different order of Sugeno polynomial, however, in several of the cases, the best performance is obtained for first-order polynomial, then, Fig. 13 illustrates the average of the evaluated datasets.

It is interesting to observe that the better performance is obtained for first-order Sugeno polynomial and exist an over fitting effect, when the performance of the test data decrease, the performance of the training data grow up. Table 2 report the best performance obtained for the different datasets and document also the corresponding Sugeno polynomial order and two metrics related with the data, the fractal dimension and the data entropy, this metrics are documented in order to find a relation with them and the order of Sugeno polynomial.

Table 2. Best results

Dataset	Average	Std. Dev	Order	Fractal Dim.	Entropy
Breast Cancer Wisconsin (Original) Data Set	96.53	1.15	3	0.993	0.93
Breast Cancer Wisconsin (Diagnostic) Data Set	96.26	1.84	3	0.9628	0.9531
Haberman's Survival Data	75.01	1.82	10	0.9301	0.8326
Mammographic Mass Data Set	84.73	1.82	1	0.917	0.9947
Pima Indians Diabetes Data Set	76.76	2.5	2	0.8884	0.9311
Fertility Data Set	85.21	5.57	1	0.8924	0.5221
Immunotherapy Data Set	100	0	1	1	0.7186
Cryotherapy Data Set	87.22	7.39	2	1	0.9964
Breast Cancer Coimbra	72.01	5.72	5	0.9196	0.9916
Heart dataset	81.29	3.33	1	0.9522	0.9913

Tables 3, 4, 5, 6, 7, 8, 9 and 10 documents some of the results reported in the literature, they are presented in order to compare the performance obtained with respect to other kinds of methods reported in the literature, for examples, decision trees, Support-Vector-Machine or Artificial Neural Networks.

Table 3. Breast Cancer Wisconsin (Original) Data Set

Author	Method	Average	Std. dev.	Reference
Yoon	SVM	97.1		[13]
Goncalves	Neuro-Fuzzy	98.26		[19]
Elyan	Random Forest	96.73		[20]
Roy	SMO	97.85		[21]
Sheng	NN	97.71		[14]
Proposed approach		96.53	1.15	

Table 4. Breast Cancer Wisconsin (Diagnostic) Data Set

Author	Method	Average	Std. dev.	Reference
Yoon	SVM	97.68		[13]
Roy	SMO	98.77		[21]
Boros	LAD	96.9		[22]
Proposed approach		96.26	1.84	

Table 5. Haberman's Survival Data

Author	Method	Average	Std. dev.	Reference
Morente-Molinera	Supervised Classification	74.2		[23]
Kahraman	ABC-based k-nn	87.28		[24]
YoungII	V-synth	79		[25]
Proposed approach		75.01	1.82	

Table 6. Mammographic Mass Data Set

Author	Method	Average	Std. dev.	Reference
Saritas	ANN	85.5		[15]
Nugroho	DT	83.3		[26]
Zadeh Shirazi	SOM	94.5		[27]
Elyan	Random Forests	83.83		[20]
Proposed approach		84.73	1.82	

Table 7. Pima Indians Diabetes Data Set

Author	Method	Average	Std. dev.	Reference
Goncalves	Neuro-Fuzzy	78.6		[19]
Kahramanli	FNN	84.2		[28]
Güneş	GDA-LS-SVM	82.05		[29]
Mansourypoor	RLEFRBS	84		[30]
Roy	SMO	79.3		[21]
Sheng	NN	79.6		[14]
Proposed approach		76.76	2.5	

Table 8. Immunotherapy Data Set

Author	Method	Average	Std. dev.	Reference
Khatri	DT	96.6		[31]
Akben	DT	90		[32]
Khozeimeh	FIS	83.33		[33]
Proposed approach		100	0	

Table 9. Cryotherapy Data Set

Author	Method	Average	Std. dev.	Reference
Khatri	DT	98.9		[31]
Akben	DT	94.4		[32]
Khozeimeh	FIS	80		[33]
Proposed approach		87.22	7.39	

Table 10. Heart dataset

Author	Method	Average	Std. dev.	Reference
Elyan	Random Forests	83.96		[34]
Ustun	SLIM	83.5		[35]
Sheng	NN	83.6		[14]
Proposed approach		81.29	3.33	

5   Conclusions

Based in the realized experiments we can conclude that the use of low order of low-order polynomial in T1 TSK FDS is recommended. This can be explained because when the order of Sugeno polynomial is increased, the system is most complex and tends to memorize the diagnosis and it is not useful for another data. As can be observed in the experiment, there not exist relationship between the accuracy obtained for training data and test data, so, optimize a FDS based in the training data produce an over fitting. Another conclusion, based in the consulted literature, is that the fuzzy logic does not show the better results in comparison with respect another method designed for classifications or most complex methods for example Artificial Neural Networks. However, the obtained results are similar than other complex methods and as future work it is possible to analyze the performance by using Type-2 Fuzzy Logic [36,37] We also could consider other application areas like in [38,39].

References

1. Caraveo, C., Valdez, F., Castillo, O.: Optimization of fuzzy controller design using a new bee colony algorithm with fuzzy dynamic parameter adaptation. Appl. Soft Comput. **43**, 131–142 (2016)
2. Castillo, O., Amador-Angulo, L., Castro, J.R., Garcia-Valdez, M.: A comparative study of type-1 fuzzy logic systems, interval type-2 fuzzy logic systems and generalized type-2 fuzzy logic systems in control problems. Inf. Sci. **354**, 257–274 (2016)
3. Castillo, O., Melin, P., Alanis, A., Montiel, O., Sepulveda, R.: Optimization of interval type-2 fuzzy logic controllers using evolutionary algorithms. Soft Comput. **15**, 1145–1160 (2011)
4. Cervantes, L., Castillo, O.: Type-2 fuzzy logic aggregation of multiple fuzzy controllers for airplane flight control. Inf. Sci. **324**, 247–256 (2015)
5. Ontiveros-Robles, E., Melin, P., Castillo, O.: Comparative analysis of noise robustness of type 2 fuzzy logic controllers. Kybernetika **54**, 175–201 (2018)
6. Roose, A.I., Yahya, S., Al-Rizzo, H.: Fuzzy-logic control of an inverted pendulum on a cart. Comput. Electr. Eng. **61**, 31–47 (2017)
7. Melin, P., Ontiveros-Robles, E., Gonzalez, C.I., Castro, J.R., Castillo, O.: An approach for parameterized shadowed type-2 fuzzy membership functions applied in control applications. Soft Comput. **23**, 3887–3901 (2018)
8. Gonzalez, C.I., Melin, P., Castro, J.R., Castillo, O., Mendoza, O.: Optimization of interval type-2 fuzzy systems for image edge detection. Appl. Soft Comput. **47**, 631–643 (2016)
9. Melin, P., Gonzalez, C.I., Castro, J.R., Mendoza, O., Castillo, O.: Edge-detection method for image processing based on generalized type-2 fuzzy logic. IEEE Trans. Fuzzy Syst. **22**, 1515–1525 (2014)
10. Khooban, M.H., Vafamand, N., Liaghat, A., Dragicevic, T.: An optimal general type-2 fuzzy controller for Urban Traffic Network. ISA Trans. **66**, 335–343 (2017)
11. Juang, C.F., Juang, K.J.: Circuit Implementation of data-driven TSK-type interval type-2 neural fuzzy system with online parameter tuning ability. IEEE Trans. Ind. Electron. **64**, 4266–4275 (2017)
12. Debnath, J., Majumder, D., Biswas, A.: Air quality assessment using weighted interval type-2 fuzzy inference system. Ecol. Inform. **46**, 133–146 (2018)
13. Wang, H., Zheng, B., Yoon, S.W., Ko, H.S.: A support vector machine-based ensemble algorithm for breast cancer diagnosis. Eur. J. Oper. Res. **267**, 687–699 (2018)
14. Sheng, W., Shan, P., Chen, S., Liu, Y., Alsaadi, F.E.: A niching evolutionary algorithm with adaptive negative correlation learning for neural network ensemble. Neurocomputing **247**, 173–182 (2017)

15. Saritas, I.: Prediction of Breast Cancer Using Artificial Neural Networks. J. Med. Syst. **36**, 2901–2907 (2012)
16. Zadeh, L.A.: Fuzzy sets. Inf. Control **8**, 338–353 (1965)
17. Castro, J.R., Castillo, O., Sanchez, M.A., Mendoza, O., Rodríguez-Diaz, A., Melin, P.: Method for higher order polynomial Sugeno Fuzzy inference systems. Inf. Sci. **351**, 76–89 (2016)
18. Jang, J.-S.R.: ANFIS: adaptive-network-based fuzzy inference system. IEEE Trans. Syst. Man Cybern. **23**, 665–685 (1993)
19. Goncalves, L.B., Vellasco, M.M.B.R., Pacheco, M.A.C., de Souza, F.J.: Inverted hierarchical neuro-fuzzy BSP system: a novel neuro-fuzzy model for pattern classification and rule extraction in databases. IEEE Trans. Syst. Man Cybern. Part C Appl. Rev. **36**, 236–248 (2006)
20. Elyan, E., Gaber, M.M.: A fine-grained random forests using class decomposition: an application to medical diagnosis. Neural Comput. Appl. **27**, 2279–2288 (2016)
21. MadhuSudana Rao, N., Kannan, K., Gao, X., Roy, D.S.: Novel classifiers for intelligent disease diagnosis with multi-objective parameter evolution. Comput. Electr. Eng. **67**, 483–496 (2018)
22. Boros, E., Hammer, P.L., Ibaraki, T., Kogan, A., Mayoraz, E., Muchnik, I.: An implementation of logical analysis of data. IEEE Trans. Knowl. Data Eng. **12**, 292–306 (2000)
23. Morente-Molinera, J.A., Mezei, J., Carlsson, C., Herrera-Viedma, E.: Improving supervised learning classification methods using multigranular linguistic modeling and fuzzy entropy. IEEE Trans. Fuzzy Syst. **25**, 1078–1089 (2017)
24. Kahraman, H.T.: A novel and powerful hybrid classifier method: development and testing of heuristic k-nn algorithm with fuzzy distance metric. Data Knowl. Eng. **103**, 44–59 (2016)
25. Young, W.A., Nykl, S.L., Weckman, G.R., Chelberg, D.M.: Using Voronoi diagrams to improve classification performances when modeling imbalanced datasets. Neural Comput. Appl. **26**, 1041–1054 (2015)
26. Nugroho, K.A., Setiawan, N.A., Adji, T.B.: Cascade generalization for breast cancer detection. In: 2013 International Conference on Information Technology and Electrical Engineering (ICITEE), pp. 57–61. IEEE, Yogyakarta, Indonesia (2013)
27. Zadeh Shirazi, A., Chabok, S.J.S.M., Mohammadi, Z.: A novel and reliable computational intelligence system for breast cancer detection. Med. Biol. Eng. Comput. **56**, 721–732 (2018)
28. Kahramanli, H., Allahverdi, N.: Design of a hybrid system for the diabetes and heart diseases. Expert Syst. Appl. **35**, 82–89 (2008)
29. Polat, K., Güneş, S., Arslan, A.: A cascade learning system for classification of diabetes disease: generalized discriminant analysis and least square support vector machine. Expert Syst. Appl. **34**, 482–487 (2008)
30. Mansourypoor, F., Asadi, S.: Development of a reinforcement learning-based evolutionary fuzzy rule-based system for diabetes diagnosis. Comput. Biol. Med. **91**, 337–352 (2017)
31. Khatri, S., Arora, D., Kumar, A.: Enhancing decision tree classification accuracy through genetically programmed attributes for wart treatment method identification. Procedia Comput. Sci. **132**, 1685–1694 (2018)
32. Akben, S.B.: Predicting the success of wart treatment methods using decision tree based fuzzy informative images. Biocybern. Biomed. Eng. **38**, 819–827 (2018)
33. Khozeimeh, F., Alizadehsani, R., Roshanzamir, M., Khosravi, A., Layegh, P., Nahavandi, S.: An expert system for selecting wart treatment method. Comput. Biol. Med. **81**, 167–175 (2017)

34. Elyan, E., Gaber, M.M.: A genetic algorithm approach to optimising random forests applied to class engineered data. Inf. Sci. **384**, 220–234 (2017)
35. Ustun, B., Rudin, C.: Supersparse linear integer models for optimized medical scoring systems. Mach. Learn. **102**, 349–391 (2016)
36. Mendez, G.M., Castillo, O.: Interval type-2 TSK fuzzy logic systems using hybrid learning algorithm. In: The 14th IEEE International Conference on Fuzzy Systems 2005, FUZZ 2005, pp. 230–235. IEEE, Reno, Nevada, USA (2005)
37. Rubio, E., Castillo, O., Valdez, F., Melin, P., Gonzalez, C.I., Martinez, G.: An extension of the fuzzy possibilistic clustering algorithm using type-2 fuzzy logic techniques. Adv. Fuzzy Syst. **2017**, 1–23 (2017)
38. Melin, P., Castillo, O.: Intelligent control of complex electrochemical systems with a neuro-fuzzy-genetic approach. IEEE Trans. Ind. Electron. **48**, 951–955 (2001)
39. Melin, P., Castillo, O.: Adaptive intelligent control of aircraft systems with a hybrid approach combining neural networks, fuzzy logic and fractal theory. Appl. Soft Comput. **3**, 353–362 (2003)

Fault Tolerant Controller Using Interval Type-2 TSK Logic Control Systems: Application to Three Interconnected Conical Tank System

Himanshukumar R. Patel$^{(\boxtimes)}$ ⓘ and Vipul Shah

Dharmsinh Desai University, Nadiad 387001, Gujarat, India
{himanshupatel.ic,vashah.ic}@ddu.ac.in

Abstract. Type-2 fuzzy logic systems have recently been utilized in many control process due to their ability to model uncertainties. This paper proposes an inference mechanism for an interval type-2 Takagi-Sugeno-Kang fuzzy logic control system (IT2 TSK FLCS) when antecedents are type-2 fuzzy sets and consequent are crisp. This paper focus on fault-tolerant control application for the three inter-connected conical tank system with following cases: (1) Servo control without fault, and, (2) Regulatory control with the actuator, system component faults, and process disturbances. In both the cases, IAE and ISE are calculated. The methods presented in this paper facilitating the design of fault-tolerant controller design using IT2 TSK FLCs with significantly improved performance over type-1 FLCs approaches.

Keywords: Actuator fault · Fault tolerant control ·
Interval type 2 fuzzy logic control · System component fault ·
Three interconnected conical tank

1 Introduction

The knowledge embedded in Rule-Base systems, derived either from human experts or from clustering algorithms, is most of the times inconsistent due to interpersonal differences on the definition of the rule's membership functions or incomplete in some regions of the input/output space as a result from operation conditions not experienced during a model's training stage. Type-2 Fuzzy Logic Systems particularly focus on the mitigation of these problems and, with the development of simpler Interval Type-2 Fuzzy Sets and computationally efficient Type-Reduction algorithms, the range of its possible application scenarios has been broadly expanded in recent years [1,2]. The Type-2 fuzzy logic system (T-2 FLS), initially developed by Zadeh [1], is unique in its ability to model and handle uncertainty while also being able to handle complex control system

Supported by Department of IC Engineering, Dharmsinh Desai University.

R. B. Kearfott et al. (Eds.): IFSA 2019/NAFIPS 2019, AISC 1000, pp. 466–482, 2019.
https://doi.org/10.1007/978-3-030-21920-8_42

and linguistic variables. Over the years, a great deal of efforts have been put into reducing the computation cost, and thus, improving the performance of the Type-2 fuzzy controllers. The focus of these efforts has been mainly on the type reduction of fuzzy type-2 sets in the last stage of control computation process in a general T-2 FLC (e.g., [2–7]). In comparison, in this paper, we propose a technique for designing a T-2 FLC which reduces the computation cost both in input fuzzification and output signal processing. For this purpose, we use Interval Type-2 fuzzy sets to capture the input space. The Interval Type-2 fuzzy sets [1, 8, 9], were initially introduced to reduce the computational complexity of the general T-2 FLSs. Interval T-2 FLSs have become even more popular over the recent years because they can be extended to general T-2 FLSs [6, 10], while still preserving the major properties of general T-2 FLSs. On the other hand, to reduce the computation cost in the output processing part of the FLCs, we employ the Takagi-Sugeno-Kang (TSK) controller. The TSK controller or Sugeno controller was originally introduced by Sugeno et al. [11].

The TSK controller allows for the modeling of more complex output functions while greatly reducing the computation cost by parallel processing of the inputs and out-puts. While some work has been done on Type-2 TSK controller design (e.g., [12–14]), this paper provides a unique theoretical approach and theorem which together, simplifies the understanding of the Type-2 TSK controller design. The pro-posed algorithm ultimately requires little information (the upper and lower bounds for the firing levels and output constants) which significantly reduces the computation cost for calculating the output control signal.

Thus, the contribution of this paper is the development of an effective fault tolerant controller using Interval Type-2 fuzzy logic controller subject to actuator, system component faults, and process disturbances. Interval Type-2 fuzzy sets are used to capture the input space, and output processing is done by employing the TSK Fuzzy Logic Controller. An illustrative example and the simulation results for a three inter-connected conical tank arc provided which show the satisfactory performance of the developed fault tolerant controller as compared to type 1 fuzzy logic controller (T1 FLC). Integrating the interval type-2 FLC with Takagi-Sugeno rules makes it possible to use parallel processing of input firing levels and rule outputs which the Mamdani approach is not able to do. This simplifies the control structure so that only the upper and lower bounds of the firing levels and rule outputs suffices to calculate the controller output. Unlike many other methods, this technique directly generates a fuzzy type-1 output which can be simply defuzzified and hence, there is no need for type reduction.

The main contribution of the paper is to present novel model base approach for Fault-tolerant control using IT2 TSK FLC. The simulation results of proposed approach is compared with the fault-tolerant control scheme using conventional T1 FLC with same fault type and magnitude. The article focuses on controlling the highly nonlinear MIMO process having problem of interaction, changing dynamics, possible faults, and unmeasured disturbances. For validation of the proposed scheme standard three inter-connected conical tank system

(TICTS) is taken as a case study. The strong reasons are listed out in conclusion section for designing FTC using Interval Type-2 Takagi-Sugeno Kang FLC for proposed nonlinear MIMO (TICTS) system.

The rest of this paper is organized as follows. In Sect. 2, we briefly discuss the necessary preliminaries and Fuzzy sets background. In Sect. 3, we detail the development of the proposed IT-2 TSK FLC and discuss its different parts including the input processing, rule sets and output processing. Section 4 simulates the developed controller. The paper is concluded in Sect. 5.

2   Preliminaries

In this section, we briefly review the necessary background on Fuzzy Set Theory which is essential to explain the proposed Type-2 fuzzy controller. Some of the notations are borrowed from Mendel et al. [9].

Definition 1: Type-1 Fuzzy Sets (T-1 FSs)
A Type-1 Fuzzy Set [4] is composed of pairs of $(x, \mu_A(x))$ in which for each member of domain, $x \in X$, a membership value $\mu_A(x) \in [0, 1]$ can be defined as follows:

$$A = \{(x, \mu_A(x))) \mid \forall x \in X, \mu_A(x) \in [0, 1]\} = \sum_{x \in X} (x, \mu_A(x)) \qquad (1)$$

Here $\sum$ denotes the collection of elements of a set. Extending from Type-1 fuzzy sets, Type-2 fuzzy sets allow for capturing more degrees of uncertainty which can be defined as follows:

Definition 2: Type-2 Fuzzy Sets (T-2 FSs)
A Type-2 Fuzzy Set [8] is composed of triples $((x, u), \mu_A(x, u))$ in which for each member of domain $x \in X$, a primary membership value, $u \in J_x$ (J_x is the range of primary membership for a given x) and a secondary membership, $\mu_{\tilde{A}}(x, u)$ can be defined as follows:

$$\tilde{A} = \{(x, u), \mu_{\tilde{A}}(x, u)) \forall x \in X, \forall u \in J_x \subseteq [0, 1], \mu_{\tilde{A}}(x, u)) \in [0, 1]\} =$$
$$\sum_{u \in J_x} \sum_{x \in X} ((x, u), \mu_{\tilde{A}}(x, u)) \qquad (2)$$

Figure 1a shows a simple Type-2 fuzzy set for a case that X and J_x are connected sets, and $\mu_{\tilde{A}}$ is a continuous function. Type-2 fuzzy sets can be seen as a set of weighted Type-1 fuzzy sets, and can model uncertainty in memberships due to imprecise measurements. Despite the capabilities of this set up, particularly in capturing uncertainties, it is found to be computationally expensive. Alternatively we can use Interval Type-2 fuzzy sets, which significantly reduce the computation costs while maintaining major advantages of Type-2 fuzzy sets.

Definition 3: Interval Type-2 Fuzzy Sets (IT2 FSs)

An Interval Type-2 Fuzzy Set [9] is a Type-2 Fuzzy Set in which the secondary grade values are always unity:

$$\tilde{A} = \{((x,u),1)\, \forall x \in X, \forall u \in J_x \subseteq [0,1]\} = \sum_{u \in J_x} \sum_{x \in X} ((x,u),1) \qquad (3)$$

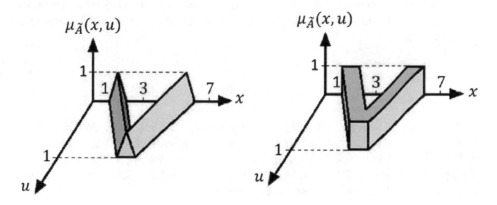

Fig. 1. (a) A type-2 FS. (b) An interval type-2 FS.

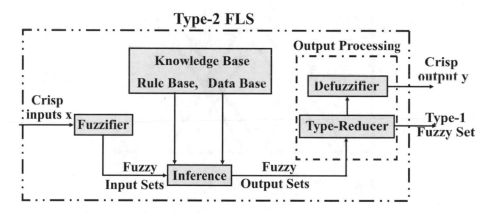

Fig. 2. IT2 TSK FLC structure [2].

3   Developing the Fault Tolerant Controller

The basic idea for developing the proposed IT2 TSK FLC is to use [19] to decompose Type-2 fuzzy sets into Type-1 fuzzy sets for which we can employ well-matured control techniques such as TSK to design the controller. Figure 2 shows the structure of the proposed IT2 TSK FLC. The blocks of this control structure are detailed in the following sections.

3.1 Interval Type-2 Fuzzification

– Membership Decomposition: The Fuzzifier block converts the control inputs (sensor readings), which are crisp values, to fuzzy values using the predefined Fuzzy Type-2 memberships. Here, we use IT2 fuzzy sets to describe the input space. Each input channel, X_i, can be captured by n_i membership functions as follows:

$$F_i = \sum_{j=1}^{n_i} F_i, j \tag{4}$$

Where F_i represents all MFs related to input channel X_i, including $F_i,1$, $F_i,2$, ..., F_i,n. The subscript j in F_i,j is used to indicate the jth individual MF in F_i. Figure 3 shows the IT2 input membership function for the input channel of a system with three IT2 member-ship sets. For simplicity, the secondary grade values are not shown as they are always unity.

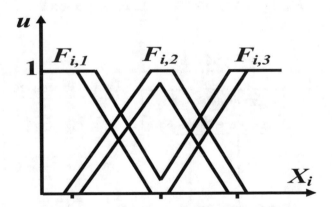

Fig. 3. IT2 membership functions for input X_i.

3.2 Inference Engine

In a fuzzy interval type-2 using the minimum or product t-norms operations, the ith activated rule $F^i (x_1,... x_n)$ produces the interval that is determined by two extremes

$\underline{f}^i (x_1,... x_n)$ and $\overline{f}^i (x_1,... x_n)$ like written below [14]:

$$F^i = (x_1, \cdots x_n) = \left[\underline{f}^i (x_1, \cdots x_n), \overline{f}^i (x_1, \cdots x_n) \right] \equiv \left[\underline{f}^i, \overline{f}^i \right] \tag{5}$$

Where $\underline{f}^i$ and $\overline{f}^i$ can be defined as follow:

$$\underline{f}^i = \underline{\mu}_{F_1^i}(x_1) * \cdots * \underline{\mu}_{F_n^i}(x_n) \tag{6}$$

$$\overline{f}^i = \overline{\mu}_{F_1^i}(x_1) * \cdots * \overline{\mu}_{F_n^i}(x_n) \tag{7}$$

3.3 Type Reducer

After definition of the rules and executing the inference, the type-2 fuzzy system resulting in type-1 fuzzy system is computed. In this part, the available methods to compute the centroid of type-2 fuzzy system using the extention principle are discussed [15]. The centroid of type-1 fuzzy system A is given by:

$$C_A = \frac{\sum_{i=1}^{n} z_i w_i}{\sum_{i=1}^{n} w_i} \tag{8}$$

Where: n represents the number of discretized domain of A, $z_i \in R$ and $w_i \in [0, 1]$. If each z_i and w_i is replaced by a type-1 fuzzy system (Z_i and W_i), with associated membership functions of $\mu_Z(z_i)$ and $\mu_W(w_i)$ respectively, and by using the extention principle, the generalized centroid for type-2 fuzzy system $\tilde{A}$ can be expressed by:

$$GC_{\tilde{A}} = \int_{z_1 \in Z_1} \cdots \int_{z_n \in Z_n} \int_{w_1 \in W_1} \cdots \int_{w_n \in W_n} [T_{i=1}^n \mu_Z(Z_i)^* T_{i=1}^n \mu_W(Z_i)] / \frac{\sum_{i=1}^{n} z_i w_i}{\sum_{i=1}^{n} w_i} \tag{9}$$

T is a t-norm and $GC_{\tilde{A}}$ is a type-1 fuzzy system. For an interval type-2 fuzzy system, it can be written:

$$GC_A = [y_l(x), y_r(x)] - \int_{y^M \in [y_l^1, y_r^1]} \cdots \int_{y^M \in [y_l^M, y_r^M]} \int_{f^1 \in [\underline{f}^1 \overline{f}^1]} \cdots \int_{f^M \in [\underline{f}^M \overline{f}^M]} 1 / \frac{\sum_{i=1}^{M} f^i y^i}{\sum_{i=1}^{M} f^i} \tag{10}$$

3.4 Defuzzifier

To get a crisp output from a type-1 fuzzy logic system, the type-reduced set must be defuzzified. The most common method to do this is to find the centroid of the type-reduced set. If the type-reduced set Y is discretized to n points, then the following expression gives the centroid of the type-reduced set:

In Fig. 4, graphical representation example is shown to computing the switching point y_l and y_r respectively. After the calculating the switching points further procedure can be done for calculating final defuzzified output.

$$y_{output}(x) = \frac{\sum_{i=1}^{n} y^i \mu(y^i)}{\sum_{i=1}^{m} \mu(y_i)} \tag{11}$$

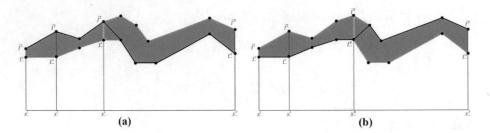

Fig. 4. (a) A computing y_l: switching from the upper bounds of the firing intervals to the lower bounds. (b) Computing y_r: switching from the lower bounds of the firing intervals to the upper bounds.

The output can be computed using the iterative Karnik Mendel Algorithms [16–19]. Therefore, the defuzzified output of an interval type-2 TSK FLC is:

$$y_{output}(x) = \frac{y_l(x) + y_r(x))}{2} \tag{12}$$

With:

$$y_l(x) = \frac{\sum_{i=1}^{M} f_l^i y_l^i}{\sum_{i=1}^{M} f_l^i} \quad y_r(x) = \frac{\sum_{i=1}^{M} f_r^i y_r^i}{\sum_{i=1}^{M} f_r^i} \tag{13}$$

This final crisp output, y represents the final response of the IT2 TSK FLC for the crisp inputs x and v. The interval $[y_l,\ y_r]$ represents its associate uncertainty range.

3.5   Interval Type-2 TSK Fuzzy Logic Controller

Model Rule i

if e_1 is $\tilde{F}_1^i$ and e_2 is $\tilde{F}_2^i$ and $\cdots e_n$ is $\tilde{F}_n^i$, Than $\dot{x} = A_i x(t) + B_i u(t))$, $i = 1, \cdots, M$ $\tag{14}$

Here M is the number of IF-THEN rules. $x(t) \in \mathbb{R}^{n \times 1}$ is state vector of the system. $u(t) \in \mathbb{R}^{m \times 1}$ is the input vector of the system. A is the state matrix of the system. B is the input matrix of the system.

Control Rule i

$if e_1$ is $\tilde{F}_1^i$ and e_2 is $\tilde{F}_2^i$ and $\cdots e_n$ is $\tilde{F}_n^i$, Than $y^i = \tilde{G}^i$ $i = 1, \cdots, M$ $\tag{15}$

where $\tilde{G}^i \in \mathbb{R}^{1 \times 1}$ is the output of the interval type-2 fuzzy system for the rule i using the centroid defuzzification. The procedure for processing the output value of the proposed control structure is described in Algorithm 1.

Algorithm 1. Interval Type-2 TSK Fuzzy Control
Input: crisp inputs, rule base & input and output MFs
Output: crisp output control signal & uncertainty range
Begin Procedure

Step 1: For the given set of input signals, compute upper and lower bound for the firing level of each rule ($\underline{f}^i$ and $\overline{f}^i$, i= 1,....,m using (6) and (7).

Step 2: For the given set of input signals, compute upper and lower bound for the rule outputs using (9) and (10).

Step 3: Compute the upper and lower bounds for the output signal (y_l and y_r) using (13).

Step 4: Compute the output control signal y using (12).

Step 5: Return output control signal, y and uncertainty range, [y_l and y_r].

End Procedure

Remark 1: Note that Step 1 and Step 2 are independent and can be processed in parallel thus contributing to reduced computation time. This is possible through the use of the TSK technique.

3.6   Proposed FTC Structure Using IT2 TSK FLC

In order to eliminate the peak overshoot, nonlinear system and model uncertainty, a continuous Interval Type-2 TSK FLC (IT2 TSK FLC) is used to approximate the discontinue control. The proposed control scheme is shown in Fig. 5.

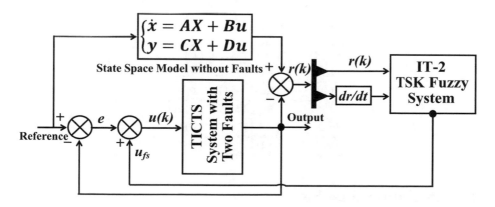

Fig. 5. Fault-tolerant controller structure using (IT2 TSK FLC) for TICTS.

The equivalent control (u_k) is calculated in such a way to have rate of change of r(k) is zero.

$$u_{fs} = IT2TSKFLC\left(r(k), r(\dot{k}))\right) \tag{16}$$

u_{fs} is the output of the IT2 TSK FLC, which depends on the normalized $r(k)$ and rate of change of $r(k)(k)$.

All the membership functions of the fuzzy input variable are chosen to be Gaussian for all upper and lower membership functions. The uses labels of the fuzzy variable residue and its derivative are: negative big (NB), negative small (NS), medium (M), Positive small (PS) and positive big (PB). Figure 6 presents the type-2 membership functions for the IT2 FLC. The corrective control is decomposed into five levels, so total rules can be 25 presented in Table 1.

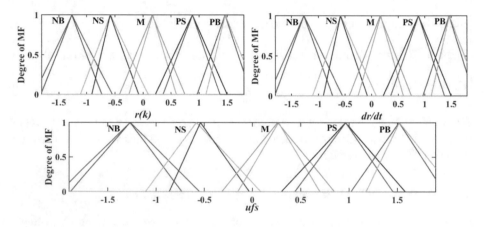

Fig. 6. Type-2 fuzzy set for input output for IT2 TSK FLC.

Table 1. Type 2 fuzzy rules

$\dot{r}(k)$					
u_{fs} & r(k)	NB	NS	M	PB	PB
NB	NB	M	M	PB	PB
NS	M	M	PS	PS	PB
M	M	PS	PS	PB	PB
PS	PS	PS	PB	PB	PB
PB	NB	NS	M	PS	PB

4   Simulation Results

To verify the proposed control algorithm, we have applied it to a Three Interconnected Conical Tank System (TICTS). In this nonlinear system, f_1 and f_2 is two inlet flow rate to that tank 1 and tank 2 respectively, which are control using CV_1 and CV_2. There are three outlet flow rate f_{o1}, f_{o2}, and f_{o3} from the conical tank 1, tank 2, and tank 3 respectively. The objective of the system is to control the tank 3 height using manipulated variable inlet flow rate f_1 and f_2. Control valve V_1 and V_2 introduce interaction into tank 3, this results nonlinear behavior to the system. Two fault introduce into the system to verify the effectiveness of the proposed fault-tolerant control algorithm, one is actuator fault and second one is system component fault (leak) faults. The Fig. 7 illustrated the prototype structure of Three Interconnected Conical Tank System (TICTS).

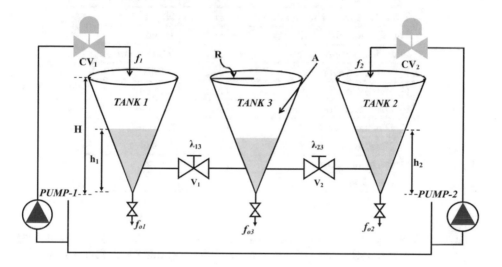

Fig. 7. Three interconnected conical tank system (TICTS) prototype.

The TICTS dynamics can be described as:

$$\frac{dh_1}{dt} = \frac{\left[f_1 - \frac{1}{3}h_1 \frac{dA(h_1)}{dt} - \lambda_1 \sqrt{h_1} - \lambda_{13}\sqrt{h_1 - h_3} \right]}{\frac{1}{3}\pi R^2 \frac{h_1^2}{H^2}} \tag{17}$$

$$\frac{dh_2}{dt} = \frac{\left[f_2 - \frac{1}{3}h_2 \frac{dA(h_2)}{dt} - \lambda_2 \sqrt{h_2} - \lambda_{23}\sqrt{h_2 - h_3} \right]}{\frac{1}{3}\pi R^2 \frac{h_2^2}{H^2}} \tag{18}$$

$$\frac{dh_3}{dt} = \frac{\left[\lambda_1 \sqrt{h_1 - h_3} + \lambda_{23}\sqrt{h_2 - h_3} - \frac{1}{3}h_3 \frac{dA(h_3)}{dt} - \lambda_3 \sqrt{h_3} \right]}{\frac{1}{3}\pi R^2 \frac{h_3^2}{H^2}} \tag{19}$$

Area of the canonical Tank 1 at height (h_1), Tank 2 at any height (h_2), and Tank 3 at any height (h_3) (Table 2)

$$A_1 = \frac{\pi R^2 h_1^2}{H^2}; A_2 = \frac{\pi R^2 h_2^2}{H^2}; A_3 = \frac{\pi R^2 h_3^2}{H^2} \tag{20}$$

Table 2. Three interconnected conical-tank system parameters.

Parameters	Symbol	Value with Unit
Total Height of the each tank	H	90 cm
Top Radius of the all three Tank	R	15 cm
Inlet flow rate of Tank 1	F_{in}	0.00278 cm^3/s
V_1 Valve Co-efficient	λ_{13}	5 cm^2/s
V_2 Valve Co-efficient	λ_{23}	4.25 cm^2/s
V_3 Discharge co-efficient	λ_1	0.85 cm^2/s
V_4 Discharge co-efficient	λ_2	1.03 cm^2/s
V_5 Discharge co-efficient	λ_3	0.1839 cm^2/s
Gravitational constant	g	9.81 m/s^2
Process Delay	τ_d	0 s

The output of the system considering $y = \begin{bmatrix} 1 & 1 & h_3 \end{bmatrix}^T$ and input vector $u = \begin{bmatrix} u_1 & u_2 \end{bmatrix}^T$. For the T-S model of the system following matrix is found form the system model:

$$A_1 = \begin{bmatrix} 0.0224 & 0 & 0.0224 \\ 0 & -0.025 & 0.0224 \\ 0.0033 & 0.0033 & -0.0132 \end{bmatrix} \quad A_2 = \begin{bmatrix} 0.0224 & 0 & 0.0224 \\ 0 & 0.025 & 0.0224 \\ 0.0033 & 0.0033 & 0.0132 \end{bmatrix}$$

$$B_1 = \begin{bmatrix} 33.8311 & 0 \\ 0 & 33.8311 \\ 0 & 0 \end{bmatrix} \quad B_2 = \begin{bmatrix} 3.8311 & 0 \\ 0 & 33.8311 \\ 0 & 0 \end{bmatrix}$$

4.1   Simulation

The proposed controller is verified with three different cases, each case represents the system component (leak) fault (f_{sys}), actuator fault (f_a), and process disturbances (d) respectively. The comparative simulation results of IT2 TSK FLC with T1 FLC is carried out in MATLAB simulink platform and demonstrated as follows:

Case 1: TICTS regulatory response subject to system component (leak) fault

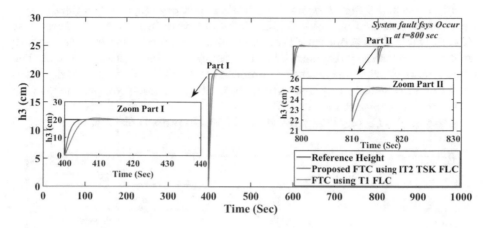

Fig. 8. Comparison of regulatory response for TICTS subject to system component fault.

Case 2: TICTS regulatory response subject to actuator fault

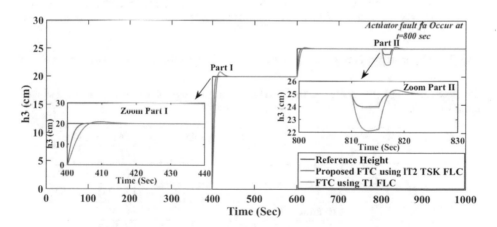

Fig. 9. Comparison of regulatory response for TICTS subject to actuator fault.

The dynamic behavior of the TICTS system has been simulated by using the integration of the nonlinear differential equations (17–19) in state space form. This section presents the results from the simulation of FTC using interval type-2 TSK FLC subject to actuator and system component (tank leak) faults. The simulation are done for regulatory and servo problem for proposed TICTS. The fault introduce into the system is abrupt form of nature. The regulatory and

servo responses of TICTS were compared to proposed FTC using IT2 TSK FLC
with T1 FLC. The efficacy of the proposed controller found using IAE and ISE.

Figures 8 and 9 demonstrate the TICTS regulatory responses subject to system component and actuator faults. The fault ware introduce into system at
time t = 800 s. Also proposed FTC tested subject to process disturbances, Fig. 10
present the regulatory response. The error comparison illustrated in Table 3. The
error results clearly shown that the proposed IT2 TSK FLC gives superior tran-
sient and steady-state response subject to faults. Also same computational time
is required as compared to T1 FLC.

Case 3: TICTS regulatory response subject to process disturbances

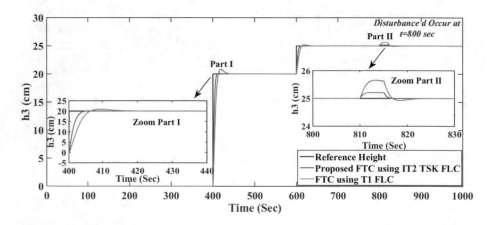

Fig. 10. Comparison of regulatory response for TICTS subject to process disturbances.

Table 3. IAE and ISE error results from simulation of regulatory responses with abrupt
form of faults.

Controller structure	Response type	Fault or Disturbance	IAE	ISE
			h_2	h_2
IT 2 TSK FLC			**0.2691**	**1.5814**
T1 FLC		f_{sys}	1.3941	2.9347
IT 2 TSK FLC	Regulatory		**0.3681**	**1.6912**
T1 FLC	Response	f_a	1.4375	2.1863
IT 2 TSK FLC			**0.2148**	**1.4291**
T1 FLC		d	1.2349	2.7885

4.2   Servo Response of TICTS

The proposed controller is also tested with servo problem without fault. Two
different trajectories r_1 and r_2 are applied on TICTS. In Figs. 11 and 12 demonstrates the effectiveness of the proposed scheme as compared to T1 FLC. Table 4
represents the error comparison for the servo responses of TICTS without fault.

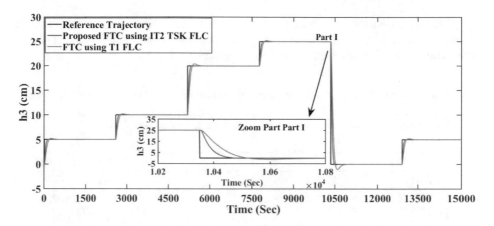

Fig. 11. Comparison of servo response for TICTS with trajectory r_1.

It is obvious that the IT2 TSK FLC outperform the T1 FLC for each faults and process disturbances as shown in Figs. 8, 9 and 10. In fact, the error when we use IT2 TSK FLS is more close to zero than the error when we use T1 FLS. The superiority of the proposed approach is due to the use of T2 FSs for the TICTS system, which offer more degrees of freedom and have an excellent capability to handle uncertainties.

Critical observing the servo response of TICTS with two different trajectories r_1 and r_2, IT2 TSK FLC will track perfectly the trajectories as compared to IT2 FLC. Also IT2 TSK FLC will tracking the reference trajectories smoothly without any overshoot.

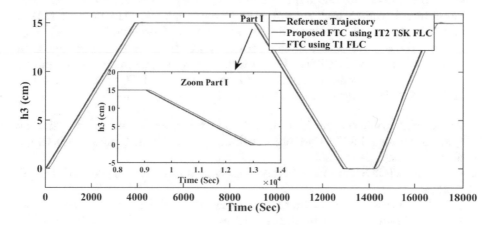

Fig. 12. Comparison of servo response for TICTS with trajectory r_2.

Table 4, presented two integral error indices for servo response of TICTS without faults. The error results clearly demonstrate the IT2 TSK FLC outperform the T1 FLC controller even without faults.

Table 4. IAE and ISE error results from simulation of servo responses without faults.

Controller structure	Response type	Fault or Disturbance	IAE h_2	ISE h_2
IT 2 TSK FLC			**0.8936**	**2.3612**
T1 FLC	Servo	NA	1.6981	3.1698
IT 2 TSK FLC	Response		**1.0336**	**2.9829**
T1 FLC		NA	2.1481	3.9492

5 Conclusion and Future Work

In this paper, we developed fault tolerant controller using IT2 TSK FLC algorithm. The proposed control algorithm, on the one hand, takes advantage of Fuzzy type-2 controller, and on the other hand, significantly reduces the computation cost by describing the inputs using IT2 FSs and integrating the control algorithm with the TSK control technique. The developed controller was applied to a three interconnected conical tank system subject to actuator and system component faults along with process disturbances and the simulation results verified the efficacy of the proposed controller as compared to T1 FLCs. As future work, we intend to develop an algorithm for tuning the parameters of the proposed control structure and apply the developed controller to a real time three interconnected conical tank system where we can compare the performance of the system compared to other existing approaches. Third possible sensor (f_a) faults in TICTS is not scope of this article, it will covered into future research work and validate the IT2 TSK FLC performance when presence of sensor faults.

The following reasons are listed out for proposed new FTC scheme using IT2 TSK FLC for MIMO level control process (TICTS):

1. Takagi-Sugeno fuzzy model is found for three-interconnected conical tank system because TICTS dynamics is changing continuously due to it's shape and interaction is present.
2. Proposed TICTS system is highly nonlinear due to interaction and MIMO process.
3. Due to process and modeling uncertainty of TICTS interval Type-2 fuzzy system is used, so mathematical model mismatch error can be eliminated.
4. Type-2 TSK fuzzy-based controller has an ability to stabilize and control TICTS with acceptable performance even though process disturbances and possible faults occurs in the system.

The following reason is listed out for degrade the control performance of T1 FLC as compared to IT2 TSK FL for MIMO level control process (TICTS):

1. Process dynamics is continuously change due to highly nonlinear system.
2. Due to changing process dynamics Type-1 FLC will not cope up that situation into account at time of FLC designing.
3. Type-1 FLC will not give optimum performance in the presence of unmeasured and abrupt disturbances.

References

1. Zadeh, L.A.: The concept of a linguistic variable and its application to approximate reasoning II. Inf. Sci. **8**(4), 301–357 (1975)
2. Patel, H., Shah, V.: Fault tolerant control using interval type-2 Takagi-Sugeno fuzzy controller for nonlinear system. In: 18th International Conference on Intelligent Systems Design and Applications, Advances in Intelligent Systems and Computing (AISC). Springer, Vellore Institute of Technology, Vellore, India (2018, to be published)
3. Patel, H., Shah, V.: Passive fault tolerant control based on interval type-2 fuzzy controller for coupled tank system. In: Proceedings of International Conference Artificial Intelligence, Smart Grid and Smart City Applications (AISGSC 2019). Springer, PSG College of Technology, Coimbatore, TamilNadu, India (2019, to be published)
4. Wu, D., Nie, M.: Comparison and practical implementation of type-reduction algorithms for type-2 fuzzy sets and systems. In: Fuzzy Systems (FUZZ), IEEE, Taipei, Taiwan, pp. 2131–2138 (2011)
5. Greenfield, S., Chiclana, F.: Type-reduction of the discretized interval type-2 fuzzy set: approaching the continuous case through progressively finer discretization. J. Artif. Intell. Soft Comput. Res. **1**(3), 183–193 (2011)
6. Liu, F.: An efficient centroid type-reduction strategy for general type 2 fuzzy logic system. Inf. Sci. **178**(9), 2224–2236 (2008)
7. Nie, M., Tan, W. W.: Towards an efficient type-reduction method for interval type-2 fuzzy logic systems. In: IEEE World Congress on Computational Intelligence, IEEE, Hong Kong, China, pp. 1425–1432 (2008)
8. Mendel, J.M., John, R.B.: Type-2 fuzzy sets made simple. IEEE Trans. Fuzzy Syst. **10**(2), 117–127 (2002)
9. Mendel, J.M., John, R.I., Liu, F.: Interval type-2 fuzzy logic systems made simple. IEEE Trans. Fuzzy Syst. **14**(6), 808–821 (2006)
10. Mendel, J.M., Rajati, M.: On computing normalized interval type-2 fuzzy sets. IEEE Trans. Fuzzy Syst. **22**(5), 1335–1340 (2014)
11. Sugeno, M.: Industrial Applications of Fuzzy Control. Elsevier Science Inc., New York (1985)
12. Ren, Q., Balazinski, M., Baron, L.: High-order interval type-2 takagi-sugeno-kang fuzzy logic system and its application in acoustic emission signal modeling in turning process. Int. J. Adv. Manuf. Technol. **63**(9–12), 1057–1063 (2012)
13. Tseng, C.L., Wang, S.Y., Lin, S.C., Chen, Y.Y.: Interval type-2 Takagi-Sugeno fuzzy controller design for a class of nonlinear singular networked control systems. In: International Conference on Fuzzy Theory and it's Applications (iFUZZY), IEEE, Taichung, Taiwan, pp. 268–272 (2012)

14. Liang, Q., Mendel, J.M.: Interval type-2 fuzzy logic systems: theory and design. IEEE Trans. Fuzzy Syst. **8**(5), 535–550 (2000)
15. Castillo, O., Melin, P.: A review on the design and optimization of interval type-2 fuzzy controllers. Appl. Soft Comput. **12**(4), 1267–1278 (2012)
16. Juan, R., Castillo, O., Melin, P., Rodríguez-Díaz, A.: A hybrid learning algorithm for a class of interval type-2 fuzzy neural networks. Inf. Sci. **179**(13), 2175–2193 (2009)
17. Martínez, R., Castillo, O., Aguilar, L.: Optimization of interval type-2 fuzzy logic controllers for a perturbed autonomous wheeled mobile robot using genetic algorithms. Inf. Sci. **179**(13), 2158–2174 (2009)
18. Wu, D., Tan, W.: A simplified type-2 fuzzy logic controller for real-time control. ISA Trans. **45**(4), 503–510 (2006)
19. Karnik, N.N., Mendel, J.M., Liang, Q.: Type-2 fuzzy logic systems. IEEE Trans. Fuzzy Syst. **7**, 643–658 (1999)

The Dominance, Covering and Supercovering Relations in the Context of Multiple Partially Specified Reciprocal Relations

Raúl Pérez-Fernández[1,2(✉)], Irene Díaz[3], Susana Montes[1],
and Bernard De Baets[2]

[1] Department of Statistics and O.R. and Mathematics Didactics,
University of Oviedo, Oviedo, Spain
{perezfernandez,montes}@uniovi.es
[2] KERMIT, Department of Data Analysis and Mathematical Modelling,
Ghent University, Ghent, Belgium
{raul.perezfernandez,bernard.debaets}@ugent.be
[3] Department of Informatics, University of Oviedo, Oviedo, Spain
sirene@uniovi.es

Abstract. The problem of ranking different candidates or alternatives according to the preferences of different voters or experts is a common study subject in the fields of social choice theory and preference modelling. Whereas the former field normally restricts its attention to preferences given in the form of rankings (with ties), the latter field embraces the use of (partially specified) reciprocal relations. In this contribution, we study the notions of dominance relation, covering relation and supercovering relation, which are widely studied in the setting in which we are dealing with rankings (with ties), and adapt them to the setting in which we are dealing with partially specified reciprocal relations by using a tool similar to stochastic dominance.

Keywords: Reciprocal relation · Dominance · Covering ·
Supercovering

1 Introduction

The aggregation of preferences (on a set of candidates) given in the form of a list of rankings has been a popular topic in the field of social choice theory for centuries [6,9]. This field has witnessed its greatest period of splendor since the appearence of Arrow's impossibility theorem [1,2], which states that there is no method for aggregating a given list of rankings into a single ranking in

© Springer Nature Switzerland AG 2019
R. B. Kearfott et al. (Eds.): IFSA 2019/NAFIPS 2019, AISC 1000, pp. 483–492, 2019.
https://doi.org/10.1007/978-3-030-21920-8_43

a way that satisfies some desirable properties. One should note that Arrow's impossibility theorem only arises in the presence of three or more candidates since in the case of two candidates the simple majority rule [11,17,21,31][1] fulfills the aforementioned desirable properties. Unfortunately, in case three or more candidates are considered, it is known that the simple majority rule might yield a cyclical relation – a phenomenon usually referred to as the Condorcet paradox – and, thus, lead to Arrow's impossibility theorem.

Once it is established there is no perfect method for the aggregation of rankings, different study subjects arise. For instance, we could take a totally different approach and abandon the aggregation of rankings by asking the voters to evaluate the candidates individually, thus obtaining an aggregated utility value for each candidate, which could be used for ranking all the candidates. This problem has been addressed by many authors, especially in the context of bargaining [23], decision making [24] and voting [3,4]. If one would like to stick to the problem of the aggregation of rankings, one could focus on the proposal of different methods fulfilling other desirable properties not listed by Arrow (e.g. the method of Kemeny [19] can be interpreted as a maximum-likelihood estimator of a true unobservable ranking [34] or the method of ranked pairs [33,35] and the method of Schulze [30] have been proved to be independent of clones). Another option is to admit as a possible outcome of the aggregation any possible binary relation. For instance, one could simply consider the dominance relation given by the simple majority rule, which amounts to computing the (smallest) median relation [5]. More elaborate options are the covering relation [22] and the super-covering relation [25], which are typically quite sparse.

Nobel-prize laureate Sen [32] expressed: "it is certainly arguable that what matters is not merely the number who prefer x to y and the number who prefer y to x, but also by how much each prefers one alternative to the other". In this direction, the field of (fuzzy) preference modelling [13,29] has also built on some similar foundations as social choice theory, while now allowing the voters to express different intensities of preference in terms of, for instance, (valued) reciprocal relations or linguistic labels. Prominent notions in social choice theory have been analysed in the setting in which voters express different intensities of preference, e.g., the simple majority rule [14], the Borda count [16], approval voting [15] and the covering relation [28].

In this paper, we follow the direction started in [27] and discuss the setting in which each voter expresses a partially specified reciprocal relation. In particular, we recall the most important notions from [27] in Sect. 2 and discuss how the covering relation and the supercovering relation could be defined in this setting in Sect. 3. We end with some conclusions in Sect. 4.

---

[1] We recall that a candidate is said to defeat another candidate by simple majority if the number of voters who prefer the former candidate to the latter one is greater than the number of voters who prefer the latter candidate to the former one.

2   Preliminaries

We consider the problem setting in which r voters/experts[2] are asked to compare k candidates/alternatives[2] and the goal is to obtain a collective comparison of the different candidates (preferably by ranking them from best to worst). In our particular problem setting, each voter expresses his/her own personal partially specified reciprocal relation [7,10,18] on the set of candidates $\mathscr{C}$. This means that, for at least one couple of candidates, but not necessarily for all possible couples of candidates, the voter expresses a value in the unit interval[3]. The value 0.5 represents indifference between two candidates and the closer to 1 (resp. 0), the stronger the degree of preference of the first (resp. second) candidate over the second (resp. first) one. It is assumed that every two candidates are compared by at least one voter. In that way, each voter expresses a relation $P^\ell : \mathscr{C}^2 \to [0,1]$ where, for any $\ell \in \{1,\ldots,r\}$ and any $(a_i, a_j) \in \mathscr{C}^2$, it holds that $P^\ell(a_i, a_j) + P^\ell(a_j, a_i) = 1$ or that both $P^\ell(a_i, a_j)$ and $P^\ell(a_j, a_i)$ are undefined. By convention, $P^\ell(a_i, a_i)$ is considered undefined for any $\ell \in \{1,\ldots,r\}$ and any $a_i \in \mathscr{C}$. The fact that an element $P^\ell(a_i, a_j)$ is undefined is denoted by $P^\ell(a_i, a_j) = \emptyset$. For any $\ell \in \{1,\ldots,r\}$ and any $a_i, a_j \in \mathscr{C}$, $P^\ell(a_i, a_j)$ is referred to as the degree of preference of the ℓ-th voter for candidate a_i over candidate a_j.

Given the partially specified reciprocal relations $(P^\ell)_{\ell=1}^r$ given by the voters, we denote by $D(a_i, a_j)$ the multi-set of all degrees of preference expressed for a candidate a_i over a candidate a_j, i.e.,

$$D(a_i, a_j) = \{\!\{ P^\ell(a_i, a_j) \mid \ell \in \{1,\ldots,r\} \wedge P^\ell(a_i, a_j) \neq \emptyset \}\!\} .$$

We refer to $D(a_i, a_j)$ as the preference distribution for candidate a_i over candidate a_j or, equivalently, as the preference distribution for the couple (a_i, a_j). Multi-sets, which are sets of elements where duplicated elements are allowed, are denoted with double curly brackets $\{\!\{\,\}\!\}$.

Example 1. Consider the set $\mathscr{C} = \{a, b, c, d\}$ of $k = 4$ candidates and the following partially specified reciprocal relations given by $r = 4$ voters, represented as matrices P^ℓ, where the element at the i-th row and the j-th column represents the degree of preference of the ℓ-th voter for the i-th candidate over the j-th candidate (candidates are indexed in alphabetical order):

$$P^1 = \begin{pmatrix} \emptyset & 0.6 & 1 & \emptyset \\ 0.4 & \emptyset & 0.4 & \emptyset \\ 0 & 0.6 & \emptyset & 0.6 \\ \emptyset & \emptyset & 0.4 & \emptyset \end{pmatrix}, \quad P^2 = \begin{pmatrix} \emptyset & 0.7 & \emptyset & 1 \\ 0.3 & \emptyset & \emptyset & 0.4 \\ \emptyset & \emptyset & \emptyset & \emptyset \\ 0 & 0.6 & \emptyset & \emptyset \end{pmatrix},$$

---

[2] The terms 'voter' and 'candidate' are commonly used in the field of social choice, whereas the terms 'expert' and 'alternative' are more extended in the field of preference modelling. We stick to the former notation throughout the remainder of this paper.

[3] Admittedly, a voter should not be expected to provide a large variety of values in the unit interval. However, a broad range of values might be a result of a prior process for making uniform the information provided by the different voters [8].

$$P^3 = \begin{pmatrix} \emptyset & \emptyset & \emptyset & 1 \\ \emptyset & \emptyset & 0.8 & \emptyset \\ \emptyset & 0.2 & \emptyset & 0.6 \\ 0 & \emptyset & 0.4 & \emptyset \end{pmatrix}, \quad P^4 = \begin{pmatrix} \emptyset & \emptyset & \emptyset & 1 \\ \emptyset & \emptyset & \emptyset & \emptyset \\ \emptyset & \emptyset & \emptyset & 0.7 \\ 0 & \emptyset & 0.3 & \emptyset \end{pmatrix}.$$

For the couple (a, b), the first voter has expressed a degree of preference of 0.6 for a over b (thus, 0.4 for b over a), whereas the second voter has expressed a degree of preference of 0.7 for a over b (thus, 0.3 for b over a). Note that the third voter and the fourth voter have not expressed any preference for a over b (nor for b over a). Therefore, the preference distribution for a over b is the multi-set $D(a, b) = \{\!\{0.6, 0.7\}\!\}$. Since we are dealing with reciprocal relations, the preference distribution for b over a is the multi-set $D(b, a) = \{\!\{0.3, 0.4\}\!\}$. We can calculate the preference distribution of any couple in $\mathscr{C}_{\neq}^2 = \{(a, a') \in \mathscr{C}^2 \mid a \neq a'\}$ and obtain:

$$
\begin{aligned}
D(a, b) &= \{\!\{0.6, 0.7\}\!\}, & D(b, a) &= \{\!\{0.3, 0.4\}\!\}, \\
D(a, c) &= \{\!\{1\}\!\}, & D(c, a) &= \{\!\{0\}\!\}, \\
D(a, d) &= \{\!\{1, 1, 1\}\!\}, & D(d, a) &= \{\!\{0, 0, 0\}\!\}, \\
D(b, c) &= \{\!\{0.4, 0.8\}\!\}, & D(c, b) &= \{\!\{0.2, 0.6\}\!\}, \\
D(b, d) &= \{\!\{0.4\}\!\}, & D(d, b) &= \{\!\{0.6\}\!\}, \\
D(c, d) &= \{\!\{0.6, 0.6, 0.7\}\!\}, & D(d, c) &= \{\!\{0.3, 0.4, 0.4\}\!\}. \triangleleft
\end{aligned}
$$

Based on the notion of stochastic dominance [20], which is a popular stochastic order for the comparison of probability distributions, we can consider a dominance relation between the preference distributions for two different couples of candidates [27].

Definition 1. Let $\mathscr{C}$ be a set of k candidates, r be the number of voters and $(P^\ell)_{\ell=1}^r$ be the partially specified reciprocal relations given by the voters with associated preference distributions $\{D(a_i, a_j)\}_{(a_i, a_j) \in \mathscr{C}_{\neq}^2}$. For any $(a_{i_1}, a_{j_1}), (a_{i_2}, a_{j_2}) \in \mathscr{C}_{\neq}^2$, we say that $D(a_{i_1}, a_{j_1})$ *weakly dominates* $D(a_{i_2}, a_{j_2})$, denoted as $D(a_{i_1}, a_{j_1}) \trianglerighteq D(a_{i_2}, a_{j_2})$, if, for any $p \in [0, 1]$, it holds that

$$F_{a_{i_1}, a_{j_1}}^*(p) \leq F_{a_{i_2}, a_{j_2}}^*(p),$$

where, for any $(a_i, a_j) \in \mathscr{C}_{\neq}^2$, F_{a_i, a_j}^* is the cumulative preference distribution defined by, for any $p \in [0, 1]$:

$$F_{a_i, a_j}^*(p) = \frac{\#\{\!\{d \in D(a_i, a_j) \mid d \leq p\}\!\}}{\# D(a_i, a_j)}.$$

If $D(a_{i_1}, a_{j_1}) \trianglerighteq D(a_{i_2}, a_{j_2})$ and $D(a_{i_2}, a_{j_2}) \ntrianglerighteq D(a_{i_1}, a_{j_1})$, then we say that $D(a_{i_1}, a_{j_1})$ *(strictly) dominates* $D(a_{i_2}, a_{j_2})$, denoted as $D(a_{i_1}, a_{j_1}) \triangleright D(a_{i_2}, a_{j_2})$.

Example 2. We continue with Example 1. Let us consider the preference distributions of (a, b) and (b, a):

$$
\begin{aligned}
D(a, b) &= \{\!\{0.6, 0.7\}\!\}, \\
D(b, a) &= \{\!\{0.3, 0.4\}\!\}.
\end{aligned}
$$

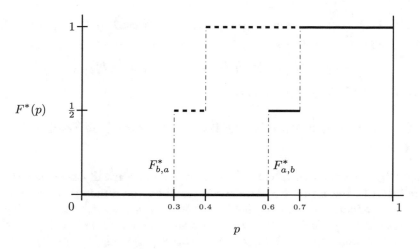

Fig. 1. Graphical representation of $F_{a,b}^*$ and $F_{b,a}^*$.

It is easily verified that $F_{a,b}^*$ and $F_{b,a}^*$ are respectively given by:

$$F_{a,b}^*(p) = \begin{cases} 0, & \text{if } p \in [0, 0.6[, \\ \frac{1}{2}, & \text{if } p \in [0.6, 0.7[, \\ 1, & \text{if } p \in [0.7, 1], \end{cases}$$

and

$$F_{b,a}^*(p) = \begin{cases} 0, & \text{if } p \in [0, 0.3[, \\ \frac{1}{2}, & \text{if } p \in [0.3, 0.4[, \\ 1, & \text{if } p \in [0.4, 1]. \end{cases}$$

As illustrated in Fig. 1, we can see that $F_{a,b}^*(p) \leq F_{b,a}^*(p)$ for any $p \in [0, 1]$ with the inequality being strict for $p \in [0.3, 0.7[$. We conclude that $D(a, b) \rhd D(b, a)$ and, thus, $D(b, a) \not\rhd D(a, b)$.   ◁

We say that a candidate a_i dominates another candidate a_j if the preference distribution for a_i over a_j dominates the preference distribution for a_j over a_i.

Definition 2. Let $\mathscr{C}$ be a set of k candidates, r be the number of voters and $(P^\ell)_{\ell=1}^r$ be the partially specified reciprocal relations given by the voters with associated preference distributions $\{D(a_i, a_j)\}_{(a_i, a_j) \in \mathscr{C}_{\neq}^2}$. We say that a candidate $a_i \in \mathscr{C}$ *dominates* another candidate $a_j \in \mathscr{C}$, denoted by $a_i \succ a_j$, if it holds that

$$D(a_i, a_j) \rhd D(a_j, a_i).$$

One should note that neither $\rhd$ defines a strict order relation on the set of preference distributions, nor $\succ$ defines a strict order relation on the set of candidates. The former is transitive but does not fulfill the antisymmetry property, whereas the latter is antisymmetric (actually asymmetric) but might be cyclical, as can be seen in the following example.

Example 3. We continue with Example 1. As discussed in Example 2, it holds that $D(a, b) \rhd D(b, a)$, thus, $a \succ b$. In general, $\succ$ is given by:

$$\succ = \{(a, b), (a, c), (a, d), (b, c), (c, d), (d, b)\}.$$

We find the cycle $b \succ c$, $c \succ d$ and $d \succ b$.   ◁

3   Introducing Some Ideas from the Field of Social Choice Theory

The setting in which each voter expresses his/her partially specified reciprocal relation in the form of a strict order relation (or ranking) on $\mathscr{C}$ has been addressed systematically in the field of social choice theory. In this setting, the dominance $a_i \succ a_j$ is equivalent to the fact that the number of voters preferring a_i over a_j is greater than the number of voters preferring a_j over a_i (for more details, we refer to Sect. 5 of [27]). The latter fact is commonly referred to as 'candidate a_i defeats candidate a_j by simple majority' [11,17,21,31]. It is known that the simple majority relation on the set of candidates might be cyclical. This phenomenon in which all voters express a strict order relation on the set of candidates but the simple majority relation turns out to be cyclical is called 'the Condorcet paradox' or 'the voting paradox' [9]. For this reason, there has been quite some interest in the study of a candidate that defeats all other candidates by simple majority. Such candidate is commonly referred to as the Condorcet candidate. This notion could be easily extended to our setting in which we are dealing with multiple partially specified reciprocal relations.

Definition 3. Let $\mathscr{C}$ be a set of k candidates, r be the number of voters and $(P^\ell)_{\ell=1}^r$ be the partially specified reciprocal relations given by the voters with associated preference distributions $\{D(a_i, a_j)\}_{(a_i, a_j) \in \mathscr{C}_{\neq}^2}$. Consider the dominance relation $\succ$ defined as in Definition 2. A candidate $a_i \in \mathscr{C}$ is called the *Condorcet winner* if, for any $a_j \in \mathscr{C} \backslash \{a_i\}$, it holds that $a_i \succ a_j$.

Example 4. We continue with Example 1. As discussed in Example 3, it holds that $a \succ b$, $a \succ c$ and $a \succ d$. Thus, candidate a is the Condorcet winner.   ◁

Along the same vein, we could also extend the notions of covering relation [22] and supercovering relation [25] to the setting of multiple partially specified relations.

Definition 4. Let $\mathscr{C}$ be a set of k candidates, r be the number of voters and $(P^\ell)_{\ell=1}^r$ be the partially specified reciprocal relations given by the voters with associated preference distributions $\{D(a_i, a_j)\}_{(a_i, a_j) \in \mathscr{C}_{\neq}^2}$. We say that a candidate $a_i \in \mathscr{C}$ *covers* another candidate $a_j \in \mathscr{C}$, denoted by $a_i \gg a_j$, if it holds that

$$D(a_i, a_j) \rhd D(a_j, a_i),$$

and, for any $a_\ell \in \mathscr{C} \backslash \{a_i, a_j\}$,

$$D(a_j, a_\ell) \rhd D(a_\ell, a_j) \Rightarrow D(a_i, a_\ell) \rhd D(a_\ell, a_i).$$

Definition 5. Let $\mathscr{C}$ be a set of k candidates, r be the number of voters and $(P^\ell)_{\ell=1}^r$ be the partially specified reciprocal relations given by the voters with associated preference distributions $\{D(a_i, a_j)\}_{(a_i,a_j)\in\mathscr{C}_{\neq}^2}$. We say that a candidate $a_i \in \mathscr{C}$ *supercovers* another candidate $a_j \in \mathscr{C}$, denoted by $a_i \gg a_j$, if it holds that

$$D(a_i, a_j) \rhd D(a_j, a_i),$$

and, for any $a_\ell \in \mathscr{C}\backslash\{a_i, a_j\}$,

$$D(a_i, a_\ell) \unrhd D(a_j, a_\ell).$$

As in the original setting of social choice theory, the supercovering relation can be proved to further restrict the covering relation.

Proposition 1. *Let $\mathscr{C}$ be a set of k candidates, r be the number of voters and $(P^\ell)_{\ell=1}^r$ be the partially specified reciprocal relations given by the voters with associated preference distributions $\{D(a_i, a_j)\}_{(a_i,a_j)\in\mathscr{C}_{\neq}^2}$. For any two candidates $a_i, a_j \in \mathscr{C}$, it holds that*

$$a_i \gg a_j \Rightarrow a_i \gg a_j.$$

Proof. Consider $a_i, a_j \in \mathscr{C}$ such that $a_i \gg a_j$. It trivially holds that $D(a_i, a_j) \rhd D(a_j, a_i)$. We only need to prove that, for any $a_\ell \in \mathscr{C}\backslash\{a_i, a_j\}$,

$$D(a_j, a_\ell) \rhd D(a_\ell, a_j) \Rightarrow D(a_i, a_\ell) \rhd D(a_\ell, a_i).$$

Consider any $a_\ell \in \mathscr{C}\backslash\{a_i, a_j\}$ and assume that $D(a_j, a_\ell) \rhd D(a_\ell, a_j)$. It thus holds that

$$D(a_i, a_\ell) \unrhd D(a_j, a_\ell) \rhd D(a_\ell, a_j) \unrhd D(a_\ell, a_i),$$

and the result finally follows from the transitivity of $\unrhd$ and the definition of $\rhd$. $\square$

Example 5. We continue with Example 1. Since a is the Condorcet winner and the other three candidates form a majority cycle, the covering relation is given by:

$$\gg= \{(a, b), (a, c), (a, d)\}.$$

It thus remains to be seen whether a supercovers any of the other candidates. For b, we have that

$$D(a, b) = \{\!\{0.6, 0.7\}\!\} \rhd \{\!\{0.3, 0.4\}\!\} = D(b, a),$$
$$D(a, c) = \{\!\{1\}\!\} \unrhd \{\!\{0.4, 0.8\}\!\} = D(b, c),$$
$$D(a, d) = \{\!\{1, 1, 1\}\!\} \unrhd \{\!\{0.4\}\!\} = D(b, d).$$

Thus, it holds that $a \gg b$. Similarly, we have that

$$D(a, c) = \{\!\{1\}\!\} \rhd \{\!\{0\}\!\} = D(c, a),$$
$$D(a, b) = \{\!\{0.6, 0.7\}\!\} \unrhd \{\!\{0.2, 0.6\}\!\} = D(c, b),$$
$$D(a, d) = \{\!\{1, 1, 1\}\!\} \unrhd \{\!\{0.6, 0.6, 0.7\}\!\} = D(c, d),$$

and

$$D(a,d) = \{\!\{1,1,1\}\!\} \rhd \{\!\{0,0,0\}\!\} = D(d,a),$$
$$D(a,b) = \{\!\{0.6,0.7\}\!\} \unrhd \{\!\{0.6\}\!\} = D(d,b),$$
$$D(a,c) = \{\!\{1\}\!\} \unrhd \{\!\{0.3,0.4,0.4\}\!\} = D(d,c).$$

Thus, it holds that $a \gg c$ and $a \gg d$. We conclude that

$$\gg = \{(a,b),(a,c),(a,d)\}.$$

As anticipated in Proposition 1, it holds that $\gg \subseteq \gg \subseteq \succ$.   ◁

Interestingly, unlike the dominance relation $\succ$, the covering relation $\gg$ defines an order relation on $\mathscr{C}$. In the sense of social choice theory, it is known that the Condorcet winner is equivalently defined by using the covering relation rather than the dominance relation. This result is also immediately extended to the setting of multiple partially specified reciprocal relations.

Proposition 2. *Let $\mathscr{C}$ be a set of k candidates, r be the number of voters and $(P^\ell)_{\ell=1}^r$ be the partially specified reciprocal relations given by the voters with associated preference distributions $\{D(a_i,a_j)\}_{(a_i,a_j)\in\mathscr{C}_{\neq}^2}$. Consider the covering relation $\gg$ defined as in Definition 4. A candidate $a_i \in \mathscr{C}$ is the Condorcet winner if and only if, for any $a_j \in \mathscr{C}\backslash\{a_i\}$, it holds that $a_i \gg a_j$.*

Proof. On the one hand, let a_i be the Condorcet winner. Consider any $a_\ell \in \mathscr{C}\backslash\{a_i,a_j\}$ such that $a_j \succ a_\ell$. By definition, it holds that $a_i \succ a_j$ for any $a_j \in \mathscr{C}\backslash\{a_i\}$, thus, $a_i \succ a_\ell$. We conclude that $a_i \gg a_j$.

On the other hand, let a_i be such that $a_i \gg a_j$ for any $a_j \in \mathscr{C}\backslash\{a_i\}$. By definition of $\gg$, it follows that $a_i \succ a_j$ for any $a_j \in \mathscr{C}\backslash\{a_i\}$, which exactly is the definition of a Condorcet winner.   □

Unlike with the dominance relation $\succ$ and the covering relation $\gg$, requiring a candidate to supercover all other candidates yields a stronger type of winner than the Condorcet winner: the pairwise winner [25].

Definition 6. Let $\mathscr{C}$ be a set of k candidates, r be the number of voters and $(P^\ell)_{\ell=1}^r$ be the partially specified reciprocal relations given by the voters with associated preference distributions $\{D(a_i,a_j)\}_{(a_i,a_j)\in\mathscr{C}_{\neq}^2}$. Consider the supercovering relation $\ggg$ defined as in Definition 5. A candidate $a_i \in \mathscr{C}$ is called the *pairwise winner* if, for any $a_j \in \mathscr{C}\backslash\{a_i\}$, it holds that $a_i \ggg a_j$.

Obviously, if a pairwise winner exists, a Condorcet winner exists and needs to coincide with this pairwise winner.

Example 6. We continue with Example 1. From Example 5, we conclude that a is both the Condorcet winner and the pairwise winner.   ◁

4   Conclusions

In this contribution, we have brought the notions of dominance relation, covering relation and supercovering relation to the field of preference modelling. In particular, we have adapted these three notions to the setting in which each voter provides a partially specified reciprocal relation. There is a clear analogy between this setting and the one in social choice theory due to the preorder nature of the considered dominance relation on the set of preference distributions for couples of candidates.

Future research is anticipated in two directions. Firstly, one could follow the direction in [27] and study how to act in case the covering and supercovering relations are sparse and, thus, there does not exist a pairwise/Condorcet winner. Secondly, one could study the notions of Borda-dominance relation [12] and superdominance relation [26], which relate to the positionalistic point of view of social choice theory (usually attributed to Borda [6]), in the context of multiple partially specified reciprocal relations, following a similar direction to that of [15, 16].

Acknowledgements. This research has been partially supported by Spanish MINECO (TIN2017-87600-P). Raúl Pérez-Fernández acknowledges the support of the Research Foundation of Flanders (FWO17/PDO/160).

References

1. Arrow, K.J.: A difficulty in the concept of social welfare. J. Polit. Econ. **58**(4), 328–346 (1950)
2. Arrow, K.J.: Social Choice and Individual Values, 2nd edn. Yale University Press, New Haven (1963)
3. Balinski, M., Laraki, R.: A theory of measuring, electing and ranking. PNAS **104**(21), 8720–8725 (2007)
4. Balinski, M., Laraki, R.: Majority Judgment: Measuring, Ranking, and Electing. MIT Press, Cambridge (2010)
5. Barthelemy, J.P., Monjardet, B.: The median procedure in cluster analysis and social choice theory. Math. Soc. Sci. **1**, 235–267 (1981)
6. Borda, J.C.: Mémoire sur les Élections au Scrutin. Histoire de l'Académie Royale des Sciences, Paris (1781)
7. Chiclana, F., Herrera, F., Herrera-Viedma, E.: Integrating three representation models in fuzzy multipurpose decision making based on fuzzy preference relations. Fuzzy Sets Syst. **97**(1), 33–48 (1998)
8. Chiclana, F., Herrera, F., Herrera-Viedma, E.: Integrating multiplicative preference relations in a multipurpose decision-making model based on fuzzy preference relations. Fuzzy Sets Syst. **122**, 277–291 (2001)
9. Condorcet, M.: Essai sur l'Application de l'Analyse à la Probabilité des Décisions Rendues à la Pluralité des Voix. De l'Imprimerie Royale, Paris (1785)
10. De Baets, B., De Meyer, H., De Schuymer, B., Jenei, S.: Cyclic evaluation of transitivity of reciprocal relations. Soc. Choice Welfare **26**(2), 217–238 (2006)
11. Fishburn, P.C.: Conditions for simple majority decision functions with intransitive individual indifference. J. Econ. Theory **2**, 354–367 (1970)

12. Fishburn, P.C.: Paradoxes of voting. Am. Polit. Sci. Rev. **68**(2), 537–546 (1974)
13. Fodor, J.C., Roubens, M.R.: Fuzzy Preference Modelling and Multicriteria Decision Support. Kluwer Academic Publishers, Dordrecht (1994)
14. García-Lapresta, J.L.: A general class of simple majority decision rules based on linguistic opinions. Inf. Sci. **176**, 352–365 (2006)
15. García-Lapresta, J.L., Martínez-Panero, M.: Borda counts versus approval voting: a fuzzy approach. Public Choice **112**, 167–184 (2002)
16. García-Lapresta, J.L., Martínez-Panero, M., Meneses, L.C.: Defining the Borda count in a linguistic decision making context. Inf. Sci. **179**, 2309–2316 (2009)
17. Inada, K.: The simple majority decision rule. Econometrica **37**(3), 490–506 (1969)
18. Kacprzyk, J., Nurmi, H., Fedrizzi, M.: Consensus Under Fuzziness, vol. 10. Kluwer Academic, Boston (1996)
19. Kemeny, J.G.: Mathematics without numbers. Daedalus **88**(4), 577–591 (1959)
20. Levy, H.: Stochastic Dominance: Investment Decision Making under Uncertainty, 3rd edn. Springer, Berlin (2016)
21. May, K.O.: A set of independent necessary and sufficient conditions for simple majority decision. Econometrica **20**, 680–684 (1952)
22. Miller, N.R.: A new solution set for tournaments and majority voting: further graph-theoretical approaches to the theory of voting. Am. J. Polit. Sci. **24**(1), 68–96 (1980)
23. Nash, J.F.: The bargaining problem. Econometrica **28**, 155–162 (1950)
24. Nguyen, H.T., Kosheleva, O., Kreinovich, V.: Decision making beyond arrow's "impossibility theorem," with the analysis of effects of collusion and mutual attraction. Int. J. Intell. Syst. **24**, 27–47 (2008)
25. Pérez-Fernández, R., De Baets, B.: The supercovering relation, the pairwise winner, and more missing links between Borda and Condorcet. Soc. Choice and Welfare **50**, 329–352 (2018)
26. Pérez-Fernández, R., De Baets, B.: The superdominance relation, the positional winner, and more missing links between Borda and Condorcet. J. Theor. Polit. **31**(1), 46–65 (2019)
27. Pérez-Fernández, R., Rademaker, M., Alonso, P., Díaz, I., Montes, S., De Baets, B.: Monotonicity-based ranking on the basis of multiple partially specified reciprocal relations. Fuzzy Sets Syst. **325**, 69–96 (2017)
28. Roubens, M.: Choice procedures in fuzzy multicriteria decision analysis based on pairwise comparisons. Fuzzy Sets Syst. **84**, 135–142 (1996)
29. Roubens, M., Vincke, P.: Preference Modelling. Springer, Berlin (1985)
30. Schulze, M.: A new monotonic, clone-independent, reversal symmetric, and Condorcet-consistent single-winner election method. Soc. Choice Welfare **36**, 267–303 (2011)
31. Sen, A.K.: A possibility theorem on majority decisions. Econometrica **34**(2), 491–499 (1966)
32. Sen, A.K.: Collective Choice and Social Welfare. Holden-Day, San Francisco (1970)
33. Tideman, T.N.: Independence of clones as a criterion for voting rules. Soc. Choice Welfare **4**(3), 185–206 (1987)
34. Young, H.P.: Condorcet's theory of voting. Am. Polit. Sci. Rev. **82**(4), 1231–1244 (1988)
35. Zavist, T.M., Tideman, T.N.: Complete independence of clones in the ranked pairs rule. Soc. Choice Welfare **6**(2), 167–173 (1989)

Hospital Data Interpretation:
A Self-Organizing Map Approach

Javid Pourkia[1]([⊠]), Shahram Rahimi[2], and Kourosh Teimouri Baghaei[2]

[1] Southern Illinois University, Carbondale, IL 62901, USA
javid@cs.siu.edu
[2] Department of Computer Science and Engineerinbg, Mississippi State University,
Starkville, MS 39759, USA
rahimi@cs.msstate.edu, kt1414@msstate.edu

Abstract. It is not feasible to attempt to interpret complex health care data with large dimensions using typical 2D or 3D charts, diagrams, or graphs. It is helpful to be able to correlate the dimensions against one another to discover new patterns and obtain fresh knowledge. Effective interpretation of the statistical data, collected from health care centers, helps physicians and clinicians to improve their efficiency and the quality of care. This work has used different types of Self-Organizing Maps (SOM) in order to provide visual interpretability of the collected data to the hospital administration. Using the approach presented in this work, existing correlations among different attributes of collected data can be discovered and utilized to uncover hidden patterns. We illustrate how Self-Organizing Maps can be effectively used in interpretation of health care and similar data.

1   Introduction

There almost always exist important hidden information and patterns in large datasets with high dimension, which is close to impossible to be discovered using rudimentary tools. This hidden information can assist us to uncover the interconnectivity of various parts of the data which might have previously been regarded as irrelevant. One of the substantial issues in data interpretation is the existence of large number of dimensions (attributes) in datasets. Naturally, humans are used to 3D illustrations which represent the physical world as they see it. However, it is not feasible to attempt to imagine a space that displays a 100 dimensional dataset.

Health care datasets are very complex and have numerous dimensions. The discovery of the correlations among the attributes of these datasets can significantly contribute to the improvement of not only the health care services provided by hospitals and medical centers but to the excavation of hidden patterns to find the causes of deficiencies or complications [1].

As of a few years ago, several federal and private organizations introduced standard surveys and data collection procedures to add consistency to health

R. B. Kearfott et al. (Eds.): IFSA 2019/NAFIPS 2019, AISC 1000, pp. 493–504, 2019.
https://doi.org/10.1007/978-3-030-21920-8_44

care data collection and provide reliable information for patients, consumers, and health care providers[1]. Having access to a standard platform, researchers have explored the possible correlations among patient survey reviews and quality measurements of hospitals [1,3]. These researches yielded noticeable findings in how hospitals' ratings in different aspects correlate with the actual observations such as patient behaviour patterns, clinical processes and outcomes, efficiency and safety [1].

In this work, we have employed Self-Organizing Map (SOM), which is a type of unsupervised Artificial Neural Network, to better represent multidimensional health care data and to help their interpretability, so as to disclose the elements that require the attention of health care authorities and other interested parties. The main objective of this research is to provide tools to explore and discover the hidden patterns from the data provided by Centers for Medicare and Medicaid Services (CMS)[1] that would provide hints in making modifications to health care procedures.

Here is the outline of the rest of this paper. Section two provides a background and literature review over the method used and its applications, as well as previous similar works. Section three explains the datasets and the pre-processing of them. Section four provides results and analysis. And in the final section, the conclusion and future works are discussed.

2   Background and Literature Review

Self-Organizing Map is one of the promising unsupervised learning methods for data clustering, introduced by Kohonen in 1980s [4]. Before going through the related literature and SOM applications, a brief overview of the way a SOM works is given next.

2.1   Self-Organizing Map

Self-Organizing Map is essentially a neural network that is able to group input data points based on their discrepancies in an unsupervised manner. Considering a finite set of weight vectors as nodes of a grid, the objective of the training process is to move around these nodes so that they cover the whole input data. Thus, each weight vector, becomes representative of the nearest data points surrounding it [4]. This grid of nodes is called *Map*.

Here are the steps that are followed for training a SOM [5]:

1. For each node w_i, the weight vector is initialized with a random value.
2. One data point x_j is randomly chosen from the input data.
3. The nearest node c from the map to the selected data point is found. It is considered as *the Winning Node* and it is commonly known as *Best Matching Unit(BMU)*.

$$c = \arg \min_i \|x_j - w_i\|$$

---

[1] For more information refer to [2]. This task is currently performed under the supervision of Centers for Medicare and Medicaid Services https://www.cms.gov.

4. BMU is moved towards the data point. A parameter called learning rate which decreases over time, determines the speed of BMU's displacement.
5. BMU's neighbouring nodes are also moved towards data point. Their update rate is proportional to their proximity to data point. Nodes within a certain radius are considered neighbours and radius is decreased over time.
6. Steps 2 to 5 are repeated until the map nodes are stabilized.

Once the training is completed, each node can represent a certain number of training points. Based on these numbers a simple 2D heat-map can be used in order to visualize the distribution of training points among different clusters [5].

There are various approaches to train a SOM [5]. Various neighbourhood functions and measurements for similarity and distance can be implemented [5]. However, the general idea is almost the same as explained, and in most applications, Euclidean distance is used for calculating the distances among data points and nodes [5].

2.2   Previous Studies

The very nature of this model in visualization of the correlations among data points, makes it a *simulatable* model in terms of *model interpretability* [6], hence, making it a suitable model for the task of visually analyzing data. This model has been successfully utilized in many applications. SOM proved to be a successful means of fraud detection as in [7] that prevents online fraudulent transactions, and in [8,9] that target financial data and fraud profiling in leasing respectively. Modified variations of SOMs were used in several papers [10–13] for analysis of network traffic in order to detect malicious network activities. Langin et. al [14] proposed a 3D model of SOM that visualizes network traffic, thereupon making detection of intrusion possible. Authors in [15] used SOM to detect Denial of Service attacks. SOM has also gained some attention in medical and health care applications. Authors in [16,17] explored breast cancer data using SOM and [18] used SOM to classify and cluster breast cancer images. A modification of SOM with an added supervised layer was proposed by [19] to analyze breast cancer alongside cardiotocography data.

SOM has also been used for detection of anomalies in biosignals. As an example, Electromyographic signals (EMG) that can be used for detecting disorders in muscle, nerve and plexus were analyzed in [20]. SOM's potential in class discovery on Human brain's electric signals (EEG) was shown in [21]. Classification of normal and abnormal beats in Electrocardiograms (ECG) was made possible by SOM in [22]. Another application of SOM in health care is the analysis of informational documents in different contexts. Analysis of multivariate social, economic, and environmental datasets with SOM led to the conclusion that there is a similarity in distribution of disease among a group of people with akin environmental characteristics [22]. [23] explored the correlation between habits and outcomes of Type-1 Diabetes patients from a self-care survey. In [24] the authors used SOM to group bio-medical documents into informative clusters that facilitate information retrieval and visualization of clusters.

Analysis of surveys of medical context, reveal some valuable information about the possible correlations among variables that cannot be concluded from the raw data [25]. This approach was implemented in [26] to gain insights into preferences and standpoints of patients in primary care and referring to specialists. In other studies, surveys were analyzed using SOM to detect and possibly modify the influential factors in order to provide better quality and service to the health care plan's customers [27,28]. In this research project which is essentially based on [25–27], a Self-Organizing Map is used in order to delve into hospitals' data to extract hidden patterns that might be helpful in improving the procedures and services.

3   Data Preprocessing and Implementation

In order to help improve the quality of care in U.S. hospitals, Centers for Medicare and Medicaid Services[2] (CMS) and a group of public and private organizations formerly known as Hospital Quality Alliance[3], collaborated in creating a credible database that provides useful information about hospitals. This online database provides consumers with various information about different hospitals and helps them in deciding on choosing care [2].

This data is publicly available online. The dataset consists of forty five different tables, from which we have mined and processed the key attributes to form three core tables that feed the generated SOMs. The complexity of the data came from the fact that each record contains information of a hospital and a measurement name and its value in a single string, so each data entry incorporates more than a single content. Therefore, for the data to be useful for SOM generation, it was reorganized so that each row represents a hospital and columns represent different measurement values. There were occasional incompleteness in data; however, after careful evaluation, it was determined that filling the missing values using estimation methods would affect the results negatively. Consequently, it was decided to omit the records that were too sparse. Fortunately, there were not too many such record in each table, and we usually had over 3000 hospitals with reliable data in which all of the values for the attribute were available.

The last step of data preparation was to normalize the data tables. The tables stored data in different formats for different attributed, some as percentages, some in real format, etc. This would almost certainly corrupt SOMs' outputs, considering the need of SOMs (as in other forms of Neural Networks) for uniform data for the purpose of distance calculation and mapping.

3.1   Categories Selected for Analysis

Among the major categories that could be selected for the purpose of this work, data was mined and processed for the following categories to be analyzed:

---

[2] https://www.cms.gov.

[3] It was formed in 2002 and later disbanded in 2011. Since then, Hospital Compare has been maintained by CMS only.

3.1.1 Process of Care, Emergency Department

This category requires data that contains average time elapsed before hospitals' staffs take action for several particular medical conditions. It also requires percentage of patients who gone under various types of procedures for certain emergency situations [2].

3.1.2 Hospital Consumer Assessment of Health Care Providers and Systems Patient Survey (HCAHPS)

A 32 item-survey is collected randomly from adult patients across the nation within the time interval of two days to six weeks after discharge. That indicates patient's overall satisfaction in various aspects of the hospitals' services such as how well patients are being communicated and informed of key information when discharged. The responsiveness of hospital staff to patients' demands and how well patients are helped in managing pain. Other factors are also considered such as how clean and quiet are the patients' rooms and whether a patient is willing to recommend that hospital to his circle of friends and relatives. A full description of the parameters of this dataset is available in [2,29].

3.1.3 30-Day Mortality and Readmission Measures

As standard measures proposed and published by a group of clinical doctors and statisticians from Harvard and Yale universities, they imply how likely a patient is to be readmitted or end up in death while in hospital. Factors like age, severe conditions or medical history of patients are also taken into account in order to provide a fair comparison. CMS calculates 30-day mortality and readmission rates of each hospital based on the Medicare claims and other administrative data. For more details take a look at [2].

3.2 Types of Generated SOMs

For the purpose of this study, and to provide the best possible visualization of the data, three different types of SOMs were generated. Each of these SOMs represent a different metric explained in the following sub-sections.

3.2.1 Neighbor Distances

The first type of SOM represents the distances among the neighboring nodes as a heat-map (Fig. 1a). The purple hexagons in the figure represent nodes of the SOM and the irregular hexagon residing in between each two pair of the nodes is meant to represent the distance among the corresponding nodes with its color. The darker the color, the further the nodes are from one another.

3.2.2 Sample Hits

Another type of SOM that is used in this study, is called "SOM Sample Hits" (Fig. 1b). Each hexagon in this map, represents one node of the SOM, and the overlaying number represents the number of nearest neighbors from input data to that node.

3.2.3   Weight Planes

As part of our analysis, we were curious to know the patterns of distances among the different features of the input data. Thereupon, we created another set of SOM maps that visualizes the neighbor distances of each of the features as in Fig. (1c). The idea behind these maps is that any possible similarities among these SOMs, would suggest possible correlations among them.

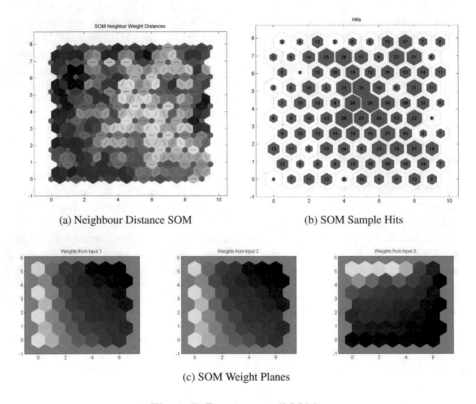

(a) Neighbour Distance SOM   (b) SOM Sample Hits

(c) SOM Weight Planes

Fig. 1. Different types of SOMs.

3.3   SOM Discover

For the purpose of this project, an application called "SOM Discover" was implemented that enables users to get the list of hospitals associated with a node from the 2D grids [25]. It also provides a more detailed visualization of measurements associated with the hospitals that reside in the same neighbourhood of the targeted node. Thus, helping experts to be able to take a closer look at hospitals' data and make more reliable comparisons about them over the SOMs. Fig. 2 is a screen shot of the SOM Discover's user interface [25].

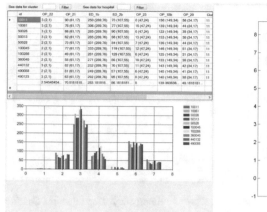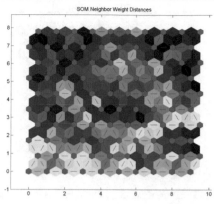

Fig. 2. SOM Discover's user interface. By clicking on any nodes of the heat- map grid on the right, the list of hospitals associated with that node will be presented in the upper left table. And the parameters of those hospitals are shown in bar diagram.

4   Case Studies and Discussion

MATLAB's NCToolbox was utilized in this project in order to generate the SOMs described above. At this point of the project, the core objective is to provide experts with visualization of the data and the statistical analysis of the SOMs, using SOM Discover [25]. In the following subsections three case studies are presented and discussed on the three different categories explained above.

4.1   Case Study I: Process of Care, Emergency Department

In this case study, the processed data of Emergency departments of various hospitals are considered. The measurements are indicative of the performance and efficiency of the hospitals.

Figures (3a) and (b) depict the SOM and the hit map of the emergency departments' data. Clearly, the figure suggests that the hospitals are grouped into six clusters. A closer examination of the data illustrates that one of the major contributing factors to this clustering has been Input 5 (OP_23)[4] which is the relative number of patients who were admitted to emergency departments having stroke symptoms, and undergone a brain scan within 45 min of arrival in percents [2]. Consulting this issue with a subject matter expert led us to conclude that the hospitals that comply to a certain "Stroke Protocol" would be clustered in right-most nodes of the SOM. The aforementioned protocol may be a simple set of rules that enforce clinical staff to take care of the stroke symptom exhibiting patients with higher priority than the other less severe patients admitted to emergency department.

---

[4] Refer to [2] for a detailed description of all of the inputs and measurements.

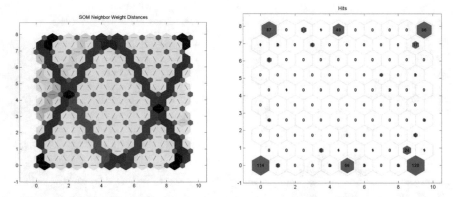

(a) Hospital Emergency Departments, SOM Heat-   (b) Hospital Emergency Departments, Sample
map   Hits

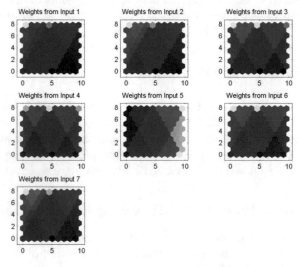

(c) Hospital Emergency Departments, Weights of inputs

Fig. 3. Hospital emergency department SOM Figures

4.2   Case Study II: Hospital Consumer Assessment of Health Care Providers and Systems Patient Survey (HCAHPS)

Due to noticeable amount of missing values for certain questions, and as explained earlier, three of the questions from our analysis were omitted. According to the weight maps of the inputs in Fig. (4a) several of the inputs have similar patterns in their SOMs. Although correlation in data does not immediately implies causation [6], these patterns are indicative of certain facts regarding the effects of patients background on their comprehension of the quality of care they have received. Closer scrutiny of Figs. (4a), (c) and (e) shows that the attributes

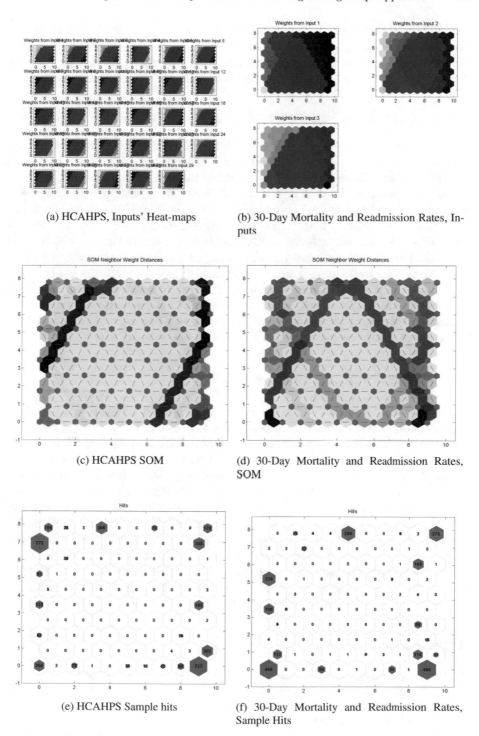

(a) HCAHPS, Inputs' Heat-maps

(b) 30-Day Mortality and Readmission Rates, Inputs

(c) HCAHPS SOM

(d) 30-Day Mortality and Readmission Rates, SOM

(e) HCAHPS Sample hits

(f) 30-Day Mortality and Readmission Rates, Sample Hits

Fig. 4. HCAHPS SOM Figures on the left and 30-Day Mortality and Readmission Rates SOMs on the right

related to age and racial profile of the patients have major affect on their reflection of their experience of the quality of care they receive. Much more can be learned and discussed using the presented visualization by the subject matter experts from the survey data which does not fit in the scope of this paper.

4.3   Case Study III: 30-Day Mortality and Readmission Measures

Upon a patient's discharge, it is possible that patient's situation aggravates and is readmitted to the hospital. The rates of mortality and readmission within 30 days of discharge is an important indicator of the quality of care.

In this study, we have only considered the rates of readmission of patients who had been admitted for Pneumonia, Heart Failure, and Heart Attack which correspond to inputs 1, 2 and 3 respectively. According to Fig. (4b), there is a correlation between readmission of Heart Attack and Heart Failure, which could mean that these hospitals are not benefiting from a very sophisticated Cardio staff. Moreover, by looking into Figs. (4d) and (f), we can clearly see that the overall performance of hospitals has been clustered into two major groups, both at the bottom of the map in the opposite corners. The small clusters are tending to stay at the top of the map. We are planning to discuss the detailed findings in a follow up publication, since they do not fit the scope of this paper.

5   Conclusions and Future Work

As an inherently simulatable model, SOM can be used to visualize certain groupings and correlations among data points [6], thus making it a suitable means of data analysis in unsupervised scenarios. In this research, we applied SOM models to hospitals' statistical and survey data in order to gain a better understanding of associations among factors that are effective in quality of the services provided by hospitals. There are a few tracks that can be followed for further research on this problem. Firstly, methods of data preprocessing and handling missing data should be improved so that less data is omitted from the dataset. Secondly, in our current approach, presence of an expert is inevitable for making proper conclusions. Therefore, implementing automated mechanisms that make use of experts' knowledge to analyze data is another track for future work [25]. Finally, making use of other unsupervised learning models might provide different results that were not possibly considered by SOM.

References

1. Price, R.A., Elliott, M.N., Zaslavsky, A.M., Hays, R.D., Lehrman, W.G., Rybowski, L., Edgman-Levitan, S., Cleary, P.D.: Examining the role of patient experience surveys in measuring health care quality. Med. Care Res. Rev. **71**(5), 522–554 (2014). PMID: 25027409
2. Centers for Medicare and Medicaid Services, Baltimore, MD. https://data.medicare.gov/data/archives/hospital-compare. Accessed 01 July 2013

3. Elliott, M.N., Cohea, C.W., Lehrman, W.G., Goldstein, E.H., Cleary, P.D., Giordano, L.A., Beckett, M.K., Zaslavsky, A.M.: Accelerating improvement and narrowing gaps: trends in patients experiences with hospital care reflected in HCAHPS public reporting. Health Serv. Res. **50**, 1850–1867 (2015)
4. Kohonen, T.: Essentials of the self-organizing map. Neural Netw. **37**, 52–65 (2013). Twenty-fifth Anniversay Commemorative Issue
5. Miljković, D.: Brief review of self-organizing maps. In: 2017 40th International Convention on Information and Communication Technology, Electronics and Microelectronics (MIPRO), pp. 1061–1066, May 2017
6. Lipton, Z.C.: The Mythos of model interpretability. CoRR, vol. abs/1606.03490 (2016)
7. Balasupramanian, N., Ephrem, B.G., Al-Barwani, I.S.: User pattern based online fraud detection and prevention using big data analytics and self organizing maps. In: 2017 International Conference on Intelligent Computing, Instrumentation and Control Technologies (ICICICT), pp. 691–694, July 2017
8. Jian, L., Ruicheng, Y., Rongrong, G.: Self-organizing map method for fraudulent financial data detection. In: 2016 3rd International Conference on Information Science and Control Engineering (ICISCE), pp. 607–610, July 2016
9. Bach, M.P., Vlahović, N., Pivar, J.: Self-organizing maps for fraud profiling in leasing. In: 2018 41st International Convention on Information and Communication Technology, Electronics and Microelectronics (MIPRO), pp. 1203–1208, May 2018
10. Langin, C., Zhou, H., Rahimi, S., Gupta, B., Zargham, M., Sayeh, M.R.: A self-organizing map and its modeling for discovering malignant network traffic. In: 2009 IEEE Symposium on Computational Intelligence in Cyber Security, pp. 122–129, March 2009
11. Yao, C., Luo, X., Zincir-Heywood, A.N.: Data analytics for modeling and visualizing attack behaviors: a case study on SSH brute force attacks. In: 2017 IEEE Symposium Series on Computational Intelligence (SSCI), pp. 1–8, November 2017
12. AlHamouz, S., Abu-Shareha, A.: Hybrid classification approach using self-organizing map and back propagation artificial neural networks for intrusion detection. In: 2017 10th International Conference on Developments in eSystems Engineering (DeSE), pp. 83–87, June 2017
13. Almi'ani, M., Ghazleh, A.A., Al-Rahayfeh, A., Razaque, A.: Intelligent intrusion detection system using clustered self organized map. In: 2018 Fifth International Conference on Software Defined Systems (SDS), pp. 138–144, April 2018
14. Langin, C., Wainer, M., Rahimi, S.: ANNaBell Island: a 3D color hexagonal SOM for visual intrusion detection. Int. J. Comput. Sci. Inf. Secur. **9**(1), 1–7 (2011)
15. Nam, T.M., Phong, P.H., Khoa, T.D., Huong, T.T., Nam, P.N., Thanh, N.H., Thang, L.X., Tuan, P.A., Dung, L.Q., Loi, V.D.: Self-organizing map-based approaches in DDoS flooding detection using SDN. In: 2018 International Conference on Information Networking (ICOIN), pp. 249–254, January 2018
16. Zribi, M., Boujelbene, Y., Abdelkafi, I., Feki, R.: The self-organizing maps of Kohonen in the medical classification. In: 2012 6th International Conference on Sciences of Electronics, Technologies of Information and Telecommunications (SETIT), pp. 852–856, March 2012
17. Markey, M.K., Lo, J.Y., Tourassi, G.D., Floyd Jr., C.E.: Self-organizing map for cluster analysis of a breast cancer database. Artif. Intell. Med. **27**, 113–127 (2003)
18. Chandra, B., Nath, S., Malhotra, A.: Classification and clustering of breast cancer images. In: The 2006 IEEE International Joint Conference on Neural Network Proceedings, pp. 3843–3847, July 2006

19. Platon, L., Zehraoui, F., Tahi, F.: Self-organizing maps with supervised layer. In: 2017 12th International Workshop on Self-Organizing Maps and Learning Vector Quantization, Clustering and Data Visualization (WSOM), pp. 1–8, June 2017
20. Ijaz, A., Choi, J.: Anomaly detection of electromyographic signals. IEEE Trans. Neural Syst. Rehabil. Eng. **26**, 770–779 (2018)
21. Ayesh, A., Arevalillo-Herráez, M., Arnau-González, P.: Class discovery from semi-structured EEG data for affective computing and personalisation. In: 2017 IEEE 16th International Conference on Cognitive Informatics Cognitive Computing (ICCI*CC), pp. 96–101, July 2017
22. Basara, H.G., Yuan, M.: Community health assessment using self-organizing maps and geographic information systems. Int. J. Health Geogr. **7**, 67 (2008)
23. Tirunagari, S., Poh, N., Hu, G., Windridge, D.: Identifying similar patients using self-organising maps: a case study on type-1 diabetes self-care survey responses. CoRR, vol. abs/1503.06316 (2015)
24. Shah, S., Luo, X.: Exploring diseases based biomedical document clustering and visualization using self-organizing maps. In: 2017 IEEE 19th International Conference on e-Health Networking, Applications and Services (Healthcom), pp. 1–6, October 2017
25. Pourkia, J.: A self-organizing map approach for hospital data analysis. MS thesis, SIUCOpenAccess (2014)
26. Grumbach, K., Selby, J.V., Damberg, C., et al.: Resolving the gatekeeper conundrum: what patients value in primary care and referrals to specialists. JAMA **282**(3), 261–266 (1999)
27. Garavaglia, S.B.: Health care customer satisfaction survey analysis using self-organizing maps and "exponentially smeared" data vectors. In: Proceedings of the IEEE-INNS-ENNS International Joint Conference on Neural Networks, IJCNN 2000. Neural Computing: New Challenges and Perspectives for the New Millennium, vol. 4, pp. 119–124, July 2000
28. Tabrizi, T.S., Khoie, M.R., Sahebkar, E., Rahimi, S., Marhamati, N.: Towards a patient satisfaction based hospital recommendation system. In: 2016 International Joint Conference on Neural Networks (IJCNN), pp. 131–138, July 2016
29. Centers for Medicare and Medicaid Services, Baltimore, MD. http://www.hcahpsonline.org. Accessed 01 July 2013

Hybrid Connection Between Fuzzy Rough Sets and Ordered Fuzzy Numbers

Piotr Prokopowicz[1]($\boxtimes$) and Marcin Szczuka[2]

[1] Institute of Mechanics and Applied Computer Science,
Kazimierz Wielki University, Kopernika 1, 85-074 Bydgoszcz, Poland
`piotrekp@ukw.edu.pl`
[2] Institute of Informatics, University of Warsaw, Banacha 2, 02-097 Warsaw, Poland
`szczuka@mimuw.edu.pl`

Abstract. Ordered Fuzzy Numbers (OFN) provide the ability of modeling data which is united with its trend. This paper presents a proposition of connecting the OFN model with the concept of information granules built as fuzzy rough sets. The procedure for gathering data and converting them into OFN is a new way of looking at transforming time series of sensor readings into granules. The introduction of the method is supported by an illustrative example. The introduced procedure for calculating similarity between OFNs allows hybridization with fuzzy rough set approach and derivation of lower and upper approximations of concepts.

Keywords: Ordered Fuzzy Number · Fuzzy rough sets ·
Hybridization · Information granulation · Time series

1 Introduction

Information granulation is one of the important ways of problem solving in Computational Intelligence. It is done through problem decomposition or modelling of cognition. Modern computations are often performed on very complex structures with plethora of compound attributes. Granules representing such complex structures often correspond to terms expressed in natural languages. They need to be approximated from data step-by-step, starting from low-level concepts derivable from raw (sensor) data. Such granules can then be treated as computational building blocks in construction of granular systems aiming at comprehension of complex situations.

In this paper we propose a new way of bridging the gap between the raw data and its granular representation. We create a hybrid approach that combines the

The hybridization concepts presented in this paper were developed as a part of the project "The hybridization of selected methods of computational intelligence for modeling non-precision data" within the program "MINIATURA 1" funded by the National Science Centre of Poland. Project No. 2017/01/X/ST6/01675.

R. B. Kearfott et al. (Eds.): IFSA 2019/NAFIPS 2019, AISC 1000, pp. 505–517, 2019.
https://doi.org/10.1007/978-3-030-21920-8_45

fuzzy extension of rough sets (fuzzy rough sets) with Ordered Fuzzy Numbers (OFNs). The strength of this hybrid is its ability to carry the information about directionality in underlying sensory data to higher levels of granular system. The use of fuzzy extension of the rough set toolbox makes it possible to express and model concepts (data objects) that are inherently imprecise, vague or incomplete.

The OFN model that is used here was formulated by Kosiński, Prokopowicz, and Ślęzak (see [10, 11, 20]). It is sometimes mistaken for the "classical", Zadeh's fuzzy numbers [27, 29] due to the very similar contexts the two notions tend to appear in. They are, however, significantly different ideas, with OFNs using an additional feature of data - a direction - taking care of the order of the characteristic parts of a fuzzy number. Another added value of the OFN model is the ability of using properly defined and well behaving arithmetic operations on OFNs. This improves on several problematic constructions in the arithmetic of conventional fuzzy numbers (see [12] for a general summary) and their extensions (see [7, 23, 26]) by eliminating some of the constraints.

The hybrid approach that we propose has as its aim building a procedure for gathering data from the sensory level in the system and converting it to a proper granular format. The advantage of OFNs in this step lays in the ability to preserve directionality in data. This may be helpful in cases where this directionality is of importance, as in time series where time precedence is crucial. We explain it in the paper using a simple example of weather data.

While the OFNs make it possible to build granular data objects from numerical data the processing of granular information, in particular derivations on granules, we perform using the rough set constructs (see [13, 14]). In order to model granule corresponding to a given imprecise concept, e.g. notion of "morning" in time domain, we employ rough approximations. In fact, since we are interested in the degree (a value in $[0, 1]$) of membership of an object in the concept, we adopt the fuzzy rough set approach, as described in [2, 5]. In order to make a good use of the capabilities of fuzzy rough sets we introduce a procedure for establishing the degree of similarity between two OFNs that takes advantages of both flexible OFN arithmetic and defuzzyfiction functional.

The main purpose of this paper is to introduce the underlying concepts and then demonstrate the potential behind their hybridization. First the granular (rough, fuzzy, and fuzzy rough) frameworks are very briefly described in Sect. 2, followed by explanation of main OFN constructs in Sect. 3. A proposal for usage of OFNs in data gathering, together with illustrative example make Sect. 4.1. Finally, the hybridization and its possible extensions are described in Sect. 5, followed by concluding remarks in Sect. 6.

2   Granular Frameworks

The fuzzy set theory [27] and the rough set theory [13] are some of the prominent frameworks for constructing information granules (cf. [25, 29]). The hybridization of the two leads to the fuzzy rough set approach, as described in, e.g. [1, 2, 14, 21]. We briefly introduce the most basic notions in these frameworks in order to match them with OFNs next.

2.1   Rough Set Theory

Following [2], data entities are represented as an *information system* $(X, \mathscr{A})$, where $X = x_1, \ldots, x_n$ and $\mathscr{A} = a_1, \ldots, a_m$ are finite, non-empty sets of objects and attributes, respectively. Each a in $\mathscr{A}$ corresponds to an $X \rightarrow V_a$ mapping, in which V_a is the value set of a over X. For every subset B of $\mathscr{A}$, the B-indiscernibility relation R_B is defined as

$$R_B = (x, y) \in X^2 \text{ and } (\forall a \in B)(a(x) = a(y)) \tag{1}$$

Clearly, R_B is an equivalence relation. Its equivalence classes $[x]_{R_B}$ can be used to approximate concepts, i.e., subsets of the universe X corresponding to basic granules. Given $A \subseteq X$, its lower and upper approximation w.r.t. R_B are defined by

$$R_B{\downarrow}A - \{x \in X | [x]_{R_B} \subseteq A\} \tag{2}$$

$$R_B{\uparrow}A = \{x \in X | [x]_{R_B} \cap A \neq \emptyset\} \tag{3}$$

2.2   Fuzzy Set Theory

The fuzzy set theory [27] allows that the objects can belong partially to a set. It is described by the degree – value from the interval $[0, 1]$, where 0 means not in the set, and 1 – full membership. The values in $(0, 1)$ correspond to partial membership. There is direct relationship between a fuzzy set and a basic information granule. If we deal with the couples of objects we can also consider the binary fuzzy relations.

We can define a fuzzy set in X as an $X \rightarrow [0, 1]$ mapping, while a fuzzy relation in X is a fuzzy set in $X \times X$ (see [2]). For all y in X, the R-foreset of y is the fuzzy set Ry defined by

$$Ry(x) - R(x, y) \tag{4}$$

for all x in X. If R is a reflexive and symmetric fuzzy relation, i.e., $R(x, x) = 1$ and $R(x, y) = R(y, x)$ hold for all x and y in X, then R is called a fuzzy tolerance relation.

2.3   Fuzzy Rough Set Theory

Following [2], research on the hybridization of fuzzy sets and rough sets emerged in the late 1980's [5] and has flourished recently. It focuses mainly on fuzzyfying the formulas (2) and (3) for lower and upper approximation.

The crisp concept (set) A may be generalized to a fuzzy set in X by allowing objects to belong to it (i.e., meet its characteristics) to varying degrees. The same can be extended to rough set approximations (granules) for the concept. In principle, rather than assessing objects' indiscernibility, we may measure their *approximate equality*, represented by a fuzzy relation R. It leads to classes or granules, with "soft" boundaries based on their similarity. As such, abrupt transitions between classes are replaced by gradual ones, allowing that an element can belong (to varying degrees) to more than one class. Typically, we assume that R is at least a fuzzy tolerance.

For the lower and upper approximation of a fuzzy set A in X by means of a fuzzy tolerance relation R, we adopt the definitions proposed by Radzikowska and Kerre in [22]. For all y in X given an implicator I and a t-norm T, they paraphrased formulas (2) and (3) to define $R{\downarrow}A$ and $R{\uparrow}A$ by

$$(R{\downarrow}A)(y) = \inf_{x\in X} I(R(x,y), A(x)) \qquad (5)$$

$$(R{\uparrow}A)(y) = \sup_{x\in X} T(R(x,y), A(x)). \qquad (6)$$

3   Ordered Fuzzy Numbers

The Ordered Fuzzy Numbers (OFNs) were introduced and developed in the series of papers [8,10,11,17,20]. The OFN model is conceptually connected with the idea of fuzzy sets, and can be used in a similar context. However, OFNs are not the classical fuzzy numbers defined by Zadeh's fuzzy sets. Thus, comparison of their operations with adequate operations on Zadeh's extension principle [28] can be done only in the context of the results of calculations, but not their definitions.

3.1   Definition of the OFN

Following the papers [9–11,19,20], a fuzzy number will be identified with the pair of functions defined on the interval $[0, 1]$.

Definition 1. An **Ordered Fuzzy Number** (**OFN**) A is an ordered pair

$$A = (f_A, g_A) \qquad (7)$$

of continuous functions $f_A, g_A : [0, 1] \to \mathbb{R}$, which are called the **up part** and the **down part** of A, respectively.

Continuity of both f_A and g_A guarantees that their images are bounded intervals. Let us denote them as $UP_A = f_A([0, 1])$ and $DOWN_A = g_A([0, 1])$ (see Fig. 1a).

Considering functions f_A and g_A as an ordered pair is the crucial difference comparing to standard fuzzy numbers that can be represented by means of so-called *L-R notation* [4]. By L and R one means left (increasing) and right (decreasing) components of a membership function of a given fuzzy number. Such components can be inverted (see Fig. 1b) to form functions assigning real values to the elements of the unit interval $[0, 1]$ (see [11]).

It is useful to mark characteristic points of an OFN. Herein, the parameters $s_A = f_A(0)$, $e_A = g_A(0)$ (s and e stand for start and end, respectively) and $1_A^- = f_A(1)$, $1_A^+ = g_A(1)$ ($-$ and $+$ stand for *reaching* and *leaving*, respectively, a precise component of a fuzzy number) will be useful (see Fig. 1b).

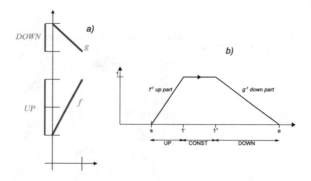

Fig. 1. (a) Ordered fuzzy number - definition. (b) Standard representation.

3.2 Arithmetic Operations

We define operations on OFNs as calculations with use of *up parts* and *down parts*.

Definition 2. Let $A = (f_A, g_A)$, $B = (f_B, g_B)$, and $C - (f_C, g_C)$ be OFNs. The sum $C = A + B$, subtraction $C = A - B$, product $C = A \cdot B$, and division $C = A \div B$ are defined by the formula:

$$f_C(y) = f_A(y) \star f_B(y) \qquad \wedge \qquad g_C(y) = g_A(y) \star g_B(y) \tag{8}$$

where "$\star$" stands for one of: "+", "−", ".", and "/". Furthermore, $A \div B$ is defined only if support of B does not contain zero. The $y \in [0, 1]$ is the argument of functions f and g.

The properties of these operations and their results were extensively discussed and analyzed in various publications (see, e.g., [11,15,20]). The fact that the opposite value is the result of the multiplying by -1 (real number - singleton) is significant and very useful. It makes a subtraction exactly the same as an addition of the opposite number. Thanks to that, the result of $A - A$ is exactly zero (real number). Also, using arithmetical operations on the OFNs, every simple equation of type $A + X = B$, where A and B are fuzzy numbers with any membership functions, can be solved. Hence, the OFN model provides the to make calculation on imprecise objects just like for (crisp, precise) real numbers at the same time retaining their fuzzy quantitative character while preventing the imprecision growth.

Figure 2 presents example of a calculation. It presents procedure for adding, however the rest of arithmetic operations is adequate.

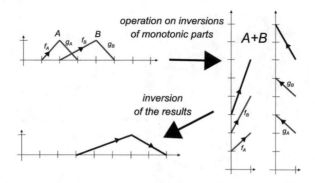

Fig. 2. Example of a calculation

3.3 Standard Representation of OFNs

As mentioned, OFNs are different mathematical objects from classical Zadeh's fuzzy numbers. However, for both concepts some intuitions and application context are similar. The transformation called *standard representation* is a tool for achieving better comparability of both models (for details see [20]).

If f_A and g_A are monotonic, then intervals UP_A and $DOWN_A$ retain the following dependencies:

$$UP_A = [\min\{s_A, 1_A^-\}, \max\{s_A, 1_A^-\}], \quad DOWN_A = [\min\{1_A^+, e_A\}, \max\{1_A^+, e_A\}]. \quad (9)$$

Furthermore, for the monotonic f_A and g_A, it is possible to determine their inverse functions from $\mathbb{R}$ to $[0, 1]$ (Fig. 1b). Inverse functions are defined within the corresponding intervals UP_A and $DOWN_A$. To obtain a continuous shape, we connect them with a plot of a constant function **1** over the interval $CONST_A = [\min\{1_A^-, 1_A^+\}, \max\{1_A^-, 1_A^+\}]$. Thus, we have three functions that can be used to represent monotonic pairs (f_A, g_A) in a form more comparable to standard (convex) fuzzy numbers. This makes it possible to define the membership function for the OFNs (see [20]) making them behave similarly to classical, convex fuzzy numbers.

3.4 Interpretation of the Direction

The direction is a key element of the OFN model. The proposition of the interpretation for this new property was presented in [8,16]. It is defined as an order on the *up part* and *down part*. It is independent of the order of real numbers. The OFNs are considered as the values representing an observation which take place in time. The time precedence intuitively corresponds to the directionality in data while the particular order of numerical values within data entities (objects) may be of lesser consequence. Hence, the direction of OFN can be used as a way of expressing trends or changes, as exemplified in [17] using the case of dynamic process.

3.5   OFN Construction

OFNs' up and down parts can be constructed from data as shown is Sect. 4 as well as arbitrarily established (see [8,16]), e.g., by using the experts' opinions about dynamic changes of the analyzed values. Here we show a very basic example of expert-based OFN building.

According to the trend-based interpretation, let OFNs A and B visible in Fig. 3a represent two statements (opinions) prepared by an expert about two units of a financial company: *A's income is at a level of 3 millions, with an upward trend* and *B's income is at a level of 6 millions, with a downward trend.*

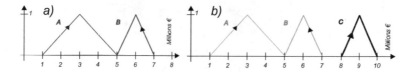

Fig. 3. (a) An income for two units of a financial company. (b) The result of adding both incomes.

By using OFNs the expert can actually describe not only the value and the trend but also an escalation of that trend. In this example we have two OFNs with different *spreads* between their up and down components – indeed, object A is *wider* than B. We can say that by making the up part of B ranging from 7 to 6 millions, the expert considers a potential of changes within 1 million whilst the up part of A ranges from 1 to 3. Hence, A could be recognized as a process that is more dynamic than B. Also, the direction of our OFNs tells us that B is a decreasing process and A is an increasing one.

One would expect here the total income, given experts' rules, at a level of 9 millions. Indeed, with OFN model by adding (ordered fuzzy) numbers A and B, we get anticipated results. We also are getting the additional information about the trend of $C = A + B$ which – as we can see in Fig. 3b – is increasing (growing). This tendency is also consistent with intuition, since the increasing process related to A is more dynamic than the decreasing process related to B. However, because of B, the overall increasing trend of C is less dynamic than in case of A.

The above example shows how interpretation of OFNs' and their direction can correspond to intuitions behind real-world observations. Such correspondence is important not only from a viewpoint of mathematical properties but becomes useful at an operational (conceptual) level.

4   Gathering Data as OFNs

Any system designed for data analysis and decision support, in particular a granular one, has to include a data preparation (gathering, cleaning, adjustment, transformation) module. Such system and its modules may be constructed in a variety of ways (see [24]), but here we want to focus on using the OFNs to help with that. We want the OFN to be a center piece of a data gathering module that stands between the raw sensor readings an analytic tools. This is conceptually shown in Fig. 4 using the example data set introduced in Sect. 4.2.

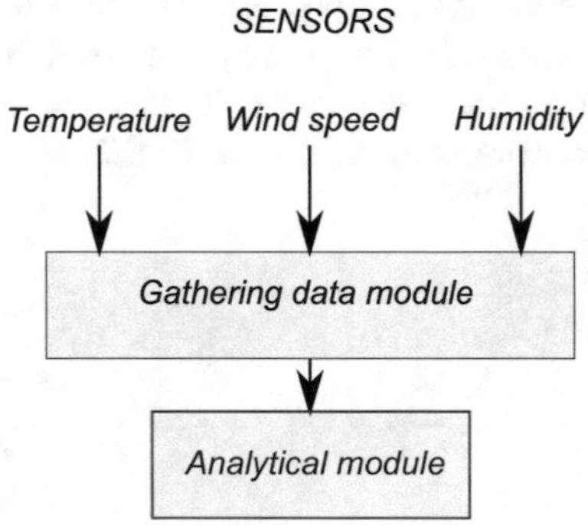

Fig. 4. The general scheme for data gathering in weather example (Sect. 4.2).

It should be mentioned, that the OFN hybridization approach is, as yet, hypothetical in the sense that no real-life applications have been implemented. Nevertheless, we use a simple example of a multi-dimensional time series (Sect. 4.2) to illustrate its potential use.

4.1   Building OFN from the Input Time Series

While gathering and processing sensor readings one can rely on widely-known methods of multivariate time series analysis [6] in the pursuit for reduction of the data dimension. Here, we propose an alternative by introducing the method for constructing OFN objects from sensor data. The main idea is to use the advantage of the direction for producing better representation of data.

We divide the whole time frame into intervals. In each interval certain number of values will be gathered. We denote the time series as $P = (p_1, p_2, ..., p_m)$ where m represents its length. The interval (or window) we will denote as subset $I \subseteq P, I = (b_k, b_{k+1}, ..., f_k)$ where $b_k, f_k \in P$ denote beginning and final data point of k-th segment in data series P, respectively. Note, that I can be also treated as time series.

Construction of the OFN will be simplified to the triangular form. The OFN $A = (s_A, 1_A, e_A)$ where $1_A = 1_{A-} = 1_{A+}$ (see Sect. 3.1).

Proposition 1. *The OFN A representing I – the k-th segment of time series P – is given by:*

$$s_A = b_k, \quad e_A = f_k, \quad 1_A = \frac{1}{n} \sum_{i=b_k}^{f_k} p_i, \tag{10}$$

where n is the number of data points in the interval I.

The middle element 1_A is an average of all data points in the given window, which consistent with natural intuition. In general, for fuzzy numbers, the kernel is a value which fully belongs to the set. Thus, in the time series context it is reasonable to use the mean of data points. Thanks to that we do not lose any initial information in the construction of OFN. Additionally, we also keep the information about the starting and ending value of I. Finally, the direction of OFN A represents general trend for the data points of I.

4.2   The Example

Let us consider a short, multi-dimensional time series that contains some weather (air) data such as: temperature, humidity and pressure. The data in this example is gathered using sensors throughout a day. If for some purposes it is useful to consider the data in the context of parts of day (morning, noon, afternoon, evening, night), then it comes quite naturally that the piece of the time series for each part of day may be replaced by a fuzzy object. In our example the morning is defined as a time interval from 6 to 10 AM. The measurements are taken every 20 min and are presented in Table 1.

From such data, using the approach detailed in Sect. 4.1 we get the OFNs representing the data for the morning measurements shown in Fig. 5. As one can notice, besides the fuzzy interpretation of values we also also retain information about the trend in measurements.

Table 1. Example of measurements - a fragment of time series for morning.

Time	Temperature [°C]	Wind [km/h]	Humidity [%]
06:00	15.1	27	100
06:20	14.8	23	100
06:40	15.0	25	95
07:00	15.3	20	85
07:20	15.3	19	90
07:40	15.8	19	92
08:00	16.2	22	85
08:20	16.0	18	88
08:40	16.4	17	85
09:00	17.1	15	89
09:20	17.9	13	82
09:40	18.5	11	80
10:00	19.2	10	83
mean	16.35	18.38	88.77

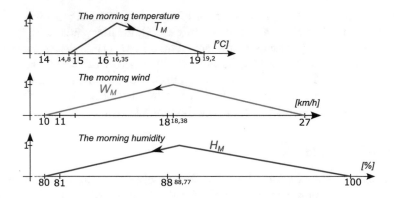

Fig. 5. The OFNs constructed from weather measurements.

5 Hybridization of OFNs and Fuzzy Rough Sets

We have demonstrated in the previous part how an OFN can be derived from raw, sensory data. Here, we want to show how to couple OFN with data represented by means of fuzzy rough sets (granules). The gist is to replace similarity calculation done with use of t-norms on sets by those performed on OFNs.

As previously was mentioned in Sect. 2.3, in fuzzy rough set theory we need a way to express the level (or degree) of similarity between objects. In case of OFNs, thanks to they nice arithmetic properties, we can revert the task of finding similarity. Instead of measuring how much in common two entities have, we concentrate on the difference between them, or rather a lack of thereof. Thanks to the flexibility of OFNs we can state that two of them are similar if there is no significant difference between them. Moreover, this difference if quite easy to derive.

To define the method of establishing similarity between ONFs we will also need a defuzzyfication method. Defuzzyfication (method) is a functional $\|.\|$ assigning a real number to a fuzzy object, in our case an OFN. There are numerous examples of such functionals in [3,9]).

Proposition 2. *Steps in calculation of similarity between two OFNs – A and B.*

1. The difference Dif between A and B is calculated by subtraction:

$$Dif_{AB} = (A - B).\tag{11}$$

2. Next, we calculate defuzzyfication of Di - crisp value:

$$Cr_{AB} = \|Dif_{AB}\|.\tag{12}$$

3. The similarity Si between OFNs A and B with the tolerance range $r \in \mathbb{R}$ is the result of:

$$
\begin{aligned}
|Cr_{AB}| \leq r: \quad & Si_{AB} = \frac{r - |Cr_{AB}|}{r} \\
|Cr_{AB}| > r: \quad & Si_{AB} = 0.
\end{aligned}
\tag{13}
$$

The resulting similarity is a function $Si : OFN^2 \times \mathbb{R} \to [0, 1]$.

The proposition maintains the basic boundary conditions. If the A is exactly the same as B, the result is equal to 1. On other hand, when there is no similarity at all the result is 0, as expected.

The parameter r represents *the range/scale* ratio important for the given situation. It can be called the *quantitative context* and needs some explanation. Usually, we have a scale relevant for actual problem. For example, dealing with the various vehicles on earth, the speed 15 km/h is not that similar to 60 km/h. But if we consider astronomical objects, then the mentioned velocities can be treated as similar. Therefore, we can think of r as the scale coefficient of the analyzed problem.

5.1 Hybridization Variants

For hybridization of fuzzy rough set concepts and the OFN model we can apply both Propositions 1 and 2 introduced earlier. Proposition 1 provides the means for gathering and representing the data while preserving its directionality (trend). Proposition 2 gives the additional ability to represent fuzzy rough approximations of a concept, namely:

- the lower approximation contains all data objects having $Si = 1$,
- the upper approximation are all the objects for which $0 > Si > 1$.

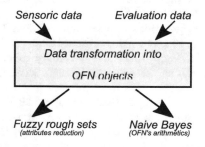

Fig. 6. The OFN hybridization potential.

Figure 6 represents the potential of OFNs hybridization. It should be mentioned that this does not exhaust the range of possibilities.

6 Summary and Perspectives for the Future

This paper presents the concept and the procedures for the hybridization of fuzzy rough sets and OFN models. It matches the granular representation of fuzzy rough sets with the possibility to represent the data with a trend as proper arithmetic of OFNs. There is even greater potential for hybridization of these two models. For instance, in [16,19] a procedure for calculating compatibility between two OFNs was presented. That procedure was used to determine the fuzzy truth value of a statement A *is* B for A, B being OFNs. We are convinced that this is a direction worthy of investigation and that with small adjustment this procedure can be adopted to determine similarity between lower and upper approximations of sets (granules).

The hybridization, as shown in this paper can also work the other way round. If we take the example from Sect. 4.2 and revisit the concept (granule) "in the morning" we can see that this concept may not be crisp/precise. To deal with this imprecision we can adopt a representation of data (a granule for morning hours) as a fuzzy rough set with lower and upper approximation described by different OFNs constructed from sensor data. This is also a topic we want to pursue in the future.

References

1. Cornelis, C., De Cock, M., Radzikowska, A.M.: Fuzzy rough sets: from theory into practice. In: Handbook of Granular Computing (Chap. 24), pp. 533–552. Wiley, Hoboken (2008). https://doi.org/10.1002/9780470724163.ch24
2. Cornelis, C., Jensen, R., Hurtado, G., Ślęzak, D.: Attribute selection with fuzzy decision reducts. Inf. Sci. **180**(2), 209–224 (2010). https://doi.org/10.1016/j.ins.2009.09.008
3. Czerniak, J.M., Dobrosielski, W.T., Filipowicz, I.: Comparing fuzzy numbers using defuzzificators on OFN shapes. In: Prokopowicz et al. [18], pp. 99–132. https://doi.org/10.1007/978-3-319-59614-3_6
4. Dubois, D., Prade, H.: Operations on fuzzy numbers. Int. J. Syst. Sci. **9**(6), 613–626 (1978). https://doi.org/10.1080/00207727808941724
5. Dubois, D., Prade, H.: Rough fuzzy sets and fuzzy rough sets. Int. J. Gen. Syst. **17**(2–3), 191–209 (1990). https://doi.org/10.1080/03081079008935107
6. Fu, T.C.: A review on time series data mining. Eng. Appl. Artif. Intell. **24**(1), 164–181 (2011). https://doi.org/10.1016/j.engappai.2010.09.007
7. Klir, G.J.: Fuzzy arithmetic with requisite constraints. Fuzzy Sets Syst. **91**(2), 165–175 (1997). https://doi.org/10.1016/S0165-0114(97)00138-3
8. Kosiński, W., Prokopowicz, P., Kacprzak, D.: Fuzziness - representation of dynamic changes by ordered fuzzy numbers. In: Seising, R. (ed.) Views on Fuzzy Sets and Systems from Different Perspectives: Philosophy and Logic, Criticisms and Applications, pp. 485–508. Springer, Heidelberg (2009). https://doi.org/10.1007/978-3-540-93802-6_24
9. Kosiński, W., Prokopowicz, P., Rosa, A.: Defuzzification functionals of ordered fuzzy numbers. IEEE Tran. Fuzzy Syst. **21**(6), 1163–1169 (2013). https://doi.org/10.1109/TFUZZ.2013.2243456
10. Kosiński, W., Prokopowicz, P., Ślęzak, D.: Ordered fuzzy numbers. Bull. Pol. Acad. Sci. Math. **51**(3), 327–338 (2003)
11. Kosiński, W., Prokopowicz, P., Ślęzak, D.: Calculus with fuzzy numbers. In: Bolc, L., et al. (eds.) Intelligent Media Technology for Communicative Intelligence. LNCS, vol. 3490, pp. 21–28. Springer, Heidelberg (2005). https://doi.org/10.1007/11558637_3
12. Mares, M.: Weak arithmetics of fuzzy numbers. Fuzzy Sets Syst. **91**(2), 143–153 (1997)

13. Pawlak, Z.: Rough Sets – Theoretical Aspects of Reasoning About Data. Springer, Heidelberg (1991). https://doi.org/10.1007/978-94-011-3534-4
14. Pawlak, Z., Skowron, A.: Rough sets: some extensions. Inf. Sci. **177**(1), 28–40 (2007). https://doi.org/10.1016/j.ins.2006.06.006
15. Prokopowicz, P.: Flexible and simple methods of calculations on fuzzy numbers with the ordered fuzzy numbers model. In: Rutkowski, L., et al. (eds.) Artificial Intelligence and Soft Computing. LNCS, vol. 7894, pp. 365–375. Springer, Heidelberg (2013). https://doi.org/10.1007/978-3-642-38658-9_33
16. Prokopowicz, P.: Processing direction with ordered fuzzy numbers. In: Prokopowicz et al. [18], pp. 81–98. https://doi.org/10.1007/978-3-319-59614-3_5
17. Prokopowicz, P.: The use of ordered fuzzy numbers for modelling changes in dynamic processes. Inf. Sci. **470**, 1–14 (2019). https://doi.org/10.1016/j.ins.2018.08.045
18. Prokopowicz, P., Czerniak, J., Mikołajewski, D., Apiecionek, Ł., Ślęzak, D. (eds.): Theory and Applications of Ordered Fuzzy Numbers: A Tribute to Professor Witold Kosiński. Studies in Fuzziness and Soft Computing, vol. 356. Springer, Heidelberg (2017). https://doi.org/10.1007/978-3-319-59614-3
19. Prokopowicz, P., Pedrycz, W.: The directed compatibility between ordered fuzzy numbers - a base tool for a direction sensitive fuzzy information processing. In: Rutkowski, L., et al. (eds.) Artificial Intelligence and Soft Computing. LNCS, vol. 9119, pp. 249–259. Springer, Heidelberg (2015). https://doi.org/10.1007/978-3-319-19324-3_23
20. Prokopowicz, P., Ślęzak, D.: Ordered fuzzy numbers: Definitions and operations. In: Prokopowicz et al. [18], pp. 57–79. https://doi.org/10.1007/978-3-319-59614-3_4
21. Qian, Y., Wang, Q., Cheng, H., Liang, J., Dang, C.: Fuzzy-rough feature selection accelerator. Fuzzy Sets Syst. **258**, 61–78 (2015). https://doi.org/10.1016/j.fss.2014.04.029
22. Radzikowska, A.M., Kerre, E.E.: A comparative study of fuzzy rough sets. Fuzzy Sets Syst. **126**(2), 137–155 (2002). https://doi.org/10.1016/S0165-0114(01)00032-X
23. Sanchez, E.: Solution of fuzzy equations with extended operations. Fuzzy Sets Syst. **12**(3), 237–248 (1984)
24. Ślęzak, D., Grzegorowski, M., Janusz, A., Kozielski, M., Nguyen, S.H., Sikora, M., Stawicki, S., Wróbel, L.: A framework for learning and embedding multi-sensor forecasting models into a decision support system: a case study of methane concentration in coal mines. Inf. Sci. **451–452**, 112–133 (2018). https://doi.org/10.1016/j.ins.2018.04.026
25. Szczuka, M., Skowron, A., Jankowski, A., Ślęzak, D.: Granular computing: from granules to systems. In: Wiley Encyclopedia of Electrical and Electronics Engineering, pp. 1–8. Wiley, Hoboken (2016). https://doi.org/10.1002/047134608X.W8293
26. Wagenknecht, M., Hampel, R., Schneider, V.: Computational aspects of fuzzy arithmetics based on archimedean t-norms. Fuzzy Sets Syst. **123**(1), 49–62 (2001). https://doi.org/10.1016/S0165-0114(00)00096-8
27. Zadeh, L.A.: Fuzzy sets. Inf. Control **8**, 338–353 (1965)
28. Zadeh, L.A.: The concept of a linguistic variable and its application to approximate reasoning I. Inf. Sci. **8**(3), 199–249 (1975). https://doi.org/10.1016/0020-0255(75)90036-5
29. Zadeh, L.A. (ed.): Computing with Words: Principal Concepts and Ideas. Studies in Fuzziness and Soft Computing, vol. 277. Springer, Heidelberg (2012)

Collaborative Evidential Clustering

Yixuan Qiao[1], Shoumei Li[1(✉)], and Thierry Denœux[2]

[1] Beijing University of Technology,
100 Pingleyuan, Chaoyang District, Beijing 100124, People's Republic of China
{yixuanqiao,lisma}@bjut.edu.cn
[2] Université de Technologie de Compiègne, UMR CNRS 7253 Heudiasyc,
Compiègne, France
tdenoeux@utc.fr

Abstract. Different companies may not be allowed to treat data together given restrictions of security, privacy or other technical reasons. In order to make better use of information from different sources, clustering algorithms based on collaboration mechanisms have been widely used. We propose the concept of collaborative evidential clustering under the framework of evidence theory. The key point is to establish collaboration among the credal partition matrices of each data site to meet the data confidentiality requirements. Considering the problems of excessive information interaction and insufficient information interaction, we design single-step and multi-step collaborative evidential clustering algorithms. Our algorithms were validated on real data sets.

1 Introduction

With the rapid development of information technology, the total amount of data is growing exponentially. Companies rely on their powerful storage capability to continuously collect, organize, and analyze data, in order to mine valuable information. A large amount of data is stored in different data sites and various types of servers. Due to security, privacy or other technical reasons, companies are reluctant to share data and only want to exchange information at non-data levels. In order to make better use of different levels of information and reveal the internal information structure at local data sites, clustering algorithms based on collaborative mechanisms have been proposed. The basic idea of collaborative clustering is to first run a clustering algorithm independently at each data site, and then interact by exchanging the local structure information of each data site to reveal the potential common underlying structure of different data sites.

The concept of collaborative fuzzy clustering (CFC) and its implementation have been introduced in [15]. The development of CFC solves many practical problems (see, e.g., [5,6,12]). Collaboration mechanisms have been extensively studied from different perspectives: collaborative frameworks based on rough-fuzzy clustering or topological maps, higher-level collaborative schemes relying on existing clustering algorithms or collaborative approaches dedicated to distributed datasets (see, e.g., [4,7,14,24]).

© Springer Nature Switzerland AG 2019
R. B. Kearfott et al. (Eds.): IFSA 2019/NAFIPS 2019, AISC 1000, pp. 518–530, 2019.
https://doi.org/10.1007/978-3-030-21920-8_46

Since the original concept of CFC only implies a single collaboration phase, Pedrycz and Rai [16,17] further refined the concept by making the collaboration an iterative process in which a specific data site would periodically use structure information from other data sites resulting from the collaboration process. In addition, experiments show that the performance of the CFC algorithm is not sensitive to the choice of the collaborative strength coefficient. In this paper, we study the specific form of the collaboration mechanism under the framework of Dempster-Shafer evidence theory.

Previous studies suggest that a direct comparison of two partition matrices after a few steps of collaboration could not be feasible as we may not have a direct correspondence between their rows (respective clusters). Consequently, several authors have proposed to rearrange the partition matrix (see e.g. [16–18]). Recently, the Hungarian algorithm [8] has been used to ensure that the same rows in the partition matrices refer to the same cluster. However, it has been found experimentally that the partition matrix reordering is not necessary, especially when facing a large number of phases of the collaboration [21].

Collaborative clustering has not been studied under the framework of evidence theory, which is considered to be a very mature theoretical system for uncertainty inference and widely used in many fields (see, e.g., [2,3,9–11]). Evidential clustering relies on the concept of credal partition [3], which uses mass functions to characterize the uncertainty in data effectively. The concept of credal partition extends those of hard, fuzzy and possibilistic partitions, and it constitutes a more general clustering framework. For each object, masses are assigned not only to single classes, but also to unions of classes. Experiments reported in [3] and [13] show that this extra flexibility allows us to have a deeper understanding of the data structure and improve the robustness to outliers.

In this paper, we study the implementation of a *collaborative evidential clustering* (CEC) algorithm, assuming the data at each site have the same sample size, the same number of clusters, and different feature spaces. Specifically, we explore the implementation of the collaboration mechanism in the ECM algorithm [13]. We consider a collaborative mechanism based on cluster structure information given existing data confidentiality requirements. Furthermore, to address the *Excessive Information Interaction* (EII) and *Insufficient Information Interaction* (III) problems, we propose a single-step CEC algorithm and a multi-step CEC algorithm separately, in which the number of multi-step collaborations is controlled based on a structural similarity index (see e.g. [16,17,21]).

This paper is organized as follows. Section 2 recalls the background notions about belief functions, the ECM algorithm and the pignistic transform. The single-step and multi-step CEC algorithms are introduced in Sects. 3 and 4, respectively. Section 5 presents experimental results and some observations. Conclusions are given in Sect. 6.

2   Preliminaries

The Dempster-Shafer theory of evidence [20,23] (or belief function theory) is a theoretical framework for representing partial and unreliable information. Let us consider a variable ω taking values in a finite set $\Omega = \{\omega_1, \ldots, \omega_j, \ldots, \omega_c\}$, called the frame of discernment. Partial knowledge regarding the actual value taken by ω can be represented by a *mass function* m, which is an application from the power set of Ω in the interval [0, 1] such that $\sum_{A \subseteq \Omega} m(A) = 1$. The subsets A of Ω such that $m(A) > 0$ are called the *focal sets* of m. The mass $m(A)$ can be interpreted as a fraction of a unit mass of belief that is allocated to A and that cannot be allocated to any subset of A. Complete ignorance is obtained when Ω is the only focal set, and full certainty when the whole mass of belief is assigned to a unique singleton of Ω. If all the focal sets of m are singletons, m is similar to a probability distribution: it is then called a *Bayesian* mass function. In the following, we use the concise notation $m_{ij} \in [0,1]$ to denote the belief of object x_i to subset A_j, and $m_{i\emptyset}$ to denote the mass assigned to the empty set. A mass function m_i such that $m_{i\emptyset} = 0$ is said to be *normalized*. Under the *open-world* assumption, the mass $m_{i\emptyset}$ is interpreted as a quantity of belief given to the hypothesis that the actual value of ω might not belong to Ω [22].

ECM is one of the algorithms proposed to derive a *credal partition* from data [13]. Deriving a credal partition implies determining, for each object x_i, the quantities $m_{ij} = m_i(A_j)$ in such a way that a low value of m_{ij} is found when the distance d_{ij} between x_i and A_j is high. In this framework, partial knowledge regarding the class membership of an object is represented by a mass function on the set of possible classes. Thus, belief mass may be given to any subset A of Ω (any set of classes), and not only to singletons of Ω. This representation makes it possible to model a wide variety of situations ranging from complete ignorance to full certainty. The ECM algorithm searches for the credal partition matrix M and cluster center matrix V that minimize the following criterion:

$$J_{ECM}(M, V) = \sum_{i=1}^{N} \sum_{\{j | A_j \neq \emptyset, A_j \subseteq \Omega\}} c_j^\alpha m_{ij}^2 d_{ij}^2 + \sum_{i=1}^{N} \delta^2 m_{i\emptyset}^2, \tag{1}$$

subject to the constraints $m_{ij} \geq 0$ for all i and j, $m_{i\emptyset} \geq 0$ for all i, and

$$\sum_{\{j | A_j \neq \emptyset, A_j \subseteq \Omega\}} m_{ij} + m_{i\emptyset} = 1, \tag{2}$$

for all i. In (1), $c_j = |A_j|$ is the cardinality of A_j and δ represents the distance of any object to the empty set.

In order to make a decision regarding the value of ω, it is possible to transform a normalized mass function m into a probability distribution using the following pignistic transformation [23]:

$$BelP(\omega) = \sum_{\{A \subseteq \Omega | \omega \in A\}} \frac{m(A)}{|A|}, \qquad \forall \omega \in \Omega. \tag{3}$$

3   A Single-Step CEC Algorithm

To meet data confidentiality requirements, we mainly consider the collaborative mechanism based on cluster structure information (including M and V) and integrate it into the objective function of the ECM algorithm.

The T data sites are denoted by $D[1], \ldots, D[t], \ldots, D[T]$. All data sites have the same number of samples denoted by N, same clusters denoted by c, but different feature spaces composed of $n[1], \ldots, n[t], \ldots, n[T]$ features, respectively. Matrix $K_{T \times T}$ is used to quantify the collaboration strength between each pair of data sites; its general term, denoted by $\kappa[t, s]$, represents the collaboration strength between sites $D[t]$ and $D[s]$. As there is no collaboration between $D[t]$ and itself, we set $\kappa[t, t] = 0$.

For a given data site $D[t]$, we combine the information provided by the local data with the cluster structure information of the collaborators to determine the cluster structure. For that purpose, the objective function is expanded into the form

$$Q[t] = \sum_{i=1}^{N} \sum_{\{j | A_j \neq \emptyset, A_j \subseteq \Omega\}} c_j^\alpha[t] m_{ij}^2[t] d_{ij}^2[t] + \sum_{i=1}^{N} \delta^2 m_{i\emptyset}^2[t]$$

$$+ \sum_{s=1, s \neq t}^{T} \kappa[t, s] \sum_{i=1}^{N} \sum_{\{j | A_j \neq \emptyset, A_j \subseteq \Omega\}} (m_{ij}[t] - m_{ij}[s])^2 d_{ij}^2[t],$$

$$(4)$$

for $t = 1, 2, \ldots, T$. The objective function $Q[t]$ contains two parts. The first part is the sum of weighted distances between the patterns in $D[t]$ and the center of the non-empty subset A_j; it is just the objective function (1) of the standard ECM applied to $D[t]$ with $\beta = 2$. The second part implements the collaborative mechanism, which makes the clustering based on the $D[t]$ aware of other collaborators. The difference between $M[t]$ and $M[s]$ represents the difference in cluster structure between data sites $D[t]$ and $D[s]$. The weight $\kappa[t, s]$ controls the balance between local data information and collaborator cluster structure information. When $\kappa[t, s] = 0$, the problem translates into a scenario where the standard ECM algorithm acts at each data site without collaboration. In general, we propose to constrain $\kappa[t, s]$ to be in the interval $[0, 1]$, so that the second term in the right-hand side of (4) does not dominate the first one.

As in the ECM algorithm, we also need $M[t]$ to satisfy constraints (2). Therefore, the single-step collaborative clustering algorithm consists in minimizing $Q[t]$ in (4) subject to (2). This optimization task splits into two problems, namely, determining the credal partition matrix $M[t]$ and the cluster center matrix $V[t]$. To determine the partition matrix, we exploit the technique of Lagrange multipliers. This leads to the new objective function that is formed separately for each data site $D[t]$, namely,

$$
L[t] = \sum_{i=1}^{N} \sum_{\{j|A_j \neq \emptyset, A_j \subseteq \Omega\}} c_j^{\alpha}[t] m_{ij}^2[t] d_{ij}^2[t] + \sum_{i=1}^{N} \delta^2 m_{i\emptyset}^2[t]
$$

$$
+ \sum_{s=1, s \neq t}^{T} \kappa[t,s] \sum_{i=1}^{N} \sum_{\{j|A_j \neq \emptyset, A_j \subseteq \Omega\}} (m_{ij}[t] - m_{ij}[s])^2 d_{ij}^2[t]
$$

$$
- \sum_{i=1}^{N} \lambda_i \left(\sum_{\{j|A_j \neq \emptyset, A_j \subseteq \Omega\}} m_{ij}[t] + m_{i\emptyset}[t] - 1 \right),
$$

$$(5)$$

where λ_i denotes a Lagrange multiplier. The necessary conditions leading to the local minimum of $M[t]$ read as follows:

$$
\frac{\partial L[t]}{\partial m_{ij}[t]} = 2c_j^{\alpha}[t] m_{ij}[t] d_{ij}^2[t] + 2 \sum_{s=1, s \neq t}^{T} \kappa[t,s](m_{ij}[t] - m_{ij}[s]) d_{ij}^2[t] - \lambda_i = 0,
$$

$$(6a)$$

$$
\frac{\partial L[t]}{\partial m_{i\emptyset}[t]} = 2\delta^2 m_{i\emptyset}[t] - \lambda_i = 0,
$$

$$(6b)$$

$$
\frac{\partial L[t]}{\lambda_i} = \sum_{\{j|A_j \neq \emptyset, A_j \subseteq \Omega\}} m_{ij}[t] + m_{i\emptyset}[t] - 1 = 0.
$$

$$(6c)$$

Introducing the notations

$$
\psi[t] = \sum_{s=1, s \neq t}^{T} \kappa[t,s], \quad \varphi_{ij}[t] = \sum_{s=1, s \neq t}^{T} \kappa[t,s] m_{ij}[s],
$$

$$(7)$$

we get the solution

$$
m_{ij}[t] = \frac{\varphi_{ij}[t]}{c_j^{\alpha}[t] + \psi[t]} + \frac{\frac{1}{d_{ij}^2[t](c_j^{\alpha}[t]+\psi[t])} \left(1 - \sum_{\{j|A_j \neq \emptyset, A_j \subseteq \Omega\}} \frac{\varphi_{ij}[t]}{c_j^{\alpha}[t]+\psi[t]}\right)}{\sum_{\{j|A_j \neq \emptyset, A_j \subseteq \Omega\}} \frac{1}{d_{ij}^2[t](c_j^{\alpha}[t]+\psi[t])} + \frac{1}{\delta^2}},
$$

$$(8a)$$

$$
m_{i\emptyset}[t] = 1 - \sum_{\{j|A_j \neq \emptyset, A_j \subseteq \Omega\}} m_{ij}[t].
$$

$$(8b)$$

In the calculations of the prototypes we confine ourselves to the weighted Euclidean distance between the sample and the centroid of the cluster, so the necessary condition for solving the local minimum of the cluster center $V[t]$ is

$$\frac{\partial L[t]}{\partial v_l[t]} = \sum_{i=1}^{N} \sum_{\{j|A_j \neq \emptyset, A_j \subseteq \Omega\}} c_j^{\alpha}[t] m_{ij}^2[t] \frac{\partial d_{ij}^2[t]}{\partial v_l[t]}$$

$$+ \sum_{s=1, s \neq t}^{T} \kappa[t,s] \sum_{i=1}^{N} \sum_{\{j|A_j \neq \emptyset, A_j \subseteq \Omega\}} (m_{ij}[t] - m_{ij}[s])^2 \frac{\partial d_{ij}^2[t]}{\partial v_l[t]}.$$

(9)

Introducing the notations

$$B_{lq}[t] = \sum_{i=1}^{N} x_{iq} \sum_{w_l \in A_j} c_j^{\alpha-1} m_{ij}^2[t], \quad B_{lq}[t,s] = \sum_{i=1}^{N} x_{iq} \sum_{w_l \in A_j} (m_{ij}[t] - m_{ij}[s])^2 \frac{1}{c_j[t]},$$

$$H_{lk}[t] = \sum_{i=1}^{N} \sum_{\{w_k, w_l\} \subseteq A_j} c_j^{\alpha-2} m_{ij}^2[t], \quad H_{lk}[t,s] = \sum_{i=1}^{N} \sum_{\{w_k, w_l\} \subseteq A_j} (m_{ij}[t] - m_{ij}[s])^2 \frac{1}{c_j^2},$$

the cluster center matrix $V[t]$ has the form

$$V[t] = \left(H[t] + \sum_{s=1, s \neq t} \kappa[t,s] H[t,s] \right)^{-1} \left(B[t] + \sum_{s=1, s \neq t} \kappa[t,s] B[t,s] \right). \quad (10)$$

More details about the derivation process can be found in [19]. The algorithm can be described in Algorithm 1. It is worth noting that for each data site, the information used by single-step collaboration comes from the cluster structure information obtained by the initial ECM algorithm, not from the improved cluster structure information of the collaborator data site through collaboration.

Termination criterion I relies on the changes to the cluster center matrices obtained in successive iterations of the single-step CEC algorithm; we chose the L_∞ norm as a measure of change in the cluster center matrices. Subsequently, the optimization is terminated when this distance is lower than an assumed threshold value $\varepsilon > 0$.

4   A Multi-step CEC Algorithm

As the single-step CEC algorithm described in Sect. 3 may face the EII and III problems, we consider a multi-step collaboration mechanism to get more information from the collaborator data site for better interaction. The original purpose of collaboration was to reconcile and optimize the differences between cluster structures of various data sites. As the reconciliation continues, we can expect that the cluster structure similarity between the data sites will gradually increase. Therefore, we can use the structural similarity index to guide the multi-step CEC algorithm.

Algorithm 1. Single-step CEC Algorithm.

Require: $D[1], \ldots, D[t], \ldots, D[T]$, c, termination criterion I, $\kappa[t,s]$, ε
1: **for** $t = 1$ **to** T **do**
2:     Use ECM to get the original $M_{original}[t]$ and $V_{original}[t]$
3:     $M_{final}[t] \leftarrow M_{original}[t]$, $V_{final}[t] \leftarrow V_{original}[t]$
4: **end for**
5: **for** $t = 1$ **to** T **do**
6:     $l \leftarrow 0$, $I^0[t] \leftarrow 1$
7:     $M^0[t] \leftarrow M_{original}[t]$ and $V^0[t] \leftarrow V_{original}[t]$
8:     **while** $I^l[t] \geq \varepsilon$ **do**
9:         **for** $q = 1$ **to** T **do**
10:             **if** $q \neq t$ **then**
11:                 $M^l[q] \leftarrow M_{original}[q]$ and $V^l[q] \leftarrow V_{original}[q]$
12:             **end if**
13:         **end for**
14:         $l \leftarrow l + 1$
15:         Compute $M^l[t]$, $V^l[t]$ using (8) and (10) with $M^{l-1}[t]$, $V^{l-1}[t]$
16:         $I^l[t] \leftarrow \max_{k \in [1,n[t]],\ j \in \{j | A_j \neq \phi, A_j \subseteq \Omega\}} (|V_{jk}^l[t] - V_{jk}^{l-1}[t]|)$
17:     **end while**
18:     $M_{final}[t] \leftarrow M^l[t]$, $V_{final}[t] \leftarrow V^l[t]$
19: **end for**
20: **return** $M_{final}[t]$ and $V_{final}[t]$

We should stress the fact that a direct comparison of two credal partition matrices could not be feasible as we may not have a direct correspondence between their rows (respective clusters). In a more general setting, we might even have different numbers of clusters at the individual data sites, and this diversity could make any attempt to form the correspondence between the partition matrices infeasible. Instead, we consider the following approach in which we test how the structure revealed at one data site performs on the remaining ones. Let us consider the following local structural similarity index:

$$W[t] = \sum_{s=1, s \neq t}^{T} \sum_{i=1}^{N} \sum_{\{j | A_j \neq \emptyset, A_j \subseteq \Omega\}} m_{ij}^2[s] \|x_i[t] - v_j[t|s]\|^2. \tag{11}$$

where

$$v_j[t|s] = \frac{\sum_{i=1}^{N} m_{ij}^2[s] x_i[t]}{\sum_{i=1}^{N} m_{ij}^2[s]}. \tag{12}$$

The rationale behind this measure is that if the structure of $D[s]$ is similar to that of $D[t]$, then the structure should also obtain a good performance on $D[t]$ (a more similar structure should lead to a lower value of $W[s]$). Finally, for all data sites, we have the following global structural similarity metric

$$W = \sum_{t=1}^{T} W[t]. \tag{13}$$

This indicator is used to control the number of iterations of the multi-step CEC algorithm, denoted as termination criterion II. So when $|W^l - W^{l-1}| < \varepsilon$, the multi-step CEC algorithm is completed.

The multi-step collaborative evidential clustering algorithm consists of three phases. It can be described in Algorithm 2.

Algorithm 2. Multi-step CEC Algorithm.

Require: $D[1], \ldots, D[t], \ldots, D[T]$, c, termination criterion I, termination criterion II,
   $\kappa[t, s]$, ε
 1: Run Steps (1)-(20) of Algorithm 1, return values are denoted as $M^0[t]$ and $V^0[t]$
 2: $l' \leftarrow 0$, II $\leftarrow W^0$
 3: **while** II $\geq \varepsilon$ **do**
 4:     **for** $t = 1$ to T **do**
 5:         $M_{original}[t] \leftarrow M^{l'}[t]$, $V_{original}[t] \leftarrow V^{l'}[t]$
 6:     **end for**
 7:     $l' \leftarrow l' + 1$
 8:     Run (5)-(20) of Algorithm 1, return values are denoted as $M^{l'}[t]$ and $V^{l'}[t]$
 9:     $II \leftarrow |W^{l'} - W^{l'-1}|$
10: **end while**
11: **return** $M^{l'}[t]$ and $V^{l'}[t]$

The multi-step collaboration process is actually a cascade of single-step collaboration processes. We can imagine that before each single-step collaboration, the structural information of all data sites enters the information interaction pool, representing all the information that can be used in this single-step CEC process. For each data site, the collaborator cluster structure information that can be utilized is constant, only its own structure is constantly changing.

The final partition is determined by assigning each object to the cluster after convergence of the algorithm with maximal pignistic probability (3). Based on the given reference partition, we use the adjusted Rand index (ARI) to characterize the local collaboration quality of each data site. The global collaborative quality assessment index is then defined as

$$AARI = \frac{1}{T} \sum_{ii=1}^{T} ARI[ii]. \tag{14}$$

5   Experimental Results

In this section, we report on experimental findings[1] for some machine learning data sets [1]. The intent is to demonstrate the effectiveness of the collaboration and get some experimental insights into the behavior of the algorithms.

---

[1] We also conducted a simulation study. Experiments with synthetic dataset can be found in [19]. The ARI index for each data site is close to 1, which demonstrates that our algorithms are very competitive. Due to space limitations, the results of simulation studies have to be omitted.

The details of the features contained in the data site in each dataset are as follows. For Iris dataset: (a) Sepal.Length, Sepal.Width; (b) Sepal.Width, Petal.Length; (c) Petal.Length, Petal.Width. For Seeds dataset: (a) area, perimeter, compactness; (b) compactness, length of kernel, width of kernel; (c) width of kernel, asymmetry coefficient, length of kernel groove. We set $c = 3$, $\kappa[t, s] = 1$, $\varepsilon = 0.0001$. The evolutions of W and $AARI$ as a function of the number of iterations are reported in Figs. 1 and 2.

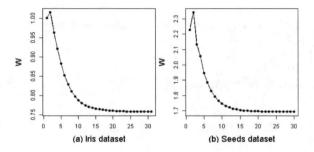

Fig. 1. With the continuous reconciliation of multi-step collaboration, the structural similarity between data sites is enhanced, and the value of W constantly declines. The EII problem emerged in the early stage, which resulted in a slight increase of W. The post-correction function of multi-step collaboration is indispensable.

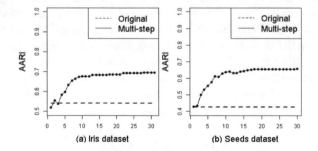

Fig. 2. EII and III led to a decrease in AARI after single step collaboration; the multi-step collaboration process performs effective information correction, making the index AARI strictly superior to the initial value after multi-step CEC. This further confirms the adequacy and necessity of the multi-step collaborative process.

At the local level, the structural similarity indicators of all data sites in each data set further confirm the existence of EII and III problems (see Fig. 3). Furthermore, we find that indices W and W[t] have roughly the same trend, which shows that W is excellent at global direction control. The ARI and its reference level for each data set are shown in Fig. 4.

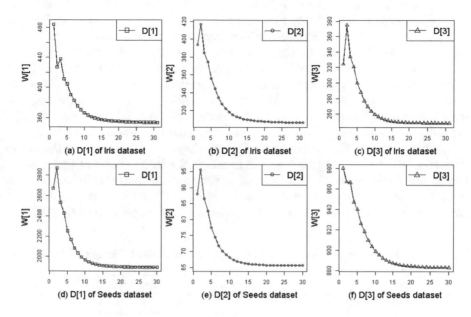

Fig. 3. The trend of local structural similarity index of Iris and Seeds dataset.

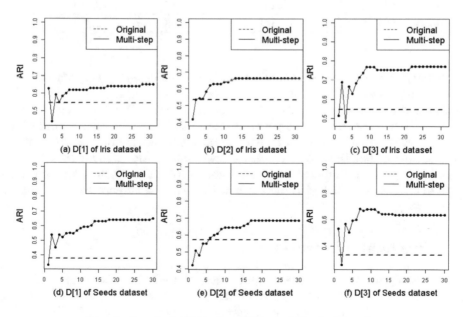

Fig. 4. Trend of ARI index of Iris and Seeds dataset.

For data sites D[1] and D[3] in the Iris dataset, we found that in the first few collaborations, the information was redistributed multiple times to find the correct direction of collaboration, which laid an important foundation for the subsequent collaboration. In the algorithm debugging phase, we find that the randomness introduced by redistribution is the key factor that determines the improvement of the final multi-step CEC algorithm. For data site D[3] in the Wine dataset, we found that ARI decreased slightly in the late stage but was still significantly higher than the baseline level, while the ARIs of data sites D[1] and D[2] were still rising. This further confirms that collaboration is a long term interaction process. The information data site D[3] lost can cause the data sites D[1] and D[2] to get a bigger boost. Therefore, the loss is beneficial in general. It also indirectly shows that our choice of the global structural similarity index W as the stopping criterion of the multi-step CEC algorithm is reasonable. The numerical results of the multi-step CEC of each dataset are shown in Table 1.

Table 1. Multi-step collaborative clustering results analysis of Iris and Wine datasets

Data site	Original ARI	Single step ARI	Multi-step ARI	Promotion (%)
$Iris - D[1]$	0.546	0.626	0.652	0.106
$Iris - D[2]$	0.534	0.416	0.664	0.130
$Iris - D[3]$	0.547	0.515	0.771	0.224
$Wine - D[1]$	0.377	0.334	0.651	0.275
$Wine - D[2]$	0.571	0.423	0.687	0.115
$Wine - D[3]$	0.333	0.521	0.636	0.303

From Table 1, we can see more clearly that the single-step cooperation algorithm has poor stability and is prone to excessive information interaction, which causes the cluster structure to change too much and affect the value of ARI. The multi-step collaboration algorithm has a good correction effect, and it can improve ARI well after the information is redistributed. By further observation, we also found that the data site with a lower initial ARI value eventually rose significantly. The information it can use comes from the original data site with a higher value of ARI and a data site with a large difference from its cluster structure.

6 Concluding Remarks

In this study we have proposed a new concept of collaborative evidential clustering. Our multi-step CEC algorithm has been validated on real data sets and the experimental results have shown competitive performances. The EII and III problems in the single step CEC algorithm play a very good role in information redistribution and lay the foundation for multi-step CEC algorithm. So far, the research we have completed is based on two assumptions: (a) all data sites have the same number of clusters, (b) the strength of collaboration between different data sites is the same. These assumptions will be relaxed in future work.

Acknowledgments. This work was supported by National Nature Science Foundation of China (No. 11571024), and by a grant to the third author as part of the Overseas Talent program from the Beijing Government.

References

1. Asuncion, A., Newman, D.: UCI machine learning repository (2007)
2. Bordes, J.B., Davoine, F., Xu, P., Denœux, T.: Evidential grammars: a compositional approach for scene understanding. Application to multimodal street data. Appl. Soft Comput. **61**, 1173–1185 (2017)
3. Denœux, T., Masson, M.H.: EVCLUS: evidential clustering of proximity data. IEEE Trans. Syst. Man Cybern. Part B (Cybern.) **34**(1), 95–109 (2004)
4. Depaire, B., Falcn, R., Vanhoof, K., Wets, G.: PSO driven collaborative clustering: a clustering algorithm for ubiquitous environments. Intell. Data Analy. **15**(1), 49–68 (2011)
5. Elhamifar, E., Vidal, R.: Sparse subspace clustering: algorithm, theory, and applications. IEEE Trans. Pattern Anal. Mach. Intell. **35**(11), 2765–2781 (2013)
6. Forestier, G., Wemmert, C., Gancarski, P.: Collaborative multi-strategical clustering for object-oriented image analysis. In: Supervised and Unsupervised Ensemble Methods and their Applications, pp. 71–88. Springer, Heidelberg (2008)
7. Ghassany, M., Grozavu, N., Bennani, Y.: Collaborative generative topographic mapping. Proceedings of International Conference on Neural Information Processing, pp. 591–598. Springer, Heidelberg (2012)
8. Kuhn, H.W.: The Hungarian method for the assignment problem. Naval Res. Logistics (NRL) **52**(1), 7–21 (2005)
9. Lelandais, B., Ruan, S., Denœux, T., Vera, P., Gardin, I.: Fusion of multi-tracer PET images for dose painting. Med. Image Anal. **18**(7), 1247–1259 (2014)
10. Lian, C., Ruan, S., Denœux, T., Jardin, F., Vera, P.: Selecting radiomic features from FDG-PET images for cancer treatment outcome prediction. Med. Image Anal. **32**, 257–268 (2016)
11. Lian, C., Ruan, S., Denœux, T., Li, H., Vera, P.: Spatial evidential clustering with adaptive distance metric for tumor segmentation in FDG-PET images. IEEE Trans. Biomed. Eng. **65**(1), 21–30 (2018)
12. Loia, V., Pedrycz, W., Senatore, S.: Semantic web content analysis: a study in proximity-based collaborative clustering. IEEE Trans. Fuzzy Syst. **15**(6), 1294–1312 (2007)
13. Masson, M.H., Denœux, T.: ECM: an evidential version of the fuzzy c-means algorithm. Pattern Recogn. **41**(4), 1384–1397 (2008)
14. Mitra, S., Banka, H., Pedrycz, W.: Rough-fuzzy collaborative clustering. IEEE Trans. Syst. Man Cybern. Part B (Cybern.) **36**(4), 795–805 (2006)
15. Pedrycz, W.: Collaborative fuzzy clustering. Pattern Recogn. Lett. **23**(14), 1675–1686 (2002)
16. Pedrycz, W., Rai, P.: A multifaceted perspective at data analysis: a study in collaborative intelligent agents. IEEE Trans. Syst. Man Cybern. Part B (Cybern.) **38**(4), 1062–1072 (2008)

17. Pedrycz, W., Rai, P.: Collaborative clustering with the use of fuzzy c-means and its quantification. Fuzzy Sets Syst. **159**(18), 2399–2427 (2008)
18. Prasad, M., Siana, L., Li, D.L., Lin, C.T., Liu, Y.T., Saxena, A.: A preprocessed induced partition matrix based collaborative fuzzy clustering for data analysis. In: Proceedings of IEEE International Conference on Fuzzy Systems (FUZZ-IEEE 2014), pp. 1553–1558 (2014)
19. Qiao, Y.: On study of collaborative evidential clustering algorithm with applications. Dissertation for M.S. degree of Beijing University of Technology under the supervision of Li S. and Denoeux T (2019)
20. Shafer, G.: A Mathematical Theory of Evidence. Princeton University Press, Princeton (1976)
21. Shen, Y., Pedrycz, W.: Collaborative fuzzy clustering algorithm: some refinements. Int. J. Approx. Reason. **86**, 41–61 (2017)
22. Smets, P.: The transferable belief model for quantified belief representation. In: Quantified Representation of Uncertainty and Imprecision, pp. 267–301. Springer, Dordrecht (1998)
23. Smets, P., Kennes, R.: The transferable belief model. Artif. Intell. **66**(2), 191–234 (1994)
24. Sublime, J., Grozavu, N., Bennani, Y., Cornujols, A.: Collaborative clustering with heterogeneous algorithms. Proceedings of IEEE International Joint Conference on Neural Networks (IJCNN-IEEE 2015), pp. 1–8 (2015)

Firefly Algorithm and Grey Wolf Optimizer for Constrained Real-Parameter Optimization

Luis Rodríguez, Oscar Castillo[✉], Mario García, and José Soria

Tijuana Institute of Technology, Tijuana, BC, Mexico
ocastillo@tectijuana.mx

Abstract. The main goal of this paper is to present the performance of two popular algorithms, the first is the Firefly Algorithm (FA) and the second one is the Grey Wolf Optimizer (GWO) algorithm for complex problems. In this case the problems that we are presenting are of the CEC 2017 Competition on Constrained Real-Parameter Optimization in order to realize a brief analysis, study and comparison between the FA and GWO algorithms respectively.

Keywords: Grey Wolf Optimizer · Firefly Algorithm · Constraints · Complex problems · Study · Optimization

1 Introduction

In the last decades technology and computer science have improved at a very fast pace and now it is more common that the computers realize arithmetic operations in a few seconds in comparisons with the last years.

A way of computational intelligence is to use methodologies and approaches that are bio-inspired and the most popular techniques are the metaheuristics, fuzzy logic and artificial neural networks that basically imitate behaviors based on the nature of humans or phenomena in the real world.

Computer Science has specifically areas to solve optimization problems when the main goal is to maximize or minimize time, costs, money or others problems that exist in the real world and that the humans always are searching new methods for improve the results of these problems.

In addition, we can note that researchers realize new metaheuristics with new bioinspired features in order to improve the performance of these new algorithms in complex problems.

Basically studied new techniques to find good results, and these techniques are based on the behavior in the nature (bio-inspired). Metaheuristics are very popular in the last years because they are: simple, flexible, stochastic and avoid the local optimal.

R. B. Kearfott et al. (Eds.): IFSA 2019/NAFIPS 2019, AISC 1000, pp. 531–541, 2019.
https://doi.org/10.1007/978-3-030-21920-8_47

Finally in this research we are presenting the FA (Firefly Algorithm) and GWO (Grey Wolf Optimizer) algorithms that are popular algorithms in the optimization area. In this work we are presenting the set of ten first problems that was used in the CEC 2017 competition for constrained functions.

This paper is organized as follows: Sect. 2 shows a brief explanation of the Firefly Algorithm, Sect. 3 describes the Grey Wolf Optimizer algorithm, in Sect. 4 we are presenting the simulation results and finally in Sect. 5 we present some conclusions.

2 Firefly Algorithm

According to the inspiration of the metaheuristics these are classified as evolutionary [1], Based on physics [2] and Swarm Intelligence [3]. In addition the No free lunch theorem [4] for optimization proved that there is no meta-heuristic appropriately suited for solving all optimization problems.

The Firefly algorithm [5] was created in 2008 by Yang and was inspired by the flashing patterns and behavior of fireflies [6]. In this algorithm we can find the two main equations, the first is the attractiveness or the light intensity of each firefly and is represented by the following equation:

$$\beta = \beta_0 e^{-\gamma r^2} \tag{1}$$

Where β_0 and γ are parameters that could be optimized according to the problem and r represents the distance among two fireflies Finally the movement of a firefly is calculated with the following equation:

$$x_i^{t+1} = x_i^t + \beta_0 e^{-\gamma r_{ij}^2}\left(x_j^t - x_i^t\right) + \alpha_t \in_i^t \tag{2}$$

Where x_i^{t+1} is the new position of the firefly, x_i^t is the current position of the firefly, where we are adding the attractiveness that exists between the other best solution (x_j^t) and the current solution (x_i^t) and finally adding the randomness parameter that is α_t being the randomization parameter, and $\in_i^t$ is a vector of random numbers drawn from a Gaussian distribution or uniform distribution at time t. Finally in Fig. 1 we are presenting a general flowchart of the Firefly Algorithm.

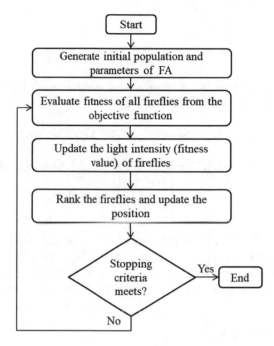

Fig. 1. General flowchart of the Firefly Algorithm

3 Grey Wolf Optimizer Algorithm

The Grey Wolf Optimizer (GWO) [7] algorithm was proposed by Mirjalili in 2014 when designed the mathematical model based on two main features of the grey wolf that are the following: the hunting process and the hierarchy in the members of the pack that Muro presented in his work [8]. Finally we can find the hierarchy pyramid [9] of the pack as a leadership strategy, in other words the best wolf is called alpha (α) and is considered as the best solution in the algorithm, the second best solution is called beta (β) and the third best solution is delta (δ) respectively. Finally, the rest of the candidate solutions are consider as the omega (ω) wolves. In order to mathematically model the two main inspirations described above, we present the following equations:

$$D = \left\| C \cdot X_p(t) - X(t) \right\| \tag{3}$$

$$X(t+1) = X_p(t) - AD \tag{4}$$

Where: Eq. 1 represents the distance between the best solution with a randomness method and the current individual that we are analyzing. Equation 2 represents the next position of the current individual based on the distance of the Eq. 1. Finally, coefficients "A" and "C" are represented by the following equations:

$$A = 2a \cdot r_1 - a \tag{5}$$

$$C = 2 \cdot r_2 \tag{6}$$

Where the "a" parameter is linearly decreasing through of the iterations. The "r_1" and "r_2" are random values between 0 and 1

$$D_\alpha = \|C_1 \cdot X_\alpha - X\|, \quad D_\beta = \|C_2 \cdot X_\beta - X\|, \quad D_\delta = \|C_3 \cdot X_\delta - X\| \tag{7}$$

$$X_1 = X_\alpha - A_1 \cdot (D_\alpha), \quad X_2 = X_\beta - A_2 \cdot (D_\beta), \quad X_3 = X_\delta - A_3 \cdot (D_\delta) \tag{8}$$

$$X(t + 1) = \frac{X_1 + X_2 + X_3}{3} \tag{9}$$

In addition Eqs. 7 and 8 are the same equations as Eqs. 3 and 4 respectively, but in this case the results are based on the leaders. Finally, Eq. 9 represents the next position of the current individual that is an average based on the results obtained in Eq. 8. In addition we can find the general flow of the GWO algorithm as in Fig. 2.

4   Simulations Results

In this paper we are presenting a set of problems that contain the presence of constraints that alter the shape of the search space making it more difficult to solve in comparison with the conventional benchmark functions [10–12]. In this section we selected the first 10 problems of the CEC 2017 Competition on Constrained Real-Parameter Optimizer [13] which can be transformed into the following format:

$$\text{Minimize: } f(X), X = (x_1, x_2, \ldots, x_n) \, and \, X \in S \tag{10}$$

$$\text{Subject to} \quad \begin{matrix} g_i(X) \leq 0, & i = 1, \ldots, p \\ h_j(X) = 0, & j = p+1, \ldots, m \end{matrix} \tag{11}$$

Usually equality constraints are transformed into inequalities of the form:

$$\left| h_j(X) \right| - \varepsilon \leq 0, \, for \, j = p+1, \ldots, m \tag{12}$$

A solution X is regarded as feasible if $g_i(X) \leq 0, \, for \, i = 1, \ldots, p$ and $\left| h_j(X) \right| - \varepsilon \leq 0$, for $j = p+1, \ldots, m$. In this competition (CEC 2017) ε is set to 0.0001.

In Table 1 we can find a brief explanation of the first 10 constraint problems, where D is the number of decision variables, I is the number of inequality constraints and E is the number of equality constraints.

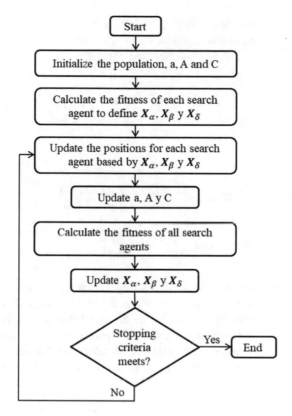

Fig. 2. General flowchart of the GWO algorithm

Table 1. Details of first 10 problems in the CEC 2017 competition

Problem	Type of objective	Number of constraints	
		E	I
C1	Non separable	0	1; Separable
C2	Non separable, rotated	0	1; Non separable, rotated
C3	Non separable	1; Separable	1; Separable
C4	Separable	0	2; Separable
C5	Non separable	0	2; Non separable, rotated
C6	Separable	6; Separable	0
C7	Separable	2; Separable	0
C8	Separable	2; Non separable	0
C9	Separable	2; Non separable	0
C10	Separable	2; Non separable	0

In addition it is important to mention that in the following tables we are presenting the mean of the violation at one solution, these violations depend on the total numbers that we presented in Table 1, and the equation for calculate this mean violations is the following:

$$v = \frac{\left(\sum_{i=1}^{p} G_i(X) + \sum_{j=p+1}^{m} H_j(X) \right)}{m} \tag{13}$$

Where

$$G_i(X) = \begin{cases} g_i(X) & \text{if } g_i(X) > 0 \\ 0 & \text{if } g_i(X) \le 0 \end{cases} \tag{14}$$

$$H_j(X) = \begin{cases} |h_j(X)| & \text{if } |h_j(X)| - \varepsilon > 0 \\ 0 & \text{if } |h_j(X)| - \varepsilon \le 0 \end{cases} \tag{15}$$

For the simulation results, we are presenting the averages and standard deviations as the result of 31 independent executions for each problem of the CEC 17 competition with 30 individuals in both algorithms. In the FA metaheuristic we are presenting tests with parameters that were used in the paper of Lagunes et al. [14]. In Table 2 we can find the averages and standard deviations for FA when the problems have 10 dimensions, also we can note the averages and standard deviations of the violations that we described above.

Table 2. Results of FA with 10 dimension in the CEC 2017 problems

FA - 10 dimensions				
Function	FA	STD	Violation	STD
F1	1.53E−04	4.75E−05	0	0
F2	1.73E−04	5.33E−05	0	0
F3	1.76E−04	4.50E−05	0.5	0
F4	7.5322	2.9612	0.2661	0.0624
F5	366.91	1158.22	0.1774	0.2432
F6	9.0410	3.6233	1	0
F7	−741.0755	54.7787	1	0
F8	−90.3668	0	1	0
F9	−90.6092	4.33E−14	0.5	0
F10	−59.7540	4.33E−14	1	0

In addition we can find in Table 3 the results of the FA when the problems have 50 dimensions.

Table 3. Results of FA with 50 dimension in the CEC 2017 problems

FA - 50 dimensions

Function	FA	STD	Violation	STD
F1	0.0564	0.0142	0	0
F2	0.0563	0.0129	0	0
F3	0.0585	0.0133	0.5	0
F4	75.4605	13.3594	0.2903	0.0935
F5	83.0312	96.7234	0.0726	0.1606
F6	78.2756	21.2956	1	0
F7	−2852.58	201.07	1	0
F8	−90.36682	0.00	1	0
F9	−90.0593	5.78E−14	0.5	0
F10	−53.1833	4.33E−14	1	0

Finally in both cases we respected the maximum number of evaluation functions allowed in the CEC 17 competition and for GWO we are using the parameters that the author recommended in the original paper [15].

Table 4. Results of GWO with 10 dimension in the CEC 2017 problems

GWO - 10 dimensions

Function	GWO	STD	Violation	STD
F1	144.0604	259.6247	0	0
F2	103.4100	129.6063	0	0
F3	88.7982	118.2894	0.5	0
F4	12.4148	8.7777	0.3333	0.1217
F5	14.3888	43.0255	0.0215	0.1197
F6	18.0462	9.8407	0.8571	3.39E−16
F7	−726.2750	69.7004	0.6667	1.13E−16
F8	−90.3668	0	0.6667	1.13E−16
F9	−90.6092	4.33E−14	0.5	0
F10	−59.7540	4.33E−14	0.6667	1.13E−16

In Table 4 we can find the averages and standard deviations of GWO algorithm when the problems have 10 dimensions, also we can note the averages and standard deviations of the violations. Finally, in Table 5 we can find the results of the GWO when the problems have 50 dimensions.

Table 5. Results of GWO with 50 dimension in the CEC 2017 problems

GWO - 50 dimensions

Function	GWO	STD	Violation	STD
F1	8657.98	2032.415	0	0
F2	8657.98	2032.42	0	0
F3	9734.57	2605.37	0.50	0
F4	297.0701	40.1308	0.2742	0.0751
F5	31557.05	23287.72	0	0
F6	513.4068	136.0203	1	0
F7	−1450.85	238.30	1	0
F8	−45.9427	40.9837	1	0
F9	−53.0457	41.4657	0.5	0
F10	−28.3182	2.2731	1	0

In this paper we are also presenting a hypothesis test [16], specifically the z-test in order to show statistically a comparison between the performance of the GWO and FA algorithm respectively as a brief explanation, we can mention that if the z-value is less than −1.645 we can conclude that FA algorithm has better performance than the GWO algorithm.

Table 6 shows the results of the hypothesis test realized between the two algorithms that we presented above, we can find the Z-value just in the last column and according of these results we can conclude that for 10 dimensions in the constrained problems the FA has better performance that the GWO algorithm in 5 of the 7 first problems.

Finally in Table 7 we can find the results of hypothesis testing among the FA and GWO algorithms respectively with 50 dimensions for the constrained problems in the CEC 17 competition and according with these results we can conclude that the Firefly Algorithm is better in all constrained functions that were analyzed in this paper.

Table 6. Comparison between FA and GWO algorithms with 10 dimensions

10 dimensions

Function	GWO	STD	FA	STD	Z - Value
F1	144.0604	259.6247	**1.53E−04**	**4.75E−05**	−3.0894
F2	103.4100	129.6063	**1.73E−04**	**5.33E−05**	−4.4424
F3	88.7982	118.2894	**1.76E−04**	**4.50E−05**	−4.1796
F4	12.4148	8.7777	**7.5322**	**2.9612**	−2.9346
F5	**14.3888**	**43.0255**	366.91	1158.22	1.6935
F6	18.0462	9.8407	**9.0410**	**3.6233**	−4.7813
F7	**−726.2750**	**69.7004**	−741.0755	54.7787	-0.9296
F8	−90.3668	0	−90.3668	0	-
F9	−90.6092	4.33E−14	−90.6092	4.33E−14	0
F10	−59.7540	4.33E−14	−59.7540	4.33E−14	0

Table 7. Comparison between FA and GWO algorithms with 50 dimensions

50 dimensions					
Function	GWO	STD	FA	STD	Z - Value
F1	8657.98	2032.415	**0.0564**	**0.0142**	−23.7182
F2	8657.98	2032.42	**0.0563**	**0.0129**	−23.7182
F3	9734.57	2605.37	**0.0585**	**0.0133**	−20.8030
F4	297.0701	40.1308	**75.4605**	**13.3594**	−29.1722
F5	31557.05	23287.72	**83.0312**	**96.7234**	−7.5249
F6	513.4068	136.0203	**78.2756**	**21.2956**	−17.5970
F7	−1450.85	238.30	**−2852.58**	**201.07**	−25.0305
F8	−45.9427	40.9837	**−90.36682**	**0.00**	−6.0352
F9	−53.0457	41.4657	**−90.0593**	**5.78E−14**	−4.9700
Γ10	−28.3182	2.2731	**−53.1833**	**4.33E−14**	−60.9058

5 Conclusions

In this work we presented a comparative study between two popular algorithms that were the Firefly Algorithm (FA) and Grey Wolf Optimizer (GWO) and these algorithms were test with a set of 10 problems of the CEC 2017 Competition on Constrained Real-Parameter Optimizer in order to study their performance in complex problems. In addition we presented a hypothesis test in order to demonstrate statistically which algorithm has better performance and for these problems we presented tests with 10 and 50 dimensions respectively and we presented these general results in tables with information according with the guidelines that the competition describes in the original paper. In addition we can conclude that for 10 dimensions the FA algorithm has better performance in approximately 70 percent of the problems analyzed and for 50 dimensions the FA performs better in all problems that we presented in this paper, so we can conclude that for these set of problems the Firefly algorithm has better performance than the GWO algorithm according with the hypothesis test that we presented above. As a future work it is important improve the results in the experiments of the constrained problems of the CEC 2017 competition using other metaheuristics, for example GSA [17], ICA [18], FWA [19, 20] and include fuzzy logic [21] in order to dynamically adjust the parameters in the other algorithms [22, 23]. In addition, instead of using type-1 fuzzy logic we could consider type-2 fuzzy as in [24–32], or consider other areas of application as in [33–38].

References

1. Maier, H.R., Kapelan, Z.: Evolutionary algorithms and other metaheuritics in water resources: current status, research challenges and future directions. Environ. Model. Softw. **62**, 271–299 (2014)
2. Can, U., Alatas, B.: Physics based metaheuristic algorithms for global optimization. Am. J. Inf. Sci. Comput. Eng. **1**, 94–106 (2015)

3. Yang, X., Karamanoglu, M.: Swarm intelligence and bio-inspired computation: an overview. In: Swarm Intelligence and Bio-Inspired Computation, pp. 3–23 (2013)
4. Wolpert, D.H., Macready, W.G.: No free lunch theorems for optimization. IEEE Trans. Evol. Comput. **1**, 67–82 (1997)
5. Yang, X-S.: Firefly Algorithm, Lévy Flights and Global Optimization arXiv:1003.1464v1 (2010)
6. Yang, X.-S.: Firefly Algorithm: Recent Advances and Applications arXiv:1308.3898v1 (2013)
7. Mirjalili, S., Mirjalili, M., Lewis, A.: Grey wolf optimizer. Adv. Eng. Softw. **69**, 46–61 (2014)
8. Muro, C., Escobedo, R., Spector, L., Coppinger, R.: Wolf-pack (Canis lupus) hunting strategies emerge from simple rules in computational simulations. Behav. Process. **88**, 192–197 (2011)
9. Rodríguez, L., Castillo, O., Valdez, M., Soria, J.: A comparative study of dynamic adaptation of parameters in the GWO algorithm using type-1 and interval type-2 fuzzy logic. In: Fuzzy Logic Augmentation of Neural and Optimization Algorithms: Theoretical Aspects and Real Applications, pp. 3–17 (2018)
10. Digalakis, J., Margaritis, K.: On benchmarking functions for genetic algorithms. Int. J. Comput. Math. **77**, 481–506 (2001)
11. Molga, M., Smutnicki, C.: Test functions for optimization needs. Test functions for optimization needs (2005)
12. Yang, X.-S.: Test problems in optimization, arXiv, preprint arXiv:1008.0549 (2010)
13. Guohua, W., Mallipeddi, R., Suganthan, P.N.: Problem Definitions and Evaluation Criteria for the CEC 2017 Competition on Constrained Real-Parameter Optimization (2017)
14. Lagunes, M., Castillo, O., Soria, J.: Optimization of membership functions parameters for fuzzy controller of an autonomous mobile robot using the firefly algorithm. In: Fuzzy Logic Augmentation of Neural and Optimization Algorithms, pp. 199–206 (2018)
15. Rodriguez, L., Castillo, O., Soria, J., Melin, P., Valdez, F., Gonzalez, C., Martinez, G., Soto, J.: A fuzzy hierarchical operator in the grey wolf optimizer algorithm. Appl. Soft. Comput. **57**, 315–328 (2017)
16. Larson, R., Farber, B.: Elementary Statistics Picturing the World, pp. 428–433. Pearson Education Inc. (2003)
17. Gonzalez, B., Melin, P., Valdez, F., Prado-Arechiga, G.: Ensemble neural network optimization using a gravitational search algorithm with interval type-1 and type-2 fuzzy parameter adaptation in pattern recognition applications. In: Fuzzy Logic Augmentation of Neural and Optimization Algorithms: Theoretical Aspects and Real Applications, pp. 17–27 (2018)
18. Bernal, E., Castillo, O., Soria, J.: Imperialist competitive algorithm with dynamic parameter adaptation applied to the optimization of mathematical functions. In: Nature-Inspired Design of Hybrid Intelligent Systems, pp. 329–341 (2017)
19. Barraza, J., Melin, P., Valdez, F., Gonzalez, C.I.: Fuzzy fireworks algorithm based on a sparks dispersion measure. Algorithms **10**, 83 (2017)
20. Barraza, J., Melin, P., Valdez, F., Gonzalez, C.: Fuzzy FWA with dynamic adaptation of parameters. In: IEEE CEC 2016, pp. 4053–4060 (2016)
21. Rodríguez, L., Castillo, O., García, M., Soria, J.: A comparative study of dynamic adaptation of parameters in the GWO algorithm using type-1 and interval type-2 fuzzy logic. In: Fuzzy Logic Augmentation of Neural and Optimization Algorithms: Theoretical Aspects and Real Applications, pp. 3–16 (2018)

22. Caraveo, C., Valdez, F., Castillo, O.: Optimization mathematical functions for multiple variables using the algorithm of self-defense of the plants. In: Nature-Inspired Design of Hybrid Intelligent Systems, pp. 631–640 (2017)
23. Guerrero, M., Castillo, O., Garcia, M.: Cuckoo search algorithm via lévy flight with dynamic adaptation of parameter using fuzzy logic for benchmark mathematical functions. In: Design of Intelligent Systems Based on Fuzzy Logic, Neural Networks and Nature-Inspired Optimization. Studies in Computational Intelligence, pp. 555–571 (2016)
24. Leal-Ramírez, C., Castillo, O., Melin, P., Rodríguez-Díaz, A.: Simulation of the bird age-structured population growth based on an interval type-2 fuzzy cellular structure. Inf. Sci. **181**(3), 519–535 (2011)
25. Cázarez-Castro, N.R., Aguilar, L.T., Castillo, O.: Designing type-1 and type-2 fuzzy logic controllers via fuzzy lyapunov synthesis for nonsmooth mechanical systems. Eng. Appl. AI **25**(5), 971–979 (2012)
26. Rubio, E., Castillo, O., Valdez, F., Melin, P., González, C.I., Martinez, G.: An extension of the fuzzy possibilistic clustering algorithm using type-2 fuzzy logic techniques. Adv. Fuzzy Syst. **2017**, 7094046:1–7094046:23 (2017)
27. Castillo, O., Melin, P.: Intelligent systems with interval type-2 fuzzy logic. Int. J. Innovative Comput. Inf. Control **4**(4), 771–783 (2008)
28. Mendez, G.M., Castillo, O.: Interval type-2 TSK fuzzy logic systems using hybrid learning algorithm. In: The 14th IEEE International Conference on Fuzzy Systems, FUZZ 2005, pp. 230–235 (2005)
29. Melin, P., González, C.I., Castro, J.R., Mendoza, O., Castillo, O.: Edge-detection method for image processing based on generalized type-2 fuzzy logic. IEEE Trans. Fuzzy Syst. **22**(6), 1515–1525 (2014)
30. González, C.I., Melin, P., Castro, J.R.: Oscar castillo, olivia mendoza: optimization of interval type-2 fuzzy systems for image edge detection. Appl. Soft Comput. **47**, 631–643 (2016)
31. González, C.I., Melin, P., Castro, J.R.: Olivia mendoza, oscar castillo: an improved sobel edge detection method based on generalized type-2 fuzzy logic. Soft. Comput. **20**(2), 773–784 (2016)
32. Ontiveros, E., Melin, P., Castillo, O.: High order α-planes integration: a new approach to computational cost reduction of general type-2 fuzzy systems. Eng. Appl. AI **74**, 186–197 (2018)
33. Melin, P., Castillo, O.: Intelligent control of complex electrochemical systems with a neuro-fuzzy-genetic approach. IEEE Trans. Ind. Electron. **48**(5), 951–955 (2001)
34. Aguilar, L., Melin, P., Castillo, O.: Intelligent control of a stepping motor drive using a hybrid neuro-fuzzy ANFIS approach. Appl. Soft Comput. **3**(3), 209–219 (2003)
35. Melin, P., Castillo, O.: Adaptive intelligent control of aircraft systems with a hybrid approach combining neural networks, fuzzy logic and fractal theory. Appl. Soft Comput. **3**(4), 353–362 (2003)
36. Melin, P., Amezcua, J., Valdez, F., Castillo, O.: A new neural network model based on the LVQ algorithm for multi-class classification of arrhythmias. Inf. Sci. **279**, 483–497 (2014)
37. Melin, P., Castillo, O.: Modelling, Simulation and Control of Non-Linear Dynamical Systems: An Intelligent Approach Using Soft Computing and Fractal Theory. CRC Press, Boca Raton (2001)
38. Melin, P., Sánchez, D., Castillo, O.: Genetic optimization of modular neural networks with fuzzy response integration for human recognition. Inf. Sci. **197**, 1–19 (2012)

Canonical Fuzzy Preference Relations

Thomas A. Runkler$^{(\boxtimes)}$

Siemens AG, Otto-Hahn-Ring 6, 81739 Munich, Germany
thomas.runkler@siemens.com

Abstract. Fuzzy preference matrices are an important tool for group decision making. Often not all group members are willing or able to provide their preferences in the form of fuzzy preference matrices but only in the form of crisp rank orders of options. In this paper we introduce the concepts of reciprocal canonical fuzzy preference matrices and preference weight vectors for additive and multiplicative fuzzy preferences. Consistency is an important property of fuzzy preference matrices, so we present two methods that allow to construct consistent reciprocal canonical additive fuzzy preference matrices and consistent reciprocal canonical multiplicative fuzzy preference matrices from crisp rank orders for arbitrary numbers of elements. These transformations allow to process fuzzy preference matrices and crisp rank orders in the same mathematical framework in group decision making.

1 Introduction

This paper is motivated by recommender systems [3,7] and group decision processes [5,12] where preferences of individuals are modeled by pairwise fuzzy preference relations [1]. Given a set of n options, a fuzzy preference relation is specified by an $n \times n$ fuzzy preference matrix

$$P = \begin{pmatrix} p_{11} & \cdots & p_{1n} \\ \vdots & \ddots & \vdots \\ p_{n1} & \cdots & p_{nn} \end{pmatrix} \qquad (1)$$

where each matrix element p_{ij} quantifies the degree of preference of option i over option j, $i, j = 1, \ldots, n$. We distinguish additive and multiplicative fuzzy preference matrices. For additive fuzzy preference matrices [1] we have $p_{ij} \in [0, 1]$, where $p_{ij} = 0$ indicates absolutely no preference, $p_{ij} = 1$ indicates complete preference, and $p_{ij} = 1/2$ indicates equivalence, so the elements of the main diagonal are $p_{ii} = 1/2$ for all $i = 1, \ldots, n$. For multiplicative fuzzy preference matrices [11], we have $p_{ij} \in (0, \infty)$, where $p_{ij} = 1$ indicates equivalence, so the elements of the main diagonal are $p_{ii} = 1$ for all $i = 1, \ldots, n$. In a group decision making process each group member may provide her or his preferences in the form of such a fuzzy prefence matrix, and many different approaches have been proposed to merge individual preferences to joint group decisions. For an overview we refer to [1,11,14].

© Springer Nature Switzerland AG 2019
R. B. Kearfott et al. (Eds.): IFSA 2019/NAFIPS 2019, AISC 1000, pp. 542–555, 2019.
https://doi.org/10.1007/978-3-030-21920-8_48

Often one or several members of the group are not willing or not able to provide such a fuzzy preference matrix, especially when the number of options is high and therefore specifying all entries of the preference matrix is tedious. Such experts may rather be willing and able to express their preferences in the form of (crisp) rank orders [4]. Given a set of n options, a rank order is specified by a permutation π of the n option indices

$$\pi : \{1, \ldots, n\} \rightarrow \{1, \ldots, n\} \tag{2}$$

so option i_1 with $\pi(i_1) = 1$ is most preferred (rank 1), option i_2 with $\pi(i_2) = 2$ is second most preferred (rank 2), and so on, and option i_n with $\pi(i_n) = n$ is least preferred (rank n). For simplicity we will not consider ties (different options with the same rank) here, but all methods presented in this paper can be easily extended to handle such cases.

In group decision processes we want to treat individual preferences expressed either by fuzzy preference matrices or by rank orders in the same framework. To do so, we want to convert rank orders into what we will call *reciprocal canonical fuzzy preference matrices*. Then the preferences of all group members will be available in the same format (as fuzzy preference matrices) and can be processed by any of the group decision making methods for fuzzy preference matrices proposed in the literature.

An important property of fuzzy preference matrices is consistency [12,13]. Therefore, the focus of this paper is to convert rank orders to consistent reciprocal canonical additive or multiplicative fuzzy preference matrices. This is related to earlier work dealing with the conversion of utility values to fuzzy preference matrices [8], to consistent additive or multiplicative fuzzy preference matrices [10], and to consistent Łukasiewicz fuzzy preference matrices [9]. Herrera–Viedma *et al.* [2] proposed a method for constructing consistent reciprocal non-canonical additive and multiplicative fuzzy preference matrices by specifying all the preferences along the first superdiagonal of P. Here, we consider the case when we do not want to specify any element of P (except for the maximum preference) but construct P entirely from the rank order π. Ma *et al.* [6] proposed an approach to construct fuzzy preference matrices and then repair the inconsistency. Here, we want to immediately construct consistent fuzzy preference matrices.

This paper is structured as follows: Sect. 2 introduces the concept of reciprocal canonical (additive and multiplicative) fuzzy preference matrices. Section 3 shows how consistent reciprocal canonical *additive* fuzzy preference matrices can be constructed. Accordingly, Sect. 4 shows how consistent reciprocal canonical *multiplicative* fuzzy preference matrices can be constructed. Section 5 summarizes our conclusions and gives an outline for future research.

2   Reciprocal Canonical Fuzzy Preference Matrices

Our goal here is to use a rank order π (2) to construct a fuzzy preference matrix P (1). In the canonical setting we assume that the degree of preference of the rank 1 option over the rank 2 option is the same as the degree of preference of the rank 2 option over the rank 3 option, and so on. So we define the preference of a rank i option over the rank $i+1$ option as w_1, $i = 1, \ldots, n-1$. For additive preference we have $w_1 \in (0.5, 1]$, and for multiplicative fuzzy preference we have $w_1 \in (1, \infty)$. More generally, we assume that the degrees of preference of any rank i option over the rank $i+k$ option are all equal to w_k, $k = 1, \ldots, n-1$, $i = 1, \ldots, n-k$, where $w_k \in (0.5, 1]$ for additive preference and $w_k \in (1, \infty)$ for multiplicative fuzzy preference. So for any $i, j = 1, \ldots, n$, $\pi(i) < \pi(j)$, we can write

$$p_{ij} = w_{\pi(j)-\pi(i)} \tag{3}$$

Moreover, it is reasonable to assume that the degree of preference of a rank i option over the rank $i+k$ option is larger than the degree of preference of a rank i option over the rank $i+l$ option if $k > l$, so we assume that the elements of w are strictly monotonically increasing,

$$w_i < w_j \tag{4}$$

for all $i, j = 1, \ldots, n-1$, $i < j$. We will call w a *preference weight vector*. Notice the similarity to *ordered weighted averaging* (OWA) operators [15]. Next, we require that the preferences should be reciprocal. An additive fuzzy preference matrix is called reciprocal if and only if

$$p_{ij} = 1 - p_{ji} \tag{5}$$

for all $i, j = 1, \ldots, n$. So for reciprocal canonical additive fuzzy preference matrices we define

$$p_{ij} = 1 - w_{\pi(i)-\pi(j)} \tag{6}$$

for $\pi(i) > \pi(j)$. A multiplicative fuzzy preference matrix is called reciprocal if and only if

$$p_{ij} = 1/p_{ji} \tag{7}$$

for all $i, j = 1, \ldots, n$. So for reciprocal canonical multiplicative fuzzy preference matrices we define

$$p_{ij} = 1/w_{\pi(i)-\pi(j)} \tag{8}$$

for $\pi(i) > \pi(j)$. This leads us to the following definitions of reciprocal canonical additive and multiplicative fuzzy preference matrices.

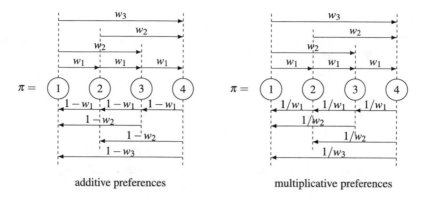

Fig. 1. Schematic view of reciprocal canonical fuzzy preference relations.

Definition 1. An $n \times n$ matrix P is called a *reciprocal canonical additive fuzzy preference matrix* if there is a permutation $\pi : \{1, \ldots, n\} \rightarrow \{1, \ldots, n\}$ and a strictly monotonically increasing vector $w = (w_1, \ldots, w_{n-1}) \subset (0.5, 1]$, $w_i < w_j$, for all $i, j = 1, \ldots, n - 1$, $i < j$, so that the elements of P can be written as

$$p_{ij} = \begin{cases} w_{\pi(j)-\pi(i)} & \text{for } \pi(i) < \pi(j) \\ 1/2 & \text{for } \pi(i) = \pi(j) \\ 1 - w_{\pi(i)-\pi(j)} & \text{for } \pi(i) > \pi(j) \end{cases} \tag{9}$$

for all $i, j = 1, \ldots, n$.

Definition 2. An $n \times n$ matrix P is called a *reciprocal canonical multiplicative fuzzy preference matrix* if there is a permutation $\pi : \{1, \ldots, n\} \rightarrow \{1, \ldots, n\}$ and a strictly monotonically increasing vector $w = (w_1, \ldots, w_{n-1}) \subset (1, \infty)$, $w_i < w_j$, for all $i, j = 1, \ldots, n - 1$, $i < j$, so that the elements of P can be written as

$$p_{ij} = \begin{cases} w_{\pi(j)-\pi(i)} & \text{for } \pi(i) < \pi(j) \\ 1 & \text{for } \pi(i) = \pi(j) \\ 1/w_{\pi(i)-\pi(j)} & \text{for } \pi(i) > \pi(j) \end{cases} \tag{10}$$

for all $i, j = 1, \ldots, n$.

We say that the reciprocal canonical fuzzy preference matrix P is *induced* by the rank order π and the preference weight vector w. Figure 1 shows an example of four options with ranks $\pi = 1, \ldots, 4$, the corresponding canonical preferences $w_1, \ldots, w_3$, the reciprocal canonical additive preferences $1 - w_1, \ldots, 1 - w_3$ (left), and the reciprocal canonical multiplicative preferences $1/w_1, \ldots, 1/w_3$ (right). We may rearrange the order of options according to the ranks, so that the permutation π turns into the identity function

$$\pi(i) = i \tag{11}$$

for all $i = 1, \ldots, n$, i.e. the first option is rank 1, the second option is rank 2, and so on. After such rearrangement a reciprocal canonical additive fuzzy preference matrix can be written as

$$
P = \begin{pmatrix}
1/2 & w_1 & w_2 & \cdots & w_{n-3} & w_{n-2} & w_{n-1} \\
1-w_1 & 1/2 & w_1 & \cdots & w_{n-4} & w_{n-3} & w_{n-2} \\
1-w_2 & 1-w_1 & 1/2 & \cdots & w_{n-5} & w_{n-4} & w_{n-3} \\
\vdots & \vdots & \vdots & \ddots & \vdots & \vdots \\
1-w_{n-3} & 1-w_{n-4} & 1-w_{n-5} & \cdots & 1/2 & w_1 & w_2 \\
1-w_{n-2} & 1-w_{n-3} & 1-w_{n-4} & \cdots & 1-w_1 & 1/2 & w_1 \\
1-w_{n-1} & 1-w_{n-2} & 1-w_{n-3} & \cdots & 1-w_2 & 1-w_1 & 1/2
\end{pmatrix}
\tag{12}
$$

with constant diagonals whose elements can be written as

$$
\mathrm{diag}_k P = \begin{cases}
w_k & \text{for } k > 0 \\
1/2 & \text{for } k = 0 \\
1 - w_{-k} & \text{for } k < 0
\end{cases}
\tag{13}
$$

for all $k = 1 - n, \ldots, n - 1$, where $k = 0$ refers to the main diagonal, $k > 0$ are the superdiagonals, and $k < 0$ are the subdiagonals. A reciprocal canonical multiplicative fuzzy preference matrix can then be written as

$$
P = \begin{pmatrix}
1 & w_1 & w_2 & \cdots & w_{n-3} & w_{n-2} & w_{n-1} \\
1/w_1 & 1 & w_1 & \cdots & w_{n-4} & w_{n-3} & w_{n-2} \\
1/w_2 & 1/w_1 & 1 & \cdots & w_{n-5} & w_{n-4} & w_{n-3} \\
\vdots & \vdots & \vdots & \ddots & \vdots & \vdots \\
1/w_{n-3} & 1/w_{n-4} & 1/w_{n-5} & \cdots & 1 & w_1 & w_2 \\
1/w_{n-2} & 1/w_{n-3} & 1/w_{n-4} & \cdots & 1/w_1 & 1 & w_1 \\
1/w_{n-1} & 1/w_{n-2} & 1/w_{n-3} & \cdots & 1/w_2 & 1/w_1 & 1
\end{pmatrix}
\tag{14}
$$

$$
\mathrm{diag}_k P = \begin{cases}
w_k & \text{for } k > 0 \\
1 & \text{for } k = 0 \\
1/w_{-k} & \text{for } k < 0
\end{cases}
\tag{15}
$$

Herrera-Viedma *et al.* [2] have considered the case of consistent reciprocal non-canonical fuzzy preference matrices where the elements of the first superdiagonal of P are specified. Here, we consider the canonical case, where the elements of each sub- and superdiagonal are the same.

An important property of fuzzy preference relations is consistency. In the following two sections we will show how we can construct consistent reciprocal canonical additive fuzzy preference matrices (Sect. 3) and consistent reciprocal canonical multiplicative fuzzy preference matrices (Sect. 4).

3   Constructing Additive Preferences from Ranks

An additive fuzzy preference relation is called *consistent* [12] if and only if

$$(p_{ij} - 0.5) + (p_{jk} - 0.5) = (p_{ik} - 0.5) \tag{16}$$

for all $i, j, k = 1, \ldots, n$.

Our goal here is to find a formula for the elements of the preference weight vector w so that the resulting reciprocal canonical additive fuzzy preference matrix P is consistent. Let us first consider a case of $i, j, k \in \{1, \ldots, n\}$ so that

$$\pi(i) = 1, \quad \pi(j) = 2, \quad \pi(k) = 3 \tag{17}$$

and with (9) we obtain

$$p_{ij} = w_1, \quad p_{jk} = w_1, \quad p_{ik} = w_2 \tag{18}$$

Inserting these into (16) yields

$$(w_1 - 0.5) + (w_1 - 0.5) = (w_2 - 0.5) \tag{19}$$

$$\Rightarrow \quad w_2 = 2w_1 - 0.5 \tag{20}$$

Next we consider the case

$$\pi(i) = 1, \quad \pi(j) = 2, \quad \pi(k) = 4 \tag{21}$$

$$\Rightarrow \quad p_{ij} = w_1, \quad p_{jk} = w_2, \quad p_{ik} = w_3 \tag{22}$$

$$\Rightarrow \quad (w_1 - 0.5) + (w_2 - 0.5) = (w_3 - 0.5) \tag{23}$$

$$\Rightarrow \quad w_3 = w_2 + w_1 - 0.5 = 2w_1 - 0.5 + w_1 - 0.5 = 3w_1 - 1 \tag{24}$$

Obviously, whenever $\pi(k)$ increases by one, then the corresponding $w_{\pi(k)-1}$ increases by $w_1 - 0.5$. So for arbitrary $i = 1, \ldots, n-1$ we obtain

$$w_i = i \cdot w_1 - \frac{i-1}{2} \tag{25}$$

For the computation of the elements of w we may not want to specify the smallest preference w_1 but the maximum preference $\hat{w} = w_{n-1}$. For this case we obtain

$$\hat{w} = w_{n-1} = (n-1) \cdot w_1 - \frac{n-2}{2} \tag{26}$$

$$\Rightarrow \quad w_1 = \frac{\hat{w}}{n-1} + \frac{n-2}{2n-2} \tag{27}$$

$$\Rightarrow \quad w_i = i \cdot \left(\frac{\hat{w}}{n-1} + \frac{n-2}{2n-2} \right) - \frac{i-1}{2} = \frac{1}{2} + \frac{i}{n-1}(\hat{w} - \frac{1}{2}) \tag{28}$$

For $\hat{w} > 1/2$ this yields $w_i > 1/2$ for all $i = 1, \ldots, n-1$. If we set the maximum preference to $\hat{w} = 1$, then we obtain the simpler formula

$$w_i = \frac{1}{2} + \frac{i}{2n-2} \tag{29}$$

We summarize these findings in

Theorem 1. *Let the maximum preference be defined as $\hat{w} \in (0.5, 1]$. An arbitrary rank order $\pi : \{1, \ldots, n\} \to \{1, \ldots, n\}$ and the preference weight vector w with the elements*

$$w_i = \frac{1}{2} + \frac{i}{n-1}\left(\hat{w} - \frac{1}{2}\right) \tag{30}$$

$i = 1, \ldots, n-1$, induce a consistent reciprocal canonical additive fuzzy preference matrix P.

Proof. Given $i, j, k \in \{1, \ldots, n\}$, there are six possibilities of orders of $\pi(i)$, $\pi(j)$, and $\pi(k)$. Consider for example the case

$$\pi(j) < \pi(k) < \pi(i) \tag{31}$$

With (9) and (30) we obtain

$$p_{ij} = 1 - \frac{1}{2} - (\pi(i) - \pi(j))\frac{2\hat{w} - 1}{2n - 2} = \frac{1}{2} + (\pi(j) - \pi(i))\frac{2\hat{w} - 1}{2n - 2} \tag{32}$$

$$p_{jk} = \frac{1}{2} + (\pi(k) - \pi(j))\frac{2\hat{w} - 1}{2n - 2} \tag{33}$$

$$p_{ik} = 1 - \frac{1}{2} - (\pi(i) - \pi(k))\frac{2\hat{w} - 1}{2n - 2} = \frac{1}{2} + (\pi(k) - \pi(i))\frac{2\hat{w} - 1}{2n - 2} \tag{34}$$

It is easy to verify that for all other five possibilities of orders of $\pi(i)$, $\pi(j)$, and $\pi(k)$ we obtain the same p_{ij}, p_{jk}, and p_{ik}. Inserting these into (16) yields

$$(\pi(j) - \pi(i)) + (\pi(k) - \pi(j)) = (\pi(k) - \pi(i)) \quad \square \tag{35}$$

Figure 2 shows the graphs of the functions $w_1(\hat{w}), \ldots, w_8(\hat{w})$ according to (30), for $n = 9$ and $\hat{w} \in (0.5, 1]$. Each function w_i, $i = 1, \ldots, n-1$, increases linearly with $\hat{w}$, starting at the point $w_i = \hat{w} = 0.5$. For a given $\hat{w}$ we find the corresponding preference values $w_1, \ldots, w_{n-1}$ along the vertical at $\hat{w}$. The dashed lines illustrate this for $\hat{w} = 0.9$ where we find $w_1 = 0.55$, $w_2 = 0.6$, $w_3 = 0.65$, $w_4 = 0.7$, $w_5 = 0.75$, $w_6 = 0.8$, $w_7 = 0.85$, and $w_8 = \hat{w} = 0.9$. These preference values and their complements $1 - w_i$, $i = 1, \ldots, n-1$, can be used to construct a consistent reciprocal canonical additive fuzzy preference matrix P according to Definition 1. Table 1 shows the consistent reciprocal canonical additive fuzzy preference matrices P for $n = 2, \ldots, 9$ for maximum preference $\hat{w} = 0.9$ and $\pi(i) = i$ for all $i = 1, \ldots, n$. The values found in Fig. 2 correspond to the elements of the first row and last column of the bottom matrix in Table 1.

Table 1. Consistent reciprocal canonical additive fuzzy preference matrices P, $\hat{w} = 0.9$, $\pi(i) = i$, $n = 2, \ldots, 9$.

$$\begin{pmatrix} 0.500 & 0.900 \\ 0.100 & 0.500 \end{pmatrix}$$

$$\begin{pmatrix} 0.500 & 0.700 & 0.900 \\ 0.300 & 0.500 & 0.700 \\ 0.100 & 0.300 & 0.500 \end{pmatrix}$$

$$\begin{pmatrix} 0.500 & 0.633 & 0.767 & 0.900 \\ 0.367 & 0.500 & 0.633 & 0.767 \\ 0.233 & 0.367 & 0.500 & 0.633 \\ 0.100 & 0.233 & 0.367 & 0.500 \end{pmatrix}$$

$$\begin{pmatrix} 0.500 & 0.600 & 0.700 & 0.800 & 0.900 \\ 0.400 & 0.500 & 0.600 & 0.700 & 0.800 \\ 0.300 & 0.400 & 0.500 & 0.600 & 0.700 \\ 0.200 & 0.300 & 0.400 & 0.500 & 0.600 \\ 0.100 & 0.200 & 0.300 & 0.400 & 0.500 \end{pmatrix}$$

$$\begin{pmatrix} 0.500 & 0.580 & 0.660 & 0.740 & 0.820 & 0.900 \\ 0.420 & 0.500 & 0.580 & 0.660 & 0.740 & 0.820 \\ 0.340 & 0.420 & 0.500 & 0.580 & 0.660 & 0.740 \\ 0.260 & 0.340 & 0.420 & 0.500 & 0.580 & 0.660 \\ 0.180 & 0.260 & 0.340 & 0.420 & 0.500 & 0.580 \\ 0.100 & 0.180 & 0.260 & 0.340 & 0.420 & 0.500 \end{pmatrix}$$

$$\begin{pmatrix} 0.500 & 0.567 & 0.633 & 0.700 & 0.767 & 0.833 & 0.900 \\ 0.433 & 0.500 & 0.567 & 0.633 & 0.700 & 0.767 & 0.833 \\ 0.367 & 0.433 & 0.500 & 0.567 & 0.633 & 0.700 & 0.767 \\ 0.300 & 0.367 & 0.433 & 0.500 & 0.567 & 0.633 & 0.700 \\ 0.233 & 0.300 & 0.367 & 0.433 & 0.500 & 0.567 & 0.633 \\ 0.167 & 0.233 & 0.300 & 0.367 & 0.433 & 0.500 & 0.567 \\ 0.100 & 0.167 & 0.233 & 0.300 & 0.367 & 0.433 & 0.500 \end{pmatrix}$$

$$\begin{pmatrix} 0.500 & 0.557 & 0.614 & 0.671 & 0.729 & 0.786 & 0.843 & 0.900 \\ 0.443 & 0.500 & 0.557 & 0.614 & 0.671 & 0.729 & 0.786 & 0.843 \\ 0.386 & 0.443 & 0.500 & 0.557 & 0.614 & 0.671 & 0.729 & 0.786 \\ 0.329 & 0.386 & 0.443 & 0.500 & 0.557 & 0.614 & 0.671 & 0.729 \\ 0.271 & 0.329 & 0.386 & 0.443 & 0.500 & 0.557 & 0.614 & 0.671 \\ 0.214 & 0.271 & 0.329 & 0.386 & 0.443 & 0.500 & 0.557 & 0.614 \\ 0.157 & 0.214 & 0.271 & 0.329 & 0.386 & 0.443 & 0.500 & 0.557 \\ 0.100 & 0.157 & 0.214 & 0.271 & 0.329 & 0.386 & 0.443 & 0.500 \end{pmatrix}$$

$$\begin{pmatrix} 0.500 & 0.550 & 0.600 & 0.650 & 0.700 & 0.750 & 0.800 & 0.850 & 0.900 \\ 0.450 & 0.500 & 0.550 & 0.600 & 0.650 & 0.700 & 0.750 & 0.800 & 0.850 \\ 0.400 & 0.450 & 0.500 & 0.550 & 0.600 & 0.650 & 0.700 & 0.750 & 0.800 \\ 0.350 & 0.400 & 0.450 & 0.500 & 0.550 & 0.600 & 0.650 & 0.700 & 0.750 \\ 0.300 & 0.350 & 0.400 & 0.450 & 0.500 & 0.550 & 0.600 & 0.650 & 0.700 \\ 0.250 & 0.300 & 0.350 & 0.400 & 0.450 & 0.500 & 0.550 & 0.600 & 0.650 \\ 0.200 & 0.250 & 0.300 & 0.350 & 0.400 & 0.450 & 0.500 & 0.550 & 0.600 \\ 0.150 & 0.200 & 0.250 & 0.300 & 0.350 & 0.400 & 0.450 & 0.500 & 0.550 \\ 0.100 & 0.150 & 0.200 & 0.250 & 0.300 & 0.350 & 0.400 & 0.450 & 0.500 \end{pmatrix}$$

Table 2. Consistent reciprocal canonical multiplicative fuzzy preference matrices P, $\hat{w} = 2$, $\pi(i) = i$, $n = 2, \ldots, 9$.

$$\begin{pmatrix} 1.000 & 2.000 \\ 0.500 & 1.000 \end{pmatrix}$$

$$\begin{pmatrix} 1.000 & 1.414 & 2.000 \\ 0.707 & 1.000 & 1.414 \\ 0.500 & 0.707 & 1.000 \end{pmatrix}$$

$$\begin{pmatrix} 1.000 & 1.260 & 1.587 & 2.000 \\ 0.794 & 1.000 & 1.260 & 1.587 \\ 0.630 & 0.794 & 1.000 & 1.260 \\ 0.500 & 0.630 & 0.794 & 1.000 \end{pmatrix}$$

$$\begin{pmatrix} 1.000 & 1.189 & 1.414 & 1.682 & 2.000 \\ 0.841 & 1.000 & 1.189 & 1.414 & 1.682 \\ 0.707 & 0.841 & 1.000 & 1.189 & 1.414 \\ 0.595 & 0.707 & 0.841 & 1.000 & 1.189 \\ 0.500 & 0.595 & 0.707 & 0.841 & 1.000 \end{pmatrix}$$

$$\begin{pmatrix} 1.000 & 1.149 & 1.320 & 1.516 & 1.741 & 2.000 \\ 0.871 & 1.000 & 1.149 & 1.320 & 1.516 & 1.741 \\ 0.758 & 0.871 & 1.000 & 1.149 & 1.320 & 1.516 \\ 0.660 & 0.758 & 0.871 & 1.000 & 1.149 & 1.320 \\ 0.574 & 0.660 & 0.758 & 0.871 & 1.000 & 1.149 \\ 0.500 & 0.574 & 0.660 & 0.758 & 0.871 & 1.000 \end{pmatrix}$$

$$\begin{pmatrix} 1.000 & 1.122 & 1.260 & 1.414 & 1.587 & 1.782 & 2.000 \\ 0.891 & 1.000 & 1.122 & 1.260 & 1.414 & 1.587 & 1.782 \\ 0.794 & 0.891 & 1.000 & 1.122 & 1.260 & 1.414 & 1.587 \\ 0.707 & 0.794 & 0.891 & 1.000 & 1.122 & 1.260 & 1.414 \\ 0.630 & 0.707 & 0.794 & 0.891 & 1.000 & 1.122 & 1.260 \\ 0.561 & 0.630 & 0.707 & 0.794 & 0.891 & 1.000 & 1.122 \\ 0.500 & 0.561 & 0.630 & 0.707 & 0.794 & 0.891 & 1.000 \end{pmatrix}$$

$$\begin{pmatrix} 1.000 & 1.104 & 1.219 & 1.346 & 1.486 & 1.641 & 1.811 & 2.000 \\ 0.906 & 1.000 & 1.104 & 1.219 & 1.346 & 1.486 & 1.641 & 1.811 \\ 0.820 & 0.906 & 1.000 & 1.104 & 1.219 & 1.346 & 1.486 & 1.641 \\ 0.743 & 0.820 & 0.906 & 1.000 & 1.104 & 1.219 & 1.346 & 1.486 \\ 0.673 & 0.743 & 0.820 & 0.906 & 1.000 & 1.104 & 1.219 & 1.346 \\ 0.610 & 0.673 & 0.743 & 0.820 & 0.906 & 1.000 & 1.104 & 1.219 \\ 0.552 & 0.610 & 0.673 & 0.743 & 0.820 & 0.906 & 1.000 & 1.104 \\ 0.500 & 0.552 & 0.610 & 0.673 & 0.743 & 0.820 & 0.906 & 1.000 \end{pmatrix}$$

$$\begin{pmatrix} 1.000 & 1.091 & 1.189 & 1.297 & 1.414 & 1.542 & 1.682 & 1.834 & 2.000 \\ 0.917 & 1.000 & 1.091 & 1.189 & 1.297 & 1.414 & 1.542 & 1.682 & 1.834 \\ 0.841 & 0.917 & 1.000 & 1.091 & 1.189 & 1.297 & 1.414 & 1.542 & 1.682 \\ 0.771 & 0.841 & 0.917 & 1.000 & 1.091 & 1.189 & 1.297 & 1.414 & 1.542 \\ 0.707 & 0.771 & 0.841 & 0.917 & 1.000 & 1.091 & 1.189 & 1.297 & 1.414 \\ 0.648 & 0.707 & 0.771 & 0.841 & 0.917 & 1.000 & 1.091 & 1.189 & 1.297 \\ 0.595 & 0.648 & 0.707 & 0.771 & 0.841 & 0.917 & 1.000 & 1.091 & 1.189 \\ 0.545 & 0.595 & 0.648 & 0.707 & 0.771 & 0.841 & 0.917 & 1.000 & 1.091 \\ 0.500 & 0.545 & 0.595 & 0.648 & 0.707 & 0.771 & 0.841 & 0.917 & 1.000 \end{pmatrix}$$

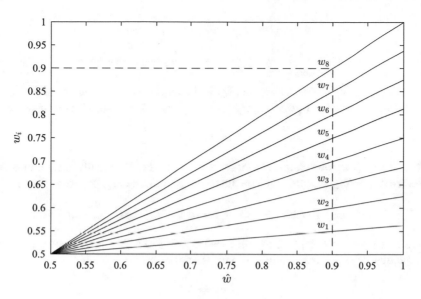

Fig. 2. Graphs of the additive preference weight functions $w_1(\hat{w}), \ldots, w_8(\hat{w})$ for $n = 9$.

4   Constructing Multiplicative Preferences from Ranks

A multiplicative fuzzy preference relation is called *consistent* [13] if and only if

$$\frac{p_{ji}}{p_{ij}} \cdot \frac{p_{kj}}{p_{jk}} = \frac{p_{ki}}{p_{ik}} \tag{36}$$

for all $i, j, k = 1, \ldots, n$. For reciprocal preferences (7) we obtain

$$\frac{1}{p_{ij}^2} \cdot \frac{1}{p_{jk}^2} = \frac{1}{p_{ik}^2} \tag{37}$$

Similar to the previous section we now want to find a formula for the elements of the preference weight vector w so that the resulting reciprocal canonical multiplicative fuzzy preference matrix P is consistent. Just as for additive preference we first consider $i, j, k \in \{1, \ldots, n\}$ with

$$\pi(i) = 1, \quad \pi(j) = 2, \quad \pi(k) = 3 \tag{38}$$

$$\Rightarrow \quad p_{ij} = w_1, \quad p_{jk} = w_1, \quad p_{ik} = w_2 \tag{39}$$

We insert these into (37) and obtain

$$\frac{1}{w_1^2} \cdot \frac{1}{w_1^2} = \frac{1}{w_2^2} \quad \Rightarrow \quad w_2 = w_1^2 \tag{40}$$

Next we consider

$$\pi(i) = 1, \quad \pi(j) = 2, \quad \pi(k) = 4 \tag{41}$$

$$\Rightarrow \quad p_{ij} = w_1, \quad p_{jk} = w_2, \quad p_{ik} = w_3 \tag{42}$$

$$\Rightarrow \quad \frac{1}{w_1^2} \cdot \frac{1}{w_2^2} = \frac{1}{w_3^2} \quad \Rightarrow \quad w_3 = w_1 \cdot w_2 = w_1^3 \tag{43}$$

So, whenever $\pi(k)$ increases by one, then the exponent in the formula for $w_{\pi(k)-1}$ also increases by one. For arbitrary $i = 1, \ldots, n-1$ we can therefore write

$$w_i = w_1^i \tag{44}$$

Just as in the previous chapter we may want to find w by specifying the maximum preference $\hat{w} = w_{n-1}$, and not the smallest preference w_1, so we write

$$\hat{w} = w_{n-1} = w_1^{n-1} \quad \Rightarrow \quad w_1 = \hat{w}^{\frac{1}{n-1}} \quad \Rightarrow \quad w_i = \hat{w}^{\frac{i}{n-1}} \tag{45}$$

For $\hat{w} > 1$ we obtain $w_i > 1$ for all $i = 1, \ldots, n-1$. If the maximum preference approaches $\hat{w} \to 1$, then we obtain the trivial case

$$w_i = 1 \tag{46}$$

for all $i = 1, \ldots, n-1$. Therefore, for non-trivial preferences we require $\hat{w} > 1$. This leads us to.

Theorem 2. *Let the maximum preference be defined as $\hat{w} \in (1, \infty)$. An arbitrary rank order $\pi : \{1, \ldots, n\} \to \{1, \ldots, n\}$ and the preference weight vector w with the elements*

$$w_i = \hat{w}^{\frac{i}{n-1}} \tag{47}$$

$i = 1, \ldots, n-1$, induce a consistent reciprocal canonical multiplicative fuzzy preference matrix P.

Proof. Given $i, j, k \in \{1, \ldots, n\}$, there are six possibilities of orders of $\pi(i)$, $\pi(j)$, and $\pi(k)$. Consider for example the case

$$\pi(j) < \pi(k) < \pi(i) \tag{48}$$

With (10) and (47) we obtain

$$p_{ij} = \hat{w}^{-\frac{\pi(i)-\pi(j)}{n-1}} \tag{49}$$

$$p_{jk} = \hat{w}^{\frac{\pi(k)-\pi(j)}{n-1}} \tag{50}$$

$$p_{ik} = \hat{w}^{-\frac{\pi(i)-\pi(k)}{n-1}} \tag{51}$$

Inserting these into (37) yields

$$\hat{w}^{2\left(\frac{\pi(i)-\pi(j)}{n-1} - \frac{\pi(k)-\pi(j)}{n-1}\right)} = \hat{w}^{2\frac{\pi(i)-\pi(k)}{n-1}} \tag{52}$$

$$\pi(i) - \pi(j) - \pi(k) + \pi(j) = \pi(i) - \pi(k) \quad \square \tag{53}$$

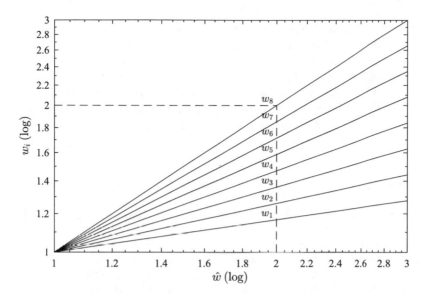

Fig. 3. Graphs of the multiplicative preference weight functions $w_1(\hat{w}), \ldots, w_8(\hat{w})$ for $n = 9$.

It is easy to verify that for all other five possibilities of orders of $\pi(i)$, $\pi(j)$, and $\pi(k)$ we obtain the same result.

Figure 3 shows a log-log plot of the graphs of the functions $w_1(\hat{w}), \ldots, w_8(\hat{w})$ according to (47), for $n = 9$ and $\hat{w} \in (1, 3]$. The functions w_i, $i = 1, \ldots, n-1$, are nonlinear, but in the log-log plot against $\hat{w}$ they appear as lines through the point $w_i = \hat{w} = 1$. Just as in the additive case, for a given $\hat{w}$ we find the corresponding preference values $w_1, \ldots, w_{n-1}$ along the vertical at $\hat{w}$. The dashed lines illustrate this for $\hat{w} = 2$ where we find $w_1 = 1.091$, $w_2 = 1.189$, $w_3 = 1.297$, $w_4 = 1.414$, $w_5 = 1.542$, $w_6 = 1.682$, $w_7 = 1.834$, $w_8 = \hat{w} = 2$. With these preference values and their complements $1/w_i$, $i = 1, \ldots, n-1$, we can construct a consistent reciprocal canonical multiplicative fuzzy preference matrix P according to Definition 2. Table 2 shows the consistent reciprocal canonical multiplicative fuzzy preference matrices P for $n = 2, \ldots, 9$ for maximum preference $\hat{w} = 2$ and $\pi(i) = i$ for all $i = 1, \ldots, n$. The values found in Fig. 3 correspond to the elements of the first row and last column of the bottom matrix in Table 2.

5   Conclusions

We have introduced the concept of canonical fuzzy preference relations which can be used to construct fuzzy preference matrices by specifying only the rank order π of the options and the maximum preference $\hat{w}$. This is useful in group decision processes, when some group members specify their preferences as fuzzy preference matrices and others as rank orders. The rank orders can then be transformed to canonical fuzzy preference matrices, and then all group preferences can be processed in the same format as (canonical and non-canonical) fuzzy preference matrices.

In the future we are planning to extend this work in the following directions: extension to Łukasiewicz preferences, consideration of other types of mathematical properties such as max-min and max-max transitivity, and the extension to fuzzy ranking, and links to other approaches for mapping between ordered labels and numerical degrees [16,17].

References

1. Fodor, J.C., Roubens, M.R.: Fuzzy Preference Modelling and Multicriteria Decision Support, vol. 14. Springer, Heidelberg (1994)
2. Herrera-Viedma, E., Herrera, F., Chiclana, F., Luque, M.: Some issues on consistency of fuzzy preference relations. Eur. J. Oper. Res. **154**(1), 98–109 (2004)
3. Hildebrandt, M., Sunder, S.S., Mogoreanu, S., Thon, I., Tresp, V., Runkler, T.: Configuration of industrial automation solutions using multi-relational recommender systems. In: Proceedings of the European Conference on Machine Learning and Principles and Practice of Knowledge Discovery in Databases, Dublin, Ireland, September 2018
4. Hüllermeier, E., Fürnkranz, J.: Comparison of ranking procedures in pairwise preference learning. In: Proceedings of the International Conference on Information Processing and Management of Uncertainty in Knowledge-Based Systems, Perugia, Italy (2004)
5. Kacprzyk, J., Fedrizzi, M., Nurmi, H.: Group decision making and consensus under fuzzy preferences and fuzzy majority. Fuzzy Sets Syst. **49**(1), 21–31 (1992)
6. Ma, J., Fan, Z.-P., Jiang, Y.-P., Mao, J.-Y., Ma, L.: A method for repairing the inconsistency of fuzzy preference relations. Fuzzy Sets Syst. **157**(1), 20–33 (2006)
7. Ricci, F., Rokach, L., Shapira, B.: Recommender Systems Handbook. Springer, New York (2015)
8. Runkler, T.A.: Constructing preference relations from utilities and vice versa. In: Carvalho, J., Lesot, M.J., Kaymak, U., Vieira, S., Bouchon-Meunier, B., Yager, R.R. (eds.) Information Processing and Management of Uncertainty in Knowledge-Based Systems. CCIS, vol. 611, pp. 547–558. Springer, Cham (2016)
9. Runkler, T.A.: Generating preference relation matrices from utility vectors using Łukasiewicz transitivity. In: Kóczy, L.T., Medina-Moreno, J., Ramírez-Poussa, E., Šostak, A. (eds.) Computational Intelligence and Mathematics for Tackling Complex Problems. Studies in Computational Intelligence, vol. 819, pp. 123–130 Springer, Cham (2020)
10. Runkler, T.A.: Mapping utilities to transitive preferences. In: Medina, J., Ojeda-Aciego, M., Verdegay, J.L., Pelta, D.A., Cabrera, I.P., Bouchon-Meunier, B., Yager, R.R. (eds.) Information Processing and Management of Uncertainty in Knowledge-Based Systems. CCIS, vol. 853, pp. 127–139. Springer, Cham (2018)
11. Saaty, T.L.: Analytic hierarchy process. In: Encyclopedia of Operations Research and Management Science, pp. 52–64. Springer, Heidelberg (2013)
12. Tanino, T.: Fuzzy preference orderings in group decision making. Fuzzy Sets Syst. **12**(2), 117–131 (1984)

13. Tanino, T.: Fuzzy preference relations in group decision making. In: Non-Conventional Preference Relations in Decision Making, pp. 54–71. Springer, Heidelberg (1988)
14. Triantaphyllou, E.: Multi-Criteria Decision Making Methods: A Comparative Study. Springer, Boston (2000)
15. Yager, R.R.: Families of OWA operators. Fuzzy Sets Syst. **59**(2), 125–148 (1993)
16. Kosheleva, O., Kreinovich, V., Escobar, M.O., Kato, K.: Towards the most robust way of assigning numerical degrees to ordered labels, with possible applications to dark matter and dark energy. In: Proceedings of the Annual Conference of the North American Fuzzy Information Processing Society, El Paso, Texas (2016)
17. Kosheleva, O., Kreinovich, V., Lorkowski, J., Osegueda, M.: How to transform partial order between degrees into numerical values. In: Proceedings of the IEEE International Conference on Systems, Man, and Cybernetics, Budapest, Hungary, pp. 2489–2494 (2016)

Discrete and Continuous Logistic
p-Fuzzy Models

Daniel Eduardo Sánchez[1(✉)], Estevão Esmi[2], and Laécio Carvalho de Barros[2]

[1] University Austral of Chile, Patagonia Campus, Coyhaique, Chile
`danielsanch@gmail.com`
[2] Institute of Mathematics, Statistics and Scientific Computing,
University of Campinas, Campinas, Brazil
`eelaureano@gmail.com`, `laeciocb@ime.unicamp.br`

Abstract. This manuscript investigates the capacity of the so-called p-fuzzy systems to model both discrete and continuous dynamic systems. Recall that one can apply a p- fuzzy system in order to combine fuzzy rule-based systems (FRBSs) and classical numerical methods to simulate the dynamics of an evolutionary system. Here, we focus on the well-known discrete and continuous Logistic models that can be used to represent several problems of Biomathematics such as dynamic population. We conduct a series of simulations using both continuous and discrete models for several growth rates. We obtain qualitative and quantitative results similar to the analytical solutions, including bifurcations in the discrete case.

1 Introduction

Logistic model, sometimes called Verhulst growth, plays an important role in biomathematics, describing many phenomena such as dynamic population. The discrete and continuous logistic models are given by difference and differential equations, respectively [3, 4].

Many phenomena can be described using differential or difference equations. Differential and difference equation theories can be used to model many phenomena from several area of knowledge [3, 4]. However, they require explicit formulas for functions that describe vector fields or variational behaviors of the dynamics. In general, such a function depends on several hypotheses and/or conditions of the phenomenon under consideration and its formulation by an expert may not be simple, specially if she/he has no previous experience or adequate training on differential or difference equations. Nevertheless, if the expert has an idea how the dynamics works in terms of rules then this expert can simulate the dynamical behaviour of the underlying phenomenon by means of a p-fuzzy system [1].

A p-fuzzy system can be viewed as a dynamic system whose the function that describes the evolution rule is given by a fuzzy rule-based systems (FRBS) [1]. In this work we use Mamdani fuzzy controllers based on fuzzy rules containing opposite semantics, such as the processes of "inhibition" of the population

© Springer Nature Switzerland AG 2019
R. B. Kearfott et al. (Eds.): IFSA 2019/NAFIPS 2019, AISC 1000, pp. 556–566, 2019.
https://doi.org/10.1007/978-3-030-21920-8_49

described by the logistic model. Based on this FRBS, we determine numerical solutions for the discrete and continuous cases of the Verhulst growth (or logistic model) via a p-fuzzy system.

Finally, we provide several examples to illustrate this methodology to simulate the logistic model for different growth rates, that include the interesting phenomena of oscillation and bifurcation in the discrete case.

2 Preliminary

This section presents the mathematical background of the classical discrete and continuous Logistic models and of fuzzy set theory.

2.1 Discrete and Continuous Logistic Models

The discrete Verhulst growth is known as the *logistic difference equation* and is given by [3,4,6]:

$$x_{n+1} = \lambda x_n \left(1 - x_n\right). \tag{1}$$

The variable x_n can be interpreted as the population of a given species in the nth generation. In [6], it was observed that this model presents a complex behaviour with respect to the parameter λ, including bifurcations and chaos.

Equation (1) consists of a nonlinear difference equation with just one parameter λ, where the number n represents the nth iteration. In particular, the Eq. (1) is restrict to the intervals $x_n \in (0,1)$ and $\lambda \in (1,4)$ since otherwise the population become extinct [3,4].

The unique non-trivial steady state of Eq. (1) is given by

$$\overline{x} = 1 - \frac{1}{\lambda}, \tag{2}$$

which is asymptotically stable if $1 < \lambda < 3$ [3,4]. For $\lambda > 3$, the discrete logistic model presents a sequence of bifurcations that lead to a very complicated dynamics [4,6].

In this work, we focus on three special cases [3,4,6]:

(i)   If $\lambda \in (1,2)$ then the steady state in Eq. (2) is globally attracting and the iterations of Eq. (1) converge monotonically to $\overline{x}$ (see the analytic solution in Fig. 3).

(ii) If $\lambda \in (2,3)$ then the steady state in Eq. (2) is still globally attracting and the iterations of Eq. (1) converge to $\overline{x}$ but not monotonically (see analytic solution in Fig. 4).

(iii) If $\lambda \in (3, 1 + \sqrt{6})$ then stable oscillations of period 2 appear to Eq. (1), that is, successive generations alternate between two fixed values of x, say $\overline{\overline{x_1}}$ and $\overline{\overline{x_2}}$, given by the roots of the following quadratic expression

$$\overline{\overline{x_1}}, \overline{\overline{x_2}} = \frac{(\lambda + 1) \pm \sqrt{(\lambda - 3)(\lambda + 1)}}{2\lambda}.$$

In this case, the fixed point $\overline{x}$ is unstable (see analytic solution in Fig. 5).

The continuous Verhulst growth is given by the following ordinary differential equation [3]:

$$\frac{dP}{dt} = \alpha P \left(1 - \frac{P}{K}\right), \tag{3}$$

where α is the intrinsic growth rate and K carrying capacity. The state variable $P = P(t)$ represents the population at the time t.

The analytic solution to Eq. (3) is given by [3]

$$P(t) = \frac{P_0 K}{P_0 + (K - P_0)e^{-\alpha t}}, \tag{4}$$

where $P_0 = P(0)$ is the initial population. In contrast, the continuous logistic model has no oscillations or bifurcations. Moreover, the unique non-trivial steady state $\overline{P} = K$ is stable for any $\alpha > 0$.

2.2 Fuzzy Rules and p-Fuzzy Systems

A *fuzzy subset* A of an universal set Z is characterized by a function $\varphi_A : Z \to [0,1]$ called membership function of A, where $\varphi_A(z)$ represents the membership degree of z in A [1]. For notation convenience, we will use the symbol $A(z)$ instead of $\varphi_A(z)$. The class of fuzzy sets of Z is denoted by $\mathscr{F}(Z)$. Here, we focus on a particular class of fuzzy sets of $\mathbb{R}$, called *fuzzy numbers*.

A trapezoidal fuzzy number A, denoted by a quadruple $(a; b_1; b_2; c)$, with $a, b_1, b_2, c \in \mathbb{R}$ and $a \leq b_1 \leq b_2 \leq c$, consists of a fuzzy number whose membership function is given by [1,7]:

$$A(z) = \begin{cases} \frac{z-a}{b_1-a} , & \text{if } z \in [a, b_1), \\ 1 , & \text{if } z \in [b_1, b_2], \\ \frac{c-z}{c-b_2} , & \text{if } z \in (b_2, c], \\ 0 , & \text{otherwise.} \end{cases}$$

In the case where $b_1 = b_2 = b$, we speak of triangular fuzzy number and it is denoted by the symbol $(a; b; c)$ instead of $(a; b; b; c)$ [1].

Fuzzy Rule-Based Systems (FRBS) is a map composed by 4 components: a fuzzification module, a fuzzy rule base, a fuzzy inference method, and a defuzzification module [1,5,8].

In the fuzzification module, real-valued inputs are translated into fuzzy numbers of their respective universes. Expert knowledge plays an important role to build the membership functions for each fuzzy number associated with the inputs [7]. Here, we use the most basic method called canonical inclusion that associates each real number with its characteristic function.

Here, we consider a fuzzy rule base given by a collection of "r" fuzzy conditional rules of the form "if u is A_j then v is B_j", where $A_j \in \mathscr{F}(U)$ and $B_j \in \mathscr{F}(V)$, for $j = 1, \ldots, r$, are fuzzy sets that represent linguistic terms and are called *antecedents* and *consequent* of jth fuzzy rule, respectively [1]. A fuzzy inference method consists of a mapping that associates fuzzy sets of U to fuzzy sets of V, based on the fuzzy rule base. In this work, we use the Mamdani inference method. Specifically, given an input $u \in \mathbb{R}$, the Mamdani inference produces as output a fuzzy set B given by [1]:

$$B(v) = \max_{j=1,\ldots,r} \min\{\, A_j(u)\,,\, B_j(v)\,\},\ \forall v \in \mathbb{R}. \tag{5}$$

Defuzzification module consist of a process that allows us to represent a fuzzy set by a real value. In this manuscript, we adopt the centroid scheme [1,5,7].

A *partially fuzzy system* or, for short, a p-fuzzy system, is a dynamical system where the direction field or variational behaviour (or function) is given by a FRBS whose corresponding fuzzy rule base is based on some partial knowledge about the phenomena under consideration [1,9].

Here, we focus on autonomous systems, that is, dynamic systems whose variations (or specific variation rates) do not depend explicitly on time. Formally, the fuzzy rule base of a continuous p-fuzzy systems is similar to the one of a discrete system. The difference lies essentially on the formulation of each rule. In the discrete systems, the variations are qualified in absolute terms. On the other hand, in the continuous case, the variation rate must have qualitative properties that are consistent with the concept of derivative. The state variables and their variations are considered linguistic variables for the FRBS. Furthermore, the state variables are associated with their variations by means of fuzzy rules where the state variables are the input and their variations are the outputs.

We consider discrete initial value problems (IVPs) of the form

$$x_{n+1} = f(x_n), \quad x(0) = x_0 \in \mathbb{R}, \tag{6}$$

where f is only partially known. Using these partially information, we estimate f by a fuzzy rule-based system $FRBS_f$. Thus, we study the system (6) by means of the following associated p-fuzzy system:

$$x_{n+1} = FRBS_f(x_n), \quad x(0) = x_0 \in \mathbb{R}, \tag{7}$$

Similarly, we also consider continuous IVPs of the form

$$\frac{dP}{dt} = g(P), \quad P(0) = P_0 \in \mathbb{R}, \tag{8}$$

where the vector field g is only partially known. Again, we can use these partially information to estimate g by means of a fuzzy rule-based system $FRBS_g$. Thus, one can investigate the qualitative and quantitative behaviour of the system (8) from the following associated p-fuzzy system:

$$\frac{dP}{dt} = FRBS_g(P), \quad P(0) = P_0 \in \mathbb{R}, \tag{9}$$

In Equations (6) and (8), the functions f and g are partially known, in the sense that we have only some qualitative and/or quantitative information about f and g, usually provided by an expert and/or extracted from a given dataset, but not explicit formulas for them. Recall that Mamdani fuzzy controller yields a function $f_r^* = FRBS_f$ (or $g_r^* = FRBS_g$) from $\mathbb{R}$ to $\mathbb{R}$, where r denotes the number of rules in the fuzzy rule base. It seems reasonable to assume that the function f_r^* (or g_r^*) approximates f (or g) when the number of data r increases since we expect that the specificity of each rule increases with the number of rules [2].

Note that the IVP (6) represents a difference equation and the IVP (8) a first order differential equation. In order to obtain solutions for the p-fuzzy systems (7) and (9), we consider the qualitative information available to design a fuzzy rule base which represent the properties that characterize the phenomenon [2].

In the discrete case, the solution of the p-fuzzy system (7) is obtained by successive iterations of the following equation:

$$x_{k+1} = x_k + \Delta x_k, \tag{10}$$

where the variation function Δx_k is the output of the fuzzy controller in the kth iteration [1].

In the continuous case, the solution $P(t)$ of the p-fuzzy system (9) is given by a sequence P_k obtained using numerical methods for solving ordinary differential equations (ODEs) such as Euler and Runge–Kutta methods [1]. Here, we use the following formula:

$$P_{k+1} = P_k + h \, \Delta P_k, \tag{11}$$

where h is the step (in time) and ΔP_k is the variation produced by $FRBS_f$ at kth iteration.

3 P-Fuzzy Logistic Models

This section presents the methodology and results of the computational simulations for discrete and continuous Logistic models using p-fuzzy systems.

In particular, for the continuous case, we consider the IVP with *specific* variations, that is

$$\frac{1}{P} \frac{dP}{dt} = g(P), \quad P(0) = P_0 \in \mathbb{R}, \tag{12}$$

Thus, Eq. (11) becomes

$$P_{k+1} = P_k + h \, P_k \, \Delta P_k. \tag{13}$$

3.1 Methodology

Recall that in a p-fuzzy system the vector fields are given by FRBSs (see in Subsect. 2.2). The antecedents and consequents of the corresponding fuzzy rules are linguistic terms associated respectively with the input and output variables. Here we use trapezoidal or triangular fuzzy numbers to represent this linguistic terms [9].

For the Logistic model, we consider as input the variable X which represents the populations x_n and $P(t)$ in the discrete and continuous cases, respectively. The variable X can assume 6 linguistic terms expressed as *"low"*, *"average low"*, *"average"*, *"average high"*, *"high"* and *"very high"*. These linguistic terms are modelled respectively by the fuzzy numbers $A_1, \ldots, A_6$ depicted in Fig. 1.

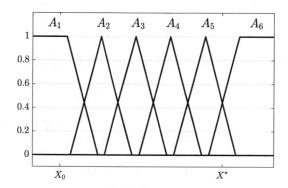

Fig. 1. Antecedents of population (X) for the Logistic p-fuzzy model, where X_0 represents the initial population and X^* the carrying capacity.

Also, we consider as output the variable Y which represents the variational and specific growths for the discrete and continuous cases, respectively. The variable Y can assume the following linguistic terms *"low negative"*, *"low positive"*, *"average positive"*, and *"high positive"* that are modelled respectively by the fuzzy numbers $B_1, \ldots, B_4$. Figure 2 illustrates these linguist terms.

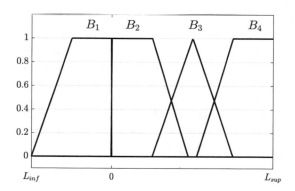

Fig. 2. Consequents of the variations or specific variations (Y) for the Logistic p-fuzzy model, where $[L_{inf}, L_{sup}]$ represents the range of possible values of Y.

Membership functions of the fuzzy terms can be adjusted by experts and/or from an available dataset about the phenomenon [7,9]. In particular for the

Logistic model, we elaborate a FRBS to model population variation based on its density, with semantic opposition [1]. These assumptions can be translated into a set of fuzzy rules that play the role of a direction field. Thus, we propose a rule base with 6 fuzzy rules given by

r_1 :   If X is "low" (A_1) then the variation is "$low\ positive$" (B_2).
r_2 :   If X is "$average\ low$" (A_2) then the variation is "$average\ positive$" (B_3).
r_3 :   If X is "$average$" (A_3) then the variation is "$high\ positive$" (B_4).
r_4 :   If X is "$average\ high$" (A_4) then the variation is "$average\ positive$" (B_3).
r_5 :   If X is "$high$" (A_5) then the variation is "$low\ positive$" (B_2).
r_6 :   If X is "$very\ high$" (A_6) then the variation is "$low\ negative$" (B_1).

3.2   Results

This section presents the solutions of p-fuzzy systems which simulate the dynamic behaviour of the variable X, that is, the population in the Logistic model. These solutions are obtained by Eqs. (10) and (13), respectively. Moreover, we provide a visual comparison between the obtained solutions and the analytical solutions. Recall that the analytical solutions for the discrete and continuous cases are given respectively by Eqs. (1) and (4).

For discrete case, we adjust the antecedents using the theory of the equilibrium and stability of one-dimensional discrete p-fuzzy systems [10]. Thus, the fuzzy terms are determined using the initial population X_0 and the carrying capacity X^*. We use the steady state $\bar{x}$ given in Eq. (2) to simulate the special cases (i) and (ii). For simulating the special case (iii), we use X^* as been the mean between $\overline{\overline{x_1}}$ and $\overline{\overline{x_2}}$. In all continuous cases, we consider $X^* = K$. With respect to the consequents given as in Fig. 2, we define $L_{sup} = 2|L_{inf}|$ for the discrete case and $L_{sup} = 3|L_{inf}|$ for the continuous case.

3.2.1   Discrete p-Fuzzy

Figures 3, 4, and 5 correspond to the solutions of the discrete p-fuzzy system given by Equation (10) for $\lambda = 1.9, 2.8, 3.05$, respectively. In the first case, one can observe in Fig. 3 a monotonic and asymptotic convergence of the solution to $0.474 \approx \bar{x} = 1 - \frac{1}{1.9}$. In the second case, Fig. 4 reveals that the solution converges to $\bar{x} = 1 - \frac{1}{2.8} \approx 0.643$ and this convergence is oscillatory and non-monotonic. As we can observe in Fig. 5, the third case presents a bifurcation phenomena with oscillations of 2-period between $0.5902 \approx \overline{\overline{x_1}}$ and $0.7377 \approx \overline{\overline{x_2}}$. For a visual comparison, the corresponding discrete analytical solutions are also depicted in these figures.

3.2.2   Continuous p-Fuzzy

Figures 6, 7, and 8 present the solutions for the Logistic p-fuzzy model given as in Eq. (13) with initial population $P_0 = 15$ and $h = 0.05$ and $\alpha = 1.9, 2.8, 3.05$, respectively. All these cases present monotonic stability at the carrying capacity $K = 200$, in agreement with the differential equation theory. For a visual comparison, we also include the corresponding analytical solutions in these figures.

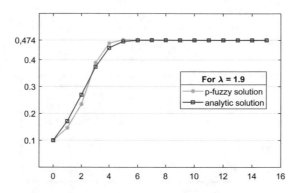

Fig. 3. Discrete p-fuzzy solution Eq. (10) and analytical logistic solution Eq. (1), for $\lambda = 1.9$ and with $\bar{x} \approx 0.474$.

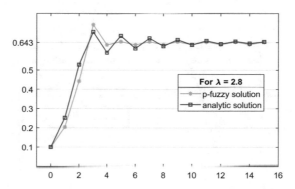

Fig. 4. Discrete p-fuzzy solution Eq. (10) and analytical logistic solution Eq. (1), for $\lambda = 2.8$ and with $\bar{x} \approx 0.643$.

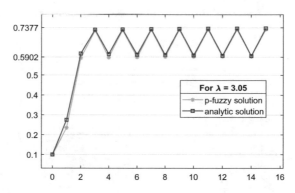

Fig. 5. Discrete p-fuzzy solution Eq. (10) and analytical logistic solution Eq. (1), for $\lambda = 3.05$ and with $\overline{x_1} \approx 0.5902$ and $\overline{x_1} \approx 0.7377$.

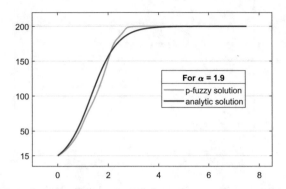

Fig. 6. Continuous p-fuzzy solution Eq. (13) and analytical logistic solution Eq. (4) for $\alpha = 1.9$.

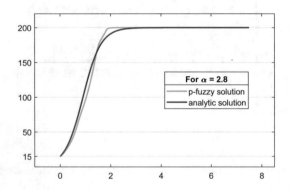

Fig. 7. Continuous p-fuzzy solution Eq. (13) and analytical logistic solution Eq. (4) for $\alpha = 2.8$.

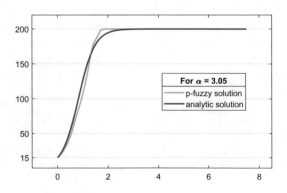

Fig. 8. Continuous p-fuzzy solution Eq. (13) and analytical logistic solution Eq. (4) for $\alpha = 3.05$.

4   Final Remarks

This paper illustrates, by means of examples, that the p-fuzzy systems are indeed useful mathematical tool to modelling evolutionary systems whose dynamics are partially known. Here, we focus on the continuous and discrete Logistic models. Although the continuous logistic model is well-behaved, depending on choice of the parameter λ in Eq. 1, the discrete logistic model may exhibit a complex behaviour including bifurcations and chaos.

As we observed in Sect. 3.2, the discrete and continuous solutions obtained via p-fuzzy systems are quantitative and qualitative similar to the ones given by analytical solutions.

In the discrete case, we provide 3 different situations considering different values for the parameter λ. The corresponding solutions of the p-fuzzy systems present similar behaviors with respect to monotony, oscillations and stability (see Figs. 3 and 4), and bifurcation with oscillations of 2-period (see Fig. 5).

In the continuous case, we also observe a similar behavior between the solutions of the p-fuzzy systems and the corresponding analytical solutions given as in Eq. (4) (see Figs. 6, 7, and 8). In our simulations, for any intrinsic growth rate α the logistic model is stable at the carrying capacity and does not present oscillations or bifurcations.

Finally, we recall that our proposal can be used by any specialist and does not require previous experience with differential equations. Moreover, the use of the p-fuzzy system is based on the universal approximation property of the FRBSs, which indicates that the p-fuzzy systems can be used to estimate the dynamics of various phenomena.

Acknowledgments. This research was partially supported by FAPESP under grants no. 2018/10946-2, and 2016/26040-7, and CNPq under grant no. 306546/2017-5.

References

1. Barros, L.C., Bassanezi, R.C., Lodwick, W.A.: A First Course in Fuzzy Logic, Fuzzy Dynamical Systems, and Biomathematics. Springer, Heidelberg (2017)
2. Dias, M.R., Barros, L.C.: Differential equations based on fuzzy rules. In: Proceedings of IFSA/EUSFLAT Conference, pp. 240–246 (2009)
3. Edelstein-Keshet, L.: Mathematical Models in Biology. Classics in Applied Mathematics. SIAM (2005)
4. Hale, J.K., Koçak, H.: Dynamics and Bifurcations. Springer, New York (1996)
5. Jafelice, R.M., Barros, L.C., Bassanezi, R.C., Gomide, F.: Fuzzy modeling in symptomatic HIV virus infected population. Bull. Math. Biol. **66**(6), 1597–1620 (2004)
6. May, R.M.: Simple mathematical models with very complicated dynamics. Nature **261**, 459–467 (1976)
7. Pedrycz, W., Gomide, F.: Fuzzy Systems Engineering Toward Human-centric Computing. IEEE Press/Wiley (2007)
8. Peixoto, M.S., Barros, L.C., Bassanezi, R.C.: Predator-prey fuzzy model. Ecol. Model. **214**, 39–44 (2008)

9. Sánchez, D., Barros, L.C., Esmi, E., Miebach, A.D.: Goodwin model via p-fuzzy system. In: Data Science and Knowledge Engineering for Sensing Decision Support. World Scientific Proceedings Series, vol. 11, pp. 977–984 (2018)
10. Silva, J.D.M., Leite, J., Bassanezi, R.C., Cecconello, M.S.: Stationary points-I: one-dimensional p-fuzzy dynamical systems. J. Appl. Math. **2013**, 1–11 (2013)

Measure of Interactivity on Fuzzy Process Autocorrelated: Malthusian Model

Francielle Santo Pedro[(✉)], Estevão Esmi, and Laécio Carvalho de Barros

Department of Applied Mathematics, University of Campinas, Campinas, Brazil
fran.stopedro@gmail.com, eelaureano@ime.unicamp.br, laeciocb@ime.unicamp.br

Abstract. In this manuscript, we study the measure of interactivity in fuzzy process, more specifically, in linearly correlated fuzzy processes. This measure is a covariance average of α-levels of two fuzzy numbers. The analysis is illustrated using the Malthusian model, for which we compute the interactivity of the solution at instants $t + h$ and t. Also, we observe its behavior when t tends to infinity and h tends to zero.

1 Introduction

Generally, dynamic evolutionary phenomena such as population dynamics and epidemiology, involve correlated data [1,12,14]. So, it is essential to study evolution of interactivity in these processes. In [3], we studied autocorrelated fuzzy processes, more particularly linearly correlated fuzzy processes, that is, processes which states are linearly locally correlated. These processes are based on ideas similar to stochastic processes with memories (time series). In this paper we study concepts of mean value, measure of interactivity and interactivity function in the linearly correlated fuzzy processes, these concerns play a fundamental role in both possibility and probability theory [8].

1.1 Preliminary

We will denote the space of the real numbers by $\mathbb{R}$, the space of strictly positive real numbers by $\mathbb{R}^+$ and the space of fuzzy number by $\mathbb{R}_{\mathscr{F}}$.

A *fuzzy* number A is a fuzzy subset in $\mathbb{R}$ that is normal, fuzzy convex (i.e. its membership function is quasiconcave), and its membership function is continuous with bounded support. The membership function of the fuzzy number A is denoted by $A(\cdot)$ and fuzzy numbers can be considered as possibility distributions [9]. The well-known α-*levels sets* of the fuzzy number A are given by

$$[A]_\alpha = \{x \in \mathbb{R} : A(x) \geq \alpha\}, \text{ for } \alpha > 0,$$

and

$$[A]_0 = \mathrm{cl}\{x \in \mathbb{R} : A(x) > 0\}, \text{ for } \alpha = 0.$$

R. B. Kearfott et al. (Eds.): IFSA 2019/NAFIPS 2019, AISC 1000, pp. 567–577, 2019.
https://doi.org/10.1007/978-3-030-21920-8_50

From definition of fuzzy number we have that its α-levels are closed and bounded intervals [2,5]. We will use

$$[A]_\alpha = [a_\alpha^-, a_\alpha^+],$$

where a_α^-, a_α^+ are the left and right end points.

A n-dimensional joint possibility distribution J is a normal fuzzy subset of $\mathbb{R}^n$ such that the support of its membership function is bounded. We denote by $\mathscr{F}(\mathbb{R}^n)$ the family of joint possibility distribution of $\mathbb{R}^n$.

Let $A_1, A_2, ..., A_n$ be fuzzy numbers and $J \in \mathscr{F}(\mathbb{R}^n)$, then [6] defines that J is a *joint possibility distribution* of $A_1, A_2, ..., A_n$ if

$$\max_{x_j \in \mathbb{R}, j \neq i} J(x_1, ..., x_n) = A_i(x_i). \tag{1}$$

A_i is called the i-th *marginal possibility distribution* of J and we denote $A_i = \pi_i(J)$, where π_i is the *projection operator* in $\mathbb{R}^n$ onto ith axis, $i = 1, \ldots n$ [9].

The interactivity between fuzzy numbers is determined from a joint possibility distribution [6].

If J is a possibility distribution of fuzzy numbers $A_1, A_2, ..., A_n$, then the following relationship holds

$$J(x_1, ..., x_n) \leq \min\{A_1(x_1), ..., A_n(x_n)\}.$$

In addition,

$$[J]_\alpha \subseteq [A_1]_\alpha \times ... \times [A_n]_\alpha, \ \forall \alpha \in [0,1].$$

We say that the fuzzy numbers $A_1, A_2, ..., A_n$ are *non interactive* when

$$J(x_1, ..., x_n) = \min\{A_1(x_1), ..., A_n(x_n)\},$$

or equivalently,

$$[J]_\alpha = [A_1]_\alpha \times ... \times [A_n]_\alpha, \ \forall \alpha \in [0,1].$$

Otherwise they are *interactive*.

The *sup-J extension principle* of f at $(A_1, \ldots, A_n)$ [6] is defined as $f_J(A_1, ..., A_n)(y) =$

$$\begin{cases} \sup_{y=f(x_1,...,x_n)} J(x_1, ..., x_n) \text{ if } y \in \text{Im}(f) \\ 0 \qquad\qquad\qquad\qquad\quad \text{ if } y \notin \text{Im}(f) \end{cases}. \tag{2}$$

where $\text{Im}(f) = \{f(x) : x \in \mathbb{R}^n\}$.

Theorem 1 [6,7]. *Let $A_1, ..., A_n$ be fuzzy numbers, J their joint possibility distribution and $f : \mathbb{R}^n \to \mathbb{R}$ a continuous function. Then,*

$$[f_J(A_1, ..., A_n)]_\alpha = f([J]_\alpha),$$

for all $\alpha \in [0,1]$.

Carlsson, Fúller and Majlender [6] introduced the concept of completely correlated fuzzy numbers using the concept of possibility distribution.

Two fuzzy numbers A_1 and A_2 are said to be *completely correlated* if there exist real numbers $q \neq 0$ and r such that their joint possibility distribution is given by [6]:

$$
\begin{aligned}
C(x_1, x_2) &= A_1(x_1)\mathscr{X}_{\{qx_1+r=x_2\}}(x_1, x_2) \\
&= A_2(x_2)\mathscr{X}_{\{qx_1+r=x_2\}}(x_1, x_2)
\end{aligned}
\tag{3}
$$

where $\mathscr{X}_{\{qx_1+r=x_2\}}(x_1, x_2)$ is the characteristic function of the line

$$
\{(x_1, x_2) \in \mathbb{R}^2 : qx_1 + r = x_2\}.
$$

By Theorem 1, we have, that for all $\alpha \in [0, 1]$, the four operations of completely correlated fuzzy numbers are given by [13, 15].

- $[B +_C A]_\alpha = (q+1)[A]_\alpha + r$;
- $[B -_C A]_\alpha = (q-1)[A]_\alpha + r$;
- $[B \cdot_C A]_\alpha = \{qx_1^2 + rx_1 \in \mathbb{R} | x_1 \in [A]_\alpha\}$;
- $[B \div_C A]_\alpha = \{qx_1 + \frac{r}{x_1} \in \mathbb{R} | x_1 \in [A]_\alpha\}$.

Barros and Pedro [3], introduced the concept of linearly correlated fuzzy numbers, in order to avoid the knowledge of the joint distribution. Two fuzzy numbers A and B are called *linearly correlated* if there exist $q, r \in \mathbb{R}$ such that

$$
[B]_\alpha = q[A]_\alpha + r,
$$

for all $\alpha \in [0, 1]$.

When $q > 0$ ($q < 0$), we say that A and B are linearly positively (negatively) correlated.

1.2   Possibilistic Measures of Central Tendency and Interactivity

Here, the concepts of central value, expected value, measure of covariance, and measure of possibilistic interactivity are similar to the ones used in statistic theory. To this end, let us recall some results from [9, 10].

The definition of these measures in the possibility theory are based on probabilistic measures for uniformly distributed random variables on the α-levels of the fuzzy sets and joint possibility distributions. The expected value and measure of possibilistic interactivity are defined in terms of a weighting function ω.

According to [8], the *weighting function* $\omega : [0, 1] \longrightarrow \mathbb{R}$ is a non-negative, monotone increasing and normalized over the unit interval function, that is,

$$
\int_0^1 \omega(s)ds = 1.
$$

Let J be a joint possibility distribution in $\mathbb{R}^n$, $p : \mathbb{R}^n \longrightarrow \mathbb{R}$ be an integrable function, and $\alpha \in [0, 1]$. The *central value* of p on $[J]_\alpha$ is given by

$$
\mathscr{C}_{[J]_\alpha}(p) = \frac{1}{\int_{[J]_\alpha} dx} \int_{[J]_\alpha} p(x)dx
\tag{4}
$$

and the *expected value* of p on J, with respect to ω, is given by

$$E_\omega(p; J) = \int_0^1 \mathscr{C}_{[J]_\alpha}(p)\omega(\alpha)d\alpha. \tag{5}$$

Notice that $E_\omega(p; J)$ computes the ω-weighted average of the central values of the function p on the α-level sets of J and, for any possibility distribution J, we have that $E_\omega(\cdot; J)$ is a linear operator.

Additionally, according to [9], if $p : \mathbb{R} \longrightarrow \mathbb{R}$ is the identity function, then, for any fuzzy number A, we have

$$\mathscr{C}_{[A]_\alpha}(id) = \frac{1}{\int_{[A]_\alpha} dx} \int_{[A]_\alpha} x dx = \frac{a_\alpha^- + a_\alpha^+}{2} \tag{6}$$

and the expected value of id on A with respect to ω is given by

$$E_\omega(id; A) = \int_0^1 \mathscr{C}_{[A]_\alpha}(id)\omega(\alpha)d\alpha. \tag{7}$$

Let J be a joint possibility distribution in $\mathbb{R}^2$ with marginal possibility distributions A and B and let $\alpha \in [0, 1]$. Then the *measure of interactivity between the α-level sets of A and B* (with respect to $[J]_\alpha$) is given by

$$\begin{aligned}
I([A]_\alpha, [B]_\alpha) &= \frac{1}{\int_{[J]_\alpha} dxdy} \int_{[J]_\alpha} xy \, dxdy \\
&- \left(\frac{1}{\int_{[A]_\alpha} dx} \int_{[A]_\alpha} x \, dx\right)\left(\frac{1}{\int_{[B]_\alpha} dy} \int_{[B]_\alpha} y \, dy\right).
\end{aligned} \tag{8}$$

It is worth noticing that, when one considers the uniform distribution, the Eq. (8) reflects the covariance between the characteristic functions of $[A]_\alpha$ and $[B]_\alpha$. The *measure of interactivity between A and B* (with respect to their joint possibility distribution J and weighting function ω) is given by

$$I_\omega(A, B) = \int_0^1 I([A]_\alpha, [B]_\alpha)\omega(\alpha)d\alpha. \tag{9}$$

In other words, Eq. (9) corresponds to ω-weighted average interactivity (or correlation) between the α-levels of A and B. Pedro et al. [13] showed that when $B = \mathscr{F}(A)$, that is, $[B]_\alpha = \mathscr{F}([A]_\alpha)$, for all $\alpha \in [0, 1]$, where $\mathscr{F}$ is a real function, then A and B are $\mathscr{F}$-*correlated*. In the case of linearly correlated fuzzy numbers, $\mathscr{F}$ is a linear function.

If $A, B \in \mathbb{R}_{\mathscr{F}} \setminus \mathbb{R}$ are $\mathscr{F}$-correlated, where $\mathscr{F}$ is a monotone continuous function and $B = \mathscr{F}(A)$, then the measure of possibilistic interactivity between A and B, becomes $I_\omega(A, B) =$

$$\int_0^1 \left(\frac{1}{a_\alpha^+ - a_\alpha^-} \int_{a_\alpha^-}^{a_\alpha^+} x\mathscr{F}(x)dx - \frac{(a_\alpha^+ + a_\alpha^-)(b_\alpha^+ + b_\alpha^-)}{4}\right)\omega(\alpha)d\alpha. \tag{10}$$

If $A, B \in \mathbb{R}_{\mathscr{F}} \setminus \mathbb{R}$ are linearly correlated and the joint possibility distribution is given by (3), then, from operator $\cdot_C$ and (10), the measure of interactivity between A and B with respect to their joint possibility distribution C is given by

$$
\begin{aligned}
I_\omega(A, B) &= \int_0^1 I([A]_\alpha, [B]_\alpha)\omega(\alpha)d\alpha \\
&= \pm\frac{1}{12} \int_0^1 (a_\alpha^+ - a_\alpha^-)(b_\alpha^+ - b_\alpha^-)w(\alpha)d\alpha,
\end{aligned}
\tag{11}
$$

where the sign is positive if A and B are positive linearly correlated and negative if A and B are negative linearly correlated [9,13].

Since A and B are linearly correlated fuzzy numbers, there exists $q > 0$ (or $q < 0$), such that $B = qA + r$ and

$$
b_\alpha^+ - b_\alpha^- = q(a_\alpha^+ - a_\alpha^-), \quad \forall \alpha \in [0, 1].
$$

Thereby,

$$
I_\omega(A, B) = \pm\frac{q}{12} \int_0^1 [a_\alpha^+ - a_\alpha^-]^2 \omega(\alpha)d\alpha.
\tag{12}
$$

1.3   Linearly Correlated Fuzzy Process and Interactive Derivative

A *fuzzy process* $F : [a, b] \longrightarrow \mathbb{R}_{\mathscr{F}}$ is a fuzzy-valued function which associates every t with a fuzzy number, such that the level sets of $F(t)$ are non-empty, compact and convex subsets of $\mathbb{R}$.

The process F is called *autocorrelated fuzzy process* when, for every h sufficiently small, there is a joint possibility distribution that relates $F(t + h)$ with $F(t)$ for all $t, t + h \in [a, b]$.

In particular, if there exist $q(h)$ and $r(h)$, for $h > 0$ sufficiently small, such that, in levels, we have

$$
[F(t + h)]_\alpha = q(h)[F(t)]_\alpha + r(h),
\tag{13}
$$

then we say that F is a *locally linearly correlated fuzzy process*.

The Eq. (13) means that the future value $F(t+h)$ is linearly correlated with the present value $F(t)$, for every h with absolute value sufficiently small.

The metric used is the *Pompieu-Hausdorff distance* $d_\infty : \mathbb{R} \times \mathbb{R} \to \mathbb{R}_+ \cup \{0\}$, and it is defined [2] by

$$
d_\infty(A, B) = \sup_{0 \leq \alpha \leq 1} \max\{|a_\alpha^- - b_\alpha^-|, |a_\alpha^+ - b_\alpha^+|\},
$$

where $A, B \in \mathbb{R}_{\mathscr{F}}$, $[A]_\alpha = [a_\alpha^-, a_\alpha^+]$ and $[B]_\alpha = [b_\alpha^-, b_\alpha^+]$.

Let $F : [a, b] \to \mathbb{R}_{\mathscr{F}}$ be an linearly correlated fuzzy process. According to [3], F is called *L-differentiable* at t_0, if there exists a fuzzy number $F_L'(t_0)$ such that the limit

$$
\lim_{h \to 0} \frac{F(t_0 + h) -_L F(t_0)}{h}
$$

exists and is equal to $F'_L(t_0)$ using the metric d_∞. In addition, $F'_L(t_0)$ is called linearly correlated fuzzy derivative of F at t_0. At the endpoints of $[a, b]$, we consider only one-sided derivative.

If $F(t) = [f_\alpha^-(t), f_\alpha^+(t)]$, according to [3,13], we have

$$[F(t)]_\alpha = \begin{cases} [(f_\alpha^-)'(t), (f_\alpha^+)'(t)] & \text{if } q(h) \geq 1 \\ [(f_\alpha^+)'(t), (f_\alpha^-)'(t)] & \text{if } 0 < q(h) \leq 1 \end{cases}. \tag{14}$$

2   Interactive Malthusian Model

2.1   Interactivity Functions

According to [16], the autocovariance function of a stochastic process provides very important information about the structure of the process in the time. Recall that $\text{Cov}(X, Y)$ is an indication of how much information random variable X provides about random variable Y. When the magnitude of the covariance is high, an observation of X provides an accurate indication of the value of Y. If the two random variables are observations of $X(t)$ taken at two different times, t_1 seconds and $t_2 = t_1 + h$ seconds, the covariance indicates how much the process is likely to change in the h seconds elapsed between t_1 and t_2. A high covariance indicates that the sample function is unlikely to change much in the h-second interval. A covariance near zero suggests a rapid change. This information is conveyed by the autocovariance function. The *autocovariance function* of the stochastic process $X(t)$ is $C_X(t, h) = \text{Cov}(X(t), X(t + h))$ [16].

In an analogous way, we will study the evolution of interactivity autocorrelated fuzzy process which is the solution $X(t)$ of FDE with interactivity. The *interactivity function* is given

$$I_\omega(X(t), X(t + h)) = \int_0^1 I([X(t)]_\alpha, [X(t + h)]_\alpha)\omega(\alpha)d\alpha,$$

From (11), for a linearly fuzzy process $X(t)$, we have

$$I_\omega(X(t), X(t + h)) = \pm \frac{q}{12} \int_0^1 [x_\alpha^+(t) - x_\alpha^-(t)]^2 \omega(\alpha)d\alpha, \tag{15}$$

where $[X(t)]_\alpha = [x_\alpha^-(t), x_\alpha^+(t)]$.

We will focus on the Malthusian growth model.

2.2   Malthusian Model

- Malthusian decay (Fig. 1)
  Consider the fuzzy initial value problem FIVP given by

$$\begin{cases} X'(t) = -\lambda X \\ X(0) = X_0 \end{cases}, \tag{16}$$

where $\lambda > 0$ and X_0 is a fuzzy number.

In [3] we can see that FIVP (16) can be written in levels by:

$$\begin{cases} [(x_\alpha^+)'(t), (x_\alpha^-)'(t)] = [-\lambda x_\alpha^+(t), -\lambda x_\alpha^-(t)] \text{ if } 0 < q(h) < 1 \\ [(x_\alpha^-)'(t), (x_\alpha^+)'(t)] = [-\lambda x_\alpha^+(t), -\lambda x_\alpha^-(t)] \text{ if } q(h) \geq 1 \end{cases},$$

and the solution is given, in levels, by:

$$[X(t)]_\alpha = [x_{0\alpha}^-, x_{0\alpha}^-]e^{-\lambda t} \text{ if } 0 < q(h) < 1. \tag{17}$$

$$\begin{cases} x_\alpha^-(t) = c_\alpha^- e^{\lambda t} + c_\alpha^+ e^{-\lambda t} \\ x_\alpha^+(t) = -c_\alpha^- e^{\lambda t} + c_\alpha^+ e^{-\lambda t} \end{cases} \text{ if } q(h) \geq 1 \tag{18}$$

where $[X_0]_\alpha = [x_{0\alpha}^-, x_{0\alpha}^+]$ and

$$c_\alpha^- = \frac{x_{0\alpha}^- - x_{0\alpha}^+}{2} \text{ and } c_\alpha^+ = \frac{x_{0\alpha}^- + x_{0\alpha}^+}{2}.$$

Therefore, by Eq. (12), we have:

– for contrative process $(0 < q(h) < 1)$

$$I_w(X(t), X(t+h)) = \frac{e^{-\lambda(2t+h)}}{12} \int_0^1 [x_{0\alpha}^+ - x_{0\alpha}^-]^2 w(\alpha) d\alpha; \tag{19}$$

– for expansive process $(q(h) > 1)$

$$I_w(X(t), X(t+h)) = \frac{e^{\lambda(2t+h)}}{12} \int_0^1 [x_{0\alpha}^+ - x_{0\alpha}^-]^2 w(\alpha) d\alpha. \tag{20}$$

Next we consider $X_0 = (u - \delta; u; u + \delta)$ the triangular fuzzy number such that $[X_0]_\alpha = [u + \delta(\alpha - 1), u + \delta(1 - \alpha)]$ and compute interactive function $I_w(X(t+h), X(t))$ with $\omega(\alpha)$ given by beta distribution.

The weighting function is given by the *beta function*

$$w(\alpha) = \beta(\alpha) = \frac{\alpha^{a-1}(1 - \alpha)^{b-1}}{B(a,b)}, \quad \alpha \subset [0,1],$$

where $a \geq 1$, $0 < b \leq 1$ and $B(a,b) = \int_0^1 \alpha^{a-1}(1 - \alpha)^{b-1} d\alpha$. According to [4,11], this weighting function is suitable for variables measure on interval $(0,1)$, for example, rates and proportions. So we have

– for contractive process $(0 < q(h) < 1)$

$$I_w(X(t), X(t+h)) = \frac{\delta^2 B(a, b+2)}{3 B(a,b)} e^{-\lambda(2t+h)}; \tag{21}$$

– for expansive process $(q(h) \geq 1)$

$$I_w(X(t), X(t+h)) = \frac{\delta^2 B(a, b+2)}{3 B(a,b)} e^{\lambda(2t+h)}. \tag{22}$$

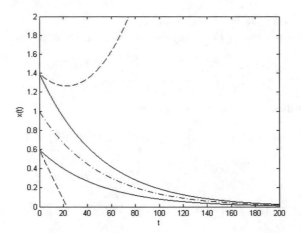

Fig. 1. Solutions of the Malthusian model (16). The 0-level (continuous curve) of the contractive solution, the 0-level (dashed curve) of the expansive solution and the 1-level (dash-point curve) of both. We consider $X_0 = (0.6; 1; 1.4)$ and $\lambda = 0.2$.

Notice that β distribution with $a = 1 = b$ is a uniform distribution on the interval $[0, 1]$ and, in this case, we have, respectively, $I_w(X(t), X(t + h)) = 0.001\overline{7}e^{-0.2(2t+h)}$ $(0 < q(h) < 1)$ and $I_w(X(t), X(t + h)) = 0.001\overline{7}e^{0.2(2t+h)}$ $(q(h) \geq 1)$.

- Malthusian growth (Fig. 2)

  If we consider the fuzzy initial value problem given by

$$\begin{cases} X'(t) = \lambda X \\ X(0) = X_0 \end{cases}, \tag{23}$$

where $\lambda > 0$ and X_0 is a fuzzy number.

The solution is given, in levels, by:

$$[X(t)]_\alpha = [x_{0\alpha}^-, x_{0\alpha}^+]e^{\lambda t} \quad \text{if} \quad q(h) \geq 1. \tag{24}$$

$$\begin{cases} x_\alpha^-(t) = c_\alpha^- e^{-\lambda t} + c_\alpha^+ e^{\lambda t} \\ x_\alpha^+(t) = -c_\alpha^- e^{-\lambda t} + c_\alpha^+ e^{\lambda t} \end{cases} \quad \text{if} \quad 0 < q(h) < 1 \tag{25}$$

where $[X_0]_\alpha = [x_{0\alpha}^-, x_{0\alpha}^+]$,

$$c_\alpha^- = \frac{x_{0\alpha}^- - x_{0\alpha}^+}{2} \quad \text{and} \quad c_\alpha^+ = \frac{x_{0\alpha}^- + x_{0\alpha}^+}{2}.$$

Then, by Eq. (12) with $[X_0]_\alpha = [u + \delta(\alpha - 1), u + \delta(1 - \alpha)]$, we have:
- for contractive process $(0 < q(h) < 1)$,

$$I_w(X(t), X(t + h)) = \frac{\delta^2 B(a, b + 2)}{3 B(a, b)} e^{-\lambda(2t+h)}; \tag{26}$$

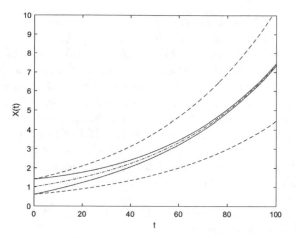

Fig. 2. Solutions of the Malthusian model (23). The 0-level (continuous curve) of the contractive solution, the 0-level (dashed curve) of the expansive solution and the 1-level (dash-point curve) of both. We consider $X_0 = (0.6; 1; 1.4)$ and $\lambda = 0.2$.

– for expansive process $(q(h) > 1)$

$$I_w(X(t), X(t + h)) = \frac{\delta^2 B(a, b + 2)}{3\, B(a, b)} e^{\lambda(2t+h)}. \tag{27}$$

So, we obtain the same interactivity function for systems (16) and (23). This result shows us that when the parameter q from L-derivative is between 0 and 1, the interactivity measure decreases and when it is bigger than 1, the interactivity measure increases over time no matter what is the signal of λ. Thus, what influences in the interactivity measure is the parameter q in the L-derivative and not the signal of parameter λ. Following, we have the final comments.

3   Conclusion

In this paper, we compute the function of interactivity between $X(t + h)$ and $X(t)$, where X is a linearly correlated fuzzy process and the solution of FIVP (16) or (23). Moreover, from t to $t + h$, the interactivity decreases in the proportion $e^{-\lambda}$, when $0 < q(h) < 1$, and it increases in the proportion e^{λ}, when $q(h) \geq 1$. By computing the function of interactivity, we conclude, by (19) and (20), that for any weighting function w, when $t \to \infty$, we have

- for $0 < q(h) < 1$: $I_w(X(t), X(t + h)) \to 0$;
- for $q(h) > 1$: $I_w(X(t), X(t + h)) \to \infty$.

In this way, we also conclude that the parameter q in L - derivative is the one that determines if the fuzziness of the dynamic system increases or decreases over time (fuzziness in the sense of solution diameter) and not the signal of the parameter λ in both (16) and (23).

Moreover, by [13], we obtain the following relation for linearly fuzzy process, as is the case of the Malthusian model, $\text{diam}([X'(t)]_\alpha) = |q'(0)|\text{diam}([X(t)]_\alpha)$, where diam is the diameter of the α-sets. Unfortunately, for the general case in which the differential equation is not linear, we do not have a formula like this last one. However, Pedro et al. [13] suggest a formula that allows us to relate the diameter of $X'(t)$ with the diameter of the fuzzy solution, even unknown. Thus, we are able to estimate the fuzziness of the solution from the fuzziness of the derivative, since this is given by the field of the differential equation in question.

Finally, in Malthus model, if one chooses gH-derivative [5], one cannot obtain the degree of interactivity of process $X(t)$, since the joint possibility distribution involved in fuzzy process is not known.

Acknowledgements. The authors would like to thank CAPES, FAPESP 2016/ 26040 − 7 and CNPq 306546/2017 − 5.

References

1. Barlow, R.: Population growth and economic growth: some more correlations. Popul. Dev. Rev. **20**, 153–165 (1994)
2. Barros, L.C., Bassanezi, R.C., Lodwick, W.A.: A First Course in Fuzzy Logic, Fuzzy Dynamical Systems, and Biomathematics. Studies in Fuzziness and Soft Computing, vol. 347, 1st edn. Springer, Heidelberg (2017)
3. Barros, L.C., Santo Pedro, F.: Fuzzy differential equations with interactive derivative. Fuzzy Sets Syst. **309**, 64–80 (2017)
4. Barros, O.A.: Estimação dos parâmetros da distribuição beta bivariada: aplicações em severidade de doenças em plantas. Master's thesis, Universidade de São Paulo (2015). (in Portuguese)
5. Bede, B.: Mathematics of Fuzzy Sets and Fuzzy Logic, vol. 295. Springer, Heidelberg (2013)
6. Carlsson, C., Fullèr, R., et al.: Additions of completely correlated fuzzy numbers. In: Proceedings of 2004 IEEE International Conference on Fuzzy Systems, vol. 1, pp. 535–539. IEEE (2004)
7. Esmi, E., Sussner, P., Ignácio, G.B.D., Barros, L.C.: A parametrized sum of fuzzy numbers with applications to fuzzy initial value problems. Fuzzy Sets Syst. **331**, 85–104 (2017)
8. Fullèr, R., Majlender, P.: On weighted possibilistic mean and variance of fuzzy numbers. Fuzzy Sets Syst. **136**(3), 363–374 (2003)
9. Fullèr, R., Majlender, P.: On interactive fuzzy numbers. Fuzzy Sets Syst. **143**(3), 355–369 (2004)
10. Fullèr, R., Majlender, P.: Correction to: "on interactive fuzzy numbers" [Fuzzy Sets and Systems 143 (2004) 355–369]. Fuzzy Sets Syst. **152**(1), 159 (2005)
11. Gupta, A.K., Nadarajah, S.: Handbook of Beta Distribution and Its Applications. CRC Press, Bosa Roca (2004)
12. Kelley, A.C., Schmidt, R.M.: Aggregate population and economic growth correlations: the role of the components of demographic change. Demography **32**(4), 543–555 (1995)
13. Pedro, F.S., de Barros, L.C., Esmi, E.: Population growth model via interactive fuzzy differential equation. Inf. Sci. **481**, 160–173 (2019). http://www.sciencedirect.com/science/article/pii/S0020025518310351

14. Robinson, W.S.: Ecological correlations and the behavior of individuals. Int. J. Epidemiol. **38**(2), 337–341 (2009)
15. Simões, F.S.P.: Sobre equações diferenciais para processos fuzzy linearmente correlacionados: aplicações em dinâmica de população. Ph.D. thesis, UNICAMP (2017). (in Portuguese)
16. Trivedi, K.S.: Probability & Statistics with Reliability, Queuing and Computer Science Applications. Wiley, Hoboken (2008)

Genetic Fuzzy System for Anticipating Athlete Decision Making in Virtual Reality

Anoop Sathyan[1]([✉]), Henry S. Harrison[2], Adam W. Kiefer[2,3], Paula L. Silva[4], Ryan MacPherson[2], and Kelly Cohen[1]

[1] Department of Aerospace Engineering, University of Cincinnati,
Cincinnati, OH, USA
{anoop.sathyan,kelly.cohen}@uc.edu
[2] Division of Sports Medicine, Cincinnati Children's Hospital, Cincinnati, OH, USA
{Henry.Harrison,Adam.Kiefer,Ryan.MacPherson}@cchmc.org
[3] Department of Pediatrics, College of Medicine, University of Cincinnati,
Cincinnati, OH, USA
[4] Department of Psychology, University of Cincinnati, Cincinnati, OH, USA
silvapa@ucmail.uc.edu

Abstract. Intercepting and impeding an opponent is a fairly common behavior in contact and collision sports such as soccer, football, lacrosse or basketball. In soccer, for example, the main objective of a defender is to intercept and impede an attacking opponent as he or she navigates toward the goal. These athlete vs. athlete interactions often lead to collisions, and the uncertainty surrounding them frequently leads to injury. A virtual reality (VR) training platform with non-player characters (NPC) that can anticipate an athletes decisions would, therefore, be a desirable tool to be used by sports trainers to safely and effectively promote the resiliency of athletes to these types of situations. Here we applied this platform to a VR task that required the athletes to run past a series of NPCs to reach a stationary virtual waypoint, or goal. Each NPC is modeled as a Genetic Fuzzy System (GFS) that is trained using a new methodology, called FuzzyBolt, that is capable of training large fuzzy logic systems efficiently to provide better predictive quality. The end result is that such an intelligent NPC is able to more accurately predict athlete movements such that it becomes more difficult for the athlete to successfully navigate around the NPC and to the virtual goal. This, in turn, forces the athlete to develop new movement and decision making strategies in order to evade the NPC, thus enhancing their resiliency and ultimately reducing the risk of collision-based injury on the field of play.

1   Introduction

Athlete-to-athlete collisions in sport are extremely common and are the most frequent cause of injury. Preventing these injuries is of course desirable but a challenge that has not yet been met. To develop resiliency against collisions, athletes

© Springer Nature Switzerland AG 2019
R. B. Kearfott et al. (Eds.): IFSA 2019/NAFIPS 2019, AISC 1000, pp. 578–588, 2019.
https://doi.org/10.1007/978-3-030-21920-8_51

must develop new perceptual and movement strategies to respond better during uncertain interactions with other athletes in the field. A crucial challenge—one we tackle here—is how to promote such strategies without exposing the athlete to the risky sport-specific situations that lead to collision and injury. Virtual reality (VR) is a platform that offers the potential for performance enhancement in these types of sport contexts without putting athletes at risk of physical harm. VR also enables the embedding of artificial intelligence models into non-player characters (NPCs), based on real-time performance data collected from the athletes, to precisely challenge the athletes and promote more resilient athletic behaviors.

Fuzzy Inference System (FIS) is one such intelligent system that can be used to model the NPCs. It uses fuzzification, rule-inference and defuzzification to make decisions. Designing a FIS involves developing the set of membership functions for each input and output variable, and defining the rules within the rulebase to define the relationship between the inputs and outputs. The designer can also choose the defuzzification method, conjunction, disjunction, implication methods, etc. Although expert knowledge can be used to build FISs and this capability is appealing to a lot of applications, it makes sense to have a mechanism to tune the parameters of the FIS automatically. Self-tuning FISs are very useful especially when there are many inputs and outputs and their relationships are not that straightforward or well known. FISs can be given such self-tuning capability by using different optimization algorithms. These include ANFIS (Adaptive Network based Fuzzy Inference System) [1], Simulated Annealing [2], Genetic Algorithm (GA) [3], etc. The process of choosing all these parameters can be automated by using an optimization approach, such as GA, that can choose a near-optimal set of parameters to minimize some pre-defined cost function. A FIS that is tuned using GA is called a Genetic Fuzzy System (GFS). Such GFSs have been developed with much success for clustering and task planning [4], simulated air-to-air combat [5], aircraft conflict resolution [6], collaborative robotics [7,8], etc. Since FISs include a set of linguistic rules that define the relationship between the inputs and outputs, it is more interpretable compared to other machine learning techniques such as neural networks, support vector machines, etc.

In this paper, we discuss GFS models that are trained using a new and efficient algorithm called FuzzyBolt, for predicting athlete movements. Since FuzzyBolt uses GA to perform the search, the model is still called a GFS. The data for training the GFS is obtained by tracking athlete movements within a VR sport navigation environment shown in Fig. 1.

Fig. 1. First person perspective of a scenario within our VR environment

2 NPC Dataset

The current dataset was collected as part of a wider study aimed at developing an AI-driven VR paradigm for training athletes in the avoidance of on-field collisions, particularly unanticipated collisions. Data were collected on 29 high-school and college soccer athletes between the ages of 14 and 21. These included 13 male and 16 female athletes. Across two visits, each athlete performed a total of 60 VR navigation trials of various complexity and their task was to run to a stationary goal marker, avoiding all obstacles.

The athletes wore a custom-fabricated wireless head-mounted full high-definition display (HMD) with built-in wireless eye gaze tracking (Tobii Pro, Stockholm, Sweden), allowing the athletes to move freely and wirelessly within our $10\,\mathrm{m} \times 20\,\mathrm{m}$ physical laboratory. Athletes positional data were captured using 3D motion capture with passive markers attached to the HMD. Positional data and eye-tracking data were recorded at 8.3 Hz, as data updates from the disparate data collection systems coincided 12 times per second, for an effective sample rate of 8.3 Hz. Virtual environments were presented with a selection of behaviorally responsible NPCs controlled via modified versions of both a steering dynamics model and an affordance-based collision model, combined with up to four state machines that dictated parameters relative to NPC capabilities. These behaviors drove the NPCs interception behaviors relative to the athlete.

From the full trial data, smaller subsets were selected to create the NPC dataset. Each NPC interaction was defined as the time at which an athlete came within 2 m of an NPC, as shown in Fig. 2. For each interaction, a time series of trial data was selected that began a maximum of 3 s prior to the start of the interaction. The data was collected in intervals of 0.2 s. Passage to the left or right of the NPC was determined by projecting both the athlete and NPC trajectories onto the line segment joining the athlete's starting position with the position of the goal marker. Passage time was determined as the first timestep at which the athlete's position on this line was closer to the goal position than the

NPC

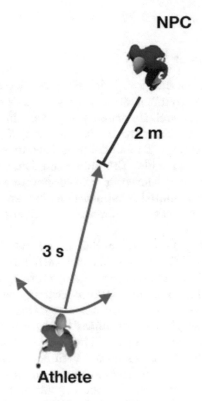

2 m

3 s

Athlete

Fig. 2. Top view of a scenario showing the athlete trying to get past the NPC. The dataset considers the data collected from 3 s prior to the athlete reaching within 2 m of the NPC.

NPCs. Passage direction was determined by the angle between the goal, NPC, and athlete at the passage time. Some NPCs were never passed before the end of the trial; these were dropped from the following analyses. Others had already been passed before the interaction began (in this case, the NPC is catching up to the athlete from behind); these were dropped as well. After ignoring the cases where the athlete collided with the NPC, a total of 545 interaction events were isolated to make up our NPC dataset. The following 12 variables were tracked: (1) Absolute NPC angle, (2) rate of change of the absolute NPC angle, (3) rate of approach of the athlete with respect to the NPC, (4) rate of change of the heading angle, (5) eye angle, (6) heading relative to the NPC angle, (7) NPC distance, (8) NPC heading, (9) NPC speed, (10) athlete's heading, (11) athlete's speed, and (12) outcome of the run listed as either "pass-left" or "pass-right". The outcome of each interaction was the variable to be predicted (i.e., whether the athlete passed the NPC on the left or right). Accurate prediction of the athlete movements can be used to improve the performance of the NPC, which in turn means that the athletes have to improve their decision making to successfully navigate around a given NPC.

3   Methodology

3.1   Steering Dynamics Model

The steering dynamics model [9–12] describes the locomotor behavior of an
agent navigating an environment as a system of differential equations based
on a simple mass-spring model. Its primary state variable is the agents heading,
which is pulled toward an attractor in the direction of goals, and pushed, via
repellers, away from obstacles. The model equations are consistent with theories
of perception-action, in particular the guidance of movement by continuously
available visual information. Moreover, the locomotor strategies implemented
by the model have been validated in empirical human-subjects research, and the
models parameters set in order to minimize the divergence between observed
and simulated trajectories.

   We used the steering dynamics model to generate baseline left-right predictions of the athlete-NPC interactions. For each interaction, a set of initial
conditions was sampled from the observed trajectories, at t_i and up to 3 s preceding, in intervals of 0.2 s. For interactions occurring within the first 3 s of the
observed trial, the set of initial conditions was reduced accordingly. For each
simulation, the NPC was placed on a linear trajectory from its initial position,
traveling forward at a constant speed. The athletes trajectory was simulated
according to the steering dynamics model with a stationary goal, treating the
NPC as a moving obstacle. Solutions to model equations were approximated
using the fourth-order Runge-Kutta method, resulting in simulated athlete trajectories. Passage to the left or right of the NPC was determined in the same
manner as in the observed dataset.

3.2   FuzzyBolt

FuzzyBolt is a GA-based algorithm that can be used for training FISs with many
inputs. Previously, GA was directly applied to tune the parameters of a FIS.
Ideally, in a FIS, the rulebase should include all possible combinations of input
membership functions in their antecedents. Therefore, if a FIS is defined using n
inputs with each input defined by m membership functions, the rulebase should
have m^n rules to include all possible combinations. This means that GA has to
tune m^n consequent parameters and along with other parameters that define
the membership function boundaries. So, as the number of inputs increase, the
number of parameters that need to be tuned using GA increases exponentially,
thus increasing the overall computational complexity of the search process.

   Genetic Fuzzy Trees (GFTs), shown in Fig. 3, mitigate this problem to some
extent by dividing the computations between several smaller FISs, each of which
only take in two or three inputs. These smaller FISs are connected together in
a tree-like architecture that outputs the desired variable(s). This reduces the
number of parameters that are tuned by GA, thus reducing the complexity.
But, the GFT architecture needs to be defined beforehand to some extent and
there is the possibility of missing some essential connections between some input
variables.

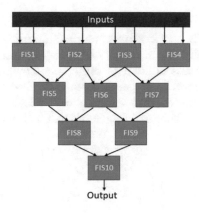

Fig. 3. Schematic of a sample GFT

FuzzyBolt also uses GA to tune the parameters. By intelligently reducing the search space, FuzzyBolt is able to define the relationship between inputs and outputs without the need to breakdown the system to a GFT format. It works on a standard single n-input-p-output system, as shown in Fig. 4, without the need to break it down to a tree architecture. Apart from reducing the search space based on the dataset, FuzzyBolt is also able to reduce the number of parameters that need to be tuned which reduces the chances of overfitting. FuzzyBolt is computationally efficient as it is able to train even large systems with tens or hundreds of inputs within minutes on a basic 8 GB CPU. The number of parameters does not increase substantially even when the number of membership functions for the inputs and outputs are increased. This provides more flexibility for the designer as they can make changes to design parameters with minimal effect on the overall training time.

Fig. 4. An n-input FIS that can be trained by FuzzyBolt

4   Results

4.1   Training the GFS Using FuzzyBolt

The steering dynamics model is used to provide a baseline prediction. The equation results for the given timepoint, t_i, was also one of the inputs to the GFS. Since the dataset already has 11 input variables, FuzzyBolt trains a 12-input-1-output system that predicts the outcome of each run. The dataset is split into 80% for training and 20% for validation. A different GFS is trained for each time-step. Accuracy of the predictions is used as the fitness function by

FuzzyBolt, as the objective is to maximize accuracy. Since this is a binary classification problem, the GFS uses 2 output membership functions, viz. left and right.
The designer can modify the number of input membership functions needed,
N_{mf}. We assume the same number of input membership functions for each of
the 12 variables. The number of parameters that need to be tuned increase as
N_{mf} is increased. We notice that an $N_{mf} = 5$ provide good results for both
training and validation. Other attributes of the GFS include:

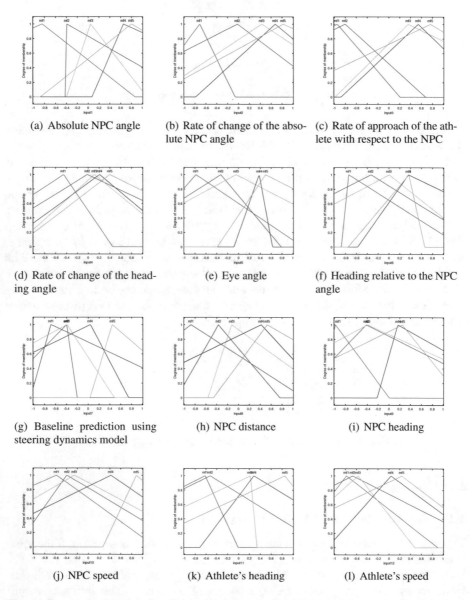

(a) Absolute NPC angle

(b) Rate of change of the absolute NPC angle

(c) Rate of approach of the athlete with respect to the NPC

(d) Rate of change of the heading angle

(e) Eye angle

(f) Heading relative to the NPC
angle

(g) Baseline prediction using
steering dynamics model

(h) NPC distance

(i) NPC heading

(j) NPC speed

(k) Athlete's heading

(l) Athlete's speed

Fig. 5. Input membership functions. Each input variable is defined using five membership functions.

1. Product AndMethod is used. This means that the firing strength of each rule is evaluated by multiplying the membership values of the inputs.
2. Implication is minimum. So, the consequent membership function of each rule is clipped at the firing strength value.
3. Mean-of-max defuzzification is used. So, in most cases, a single rule dictates the decision made by the GFS.

The NPC dataset provides information for time-steps starting at $t_s = -3\,$s to $t_s = 0\,$s at increments of 0.2 s. Thus, we have 16 different datasets, each pertaining to a single timestep, t_s, that can be used to train 16 different GFSs. The membership functions associated with each input pertaining to the GFS trained for $t = 0\,$s are shown in Fig. 5. The inputs are normalized using min-max normalization to the range $[-1, 1]$. The membership functions associated with the Left and Right classes for the predicted variable are shown in Fig. 6. The predicted class is obtained using the membership values associated with the defuzzified output. The class with the higher membership value is determined as output from the GFS.

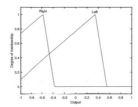

Fig. 6. The membership functions associated with the left and right classes of the output variable.

Figure 7 shows a comparison of the accuracy obtained by the GFS and the steering model for each time step. The shaded region shows the 95% confidence interval for the predictions. The accuracy of the GFSs are measured on the validation set which constitutes 20% of the entire dataset. As expected, both models have better accuracy of prediction as the athlete moves closer to the NPC. Although the two models have similar accuracies at $t = -3\,$s, GFSs have much better accuracy as the athlete moves closer to the NPC, or as t approaches zero.

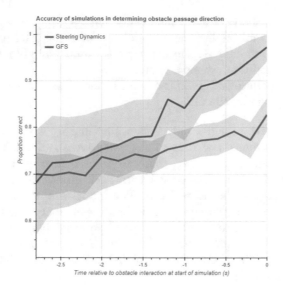

Fig. 7. A comparison of the performance between the GFS and the steering dynamics model for each time-step

4.2   Explainability

The linguistic nature of the rules provide a level of explainability to fuzzy logic systems. The trained GFS for $t = 0$ has a total of 108 rules. Since the GFS models trained for this research used mean-of-max defuzzification, the output class, in most cases, will be dependent on a single rule. This rule will have the highest firing strength. The firing strength of each rule is obtained by multiplying the membership values of each input in the antecedent part of a rule, as *prod* And-Method is used. Figure 8 illustrates this aspect by showing the most dominant rule for three sample datapoints from the validation set. For Case 1, the most dominant rule is

If X_1 is mf3 AND X_2 is mf4 AND X_3 is mf4 AND X_4 is mf3 AND X_5 is mf3 AND X_6 is mf3 AND X_7 is mf4 AND X_8 is mf5 AND X_9 is mf5 AND X_{10} is mf1 AND X_{11} is mf5 AND X_{12} is mf3 Then Y is RIGHT.

The membership values associated with the antecedent part of each input is shown underneath the rule. For example, the membership value of X_1 in $mf3$ is 0.965. The firing strength of the rule is obtained by multiplying these membership values. For each datapoint in Fig. 8, the rule shown had the highest firing strength. It is to be noted that the reason for a particular rule to have the highest firing strength is also the result of the other rules having lower firing strengths. Hence, we cannot say that one single rule dictates the output. Nevertheless, since we have a dominant rule that directly defines the output class, we can say that the particular rule played the most part in making the particular prediction. It can be noted that the dominant rules for Cases 2 and 3 look very similar.

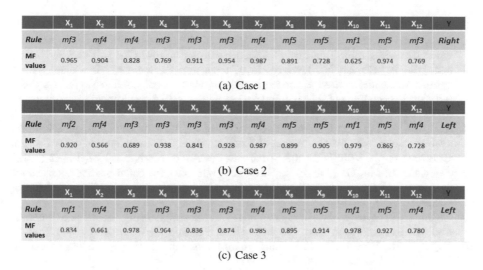

(a) Case 1

(b) Case 2

(c) Case 3

Fig. 8. The dominant rules for three different datapoints in the validation set

5 Conclusions

In this paper, we presented the performance of a set of GFSs for anticipating athlete decision making in soccer. We showed the applicability of FuzzyBolt for such a supervised learning problem involving 12 inputs. The GFSs were trained and validated using our NPC dataset. The GFS methodology was compared with the steering dynamics model, that was also used to obtain baseline predictions for the GFS. We observe that the predictive accuracy increases as the athlete moves closer to the NPC, i.e. as $t \rightarrow 0$. The explainability of the trained GFS has also been discussed. The linguistic nature of the rulebase provides a certain level of understanding the basis of the decisions made by the GFS.

Acknowledgements. This work was supported by a GAP award and Innovation award from the Cincinnati Childrens Hospital Research Foundation (Kiefer).

References

1. Jang, J.-S.R.: ANFIS: adaptive-network-based fuzzy inference system. IEEE Trans. Syst. Man Cybern. **23**(3), 665–685 (1993)
2. Jain, R., Sivakumaran, N., Radhakrishnan, T.K.: Design of self tuning fuzzy controllers for nonlinear systems. Expert Syst. Appl. **38**(4), 4466–4476 (2011)
3. Cordón, O., Gomide, F., Herrera, F., Hoffmann, F., Magdalena, L.: Ten years of genetic fuzzy systems: current framework and new trends. Fuzzy Sets Syst. **141**(1), 5–31 (2004)
4. Sathyan, A., Ernest, N., Cohen, K.: An efficient genetic fuzzy approach to UAV swarm routing. Unmanned Syst. **4**(02), 117–127 (2016)

5. Ernest, N., Carroll, D., Schumacher, C., Clark, M., Cohen, K., Lee, G.: Genetic fuzzy based artificial intelligence for unmanned combat aerial vehicle control in simulated air combat missions. J. Defense Manag. **6**(144) (2016). ISSN 2167-0374
6. Sathyan, A., Ernest, N., Lavigne, L., Cazaurang, F., Kumar, M., Cohen, K.: A genetic fuzzy logic based approach to solving the aircraft conflict resolution problem. In: Proceedings of AIAA Information Systems-AIAA Infotech@ Aerospace, p. 1751 (2017)
7. Sathyan, A., Ma, O.: Collaborative control of multiple robots using genetic fuzzy systems approach. In: Proceedings of ASME 2018 Dynamic Systems and Control Conference, pp. V001T03A002. American Society of Mechanical Engineers (2018)
8. Sathyan, A., Ma, O., Cohen, K.: Intelligent approach for collaborative space robot systems. In: Proceedings of 2018 AIAA SPACE and Astronautics Forum and Exposition, p. 5119 (2018)
9. Fajen, B.R., Warren, W.H.: Behavioral dynamics of steering, obstacle avoidance, and route selection. J. Exp. Psychol.: Hum. Percept. Perform. **29**(2), 343 (2003)
10. Fajen, B.R., Warren, W.H.: Visual guidance of intercepting a moving target on foot. Perception **33**(6), 689–715 (2004)
11. Fajen, B.R., Warren, W.H.: Behavioral dynamics of intercepting a moving target. Exp. Brain Res. **180**(2), 303–319 (2007)
12. Warren, W.H., Fajen, B.R.: Behavioral dynamics of visually guided locomotion. In: Coordination: Neural, Behavioral and Social Dynamics, pp. 45–75. Springer, Heidelberg (2008)

Modeling Probabilistic Data with Fuzzy Probability Measures in UML Class Diagrams

Jie Sheng, Li Yan$^{(\boxtimes)}$, and Zongmin Ma

College of Computer Science and Technology,
Nanjing University of Aeronautics and Astronautics, Nanjing 211106, China
yanli@nuaa.edu.cn

Abstract. Being a standard of the Object Management Group (OMG), the Unified Modeling Language (UML) has been applied to diverse domains. UML class diagram model is a conceptual data model and has been widely used for database design and information modeling. Information in real-world applications is often uncertain. To model and deal with uncertain data, various uncertain databases are pro-posed, including fuzzy ones and probabilistic ones. Also, there are few efforts in modeling fuzzy and probabilistic data in databases. But few efforts have been made on modeling uncertainty in conceptual data models. In this paper, we concentrate on modeling probabilistic data with fuzzy probability measures in the UML class diagram model. We introduce the semantics of fuzzy and probabilistic information into the UML class diagram model and extend several major constructs of UML class diagrams accordingly. We present the corresponding graph-ical representations of the extended UML class diagram model in the paper.

1 Introduction

UML (Unified Modeling Language) is a modeling language for object-oriented software engineering [1, 2]. Being a standard of the OMG (Object Management Group), UML combines multiple techniques from data modeling, business modeling, object modeling and component modeling, and hereby includes elements such as *activities, actors, business processes, database schemas, (logical) components, programming language statements* and *reusable software components*. UML can cover the whole software development life cycle and has been applied to many areas of software engineering and knowledge engineering [3]. Viewed from traditional data modeling, UML describes the complete development of databases for business requirements in [4] and the UML class diagrams, which are a kind of conceptual data model, have been extensively applied in database conceptual design [5]. Actually, the design of large and complex databases in applications generally starts with designing the conceptual data models, which are then mapped into the database models [39]. Recently, with the wide utilization of the Web and the availability of huge amounts of resources, the UML class

© Springer Nature Switzerland AG 2019
R. B. Kearfott et al. (Eds.): IFSA 2019/NAFIPS 2019, AISC 1000, pp. 589–600, 2019.
https://doi.org/10.1007/978-3-030-21920-8_52

diagram model has also been used to conceptual design of XML [6] and OWL in the Semantic Web [7]. In this paper, we concentrate on the UML class diagrams as a conceptual data model.

We argue that UML has the capability of object-oriented data modeling, but it suffers from some inadequacy of necessary semantics that are needed by real-world applications. Information in real-world applications is not always perfect and may be uncertain [8]. Fuzzy sets [9] and probability theory are two major foundations for uncertain data representation, which are applied to deal with objective uncertainty by means of objective statistics and subjective uncertainty by means of subjective estimation and judgment, respectively. To represent and handle probabilistic data in databases, the probabilistic relational databases (e.g., [14]) and probabilistic object-oriented databases (e.g., [15]) are proposed. Also, fuzzy data modeling has been extensively investigated in various database mod-els, including the fuzzy relational databases (e.g., [10]), fuzzy object-oriented databases (e.g., [11]) and fuzzy object-relational databases (e.g., [12]). More recently, to model uncertain data on the Web, probabilistic XML model [16] and fuzzy XML model [13] are widely investigated. In the context of the UML class diagrams, the fuzzy UML class diagram model is proposed, which can be typically applied for conceptually designing fuzzy databases, say conceptual design of the fuzzy relational databases [34], conceptual design of the fuzzy object-oriented databases [35] and conceptual design of the fuzzy XML [36].

We note that probabilistic and fuzzy data modeling is mostly investigated separately to handle objective and subjective uncertainties, respectively. In the real life, some data contain both probabilistic and fuzzy uncertainties and there are a kind of hybrid uncertain data called *fuzzy probability* [17–21]. Basically, we can identify two major forms for fuzzy probabilistic data in databases [25]: the first one is that fuzzy sets are applied to describe fuzzy attribute values of probabilistic entities; the second one is that fuzzy sets are applied to describe probability measures of probabilistic entities. For the former, a deductive probabilistic and fuzzy object-oriented database language as well as a probabilistic and fuzzy object base model are proposed in [22] and [23], respectively. For the latter, a probabilistic relational model and a probabilistic object-oriented model both with fuzzy probability measures are proposed in [24] and [25], respectively. The probabilistic relational model proposed in [24] is reengineered into XML model in [26].

Like the classical databases, the conceptual designs of uncertain databases need to apply uncertain conceptual data models also. Although many efforts have been carried out to propose the uncertain (fuzzy, probabilistic, or fuzzy and probabilistic) relational and object-oriented databases models as well as the un-certain (fuzzy or probabilistic) XML data models, there are only few works on modeling uncertain data in the conceptual data models. Fuzzy sets are originally introduced into the ER (entity-relationship) model by Zvieli and Chen [27]. The fuzzy ER model is used for the design and development of fuzzy relational databases in [28]. Several major EER (enhanced/extended ER) concepts are fuzzily extended and the fuzzy EER data model is proposed [29,30]. A fuzzy EER model and the corresponding graphical representations are proposed in [31]. The formal mappings from the fuzzy EER model to the fuzzy databases [31,32] and fuzzy XML model [33] are investigated. In addition to the fuzzy ER/EER models, the fuzzy UML data model is proposed and further applied for mapping to the fuzzy databases [34,35] and fuzzy XML model [36]. As to the fuzzy conceptual data modeling, little research has been done in propose the probabilistic conceptual data models.

Conceptual data models can represent rich and complex semantics at a highly abstract level. Viewed from conceptual design of databases, design requirements are generally full of uncertainties and it is especially true in the preliminary design phrase. In the preliminary design phrase of product Car, for example, we may express *"the probability that vehicle navigation system will be installed in car is high"* instead of *"the probability that vehicle navigation system will be installed in car is 0.8"* because the former is more natural and is easily understood by human beings. In order to capture and represent such uncertainties in the UML class diagrams and further serve as the conceptual design of fuzzy probabilistic databases, in this paper, we devote to extend the UML class diagrams in order to model a kind of hybrid uncertain data, where probabilistic events are associated with fuzzy probability measures. For this purpose, we extend several major concepts of the UML class model (e.g., association and dependency), which allow for the full consideration of different uncertainties in classes. We present the graphical representations of the probabilistic UML class diagrams with fuzzy probability measures in the paper. To the best of our knowledge, there is not any report on modeling both fuzzy and probabilistic data in the UML class diagrams.

The remainder of this paper is organized as follows. Section 2 provides preliminaries of fuzzy sets and fuzzy probability measures. Probabilistic UML class model with fuzzy probability measures is proposed in Sect. 3. Section 4 concludes this paper.

2   Preliminaries

This section provides preliminaries of fuzzy sets and probability theory as well as imprecise probability measures.

2.1   Fuzzy Sets

Fuzzy sets are originally proposed by Zadeh [9]. A fuzzy set is a set of elements, in which an element in the universe of discourse belongs to the fuzzy set with a membership degree in [0, 1]. In a conventional set, an element in the universe of discourse belongs to the set with a membership degree in {0, 1}.

Let F be a fuzzy set in a universe of discourse U. A membership function $\mu_F : U \to [0,1]$ is defined for F, in which, for each $u \in U$, $\mu_F(u)$ denotes the membership degree of u in F. Then the fuzzy set F is formally described as follows.

$$F = \{(u_1, \mu_F(u_1)), (u_2, \mu_F(u_1)), ..., (u_n, \mu_F(u_n))\}$$

It is shown that an element in the fuzzy set F (say (u_i)) is associated with its membership degree (say $\mu_F(u_i)$), which must be explicitly indicated. When the membership degrees of all elements in a fuzzy set are exactly 1, the fuzzy set reduces to a conventional set.

With the fuzzy set $F = \{(u_1, \mu_F(u_1)), (u_2, \mu_F(u_1)), ..., (u_n, \mu_F(u_n))\}$, we can obtain a conventional set, which are consisted of the the the elements in F with nonzero membership degrees. Such a set is called the support of F, denoted by

$$\text{supp}(F) = \{u | u \in U \text{ and } \mu_F(u) > 0\}.$$

Zadehs extension principle [37], which is also referred as maximum-minimum principle, can be used to extend nonfuzzy mathematical concepts. Its basic idea is to induce a fuzzy set from a number of given fuzzy sets using a mapping. Formally, let $A = \{(u_i, \mu_A(u_i)) | u_i \in U \wedge 1 \leq i \leq n\}$ and $B = \{(v_j, \mu_B(v_j)) | v_j \in U \wedge 1 \leq j \leq n\}$ be two fuzzy sets on the universe of discourse $U = \{u_1, u_2, ..., u_n\}$. Then, following the Zadehs extension principle, the operation with an infix operator "θ" on A and B is defined as follows.

$$A\theta B = \{max(u_i \theta v_j, min(\mu_A(u_i), \mu_B(v_j))) | u_i, v_j \in U \wedge 1 \leq i, j \leq n\}$$

2.2   Imprecise Probability Distributions

A stochastic event corresponds to an entity with uncertain status. To represent the stochastic event described by the entity with uncertainty, a probability measure is used to be associated with the entity.

Formally, let e be an entity describing a stochastic event and $p(e)$ be its probability measure $(0 \leq p(e) \leq 1)$. Then the uncertain status of e is described by a pair $(e, p(e))$. Let S be a stochastic phenomenon, which contains multiple stochastic events described by set of entities $\{e_1, e_2, ..., e_k\}$. As a result, S can be described by a probability distribution $P_S = \{(e_1, p_S(e_1)), (e_2, p_S(e_2)), ..., (e_k, p_S(e_k))\}$, where $p_S(e_i)$ is the probability measure of stochastic event $e_i (1 \leq i \leq k)$ with respect to stochastic phenomenon S. Here we have a probabilistic constraint on P_S : the sum of probability measures of all stochastic events with respect to S must be less than or equal to 1, that is, $\sum p_S(e_i) \leq 1 (1 \leq i \leq k)$.

For $(e_i, p_S(e_i))$, the classical probability theory assumes that $p_S(e_i)$ is a crisp value in [0, 1]. In the real-world applications, however, it is possible that $p_S(e_i)$ is known imprecisely. To deal with such a problem, *interval probability measures instead of crisp probability measures* are applied in [14,15]. An interval probability measure is described by two values in [0, 1], which represent the lower and upper boundaries of the imprecise probability measures, respectively. Note that each value in the interval has the same possibility being the final probability measure of the stochastic event. It is a common case in the real life that values in intervals may have different possibility being true. At this point, fuzzy sets can explicitly represent such different possibility degrees. Compared with intervals, fuzzy sets are more informative.

An imprecise probability measure can be described by a fuzzy set instead of a crisp value or an interval, which is called *fuzzy probability measure* in [24, 25]. Formally, let $p(e)$ be the fuzzy probability measure of stochastic event e. Then it is described by $\{(p_1, \mu_e(p_1)), (p_2, \mu_e(p_2)), ..., (p_m, \mu_e(p_m))\}$. Here $p_j \in [0, 1](1 \leq j \leq m)$ is a possible probability measure of $p(e)$ with a membership degree $\mu_e(p_j)$. For the probability distribution $P_S = \{(e_1, p_S(e_1)), (e_2, p_S(e_{21})), ..., (e_k, p_S(e_k))\}$, where $p_S(e_i)$ with fuzzy probability measures, where $p_S(e_i) = \{(p_{i1}, \mu_{ei}(p_{i1})), (p_{i2}, \mu_{ei}(p_{i2})), ..., (p_{im}, \mu_{ei}(p_{im}))\}(1 \leq i \leq k)$, the probabilistic constraint called fuzzy probabilistic constraint is represented as follows.

$$\sum max(supp(p_S(e_i))) \leq 1 (1 \leq i \leq k)$$

Here for $p_S(e_i) = (p_{i1}, \mu_{ei}(p_{i1})), (p_{i2}, \mu_{ei}(p_{i2})), ..., (p_{im}, \mu_{ei}(p_{im}))(1 \leq i \leq k)$, supp $(p_S(e_i))$ represents the support of $p_S(e_i)$, and max $(supp (p_S(e_i)))$ represents a value p_i with maximum $\mu_{ei}(p_i)$ in this support.

3   UML Modeling of Probabilistic Data with Fuzzy Probability Measures

The constructs of UML class diagram mainly contain *class* and *relationships*. In the following, we present a formal description of the UML class diagram [34].

A UML class diagram can be formally represented by $D = (C, A, R, O, M, S)$, where C is a finite set of classes, A is a finite set of attributes, R is a finite set of relationships, O is a finite set of objects, M is a finite set of methods, and S is a finite set of constraints. In this paper, we pay attention on the classes, attributes, relationships and objects and use a simplified UML class diagram model $D = (C, A, R, O)$, in which $C = \{c_1, c_2, ..., c_k\}$, $A = \{a_1, a_2, ..., a_l\}$, $R = \{r_1, r_2, ..., r_m\}$, and $O = \{o_1, o_2, ..., o_n\}$.

(1) $R \subseteq C \times C$ is a binary relation that represents the generalization, aggregation, association or dependency.
(2) For $c_i \in C(1 \leq i \leq k)$, $A(c_i)$ represents a set of attributes of c_i. Clearly $A(c_i) \subseteq \{a_{1i}, a_2, ..., a_l\}$, i.e., $A(c_i) \subseteq A$. For $a_j \in A(1 \leq j \leq l)$, $a_j(c_i)$ denotes the attribute a_j of c_i. In the context of the given c_i, a_j is used instead of $a_j(c_i)$.
(3) For $c_i \in C(1 \leq i \leq k)$, $O(c_i)$ means a set of objects that c_i contains. Here, $O(c_i) \subseteq \{o_1, o_2, ..., o_n\}$, i.e., $O(c_i) \subseteq O$. For $o_p \in O(1 \leq p \leq n)$ and $a_j \in A(1 \leq j \leq l)$, $o_p(c_i)$ denotes the object o_p of c_i, and $o_p(a_j(c_i))$ denotes the value of object o_p on attribute a_j. In the context of the given c_i, o_p is used instead of $o_p(c_i)$ and $o_p(a_j)$ is used instead of $o_p(a_j(c_i))$.

For the purpose of modeling probabilistic data with fuzzy probability measures in the UML class diagram, the above-mentioned four constructs of UML class diagrams should be extended according to the semantics of probabil-ity distributions and fuzzy probability measures. A probabilistic UML class dia-gram is a quadruple $D_S = (C_S, A_S, R_S, O_S)$, in which C_S is a probabilistic set of classes, A_S is a probabilistic set of attributes, R_S is a probabilistic set of relationships, and O_S is a probabilistic set of objects.

3.1   Probabilistic Classes with Fuzzy Probability Measures

The objects with the same properties constitute a class. A stochastic event with uncertain status corresponds to an object which is associated with a probability measure. Here the probability measure may be a crisp value of $[0, 1]$ or a fuzzy set in the universe of discourse $[0, 1]$ (i.e., a fuzzy probability measure). The object with probability measure is called a probabilistic object. The probabilistic objects with the same properties (excluding the property of probability measure) constitute a probabilistic class. As a result, the probabilistic objects belong to the probabilistic class with probability measures (crisp ones or fuzzy ones).

Formally, for class $O(c_i)$ and object $o_p(c_i)$, we have that $o_p(c_i)$ is the object of $O(c_i)$ with a fuzzy probability measure. To explicitly describe the fuzzy probability measure, in addition to $A(c_i)$, an additional attribute called fuzzy probabilistic attribute is needed. We use notation Λ to denote this additional attribute. Then, we have $A(c_i) \leftarrow A(c_i) \bigcup \{\Lambda\}$. Here $o_p(\Lambda(c_i))$ is a fuzzy set in the universe of discourse $[0, 1]$, meaning a fuzzy probability measure associated with $o_p(c_i)$.

To graphically represent a probabilistic class with a fuzzy probability measure in the UML class diagram, a dashed rectangle with two compartments separated by horizontal lines. The top compartment holds the class name and the bottom compartment holds a list of attributes (including the fuzzy probabilistic attribute). Figure 1 shows a probabilistic class with a fuzzy probability measure.

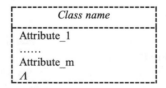

Fig. 1. The probabilistic class icon of the UML class diagram.

3.2   Probabilistic Generalization with Fuzzy Probability Measures

Generalization is used to build a hierarchical relationship between classes, in which a class referred to as the superclass is a more general description of a set of other classes called the subclasses. The subclasses inherit the attributes and methods of the superclass. As a result, each object belonging to a subclass must belong to its superclass and this is usually applied to determine if two classes have a relationship if subclass/superclass.

Being different from the classical subclass-superclass relationship, the subclass-superclass relationship is uncertain due to probabilistic classes with fuzzy probability measures. At this moment, a (probabilistic) class is the subclass of another (probabilistic) class and a (probabilistic) class is the superclass of one or more other (probabilistic) classes all with a probability measure, and this probability measure may be a crisp value of $[0, 1]$ or a fuzzy set in the universe of discourse $[0, 1]$ (i.e., a fuzzy probability measure).

Let $O(c_i)$ and $O(c_j)$ be two probabilistic classes with fuzzy probabilistic attributes $\Lambda(c_i)$ and $\Lambda(c_j)$, respectively. Then $O(c_i)$ is the subclass of $O(c_j)$ and $O(c_j)$ is the superclass of $O(c_i)$ if and only if

$$(\forall o)(o \in O \wedge o(c_i) \wedge o(c_j) \wedge o(\Lambda(c_i)) \leq o(\Lambda(c_j)))$$

It means that, for any object, the probability measure to which it belongs to the subclass $O(c_i)$ must be less than or equal to the probability measure to which it belongs to the superclass $O(c_j)$.

Note that the probability measures may be crisp values of $[0, 1]$ or fuzzy sets in the universe of discourse $[0, 1]$. In case that $o(\Lambda(c_i))$ and $o(\Lambda(c_j))$ are two fuzzy sets, we need to compare these two fuzzy sets. For this purpose, we apply the concept of semantic inclusion proposed in [38] to calculate how $o(\Lambda(c_i))$ is included by $o(\Lambda(c_j))$. It is required that the inclusion degree that $o(\Lambda(c_i))$ is included by $o(\Lambda(c_j))$ is greater than 0.

A dashed triangular arrowhead is applied to describe a probabilistic generalization relationship, where the arrowhead points to the superclass and there are one or more lines that proceed from the superclass of the arrowhead connecting it to the subclasses. Figure 2 shows a probabilistic generalization relationship with a fuzzy probability measure.

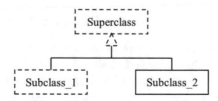

Fig. 2. A probabilistic generalization relation in the UML class diagram.

3.3 Probabilistic Aggregation with Fuzzy Probability Measures

An aggregation is used to represent a whole-part relationship between an aggregate, which is a class representing the whole, and several constituent parts, which are classes representing the part. As a result, an object of the aggregate can be projected into a set of objects, and each object belongs to a constituent part.

Suppose that a class is aggregated from several probabilistic classes. Then the aggregate is a probabilistic class also and we have a probabilistic aggregation with a probability measure. Here this probability measure may be a crisp value of $[0, 1]$ or a fuzzy set in the universe of discourse $[0, 1]$ (i.e., a fuzzy probability measure).

Let $O(c_i)$ be a probabilistic aggregation of probabilistic classes $O(c_{j1})$, $O(c_{j2})$, ..., $O(c_{jn})$. These probabilistic classes have fuzzy probabilistic attributes $\Lambda(c_i)$, $\Lambda(c_{j1})$, $\Lambda(c_{j2})$, ..., $\Lambda(c_{jn})$, respectively. Then we have

$$(\forall o)(o \in O \wedge o(c_i) \wedge o_1(c_{j1}) \wedge o_2(c_{j2}) \wedge ... \wedge o_n(c_{jn}) \wedge o_1(c_{j1}) \times o_2(c_{j2}) \times ... \times$$
$$o_n(c_{jn}) = o(c_i) \wedge o(\Lambda(c_i)) \leq min(o_1(\Lambda(c_{j1})), o_2(\Lambda(c_{j2})), ..., o_n(\Lambda(c_{jn})))).$$

It means that, for any object, the probability measure to which it belongs to the aggregate $O(c_i)$ must be less than or equal to the probability measure to which its projection to each constituent part belongs to the corresponding constituent part. Here $o_1(c_{j1}) \times o_2(c_{j2}) \times ... \times o_n(c_{jn}) = o(c_i)$ means that the projects of $o(c_i)$ to each constituent part.

A dashed open diamond is applied to describe a probabilistic aggregation relationship, in which the class touched with the dashed diamond is the aggregate. Figure 3 shows a probabilistic aggregation relationship.

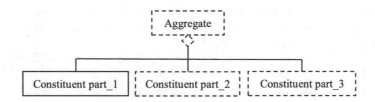

Fig. 3. A probabilistic aggregation relationship in the UML class diagram.

3.4 Probabilistic Association with Fuzzy Probability Measures

A named association relationship is used to connect two classes, specifying that the objects of one class are connected to the objects of another class with a role (i.e., the name of association). Actually, an association relationship is a more general relationship than aggregation or generalization.

When two probabilistic classes are associated, we may have a probabilistic association relationship. For a probabilistic class, the objects belong to the class with probability measure. Then two objects that respectively belong two associated classes have the given association relationship with a probability measure, and this probability measure may be a crisp value of $[0, 1]$ or a fuzzy set in the universe of discourse $[0, 1]$.

Let $O(c_i)$ and $O(c_j)$ be two probabilistic classes with fuzzy probabilistic attributes $\Lambda(c_i)$ and $\Lambda(c_j)$, respectively. Also, let $O(c_i)$ is associated with $O(c_j)$.

Then, for object of $O(c_i)$ and $o_j(c_j)$ of $O(c_j)$, the association relationship between $o_i(c_i)$ and $o_j(c_j)$ has a probability measure, which can be calculated as follows.

$$min(o_i(\Lambda(c_i)), o_j(\Lambda(c_j))).$$

It can be seen that, for a probabilistic association, different pairs of objects from the associated probabilistic classes may have different probability measures with respect to the probabilistic association relationship.

A probabilistic association is represented by a dashed double arrow between two (probabilistic) classes. Figure 4 shows a probabilistic association relationship.

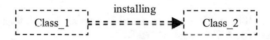

Fig. 4. A probabilistic association relation in UML class diagram.

3.5   Probabilistic Dependency with Fuzzy Probability Measures

A dependency is used to indicate a semantic relationship between two classes: one class is dependent on another class, in which the dependant is not complete without the provider. A dependency relationship between the provider class and the dependant class is established on the basis of classes instead of objects of classes.

For two probabilistic classes, the dependency relationship between them is not affected by such probabilistic classes because the dependency relationship is de-fined over classes rather than objects. The objects belong to the corresponding classes with probability measures, and two objects that respectively belong to the related classes may have a dependency relationship with a probability measure.

A dashed arrow can still be used to denote the probabilistic dependency relationship, where the class at the tail of the arrow depends on the class at the arrowhead. Figure 5 shows a probabilistic dependency relationship.

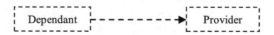

Fig. 5. A probabilistic dependency relationship in the UML class diagram.

4   Conclusions

This paper presents an extended UML data model, which can model probabilistic as well as complex objects in the real world at a conceptual level. The probabilistic objects correspond to stochastic event with fuzzy probability measures.

To model probabilistic data with fuzzy probability measures in the UML class diagrams, we extend the concepts of objects and classes and investigate several major relationships between probabilistic classes, including probabilistic generalizations, probabilistic aggregations, probabilistic associations and probabilistic dependencies. We also develop the corresponding graphical representations of probabilistic classes and these probabilistic relationships.

The UML class diagrams have been extensively applied for data and knowledge modeling. In our future work, we will concentrate on the studies of class operations, constraints and rules in the probabilistic UML class diagrams with fuzzy probability measures. In addition, we will apply the probabilistic UML data model with fuzzy probability measures for mapping from or to data-bases and ontology knowledge bases.

Acknowledgment. The work was supported in part by the National Natural Science Foundation of China (61772269 and 61370075).

References

1. Booch, G., Rumbaugh, J., Jacobson, I.: The Unified Modeling Language User Guide. Addison-Welsley Longman, Inc. (1998)
2. Object Management Group (OMG), Unified Modeling Language (UML), version 1.5, Technical report, OMG (2003). www.omg.org
3. Berardi, D., Calvanese, D., De Giacomo, G.: Reasoning on UML class diagrams. Artif. Intell. **168**(1–2), 70–118 (2005)
4. Marcos, E., Vela, B., Cavero, J.M.: Extending UML for object-relational database design. In: Proceedings of the 4th International Conference on the Unified Modeling Language, Modeling Languages, Concepts, and Tools, pp. 225–239 (2001)
5. Ambler, S.W.: The Design of a Robust Persistence Layer for Relational Databases (2000). http://www.ambysoft.com/persistenceLayer.pdf
6. Conrad, R., Scheffiner, D., Freytag, J.C.: XML conceptual modeling using UML. In: Proceeding of the 19th International Conference on Conceptual Modeling, pp. 558–571 (2000)
7. Falkovych, K., Sabou, M., Stuckenschmidt, H.: UML for the semantic web: transformation-based approaches. In: Knowledge Transformation for the Semantic Web. IOS Press (2003)
8. Parsons, S.: Current approaches to handling imperfect information in data and knowledge Bases. IEEE Trans. Knowl. Data Eng. **8**(3), 353–372 (1996)
9. Zadeh, L.A.: Fuzzy sets. Inf. Control **8**(3), 338–353 (1965)
10. de Tr, G., de Caluwe, R., Prade, H.: Null values in fuzzy databases. J. Intell. Inf. Syst. **30**(2), 93–114 (2008)
11. Ma, Z.M., Zhang, W.J., Ma, W.Y.: Extending object-oriented databases for fuzzy information modeling. Inf. Syst. **29**(5), 421–435 (2004)
12. Cuevas, L., et al.: pg4DB: a fuzzy object-relational system. Fuzzy Sets Syst. **159**(12), 1500–1514 (2008)
13. Ma, Z., Yan, L.: Modeling fuzzy data with XML: a survey. Fuzzy Sets Syst. **301**, 146–159 (2016)
14. Lakshmanan, L.V.S., et al.: ProbView: a flexible probabilistic database system. ACM Trans. Database Syst. **22**(3), 419–469 (1997)

15. Eiter, T., et al.: Probabilistic object bases. ACM Trans. Database Syst. **26**(3), 264–312 (2001)
16. Kimelfeld, B., Senellart, P.: Probabilistic XML: models and complexity. In: Advances in Probabilistic Databases for Uncertain Information Management, pp. 39–66. Springer, Heidelberg (2013)
17. Baldwin, J.M., Lawry, J., Martin, T.P.: A note on probability/possibility consistency for fuzzy events. In: Proceedings of the 6th International Conference on Information Processing and Management of Uncertainty in Knowledge-Based Systems, Granada, Spain, pp. 521–525, July 1996
18. Buckley, J.J.: Fuzzy Probabilities: New Approach and Applications. Springer, Heidelberg (2005)
19. Ralescu, A.: Fuzzy probabilities and their applications to statistical inference. In: Proceedings of the 5th International Conference on Processing and Management of Un-certainty in Knowledge-Based Systems, pp. 217–222 (1994)
20. Rebiasz, B.: New methods of probabilistic and possibilistic interactive data processing. J. Intell. Fuzzy Syst. **30**(5), 2639–2656 (2016)
21. Zadeh, L.A.: Fuzzy pobabilities. Inf. Process. Manag. **20**(3), 363–372 (1984)
22. Cao, T.H., Nguyen, H.: Uncertain and fuzzy object bases: a data model and algebraic operations. Int. J. Uncertainty Fuzziness Knowl.-Based Syst. **19**(2), 275–305 (2011)
23. Cao, T.H., Rossiter, J.M.: A deductive probabilistic and fuzzy object-oriented database language. Fuzzy Sets Syst. **140**(1), 129–150 (2003)
24. Yan, L., Ma, Z.M.: A fuzzy probabilistic relational database model and algebra. Int. J. Fuzzy Syst. **15**(2), 244–253 (2013)
25. Yan, L., Ma, Z.: A probabilistic object-oriented database model with fuzzy measures and its algebraic operations. J. Intell. Fuzzy Syst. **28**(5), 1969–1984 (2015)
26. Ma, Z., Li, C., Yan, L.: Reengineering probabilistic relational da-tabases with fuzzy probability measures into XML model. J. Database Manag. **28**(3), 26–47 (2017)
27. Zvieli, A., Chen, P.P.: Entity-relationship modeling and fuzzy databases. In: Proceedings of the 2nd IEEE International Conference on Data Engineering, pp. 320–327 (1986)
28. Chaudhry, N.A., Moyne, J.R., Rundensteiner, E.A.: An extended database design methodology for uncertain data management. Inf. Sci. **121**(1–2), 83–112 (1999)
29. Chen, G.Q., Kerre, E.E.: Extending ER/EER concepts towards fuzzy conceptual data modeling. In: Proceedings of the 7th IEEE International Conference on Fuzzy Systems, pp. 1320–1325 (1998)
30. Galindo, J., et al.: Relaxing constraints in enhanced entity-relationship models using fuzzy quantifiers. IEEE Trans. Fuzzy Syst. **12**(6), 780–796 (2004)
31. Ma, Z.M., et al.: Conceptual design of fuzzy object-oriented databases using extended entity-relationship model. Int. J. Intell. Syst. **16**(6), 697–711 (2001)
32. Yan, L., Ma, Z.M.: Modeling fuzzy information in fuzzy extended entity-relationship model and fuzzy relational databases. J. Intell. Fuzzy Syst. **27**(4), 1881–1896 (2014)
33. Yan, L., Ma, Z.M.: Formal translation from fuzzy EER model to fuzzy XML model. Expert Syst. Appl. **41**(8), 3615–3627 (2014)
34. Ma, Z.M., Zhang, F., Yan, L.: Fuzzy information modeling in UML class diagram and relational database models. Appl. Soft Comput. **11**(6), 4236–4245 (2011)
35. Ma, Z.M., Yan, L., Zhang, F.: Modeling fuzzy information in UML class diagrams and object-oriented database models. Fuzzy Sets Syst. **186**(1), 26–46 (2012)
36. Ma, Z.M., Yan, L.: Fuzzy XML data modeling with the UML and relational data models. Data Knowl. Eng. **63**(3), 970–994 (2007)

37. Zadeh, L.A.: The concept of a linguistic variable and its application to approximate reasoning. Inf. Sci. **8**(3), 199–249 (1975); **8**(4), 301–357 (1975); **9**(1), 43–80 (1975)
38. Ma, Z.M., Zhang, W.J., Ma, W.Y.: Semantic measure of fuzzy data in extended possibility-based fuzzy relational databases. Int. J. Intell. Syst. **15**(8), 705–716 (2000)
39. Teorey, T.J., Yang, D.Q., Fry, J.P.: A logical design methodology for relational databases using the extended entity-relationship model. ACM Comput. Surv. **18**(2), 197–222 (1986)

Fuzzy Bi-implications Generated by t-norms and Fuzzy Negations

Antonio Diego S. Farias[1]([✉]), Claudio Callejas[1], João Marcos[2],
Benjamín Bedregal[2], and Regivan Santiago[2]

[1] Centro Multidisciplinar de Pau dos Ferros,
Universidade Federal Rural do Semi-Árido - UFERSA, Pau dos Ferros, RN, Brazil
`{antonio.diego,claudio.callejas}@ufersa.edu.br`
[2] Departamento de Informática e Matemática Aplicada,
Universidade Federal do Rio Grande do Norte - UFRN, Natal, RN, Brazil
`{jmarcos,bedregal,regivan}@dimap.ufrn.br`

Abstract. In the literature, there are several forms of extensions of the classical bi-implication for the fuzzy logic, as for example, the axiomatization proposed by Fodor and Roubens [1]. Another way to obtain a generalization is to provide a definition based on the classical equivalence $\phi \iff \psi \equiv (\phi \Rightarrow \psi) \wedge (\psi \Rightarrow \phi)$, in which the classical operators of conjunction and implication are replaced, respectively, by a t-norm (T) and a fuzzy implication (I). In this paper, we investigate a particular class of fuzzy bi-implications $B(x,y) = T(I(x,y), I(y,x))$, in which I is a fuzzy (T, N)-implication introduced by Bedregal [2]. We study several properties satisfied by (T, N)-bi-implications, such as the sufficient conditions that they must satisfy in order to be a f-bi-implication.

1 Introduction

Since the introduction of fuzzy set theory [3], where the crisp membership functions valued in $\{0, 1\}$ were generalized to allow degrees of membership valued in [0,1], the investigation of fuzzy logic began as a family of multivalued logics, referred by Petr Hájek as fuzzy logic in a narrow sense [4, p. 2], which is the object of investigation of the mathematical fuzzy logic community. What differentiates fuzzy logics in a narrow sense from other multivalued logics, is that the former has both truth-functionality and truth degrees in [0,1] as fundamental assumptions (see [5]).

Several generalizations of the classical boolean connectives to the fuzzy setting have been introduced and studied. In particular the classical conjunction was extended in fuzzy logic by the triangular norms (see for instance [6–9]), the disjunction by the triangular conorms (see for instance [6–9]), the negation by the fuzzy negation [10] and the implication by the fuzzy implication (see for instance [11]). All these together have been used in several applications, for example the fuzzy implications have been useful to implement automated decision support systems with "if-then" rules, where depending on the context a suitable fuzzy implication is selected to implement such rules (see for instance [12]).

© Springer Nature Switzerland AG 2019
R. B. Kearfott et al. (Eds.): IFSA 2019/NAFIPS 2019, AISC 1000, pp. 601–612, 2019.
https://doi.org/10.1007/978-3-030-21920-8_53

These fuzzy operators have been used as truth-functional interpretations of formulas over the unit interval. For instance, Hájek [4] proposed a class of logics, called Basic Logics (BL), based on continuous triangular norms and their residua, were further to be extended to logics based on left-continuous triangular norms [5].

In classical logic the binary operator that semantically is true only when the truth-value of its operands are equal, is called equivalence or bi-implication [13, p. 7] or biconditional when the implication is called conditional [14, p. 70]. Since equivalences are reflexive, symmetric and transitive relations [13, p. 22] and the class of operators introduced in this paper are generated by two occurrences of an implication we prefer the name fuzzy bi-implication, just as in [6, p. 235].

In the literature there is not a consensus upon what a fuzzy bi-implication should be and one may find it under the names of T-indistinguishability operator [15, p. 18], fuzzy bi-implication [16,17], fuzzy equality [18], fuzzy bi-residuation [19], fuzzy equivalence [1,20], T-equivalence [21], fuzzy similarity [4, p. 123] and restricted equivalence function [22].

The first steps made in order to study the relations in between several of these extensions and to provide a few novel extensions, were made in [23–25]. In this paper we propose a novel class of fuzzy bi-implications obtained by the composition of a (T, N)-implication [2,26–28] and a fuzzy negation. We study several of its properties and, determine the sufficient conditions for such a fuzzy bi-implication to constitute a sub-class of the well-known axiomatization proposed by Fodor and Roubens in [1].

Several among these fuzzy bi-implications, and, in particular, those proposed and studied in this paper, can be applied, for example, for image comparison (see for instance [29]) as well as, be used as a truth-functional interpretation of formulas with occurrences of bi-implications, just as the Goedel logic has the conjunction interpreted as the minimum t-norm, probably in other fuzzy logics the bi-implication could be interpreted as a particular fuzzy bi-implication in between those proposed in this paper.

This paper is organized in the following manner: in Sect. 2 we provide the basic concepts needed in order to make this paper self-contained; in Sect. 3 we propose the class of (T, N)-bi-implications, study several of its properties and relate it with the Fodor-Roubens axiomatization; finally in Sect. 4 we provide some conclusions and propose directions for future works.

2   Preliminaries

In this part of the paper, we present some important preliminary notions for the development of this work.

2.1   t-Norms and Fuzzy Negations

In the literature, there are several operators that extend the classical conjunction for fuzzy logic, as for example the t-norms defined below:

Definition 1 ([6] **p. 4**). *A function $T : [0,1]^2 \to [0,1]$ is a* **t-norm** *if for all $x, y, z \in [0,1]$, the following axioms are satisfied:*

1. *(T1) Commutativity: $T(x,y) = T(y,x)$;*
2. *(T2) Associativity: $T(x, T(y,z)) = T(T(x,y), z)$;*
3. *(T3) Monotonicity: $T(x,y) \leq T(x,z)$, whenever $y \leq z$;*
4. *(T4) 1-ident: $T(x,1) = x$.*

Some classical examples of t-norms are:

Example 1 ([6] **p. 4**).

1. *The minimum $T_{\mathbf{M}}(x,y) = min(x,y)$;*
2. *The product $T_{\mathbf{P}}(x,y) = x \cdot y$;*
3. *The drastic product $T_{\mathbf{D}}(x,y) = \begin{cases} 0, & if\ (x,y) \in [0,1)^2 \\ min(x,y), & otherwise \end{cases}$*

t-norms satisfy the following properties:

Proposition 1 ([6] **pp. 5–6**). *If $T : [0,1]^2 \to [0,1]$ is a t-norm, then:*

1. *$T(0,x) = T(x,0) = 0$, for all $x \in [0,1]$;*
2. *$T(1,x) = x$, for all $x \in [0,1]$;*
3. *$T(x_1, y_1) \leq T(x_2, y_2)$, whenever $x_1 \leq x_2$ and $y_1 \leq y_2$;*
4. *$T_{\mathbf{D}}(x,y) \leq T(x,y) \leq T_{\mathbf{M}}(x,y)$, for all $x,y \in [0,1]$;*

The following Remark is a direct consequence of Definition 1 and Proposition 1.

Remark 1.

1. *$T(x,y) = 1 \iff x = 1$ and $y = 1$, for all $x,y \in [0,1]$;*
2. *$T(x,x) = 1 \Rightarrow x = 1$.*

Another important connective of fuzzy logic is the negation, as below:

Definition 2 ([11] **pp. 13–14 and** [30]).

1. *A* **fuzzy negation** *is a non-increasing function $N : [0,1] \to [0,1]$ such that $N(1) = 0$ and $N(0) = 1$;*
2. *When a fuzzy negation is involutive, i.e., satisfies the property $N(N(x)) = x$ for each $x \in [0,1]$, we say that N is a* **strong fuzzy negation**;
3. *A fuzzy negation is* **strict** *if it is continuous and for each $x,y \in [0,1]$ satisfies $N(x) > N(y)$ whenever $x < y$;*
4. *A fuzzy negation is* **non-filling** *if $N(x) = 1 \iff x = 0$;*
5. *A fuzzy negation is* **crisp** *if $N(x) \in \{0,1\}$ for any $x \in [0,1]$.*

Remark 2. *It is relevant for this work to emphasize that every strong fuzzy negation is strict [11, p. 15].*

In the example below we present some fuzzy negations.

Example 2 ([11] pp. 14–15).

1. *The Zadeh's negation $N_C(x) = 1 - x$ is strong and non-filling, but is not crisp;*
2. *$N_R(x) = 1 - \sqrt{x}$ is strict, non-filling, but is not crisp and strong;*
3. *The threshold $N^t(x) = \begin{cases} 1, & if\ x < t \\ 1\ or\ 0, & if\ x = t, t \in (0,1) \\ 0, & if\ x > t \end{cases}$ is crisp, but is neither strict nor non-filling.*

2.2 Fuzzy Implications

The notion of implication in Fuzzy Logic has many non equivalent extensions. In this paper we are going to use the next one.

Definition 3 ([11] p. 2). *A **fuzzy implication** is a binary operator $I : [0,1]^2 \to [0,1]$ that satisfies:*

(I1) $I(x_1, y) \geq I(x_2, y)$, whenever $x_1, x_2, y \in [0,1]$ and $x_1 \leq x_2$;
(I2) $I(x, y_1) \leq I(x, y_2)$, whenever $x, y_1, y_2 \in [0,1]$ and $y_1 \leq y_2$;
(I3) $I(0,0) = 1$;
(I4) $I(1,1) = 1$;
(I5) $I(1,0) = 0$.

Properties $(I1)$ and $(I2)$ are called *first place antitonicity* and *second place isotonicity*, respectively The remaining properties together with

(I6) $I(0,1) = 1$,

are called boundary conditions. The boundary conditions guarantee that the class of fuzzy implications extend the classical implication. The next result has $(I6)$ as a particular case. Hence, $(I6)$ is unnecessary in the Definition 3.

Proposition 2 ([11] p. 2). *If $I : [0,1]^2 \to [0,1]$ is a fuzzy implication, then*

$$I(x,1) = 1, \ for\ all\ x \in [0,1]$$

The properties considered in the next two definitions were stated for fuzzy implications in [11, pp. 9,20] and [24], but we will study them for fuzzy bi-implications, which will be presented in Sect. 3.

Definition 4. *A fuzzy operator $F : [0,1]^2 \to [0,1]$ is said to satisfy the property of:*

*(LNP) **left neutrality** if:*

$$F(1,y) = y, \ for\ all\ y \in [0,1]$$

*(IP) **identity** if:*

$$F(x,x) = 1, \ for\ all\ x \in [0,1]$$

(LOP) **left-ordering** *if:*

$$F(x, y) = 1, \quad whenever \ x \le y$$

Definition 5. *Let* $N : [0,1] \to [0,1]$ *be a fuzzy negation. A fuzzy operator* $F :$ $[0,1] \to [0,1]$ *is said to satisfy:*

(CP) the **contraposition law** *with respect to* N, *if:*

$$F(x, y) = F(N(y), N(x)), \quad for \ all \ x, y \in [0,1]$$

(LCP) the **left contraposition law** *with respect to* N, *if:*

$$F(N(x), y) = F(N(y), x), \quad for \ all \ x, y \in [0,1]$$

(RCP) the **right contraposition law** *with respect to* N, *if:*

$$F(x, N(y)) = F(y, N(x)), \quad for \ all \ x, y \in [0,1]$$

If F satisfies the contraposition law (or left contraposition or right contraposition) with respect to N, then we denote these properties by $CP(N)$ (respectively, by $LCP(N)$ or $RCP(N)$).

2.2.1   (T,N)-implications

In [2], the author introduced a class of fuzzy implications obtained by the defining standard based on the classical equivalence $\phi \Rightarrow \psi \equiv \neg(\phi \wedge \neg\psi)$

Definition 6 ([2]). *Let* $T : [0,1]^2 \to [0,1]$ *be a t-norm and* $N : [0,1] \to [0,1]$ *be a fuzzy negation. The function defined by:*

$$I_T^N(x, y) = N(T(x, N(y))), \quad for \ every \ x, y \in [0,1]$$

is called N-dual fuzzy implication of T.

In [26–28], the authors called the N-dual fuzzy implications simply as (T, N)-implications and studied properties of these fuzzy operators. For example, they studied the conditions for (T, N)-implications to satisfy (EP), (CP), (LCP), (RCP) and (LNP). For this work, it is important to mention the following result:

Proposition 3 ([27]). *Let* $T : [0,1]^2 \to [0,1]$ *be a t-norm. If* $N : [0,1] \to [0,1]$ *is a crisp fuzzy negation, then* I_T^N *satisfies (LOP).*

2.3   Fuzzy Bi-implications

In what follows we show an axiomatization of a class of fuzzy bi-implications proposed by Fodor and Roubens.

Definition 7 ([1] p. 33). *A function* $B : [0,1]^2 \to [0,1]$ *is called* ***f-bi-implication*** *if it satisfies the following axioms:*

(B1) $B(x,y) = B(y,x)$, for all $x, y \in [0,1]$; (commutativity)
(B2) $B(0,1) = B(1,0) = 0$; (boundary condition)
(B3) $B(x,x) = 1$, for all $x \in [0,1]$; (identity principle)
(B4) $B(x,y) \leq B(x',y')$, whenever $x \leq x' \leq y' \leq y$.

It is important to note that other two further boundary conditions that an extension of the classical bi-implication needs to satisfy are $B(0,0) = B(1,1) = 1$ which are immediate consequences of *(B3)*.

Example 3. *Examples of f-bi-implications are:*

1. $B_{\mathbf{M}}(x,y) = \begin{cases} 1, & \text{if } x = y \\ min(x,y), & \text{otherwise} \end{cases}$, that satisfies (LNP), (IP), but does not satisfy (LOP);

2. $B_{\mathbf{P}}(x,y) = \begin{cases} 1, & \text{if } x = y \\ \frac{min(x,y)}{max(x,y)}, & \text{otherwise} \end{cases}$, that satisfies (LNP), (IP), but does not satisfy (LOP);

3. $B_{\mathbf{KP}}(x,y) = 1 - max(x^2, y^2) + xy$, that satisfies (IP), but does not satisfy (LNP) and (LOP).

Note that, by *(B2)* their does not exist a f-bi-implication that satisfies (LOP).

3   Fuzzy Bi-implications Generated by t-Norms and Fuzzy Negations

In this section, we investigate a special type of fuzzy bi-implications obtained by the defining standard based on the classical logical equivalence $\phi \iff \psi \equiv (\phi \Rightarrow \psi) \wedge (\psi \Rightarrow \phi)$ Considering the t-norms and fuzzy implications as a generalizations, respectively, of the classical conjunction and classical implication, we have the following function $B(x,y) = T(I(x,y), I(y,x))$, where T is a t-norm and I is a fuzzy implication. In this paper, we study the case in which I is a (T,N)-implication.

Definition 8. *Let $N : [0,1] \rightarrow [0,1]$ be a fuzzy negation and $T : [0,1]^2 \rightarrow [0,1]$ be a t-norm. The N dual fuzzy bi-implication of T (or fuzzy (T,N)-bi-implication or simply (T,N)-bi-implication) is a function $B_T^N : [0,1]^2 \rightarrow [0,1]$ of the form:*

$$B_T^N(x,y) = T(I_T^N(x,y), I_T^N(y,x))$$
$$= T(N(T(x,N(y))), N(T(y,N(x))))$$

Example 4. *Let $T_{\mathbf{M}}$ and N_C be the operators defined, respectively, in Examples 1 and 2, then:*

$$B_{T_{\mathbf{M}}}^{N_C}(x,y) = min(1 - min(x, 1-y), 1 - min(y, 1-x))$$
$$= min(max(1-x, y), max(1-y, x))$$

Note that $B_{T_{\mathbf{M}}}^{N_C}$ is not a f-bi-implication, since it does not satisfy *(B3)*, for example, $B_{T_{\mathbf{M}}}^{N_C}(0.7, 0.7) = 0.7$. Thus, (T,N)-bi-implications fail to constitute a subclass of the class of f-bi-implications.

3.1   Properties of (T, N)-bi-implications

In this subsection we investigate some properties of (T, N)-bi-implications. In the next result we show that there exist an unique fuzzy negation that generates a given (T, N)-bi-implication.

Proposition 4. *Let $T : [0, 1]^2 \to [0, 1]$ be a t-norm and $N : [0, 1] \to [0, 1]$ be a fuzzy negation. Then,*

$$B_T^N(x, 0) = N(x), \text{ for any } x \in [0, 1]$$

Proof. Just note that:

$$
\begin{aligned}
B_T^N(x, 0) &= T(N(T(x, N(0))), N(T(0, N(x)))) \\
&= T(N(T(x, 1)), N(0)) \quad - \quad \text{by 1 of Proposition 1} \\
&= T(N(x), 1) \quad\quad\quad - \quad \text{by (T4) and Definition 2} \\
&= N(x) \quad\quad\quad\quad\quad - \quad \text{by (T4)}
\end{aligned}
$$

Corollary 1. *If $T : [0, 1]^2 \to [0, 1]$ is a t-norm and $N : [0, 1] \to [0, 1]$ is a fuzzy negation, then $B_T^N(\cdot, 0) : [0, 1] \to [0, 1]$ is a fuzzy negation.*

A natural question is, does there exist an unique t-norm that generates a given (T, N)-bi-implication. For a (T, N)-bi-implication generated by strong fuzzy negations, the following Proposition proves a positive answer for this question.

Proposition 5. *If N is a strong fuzzy negation and T is t-norm, then it does not exist a t-norm $T' \neq T$ such that $B_T^N = B_{T'}^N$.*

Proof. Just note that, for any strong negation N and $x, y \in [0, 1]$

$$
\begin{aligned}
I(x, y) = N(T(x, N(y))) &\Rightarrow N(I(x, y)) = T(x, N(y)) \\
&\Rightarrow T(x, y) = N(I(x, N(y)))
\end{aligned}
$$

Note that when a (T, N)-bi-implication is generated by a strong fuzzy negation, both the fuzzy negation and the t-norm, are unique. The next two results are immediately obtained by the Definition 8 and the Corollary 1.

Proposition 6. *If $T : [0, 1]^2 \to [0, 1]$ is a t-norm and $N : [0, 1] \to [0, 1]$ is a fuzzy negation, then B_T^N satisfies (B1).*

Corollary 2. *If $T : [0, 1]^2 \to [0, 1]$ is a t-norm and $N : [0, 1] \to [0, 1]$ is a fuzzy negation, then $B_T^N(0, \cdot) : [0, 1] \to [0, 1]$ is a fuzzy negation.*

Since there is a unique fuzzy negation that generates a (T, N)-bi-implication, the fuzzy negation of Corollaries 1 and 2 coincide. In the following results we investigate the sufficient conditions for a (T, N)-bi-implication to be a f-bi-implication.

Proposition 7. *If $T : [0,1]^2 \to [0,1]$ is a t-norm and $N : [0,1] \to [0,1]$ is a fuzzy negation, then B_T^N satisfies (B2).*

Proof. Follows from (B1) and by the equality $B_T^N(1,0) = N(1) = 0$ of Proposition 4.

Proposition 8. *If N is a non-filling fuzzy negation, then B_T^N satisfies (B3) if, and only if, the pair (T, N) satisfies the law of non-contradiction.*[1]

Proof. Since,

$$B_T^N(x,x) = 1 \iff T(N(T(x, N(x))), N(T(x, N(x)))) = 1 \; - \; \text{by Definition 8}$$
$$\iff N(T(x, N(x))) = 1 \qquad\qquad - \; \text{by Remark 1}$$
$$\iff T(x, N(x)) = 0 \qquad\qquad - \; \text{by non-filling condition}$$

Corollary 3. *If N is a strict fuzzy negation, then B_T^N satisfies (B3) if, and only if, the pair (T, N) satisfies the law of non-contradiction.*

Proof. It follows from the fact that all strict fuzzy negation are injective hence, non-filling.

Example 5. *If $T = T_\mathbf{P}$ and $N_\perp(x) = \begin{cases} 1, & \text{if } x = 0 \\ 0, & \text{otherwise} \end{cases}$, then (T, N) satisfies the law of non-contradiction, $N_\perp$ is a non-filling fuzzy negation that fails to be strict and $B_{T_\mathbf{P}}^{N_\perp}$ satisfies (B3), by Proposition 7.*

Theorem 1. *Let $T : [0,1]^2 \to [0,1]$ be a t-norm and $N : [0,1] \to [0,1]$ be a fuzzy negation. If I_T^N satisfies (LOP), then B_T^N satisfies (B4).*

Proof. First let's see that:

$$B_T^N(x,y) = \begin{cases} N(T(x, N(y))), & \text{if } x \geq y \\ N(T(y, N(x))), & \text{if } x \leq y \end{cases} \tag{1}$$

Indeed, as for all $x, y \in [0,1]$ we have

$$x \leq y \text{ or } y \leq x$$

Then, by (LOP) of I_T^N, we have

$$I_T^N(x,y) = 1 \text{ or } I_T^N(y,x) = 1$$

Thus, the equality

$$B_T^N(x,y) = T(N(T(x, N(y))), N(T(y, N(x))))$$
$$= T(I_T^N(x,y), I_T^N(y,x))$$

---

[1] If $T : [0,1]^2 \to [0,1]$ is a t-norm and $N : [0,1] \to [0,1]$ is a fuzzy negation, then we say that the pair (T, N) satisfies the law of non-contradiction if $T(x, N(x)) = 0$, for all $x \in [0,1]$ (this law is equivalently stated in [11, p. 55]).

ensures the Eq. (1) is valid.

Now, given $x \leq x' \leq y' \leq y$, then

$$B_T^N(x, y) = N(T(y, N(x))) \text{ and } B_T^N(x', y') = N(T(y', N(x')))$$

Therefore, by the monotonicity conditions of T and N, we conclude that

$$B_T^N(x, y) \leq B_T^N(x', y')$$

Corollary 4. *If $T : [0, 1]^2 \to [0, 1]$ is a t-norm and $N : [0, 1] \to [0, 1]$ is a non-filling fuzzy negation such that the pair (T, N) satisfies the law of non-contradiction and I_T^N satisfies (LOP), then B_T^N is a f-bi-implication.*

Proof. Follows from Propositions 5, 6 and 7, and Theorem 1.

Corollary 5. *If N is a crisp fuzzy negation and T is a t-norm, then B_T^N satisfies $(B4)$.*

Proof. By Proposition 2, I_T^N satisfies (LOP). Therefore, by Theorem 1, B_T^N satisfies $(B4)$.

There is only one crisp fuzzy negation that is non-filling, that is $N_\perp$, which in turn also satisfies the law of non-contradiction with any t-norm. Consequently, by Corollaries 4 and 5, every (T, N)-bi-implication generated by $N_\perp$ and a t-norm is a f-bi-implication.

There are (T, N)-implications that fail to satisfy (LOP). For example, for $T_\mathbf{M}$ and N_C we have that $I_{T_\mathbf{M}}^{N_C}(0.3, 0.5) = 0.7$. But in the next result the (T, N)-bi-implication generated by $T_\mathbf{M}$ and a strong fuzzy negation satisfies $(B4)$.

Theorem 2. *If $T = T_\mathbf{M}$ and N is a strong fuzzy negation, then $B_{T_\mathbf{M}}^N$ satisfies $(B4)$.*

Proof. For any $x, y \in [0, 1]$, either $x < N(y)$ or $x \geq N(y)$. In addition, as N is a strong fuzzy negation:

$$x \leq N(y) \iff y \leq N(x)$$

and

$$x \geq N(y) \iff y \geq N(x)$$

If $x \leq N(y)$, then $y \leq N(x)$ and so,

$$\begin{aligned}
B_{T_\mathbf{M}}^N(x, y) &= T_\mathbf{M}(N(T_\mathbf{M}(x, N(y))), N(T_\mathbf{M}(y, N(x)))) \\
&= min(N(min(x, N(y))), N(min(y, N(x)))) \\
&= min(N(x), N(y))
\end{aligned}$$

If $x \geq N(y)$, then $y \geq N(x)$ and so

$$\begin{aligned}
B_{T_\mathbf{M}}^N(x, y) &= T_\mathbf{M}(N(T_\mathbf{M}(x, N(y))), N(T_\mathbf{M}(y, N(x)))) \\
&= min(N(min(x, N(y))), N(min(y, N(x)))) \\
&= min(N(N(y)), N(N(x))) \\
&= min(x, y) \qquad - \quad \text{because } N \text{ is strong}
\end{aligned}$$

Thus,

$$B_{T_M}^N(x,y) = \begin{cases} min(N(x), N(y)), & \text{if } x \leq N(y) \\ min(x,y), & \text{if } x \geq N(y) \end{cases}$$

Let $x \leq x' \leq y' \leq y$. Then,

- for $x \leq N(y)$, we have

$$B_{T_M}^N(x,y) = N(y)$$

Thus, if $x' \leq N(y')$, then $B_{T_M}^N(x',y') = N(y') \geq N(y) = B_{T_M}^N(x,y)$. If $x' \geq N(y')$, then $B_{T_M}^N(x',y') = x' \geq N(y') \geq N(y) = B_{T_M}^N(x,y)$. So, $B_{T_M}^N(x,y) \leq B_{T_M}^N(x',y')$.

- for $x \geq N(y)$, we have

$$B_{T_M}^N(x,y) = x$$

Thus, if $x' \leq N(y')$, then $B_{T_M}^N(x',y') = N(y') \geq x' \geq x = B_{T_M}^N(x,y)$. If $x' \geq N(y')$, then $B_{T_M}^N(x',y') = x' \geq x = B_{T_M}^N(x,y)$. So, $B_{T_M}^N(x,y) \leq B_{T_M}^N(x',y')$.

Therefore, $B_{T_M}^N$ satisfies $(B4)$.

Even though $B_{T_M}^N$ of Theorem 2 satisfies $(B4)$ it is worthy to mention that the pair (T_M, N), where N is a strong fuzzy negation, does not satisfy the law of non-contradiction, because $min(x, N(x)) = 0 \iff x = 0$ or $N(x) = 0$. Hence $B_{T_M}^N$ fails to be a f-implication. The next Propositions show sufficient conditions for a (T, N)-bi-implication to satisfy (LNP), $LCP(N)$, $RCP(N)$ and $CP(N)$.

Proposition 9. *Let $T : [0,1]^2 \to [0,1]$ be a t-norm and $N : [0,1] \to [0,1]$ be a fuzzy negation. If N is strong, then B_T^N satisfies (LNP).*

Proof. For (LNP) we have:

$$\begin{aligned} B_T^N(1,x) &= T(N(T(1, N(x))), N(T(x, N(1))) && \text{– by Definition 8} \\ &= T(N(N(x)), N(T(x,0))) && \text{– by (T1) and (T4)} \\ &= T(x, N(0)) && \text{– by Proposition 1 and because } N \text{ is strong} \\ &= x && \text{– by Definition 2 and (T4)} \end{aligned}$$

Proposition 10. *If N is a strong fuzzy negation and T is a t-norm, then B_T^N satisfies $LCP(N)$, $RCP(N)$ and $CP(N)$.*

Proof. For $LCP(N)$, just see that for all $x, y \in [0,1]$:

$$\begin{aligned} B_T^N(N(x), y) &= T(N(T(N(x), N(y))), N(T(y, N(N(x))))) \\ &= T(N(T(N(x), N(y))), N(T(y,x))) \\ &= T(N(T(N(y), N(x))), N(T(x, N(N(y))))) \\ &= B_T^N(N(y), x) \end{aligned}$$

The property $RCP(N)$ follows from Proposition 9 and $(B1)$, and for $CP(N)$, just note that for all $x, y \in [0, 1]$:

$$
\begin{aligned}
B_T^N(N(x), N(y)) &= T(N(T(N(x), N(N(y)))), N(T(N(y), N(N(x))))) \\
&= T(N(T(N(x), y)), N(T(N(y), x))) \\
&= T(N(T(x, N(y))), N(T(y, N(x)))) \\
&= B_T^N(x, y)
\end{aligned}
$$

4   Conclusions and Future Works

In this paper, we introduce a new class of binary operators that extend the classical bi-implications, called (T, N)-bi-implications. We show that the class of (T, N)-bi-implications is not contained in the class of f-bi-implications and that these two classes of functions have a non-empty intersection. We also obtain sufficient conditions for a (T, N)-bi-implication to be a f-bi-implication. Some open questions are: if the class of f-bi-implications is a subclass of the class of (T, N)-bi-implications and investigate other properties satisfied by the (T, N)-bi-implications, as the exchange principle.

Acknowledgement. This work is partially supported by Universidade Federal Rural do Semi-Árido - UFERSA (Project PIH10002-2018).

References

1. Fodor, J.C., Roubens, M.: Fuzzy Preference Modelling and Multicriteria Decision Support, vol. 14. Springer, Heidelberg (1994)
2. Bedregal, B.C.: A normal form which preserves tautologies and contradictions in a class of fuzzy logics. J. Algorithms **62**(3), 135–147 (2007)
3. Zadeh, L.A.: Fuzzy sets. Inf. Control **8**(3), 338–353 (1965)
4. Hájek, P.: Metamathematics of Fuzzy Logic, vol. 4. Springer, Heidelberg (2013)
5. Behounek, L., Cintula, P., Hájek, P.: Introduction to Mathematical Fuzzy Logic. College Publications (2011). Ch. 1
6. Klement, E.P., Mesiar, R., Pap, E.: Triangular Norms, vol. 8. Springer, Heidelberg (2000)
7. Klement, E., Mesiar, R., Pap, E.: Triangular norms. Position paper I: basic analytical and algebraic properties. Fuzzy Sets Syst. **143**(1), 5–26 (2004)
8. Klement, E., Mesiar, R., Pap, E.: Triangular norms. Position paper II: general constructions and parameterized families. Fuzzy Sets Syst. **145**(3), 411–438 (2004)
9. Klement, E., Mesiar, R., Pap, E.: Triangular norms. Position paper III: continuous t-norms. Fuzzy Sets Syst. **145**(3), 439–454 (2004)
10. Bertei, A., Zanotelli, R., Cardoso, W., Reiser, R., Foss, L., Bedregal, B.: Correlation coefficient analysis based on fuzzy negations and representable automorphisms. In: 2016 IEEE International Conference on Fuzzy Systems, FUZZ-IEEE 2016, 24–29 July 2016, Vancouver, BC, Canada, pp. 127–132 (2016)
11. Baczyński, M., Jayaram, B.: Fuzzy Implications. Studies in Fuzziness and Soft Computing, vol. 231. Springer, Berlin (2008)

12. Mizumoto, M.: Fuzzy controls under various fuzzy reasoning methods. Inf. Sci. **45**(2), 129–151 (1988)
13. Dalen, D.: Logic and Structure, 5th edn. Springer, Heidelberg (2013). Universitext
14. Kleene, S.: Mathematical Logic. Dover Books on Mathematics. Dover Publications, Mineola (2002)
15. Recasens, J.: Indistinguishability Operators: Modelling Fuzzy Equalities and Fuzzy Equivalence Relations, vol. 260. Springer, Heidelberg (2010)
16. Bodenhofer, U.: A compendium of fuzzy weak orders: representations and constructions. Fuzzy Sets Syst. **158**(8), 811–829 (2007)
17. Bedregal, B.C., Cruz, A.P.: A characterization of classic-like fuzzy semantics. Logic J. IGPL **16**(4), 357–370 (2008)
18. Novák, V., De Baets, B.: Eq-algebras. Fuzzy Sets Syst. **160**(20), 2956–2978 (2009)
19. Mesiar, R., Novák, V.: Operations fitting triangular-norm-based biresiduation. Fuzzy Sets Syst. **104**(1), 77–84 (1999)
20. Ćirić, M., Ignjatović, J., Bogdanović, S.: Fuzzy equivalence relations and their equivalence classes. Fuzzy Sets Syst. **158**(12), 1295–1313 (2007)
21. Moser, B.: On the t-transitivity of kernels. Fuzzy Sets Syst. **157**(13), 1787–1796 (2006)
22. Bustince, H., Barrenechea, E., Pagola, M.: Restricted equivalence functions. Fuzzy Sets Syst. **157**(17), 2333–2346 (2006)
23. Callejas, C.: What is a fuzzy bi-implication? Master's thesis, Universidade Federal do Rio Grande do Norte (2012)
24. Callejas, C., Marcos, J., Bedregal, B.R.C.: On some subclasses of the Fodor-Roubens fuzzy bi-implication. In: Proceedings of Logic, Language, Information and Computation - 19th International Workshop, WoLLIC 2012, 3–6 September 2012, Buenos Aires, Argentina, pp. 206–215 (2012)
25. Callejas, C., Marcos, J., Bedregal, B.: Actions of automorphisms on some classes of fuzzy bi-implications. Mathware Soft Comput. Mag. **20**, 94–97 (2013)
26. Pinheiro, J., Bedregal, B., Santiago, R., Santos, H., Dimuro, G.P.: (T,N)-implications and some functional equations. In: Barreto, G.A., Coelho, R. (eds.) Fuzzy Information Processing, pp. 302–313. Springer, Cham (2018)
27. Pinheiro, J., Bedregal, B., Santiago, R.H., Santos, H.: A study of (T,N)-implications and its use to construct a new class of fuzzy subsethood measure. Int. J. Approximate Reason. **97**, 1–16 (2018)
28. Pinheiro, J., Bedregal, B., Santiago, R.H.N., Santos, H.: (T, N)-implications. In: 2017 IEEE International Conference on Fuzzy Systems (FUZZ-IEEE), pp. 1–6 (2017)
29. Bustince, H., Barrenechea, E., Pagola, M.: Image thresholding using restricted equivalence functions and maximizing the measures of similarity. Fuzzy Sets Syst. **158**(5), 496–516 (2007)
30. Dimuro, G.P., Bedregal, B., Bustince, H., Jurio, A., Baczynski, M., Mis, K.: QL-operations and QL-implication functions constructed from tuples (O, G, N) and the generation of fuzzy subsethood and entropy measures. Int. J. Approximate Reason. **82**, 170–192 (2017)

Calculations of Zadeh's Extension of Piecewise Linear Functions

Jiří Kupka and Nicole Škorupová[(✉)]

Centre of Excellence IT4Innovations, Division University of Ostrava, IRAFM,
University of Ostrava, 30. dubna 22, 701 03 Ostrava 1, Czech Republic
{Jiri.Kupka,Nicole.Skorupova}@osu.cz

Abstract. Zadeh's extension principle is one of the most classical techniques in fuzzy set theory. It is a tool which, for example, can naturally extend a real-valued continuous map to a map having fuzzy sets as its arguments. Theoretically, it is a nice mathematical tool used in many theories, e.g. in studies on fuzzy dynamical systems. However, concrete calculations or even approximations can be very difficult in general and, consequently, many approaches trying to solve this problem appeared. In this work we present a novel algorithm which can compute Zadeh's extension of given continuous piecewise linear functions. Among other things, an advantage of this approach is that, unlike almost all former approaches, it can deal with discontinuities which naturally appear in simulations of fuzzy dynamical systems.

1 Introduction

So-called Zadeh's extension principle plays an important role in fuzzy mathematics because it can be naturally used to extend many classical (non-fuzzy) operations into fuzzy ones, i.e. those working with fuzzy sets [21]. This extension principle can also be used for extending a single valued map. Namely, by using the following formula

$$(z_f(A))(x) = \sup_{y \in f^{-1}(x)} \{A(y)\}, \tag{1}$$

each map $f \colon X \to Y$ uniquely induces a map $z_f \colon \mathbb{F}(X) \to \mathbb{F}(Y)$, where $\mathbb{F}(X)$ (resp. $\mathbb{F}(Y)$) is an appropriate class of fuzzy sets defined on X (resp. Y). For definitions of other notions used in this introduction we refer to the next section.

There exists a very tight relation of the expression (1) to theory of dynamical systems. Namely, a pair (X, f), where X is a topological space and $f \colon X \to X$ is continuous, is called a *(discrete) dynamical system*. Kloeden in [14] elaborated a nice formal mathematical model which allowed to connect theory of (non-fuzzy) dynamical systems to theory of fuzzy dynamical systems. By using this model and (1), for every dynamical system (X, f) we can create induced fuzzy dynamical system $(\mathbb{F}(X), z_f)$, and, consequently, mathematicians can study properties of fuzzy dynamical systems with the help of classic theory which has been

© Springer Nature Switzerland AG 2019
R. B. Kearfott et al. (Eds.): IFSA 2019/NAFIPS 2019, AISC 1000, pp. 613–624, 2019.
https://doi.org/10.1007/978-3-030-21920-8_54

elaborated for decades for non-fuzzy systems. There are many papers contributing to this topic, e.g. [4,5,14,16] and references therein.

However the problem often appears when the users try to calculate (1) in particular applications. In general, calculation of Zadeh's extension can be a difficult task [21]. The difficulty comes from computing the inverse map f^{-1} at particular points. Therefore, many authors proposed various approaches and tried to find approximations of images of given fuzzy sets when using Zadeh's extension principle (1). Below we discuss some of them.

For example in [8], a method approximating $z_f(A)$ via decomposition into fuzzy intervals and multilinearization of f was introduced. In particular, the proposed algorithm decomposed a given fuzzy interval A into n compact fuzzy sets $A_1, A_2, \ldots, A_n$, such that $A = \sup A_i$ and each A_i has an uncertainty interval smaller than the original fuzzy set A. Then, using the differentiability of f a multilinearization of f is obtained and used for approximation. A similar technique approximating $z_f(A)$ with help of the decomposition of the fuzzy interval and linear spline functions was introduced in [7]. This method was also extended in [6] for the case when the approximated function has more than two arguments.

In [1], the authors dealt with non-monotone functions and used optimization techniques (based on Brent's method [3]) to find global maximum and minimum values. As a consequence, a method using an optimization over α-cuts of a given fuzzy set was proposed. This approach ensures convexity of the solution $(z_f(A))$ however it is computationally efficient.

There are also approaches using different parametric representations of a fuzzy set A. The use of the parametric LU-representation of fuzzy numbers was suggested for example in [12] and [20]. Note that this specific LU-fuzzy representation allows the user a relatively easy and fast simulation of fuzzy dynamical systems and, consequently, to avoid usual massive computational work which is an obstacle in numerical applications. However, there is a need of reconstruction of calculated fuzzy sets and the reconstructed image is always a continuous fuzzy number. There is one more paper we would like to comment. In most papers dealing with approximations of Zadeh's extension the authors used so-called levelwise metric d_∞ and only rarely provided results on quality of approximation of proposed algorithms. Related to this, the author of [17] studied the role of chosen metric and its effect to quality of the approximation. It was shown that the choice of the levelwise metric d_∞ need not be the most appropriate one, and that the approximation could be simplified by choosing another metric, namely the endograph one.

Let us comment the last three papers. In [13] the author introduced a specific implementation of α-cut based fuzzy arithmetic that avoided the well-known overestimation effect which usually arises whenever fuzzy arithmetic is based on interval calculus. This approach eliminates potential artificial widening of simulation results. In [11] an efficient algorithm for the computations of extended algebraic operations on fuzzy numbers is described. The idea again uses α-cut based representation of fuzzy numbers and, additionally, non-linear programming implementation of the extension principle. To finish our brief overview, a new optimization derivative-free method was recently published in [9]. It is based on novel low-rank tensor methods to the problem of propagating fuzzy uncertainty through a continuous real-valued function. The idea is to reformulate the Zadeh's extension as a sequence of optimization problems over nested search spaces. In this approach, the authors designed the algorithm for the use with continuous real valued functions and fuzzy numbers, resp. a vector of fuzzy numbers. The three papers mentioned above solved mainly problems related to fuzzy arithmetics, not those related to simulations of fuzzy dynamical systems.

As we could see most of existing approaches only partly coped with difficulty of calculation or approximation of Zadeh's extension, even in one-dimensional case. They either contributed to problems with fuzzy arithmetics or fuzzy numbers, resp. intervals (and their specific representations), or require the map f to satisfy some additional assumptions like differentiability.

We contribute to this problem by restricting our attention to piecewise linear fuzzy sets and continuous piecewise linear functions. Our algorithm allows us to calculate Zadeh's extension for the class of piecewise linear fuzzy sets, for which we do not assume even the continuity. On the contrary to existing approaches, our approach is not in fact an approximation but a precise calculation of fuzzy dynamical system at given fuzzy sets. This approach provides an easy-to-implement and easy-to-compute solution because the problem behind the algorithm is restricted to linear maps. Although we did not consider the class of all continuous maps but a smaller class of continuous piecewise linear functions, which is however dense in the class of continuous maps, our proposal provides one significant feature which is not involved in previous approaches. Namely, our proposal can deal with discontinuous fuzzy sets which naturally appear as images of continuous fuzzy sets – even in very simple cases as it is demonstrated in Example 2. This manuscript is intended as the first step of this direction, The fact that piecewise linear functions are dense in the space of continuous functions indicates that our approach can be used to approximate even more complex dynamical systems. Due to the page limit we have for this contribution, we have no space to describe details of our further ideas.

This manuscript has the following structure. In the next section, we introduce the most elementary terms which have to be provided when working with fuzzy dynamical systems. In Sect. 3, the above mentioned algorithm for piecewise linear functions is described. In Sect. 4 we then demonstrate our algorithm by various examples and comment some of its features and we also briefly discuss computational complexity of the proposed algorithm. And finally, some concluding remarks are formulated in Sect. 5.

2 Preliminaries

2.1 Basic Notation

Let X be a nonempty set and (X, d_X) be a compact metric space. A *fuzzy set* A on X is defined as a map $A \colon X \to [0, 1]$ and, for a given point $x \in X$, the value $A(x)$ is called a *membership degree* of x in the fuzzy set A. By a *support* of A we mean the topological closure of the set $\{x \in X \mid A(x) > 0\}$. By $\mathbb{F}(X)$ we denote a family of maps $A\colon X \to [0,1]$ such that each A is upper semi-continuous and has a compact support. Such assumptions are needed to define a metric on $\mathbb{F}(X)$. Further, for $\alpha \in (0,1]$, an α-cut of A is the set $A_\alpha = \{x \in X \mid A(x) \geq \alpha\}$. We say that a fuzzy set A is *normal* if $A(x) = 1$ for some $x \in X$. Note that if a fuzzy set A is upper semi-continuous then every α-cut is a closed subset of X.

The family of *normal* upper semi-continuous fuzzy sets on X with compact supports will be denoted by $\mathbb{F}^1(X)$ and by $\mathbb{F}^1_1(X)$ we denote a family of *fuzzy numbers* on X, i.e. a family $\mathbb{F}^1_1(X) \subset \mathbb{F}^1(X)$ of fuzzy sets for which each α-cut is topologically connected in X. Thus, if $X \subset \mathbb{R}$ and $A \in \mathbb{F}^1_1(X)$ then $[A]_\alpha$ is a (possibly even degenerated) interval for any $\alpha > 0$ and our definition of fuzzy numbers coincides with other definition of fuzzy numbers on $\mathbb{R}$, where fuzzy numbers are considered as complex normal fuzzy sets. In our work we deal with topological connectedness instead of convexity, because connectedness is preserved by continuous maps, i.e. by iterations in dynamical systems, whereas convexity is not. For further details we refer to [16]. Analogously, by $\mathbb{F}^1_k(X)$ we denote the family of fuzzy sets from $\mathbb{F}^1(X)$ for which every α-cut consists of at most k topologically connected components [17].

2.2   Metrics

In this subsection we introduce a metric for the family $\mathbb{F}(X)$ of fuzzy sets. Commonly used metrics on $\mathbb{F}(X)$ are usually based on well-known Hausdorff metric D_X between $E, F \in \mathbb{K}(X)$ [10,14], where $\mathbb{K}(X)$ denotes a space of nonempty closed subsets of X. This is one of the reasons for the upper semicontinuity of each element from $\mathbb{F}(X)$, because every α-cut of any upper semicontinuous fuzzy set on X is a closed subset of X. So, for any $E, F \in \mathbb{K}(X)$, the *Hausdorff metric* D_X is defined by the following expression

$$D_X(E, F) = \inf\{\varepsilon > 0 \mid E \subseteq U_\varepsilon(F) \text{ and } F \subseteq U_\varepsilon(E)\},$$

where

$$U_\varepsilon(E) = \{x \in X \mid D(x, E) < \varepsilon\}$$

and

$$D(x, E) = \inf_{e \in E} d_X(x, e).$$

Probably the most often used metric on $\mathbb{F}(X)$ is so-called *supremum metric* d_∞ defined as

$$d_\infty(E, F) = \sup_{\alpha \in (0,1]} D_X([E]_\alpha, [F]_\alpha),$$

for $E, F \in \mathbb{F}^1(X)$. However, very often d_∞ need not be the best metric, especially when approximating Zadeh's extension, for details we refer to [17]. Naturally, the family $\mathbb{F}^1(X)$ is equipped with the metric topology τ_∞ induced by the metric d_∞.

2.3   Dynamical Systems

By a dynamical system we denote a mathematical model of a system in which a function describes the time dependence of a point representing an initial state in a given space [18]. Formally, let X is a metric space and $f: X \rightarrow X$ be a continuous map. Then a pair (X, f) form a *(discrete) dynamical system*.

For a fixed point $x \in X$ we consider a sequence $\{f^n(x)\}_{n \in \mathbb{N}}$ of forward iterates defined inductively by $f^0(x) = x$, $f^1(x) = f(x)$ and $f^{n+1}(x) = f(f^n(x))$ for every $n \in \mathbb{N}$. This sequence is called a *forward trajectory* of the point x under the map f. Points from X can be naturally described by properties of their trajectories. For example, a point $x \in X$ is called a *fixed point* of the map f if $f(x) = x$. A point $x \in X$ is called a *periodic point* of the map f if there exists $n \in \mathbb{N}$, $n > 0$, such that $f^n(x) = x$. The example below demonstrates the fact that a very small change of the initial state may significantly change behaviour of given discrete dynamical system.

Example 1. Consider a *tent map* T, where $T: [0, 1] \rightarrow [0, 1]$ is defined by $T(x) = 2x$ for $x \in [0, 1/2)$ and $T(x) = 2(1 - x)$ otherwise.

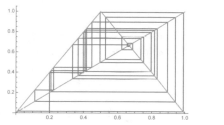 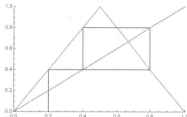

Fig. 1. 50 iterations, $x_0 = 0.199$ (the left picture), 50 iterations, $x_0 = 0.2$ (the right picture).

In Fig. 1, if we take $x_0 = 0.199$ as an initial point, this initial state generates rather complex dynamics, i.e. $x_0 = 0.199 \mapsto f(x_0) = 0.398 \mapsto f(f(x_0)) = 0.796 \mapsto f(f(f(x_0))) = 0.408 \mapsto \ldots$. However a trajectory of an initial state $x_0 = 0.2$ is depicted. Clearly, x_0 is eventually periodic, i.e. x_0 is mapped to a periodic point 0.4. More precisely, $0.2 \mapsto 0.4 \mapsto 0.8 \mapsto 0.4 \mapsto \ldots$.

2.4   Fuzzy Dynamical Systems

We are ready to define a fuzzy dynamical system with the help of so-called Zadeh's extension (fuzzification). We would like to emphasize that this notions was introduced by Kloeden in [14], further elaborated by Kloeden in [15] and other mathematicians (e.g. see [16] and references therein).

Let (X, f) be a fixed discrete dynamical system. Then the expression

$$(z_f(A))(x) = \sup_{y \in f^{-1}(x)} \{A(y)\}$$

uniquely induces a map $z_f \colon \mathbb{F}^1(X) \to \mathbb{F}^1(X)$. The map z_f is called a *fuzzification* (or *Zadeh's extension*) of the map $f \colon X \to X$. Moreover, z_f is indeed continuous (not only) in $(\mathbb{F}^1(X), \tau_\infty)$ if and only if f is continuous. It is worth mentioning, Zadeh's extension is strongly related to another natural extension $(\mathbb{K}(X), s_f)$ of (X, f), where $s_f \colon \mathbb{K}(X) \to \mathbb{K}(X)$ is defined by $s_f(C) = f(C)$ for any $C \in \mathbb{K}(X)$. Indeed, $[z_f(A)]_\alpha = s_f([A]_\alpha) = f([A]_\alpha)$ for any $A \in \mathbb{F}^1(X)$ and $\alpha \in (0, 1]$.

2.5   Piecewise Linear Functions

Below we describe our algorithm for so-called piecewise linear functions defined on $X = [0, 1]$, so let us define them. A *piecewise linear function* $f \colon [0, 1] \to [0, 1]$ given by finitely many points $(c_i, s_i) \in [0, 1] \times [0, 1]$ for $i = 1, \ldots, \ell$, is a function $f \colon [0, 1] \to [0, 1]$ such that $c_1 = 0, c_\ell = 1$, and $f(c_i) = s_i$ for each $i = 1, 2, \ldots, \ell$, and $f|_{[c_i, c_{i+1}]}$ is linear for every $i = 1, 2, \ldots, \ell - 1$. The points c_i are called *turning points*. It follows from the definition that f is continuous. We emphasize here that fuzzy sets in our algorithm need not be continuous, although they are called piecewise linear. The only difference in their definition is that they are given by pairs of pairs (c_i, s_i), which define linear segments of graphs of fuzzy sets.

There are several reasons why we choose this class of interval maps and that we consider the interval $[0, 1]$ only. The first reason is that considering another closed interval would lead to the same mathematical problem. The second reason is that we do not approximate Zadeh's extension but we exactly compute it. This is due to using piecewise linear functions and fuzzy sets. The next reason is that calculations of Zadeh's extension are restricted to computation with linear maps, which is significantly simpler task than working with continuous maps in general. The fact that continuous piecewise linear interval maps are dense in the class of continuous interval maps will help us with further generalizations of this algorithm. And finally the class of piecewise linear maps, as well as classes of piecewise monotone or simply unimodal maps, are interesting from the dynamical point of view as they provide a huge variety of dynamical properties – for some details we refer to [2] and [19].

3   Algorithm

In this section we present our algorithm. In the first part of this section we comment our algorithm in order to make the second part more legible. It should be emphasized that our algorithm is proposed for continuous piecewise linear functions and we assume $X = [0, 1]$, although our algorithm could be easily adapted to arbitrary closed subinterval of the real line $\mathbb{R}$.

The purpose of the following algorithm is a computation of a trajectory of given discrete fuzzy dynamical subsystem $(\mathbb{F}^1(X), z_f)$, which is a unique and natural extension of a discrete dynamical system (X, f). In other words, it is used for a calculation of Zadeh's extension of a continuous piecewise linear function f and a given fuzzy set A which is also defined as a piecewise linear function $A \colon X \to [0, 1]$ (not necessarily continuous). The main idea behind our algorithm is to calculate Zadeh's extension on smaller intervals, on which graphs of both f and A are linear. Due to appropriate decomposition of X into finitely many intervals, the algorithm calculates Zadeh's extension in a limited number of points, namely in starting and ending points of elements of the decomposition. This way we do not approximate the trajectory of A, but we exactly compute elements of this trajectory.

The process described in STEP 1 shows how to decompose $X = [0,1]$ into finitely many subintervals and which points to choose for calculation of Zadeh's extension. Naturally, it is necessary to consider turning points of the function f and also end-points of linear segments of a given fuzzy set A. As a result of this part we obtain a finite decomposition J' of $[0,1]$ such that $K \in J'$ if there is no bigger interval containing K for which both A and f are linear on the interior of K. This simple assumption provides us with the smallest number of points needed for calculating of Zadeh's extension.

In STEP 2 we use the fact that if both A and f are linear on some interval K then the Zadeh's extension of f and $A|_K$ is also linear on $f(K)$. As a partial outcome we obtain a finite set of linear segments given by elements from J' and, by taking the supremum of such linear segments, we obtain a graph of the image $z_f(A)$ of the fuzzy set A. Of course, the proposed algorithm needs to work for a certain number of iterations. Thus, the process described in STEPS 1 and 2 is repeated until we reach the chosen number M of iterations. To demonstrate the evolution of the fuzzy set A in a given dynamical system $([0,1], f)$ as a final result of this algorithm is the 3D plot of all calculated iterations, i.e. of $z_f(A)$, $z_f^2(A)$, ..., $z_f^M(A)$. Now the formal description of our algorithm follows. As an input we have a continuous piecewise linear function $f \colon [0,1] \to [0,1]$, a piecewise linear fuzzy set $A \colon [0,1] \to [0,1]$ and a natural number M of iterations. As an output we obtain a 3D plot containing graphs of M picewise linear fuzzy sets $z_f(A)$, $z_f^2(A)$, ..., $z_f^M(A)$.

INPUT:

- A continuous piecewise linear function $f \colon [0,1] \to [0,1]$ given by points (c_i, s_i), and $i = 1, 2, \ldots, \ell$. More precisely,

$$
f(x) = \begin{cases}
s_1 + (s_2 - s_1)\frac{(x - c_1)}{(c_2 - c_1)}, & c_1 \leq x \leq c_2, \\
s_2 + (s_3 - s_2)\frac{(x - c_2)}{(c_3 - c_2)}, & c_2 \leq x \leq c_3, \\
\vdots & \\
s_{\ell-1} + (s_\ell - s_{\ell-1})\frac{(x - c_{\ell-1})}{(c_l - c_{\ell-1})}, & c_{\ell-1} \leq x \leq c_\ell.
\end{cases}
$$

- A piecewise linear fuzzy set $A \colon [0,1] \to [0,1]$ given by n *linear segments*, i.e. by pairs of pairs $(a_i, b_i) \in [0,1] \times [0,1]$, where $i = 1, 2, \ldots, n$. This means, for every $i = 1, 2, \ldots, n$, each $((a_i, b_i), (a_i', b_i'))$ defines a linear segment of the graph of A, i.e. $A(\lambda a_i + (1 - \lambda)a_i') = \lambda b_i + (1 - \lambda)b_i'$ for every $\lambda \in (0, 1)$.
- A number of iterations $M \in \mathbb{N}$. Let $k = 1$.

STEP 1

- Create a set J of s linear segments L_i obtained from linear segments of A by the following inductive procedure. Whenever $c_j \in (a_i, a_i')$ for some

$L'_i := L_i((a_i, b_i), (a'_i, b'_i)) \in J$ and m is the number of elements in J, then J gets two new linear segments L_i, L_{m+1} instead of L'_i, where $L_i = ((a_i, b_i), (c_j, A(c_j)))$ and $L_{m+1} = ((c_j, A(c_j)), (a'_i, b'_i))$.

STEP 2

- For $i = 1, 2, \ldots, s$ and for every linear segment $L_i = ((a_i, b_i), (a'_i, b'_i))$ from J we compute its image $z_f(L_i)$ under Zadeh's extension of the map f restricted to $[a_i, a'_i]$. More precisely, for every L_i we obtain a linear segment $z_f(L_i) = ((f(a_i), b_i), (f(a'_i), b'_i))$.
- Take the supremum of obtained linear segments of $z_f(L_i)$, $i = 1, 2, \ldots, s$, in order to get the graph of $z_f^k(A)$.

STEP 3

- If $k = M$, then algorithm is finished.
- If $k < M$, then put $k = k + 1$ and continue by STEP 1.

OUTPUT

- Depict images of $z_f(A), \ldots, z_f^M(A)$ in a 3D plot.

Note that the proposed algorithm is somewhat continuation of the approximation method proposed in [17], because you can avoid computing inverses of continuous piecewise linear maps f by using properly chosen points in the space X. In the proposed algorithm we took the smallest number of points necessary for correct computation, not approximation, of Zadeh's extension.

4   Examples

In this section, we briefly demonstrate behaviour and features of our algorithm. As a simple application we show how our algorithm can be used to obtain trajectories of fuzzy dynamical subsystem $(\mathbb{F}^1_k([0, 1]), z_f)$ induced by Zadeh's extension of a given one-dimensional dynamical system $([0, 1], f)$.

Example 2. The first example calculates Zadeh's extension of a function f given by $(0,0),(1/4,9/10),(1,0)$ and of a fuzzy number A given by $(0,0),(3/4,1),(1,0)$ (Fig. 2).

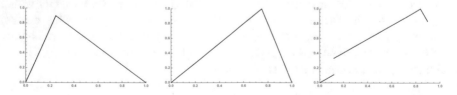

Fig. 2. The graphs of the map f, of the fuzzy set A and of $z_f^2(A)$, respectively.

The picture above shows that, even for very trivial graphs of continuous linear functions f and A, Zadeh's extension can generate discontinuous images of A, in our example $z_f^2(A)$ is discontinuous. Now we demonstrate that we can further handle with such discontinuities and we depict the first 20 elements of the trajectory of A (Fig. 3).

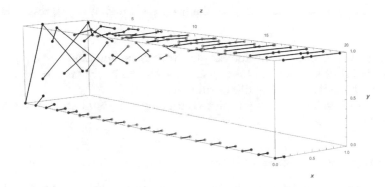

Fig. 3. The graphs of $z_f(A), \ldots, z_f^{20}(A)$.

Example 3. In the second example we consider a continuous piecewise linear function f and a continuous piecewise linear fuzzy set A depicted on Fig. 4. The first 30 iterations of the fuzzy set A induced by these maps are depicted on Fig. 5. It can be easily seen that the trajectory tends to be periodic.

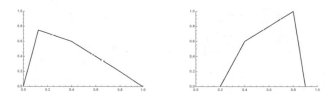

Fig. 4. The graphs of the map f and of the fuzzy set A, respectively.

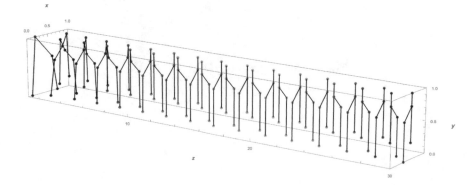

Fig. 5. The graphs of $z_f(A), \ldots, z_f^{30}(A)$.

Fig. 6. The graphs of the map f, of the fuzzy set A and of $z_f(A)$, respectively.

Example 4. We consider a continuous piecewise linear function f and a piecewise linear fuzzy set A as depicted on Fig. 6. One can again see on the same picture that the first iteration of A is discontinuous.

Finally in Fig. 7 we can see a plot containing images of the fuzzy set A for the first 20 iterations.

4.1 Computation Complexity

Here we mention a brief review of our experiences with computational costs. Naturally, requested computational time depends on several factors, i.e. on the chosen number of iterations, the number of points defining a function $f\colon [0,1] \to [0,1]$ and a fuzzy set $A\colon [0,1] \to [0,1]$ and on computer which is used for computing. In Table 1, we can see approximate computing time for the examples mentioned above.

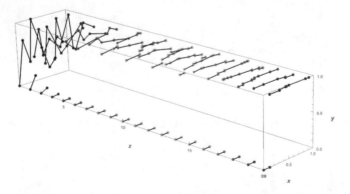

Fig. 7. The graphs of $z_f(A), \ldots, z_f^{20}(A)$.

Table 1. Computing time in seconds

	10 iterations	100 iterations	1000 iterations	10000 iterations
Example 1	0.30216	6.3155	65.4623	202.677
Example 2	0.280272	5.73293	32.3179	249.809
Example 3	0.435848	7.66259	50.2349	377.948

[a] Compiled in *Mathematica* 11.3 on a laptop with processor 1,8 GHz Intel Core i5.

5 Conclusion

In this manuscript we presented some preliminary results related to approximation of Zadeh's extension of a given continuous function f. The main idea of this approach is that we are focused on one-dimensional continuous piecewise linear functions and piecewise linear fuzzy sets. As we mentioned in the introduction, our approach is different, does not require special assumptions on calculated fuzzy sets, covers dense class of interval maps and, unlike almost all previous approaches by other authors, can deal with discontinuities which naturally appear in computations. This fact is demonstrated by Examples 2 and 4.

This is the first step of our future work, due to page limit we cannot write more. Our future work covers extending our approach to the class of continuous piecewise linear maps on spaces of dimension strictly higher than one and, with the help of obtained results, preparing algorithms for more general classes of maps, not only for piecewise linear ones.

Acknowledgements. The second author acknowledges funding from the project "Support of talented PhD students at the University of Ostrava" from the programme RRC/10/2018 "Support for Science and Research in the Moravian Silesian Region 2018".

References

1. Ahmad, M.Z., Hasan, M.K.: A new approach for computing Zadeh's extension principle. Matematika **26**, 71–81 (2010)
2. Block, L.S., Coppel, W.A.: Dynamics in One Dimension. Springer, Heidelberg (2006)
3. Brent, R.P.: Algorithms for Minimization Without Derivatives. Courier Corporation, North Chelmsford (2013)
4. Cánovas, J.S., Kupka, J.: On the topological entropy on the space of fuzzy numbers. Fuzzy Sets Syst. **257**, 132–145 (2014)
5. Cecconello, M., Bassanezi, R.C., Brandão, A.J., Leite, J.: On the stability of fuzzy dynamical systems. Fuzzy Sets Syst. **248**, 106–121 (2014)
6. Chalco-Cano, Y., Jiménez-Gamero, M., Román-Flores, H., Rojas-Medar, M.A.: An approximation to the extension principle using decomposition of fuzzy intervals. Fuzzy Sets Syst. **159**(24), 3245–3258 (2008)

7. Chalco-Cano, Y., Misukoshi, M.T., Román-Flores, H., Flores-Franulic, A.: Spline approximation for Zadeh's extensions. Int. J. Uncertainty Fuzziness Knowl.-Based Syst. **17**(02), 269–280 (2009)
8. Chalco-Cano, Y., Román-Flores, H., Rojas-Medar, M., Saavedra, O., Jiménez-Gamero, M.D.: The extension principle and a decomposition of fuzzy sets. Inf. Sci. **177**(23), 5394–5403 (2007)
9. Corveleyn, S., Vandewalle, S.: Computation of the output of a function with fuzzy inputs based on a low-rank tensor approximation. Fuzzy Sets Syst. **310**, 74–89 (2017)
10. Diamond, P., Kloeden, P.E.: Metric Spaces of Fuzzy Sets: Theory and Applications. World Scientific, Singapore (1994)
11. Dong, W.M., Wong, F.S.: Fuzzy weighted averages and implementation of the extension principle. Fuzzy Sets Syst. **21**(2), 183–199 (1987)
12. Guerra, M.L., Stefanini, L.: Approximate fuzzy arithmetic operations using monotonic interpolations. Fuzzy Sets Syst. **150**(1), 5–33 (2005)
13. Hanss, M.: The transformation method for the simulation and analysis of systems with uncertain parameters. Fuzzy Sets Syst. **130**(3), 277–289 (2002)
14. Kloeden, P.: Fuzzy dynamical systems. Fuzzy Sets Syst. **7**(3), 275–296 (1982)
15. Kloeden, P.: Chaotic iterations of fuzzy sets. Fuzzy Sets Syst. **42**(1), 37–42 (1991)
16. Kupka, J.: On fuzzifications of discrete dynamical systems. Inf. Sci. **181**(13), 2858–2872 (2011)
17. Kupka, J.: A note on the extension principle for fuzzy sets. Fuzzy Sets Syst. **283**, 26–39 (2016)
18. Lynch, S.: Dynamical Systems with Applications Using MATLAB. Springer, Heidelberg (2004)
19. Sharkovsky, A., Kolyada, S., Sivak, A., Fedorenko, V.: Dynamics of One-Dimensional Maps, vol. 407. Springer, Heidelberg (2013)
20. Stefanini, L., Sorini, L., Guerra, M.L.: Simulation of fuzzy dynamical systems using the LU-representation of fuzzy numbers. Chaos, Solitons Fractals **29**(3), 638–652 (2006)
21. Zadeh, L.A.: Fuzzy sets. Inf. Control **8**, 338–353 (1965)

A New Method to Design a Granular Hierarchy Model with Information Granularity

Mingli Song[(⊠)] and Yukai Jing

School of Computer Science, Communication University of China,
Beijing 100024, China
{songmingli, jyk0715}@cuc.edu.cn

Abstract. In this paper, we propose a new modeling strategy to deal with a complex system which has various sources of input data sets. Through a thorough study of existing aggregating methods of models, we found that most works only consider aggregation of local models with the same level of granularity. With this motivation, we construct a granular hierarchy model by local models with different levels of granularities. The constructed hierarchy model includes two levels: a lower level and a higher level. The global model in the higher level is aggregated with local models and information granularities. In this case, more abstract data (information) are used as the input of global model which assures the general output of the whole model. Knowledge reconciliation and interaction are crucial issues during the whole process. Hence we propose a method by analyzing prototypes of local models with the principle of justifiable granularity.

Keywords: Granular model · Information granularity · Aggregate

1 Introduction

In system modeling, we definitely will encounter the situation that training data come from different sources. The term "different sources" mean that data may be observed over some time periods, some different locations or analyzed at different levels of details. In this case, most researchers in literature consider to build local models first and then aggregate those local models. No matter which type of model will be selected, an essential issue is how to realize aggregation. Depending on whether the local models have same level of information granularity or not, we can discuss this issue from two aspects: (1) the local models are aggregated with the same level of granularity; (2) the local models are aggregated with different levels of granularity. Most studies in literature focus on the first case.

For a model with numeric parameters and structure, system modeling is realized in a more or less local way: the model is suitable for the data provided so far. Owing to the huge size of data stored in the database over time, an abstract (general) system model becomes an overwhelming alternative. An appealing way to solve this problem is to assign a level of granularity to the entire system. The concept of granularity mentioned here is closely concerned with the "size", "capacity" or "dimension" of the information granule [1]. Granularity brings a flexibility to view the system [14].

© Springer Nature Switzerland AG 2019
R. B. Kearfott et al. (Eds.): IFSA 2019/NAFIPS 2019, AISC 1000, pp. 625–630, 2019.
https://doi.org/10.1007/978-3-030-21920-8_55

Moreover, the level of granularity provides an opportunity to avoid the noise and disturbance on components (parameters and the structure) of the system.

In this paper, we propose a new system modeling method in which information granularity is added for different local models. When developing the global model, local models are aggregated with granularities. In this case, more abstract data (information) are used as the input of global model which assures the general output of the whole model.

In what follows, we firstly go over some popular studies of aggregation of local models with the same level of granularity. Then we propose an algorithm that considers constructing a granular hierarchy model by local models with different levels of granularities. The main contribution of this study is: (1) A new algorithm to design a more general model with the aid of information granularity is proposed. (2) The algorithm considers different effects of local models and shows the diversity of the local models.

Section 2 reviews some popular methods of aggregation of local models. Section 3 illustrates our new modeling method. Section 4 concludes the contribution of this paper.

2   Aggregation of Local Models with the Same Level of Granularity

The studies of a collection of models have been around for some time. In machine learning, it is commonly under the term "ensemble". It has been demonstrated from theory and real application (experiments) aspects that the performance of such ensembles is substantially better that a single model. There has been a plenty of contributions about ensemble approaches in literature. To give a glimpse of the existing works, we select some popular methods to show the detail description.

A very well known method is OWA (ordered weighted averaging) in [2]. In many cases, the type of aggregation operator lies between the pure "adding" of t-norm, and the pure "oring" of S operator. To solve this problem, Yager proposed a new operator which allows to easily adjust the degree of "anding" and "oring" implicit in the aggregation. Given n numeric values $a_1, a_2, \ldots, a_n$, the aggregation function is as follows:

$$F(a_1, a_2, \ldots, a_n) = W_1 b_1 + W_2 b_2 + \ldots + W_n b_n \tag{1}$$

where b_i is the ith largest element in the collection of $a_1, a_2, \ldots, a_n$. W_i is called a weight value and satisfies $\sum W_i = 1$. The aggregation result is in the format of numeric value.

In [3], Wu et al. present the linguistic weighted average (LWA), where the weights are always words modeled as interval-type 2 fuzzy sets.

$$\tilde{Y}_{LWA} = \frac{\sum_{i=1}^{n} \tilde{X}_i \tilde{W}_i}{\sum_{i=1}^{n} \tilde{W}_i} \tag{2}$$

Where $\tilde{X}_i$ and $\tilde{X}_i$ are words modeled by interval type 2 fuzzy sets. The result of LWA is also an interval type-2 fuzzy set after the entire calculation formula. The aggregation result is in the format of interval- valued fuzzy set.

The above two contributions use simple framework as the input of aggregation formula. Chen et al. [4] proposed EP (Expectation propagation) pruning algorithm to reduce the computational resource and increase the generalization performance. It will find a better subset of ensemble members with a small ensemble. In the experimental part, classification and regression trees (CART) are utilized as ensemble members. The aggregation result is in the format of a subset of ensemble members.

Zhou et al. explore the way to generate effective accurate but diverse ensemble component learners. Please refer to literature [5]. Since the popular resampling methods such as boostrap are not very effective on local learners (such as nearest neighbors). It uses majority voting as the aggregation mechanism for n nearest neighbor classifiers. The aggregation result is numeric class label.

In [6], n MLP classifiers are studied as local models. The authors analyze linear function combiner with optimal values of the classifier weights from theoretical and experimental aspect. The aggregation result is numeric class label. Kuncheva et al. [7] propose a combined fusion-selection approach to design an ensemble with n decision trees. Each classifier in the ensemble is replaced by a miniensemble of a pair of sub-classifiers with a random linear oracle to choose between the two. The output is an ensemble with numeric output. In [8], two cooperative learning algorithms are proposed. n neural network (NN) classifiers are used as local models and each NN is an element of the ensemble. It uses the validation error to stop ensemble training. The output is an ensemble with numeric output. Other studies about the ensemble learning method are listed in [9–13].

In spite of the visible diversity present above, there is a clear similarity among them: the result of aggregation is positioned at the same level of granularity as the results produced by the individual models involved in the aggregation. Every component in the local models owns the same preference. Then what if we consider the concept of information granularity when building the model for the entire system? In what follows, we elaborate on the design of a granular hierarchy model.

3   Construction of a Granular Hierarchy Model

In this section, we illustrate the process of constructing a granular hierarchy model by local models with different levels of granularities.

3.1   The Principle of Justifiable Granularity

The principle of justifiable granularity is concerned with a formation of a meaningful representation of a collection of numeric values. The numeric values can be the outputs of a model, the weights, or the different levels of granularities. The issue is how to find a proper way to represent those numeric values. The commonly used frameworks include intervals, fuzzy sets, rough sets, shadow sets, et al. In this paper, we adopt intervals as our starting point. The other formats can be applied in a similar way.

A numeric value can be expanded into an interval Ψ (= [a, b]) in such a way that the interval satisfies: (i) the numeric evidence accumulated within the bounds (a, b) of Ψ is as high as possible. It can be quantified by counting the number of data falling within the bounds of Ψ (which can be used by card). The card count has to be maximized; (ii) the support of Ψ is as low as possible, which makes the interval specific (detailed) enough. The two conditions are in conflict. To come up with a single performance index Q taken as a product of two functions f_1 and f_2, $Q = f_1 \times f_2$ where f_1 is an increasing function of the cardinality of the elements falling within the bounds of the interval and f_2 is a decreasing function of the support (length) of $|b–a|$. To address the requirements formulated above, our intent is to maximize Q with respect to the boundaries of the interval. This usually invokes the usage of an optimization algorithm, such as genetic algorithm, particle swarm optimization, ant colony, et al.

3.2   Build the Higher Level Model

The underlying concept is concisely illustrated in Fig. 1. A certain system is perceived from different points of perspectives. The local models emerging there, denoted as LM-1, LM-2, …, LM-m use locally available data D-1, D-2, …, D-m. In general, the local models can be constructed with different architectures, such as decision trees, Bayes classifier, fuzzy models and others. However, for the sake of clearly illustrating the whole design process, we may consider a single topology for all local models. The model constructed on the local models will be definitely more abstract and general than a single local model. How to construct the global model becomes a crucial issue.

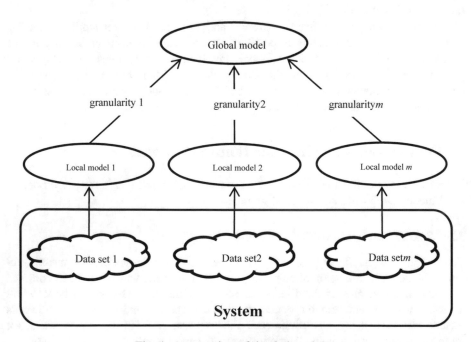

Fig. 1 An overview of the design picture

Since we assume that the local models are from different views (or using different data resources), the outputs of different local models are very likely different. What we know for the construction of the global model is the information about local models, such as the topologies and the parameters. The data used to develop the model are unknown due to the unavailable nature. If the objective of model at the higher level is to capture the property of diversity of knowledge, for the same input, it should return an output that is more abstract. This is the meaning of adopting information granules. Hence the global model can be regarded as granular global model.

Actually the global model will be constructed under two different situations:

(a) Local models are firstly built individually. Then the new data appear for all local models and for the global model. In this case, we can select some of the new data to develop the global model or using clustering techniques to generate prototypes to create the global model. The two ways naturally generate more abstract data and it promises building a more general model.

(b) Local models are built individually. However, partial local data are available for the global model. In this case, when the new data are coming, it is necessary to select a method to determine which data will be used for global model. Then we may refer to the same handling method in situation (a)

No matter which situation is faced, the core issue is how to obtain abstract information through the global model. Thus we assume there is already an approach AP (which can explored in our future research) to determine the input data of the global model.

The major steps of the algorithm can be summarized as follows:

Step 1: Create m local models with m different data sets. The architecture of local models can be neural networks (MLP) for regression problem and decision tree (CART) for classification problem. The detail design of the local models is defined by the problem. In general, the m local data sets are unseen between each other. Hence the outputs of each local model can hardly be the same.

Step 2: Adopt clustering technique (Fuzzy C Means) to create m prototype sets for the local data sets. The size of each prototype set is determined by the problem.

Step 3: Use the AP approach to obtain the training data set T for the higher level of model. Use FCM method on the T data set and returned n prototypes.

Step 4: Utilize the n prototypes on the m local models and each local model will generate n outputs. Create a vector V ($m \times n$ components) to save all the outputs.

Step 5: Use the principle of justifiable granularity to create an information granule (an interval) on the vector V. The information granule represents how abstract the hierarchy model is. The objective function here can be the combination of performance of local models. The obtained result may reflect the levels of granularity of each local model.

4   Conclusions

In this paper, a new modeling method is proposed which is designed for a granular hierarchy model. This algorithm works through using abstract prototypes to build higher level model. Before this, local models are constructed by local data sets individually. This is a starting point for our research. In the future, we will concentrate on the experimental studies to verify the effectiveness of this algorithm.

Acknowledgement. Support from the National Natural Science Foundation of China (NSFC) 61773352 and the Fundamental Research Funds for the Central Universities (Grant No. CUC2019B022) are gratefully appreciated.

References

1. Pedrycz, W., Gomide, F.: Fuzzy Systems Engineering: Toward Human-Centric Computing. Wiley, Hoboken (2007)
2. Yager, R.R.: On ordered weighted averaging aggregation operators in multicriteria decision making. Readings. Fuzzy Sets. Intell. Syst. **18**(1), 80–87 (1993)
3. Wu, D.R., Mendel, J.M.: Aggregation using the linguistic weighted average and interval type-2 fuzzy sets. IEEE Trans. Fuzzy Syst. **15**(6), 1145–1161 (2007)
4. Chen, H.H., Tiňo, P., Yao, X.: Predictive ensemble pruning by expectation propagation. IEEE Trans. Knowl. Data Eng. **25**(7), 999–1013 (2009)
5. Zhou, Z.H., Yu, Y.: Ensembling local learners through multimodal perturbation. IEEE Trans. Syst. Man Cybern. Part B **35**(4), 725–735 (2005)
6. Giorgio, F., Roli, F.: A theoretical and experimental analysis of linear combiners for multiple classifier systems. IEEE Trans. Pattern Anal. Mach. Intell. **27**(6), 942–956 (2005)
7. Kunchev, L.I., Rodriguez, J.J.: Classifier ensembles with a random linear oracle. IEEE Trans. Knowl. Data Eng. **19**(4), 500–508 (2007)
8. Islam, M.M., et al.: Bagging and boosting negatively correlated neural networks. IEEE Trans. Syst. Man Cybern. Part B Cybern. **38**(3), 84–771 (2008). A Publication of the IEEE Systems Man & Cybernetics Society
9. Wu, Q.Y., et al.: ML-Forest: A multi-label tree ensemble method for multi-label classification. IEEE Trans. Knowl. Data Eng. **28**(10), 1 (2016)
10. Luca, C., Zhang, Y., Mihaela, V.D.S.: ensemble of distributed learners for online classification of dynamic data streams. IEEE Trans. Sign. Inf. Process. Over Netw. **1**(3), 180–194 (2013)
11. Windeatt, T., Zor, C.: Ensemble pruning using spectral coefficients. IEEE Trans. Neural Netw. Learn. Syst. **24**(4), 673–678 (2013)
12. Song, M.L., Pedrycz, W.: Fuzzy modeling with genetically optimized feature space reduction. LNCS, vol. 2186, no. 1 (2010)
13. Song, M.L., Pedrycz, W.: Granular neural networks: concepts and development schemes. IEEE Trans. Neural Netw. Learn. Syst. **24**(4), 542–553 (2013)
14. Zadeh, L.A.: Toward a theory of fuzzy information granulation and its centrality in human reasoning and fuzzy logic. Fuzzy Sets Syst. **90**(90), 111–127 (1997)

Typicality of Features in Fuzzy Relational Compositions

Martin Štěpnička[1]([⊠]), Nhung Cao[1], Michal Burda[1], and Aleš Dolný[2]

[1] CE IT4I - IRAFM, University of Ostrava,
30. dubna 22, 701 03 Ostrava, Czech Republic
{martin.stepnicka,nhung.cao,michal.burda}@osu.cz
[2] Department of Biology and Ecology, Faculty of Sciences,
University of Ostrava, Chittussiho 10, 710 00 Ostrava, Czech Republic
ales.dolny@osu.cz

Abstract. Fuzzy relations and their compositions have the same crucial importance for the fuzzy mathematics that is provided to mathematics by relations and their composition. The topic, since it attracted many scholars and influenced many areas in fuzzy modeling, has been extended on distinct directions including the recent one on the incorporation of excluding features. This article brings a mathematically similar yet semantically opposite extension, in particular, the concept of typical features. We show the appropriateness of such a new concept and investigate some of its properties. Furthermore, we discuss how the concept of typical features incorporated in fuzzy relational compositions may bring significant improvement of results of some applications. This fact is demonstrated on a real example of biological species classification.

1 Introduction

Fuzzy relational compositions have a rich origin stemming from the investigations by Bandler and Kohout in early 1980s and later on by a wide scientific community. This fact can be supported by numerous developments of the related theory [3–5, 12] and as well as of the connected applications, such as architectures of information processing, modeling fuzzy rule bases and the related systems of fuzzy relational equations. Here we refer readers only to a few sources being aware that the full list only of the most important ones would outreach the abilities of this article [11, 13].

> Son: *Dad, let us play a guessing game.*
> Dad: *Ok, son. I am thinking of an animal that is hairy and it eats grass.*
> Son: *A sheep!*
> Dad: *Correct. But how did you know it is neither a horse nor a cow?*
> Son: *Being hairy is so typical for a sheep, you would never ignore it.*

Several extensions were proposed in the recent years. We recall the one [7] that proposed the concept of excluding features and that was successfully applied

R. B. Kearfott et al. (Eds.): IFSA 2019/NAFIPS 2019, AISC 1000, pp. 631–642, 2019.
https://doi.org/10.1007/978-3-030-21920-8_56

to a classification of biological samples (in particular dragonflies) into species. This choice is first of all, we will directly use this extension and secondly, our extension, namely the concept of typical features, will be strongly motivated and will actually preserve a contrary position to the recalled one. As we will show, both extensions, when jointly used, will bring significant shift in the application potential of the fuzzy relational compositions.

2 Preliminaries

2.1 Algebraic Background

Let us recall the basic definitions of the used underlying algebraic structures. Let us also recall some properties preserved in these structures and that will be used in the sequel.

Definition 1. An algebra $\mathscr{L} = \langle L, \wedge, \vee, \otimes, \rightarrow, 0, 1 \rangle$ is a *residuated lattice* if

1. $\langle L, \wedge, \vee, 0, 1 \rangle$ is a lattice with the least and the greatest element
2. $\langle L, \otimes, 0, 1 \rangle$ is a commutative monoid such that $\otimes$ is isotone in both arguments
3. the operation $\rightarrow$ is a residuation with respect to $\otimes$, i.e.

$$a \otimes b \leq c \quad \text{iff} \quad a \rightarrow c \geq b. \tag{1}$$

We can define additional operations for all $a, b \in L$, namely: biresiduation (biimplication, residual equivalence), negation, and addition, respectively:

$$a \leftrightarrow b = (a \rightarrow b) \wedge (b \rightarrow a),$$
$$\neg a = a \rightarrow 0,$$
$$a \oplus b = \neg(\neg a \otimes \neg b).$$

Furthermore, let us recall the MV-algebra.

Definition 2. An MV-algebra is an algebra $\mathscr{L} = \langle L, \oplus, \otimes, \neg, 0, 1 \rangle$ with two binary operations $\oplus, \otimes$, a unary operation $\neg$ and two constants such that $\langle L, \oplus, 0 \rangle$ and $\langle L, \otimes, 1 \rangle$ are commutative monoids and the following identities hold:

$$a \oplus \neg a = 1, \qquad\qquad a \otimes \neg a = 0,$$
$$\neg(a \oplus b) = \neg a \otimes \neg b, \qquad \neg(a \otimes b) = \neg a \oplus \neg b,$$
$$a = \neg\neg a, \qquad\qquad \neg 0 = 1,$$
$$\neg(\neg a \oplus b) \oplus b = \neg(\neg b \oplus a) \oplus a.$$

Note, that every MV-algebra $\langle L, \oplus, \otimes, \neg, 0, 1 \rangle$ is a residuated lattice [14] by putting

$$a \vee b = \neg(\neg a \oplus b) \oplus b = (a \otimes \neg b) \oplus b,$$
$$a \wedge b = \neg(\neg a \vee \neg b) = (a \oplus \neg b) \otimes b,$$
$$a \rightarrow b = \neg a \oplus b,$$

where $\rightarrow$ is a residuation operation with respect to $\otimes$. An MV-algebra or residuated lattice is complete if the underlying lattice is a complete lattice. Let us recall [14], that in a complete residuated lattice, the following holds for any index set I:

$$\bigvee_{i \in I} a_i \rightarrow b = \bigwedge_{i \in I} (a_i \rightarrow b) \quad a_i, b \in L . \tag{2}$$

As a consequence of (2) and the definition of the negation, we immediately get the following equality:

$$\neg \bigvee_{i \in I} a_i = \bigwedge_{i \in I} \neg a_i .$$

2.2   Fuzzy Relational Compositions

Let us recall the standard definitions of the fuzzy relational compositions as introduced by Bandler and Kohout [1,2]. Before doing so, let us fix the denotation of a set of all fuzzy sets on a given universe U by $\mathscr{F}(U)$.

Definition 3. Let X, Y, Z be finite non-empty universes and let $R \in \mathscr{F}(X \times Y)$ and $S \in \mathscr{F}(Y \times Z)$. Then compositions $R \circ S, R \triangleleft S, R \triangleright S, R \square S$ from $\mathscr{F}(X \times Z)$ are given as follows

$$(R \circ S)(x, z) = \bigvee_{y \in Y} (R(x, y) \otimes S(y, z)) , \tag{3}$$

$$(R \triangleleft S)(x, z) = \bigwedge_{y \in Y} (R(x, y) \rightarrow S(y, z)) , \tag{4}$$

$$(R \triangleright S)(x, z) = \bigwedge_{y \in Y} (R(x, y) \leftarrow S(y, z)) , \tag{5}$$

$$(R \square S)(x, z) = \bigwedge_{y \in Y} (R(x, y) \leftrightarrow S(y, z)) . \tag{6}$$

The semantics of the compositions may be explained on the above-mentioned classification of biological species. Indeed, consider X to be a set of samples, Y to be set of characteristic features, and finally let Z stands for a set of classes (or species in this specific example). The value $(R \circ S)(x, z)$ then represents the truth degree of the statement *"there exists a feature $y \in Y$ carried by sample $x \in X$ and related to class z"* and the value $(R \triangleleft S)(x, z)$ represents the truth degree of the statement *"all features $y \in Y$ carried by sample $x \in X$ are connected to class z"*. The other compositions, namely $R \triangleright S$ and $R \square S$, could be interpreted analogously again using the semantics of the involved operations $\leftarrow$ and $\leftrightarrow$ and the universal quantifier that is encoded in the formulas (4)–(6) the supremum.

For the sake of completeness, we may also recall the so-called "inf-S" composition[1] [10,16]. We denote this composition by the symbol ∇ used already in [7].

---

[1] The "S" in the name of the compositions implies the use of a t-conorm as this letter is often used to denote such an operation however, it has nothing common with our fuzzy relation S.

Definition 4. Let X, Y, Z be non-empty universes, let $R \in \mathcal{F}(X \times Y)$, $S \in \mathcal{F}(Y \times Z)$. Then the *composition* $\triangledown$ *of fuzzy relations* R and S is a fuzzy relation on $X \times Z$ defined as follows:

$$(R \triangledown S)(x, z) = \bigwedge_{y \in Y} (R(x, y) \oplus S(y, z)) \ , \quad (x, z) \in X \times Z \ .$$

Remark 1. Note, that Bandler-Kohout products $\triangleleft, \triangleright$, and $\square$ were re-defined by De Baets and Kerre [12]. The motivation for this step was to avoid fake suspicions that occur when the value 0 appears in all antecedents of the implications gathered in the Bandler-Kohout products. Consider, a sample $x \in X$ with no observed feature belonging to the set Y. Then for all $y \in Y$, $R(x, y) = 0$ and thus $(R \triangleleft S)(x, z) = 1$ for any z which can hardly be interpreted that the given sample x belongs to all species from Z but rather as fake suspicions. Similar fake suspicions can be also generated by $\triangleright$ and $\square$. Therefore, De Baets and Kerre [12] have added the assumption of the necessity of the existence of a connecting feature in the definitions of the three Bandler-Kohout products. For example, the Bandler-Kohout subproduct is re-defined as follows $(R \triangleleft_K S) = (R \triangleleft S) \cap (R \circ S)$ which ensures that $(R \triangleleft_K S)(x, z) = 0$ in the case described above. On the other hand, the original Bandler-Kohout definitions make theoretical sense and there are strong reasons to preserve them as defined [4]. This article will also add a clear reason to stay with the original definitions and to consider the modified definitions by De Baets and Kerre as complementing ones.

The concept of excluding features was motivated by the existence of excluding symptoms for particular diseases, i.e., symptoms directly excluding the possibility of having a particular disease, no matter how many symptoms connecting the given patient with the given diseases exist. This idea has been incorporated in the fuzzy relational compositions and demonstrated to be useful [7] on a real problem focusing on the classification of dragonfly species.

Firstly, let us recall the concept of the excluding features on the compositions of classical relations. Consider binary relations $R \subseteq X \times Y$ and $S \subseteq Y \times Z$. Furthermore, consider a crisp relation $E \subseteq Y \times Z$ with $(y, z) \in E$ expressing that y is an excluding feature for class z. Then the relation $R \circ S^{\backprime}E \subseteq X \times Z$ will be defined as follows

$$R \circ S^{\backprime}E = \{(x, z) \in X \times Z \mid (\exists \, y \in Y : (x, y) \in R \ \& \ (y, z) \in S) \ \& $$
$$(\nexists y \in Y : (x, y) \in R \ \& \ (y, z) \in E)\} \ . \tag{7}$$

The proposed composition provides the desirable meanings. Indeed, the fact that $(x, z) \in R \circ S^{\backprime}E$ expresses the following meaning: *sample x has at least one feature belonging to class z and there is no excluding feature related to class z carried by this sample.* The direct fuzzification of formula (7) leads to the following definition.

Definition 5. Let X, Y, Z be non-empty universes, let $R \in \mathscr{F}(X \times Y)$, and let $S, E \in \mathscr{F}(Y \times Z)$. Then $R \circ S^{\backprime}E$ is a fuzzy relations on $X \times Z$ defined as follows:

$$(R \circ S^{\backprime}E)(x, z) = \bigvee_{y \in Y} (R(x, y) \otimes S(y, z)) \otimes \neg \bigvee_{y \in Y} (R(x, y) \otimes E(y, z)) . \quad (8)$$

3   The Concept of Typicality

3.1   Typicality as the Concept of Unavoidable Features

As it has been shown in [7], the simple idea of excluding features, that was straightforwardly employed into the fuzzy relational compositions, brought more significant improvement than theoretically much more complicated incorporation of generalized quantifiers [6,18] or well-tuned data driven machine learning methods such as random forest. Of course, we cannot generalize the conclusions made in [7] on a single classification problem and we are fully aware of the fact that on other problems, data-driven approaches might dominate as well as there might be a plenty of approaches were the incorporation of fuzzy quantifiers could bring more significant improvement than the concept of excluding features. The message of the previous investigation may be understood as pointing out that even simple and straightforward ideas are worth of elaborating in the context of fuzzy relational compositions with a huge potential to contribute to their application potential.

In this Section, we will introduce a similarly straightforward approach, namely the concept of *typicality* or, the concept of *typical features*. By the typical features, we may denote those that are somehow unavoidable in any description of an object (sample) of the given class and we can assume that they may never be omitted. Let this relationship is encoded in a binary fuzzy relation $T \in \mathscr{F}(Y \times Z)$. So, the value of $T(y, z)$ expresses a truth degree up to which y is a typical (unavoidable) feature of class z. Indeed, one could assume some consequent natural properties, such as $T \subseteq S$ or even T being a crisp relation as considering a feature to be unavoidable in a degree sounds very unusual. However, the formal definition may remain as general as possible and these practical aspects may come alive in the practical settings of real applications.

Definition 6. Let X, Y, Z be non-empty universes, let $R \in \mathscr{F}(X \times Y)$, and let $S, E, T \in \mathscr{F}(Y \times Z)$. Then $(R \circ S^{\backprime}E)^{\triangleright T}$ is a fuzzy relations on $X \times Z$ defined as follows:

$$(R \circ S^{\backprime}E)^{\triangleright T}(x, z) = \bigvee_{y \in Y} (R(x, y) \otimes S(y, z)) \otimes \neg \bigvee_{y \in Y} (R(x, y) \otimes E(y, z))$$

$$\otimes \bigwedge_{y \in Y} (R(x, y) \leftarrow T(y, z)) . \quad (9)$$

Definition 6 assumes that both concepts, not only the typicality, but also excluding features are employed. Naturally, one could easily define the basic fuzzy relational composition involving the concept of typicality without the excluding features as $(R \circ S)^{\triangleright T}$. However, for the sake of generality, we view this step redundant as this form can be easily obtained from $(R \circ S^{\backprime} E)^{\triangleright T}$ by putting $E(y, z) = 0$ for all pairs $(y, z) \in Y \times Z$.

The meaning of $(R \circ S^{\backprime} E)^{\triangleright T}(x, z)$ can be formulated as follows: *sample x has at least one feature belonging to class z and there is no excluding feature related to class z carried by this sample and moreover, all features that are typical for z are carried by x.*

Formula (9) can be rewritten in a concise form as follows

$$(R \circ S^{\backprime} E)^{\triangleright T}(x, z) = (R \circ S)(x, z) \otimes \neg(R \circ E)(x, z) \otimes (R \triangleright T)(x, z) \qquad (10)$$

or even in a more abbreviated form as follows

$$(R \circ S^{\backprime} E)^{\triangleright T}(x, z) = (R \circ S^{\backprime} E)(x, z) \otimes (R \triangleright T)(x, z) . \qquad (11)$$

One can intuitively yet wrongly assume, that the $(R \circ S)$ part of formula (10) is practically redundant as the existence of a connecting feature should be ensured by the in practice implicit assumption on $T \subseteq S$. Thus, if all typical features of some class z are carried by a sample x, then necessarily there are some connecting features and consequently $(R \circ S)$ requiring their existence can be omitted. However, this would be true only in the case of the modified products as introduced by De Baets and Kerre [12], see Remark 1, but it is not the case of the original definitions by Bandler and Kohout. Note, that this makes a very good sense as in general we cannot make such a strong assumption that each class has its typical and unavoidable features. Indeed, even the most intuitive features for certain classes, such as being hairy or having a tail when talking about dogs, do not have to be unavoidable. And if we admit, that some of the classes may have no typical features, then $(R \triangleright T)(x, z) = 1$ for such classes and for all samples and the whole formula (9) will detect suspicious classes based on its first part that equals to $(R \circ S^{\backprime} E)$. In other words, if a class has no unavoidable feature, the classification will be based on on the existence of connecting features and non-existence of excluding features as before. If we omitted the $(R \circ S)$ part, formula (10) would collapse to $\neg(R \circ E)(x, z)$ as the last part $(R \triangleright S)$ would be equal to 1 for such cases. And this is surely not a good choice.

We may also shortly comment on using the modified version $(R \triangleright_K T)$ as introduced in [12] and as recalled in Remark 1. In the cases of classes without typical features that have been discussed above, $(R \triangleright_K T)(x, z)$ would equal to 0 as the assumption on the existence typical features wold not be met and thus, the whole formula (10) would give us $(R \circ S^{\backprime} E)^{\triangleright T}(x, z) = 0$ no matter how many connecting features would be present. This is a strong difference to the proposed formula (10) that treats such cases more carefully based on the first parts of the definition only.

Let this is not viewed as a criticism of the modified approach presented in [12] but rather as a demonstration why both approaches make their sense. Indeed, the modified one is a very useful construction using the original one. A similar approach in building more complicated compositions out of the fundamental ones has been later on adopted for the excluding features [7] and the concept of typicality also mimics the procedure adopted in [12]. For such constructive ways of the determination of new compositions the existence of the fundamental "building blocks" as defined by Bandler and Kohout seems to be crucial.

3.2 Properties

Let the complete linearly ordered residuated lattice $\mathscr{L} = \langle [0,1], \wedge, \vee, \otimes, \rightarrow, 0, 1 \rangle$ serves as the underlaying algebraic structure for the whole Section. Thus, all the results presented below are valid in this structure and all the used operations are taken from this structure. Exceptions will be explicitly mentioned. Let X, Y, Z be non-empty universes and let $R \in \mathscr{F}(X \times Y)$, $S, E, T, T_1, T_2 \in \mathscr{F}(Y \times Z)$.

Lemma 1.

$$(R \circ S^\backprime E)^{\triangleright T} \subseteq (R \circ S^\backprime E) \ . \tag{12}$$

Sketch of the proof: Direct application of the property $a \otimes b \leq b$ valid in $\mathscr{L}$

$\square$

Lemma 2.

$$(R \circ S^\backprime E)^{\triangleright (T_1 \cup T_2)} = (R \circ S^\backprime E)^{\triangleright T_1} \cap (R \circ S^\backprime E)^{\triangleright T_2} \tag{13}$$

Sketch of the proof: Due to the property

$$(a \vee b) \rightarrow c = (a \rightarrow c) \wedge (b \rightarrow c)$$

that is valid in $\mathscr{L}$ it is easy to prove that $R \triangleright (T_1 \cup T_2) = (R \triangleright T_1) \cap (R \triangleright T_2)$. This property jointly with the associativity of $\otimes$ is sufficient to prove equality (13).

$\square$

Lemma 3.

$$(R \circ S^\backprime E)^{\triangleright (T_1 \cap T_2)} = (R \circ S^\backprime E)^{\triangleright T_1} \cup (R \circ S^\backprime E)^{\triangleright T_2} \tag{14}$$

Sketch of the proof: Due to the property

$$(a \wedge b) \rightarrow c = (a \rightarrow c) \vee (b \rightarrow c)$$

that is valid in $\mathscr{L}$ it is easy to prove that $R \triangleright (T_1 \cap T_2) = (R \triangleright T_1) \cup (R \triangleright T_2)$. This property jointly with the associativity of $\otimes$ is sufficient to prove equality (14).

$\square$

Lemma 4.

$$T_1 \subseteq T_2 \Rightarrow (R \circ S^\backprime E)^{\triangleright T_1} \supseteq (R \circ S^\backprime E)^{\triangleright T_2} \tag{15}$$

Sketch of the proof: Based no the antitonicity of $\rightarrow$ in the first argument. $\square$

Now, let us downgrade a bit and focus on the binary boolean case only. Then there are three equivalent forms of the description of the same meaning that has been defined in the typicality concept by the construction of the relation $(R \circ S\backslash E)^{\rhd T}$.

Lemma 5. *Let $R, S, E,$ and T are crisp relations. Then the following three definitions of $(R \circ S\backslash E)^{\rhd T}$ are equivalent*

$$(R \circ S\backslash E)^{\rhd T} = (R \circ S) \otimes \neg(R \circ E) \otimes (R \rhd T) \,,$$

$$(R \circ S\backslash E)^{\rhd T} = (R \circ S) \otimes \neg(R \circ E) \otimes (R \triangledown \neg T) \,,$$

$$(R \circ S\backslash E)^{\rhd T} = (R \circ S) \otimes \neg(R \circ E) \otimes \neg(\neg R \circ T) \,.$$

Sketch of the proof: As in all three forms, the first two parts out of the three parts that appear on the right hands sides are equivalent, the proof consist in checking the equality of the last parts, that is, checking that

$$(R \rhd T) = (R \triangledown \neg T) \ \text{and} \ (R \rhd T) = \neg(\neg R \circ T) \,.$$

Indeed, in the Boolean algebra

$$(R \rhd T) = \{(x, z) \mid \forall y : (y, z) \in T \Rightarrow (x, y) \in R\}$$

and as the following equality $a \Rightarrow b \equiv \neg a \vee b$ holds it leads to

$$(R \rhd T) = \{(x, z) \mid \forall y : \neg((y, z) \in T) \vee (x, y) \in R\}$$

which is nothing else but the definition of $(R \triangledown \neg T)$.

Furthermore, using the double negation law and the following relationship between the conjunction and the disjunction $a \vee b = \neg(\neg a \mathbin{\&} \neg b)$ that is valid in the Boolean algebra, the proof goes as follows:

$$\forall y \in Y : \neg((y, z) \in T) \vee (x, y) \in R \equiv \forall y \in Y : \neg(\neg\neg((y, z) \in T) \wedge \neg((x, y) \in R))$$
$$\equiv \forall y \in Y : \neg((y, z) \in T \wedge \neg((x, y) \in R))$$
$$\equiv \not\exists y \in Y : (y, z) \in T \wedge \neg((x, y) \in R)$$

where the last line defines the set $\neg(\neg R \circ T)$. Thus, $(R \rhd T) = \neg(\neg R \circ T)$ and consequently also $(R \triangledown \neg T) = \neg(\neg R \circ T)$ holds in the Boolean algebra. As the residuated lattice operations on $\{0, 1\}$ coincide with operations of the Boolean algebra, it completes the proof. $\square$

Now, the natural question that can be posed is, whether a similar equivalence between the three forms is preserved in the residuated lattice generally for fuzzy (non-crisp) fuzzy relations. It is easy to build a counter-example that proves the negative answer to the question. However, if we focus on a narrower set of residuated lattices, in particular, on MV-algebras, the answer turns to a positive one.

Proposition 1. *Let the underlying algebraic structure be an MV-algebra* $\mathscr{L} = \langle L, \oplus, \otimes, \neg, 0, 1 \rangle$. *Then the following three definitions of* $(R \circ S^\backprime E)^{\triangleright T}$ *are equivalent*

$$(R \circ S^\backprime E)^{\triangleright T} = (R \circ S) \otimes \neg (R \circ E) \otimes (R \triangleright T) \ ,$$

$$(R \circ S^\backprime E)^{\triangleright T} = (R \circ S) \otimes \neg (R \circ E) \otimes (R \triangledown \neg T) \ ,$$

$$(R \circ S^\backprime E)^{\triangleright T} = (R \circ S) \otimes \neg (R \circ E) \otimes \neg (\neg R \circ T) \ .$$

Sketch of the proof: Based on the fact, that all the properties used in the proof of Lemma 5 are preserved in the MV-algebra, namely $a \to b = \neg a \oplus b$, $\neg \neg a = a$, and $a \oplus b = \neg (\neg a \otimes \neg b)$.   □

4   Experimental Justification and Conclusions

4.1   Experiment with Dragonfly Classification

In this Section, we provide a short experimental justification of the proposed concept of typical features employed in the fuzzy relational compositions. Note, that the demonstrative example is not artificial and it is a part of a long-term project connecting citizen-science, biodiversity and modern technologies in a particular cell-phone application, see [15].

The dragonfly classification problem stems from the real dataset that contains 52940 testing records (X) of dragonfly observations. Each record is described with up to 60 features (Y), including colours, intervals of altitudes, ten-day periods in the year, and morphological categories. The objective is to classify each sample with a correct class from 140 possible classes (Z). The expert knowledge provided by an odonatologist is encoded in the matrices $S, E, T \subset \mathscr{F}(Y \times Z)$, which are the matrix of features, excluding features, and typical features. Testing data samples are encoded in the matrix $R \in \mathscr{F}(X \times Y)$.

Since not all features were collected in the past (e.g. colours of observed dragonflies were not inquired originally), these features were randomly generated in the matrix R to simulate a real input. Originally [7], such random generation did not mind the typicality of some features. Therefore, the generation process was updated for this paper and therefore, the results were recomputed again even for the standard compositions without the concept of typical features.

Nevertheless, for the comparison purposes, we have recalculated the following fuzzy relational compositions: basic composition $R \circ S$, Bandler-Kohout subproduct $R \triangleleft S$, Bandler-Kohout superproduct $R \triangleright S$, and finally the basic compositions with excluding features $R \circ S^\backprime E$. Note, that due to the extreme result provided by $R \triangleright S$ and due to the fact that the square product $R \square S$ is an intersection of both triangle products, we implicitly obtain also the result provided by $R \square S$, see the discussion below.

Then we have calculated the pure typicality composition $R \triangleright T$ in order to get an intuitive insight of its influence. Furthermore, this part can be conjunctively aggregated with with other compositions, namely the following ones:

$$(R \circ S)^{\triangleright T} = (R \circ S) \otimes (R \triangleright T) \,, \quad (R \triangleleft S)^{\triangleright T} = (R \triangleleft S) \otimes (R \triangleright T) \,.$$

Due to the results provided by the Bandler-Kohout superproduct and consequently also the square product, calculating the extensions of these compositions by the typicality concept is meaningless and would provide us with an unchanged results. Finally, the composition proposed and studied in this paper, i.e., the product $(R \circ S^{\backslash} E)^{\triangleright T}$ is also constructed and its performance presented jointly with the others in Table 1. For the evaluation purposes, the following measures from [7] were adopted:

- rankM – the arithmetic mean (over the sample set X) of numbers of dragonfly species with assigned membership degree greater or equal than the membership value assigned to the correct class;
- rankGrM – the arithmetic mean (over the sample set X) of numbers of dragonfly species with assigned membership degree strictly greater than the membership value assigned to the correct class;
- #corrMax – the number samples with correct class being assigned the maximal membership degree among all other classes (not necessarily uniquely).

Table 1. Results of the dragonfly classification problem.

Composition	rankMean	rankGrMean	#correctMax
$R \circ S$	132.83	0	52940 (100.00%)
$R \triangleleft S$	9.88	1.44	36628 (69.19%)
$R \triangleright S$ (also $R \square S$)	140	0	52940 (100.00%)
$R \circ S^{\backslash} E$	18.37	0.07	52396 (98.97%)
$R \triangleright T$	26.48	0	52940 (100.00%)
$(R \circ S)^{\triangleright T}$	26.29	0	52940 (100.00%)
$(R \triangleleft S)^{\triangleright T}$	7.67	0.96	38022 (71.82%)
$(R \circ S^{\backslash} E)^{\triangleright T}$	12.27	0.04	52465 (99.10%)

4.2 Results, Discussion and Future Directions

Table 1 brings several interesting observations giving us an insight into the potential of distinct fuzzy relational based approaches. The basic composition always assigns the highest number to the correct species (#correctMax equal to 100%) however, there are on average 132.83 species that are assigned the highest number, which is only a negligible narrowing from the original list of 140 species. Using the Bandler-Kohout subproduct helps to narrow the "guessed set" to 9.88 species on average however, in more than 30% case, the correct species is not contained in it. Note, that this result is not so bad although higher accuracy is surely desirable, see [7]. The Bandler-Kohout superproduct provided full matrix of zero, i.e., $(R \triangleright S)(x, z) = 0$ for any x and for any z. This is caused by the fact that each species z is related to several ten-day periods and several altitudes however, it is impossible that a given sample x would be connected to more than a single ten-day period in the year and a single altitude. Due to the construction of the square product, it provides the same results. Let us note that this drawback may be successfully dealt with by using so-called grouping of features [8,9]. Thus, the basic composition with the concept of excluding features brought by far the best combination of the accuracy (98.87%) and the narrowness (18.37) of the results.

The performance of the pure application of the typicality provided by $R \triangleright T$ already foreshadows the positive potential in improving the results. Indeed, this approach narrows the guessed set to 26.48 species on average with the preservation of the 100% accuracy. This provides us with an interesting observation that focusing on the typical features only might be a better strategy than taking into account all possible features where the basic composition provides too wide results and the Bandler-Kohout products an unavoidable loss of accuracy. The latter is, however, not a drawback of the Bandler-Kohout products but of the width and generality of the features set Y and as soon as the number of considered features set is narrowed by the use of T instead of S, the same (Bandler-Kohout superproduct) composition suddenly brings a great potential. If the typical features are combined with the basic compositions as the change may occur only in such rare cases (species) that have no unavoidable features. The biggest potential is, indeed, demonstrated by the use of $(R \circ S^{\backprime}E)^{\triangleright T}$. There, the rankMean is lowered to 12.27 accompanied by the #corrMax accuracy equal to 99.10% which is an impressive result providing a combination of narrowness and accuracy so far never obtained by any approach.

For the sake of completeness, let us mention that the composition $(R \triangleleft S)^{\triangleright T}$ noticeably narrowed to guessed set provided by $R \triangleleft S$ (from 9.88 to 7.67) without any negative influence on the accuracy (increase from 69.19% to 71.82%). Although this composition is not the best one for the given application, even here the experiment confirmed the positive influence of the concept of typicality.

The future directions will lead to the combination with generalized quantifiers, comparison with recent results provided by these combinations, further theoretical research including the modification for partial fuzzy logics [17] and the applications. The applications will aim at other biological species classifications, medical diagnosis, and social work intelligent software support, where in the last one, typical features of communities may play an essential role in their classification and the consequent determination of the type of the social support by community projects applied in the given communities.

Acknowledgements. Supported by LQ1602 NPU II "IT4Innovations in science" by the MŠMT.

References

1. Bandler, W., Kohout, L.: Fuzzy relational products and fuzzy implication operators. In: Proceedings of the International Workshop on Fuzzy Reasoning Theory and Applications, pp. 239–244. Queen Mary College, London (1978)
2. Bandler, W., Kohout, L.: Semantics of implication operators and fuzzy relational products. Int. J. Man-Mach. Stud. **12**(1), 89–116 (1980)
3. Belohlavek, R.: Sup-t-norm and inf-residuum are one type of relational product: unifying framework and consequences. Fuzzy Sets Syst. **197**, 45–58 (2012)
4. Běhounek, L., Daňková, M.: Relational compositions in fuzzy class theory. Fuzzy Sets Syst. **160**(8), 1005–1036 (2009)
5. Bělohlávek, R.: Fuzzy Relational Systems: Foundations and Principles. Kluwer Academic; Plenum Press, Dordrecht; New York (2002)
6. Cao, N., Holčapek, M., Štěpnička, M.: Extensions of fuzzy relational compositions based on generalized quantifiers. Fuzzy Sets Syst. **339**, 73–98 (2018)
7. Cao, N., Štěpnička, M., Burda, M., Dolný, A.: Excluding features in fuzzy relational compositions. Expert Syst. Appl. **81**, 1–11 (2017)
8. Cao, N., Štěpnička, M., Burda, M., Dolný, A.: Fuzzy relational compositions based on grouping features. In: 9th International Conference on Knowledge and Systems Engineering (KSE), pp. 94–99 (2017)
9. Cao, N., Štěpnička, M., Burda, M., Dolný, A.: On the use of subproduct in fuzzy relational compositions based on grouping features. In: 2018 Information Processing and Management of Uncertainty in Knowledge-Based Systems, Applications. (Communications in Computer and Information Science), vol. 855, pp. 175–186 (2018)
10. Chung, F., Lee, T.: Analytical resolution and numerical identification of fuzzy relational systems. IEEE Trans. Syst. Man Cybern. **28**, 919–924 (1998)
11. De Baets, B.: Analytical solution methods for fuzzy relational equations. In: Dubois, D., Prade, H. (eds.) The Handbook of Fuzzy Set Series, vol. 1, pp. 291–340. Academic Kluwer Publishers, Boston (2000)
12. De Baets, B., Kerre, E.: Fuzzy relational compositions. Fuzzy Sets Syst. **60**, 109–120 (1993)
13. Di Nola, A., Sessa, S., Pedrycz, W., Sanchez, E.: Fuzzy Relation Equations and Their Applications to Knowledge Engineering. Kluwer, Boston (1989)
14. Novák, V., Perfilieva, I., Močkoř, J.: Mathematical Principles of Fuzzy Logic. Kluwer Academic Publishers, Boston (1999)
15. Ožana, S., Burda, M., Hykel, M., Malina, M., Prášek, M., Bárta, D., Dolný, A.: Dragonfly hunter CZ: mobile application for biological species recognition in citizen science. PLoS ONE **14**(1), e0210370 (2019)
16. Pedrycz, W.: Fuzzy relational equations with generalized connectives and their applications. Fuzzy Sets Syst. **10**, 185–201 (1983)
17. Štěpnička, M., Cao, N., Běhounek, L., Burda, M., Dolný, A.: Missing values and dragonfly operations in fuzzy relational compositions. Int. J. Approximate Reasoning (submitted)
18. Štěpnička, M., Holčapek, M.: Fuzzy relational compositions based on generalized quantifiers. In: Information Processing and Management of Uncertainty in Knowledge-Based Systems. PT II (IPMU 2014), Communications in Computer and Information Science, vol. 443, pp. 224–233. Springer, Berlin (2014)

A Subsethod Interval Associative Memory with Competitive Learning

Peter Sussner$^{(\boxtimes)}$, Estevão Esmi, and Luis Gustavo Jardim

IMECC, Department of Applied Mathematics, University of Campinas,
Campinas, SP, Brazil
{sussner,eelaureano,ra207576}@ime.unicamp.br

Abstract. Morphological perceptrons with competitive learning (MP/CLs) are constructive artificial neural network models having a modular architecture. Not only the weights but also the architecture of the MP/CL is automatically generated by the MP/CL training algorithm. The resulting architecture is determined by hyperboxes, i.e., closed intervals, contained in $\mathbb{F}^n$, where $\mathbb{F}$ is a totally ordered group. The group operations and the total ordering are used to perform a competition among the outputs of each module. In this paper, we present an interval subsethood associative memory whose hidden nodes compute degrees of subsethood of the input pattern in each of the closed intervals generated by the MP/CL training algorithm. We show that the resulting interval subsethood associative memory with competitive learning (S-IAM/CL) can be viewed as a Θ-fuzzy associative memory (Θ-FAM) model. We compare the performance of S-IAM/CLs for a particular choice of interval subsethood measure with the ones of the original MP/CL model and several competitive models in a number of classification problems.

1 Introduction

The class of fuzzy associative memories known as Θ-FAMs grew out of morphological associative memories, that – along with morphological perceptrons – represent one of the first classes of morphological neural networks to appear in the literature [16]. Specifically, Sussner observed that the outputs of binary autoassociative morphological memories can be described in terms of crisp subsethood and supersethood operations and proposed autoassociative memories based on Kosko's fuzzy subsethood and supersethood operations [18]. These considerations can be linked to the general theory of mathematical morphology from the geometrical or topological perspective, that is concerned with the processing and analysis of images using structuring elements (SEs) [17]. In particular, recall that the binary erosion of an image by a structuring element S yields the set of points for which certain translated versions of S are contained in the image. Similarly, a fuzzy erosion of a fuzzy image by a fuzzy structuring element $\mathbf{s}$ at a point $\mathbf{x}$ is given by a degree of fuzzy inclusion of the translated

© Springer Nature Switzerland AG 2019
R. B. Kearfott et al. (Eds.): IFSA 2019/NAFIPS 2019, AISC 1000, pp. 643–654, 2019.
https://doi.org/10.1007/978-3-030-21920-8_57

version of **s** centered at **x** in the fuzzy image [22]. The latter depends on the choice of a fuzzy inclusion or subsethood measure [10,24].

The initial idea underlying the development of morphological perceptrons (MPs) was to evaluate the crisp inclusion of an input pattern in each of the hyperboxes generated by the MP training algorithm. Morphological neural networks with competitive learning also generate a finite number of hyperboxes but their outputs are computed using lattice-algebraic operations in $\mathbb{R}_{\pm\infty}^n = (\mathbb{R}_{\pm\infty})^n$, where $\mathbb{R}_{\pm\infty}$ are the extended real numbers, i.e., $\mathbb{R} \cup \{+\infty, -\infty\}$. Since we are dealing with real-valued patterns in practice, these hyperboxes can be viewed as closed subintervals contained in $[0,1]^n$ after normalization. Therefore, any subsethood measure defined on the bounded lattice of the closed subintervals (including $\emptyset$) of $[0,1]^n$, i. e., $\mathbb{I} = \{[\mathbf{a},\mathbf{b}] : \mathbf{a},\mathbf{b} \in [0,1]^n \text{ and } \mathbf{a} \leq \mathbf{b}\} \cup \emptyset$, where $[\mathbf{a},\mathbf{b}] = \{\mathbf{x} : a_i \leq x_i \leq b_i, i = 1, \ldots, n\}$, can be applied to the input pattern, represented as an interval, and one of the hyperboxes. In this paper, we introduce a method for constructing subsethood measures on an arbitrary bounded lattice $\mathbb{L}$ using an increasing function $v : \mathbb{L} \to [0,1]$ that satisfies certain boundary conditions. Here, it suffices to consider an appropriate function $v_{\mathbb{I}}$ on $\mathbb{I}$. We show how to generate an infinite number of interval subsethood measures based on $v_{\mathbb{I}}$ and how to incorporate these interval subsethood measures as well as the result of the MP/CL training algorithm into subsethood interval associative memories that comply with the general definition of Θ-FAMs in bounded lattices [9]. We apply these S-IAMs to a number of benchmark classification problems and compare the classification rates with the ones produced by the MP/CL and other models that recently appeared in the literature.

The paper is organized as follows. Section 2 provides some mathematical background on lattice theory and, more importantly, introduces a general type of subsethood measure on bounded lattices. Section 3 briefly reviews the MP/CL model and its training algorithm as well as the Θ-FAM model. Section 4 describes the proposed S-IAM/CL model that incorporates the result of the MP/CL training algorithm into a subsethood interval associative memory. In Sect. 5, we apply the new model to a number of classification problems and compare the results with the ones produced by some other classifiers. We finish with some concluding remarks.

2   Some Mathematical Background

2.1   Some Lattice-Theoretical Concepts

Lattice theory has its origins in Boolean algebra [5] and in certain number-theoretic problems that were studied by Dedekind at the turn of the twentieth century [7]. In the following decades, lattice theory was almost forgotten before Birkhoff, Ore, and others rediscovered it in the 1930s [4]. Although lattice theory grew out of purely mathematical considerations, it has proved to be very useful in several application-oriented areas [12,13,15,20,21,23].

A *partially ordered set*, also called *poset*, is a pair $(P, \leq)$ where P is a non-empty set and $\leq$ is a partial order relation, i.e., a reflexive, anti-symmetric and transitive relation, denoted $\leq$. If the partial order relation $\leq$ clearly arises from the context, then we simply use the symbol P instead of $(P, \leq)$. If X is an arbitrary, non-empty set, then its power set $\mathscr{P}(X)$ together with the crisp set inclusion yields a poset. If P is a poset and $x \leq y$ or $y \leq x$ for all $x, y \in P$, then one refers to $\leq$ as a total order and to P as a *totally ordered set* or *chain*. The sets $[0, 1]$, $\mathbb{R}$ and $\mathbb{R}_{\pm\infty} = \mathbb{R} \cup \{-\infty, +\infty\}$ with the usual ordering represent chains.

A partial order on P gives rise to the notions of lower bound and upper bound. An element $l \in P$ is called a *lower bound* of $X \subseteq P$ if $l \leq x$ for all $x \in X$. A lower bound l of $X \subseteq P$ is called greatest lower bound or *infimum* of X if there is no other lower bound m of X such that $l \leq m$. The notion of least upper bound or *supremum* is defined similarly. One uses the symbols $\bigwedge X$ annd $\bigvee X$ to denote respectively the infimum and the supremum of X. If $X = \{x, y\}$, then one simply writes $x \wedge y$ and $x \vee y$. A poset $\mathbb{L}$ is called a *lattice* if $x \wedge y$ and $x \vee y$ exist in $\mathbb{L}$ for all $x, y \in \mathbb{L}$. If $\mathbb{M}$ is a subset of a lattice $\mathbb{L}$ and if $x \wedge y, x \vee y$, as defined in $\mathbb{L}$, are elements of $\mathbb{M}$ for all $x, y \in \mathbb{L}$, then $\mathbb{M}$ is called a sublattice of $\mathbb{L}$.

Let $\mathbb{L}$ be a lattice. If $\bigwedge \mathbb{L}, \bigvee \mathbb{L} \in \mathbb{L}$, then $\mathbb{L}$ is called a *bounded lattice* and $\bigwedge \mathbb{L}, \bigvee \mathbb{L}$ are respectively denoted $0_{\mathbb{L}}$ and $1_{\mathbb{L}}$. If $\bigwedge X, \bigvee X \in \mathbb{L}$ for all $X \subseteq \mathbb{L}$, then $\mathbb{L}$ is called a *complete lattice*. For example, $\mathscr{P}(X)$ is a complete lattice for every set $X \neq \emptyset$. The chains $[0, 1]$ and $\mathbb{R}_{\pm\infty}$ are also complete lattices. Therefore, we refer to them as complete chains.

Let $\mathbb{L}$ be a bounded lattice and let

$$\mathbb{I}_{\mathbb{L}} = \{[a, b] : a, b \in \mathbb{L} \text{ and } a \leq b\} \cup \emptyset, \tag{1}$$

where $[a, b] \subseteq \mathbb{L}$ represents the closed interval $[a, b] = \{x : a \leq x \leq b\}$ for $a \leq b$. Together with the partial ordering of set inclusion, $\mathbb{I}_{\mathbb{L}}$ yields a bounded lattice. Note that the infimum of two elements of $\mathbb{I}_{\mathbb{L}} \subseteq \mathscr{P}(\mathbb{L})$ is given by their intersection, but the supremum operation in $\mathbb{I}_{\mathbb{L}}$ differs from the union of sets. Note that $[a, b] \vee [c, d] = [a \wedge c, b \vee d] \in \mathbb{I}_{\mathbb{L}}$ for $a, b, c, d \in \mathbb{L}$ such that $a \leq b$ and $c \leq d$. Therefore, we have that $\mathbb{I}_{\mathbb{L}}$ is not a sublattice of the complete lattice $\mathscr{P}(\mathbb{L})$. If $\mathbb{L}$ is a complete lattice, then $\mathscr{P}(\mathbb{L})$ is a complete lattice as well.

Isomorphisms allow us to identify one mathematical structure with another one. For example, consider two complete lattices $(\mathbb{L}, \leq_{\mathbb{L}})$ and $(\mathbb{M}, \leq_{\mathbb{M}})$. A bijection $\phi : \mathbb{L} \to \mathbb{M}$ is called a *complete lattice isomorphism* if we have $x \leq_{\mathbb{L}} y$ if and only if $\phi(x) \leq_{\mathbb{M}} \phi(x)$ for all $x, y \in \mathbb{L}$. If there is a complete lattice isomorphism $\phi : \mathbb{L} \to \mathbb{M}$, then the complete lattices $\mathbb{L}$ and $\mathbb{M}$ are called isomorphic and one writes $\mathbb{L} \simeq \mathbb{M}$.

If $\mathbb{L}_1, \ldots \mathbb{L}_n$ are posets, then a partial order on the direct product $\mathbb{L}_1 \times \ldots \times \mathbb{L}_n$ is given by

$$(a_1, \ldots, a_n) \leq (b_1, \ldots, b_n) \Leftrightarrow a_i \leq b_i, \ i = 1, \ldots, n. \tag{2}$$

The direct product of n (> 0) copies of $\mathbb{L}$ is denoted using the symbol $\mathbb{L}^n$. If $\mathbb{L}$ is a lattice, then the direct product $\mathbb{L}^n$ is also a lattice and if $\mathbb{L}$ is bounded or complete, then $\mathbb{L}^n$ is respectively bounded or complete as well. Similarly, if $\mathbb{L}^X$

stands for the class of all functions $X \to \mathbb{L}$, then a partial order $\leq$ on $\mathbb{L}$ induces a partial order on $\mathbb{L}^X$, also denoted by $\leq$, which is defined as follows, for all $f, g \in \mathbb{L}^X$:

$$f \leq g \Leftrightarrow f(x) \leq g(x) \forall x \in X. \tag{3}$$

This definition of a partial order on $\mathbb{L}^X$ implies that a $\mathbb{L}^X$ is a lattice if $\mathbb{L}$ is a lattice. We have that the lattice $\mathbb{L}^X$ is bounded if $\mathbb{L}$ is bounded and $\mathbb{L}^X$ is complete if $\mathbb{L}$ is complete.

Mathematical morphology (MM) is usually conducted in a complete lattice framework [12]. In this framework, there are four elementary operators that are defined as follows:

Definition 1. Let $\varepsilon, \delta, \bar{\varepsilon}, \bar{\delta} : \mathbb{L} \to \mathbb{M}$, where $\mathbb{L}$ and $\mathbb{M}$ are complete lattice. The operators $\varepsilon, \delta, \bar{\varepsilon}$, and $\bar{\delta}$ are respectively called an *erosion*, a *dilation*, *anti-erosion*, and an *anti-dilation* if and only if the following equations are satisfied for all $Y \subseteq \mathbb{L}$:

$$\varepsilon\left(\bigwedge Y\right) = \bigwedge_{y \in Y} \varepsilon(y), \ \delta\left(\bigvee Y\right) = \bigvee_{y \in Y} \delta(y), \ \bar{\varepsilon}\left(\bigwedge Y\right) = \bigvee_{y \in Y} \bar{\varepsilon}(y), \ \bar{\delta}\left(\bigvee Y\right) = \bigwedge_{y \in Y} \bar{\delta}(y). \tag{4}$$

Erosions and dilations are increasing operators. Recall that a mapping $\phi : \mathbb{L} \to \mathbb{M}$ is *increasing* if $x \leq y$ implies that $\phi(x) \leq \phi(y)$. The following theorem, which is an immediate consequence of Theorem 6.1 of [3], shows that every mapping can be written in terms of erosions and anti-dilations or dilations and anti-erosions.

Theorem 1. *Let $\mathbb{L}$ and $\mathbb{M}$ be complete lattices and let $\psi : \mathbb{L} \to \mathbb{M}$ be arbitrary. There exists an index set I and erosions ε^i and anti-dilations $\bar{\delta}^i$ for $i \in I$ such that*

$$\psi = \bigvee_{i \in I} \left(\varepsilon_i \wedge \bar{\delta}_i\right). \tag{5}$$

Similarly, there exists an index set J and dilations δ_j and anti-erosions $\bar{\varepsilon}^j$ for $j \in J$ such that

$$\psi = \bigwedge_{j \in J} \left(\delta_j \vee \bar{\varepsilon}_j\right). \tag{6}$$

For the purposes of this paper, it suffices to recall very specific examples of erosions and anti-dilations. First note that $\mathbb{R}_{\pm\infty} = \mathbb{R} \cup \{+\infty, -\infty\}$ is a complete chain. Let $\mathbb{R}_{\pm\infty}^n$ denote $(\mathbb{R}_{\pm\infty})^n$. Consider the following commutative and associative operator $+' : \mathbb{R}_{\pm\infty} \times \mathbb{R}_{\pm\infty} \to \mathbb{R}_{\pm\infty}$ that satisfies

1. $a +' b = a + b \ \forall a, b \in \mathbb{R}$;
2. $a +' (+\infty) = +\infty, \forall a \in \mathbb{R}_{\pm\infty}$;
3. $a +' (-\infty) = -\infty, \forall a \in \mathbb{R} \cup \{-\infty\}$.

The *conjugate* of an element $x \in \mathbb{R}_{\pm\infty}$ is given by x^*, where [6]

$$x^* = \begin{cases} -x & \text{if } x \in \mathbb{R}, \\ -\infty & \text{if } x = +\infty, \\ +\infty & \text{if } x = -\infty. \end{cases} \tag{7}$$

For every $\mathbf{w} \in \mathbb{R}^n_{\pm\infty}$, the following operators $\varepsilon_{\mathbf{w}}, \bar{\delta}_W : \mathbb{R}^n_{\pm\infty} \to \mathbb{R}_{\pm\infty}$ represent respectively an erosion and an anti-dilation.

$$\varepsilon_{\mathbf{w}}(\mathbf{x}) = \bigwedge_{i=1}^{n} (x_i +' w_i), \quad \bar{\delta}_{\mathbf{w}}(\mathbf{x}) = \bigwedge_{i=1}^{n} (x_i^* +' w_i). \tag{8}$$

Here x_i^* stands for $(x_i)^*$. Operators of the type given in Eq. (8) are employed in MP/CLs.

2.2 Subsethood Measures on Bounded Lattices

Subsethood measures were previously defined on $\mathscr{F}(X)$, the class of fuzzy sets of $X \neq \emptyset$. Recall that a fuzzy set A on the universe X is the graph of its membership function $\mu_A : X \to [0, 1]$. Together with the partial order given by the inclusion of fuzzy sets, $\mathscr{F}(X)$ is a complete lattice that is isomorphic to $[0, 1]^X$ with the product partial order given by Eq. (2). Recall that $A \subseteq B$ for $A, B \in \mathscr{F}(X)$ if and only if $\mu_A(x) \leq \mu_B(x) \forall x \in X$. Note that fuzzy set inclusion is a crisp measure, that is, for two fuzzy sets in $\mathscr{F}(X)$, we either have that one of them is included in the other one or not. In contrast, evaluating the fuzzy inclusion or fuzzy subsethood of $A \in \mathscr{F}(X)$ in $B \in \mathscr{F}(X)$ yields an element of $[0, 1]$. We recently generalized the axiomatic characterization of a fuzzy subsethood measure according to Fan et al. as follows [9,10]:

Definition 2. A function $S : \mathbb{L} \times \mathbb{L} \to [0, 1]$ satisfying the following properties for all $\mathbf{x}, \mathbf{y}, \mathbf{z} \subset \mathbb{L}$ is said to be a *subsethood measure* on $\mathbb{L}$:

1. If $\mathbf{x} \leq \mathbf{y}$, then $S(\mathbf{x}, \mathbf{y}) = 1$;
2. $S(1_{\mathbb{L}}, 0_{\mathbb{L}}) = 0$;
3. If $\mathbf{x} \leq \mathbf{y} \leq \mathbf{z}$ then $S(\mathbf{z}, \mathbf{x}) \leq S(\mathbf{y}, \mathbf{x})$ and $S(\mathbf{z}, \mathbf{x}) \leq S(\mathbf{z}, \mathbf{y})$.

Fan et al.'s definition of a fuzzy subsethood measure arises by setting $\mathbb{L} = \mathscr{F}(X)$ for some non-empty universe X. Let us recall a useful recipe for constructing fuzzy subsethood measures [9]. Specifically, the following operators $S^{\cap}$ and $S^{\cup}$ represent fuzzy subsethood measures if I is a fuzzy implication that satisfies

$$I(a, b) = 1, \forall a \leq b \in [0, 1] \tag{9}$$

and if $v : \mathscr{F}(X) \to [0, 1]$ is an increasing function such that $v(\emptyset) = 0$ and $v(X) = 1$:

(a) $$S^{\cap}(A, B) = I(v(A), v(A \cap B)) \tag{10}$$

(b) $$S^{\cup}(A, B) = I(v(A \cup B), v(B)) \tag{11}$$

Let us generalize this construction method as follows:

Proposition 1. *Let* $\mathbb{L}$ *be a bounded lattice, I be a fuzzy implication that satisfies Eq. (9), and let $v : \mathbb{L} \to [0,1]$ be increasing with $v(0_\mathbb{L}) = 0$ and $v(1_\mathbb{L}) = 1$. The following formulas yield subsethood measures on* $\mathbb{L}$:

$$(a) \qquad\qquad S^\wedge(\mathbf{x}, \mathbf{y}) = I(v(\mathbf{x}), v(\mathbf{x} \wedge \mathbf{y})) \qquad\qquad (12)$$

$$(b) \qquad\qquad S^\vee(\mathbf{x}, \mathbf{y}) = I(v(\mathbf{x} \vee \mathbf{y}), v(\mathbf{y})) \qquad\qquad (13)$$

Proof. Let us restrict ourselves to showing that $S^\vee$ is a subsethood measure on $\mathbb{L}$. The proof of the other part is similar.

First, suppose that $\mathbf{x} \leq \mathbf{y}$. We obtain $S^\vee(\mathbf{x}, \mathbf{y}) = I(v(\mathbf{x} \vee \mathbf{y}), v(\mathbf{y})) = I(v(\mathbf{y}), v(\mathbf{y}))$. By Eq. (9), the latter equals 1.

Secondly, we have $S^\wedge(1_\mathbb{L}, 0_\mathbb{L}) = I(v(1_\mathbb{L}), v(0_\mathbb{L} \wedge 1_\mathbb{L})) = I(v(1_\mathbb{L}), v(0_\mathbb{L})) = I(1,0) = 0$ and, similarly, $S^\vee(1_\mathbb{L}, 0_\mathbb{L}) = 0$.

Thirdly, let $\mathbf{x} \leq \mathbf{y} \leq \mathbf{z}$. We have $S^\vee(\mathbf{z}, \mathbf{x}) = I(v(\mathbf{z} \vee \mathbf{x}), v(\mathbf{x})) = I(v(\mathbf{z}), v(\mathbf{x})) \leq I(v(\mathbf{y}), v(\mathbf{x})) = I(v(\mathbf{y}) \vee v(\mathbf{x}), v(\mathbf{x})) = S^\vee(\mathbf{y}, \mathbf{x})$. Additionally, we have $S^\vee(\mathbf{z}, \mathbf{x}) = I(v(\mathbf{z} \vee \mathbf{x}), v(\mathbf{x})) = I(v(\mathbf{z}), v(\mathbf{x})) \leq I(v(\mathbf{z}), v(\mathbf{y})) = I(v(\mathbf{z}) \vee v(\mathbf{y}), v(\mathbf{y})) = S^\vee(\mathbf{z}, \mathbf{y})$.

Note that every residual implication of a t-norm satisfies Eq. (9). For the purposes of this paper, it suffices to focus on interval subsethood measures from $\mathbb{I} \times \mathbb{I}$ to $[0,1]$. where $\mathbb{I}$ equals the complete lattice $\mathbb{I}_\mathbb{L}$ (cf. Eq. (1)) for $\mathbb{L} = [0,1]^n$ with the partial order given by $\subseteq$. For clarity, let us point out what $\mathbf{x} \wedge \mathbf{y}$ and $\mathbf{x} \vee \mathbf{y}$ are for $\mathbf{x} = [\underline{\mathbf{x}}, \overline{\mathbf{x}}] = ([\underline{x}_1, \overline{x}_1], \ldots, [\underline{x}_n, \overline{x}_n])^T$, $\mathbf{y} = [\underline{\mathbf{y}}, \overline{\mathbf{y}}] = ([\underline{y}_1, \overline{y}_1], \ldots, [\underline{y}_n, \overline{y}_n])^T \subseteq [0,1]^n$.

$$\mathbf{x} \wedge \mathbf{y} = \begin{cases} ([\underline{x}_1 \vee \underline{y}_1, \overline{x}_1 \wedge \overline{y}_1], \ldots, [\underline{x}_n \vee \underline{y}_n, \overline{x}_n \wedge \overline{y}_n])^T & \text{if } \underline{x}_i \vee \underline{y}_i \leq \overline{x}_i \wedge \overline{y}_i \; \forall i \\ \emptyset & \text{otherwise.} \end{cases} \qquad (14)$$

$$\mathbf{x} \vee \mathbf{y} = ([\underline{x}_1 \wedge \underline{y}_1, \overline{x}_1 \vee \overline{y}_1], \ldots, [\underline{x}_n \wedge \underline{y}_n, \overline{x}_n \vee \overline{y}_n])^T. \qquad (15)$$

Moreover, we have

$$\mathbf{x} \wedge \emptyset = \emptyset \text{ and } \mathbf{x} \vee \emptyset = \mathbf{x} \; \forall \mathbf{x} \in \mathbb{I}. \qquad (16)$$

An increasing function $v_\mathbb{I} : \mathbb{I} \to [0,1]$ that satisfies $v_\mathbb{I}(\emptyset) = 0$ and $v_\mathbb{I}([0,1]) = 1$ is given by

$$v_\mathbb{I}(\mathbf{x}) = \begin{cases} \dfrac{1}{n} \displaystyle\sum_{i=1}^{n} (\overline{x}_i - \underline{x}_i) & \text{if } \mathbf{x} = ([\underline{x}_1, \overline{x}_1], \ldots, [\underline{x}_n, \overline{x}_n])^T \in \mathbb{I}, \\ 0 & \text{if } \mathbf{x} = \emptyset \in \mathbb{I}. \end{cases} \qquad (17)$$

Using Eq. (12) or (13), one can generate a subsethood measure $\mathbb{I} \times \mathbb{I} \to [0,1]$ based on an R-implication and $v_\mathbb{I}$. In this paper, the input pattern $\mathbf{x} \in \mathbb{I}$ presented to the S-IAM model will be of the form $\mathbf{x} = [\underline{\mathbf{x}}, \overline{\mathbf{x}}]$, where $\underline{\mathbf{x}} = \overline{\mathbf{x}} \in [0,1]^n$. This implies that $v_\mathbb{I}(\mathbf{x}) = 0$ and, since $v_\mathbb{I}$ is increasing, $v_\mathbb{I}(\mathbf{x} \wedge \mathbf{y}) = 0$ for all $\mathbf{y} \in \mathbb{I}$. Hence, $S^\wedge(\mathbf{x}, \cdot)$ equals the constant function 1 if $\underline{\mathbf{x}} = \overline{\mathbf{x}}$ and if $v = v_\mathbb{I}$. Therefore, we will focus on interval subsethood measures of the form $S^\vee$. In fact, our S-IAM model will compute degrees of subsethood in terms of $S^\vee$ of the input pattern $\mathbf{x} \in [0,1]^n$ in closed subintervals of $[0,1]^n$ that were generated by the MP/CL training algorithm.

These observations also reveal that neither $S^\wedge$ nor $S^\vee$ based on $v_{\mathbb{I}}$ are inclusion measures in the sense of Kaburlasos et al. [13–15] because $S^\wedge(\mathbf{x}, \emptyset) = S^\vee(\mathbf{x}, \emptyset) = 1$ for every $\mathbf{x}$ such that $v_{\mathbb{I}}(\mathbf{x}) = 0$.

3   A Very Brief Review of the MP/CL and Θ-FAM Models

3.1   Some Comments of the MP/CL

Morphological perceptrons with competitive learning are constructive neural networks for solving classification problems. The training algorithm for morphological perceptrons with competitive learning (MP/CLs) was proven to converge in a finite number of steps and the result is independent of the sequence of the training data [19]. The MP/CL training algorithm automatically generates the modular network architecture that is visualized in Figs. 1 and 2.

The sth module has weight vectors that are denoted $\mathbf{v}_j^s$ and $\mathbf{w}_j^s$, where $j = 1, \ldots, m_s$, and produces the following output $y_s \in \mathbb{R}$ upon presentation of an input pattern $\mathbf{x} \in \mathbb{R}^n$:

$$y_s = \bigvee_{j=1}^{m_s} (\varepsilon_{\mathbf{v}_j^s}(\mathbf{x}) \wedge \bar{\delta}_{\mathbf{w}_j^s}(\mathbf{x})) \tag{18}$$

Hence, the MP/CL is equipped with $\sum_{s=1,\ldots,S} m_s$ morphological components that compute $\varepsilon_{\mathbf{v}_j^s}(\mathbf{x}) \wedge \bar{\delta}_{\mathbf{w}_j^s}(\mathbf{x})$ upon presentation of an input pattern $\mathbf{x}$. Each of these components determines a hyperbox $\mathbf{h}_j^s = [\underline{\mathbf{h}}_j^s, \overline{\mathbf{h}}_j^s]$ that is given by

$$[\underline{\mathbf{h}}_j^s, \overline{\mathbf{h}}_j^s] = \{\mathbf{x} \in \mathbb{R}_{\pm\infty}^n : 0 \le \varepsilon_{\mathbf{v}_j^s}(\mathbf{x}) \wedge \bar{\delta}_{\mathbf{w}_j^s}(\mathbf{x})\}, \text{ where } \underline{\mathbf{h}}_j^s = (\mathbf{v}_j^s)^* \text{ and } \overline{\mathbf{h}}_j^s = \mathbf{w}_j^s \tag{19}$$

If the number of classes equals S, then the final output is $\text{argmax}_{s=1,\ldots S} y_s$. If $S > 2$, then Sussner and Esmi suggest to execute the MP/CL training algorithm following a one-against-one strategy

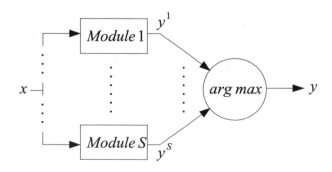

Fig. 1. Architecture of the MP/CL.

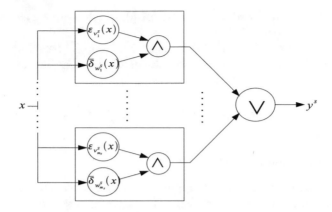

Fig. 2. Module of an MP/CL corresponding to the sth class.

3.2 Θ-FAM Model on Bounded Lattices

The associative memory models called Θ-FAMs were originally proposed for fuzzy set valued inputs and later generalized to inputs in an arbitrary bounded lattice $\mathbb{L}$. The outputs of a Θ-FAM model are fuzzy sets in $\mathscr{F}(Y)$, where Y is an arbitrary universe. The purpose of a Θ-FAM is to store a set of pattern associations, also known as the *fundamental memory set* or the set of *fundamental memories* [11], of the form $\mathscr{M} = \{(\mathbf{x}^\xi, B^\xi) \in \mathbb{L} \times \mathscr{F}(Y) \,|\, \xi \in P\}$, where P is a finite set of indices, say $P = \{1, \dots, p\}$. Let us recall the general definition of a Θ-FAM below [9].

Definition 3. Given operators $\Theta^\xi : \mathbb{L} \to [0,1]$ satisfying $\Theta^\xi(\mathbf{x}^\xi) = 1 \; \forall \, \xi \in P$, a weight vector $\mathbf{v} \in \mathbb{R}^p$, a function $F : \mathbb{R}^p \to [0,1]^p$, and an arbitrary t-norm, the following mapping $\mathscr{O} : \mathbb{L} \to \mathscr{F}(Y)$ yields a Θ-fuzzy associative memory, for short, a Θ-FAM:

$$\mathscr{O}(\mathbf{x}) = R \circ_t F(v_1 \Theta^1(\mathbf{x}), \dots, v_p \Theta^p(\mathbf{x})), \tag{20}$$

where R denotes the fuzzy relation in $Y \times P$ given by $R(y, \xi) = B^\xi(y)$ for all $y \in Y$ and $\xi \in P$.

The Θ-FAM of Definition 3 is said to exhibit perfect recall if $\mathscr{O}(\mathbf{x}^\xi) = B^\xi$ $\forall \xi = 1, \dots, P$. From now on, let $F : \mathbb{R}^p \to \{0,1\}^p \subset [0,1]^p$ be such that $F((a_1, \dots, a_p)^T)_i = 1$ if and only if $a_i = \vee_{j=1}^p a_j$. Given this choice of F and an input pattern $\mathbf{x} \in \mathbb{L}$, the Θ-FAM produces as an output the fuzzy set $\mathscr{O}(\mathbf{x}) = \bigcup_{j \in I_\mathbf{v}(\mathbf{x})} B^j$, where

$$I_\mathbf{v}(\mathbf{x}) = \left\{ j \in \{1, \dots, p\} : v_j \Theta^j(\mathbf{x}) = \bigvee_{\xi = 1, \dots, p} v_\xi \Theta^\xi(\mathbf{x}) \right\}. \tag{21}$$

Figure 3 displays the topology of a Θ-FAM for inputs $\mathbf{x}$ in a product of bounded lattices and $Y = \{y_1, \dots, y_m\}$.

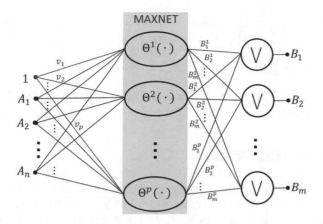

Fig. 3. Topology of a Θ-FAM.

4 S-IAM/CL: A Θ-FAM Model with Competitive Learning

As mentioned before, an associative memory model such as a Θ-FAM requires the specification of a finite fundamental memory set $\mathcal{M} \subset \mathbb{L} \times \mathscr{F}(Y)$. Let us address the use of a Θ-FAM for classification. The training and testing sets are assumed to be finite subsets of $(X_1 \times \ldots \times X_n) \times Y$, where each X_i denotes either a set of numerical values, i.e., $X_i = [v^i_{min}, v^i_{max}] \subset \mathbb{R}$, or a set of categorical values and $Y = \{1, \ldots, S\}$ is a set of class labels. One can associate each element $\mathbf{x} \in X_1 \times \ldots \times X_n$ with an element of $[0,1]^n$ by means of a normalization procedure such as the one that maps $x_i \in v^i_{min}, v^i_{max}]$ to $(x_i - v^i_{min})/(v^i_{max} - v^i_{min}) \in [0,1]$.

Suppose that $\mathbf{x}^\xi \in [0,1]^n$ is the ξth normalized pattern key in the training set and that $l^\xi \in \mathscr{C}$ is its class label. Let us associate $\mathbf{x}^\xi$ with the fuzzy set $B^\xi \in \mathscr{F}(Y)$ that is defined as follows:

$$\mu_{B^\xi}(i) = \begin{cases} 1, & \text{if } i = l^\xi \\ 0, & \text{otherwise}. \end{cases} \tag{22}$$

To conclude, we may assume that we are provided with a training set $\mathscr{T}$ of the form $\mathscr{T} = \{(\mathbf{x}^\xi, B^\xi) \in \mathbb{L} \times \mathscr{F}(Y) : \xi = 1, \ldots, k\}$, where $\mathbb{L} = [0,1]^n$.

In previous publications on Θ-FAMS, the fundamental memory set was obtained by setting $\mathcal{M} = \mathscr{T}$ [8] or by selecting an appropriate subset of $\mathscr{T}$ via the fundamental memory extraction algorithm (Algorithm 1 of [9]). In this paper, we propose a completely different strategy:

1. First, we apply the MP/CL training algorithm to the patterns $\mathbf{x}^\xi \in [0,1]^n$, having class labels l^ξ for $\xi = 1, \ldots, k$. For each class $s \in Y$, we produce m_s hyperboxes $\mathbf{h}^s_j$, where $j = 1, \ldots, S$. Let $p = \sum_{s=1,\ldots,S} m_s$ and let us rename the hyperboxes $\mathbf{h}^s_j$, where $s = 1, \ldots, S$ and $j = 1, \ldots m_s$ by referring to them

as $\mathbf{h}^\gamma$, where $\gamma = 1, \ldots, p$. Let l^γ be the class label of the patterns contained in $\mathbf{h}^\gamma$ and let $B^\gamma \in \mathscr{F}(Y)$ be the corresponding fuzzy set. Define

$$\mathscr{M} = \{(\mathbf{h}^\gamma, B^\gamma) \in \mathbb{I} \times \mathscr{F}(Y) : \gamma = 1, \ldots p\} \subset \mathbb{I} \times \mathscr{F}(Y). \qquad (23)$$

2. Define $\mathscr{M} \subset \mathbb{I} \times \mathscr{F}(Y)$ to be the fundamental memory set of a Θ-FAM model whose functions $\Theta^\gamma : \mathbb{I} \to [0, 1]$, where $\gamma = 1, \ldots, p$, are given by $S^\vee(\cdot, \mathbf{h}^\gamma)$.

5   Experimental Results

Let us present some preliminary experimental results in this section. Specifically, we performed some simulations using the S-IAM/CL model following the strategy described in the previous section. As a subsethood measure, we employed $S^\vee$ of Eq. (13). with the Lukasiewicz implication and the function $v_\mathbb{I}$ of Eq. (17). Table 1 lists ten datasets that are available at the Knowledge Extraction Based on Evolutionary Learning (KEEL)-Dataset Repository [2].

We compared the classification results produced by the proposed S-IAM/CL model with the ones produced by the MP/CL [19] and the Θ-FAM model of [8] and several other competitive classifiers that were discussed in [1]. To be more precise, the authors of [1] performed a series of computational experiments using the following models: C4.5 decision tree, structural learning algorithm on vague environment (2SLAVE), classification based on associations (CBA), an improved version of the CBA method (CBA2), classification based on multiple association rules (CMAR), classification based on predictive association rules (CPAR), fuzzy hybrid genetic based machine learning algorithm (FH-GBML), steady-state genetic algorithm for extracting fuzzy classification rules from data (SGERD), and fuzzy association rule-based classification method for high-dimensional problems (FARC-HD).

Table 1. Description of the Datasets.

Dataset	Instances	Categorical features	Numerical features	Classes
Appendicitis	106	0	7	2
Cleveland	297	0	13	5
Heart	270	0	13	2
Iris	150	0	4	3
Monks	432	0	6	2
Pima	768	0	8	2
Spectfheart	267	0	44	2
Wdbc	569	0	30	2
Wine	178	0	13	3

We conducted our simulations following the same approach suggested in [1] and we evaluated the classification performance of the S-IAM/CL using 10-fold cross validation, partitioning the data into the same folds as in [1,2]. A brief glance at Table 2 reveals that the results of our S-IAM/CL approach are competitive. Let us finish by emphasizing that – in contrast to [8,9] – we neither optimized the weights $v_1, \ldots, v_p$ shown in Fig. 1 (we set the weights $v_1, \ldots, v_p$ equal to 1 and left them unchanged) nor the functions Θ^ξ shown in Fig. 1. This issue will be investigated in the near future.

Table 2. Classification rates achieved by our S-IAM/CL approach and other models [1,8,19].

Dataset	2SLAVE	FH-GBML	SGERD	CBA	CBA2	CMAR	CPAR	C4.5	FARC-HD	Θ-FAM	MP/CL	S IAM/CL
Appendicitis	82.91	86	84.48	89.6	89.6	**89.7**	87.8	83.3	84.2	81.18	85.09	84.18
Cleveland	48.82	53.51	51.59	**56.9**	54.9	53.9	54.9	54.5	55.2	51.17	55.17	55.16
Heart	71.36	75.93	73.21	83	81.5	82.2	80.7	78.5	**84.4**	78.15	70.74	72.22
Iris	94.44	94	94.89	93.3	93.3	94	**96**	**96**	**96**	**96**	**96**	**96**
Monks	97.26	98.18	80.65	**100**	**100**	**100**	**100**	**100**	99.8	98.63	**100**	**100**
Pima	73.71	75.26	73.37	72.7	72.5	75.1	74.5	74	**75.7**	67.44	72.67	73.33
Spectfheart	79.17	72.36	78.16	79.8	79.8	79.4	78.3	76.5	79.8	**81.3**	81.29	80.91
Wdbc	92.33	92.26	90.68	94.7	95.1	94.9	95.1	95.2	95.3	**96.14**	94.55	94.03
Wine	89.47	92.61	91.88	93.8	93.8	96.7	95.6	93.3	94.3	**97.24**	94.38	95.52

Acknowledgements. This work was partially supported by CNPq under grant no. 313145/2017-2 and by FAPESP under grants nos. 2018/13657-1 and 2016/26040-7.

References

1. Alcala-Fdez, J., Alcala, R., Herrera, F.: A fuzzy association rule-based classification model for high-dimensional problems with genetic rule selection and lateral tuning. IEEE Trans. Fuzzy Syst. **19**(5), 857–872 (2011)
2. Alcalá-Fdez, J., Fernandez, A., Luengo, J., Derrac, J., García, S., Sánchez, L., Herrera, F.: KEEL data-mining software tool: data set repository, integration of algorithms and experimental analysis framework. Multiple-Valued Logic Soft. Comput. **17**(2–3), 255–287 (2011)
3. Banon, G.J.F., Barrera, J.: Decomposition of mappings between complete lattices by mathematical morphology, part 1. general lattices. Sig. Process. **30**(3), 299–327 (1993)
4. Birkhoff, G.: Lattice Theory, 3rd edn. American Mathematical Society, Providence (1993)
5. Boole, G.: An Investigation of the Laws of Thought, on Which are Founded the Mathematical Theories of Logic and Probabilities. Walton and Maberley, London (1854)
6. Cuninghame-Green, R.: Minimax Algebra. Lecture Notes in Economics and Mathematical Systems, vol. 166. Springer, New York (1979)
7. Dedekind, R.: Ueber die von drei Moduln erzeugte Dualgruppe. Math. Ann. **53**, 371–403 (1900)

8. Esmi, E., Sussner, P., Bustince, H., Fernández, J.: Theta-fuzzy associative memories (Theta-FAMs). IEEE Trans. Fuzzy Syst. **23**(2), 313–326 (2015)
9. Esmi, E., Sussner, P., Sandri, S.: Tunable equivalence fuzzy associative memories. Fuzzy Sets Syst. **292**, 242–260 (2016). Special Issue in Honor of Francesc Esteva on the Occasion of his 70th Birthday
10. Fan, J., Xie, W., Pei, J.: Subsethood measure: new definitions. Fuzzy Sets Syst. **106**(2), 201–209 (1999)
11. Hassoun, M.H. (ed.): Associative Neural Memories: Theory and Implementation. Oxford University Press, Oxford (1993)
12. Heijmans, H.J.A.M.: Morphological Image Operators. Academic Press, New York (1994)
13. Kaburlasos, V., Kehagias, A.: Fuzzy inference system (FIS) extensions based on lattice theory. IEEE Trans. Fuzzy Syst. **22**(3), 531–546 (2014)
14. Kaburlasos, V.G., Athanasiadis, I.N., Mitkas, P.A.: Fuzzy lattice reasoning (FLR) classifier and its application for ambient ozone estimation. Int. J. Approximate Reasoning **45**(1), 152–188 (2007)
15. Kaburlasos, V.G., Papakostas, G.: Learning distributions of image features by interactive fuzzy lattice reasoning in pattern recognition applications. Comput. Intell. Mag. IEEE **10**(3), 42–51 (2015)
16. Ritter, G.X., Sussner, P.: An introduction to morphological neural networks. In: Proceedings of the 13th International Conference on Pattern Recognition, pp. 709–717, Vienna, Austria (1996)
17. Serra, J.: Image Analysis and Mathematical Morphology. Academic Press, London (1982)
18. Sussner, P.: Generalizing operations of binary autoassociative morphological memories using fuzzy set theory. J. Math. Imaging Vis. **9**(2), 81–93 (2003). Special Issue on Morphological Neural Networks
19. Sussner, P., Esmi, E.: Morphological perceptrons with competitive learning: Lattice-theoretical framework and constructive learning algorithm. Inf. Sci. **181**(10), 1929–1950 (2011)
20. Sussner, P., Nachtegael, M., Mélange, T., Deschrijver, G., Esmi, E., Kerre, E.: Interval-valued and intuitionistic fuzzy mathematical morphologies as special cases of L-fuzzy mathematical morphology. J. Math. Imaging Vis. **43**(1), 50–71 (2012)
21. Sussner, P., Schuster, T.: Interval-valued fuzzy morphological associative memories: some theoretical aspects and applications. Inf. Sci. **438**, 127–144 (2018)
22. Sussner, P., Valle, M.E.: Classification of fuzzy mathematical morphologies based on concepts of inclusion measure and duality. J. Math. Imaging Vis. **32**(2), 139–159 (2008)
23. Valle, M.E., Sussner, P.: Storage and recall capabilities of fuzzy morphological associative memories with adjunction-based learning. Neural Netw. **24**(1), 75–90 (2011)
24. Young, V.R.: Fuzzy subsethood. Fuzzy Sets Syst. **77**(3), 371–384 (1996)

Fuzziness and Humor: Aspects of Interaction and Computation

Julia Taylor Rayz$^{(\boxtimes)}$ and Victor Raskin

Purdue University, West Lafayette, IN 47907, USA
{jtaylor1, vraskin}@purdue.edu

Abstract. The paper addresses the fuzzy status of verbal humor for the first time in studies of fuzziness as well as in humor research. After a brief introduction to a dominant class of linguistic theories of humor, it focuses on the ontological semantic theory of humor, where the relationship is made obvious and reaches the computational level of formality. A couple of ordinary jokes are analyzed both within ontological semantics and from the point of view of fuzziness. The final section illustrates how native speakers manipulate fuzziness in humor by maximizing the membership functions.

Keywords: Computational humor · Linguistic theories of humor · Scripts · Fuzziness of natural language · Fuzziness of humor

1 Introduction

The scientific study of humor that we have undertaken for close to 20 and over 40 years, respectively, and kept expanding lately, is not fun. It was Bergson [1, p. 62] who said famously that hoping to enjoy the analysis of a joke as much as the joke itself is similar to expecting to enjoy the recipe of a dish as much as the dish itself. Offered decades ago almost as a jest and a test for the developing linguistic semantic theory—checking if it was powerful enough to handle humor—the effort, the Script-based Semantic Theory of Humor (SSTH—[2, 3]) turned out to be the first ever full-fledged linguistic theory of humor and, as such and due in part to a formalism-free, or almost free, presentation, gained continuing currency.

The theory is based on the then developing notion that linguistic meaning was not primarily lexical, that it was larger chunks of information that mattered, that it was scripts or frames exploited and realized in text. The first and still most legitimate subject of the theory was a short verbal joke, and the main hypothesis was that its text was fully or partially compatible (overlap) with two different scripts which were opposed in a certain clever way, usually to offer and impose one script while concealing the other and then using a trigger, a punch line to switch to the other script that is recognized largely retroactively.

Thus in (1), analyzed *ad nauseam* in [3] and hundreds of times in the literature that has followed, the doctor script is opposed to the lover script by the doctor's wife's invitation to the patient to come in while her husband is not in.

© Springer Nature Switzerland AG 2019
R. B. Kearfott et al. (Eds.): IFSA 2019/NAFIPS 2019, AISC 1000, pp. 655–666, 2019.
https://doi.org/10.1007/978-3-030-21920-8_58

(1) "Is the doctor at home?" the patient asked in his bronchial whisper. "No," the doctor's young and pretty wife replied. "Come right in!"

It is the last sentence of the wife's cue that makes the obvious script of a patient with a health problem seeking to see a doctor incomprehensible and necessitates a search for a compatible interpretation. Then and only then does the hearer realize that the wife is young and pretty and that she invited him to come in while the husband is not home and they would be alone, probably mistaking his whisper for intimacy rather than illness.

The joke, selected from a 1930s American joke collection was selected for being ordinary and hackneyed (and not terribly funny even then), and it has outdated features: the doctor sees patients where he lives (and patients actually know where it is!) and his wife helps him with admitting and probably registering and billing the patients. The text also is medically incorrect: people lose their voice because of a problem with their pharynx, not their bronchi but the text will not bear the technical term 'pharyngeal,' will it?

The simplified but easily formalizable and computable (crisp) presentations of scripts were offered ([3], p. 85), with '>' standing for 'past' and '='for present, and slightly updated in (2–3).

(2) doctor

agent	+human +adult	
event	> study medicine	
	= receive patients:	patient comes or doctor visits doctor listens to complaints doctor examines patient
	= cure disease:	doctor diagnoses doctor prescribes medication
	= take payment	
time	> many years	
	= every day	
	= immediately	
condition:	physical contact	

(3) lover

agent   + human +adult gender: x
co-agent + human + adult gender: not x
activity have sex
place   secluded
time   > once
     = regularly
Condition: if married spouse(s) should not
      know

Such scripts were later developed within Ontological Semantics into collections of statements of events corresponding to such events as approaching bankruptcy [4]. They are still developed in our current Ontological Semantic Technology [5–7]. They are actively pursued in the current wave of NLP [8]. The scripts in (2–3) are definitely more complex than Minsky's [9] frames because they clearly embed complete statements/sentences with their full complement of linguistic arguments.

Rayz [10] analyzed (1) in terms of Ontological Semantics, using labeled scripts SEEK-MEDICAL-HELP and VISIT, showing that computationally evoking every component necessary for joke analysis is not as straight forward as one would wish. Moreover, while [10] did not pursue the degree of activation of every component, it should definitely be explored, and is, in part, what this paper is about.

Within a few years, SSTH was developed into the General Theory of Verbal Humor [11], and this is what is practiced now by the humor researchers. However, outside of linguistics and some adjacent disciplines, it is still SSTH, a simpler and much less adequate version that is commonly cited as definitive. GTVH has kept the script opposition (SO) and language (LA) as the actual text of the joke as two of the six components, or knowledge resources, of the joke, adding logical mechanisms (LM), situation (SI), target (TA) and narrative structure (NS) as the additional and, actually, multidisciplinary elements.

A bold attempt into the philosophy of science established a hierarchy of KRs, from the deepest SO to LM to SI to TA to NS and, finally, to the surface LA. The hierarchy was soon confirmed empirically [12] in a large psychological experiment based on the premise that jokes differing in a deeper KR are perceived as less similar than those that differ in a less deep KR (LM fell out of the experiment and has been debated ever since).

Still later, GTVH has migrated into the Ontological Semantic Theory of Humor (OSTH: [13]—see also next section) but it is used primarily within the school of Ontological Semantics and is clearly under development. It is within Ontological Semantic Technology, the latest product in this school, that the importance of humor research, especially computational humor, has been clearly established. In real Natural Language Processing, understood here as transforming text into its meaning, the actual problem is with the open, non-structured text, as in news, military and economic reports and other typical kinds of texts that need computational processing. The easiest ones are structured databases. Humor is an actual natural activity by native speakers but it is semi-structured. Almost universally, you need to discover the two opposed scripts and the punch line triggering the transition from the one script to the other. In a number of well-received papers and keynote addresses as well as well-attended and -received tutorials at major computational conferences, we have managed to make that point.

So far, however, fuzziness has not found its way explicitly into humor research, with the exception of [14] that hints at the fuzziness of Targets of the GTVH. In the following sections, we attempt to correct that.

2 Ontological Semantic Theory of Humor

Ontological Semantic Theory of Humor (OSTH: [13]) is a computational version of the Script-Based Semantic Theory of Humor and adapted to the General Theory of Verbal Humor. The underlying foundation of the OSTH is humor-free Ontological Semantic Technology [5–7, 15], which using the meaning for natural language words, produces text meaning representation per sentence or paragraph that is supposed to capture its intended meaning. It disambiguates what can be disambiguated, and leaves as many interpretations as needed for sentences that require it. Since natural language is inherently fuzzy, the ontology that drives meaning calculation is fuzzy as well [16]. We have shown that unknown words can be guessed using OST with the help of membership functions that capture both syntax and semantics of a sentence in question [17]. While humor does not require guessing of unknown words, it still requires a lot of inferencing, that may be done within the paradigm.

Since OSTH follows the notions of Script Overlap and Script Opposition, it is worth formally introducing them. Most humor scholars that accept the SSTH/GTVH/OSTH family agree that script overlap can be defined as an intersection of two scripts. While most treat it as a crisp intersection, it can be naturally extended to a fuzzy set intersection:

$$\mu_{S1\ overlap\ S2}(x) = \min(\mu_{s1}(x), \mu_{s2}(x))$$

Script Opposition presents a more challenging aspect for formal processing as its meaning has not been agreed by humor scholars. Some proposals have been made to treat it as local antonyms [18], which could be seen as exclusive OR:

$$\mu_{S1\ oppose\ S2}(y) = \max\{\min[\mu_{s1}(y), 1 - \mu_{s2}(y)],\ \min[1 - \mu_{s1}(y), \mu_{s2}(y)]\}$$

This may seem counterintuitive as it is impossible, according to these definitions, for the same elements to overlap and oppose at the same time. Indeed, this argument surfaces quite frequently, and deserves a correction: while the elements of the script overlap and oppose, they do not do so on the same semantic axes. For instance, in the example of joke (1), the participants of the joke are the same: a man and a woman. In one script, this man seems to be playing the role of a patient, and in another script this man seems to be playing the role of a lover, but, it is still the same man. Similarly, the woman is the same woman, no matter what role she plays (wife, mistress, or hostess). Thus, the overlap of the scripts in the joke is its male and female participants. It is useful to limit overlap to the objects (both animate and inanimate) that are described in a joke. The opposition is a bit trickier. Due to computational complexity of analysis of every property of a script, Taylor [19] proposed to limit them to goals or paths to these goals. Thus, the semantic axis of the opposition is very different that the participants of the joke. In the case of the first script, where the patient is seeking medical assistance, the goal is to get medically treated; in the case of the second script, the goal is a romantic meeting. However, these scripts are not the only ones that can be activated by the text of the joke, albeit these are the most frequent ones. We will address various script activation in the next sections of the paper.

3   Fuzzy? Script Activation

Most (canned) jokes are processed sequentially: every sentence is read, one by one, and, unlike a normal text where each sentence in a paragraph adds something to the previous knowledge, there comes a point where text doesn't meaningfully add anything to the previously described scenario. We can refer to it as incongruity, as a generally accepted term, and we could refer to (partial) resolution as a point when we find connections between the early portion of the text and originally-thought-of unrelated information, that we can now consider as punchline of the joke. However, in order to get to the point of recognizing that, perhaps, we could consider something to be a punchline, the scripts of the joke have to be resolved.

We will follow the doctor-lover joke as an illustration of script activation, a crisp version of which was described in [10]. The first sentence, *"Is the doctor at home?" the patient asked in his bronchial whisper,* activates several ontological concepts (capitalized), with the italicized words serving as variables to the concept membership functions:

$\mu_{DOCTOR\_MD}(doctor) = \{high\};$
$\mu_{PATIENT}(patient) = \{high\};$
$\mu_{ILLNESS}(bronchial\ whisper) = \{high\};$
$\mu_{RESIDENCE}(home) = \{high\}.$

Using ontological knowledge, we access to the following relevant information, with the terms DEFAULT, SEM, RELAXABLE-TO used as indicators of membership function values:

TREAT-ILLNESS
      AGENT   DEFAULT DOCTOR_MD
               SEM MEDICAL-PROFESSIONAL-OCCUPATION
      BENEFICIARY DEFAULT PATIENT
               SEM HUMAN
               RELAXABLE-TO ANIMAL
      LOCATION DEFAULT MEDICAL-INSTITUTION
               SEM BUILDING
               RELAXABLE-TO PHYSICAL-LOCATION
      THEME   DEFAULT ILLNESS

SEEK-MEDICAL-HELP
      AGENT DEFAULT PATIENT
               SEM HUMAN
               RELAXABLE-TO ANIMAL
      PARTICIPANT   DEFAULT DOCTOR_MD
               SEM MEDICAL-PROFESSIONAL-OCCUPATION
      LOCATION DEFAULT MEDICAL-INSTITUTION
               SEM BUILDING
               RELAXABLE-TO PHYSICAL-LOCATION
      THEME   DEFAULT ILLNESS

One can look at components of a sentence as rules that contribute to interpretation of a sentence. The rules that we are interested are these:

- If an agent of the event is doctor AND beneficiary is patient AND theme is illness AND location is medical-institution, then event being treat-illness is very high.
- If an agent of the event is doctor AND beneficiary is patient AND theme is illness AND location is building, then event being treat-illness is high.
- If an agent of the event is patient AND participant is doctor AND theme is illness AND location is medical-institution, then event being seek-medical-help is very high.
- If an agent of the event is patient AND participant is doctor AND theme is illness AND location is building, then event being seek-medical-help is high.

As a general rule, whenever default concepts are activated by explicitly stated words in the sentence, and these concepts can serve as fillers on the properties that an event needs, a membership function of the event in question is high. As a note, there are some properties that affect such rules more than others, and there are some properties that can be omitted, but such an explanation deserves a full paper by itself. The part that we are concerned with is that based on membership values of the words *doctor, patient, bronchial whisper, and home*:

$$\mu_{\text{TREAT-ILLNESS}}(1^{st}\ sentence) = \{high\};$$
$$\mu_{\text{SEEK-MEDICAL-HELP}}(1^{st}\ sentence) = \{high\}.$$

It should be noted that there are many other scripts that are activated with lower corresponding membership values. This follows from activations of other concepts, by the same words, but with lower membership values as well, examples of which are shown below:

$\mu_{DOCTOR\_MD}(doctor) = \{high\}$;
$\mu_{HUMAN}(doctor) = \{medium\}$;
$\mu_{OTHER\text{-}SOCIAL\text{-}ROLE}(doctor) = \{low\}$;
$\mu_{HUSBAND}(doctor) = \{low\}$;
$\mu_{PATIENT}(patient) = \{high\}$;
$\mu_{MAN}(patient) = \{high\}$;
$\mu_{HUMAN}(patient) = \{medium\}$;
$\mu_{OTHER\text{-}SOCIAL\text{-}ROLE}(patient) = \{low\}$;
$\mu_{PATIENT}(patient\ with\ bronchial\ whisper) = \{very\ high\}$;
$\mu_{ILL\text{-}HUMAN}(patient\ with\ bronchial\ whisper) = \{high\}$;
$\mu_{HUMAN}(patient\ with\ bronchial\ whisper) = \{medium\}$;
$\mu_{OTHER\text{-}SOCIAL\text{-}ROLE}(patient\ with\ bronchial\ whisper) = \{low\}$;

In other words, it is perfectly possible for either patient or doctor to (implicitly) participate in other activities that would normally be hinted by a use of another 'social-role' such a visitor, father, husband, lover. However, because they were not emphasized, they are only activated to a low degree of membership. There is a fairly large number of such events that could be activated based on the following criteria:

- One person is asking for another person at the latter person's residence.

One of the examples is an event VISIT that requires the following components can could be activated by the first sentence:

```
        SOCIAL-VISIT
                AGENT   DEFAULT VISITOR
                        SEM HUMAN
                        RELAXABLE-TO ANIMAL
                BENEFICIARY SEM HUMAN
                        RELAXABLE-TO ANIMAL
                LOCATION DEFAULT BUILDING
                        SEM PHYSICAL-LOCATION
```

If people in question had a high membership degree of HUMAN, activation of SOCIAL-VISIT would have medium-high value. However, both *doctor* and *patient* (with or without bronchial whisper) have a medium membership value of the concept HUMAN, which decreases SOCIAL-VISIT value activation further.

We are now ready to proceed to the next sentence: *"No," the doctor's young and pretty wife whispered in reply.* Several new concepts are activated right away:

$\mu_{WIFE}(wife) = \{high\}$;
$\mu_{WOMAN}(wife) = \{high\}$;
$\mu_{HUMAN}(wife) = \{medium\}$;

$\mu_{\text{OTHER-SOCIAL-ROLE}}(wife) = \{\text{low}\};$
$\mu_{\text{WIFE}}(young \ and \ pretty \ wife) = \{\text{high}\};$
$\mu_{\text{WOMAN}}(young \ and \ pretty \ wife) = \{\text{high}\};$
$\mu_{\text{HUMAN}}(young \ and \ pretty \ wife) = \{\text{medium}\};$
$\mu_{\text{OTHER-SOCIAL-ROLE}}(young \ and \ pretty \ wife) = \{\text{low-medium}\};$

The concept of the DOCTOR_MD remains the same, but now we have a connection to make between him and his wife, which adds the following information:

$\mu_{\text{MAN}}(doctor) = \{\text{high}\};$
$\mu_{\text{HUSBAND}}(doctor) = \{\text{high}\};$

It is still possible for the SOCIAL-VISIT to be compatible with the two sentences to a low degree, and SEEK-MEDICAL-HELP and TREAT-ILLNESS to be compatible to a high degree. What remains unresolved is why the doctor's wife whispers. Again, there could be several explanations for it, suggested in [10], one of which is that there is a baby sleeping at home (this is a young wife!) and she does not want to wake up the baby. There is just enough plausibility for it that the event could activate, with a low value. It is not clear why *she whispers in reply*, which will further reduce the activation value.

It may be the right time to bring up the property of goal or a purpose to the discussion. The goals of events are rarely explicitly stated, and thus can serve as a confirmation of an event, whenever they are present. They can also serve as a flagging mechanism to lower the activation value or even to reject the event, if the goals make no sense. The goals are usually specified in the ontology. For example, the goal of SEEK-MEDICAL-HELP is to get treatment for an illness; the goal of a SOCIAL-VISIT is social interaction. As with the first sentence, there are many events that can be activated, and all of them are now waiting for a confirmation or rejection based on the goals, if they ever become clearer.

The third sentence, *"Come right in,"* provides the necessary explanation and the punchline. However, at first, it provides the incompatibility for the highly activated scripts: if the goal is to treat illness, why is the patient coming in when the doctor is not home? Thus, we are looking for other scripts that have low activation so far, based on what we know, but the same participants – the overlap remains. A SOCIAL-VISIT or a variation on the topic does not require a doctor to be present. The only thing that needs to be calculated is what can bring its value high enough to be the second script in question.

We propose to start with entries that brought down its values: other social roles of the participants, and whispering in response. A search of various combinations of when both of these are high is when the participants are taking part in an event that should not be made publicly known. If we also assume the knowledge that two opposing and overlapping scripts, if found, is a sufficient condition for a text to be humorous, we can start looking at humor stereotypes, which a young and pretty wife can provide.

What deserves a special section, but may overwhelm the reader, is that while the scripts of NOT-SO-SOCIAL-VISIT and SEEK-MEDICAL-HELP are perfectly reasonable for an analysis of this joke, they are not as useful in the treatment of the GTVH. The General Theory of Verbal Humor compares jokes based on 6 knowledge resources, as already mentioned. However, in order to compare each of these resources one must know how similar they are. In other words, is the previously analyzed joke to this one:

> An aristocratic Bostonian lady hired a new chauffer. As they started out on their first drive, she inquired:
> "What is your name?"
> "Thomas, ma'am," he answered.
> "What is your last name?" she said. "I never call chauffeurs by their first names."
> "Darling, ma'am," he replied.
> "Drive on – Thomas," she said.

This paper addresses the script activation of jokes, and thus we can look at this component of GTVH. Both jokes are dealing with social interaction: at least one of the scripts is activated with two very district social roles: doctor and patient vs. employee and employer. While they are somewhat far apart, they are closer together than animal and food. The second script is also based on inference about romantic relationship. Thus, these jokes are much closer than:

- How do you know that an elephant was in your fridge?
- By the footsteps in the butter.

This leaves a discussion of similarity of scripts on various levels of the hierarchy – a very uncrisp business.

4 Maximizing Membership Function in Humor

As we mentioned previously [20], "[w]e have multiply explored the fuzziness of explicit natural language with regard to various parts of speech and other linguistic phenomena [16, 17, 20–27], starting out from and massively expanding Zadeh's entry into, essentially, NLP in Computing With Words (CWW—[28])—while it was substantively extended in Mendel's work [29–33], linguistically, it had remained solely focused on scalar adjectives, such as Zadeh's favorite *tall,* usually a step away from quantifiers." We have, thus, established that all text in natural languages is essentially fuzzy. It is probably less so in scientific expository text and higher in political discourse, maximizing in gossip.

We do not know for sure, talking about a man and a woman meeting for the first time at the door to a doctor's office that a proposition for sex will ensue. It is not impossible and possibly has a pretty minimal membership function. What routinely happens in jokes is that the membership function for some events is maximized, and most uncertainty is eliminated. Similarly, in the real world, no Polish American is as routinely stupid as in the notorious "Polish" jokes: what Americans do know factually is that the Polish immigration was the third most successful one by the usual standards [34]. Nor are the young and attractive blonde girls all dumbheads who wake up in the back seats of somebody's cars or worry if their babies are actually theirs.

The easy proliferation of these and many other stereotypes do not contradict what the speakers and hearers in jokes know about the real world—cf. [3, Chs. 5–6]. What happens is that they cooperate in shifting the fuzzy membership functions of features and events to the almost certain maximum, removing any doubts that would interfere with the comprehension of humor. Interestingly, it is not really the "willful suspension of disbelief" in fiction, especially the mythical one, where people fly, live forever, or travel through time. It is also not like in realistic prose about fictitious characters, where the only deviation from the truth is they are fictional characters.

In jokes, most of the ordinary ones, it is clearly a natural linguistic activity in which the fuzzy membership functions are maximized to almost absolute certainty without changing the participants view of the real world. It happens both in canned prepared, full-situation jokes and in casual, situational joking-around cues. In a sense, people are fully aware of the artificial crispening shifts, and yet it remains unmentioned and largely unconscious—even though it can be stated clearly and obviously by those willing to "spoil" the joke.

5   Conclusion

The paper has reviewed the linguistic theories of humor and revealed its fuzzy status. It showed how, with progress in the class od linguistic theories of humor, it was OSTH that reflected its fuzzy nature. That was confirmed in a detailed analysis of a couple of standard jokes ontologically. In the last section, We demonstrated how native speakers, routinely and naively, crispen humor by manipulating the membership functions of properties and events to the maximum values.

References

1. Bergson, H.: Le rire: Essai sur la signification du comique, Revue de Paris, 1–15 February, 1 March 1899 issue. English translation: Laughter, om Sypher, W. (ed.), pp. 59–190. Doubleday, Comedy, Garden City, NY, 16–17 1956
2. Raskin, V.: Semantic mechanisms of humor. In: Chiarello, C., et al. (eds.) Proceedings of the Fifth Annual Meeting of the Berkeley Linguistics Society, pp. 325–335, University of California, Berkeley (1979)
3. Raskin, V.: Semantic Mechanisms of Humor. D. Reidel, Dordrecht (1985)
4. Raskin, V., Nirenburg, S., Nirenburg, I., Hempelmann, C.F., Triezenberg, K.E.: The genesis of a script for bankruptcy in ontological semantics. In: Hirst, G., Nirenburg, S. (eds.) Proceedings of the Text Meaning Workshop, HLT/NAACL 2003: Human Language Technology and North American Chapter of the Association of Computational Linguistics Conference, ACL, Edmonton, Alberta, Canada (2003)
5. Taylor, J.M., Hempelmann, C.F., Raskin, V.: On an automatic acquisition toolbox for ontologies and lexicons in ontological semantics. In: Proceedings of International Conference on Artificial Intelligence, Las Vegas, NE (2010)
6. Hempelmann, C.F., Taylor, J.M., Raskin, V.: Application-guided ontological engineering. In: Proceedings of International Conference on Artificial Intelligence, Las Vegas, Nevada (2010)
7. Raskin, V., Hempelmann, C.F., Taylor, J.M.: Guessing vs. knowing: the two approaches to semantics in natural language processing. In: Proceedings of Annual International Conference on Artificial Intelligence Dialogue 2010, Bekasovo (Moscow), Russia (2010)

8. Launchbury, J.: A DARPA Perspective on Artificial Intelligence (2019). https://www.darpa.mil/attachments/AIFull.pdf
9. Minsky, M.: A framework for representing knowledge. In: Winston, P.H. (ed.) The Psychology of Computer Vision, pp. 211–277. McGraw Hill, New York (1975)
10. Rayz, J.T.: Scripts in the ontological semantic theory of humor. In: Attardo, S. (ed.) Festschrift for Victor Raskin. Mouton de Gruyter, Berlin (in press)
11. Attardo, S., Raskin, V.: Script theory revis(it)ed: joke similarity and joke representation model. HUMOR: Int. J. Humor Res. **4**(3–4), 293–347 (1991)
12. Ruch, W., Attardo, S., Raskin, V.: Towards an empirical verification of the general theory of verbal humor. HUMOR: Int. J. Humor Res. **6**(2), 123–136 (1993)
13. Raskin, V., Hempelmann, C.F., Taylor, J.M.: How to understand and assess a theory: the evolution of the SSTH into GTVH and now into the OSTH. J. Literary Theory **3**, 285–311 (2009). (published in 2010)
14. Taylor, J.M., Raskin, V.: Towards the cognitive informatics of natural langauge: the case of computational humor. IJCINI **7**, 25–45 (2013)
15. Raskin, V., Taylor, J.M., Hempelmann, C.F.: From disambiguation failures to common-sense knowledge acquisition: a day in the life of an ontological semantic system. In: Proceedings of Web Intelligence Conference, Lyon, France (2011)
16. Taylor, J.M., Raskin, V.: Fuzzy ontology in natural language. In: Proceedings of NAFIPS 2010 (2010)
17. Taylor, J.M., Raskin, V.: Understanding the unknown: unattested input processing in natural language. In: Proceedings of FUZZ IEEE 2011 (2011)
18. Attardo, S.: The semantic foundations of cognitive theories of humor. HUMOR: Int. J. Humor Res. **10**(4), 395–420 (1997)
19. Taylor, J.M.: Towards information computer human communication: detecting humor in restricting domain. Ph D Dissertation, University of Cincinnati (2008)
20. Taylor, J.M., Raskin, V.: Conceptual defaults in fuzzy ontology. In: Proceedings of NAFIPS 2016, El Paso, TX (2016)
21. Taylor, J.M., Mazlack, L.J.: On perception of size: comparing gigantic mice and tine elephants. In: Proceedings of NAFIPS 2008 (2008)
22. Raskin, V., Taylor, J.M.: The (not so) unbearable fuzziness of natural language: the ontological semantic way of computing with words. In: Proceedings of NAFIPS 2009 (2009)
23. Taylor, J.M., Raskin, V.: Computing with nouns and verbs. In: Proceedings of FUZZ IEEE 2012 (2012)
24. Raskin, V., Taylor, J.M., Stuart, L.M.: Is natural language ever really vague: a computational semantic view. In: Proceedings of IEEE International Conference on Cybernetics (2013)
25. Taylor, J.M., Raskin, V., Stuart, L.M.: Computing with prepositions: fuzzy semantics. In: Proceedings of NAFIPS 2014 (2014)
26. Raskin, V., Taylor, J.M.: Fuzziness, uncertainty, vagueness, possibility, and probability in natural language. In: Proceedings of NAFIPS 2014 (2014)
27. Hickman, L.C., Taylor, J.M., Raskin, V.: Fuzzy lexical acquisition of adjectives. In: Proceedings of NAFIPS 2015 (2015)
28. Zadeh, L.A.: From computing with numbers to computing with words—from manipulation of measurements to manipulation of perceptions. IEEE Trans. Circ. Syst.–I: Fundam. Theory Appl. **4**, 105–119 (1999)
29. Mendel, J.M.: Computing with words, when words can mean different things to different people. In: Proceedings of Third International ICSC Symposium on Fuzzy Logic and Applications, Rochester Univ., Rochester, NY, June 1999
30. Mendel, J.M.: An architecture for making judgments using computing with words. Int. J. Appl. Math. Comput. Sci. **12**(3), 325–335 (2002)

31. Mendel, J.M.: Fuzzy sets for words: a new beginning. In: Proceedings of FUZZ-IEEE 2003, St. Louis, MO, pp. 37–42 (2003)
32. Mendel, J.M.: Computing with words and its relationships with fuzzistics. Inf. Sci. **177**, 988–1006 (2007)
33. Mendel, J.M.: Computing with words: Zadeh, Turing, Popper and Occam. IEEE Comput. Intell. Mag. **2**, 10–17 (2007)
34. Davies, C.: On humor. In: The Inaugural Lecture, 17th Annual International Summer School on Humour and Laughter, Purdue University, W. Lafayette, Indiana (2017)

Cooperative FSEIF SLAM of Omnidirectional Mobile Multirobots

Ching-Chih Tsai[(⊠)] and Ying-Che Lai

Department of Electrical Engineering, National Chung Hsing University,
Taichung 40277, Taiwan
cctsai@nchu.edu.tw

Abstract. The paper presents a cooperative fuzzy sparse extended information filtering (FSEIF) method for simultaneous localization and mapping SLAM) of multiple three-wheeled omnidirectional mobile multirobots in a given indoor environment. After brief description of our previous fuzzy SEIF SLAM (FSEIF SLAM) algorithm for a single mobile robot, a cooperative Fuzzy SEIF SLAM approach is presented for a group of omnidirectional mobile multirobots, where the optimal path searching method is devised by incorporating with K-means and Dijkstra algorithm under the assumption of known map and correspondence conditions. The effectiveness and merits of the proposed cooperative FSEIF SLAM in a large-scale environment are well illustrated by carrying out comparative simulations for multiple mobile robots.

Keywords: Cooperative · Fuzzy logics ·
Sparse extended information filtering (SEIF) · Omnidirectional mobile robot ·
Simultaneous localization and mapping (SLAM) · Swedish wheel

1 Introduction

Over past and present decades, simultaneous localization and mapping (SLAM) problems for single mobile robots have been extensively investigated by many researchers [1–4]. Those proposed SLAM techniques can be roughly divided into at least four paradigms: Kalman filtering (KF) approach and its variants based on the Gaussian noise models, particle filtering method and its variants based on the noise models with arbitrary probability- density functions, least squares method using graph-based ideas and visual SLAM (ORB SLAM) [2]. In general, those methodologies in each category have their own strengths and advantages, but have their weak points and disadvantages. For KF-based SLAM, the extended information filter (EIF) and EKF algorithms are mathematically equivalent, but the major difference between them is that the EKF is done by using the covariance matrix of the uncertain feature points, whereas the EIF is executed by utilizing the information state and matrix. As a consequence, EIF SLAM with sparsity, called sparse SEIF SLAM [2–5], has been shown to outperform EKF SLAM for a large-scale environment and large landmarks [2, 3]. The authors in [2, 3] provided simulation results to show the high accuracy of the resulting maps using SEIF SLAM in comparison to the computationally more cumbersome EKF solutions.

Fuzzy logics has been used to improve the performance of the EIF and SEIF SLAM. For example, Tsai *et al.* [6, 7] showed that the fuzzy EIF is effective and useful in the

© Springer Nature Switzerland AG 2019
R. B. Kearfott et al. (Eds.): IFSA 2019/NAFIPS 2019, AISC 1000, pp. 667–680, 2019.
https://doi.org/10.1007/978-3-030-21920-8_59

kind of multiple sensor measurement system, and Begum *et al.* [6, 7] applied fuzzy logics and genetic algorithm to improve the performance of a SLAM [8]. Lai and Tsai in [9, 10] combine fuzzy logics and SEIF SLAM to devise a fuzzy SEIF SLAM approach with its application to an omnidirectional mobile robot with three Swedish wheels. The results in [0] showed that the proposed fuzzy SEIF SLAM outperformed SEIF SLAM in terms of accuracy of robot's pose estimation and estimated features.

Inspired by the multi-robot SLAM problem addressed in [2] and the FSIEF SLAM method presented in [10], this paper is aimed to propose a cooperative FSIEF SLAM method for a team of multiple omnidirectional mobile robots with three Swedish wheels in a given indoor environment. The proposed cooperative fuzzy SEIF SLAM will be shown effective in finding accurate pose estimation and estimated features such that the proposed method may provide useful references for related researchers and engineers.

The rest of the paper is organized as follows. Section 2 revisits the FSEIF SLAM method. Section 3 introduces the kinematic model and odometry errors of the omnidirectional mobile robot. In Sect. 4, four simulations are conducted to show the effectiveness and superiority of the proposed method in comparison with SEIF SLAM and EKF SLAM. Section 5 concludes the paper.

2   FSEIF SLAM

Since this paper is concerned with a fuzzy SEIF SLAM method for cooperative SLAM of multiple omnidirectional mobile robots, this section will recall the basic ideas behind the fuzzy SEIF SLAM method for single mobile robots. Below is the brief description of the fuzzy SEIF SLAM method in [10].

2.1   Introduction to FSEIF SLAM

In the FSIEF SLAM in [10], a simple fuzzy logic method is used to adjust the measured distance ranges and angles of the on-board ranging sensors. With these fuzzified measurement data, the conventional SEIF SLAM algorithm has been employed to address the simultaneous localization and mapping problem of the omnidirectional mobile robot with three Swedish wheels.

2.2   FSEIF-SLAM

This subsection is devoted to summarizing the proposed FSEIF SLAM algorithm. In Fig. 1, the proposed FSEIF SLAM focuses on the matching rate α of the observed features and the RMSE β of the azimuth by adding one fuzzification coverage step to dynamically adjust the measure range. Below are the key five steps of the proposed FSEIF SLAM algorithm.

1. Motion update: this step includes the control u_t in the estimation process. It only updates the components of the information vector $\bar{\xi}_t$/matrix $\bar{\Omega}_t$ that are modified in motion are those of the robot pose and the selected active features.
2. Fuzzification coverage: In Fig. 2, the step includes the previous matching rate α and the previous the RMSE β of robot azimuth to correct the sensor coverage. Therefore, measure vector $\tilde{z}_t$ will be modified by the fuzzification measurement range.

3. Measurement update: In Fig. 3, the step includes the modified measurement vector $\tilde{z}_t$ under known correspondence c_t. It only updates the information values of the robot pose and the observed features in the map.

4. Update state estimate: This step applies an amortized coordinate descent technique to recover the state estimate μ_t, updating a small number of other state vectors.

5. Sparsification: This step includes the projection matrix F_{m0}, F_{x,m_0}, F_x to modify the information matrix and information vector to remove the link of the robot pose and the past active feature.

Algorithm FSEIF_SLAM_known_correspondences $\left(\xi_{t-1}, \Omega_{t-1}, \mu_{t-1}, u_t, z_t \right)$

1. $\overline{\xi}_t, \overline{\Omega}_t, \overline{\mu}_t = $ SEIF_motion_update $\left(\xi_{t-1}, \Omega_{t-1}, \mu_{t-1}, u_t \right)$

2. $\tilde{z}_t = $ SEIF_fuzzification_coverage $\left(z_t, \alpha_{t-1}, \beta_{t-1} \right)$

3. $\xi_t, \Omega_t, \alpha_t, \beta_t = $ SEIF_measurment_update $\left(\overline{\xi}_t, \overline{\Omega}_t, \overline{\mu}_t, \tilde{z}_t \right)$

4. $\mu_t = $ SEIF_update_state_estimate $\left(\xi_t, \Omega_t, \overline{\mu}_t \right)$

5. $\tilde{\xi}_t, \tilde{\mu}_t = $ SEIF_sparsification $\left(\xi_t, \Omega_t, \mu_t \right)$

6. return $\tilde{\xi}_t, \tilde{\Omega}_t, \mu_t, \alpha_t, \beta_t$

Fig. 1. FSEIF SLAM algorithm process.

Algorithm FSEIF_fuzzification_coverage $\left(z_t, \alpha_{t-1}, \beta_{t-1} \right)$

1. rangefactor = fuzzifer(α_{t-1})

2. if $\alpha_{t-1} < $ RANGE$_{THRESHOLD}$

3.   sensor.range = sensor.ranger + rangefactor

4.   else

5.   sensor.range = sensor.ranger - rangefactor

6. endif

7. anglefactor = constant.

8. if $\beta_{t-1} < $ ANGLE$_{THRESHOLD}$

9.   sensor.angle = sensor.angle + anglefactor

10.   else

11.   sensor.angle = sensor.angle - anglefactor

12. endif

13. Return $\tilde{z}_t$

Fig. 2. Fuzzificaion coverage algorithm of the FSEIF SLAM.

3   Cooperative FSEISF SLAM Multirobots

This section will propose a cooperative FSEIF SLAM method for the multi-robot system. For cooperative SLAM, a sub-map merging method is presented to merge many sub-maps for numerous omnidirectional mobile robots whose kinematics model is briefly introduced in Sect. 3.1. Detailed descriptions of the sub-map merging method are detailed in the remaining sections.

Algorithm FSEIF_measument_update$\left(\bar{\xi}_t,\bar{\Omega}_t,\bar{\mu}_t,\tilde{z}_t\right)$

1. $Q_t=\begin{pmatrix} \sigma_r & 0 \\ 0 & \sigma_\phi \end{pmatrix}$

2. for all observed features $\tilde{z}_t^i=\left(r_t^i,\phi_t^i\right)$

3. $j=c_t^i$

4.   if landmark j never seen before

5. $\begin{pmatrix} \bar{\mu}_{j,x} \\ \bar{\mu}_{j,y} \end{pmatrix}=\begin{pmatrix} \bar{\mu}_{t,x} \\ \bar{\mu}_{t,y} \end{pmatrix}+\begin{pmatrix} r_t^i\cos(\phi_t^i+\bar{\mu}_{t,\theta}) \\ r_t^i\sin(\phi_t^i+\bar{\mu}_{t,\theta}) \end{pmatrix}$

6.   endif

7. $\delta=\begin{pmatrix} \delta_x \\ \delta_y \end{pmatrix}=\begin{pmatrix} \bar{\mu}_{j,x}-\bar{\mu}_{t,x} \\ \bar{\mu}_{j,y}-\bar{\mu}_{t,y} \end{pmatrix}$

8. $q=\delta^T\delta$

9. $\hat{z}_t^i=\begin{pmatrix} \sqrt{q} \\ \text{atan2}\left(\delta_y,\delta_x\right)-\bar{\mu}_{t,\theta} \end{pmatrix}$

10. $H_t^i=\begin{pmatrix} -\sqrt{q}\delta_x & -\sqrt{q}\delta_y & 0 & 0 & ... & 0 & +\sqrt{q}\delta_x & \sqrt{q}\delta_y & 0 & ... & 0 \\ \delta_x & -\delta_x & -q & 0 & ... & 0 & -\delta_y & +\delta_x & 0 & ... & 0 \end{pmatrix}$

11. $\beta_t^i=\text{RMSE}(\hat{z}_{\theta,t}^i,\tilde{z}_{\theta,t}^i)$

12. endfor

13. $\xi_t=\bar{\xi}_t+\sum_i H_t^i Q_t^{-1}\left[\tilde{z}_t^i-\hat{z}_t^i+H_t^i\mu_t\right]$

14. $\Omega_t=\bar{\Omega}_t+\sum_i H_t^{iT} Q_t^{-1}H_t^i$

15. $\alpha_t=(\tilde{z}_t-\hat{z}_t)/\tilde{z}_t$

16. return $\xi_t,\Omega_t,\alpha_t,\beta_t$

Fig. 3. Measurement update algorithm of the proposed SEIF SLAM.

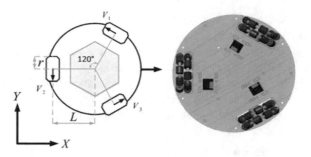

Fig. 4. Structure and geometry of the omnidirectional mobile robot in polar coordinates.

3.1 Omnidirectional Mobile Robot Model

Before recalling the kinematic model of the robot as shown in Fig. 4, we define the pose of the robot as $[x(t) \quad y(t) \quad \theta(t)]^T$ by assuming that no slips occur. Hence, the inverse kinematics of the robot in the Cartesian coordinates is given by [9]. Conversely, the forward kinematic model of the robot is expressed

$$
\begin{bmatrix} V_1(t) \\ V_2(t) \\ V_3(t) \end{bmatrix} = \begin{bmatrix} R\omega_1(t) \\ R\omega_2(t) \\ R\omega_3(t) \end{bmatrix} = P(\theta(t)) \begin{bmatrix} dx(t)/dt \\ dy(t)/dt \\ dz(t)/dt \end{bmatrix} \tag{1}
$$

$$
\begin{bmatrix} dx(t)/dt \\ dy(t)/dt \\ dz(t)/dt \end{bmatrix} = P^{-1}(\theta(t)) \begin{bmatrix} V_1(t) \\ V_2(t) \\ V_3(t) \end{bmatrix} = P^{-1}(\theta(t)) \begin{bmatrix} R\omega_1(t) \\ R\omega_2(t) \\ R\omega_3(t) \end{bmatrix} \tag{2}
$$

where

$$
P(\theta(t)) = \begin{bmatrix} -\sin\theta(t) & \cos\theta(t) & L \\ -\sin(\frac{\pi}{3} - \theta(t)) & -\cos(\frac{\pi}{3} - \theta(t)) & L \\ \sin(\frac{\pi}{3} + \theta(t)) & -\cos(\frac{\pi}{3} + \theta(t)) & L \end{bmatrix} \tag{3}
$$

and

$$
P^{-1}(\theta(t)) = \begin{bmatrix} -\frac{2}{3}\sin\theta(t) & -\sin(\frac{\pi}{3} - \theta(t)) & \sin(\frac{\pi}{3} + \theta(t)) \\ \frac{2}{3}\cos\theta(t) & -\frac{2}{3}\cos(\frac{\pi}{3} - \theta(t)) & -\frac{2}{3}\cos(\frac{\pi}{3} + \theta(t)) \\ \frac{1}{3L} & \frac{1}{3L} & \frac{1}{3L} \end{bmatrix} \tag{4}
$$

The dead-reckoning system in the vehicle simply compounds these small changes in the position and orientation to obtain a global position estimate.

3.2 Optimal Path Planning

This subsection is dedicated to describe the strategy of the initial position for the known feature marks. The proposed the optimal path plan algorithm is shown in Fig. 5, and described as in the following steps. The first step of the proposed algorithm is to cluster the feature landmarks in order to reduce the feature searching time in the loose clustering. The second step is, on basis of the clustering results, to assign one robot for each clustering. The third step is to find out the starting point of the clustering, and ends up by calculating the shortest path for the least exploration time. The fourth step sorts the visiting features order for path planning. The merge process will be proposed in next section that will be to merge the sub-map created by each robot.

3.3   K-Means Clustering

K-means clustering is a method of classifying/grouping items into k groups (where k is the number of pre-chosen groups). The grouping is done by minimizing the sum of squared distances (Euclidean distances) between items and the corresponding centroid. Given a set of observations $(x_1, x_2, \ldots, x_n)$, where each observation is a d-dimensional real vector, k-means clustering aims to partition the n observations into k ($\leq n$) sets $S = \{S_1, S_2, \ldots, S_k\}$ so as to minimize the within-cluster sum of squares (WCSS) (sum

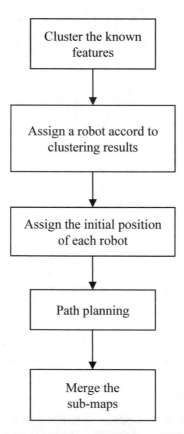

Fig. 5. The flowchart of the optimal path plan algorithm.

of distance functions of each point in the cluster to the K center). In other words, its objective is to find

$$\arg\min_{s} \sum_{i=1}^{k} \sum_{x \in S_i} \|x - \mu_i\|^2 \tag{5}$$

where μ_i is the mean of points in S_i.

3.4   Shortest Path – Dijkstra's Algorithm

The Dijkstra algorithm that is possible to determine the shortest distance between a start node and any other node in a graph. The idea of the algorithm is to continuously calculate the shortest distance beginning from a starting point, and to exclude longer distances when making an update the pseudo code is presented in Fig. 6.

3.5   Shortest Path Planning

The subsection describes the path planning strategy. The feature marks are divided into two parts in according to the K-meanings algorithm, and then the starting point of each cluster is found out as the two circles points that are presented in Fig. 7.

```
Function dijkstra(Graph, Source)
        dist[source]:=0
    for each vertex V in Graph
        if V NOT S
            dist[V] := infinite
            previous[V] := NULL
        ADD V to Q
            Q := the set of whole nodes in Graph
    while Q IS NOT EMPTY:
        U := the node in Q with nearest dist[]
        for each neighbor V of U:
            altDist := dist[U] + edge_distance(U, V)
            if altDist < dist[V]
                dist[V] := altDist
                previous[V] := U
    return dist[], previous[]
```

Fig. 6. The pseudo code of the Dijkstra algorithm.

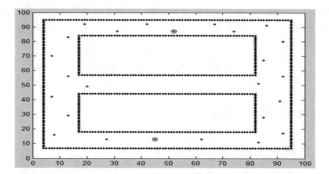

Fig. 7. The diagram of the initial and destination positions.

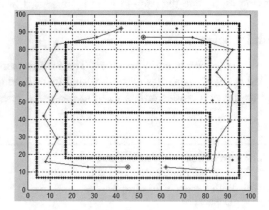

Fig. 8. Simulation results by path planning.

For optimal path planning, we traversal the whole points of clusters from the starting point, find out a point as the ending point whose distance between the starting and ending point is maximized by using Dijkstra's algorithm, and, finally, find both starting and ending points presented in Fig. 8. Therefore, the weighted graph is obtained from the Dijkstra algorithm, and the planned path is considered as the shortest path.

3.6 Sub-map Merging Method

Figure 9 shows the flowchart of the sub-map merging method for multirobots. In the beginning of the method, the initial positions of the multirobots are randomly on the planned navigation. Each robot reports the estimated features during the operating time, in order to compare the correct rate with ground truth and see if they exceed the given threshold. If the results exceed the threshold, then these repetitive estimated features are grouped. Afterwards, the centroid of the grouping features is computed to present the repetitive estimated features. Finally, the estimated features are shown on the global map.

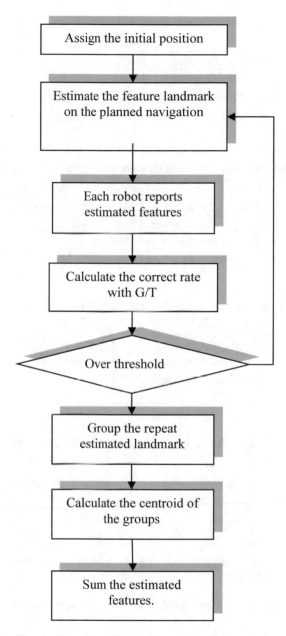

Fig. 9. Flowchart of the sub-map merging method.

Table 1. Results of the single robot with and without fuzzified range adjustment.

	R = 1, F = 0	R = 1, F = 1
RMSE_feature_map	1.7032	0.3441
RMSE_robot_position	1.9884	0.8899
RMSE_robot_Azimuth	5.4039	2.5181
Time_cost	12.79147	14.4719

Table 2. Results of the triple robots without fuzzy range adjustment.

R = 3, F = 0	R1	R2	R3
RMSE_robot_position	1.3623	0.1736	0.1306
RMSE_robot_Azimuth	4.1338	1.1671	0.5513
RMSE_feature_map	0.1898		
Time_cost	11.57		

Table 3. Results of the triple robots with fuzzy range adjustment.

R = 3, F = 1	R1	R2	R3
RMSE_robot_position	0.7317	0.2209	0.1738
RMSE_robot_Azimuth	2.1567	1.1296	0.6526
RMSE_feature_map	0.1563		
Time_cost	12.20481		

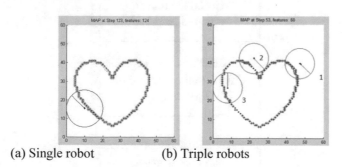

(a) Single robot   (b) Triple robots

Fig. 10. Illustration of the star simulations using (a) single robot; (b) triple robots.

4 Simulation Results and Discussion

This section is devoted to compare the FSEIF SLAM results made by the single robot and multirobots in the known feature merging. The following comparative simulations will examine the efficiency and accuracy of the FSEIF SLAM method by using the single robot or multirobots. The mentioned sub-map merging method is used to carry out the simulation.

4.1 Cooperation Simulation 1 - Heart Diagram

Figure 10 displays the moving path and features of the heart simulation using the single robot. In the simulation, the correspondences between the measured data and features are known, and two fuzzy algorithms are again employed to influence the construction quality of the FSEIF SLAM. The results in Table 1 reveal that the accuracy of the map and robot pose using the FSEIF SLAM is improved. The main reason behind it is caused by a few of features lying within the sensor range such that the used parameters are relatively small such that the fuzzy range adjustment is not affected. Therefore, the use of the *Fuzzifiedrange* algorithm still expand the measuring range, thus improving the accuracy of the map and robot pose, and increasing the cost time.

Figure 10(b) shows the build map of the FSEIF SLAM by using three mobile robots, and Fig. 11 displays the built map by multirobots in cooperative process. In Tables 2 and 3, no matter whether the two fuzzy algorithms are applied, the results are quite similar, that is to say, the accuracy of the map building and robot pose estimation is improved slightly. From the simulation results, it is easy to find that, in multirobots scenario, if the cooperative SLAM is not significantly improved.

4.2 Cooperative Fuzzy SEIF SLAM

To avoid the overlapping for cooperative SLAM using multirobot, the subsection integrates the proposed map grouping method, the optimal path planning method, the sub-map merging method and proposed FSEIF SLAM method. Let us compare the simulation results made by a single robot and multirobots as Fig. 11 show. The first simulation is that a single robot runs in the planed navigation as Fig. 12 shows. The red line means the moving trajectory of the single robot. Figures 13 and 14 depict the simulation results of the map grouping and the sub-map merging method using the two mobile robots. Table 4 summarizes the simulation results for the simulation. As can be seen in Table 4, the RMSE accuracy of the feature map built by multirobots is almost identical to that built by the single robot, but for the total spending time, multirobots have a relative smaller searching time than the single robot does. The main reason behind it is that the path is chosen optimally and the cooperative robots using the FSEIF SLAM fuse their estimated maps by using the sub-map procedure.

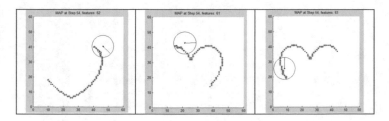

Fig. 11. Built map by multirobots in cooperation process.

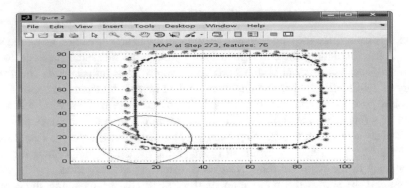

Fig. 12. Simulating the planned navigation of single robot in the known map.

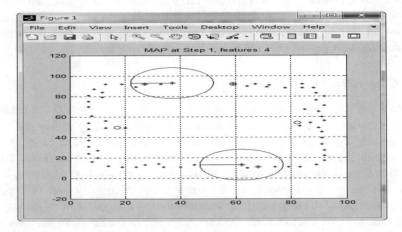

Fig. 13. Simulating the proposed map grouping method in the known map.

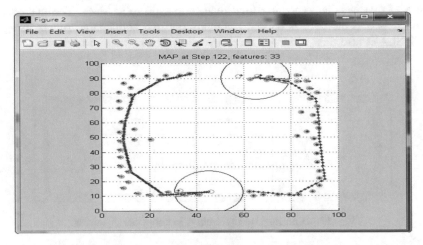

Fig. 14. Simulation results of the proposed optimal planned paths for the two robots in the known map.

Table 4. Performance comparison of the single robot using FEIF SLAM and multirobots using cooperative FEIF SLAM.

	Single	Multirobots
RMSE_feature_map	2.5633	2.1565
Time_cost	52.970	10.068

4.3   Discussion

In summary, for simple and complicated maps, that accuracy and total spending time of the estimated features and robot poses are improved by using the cooperative FSEIF SLAM via multirobots. Obviously, the use of the *Fuzzifiedrange* algorithm will result in improving the accuracy of robots' localization and robotic mapping. It is worthwhile to mention that when multirobot builds their sub-maps independently, the starting point of each robot will affect the efficiency of the mapping process.

5   Conclusions

This paper has presented a cooperative Fuzzy SEIF SLAM approach for addressing simultaneous localization and mapping problem of a group of omnidirectional mobile multirobots. The optimal path searching method via K-means and Dijkstra algorithm has been proposed under the assumption of known map and correspondence conditions. Once all the known landmarks have been detected, the sub-map merging procedure has been employed to build feature maps cooperatively. The effectiveness and merits of the proposed cooperative Fuzzy SEIF SLAM in a large-scale environment have been well illustrated by carrying out comparative simulations made by single and multiple mobile robots. A future research topic would be to conducts experiments to show the applicability of the proposed cooperative Fuzzy SEIF SLAM approach.

Acknowledgment. The authors gratefully acknowledge financial support from the Ministry of Science and Technology (MOST), Taiwan, ROC, under contract MOST 107-2221-E-005 -073-MY2.

References

1. Dissanayake, G., Newman, P., Clark, S., et al.: A solution to the simultaneous localization and map building (SLAM) problem. IEEE Transact. Robot. Autom. **17**(3), 229–241 (2001)
2. Thrun, S., Burgard, W., Fox, D.: Probabilistic Robotics. The MIT Press, Cambridge (2006)
3. Thrun, S., Liu, Y., Koller, D., Ng, A.Y., Ghahramani, Z., Durrant-Whyte, H.: Simultaneous localization and mapping with sparse extended information filters. Int. J. Robot. Res. **23**(7), 693–716 (2004)
4. Gauo, J.H., Zhao, C.X.: An improved algorithm with sparse extended information filters. Pattern Recogn. Artif. Intell. **22**(2), 269 (2009)

5. Eustice, R., Walter, M., Leonard, J.: Sparse extended information filters insights into sparsification. In Proceedings of the IEEE/RSJ Interactional Conference on Intelligent Robots and Systems, Edmonton, Canada, pp. 3281–3288 (2005)
6. Lin, H.H., Tsai, C.C.: Ultrasonic localization and pose tracking of an autonomous mobile robot via fuzzy adaptive extended information filtering. IEEE Transact. Instrum. Meas. **57** (9), 2024–2034 (2008)
7. Tai, F.C., Tsai, C.C.: Decentralized EIF-based global localization using dead-reckoning, KINECT and laser scanning for autonomous omnidirectional mobile robot. In: Proceedings of 2014 International Conference on Advanced robotics and intelligent systems, ARIS 2014, Taipei, Taiwan, pp. 85–90 (2014)
8. Begum, M., Mann, G.K.I., Gosine, R.G.: Integrated fuzzy logic and genetic algorithmic approach for simultaneous localization and mapping of mobile robots. Appl. Soft Comput. **8** (1), 15–65 (2008)
9. Lai, Y.-C., Tsai, C.C.: Fuzzy sparse EIF simultaneous localization and mapping of omnidirectional mobile robots. In: Proceedings of 2016 International Conference on Advanced Robotics and Intelligent Systems, ARIS 2016. Taipei Nangang Exhibition Center, Taipei, Taiwan (2016)
10. Tsai, C.C., Lai, Y.C.: Fuzzy SEIF SLAM for omnidirectional mobile robots. In: Proceedings of 2018 International Conference on Fuzzy Theory with Its Applications, iFuzzy 2018. EXCO, Daegu, Republic of Korea (2018)

Granular Rules for Medical Diagnosis

Shusaku Tsumoto$^{(\boxtimes)}$ and Shoji Hirano

Department of Medical Informatics, Faculty of Medicine, Shimane University,
89-1 Enya-cho, Izumo 693-8501, Japan
{tsumoto,hirano}@med.shimane-u.ac.jp
http://www.med.shimane-u.ac.jp/med_info/tsumoto/

Abstract. This paper discusses granular models of medical diagnostic rules which is an extension of rough set rule model. Medical diagnostic reasoning is characterized by three processes: focusing mechanism, differential diagnosis and detection of complications. First, focusing mechanism uses a set of symptoms which are always observed by almost all the cases of a candidate and if a case does not include any one of them, the candidate will be rejected. Second, from selected candidates, a set of symptoms which are highly observed in the cases are used for confirming the differential diagnosis. Finally, detection of complications is a set of symptoms whose occurrence of a candidate is very low but are very important for diagnosis of other diseases. These rule models can be easily described by an extension of rough set model: supporting sets of the first two sets of symptoms correspond to upper and lower approximations of a target concept. The final one is described by interrelations between a target concept and other concepts, which will be a new type of information granules.

1 Introduction

Classical medical diagnosis of a disease assumes that a disease is defined as a set of symptoms, in which the basic idea is *symptomatically*. Symptomatically had been a major diagnostic rules before laboratory and radio-logical examinations. Although the power of symptomatology for differential diagnosis is now lower, it is true that change of symptoms are very important to evaluate the status of chronic status. Even when laboratory examinations cannot detect the change of patient status, the set of symptoms may give important information to doctors.

Symptomatological diagnostic reasoning is conducted as follows. First, doctors make physical examinations to a patient and collect the observed symptoms. If symptoms are observed enough, a set of symptoms give some confidence to diagnosis of a corresponding disease. Thus, correspondence between a set of manifestations and a disease will be useful for differential diagnosis. Moreover, similarity of diseases will be inferred by sets of symptoms.

The author has been discussed modeling of symptomatological diagnostic reasoning by using the core ideas of rough sets since [12]: selection of candidates (screening) and differential diagnosis are closely related with diagnostic rules

© Springer Nature Switzerland AG 2019
R. B. Kearfott et al. (Eds.): IFSA 2019/NAFIPS 2019, AISC 1000, pp. 681–691, 2019.
https://doi.org/10.1007/978-3-030-21920-8_60

obtained by upper and lower approximations of a given concept. Thus, this paper discusses formalization of medical diagnostic rules which is closely related with rough set rule model. The important point is that medical diagnostic reasoning is characterized by focusing mechanism, composed of screening and differential diagnosis, which corresponds to upper approximation and lower approximation of a target concept. Furthermore, this paper focuses on detection of complications, which can be viewed as relations between rules of different diseases.

The paper is organized as follows. Section 2 shows characteristics of medical diagnostic process. Section 3 introduces rough sets and basic definition of probabilistic rules. Section 4 gives two style of formalization of medical diagnostic rules. The first one is a deterministic model, which corresponds to Pawlak's rough set model. And the other one gives an extension of the above ideas in probabilistic domain, which can be viewed as application of variable precision rough set model [14]. Section 5 proposes a new rule induction model, which includes formalization of rules for detection of complications. Finally, Sect. 6 concludes this chapter

2   Background: Medical Diagnostic Process

This section focuses on medical diagnostic process as rule-based reasoning. The fundamental discussion of medical diagnostic reasoning related with rough sets is given in [11].

2.1   RHINOS

RHINOS is an expert system which diagnoses clinical cases on headache or facial pain from manifestations. In this system, a diagnostic model proposed by Matsumura [1] is applied to the domain, which consists of the following three kinds of reasoning processes: exclusive reasoning, inclusive reasoning, and reasoning about complications.

First, exclusive reasoning excludes a disease from candidates when a patient does not have a symptom which is necessary to diagnose that disease. Secondly, inclusive reasoning suspects a disease in the output of the exclusive process when a patient has symptoms specific to a disease. Finally, reasoning about complications suspects complications of other diseases when some symptoms which cannot be explained by the diagnostic conclusion are obtained.

Each reasoning is rule-based and all the rules needed for diagnostic processes are acquired from medical experts in the following way.

Exclusive Rules

These rule correspond to exclusive reasoning. In other words, the premise of this rule is equivalent to the necessity condition of a diagnostic conclusion. From the discussion with medical experts, the following six basic attributes are selected which are minimally indispensable for defining the necessity condition: *1. Age, 2. Pain location, 3. Nature of the pain, 4. Severity of the pain, 5. History since onset, 6. Existence of jolt headache.* For example, the exclusive rule of common migraine is defined as:

```
In order to suspect common migraine,
the following symptoms are required:
pain location: not eyes,
nature :throbbing or persistent or radiating,
history: paroxysmal or sudden and
jolt headache: positive.
```

One of the reasons why the six attributes are selected is to solve an interface problem of expert systems: if all attributes are considered, all the symptoms should be input, including symptoms which are not needed for diagnosis. To make exclusive reasoning compact, we chose the minimal requirements only. It is notable that this kind of selection can be viewed as the ordering of given attributes, which is expected to be induced from databases. This issue is discussed later in Sect. 6

Inclusive Rules

The premises of inclusive rules are composed of a set of manifestations specific to a disease to be included. If a patient satisfies one set, this disease should be suspected with some probability. This rule is derived by asking the medical experts about the following items for each disease: *1. a set of manifestations by which we strongly suspect a disease. 2. the probability that a patient has the disease with this set of manifestations: SI (Satisfactory Index) 3. the ratio of the patients who satisfy the set to all the patients of this disease: CI (Covering Index) 4. If the total sum of the derived CI (tCI) is equal to 1.0 then end. Otherwise, goto 5. 5. For the patients with this disease who do not satisfy all the collected set of manifestations, goto 1.* Therefore a positive rule is described by a set of manifestations, its satisfactory index (SI), which corresponds to *accuracy measure*, and its covering index (CI), which corresponds to *total positive rate*. Note that SI and CI are given empirically by medical experts.

For example, one of three positive rules for common migraine is given as follows.

```
If history: paroxysmal, jolt headache: yes,
nature: throbbing or persistent,
prodrome: no, intermittent symptom: no,
persistent time: more than 6 hours,
and location: not eye,
then common migraine is suspected with
accuracy 0.9 (SI=0.9) and this rule covers
60 percent of the total cases (CI=0.6).
```

Disease Image: Complications Detection

This rule is used to detect complications of multiple diseases, acquired by all the possible manifestations of the disease. By the use of this rule, the manifestations which cannot be explained by the conclusions will be checked, which suggest complications of other diseases. For example, the disease image of common migraine is:

```
The following symptoms can be explained by
common migraine: pain location: any or
depressing: not or jolt headache: yes or ...
```

Therefore, when a patient who suffers from common migraine is depressing, it is suspected that he or she may also have other disease.

2.2   Focusing Mechanism

The most important process in medical differential diagnosis shown above is called a focusing mechanism [7,13]. Even in differential diagnosis of headache, medical experts should check possibilities of more than 100 candidates, though frequent diseases are 5 or 6. These candidates will be checked by past and present history, physical examinations, and laboratory examinations. In diagnostic procedures, a candidate is excluded one by one if symptoms necessary for diagnosis are not observed.

Focusing mechanism consists of the following two styles: exclusive reasoning and inclusive reasoning. Relations of this diagnostic model with another diagnostic model are discussed in [5,11], which is summarized in Fig. 1: First, exclusive reasoning excludes a disease from candidates when a patient does not have symptoms that is necessary to diagnose that disease. Second, inclusive reasoning suspects a disease in the output of the exclusive process when a patient has symptoms specific to a disease. Based on the discussion with medical experts, these reasoning processes are modeled as two kinds of rules, negative rules (or exclusive rules) and positive rules; the former corresponds to exclusive reasoning, the latter to inclusive reasoning [1]. [1]

3   Basics of Rule Definitions

3.1   Rough Sets

In the following sections, we use the following notation introduced by Grzymala-Busse and Skowron [4], based on rough set theory [2]. Let U denote a nonempty finite set called the universe and A denote a nonempty, finite set of attributes,

---

[1] Implementation of detection of complications is not discussed here because it is derived after main two process, exclusive and inclusive reasoning. The way to deal with detection of complications is discussed in Sect. 5.

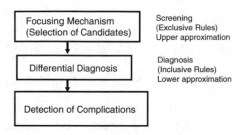

Fig. 1. Focusing mechanism.

i.e., $a : U \rightarrow V_a$ for $a \in A$, where V_a is called the domain of a, respectively. Then a decision table is defined as an information system, $A = (U, A \cup \{d\})$. The atomic formulas over $B \subseteq A \cup \{d\}$ and V are expressions of the form $[a = v]$, called descriptors over B, where $a \in B$ and $v \in V_a$. The set $F(B,V)$ of formulas over B is the least set containing all atomic formulas over B and closed with respect to disjunction, conjunction, and negation.

For each $f \in F(B,V)$, f_A denotes the meaning of f in A, i.e., the set of all objects in U with property f, defined inductively as follows:

1. If f is of the form $[a = v]$, then $f_A = \{s \in U | a(s) = v\}$.
2. $(f \wedge g)_A = f_A \cap g_A$; $(f \vee g)_A = f_A \vee g_A$; $(\neg f)_A = U - f_a$.

3.2   Classification Accuracy and Coverage

3.2.1   Definition of Accuracy and Coverage
By use of the preceding framework, classification accuracy and coverage, or true positive rate are defined as follows.

Definition 1. Let R and D denote a formula in $F(B,V)$ and a set of objects that belong to a decision d. Classification accuracy and coverage(true positive rate) for $R \rightarrow d$ is defined as:

$$\alpha_R(D) = \frac{|R_A \cap D|}{|R_A|} (= P(D|R)), \tag{1}$$

$$\kappa_R(D) = \frac{|R_A \cap D|}{|D|} (= P(R|D)), \tag{2}$$

where $|S|$, $\alpha_R(D)$, $\kappa_R(D)$, and $P(S)$ denote the cardinality of a set S, a classification accuracy of R as to classification of D, and coverage (a true positive rate of R to D), and probability of S, respectively.

It is notable that $\alpha_R(D)$ measures the degree of the sufficiency of a proposition, $R \rightarrow D$, and that $\kappa_R(D)$ measures the degree of its necessity. For example, if $\alpha_R(D)$ is equal to 1.0, then $R \rightarrow D$ is true. On the other hand, if $\kappa_R(D)$ is equal to 1.0, then $D \rightarrow R$ is true. Thus, if both measures are 1.0, then $R \leftrightarrow D$.

3.3   Probabilistic Rules

By use of accuracy and coverage, a probabilistic rule is defined as:

$$R \xrightarrow{\alpha,\kappa} d \quad s.t. \ R = \wedge_j [a_j = v_k], \alpha_R(D) \ \delta_\alpha \text{ and } \kappa_R(D) \ \delta_\kappa, \tag{3}$$

where D denotes a set of samples that belong to a class d. If the thresholds for accuracy and coverage are set to high values, the meaning of the conditional part of probabilistic rules corresponds to the highly overlapped region. This rule is a kind of probabilistic proposition with two statistical measures, which is an extension of Ziarko's variable precision model (VPRS) [14].[2]

It is also notable that both a positive rule and a negative rule are defined as special cases of this rule, as shown in the next sections.

4   Formalization of Medical Diagnostic Rules

4.1   Deterministic Model

4.1.1   Positive Rules

A positive rule is defined as a rule supported by only positive examples. Thus, the accuracy of its conditional part to a disease is equal to 1.0. Each disease may have many positive rules. If we focus on the supporting set of a rule, it corresponds to a subset of the lower approximation of a target concept, which is introduced in rough sets [2]. Thus, a positive rule is defined as:

$$R \to d \quad s.t. \quad R = \wedge_j [a_j = v_k], \quad \alpha_R(D) = 1.0 \tag{4}$$

where D denotes a set of samples that belong to a class d.

This positive rule is often called a deterministic rule. However, we use the term, positive (deterministic) rules, because a deterministic rule supported only by negative examples, called a negative rule, is introduced below.

4.1.2   Negative Rules

The important point is that a negative rule can be represented as the contra-positive of an exclusive rule [13]. An exclusive rule is defined as a rule whose supporting set covers all the positive examples. That is, the coverage of the rule to a disease is equal to 1.0. That is, an exclusive rule represents the necessity condition of a decision. The supporting set of an exclusive rule corresponds to the upper approximation of a target concept, which is introduced in rough sets [2]. Thus, an exclusive rule is defined as:

$$R \to d \quad s.t. \quad R = \vee_j [a_j = v_k], \quad \kappa_R(D) = 1.0, \tag{5}$$

where D denotes a set of samples that belong to a class d.

---

[2] This probabilistic rule is also a kind of *rough modus ponens* [3].

Next, let us consider the corresponding negative rules in the following way. An exclusive rule should be described as:

$$d \to \vee_j [a_j = v_k],$$

because the condition of an exclusive rule corresponds to the necessity condition of conclusion d. Since a negative rule is equivalent to the contrapositive of an exclusive rule, it is obtained as:

$$\wedge_j \neg [a_j = v_k] \to \neg d,$$

which means that if a case does not satisfy any attribute value pairs in the condition of a negative rule, then we can exclude a decision d from candidates.

Thus, a negative rule is represented as:

$$\wedge_j \neg [a_j = v_k] \to \neg d \quad s.t. \quad \forall [a_j = v_k] \kappa_{[a_j = v_k]}(D) = 1.0, \tag{6}$$

where D denotes a set of samples that belong to a class d.

Negative rules should also be included in a category of deterministic rules, because their coverage, a measure of negative concepts, is equal to 1.0. It is also notable that the set supporting a negative rule corresponds to a subset of negative region, which is introduced in rough sets [2].

In summary, positive and negative rules correspond to positive and negative regions defined in rough sets. Figure 2 shows the Venn diagram of those rules.

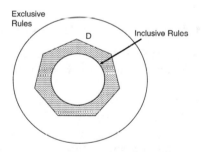

Fig. 2. Venn diagram of exclusive and positive rules.

4.2   Probabilistic Model

Although the above deterministic model exactly corresponds to original Pawlak rough set model, rules for differential diagnosis is strict for clinical setting, because clinical diagnosis may include elements of uncertainty.[3] Tsumoto [5]

---

[3] However, deterministic rule induction model is still powerful in knowledge discovery context as shown in [8].

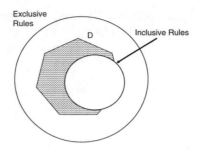

Fig. 3. Venn diagram of exclusive and inclusive rules.

relaxes the condition of positive rules and defines an inclusive rules, which models the inclusive rules of RHINOS model. The definition is almost the same as probabilistic rules defined in Sect. 3, except for the constraints for accuracy: the threshold for accuracy is sufficiently high. Thus, the definitions of rules are summarized as follows.

4.2.1   Exclusive Rules

$$R \to d \quad s.t. \ R = \vee_j [a_j = v_k], \quad (s.t. \quad \kappa[a_j = v_k](D) > \delta_\kappa) \quad \kappa_R(D) = 1.0. \quad (7)$$

4.2.2   Inclusive Rules

$$R \xrightarrow{\alpha,\kappa} d \quad s.t. \ R = \wedge_j [a_j = v_k], \quad \alpha_R(D) > \delta_\alpha \text{ and } \kappa_R(D) > \delta_\kappa. \quad (8)$$

In summary, positive and negative rules correspond to positive and negative regions defined in variable rough set model [14]. Figure 3 shows the Venn diagram of those rules.

Tsumoto introduces an algorithm for induction of exclusive and inclusive rules as PRIMEROSE-REX and conducted experimental validation and compared induced results with rules manually acquired from medical experts [5]. The results show that the rules do not include components of hierarchical diagnostic reasoning. Medical experts classify a set of diseases into groups of similar diseases and their diagnostic reasoning is multi-staged: first, different groups of diseases are checked, then final differential diagnosis is performed with the selected group of diseases. In order to extend the method into induction of hierarchical diagnostic rules, one of the authors proposes several approach to mining taxonomy from a dataset in [6,9,10].

5   Detection of Complications

The former rule induction models do not include reasoning about detection of complications, which is introduced as *disease image* as shown in Sect. 1. The core idea is that medical experts detect the symptoms which cannot be frequently occurred in the final diagnostic candidates. For example, let us assume that a patient suffering from muscle contraction headache, who usually complains of persistent pain, also complains of paroxysmal pain, say he/she feels a strong pain every one month. The situation is unusual and since paroxysmal pain is frequently observed by migraine, medical experts suspect that he/she suffers from muscle contraction headache and common migraine. Thus, a set of symptoms which are not useful for diagnosis of a disease may be important if they belong to the set of symptoms frequently manifested in other diseases. In other means, such the set of symptoms will be elements of detection of complications. Based on these observations, complications detection rules can be defined as follows:

5.1   Rules for Detection of Complications

Complications detection rule of diseases are defined as a set of rules each of which is included into inclusive rules of other diseases.[4]

$$\{R \rightarrow d \quad s.t. \ R = [a_i = v_j], \ \alpha_R(D) > \delta_\alpha, \kappa_R(D) > \delta_\kappa\} \tag{9}$$

Figure 4 depicts the relations between exclusive, inclusive and complications detection rules.

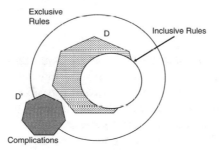

Fig. 4. Venn diagram of exclusive, inclusive and complications detection rules.

5.2   Classification of Observations

The relations between three types of rules can be visualized in a two dimensional plane, called $(\alpha, \kappa) - plane$, as shown in Fig. 5. The vertical and horizontal axis denotes the values of accuracy and coverage, respectively. Then, each rule can be plotted in the plane with its accuracy and coverage values. The region for

---

[4] The first term $R = [a_i = v_j]$ may not be needed theoretically. However, since deriving conjunction in an exhaustive way is sometimes computationally expensive, here this constraint is imposed for computational efficiency.

inclusive rules is shown in upper right, whereas the region for candidates of detection of complications is in lower left. When a rule of that region belongs to an inclusive rule of other disease, it is included into complications detection rule of the target diseases.

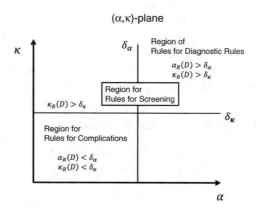

Fig. 5. Two dimensional plot: $(\alpha, \kappa) - plane$

6   Conclusion

Formalization of medical diagnostic reasoning based on symptomatology is discussed. Reasoning consists of three processes, exclusive reasoning, inclusive reasoning and complications detection, the former two of which belongs to a focusing mechanism. In exclusive reasoning, a disease is ruled out from diagnostic candidates when a patient does not have symptoms necessary for diagnosis. The process corresponds to screening. Second, in inclusive reasoning, a disease out of selected candidates is suspected when a patient has symptoms specific to a disease, which corresponds to differential diagnosis. Finally, if symptoms which are rarely observed in the final candidate, complication of other diseases will be suspected.

Previous studies are surveyed: one of the author concentrate on the focusing mechanism. First, in a deterministic version, two steps are modeled as two kinds of rules obtained from representations of upper and lower approximation of a given disease. Then, he extends it into probabilistic rule induction, which can be viewed as an application of VPRS.

Then, the authors formalize complications detection rules in this paper. The core idea is that the rules are not simply formalized by the relations between a set of symptoms and a disease, but by those between a symptoms, a target disease and other diseases. The next step will be to introduce an efficient algorithm to generate complication detection rules from data.

Acknowledgments. The author would like to thank past Professor Pawlak for all the comments on my research and his encouragement. Without his influence, one of the authors would neither have received Ph.D on computer science, nor become a professor of medical informatics. The author also would like to thank Professor Jerzy Grzymala-Busse, Andrezj Skowron, Roman Slowinski, Yiyu Yao, Guoyin Wang, Wojciech Ziarko for their insightful comments.

References

1. Matsumura, Y., Matsunaga, T., Maeda, Y., Tsumoto, S., Matsumura, H., Kimura, M.: Consultation system for diagnosis of headache and facial pain: "rhinos". In: Wada, E. (ed.) LP. Lecture Notes in Computer Science, vol. 221, pp. 287–298. Springer (1985)
2. Pawlak, Z.: Rough Scts. Kluwer Academic Publishers, Dordrecht (1991)
3. Pawlak, Z.: Rough modus ponens. In: Proceedings of International Conference on Information Processing and Management of Uncertainty in Knowledge-Based Systems 98. Paris (1998)
4. Skowron, A., Grzymala-Busse, J.: From rough set theory to evidence theory. In: Yager, R., Fedrizzi, M., Kacprzyk, J. (eds.) Advances in the Dempster-Shafer Theory of Evidence, pp. 193–236. Wiley, New York (1994)
5. Tsumoto, S.: Automated induction of medical expert system rules from clinical databases based on rough set theory. Inf. Sci. **112**, 67–84 (1998)
6. Tsumoto, S.: Extraction of experts' decision rules from clinical databases using rough set model. Intell. Data Anal. **2**(3), 215–227 (1998)
7. Tsumoto, S.: Modelling medical diagnostic rules based on rough sets. In: Polkowski, L., Skowron, A. (eds.) Rough Sets and Current Trends in Computing. Lecture Notes in Computer Science, vol. 1424, pp. 475–482. Springer, Heidelberg (1998)
8. Tsumoto, S.: Automated discovery of positive and negative knowledge in clinical databases based on rough set model (2000)
9. Tsumoto, S.: Extraction of hierarchical decision rules from clinical databases using rough sets. Inf. Sci. (2003)
10. Tsumoto, S.: Extraction of structure of medical diagnosis from clinical data. Fundam. Inform. **59**(2–3), 271–285 (2004)
11. Tsumoto, S.: Rough sets and medical differential diagnosis. In: Skowron, A., Suraj, Z. (eds.) Rough Sets and Intelligent Systems. Intelligent Systems Reference Library, vol. 42, pp. 605–621. Springer (2013)
12. Tsumoto, S., Tanaka, H.: Induction of probabilistic rules based on rough set theory. In: Jantke, K.P., Kobayashi, S., Tomita, E., Yokomori, T. (eds.) Algorithmic Learning Theory, 4th International Workshop, ALT 1993, Tokyo, Japan, 8-10 November 1993, Proceedings. *Lecture Notes in Computer Science*, vol. 744, pp. 410–423. Springer (1993). https://doi.org/10.1007/3-540-57370-4
13. Tsumoto, S., Tanaka, H.: Automated discovery of medical expert system rules from clinical databases based on rough sets. In: Proceedings of the Second International Conference on Knowledge Discovery and Data Mining 96, pp. 63–69. AAAI Press, Palo Alto (1996)
14. Ziarko, W.: Variable precision rough set model. J. Comput. Syst. Sci. **46**, 39–59 (1993)

Personalized Promotion Recommendation Through Consumer Experience Evolution Modeling

Cong Wang[1], Guoqing Chen[1], Qiang Wei[1(✉)], Guannan Liu[2], and Xunhua Guo[1]

[1] China Retail Research Center, School of Economics and Management, Tsinghua University, Beijing 100084, China
{wangc3.14,chengq,weiq,guoxh}@sem.tsinghua.edu.cn
[2] Beihang University, Beijing 100083, China
liugn@buaa.edu.cn

Abstract. Recent years have witnessed the great passion of shoppers to purchase products at promotion, resulting in "smarter" consumers with growing price sensitivity towards promotion. In order to provide such price sensitive consumers with personalized promotion recommendations, it is important to take account of the temporal uncertainty of consumer preference as well as price sensitivity simultaneously. Although consumer preference has been richly studied in recommender system, little attention has been paid to exploring the uncertainty in consumers' growing price sensitivity. In this regard, this paper seeks to bridge the gap by modeling the temporal dynamics of consumer preference and price sensitivity in a combined manner through the lens of consumer experience evolution. Given the commonly implicit nature of consumer behavior, a pairwise learning framework built on feature-based latent factor model and enhanced Bayesian personalized ranking (exFBPR) is proposed along with a corresponding learning algorithm tailored for experience evolution and multiple feedbacks is developed accordingly. Furthermore, extensive empirical experiments a real-world dataset show the superiority of the proposed framework.

Keywords: Recommender systems · Price sensitivity · Temporal uncertainty · Personalized recommendation

1 Introduction

With the proliferation of products and promotion campaigns on e-commerce platforms, consumers are becoming "smarter" through sifting various deals. According to a survey by *eMarketer* in 2017, one of the greatest challenges mentioned by retailers nowadays is the ever growing price sensitivity of consumers [8]. The problem might become more severe in the online shopping context, as information search and price comparisons are significantly easier online

© Springer Nature Switzerland AG 2019
R. B. Kearfott et al. (Eds.): IFSA 2019/NAFIPS 2019, AISC 1000, pp. 692–703, 2019.
https://doi.org/10.1007/978-3-030-21920-8_61

[1]. Thus, recommender systems could be used to match consumers with their desired products at appropriate promotion so as to achieve overall satisfaction of both retailers and consumers.

Developing such a promotion recommender system, however, is a nontrivial task. While consumer preference modeling has been widely studied in recommender system research, little attention has been paid to product price and consumers' price sensitivity. Some recent efforts have modeled product price in a discrete way [19] and address consumers' price sensitivity in a static form [20]. Yet, according to practical observations, product price is usually in numerical form and consumer price sensitivity is with temporal uncertainty rather than being static. For example, in our data collected from an online store in Taobao, 79.62% of consumers only bought products at promotion after they first encountered sales promotion in their order. Thus, it is important to model the evolution of consumers' price sensitivity when building the recommender system so as to meet the dynamically evolving needs of consumers. Concretely, the challenges for developing the promotion recommender system are summarized as follows.

Temporal Uncertainty. As mentioned above, consumers' price sensitivity is of uncertainty as evolves with time. In other words, consumers would grow more experienced as exposed to the store promotion information. For example, initially a consumer might be unaware of the marketing campaigns of the store until she occasionally encounters a price discount. After receiving the products bought at discount, she would adjust her price sensitivity according to the product quality and price paid. Gradually through exposure to store promotions, she would be more experienced in evaluating whether the price discount is appealing or not, which will affect her subsequent purchasing decisions. Likewise, the consumers' preference also evolves with time. Therefore, it is significant and necessary to model the evolvement of consumer price sensitivity as well as temporal dynamics of preference simultaneously when developing an effective recommender system.

Implicit Feedback. With the growing convenience of online shopping, consumers buy lots of things online and usually neglect to provide explicit feedbacks like ratings and reviews, giving rise to a challenge with respect to so called implicit feedback. Unlike explicit feedbacks provided by consumers themselves, implicit feedbacks are hidden reflections/expressions of the consumers that are usually not available but buried in consumer behaviors, which often needs to be captured with robust statistics. Generally speaking, purchase information (such as buying or not) is considered as a standard form of implicit feedback to infer consumer preference. However, as a common consumer only buys a small number of products, the sparsity problem pertains and needs to be solved.

Sparsity. Compared with the total number of products in the store, the number of products bought by one consumer is usually limited. In addition, taking account of the temporal evolution of consumers' preference would intensify the sparsity concern as the implicit feedback information would be scattered in a longer time frame. Due to the fact that sparsity will largely affect the performance of classic recommendation methods, a finer-grained investigation of the data is often conducted to manually develop multiple feedbacks to attenuate the degree of sparsity.

In this paper, several strategies are designed to cope with the above-mentioned challenges. First, an experience-aware latent factor model is developed to reflect the temporal dynamics of preference and price sensitivity simultaneously. Inspired by [14], the evolvement is modeled through a proxy of user experience. Next, an enhanced Bayesian Personalized Ranking (BPR) framework is developed to model the pairwise preferences and generate personalized item ranking lists for consumers. In doing so, multiple feedback information (i.e., bought and kept, bought but refunded, not bought) is incorporated to derive more preference pairs. Then, the experience-aware dynamic feature-based latent factor model is optimized by enhanced BPR framework to form an integrated method, denoted as exFBPR, so as to provide consumers with personalized promotion recommendation.

The remainder of the paper is organized as follows. Section 2 reviews the literature for extant studies. Section 3 describes preliminaries. Section 4 elaborates the proposed method. Section 5 shows the experimental results. Finally, Sect. 6 concludes the paper with future research.

2   Related Work

Our work is related to three streams of research efforts, i.e., recommender systems considering product price, temporal dynamics in recommender systems, and recommendations with implicit feedbacks.

In recommender systems research, explicitly considering product price and consumer price sensitivity has not received sufficient attention. In recent years, some emerging research effort have turned to this perspective. Ge et al. [10] incorporate cost factors into different latent factor models and evaluates the performance using real-world travel tour data. Chen et al. [4] shows that product price information and consumer price preference could be used to improve the accuracy in recommending products in un-explored categories. In [19], a price-sensitive recommender system is developed. Yet, product price is processed in a discrete way, and personalization is not fully addressed. A recent research effort [20] shares the same scenario with ours. However, consumers to match shoppers with desired promotions. Here, however, the price sensitivity is assumed to be static in [20], which does not reflect the temporal dynamics of concern.

In regard to the studies for dealing with the temporal dynamics in recommender systems, some prominent ones define such a problem as *concept drift* [18,21], which tracks a concept that is not stable and often changes with time. Two conventional strategies are commonly adopted to handle the problem, one is instance selection based on time-window partition [5,21], the other is instance weighting according to the estimated relevance [3,7,11]. In addition, some research effort focus on the finer-grained user-level modeling, e.g., taking account of both community and individual user evolution [14], incorporating textual information [15]. Notably, all these research efforts are on scenarios where users provide explicit feedbacks like ratings and reviews. However, in real-world applications, most feedbacks are tracked automatically by platforms, such as clicks, views, or purchases [16], which are destined to be implicit. Hence, our work is dedicated to addressing the issues of temporal dynamics from the implicit feedback perspective.

For the recommendations using implicit feedbacks, a dominant pairwise modeling framework is proposed by Rendle et al. [17], in which they first argued that item recommendation with implicit feedbacks should be recognized as a personalized ranking task instead of a regression task. A Bayesian personalized ranking (BPR) framework and a corresponding stochastic gradient descent (SGD)-based optimization algorithm were proposed and the framework was further developed to work with a wide range of recommendation methods like latent factor models. Some follow-up research attempts have concentrated on the sampling refinements in the BPR framework according to data characteristics [6,13]. In a similar spirit, we combine the BPR framework with an experience-aware latent factor model and further enhance it by incorporating multiple feedbacks.

3 Preliminaries

This section elaborates the preliminaries of our proposed method. Following an introduction of the basic notations in Table 1, the latent factor model and Bayesian personalized ranking framework are briefly recapitulated.

3.1 Latent Factor Model

Latent factor model is a basic yet effective model in recommender systems. It maps the representations of users and items to low dimensional latent spaces

Table 1. Basic notations

Symbol	Description
I, U	Item set, user set
$T = \{(u, i, t)\}$	Transaction log
$I_{S_u}, I_{R_u}, I_{N_u}$	Set of items bought, refunded, not bought by user u
$\hat{y}_{u,i}$	Predicted preference of user u towards item i
Φ, Ψ	Latent factor matrix of item and user, respectively
b_0, b_i, b_u	Global, item, user bias
$p_i(t), d_i(t)$	Price and discount of item i at time t
α_i, β_i	Coefficient of item i's price and discount
$\tilde{\phi}_i^{(o)}, \tilde{\psi}_u^{(o)}$	Coefficient of item i and user u
$\phi_i^{(o)}, \psi_u^{(o)}$	Coefficient of explicit item, user features
$\phi_i^{(l)}, \psi_u^{(l)}$	Remaining latent factors of item and user
$e_{u,t}$	User u's experience level at time t

through factorizing the rating matrix. The standard latent factor model predicts the "ratings" of user-item pairs (u, i) according to the bias form shown in Eq. (1).

$$\hat{y}_{u,i} = b_0 + b_i + b_u + \langle \phi_i, \psi_u \rangle \tag{1}$$

where b_0 is the global offset, b_i and b_u are item and user biases respectively, $\langle \phi_i, \psi_u \rangle$ is the inner product of ϕ_i and ψ_u that captures the interaction between user u and item i. Although the standard model could generally yield good performance, it fails to capture the temporal uncertainty embedded in the sequential user behavior data. Further, incorporating all factors in the latent variables does not help the interpretability of various user/item features. Hence we extend the latent factor model in a feature-based dynamic factorization manner, the details of which are discussed in Sect. 4.

3.2   Bayesian Personalized Ranking

BPR is a pairwise modeling framework that could learn personalized rankings from implicit feedbacks [17]. In BPR settings, in order to get the correct personalized rankings for all items, it is necessary to minimize the loss function shown in Eq. (2), where Θ represents the parameter vector, $\sigma(\cdot)$ is sigmoid function, D_S is the pairwise training instances $D_S := \{(u, i, j) | i \in I_u^+ \wedge j \in I \setminus I_u^+\}$, where I_u^+ is the set of items that users have interacted with and I is whole item set.

$$L_{BPR} = \sum_{(u,i,j) \in D_S} -\ln \sigma(\hat{y}_{u,i}(\Theta) - \hat{y}_{u,j}(\Theta)) + \lambda_\Theta \|\Theta\|^2 \tag{2}$$

The parameter learning process of BPR is usually done through SGD procedure. With the parameters learned, users' preference towards each product $\hat{y}_{u,i}$ could be calculated and thus generating ranking lists of items.

Although the BPR framework is generally well performed, it still suffers from the sparsity issue as the number of items interacted with the users are limited compared with all items. The framework is enhanced in our proposed method by leveraging multiple feedbacks to alleviate the sparsity issue to some extent.

4   Experience-Aware Feature-Based Bayesian Personalized Ranking

In this section, our proposed experience-aware feature-based Bayesian personalized ranking (exFBPR) method is presented. A parameter learning algorithm based on SGD and dynamic programming(DP) is developed for our method.

4.1   Experience-Aware Feature-Based Latent Factor Model

In order to model the temporal uncertainty of consumers' preference and price sensitivity simultaneously, we extend the standard model from an experience-aware feature-based matrix factorization perspective. First, all terms in Eq. (1)

are extended to an experience-aware manner, i.e., all parameters are expressed through the lens of user experience level $e_{u,i}$ as shown in Eq. (3).

$$\hat{y}_{u,i,t} = b_0(e_{u,t}) + b_i(e_{u,t}) + b_u(e_{u,t}) + \langle \phi_i(e_{u,t}), \psi_u(e_{u,t}) \rangle \tag{3}$$

Concretely, $e_{u,t}$ is a latent factor denoting the experience level of user u at transaction time t. Similar to [14], it is assumed that e is a categorical variable ranging from 1 to e_{max}, i.e., $e \in \{1 \ldots e_{max}\}$. As shown in Eq. (4), e is monotone non-decreasing as consumers gradually purchase products at promotion, where $\tau_k \in D$ indicates purchase at promotion.

$$\begin{cases} e_{u,\tau_k} \geq e_{u,\tau_{k-1}} & \text{if } \tau_k \in D \\ e_{u,\tau_k} = e_{u,\tau_{k-1}} & \text{otherwise} \end{cases} \tag{4}$$

To explicitly model consumers' preference and price sensitivity, the model is further extended to incorporate item and user features, which could be categorized into observed and latent types. Hence the factorization model could be further derived as shown in Eq. (5), where $\langle \psi_u^{(o)}(e_{u,t}), \tilde{\phi}_i^{(o)}(e_{u,t}) \rangle$ and $\langle \phi_i^{(o)}(e_{u,t}), \tilde{\psi}_u^{(o)}(e_{u,t}) \rangle$ are the explicit user and item effects respectively, $\langle \phi_i^{(l)}(e_{u,t}), \psi_u^{(l)}(e_{u,t}) \rangle$ is the latent interaction, $\tilde{b}_u(e_{u,t})$ and $\tilde{b}_i(e_{u,t})$ are the remaining user and item biases respectively, $\log p$ and $\log d$ are the log form of product price and discount, with $\alpha_{u,i}$ and $\beta_{u,i}$ as their coefficients.

$$\begin{aligned} \hat{y}_{u,i,t} = {} & \tilde{b}_u(e_{u,t}) + \tilde{b}_i(e_{u,t}) + b_0(e_{u,t}) + \alpha_{u,i} \log p_i(e_{u,t}) + \beta_{u,i} \log d_i(e_{u,t}) + \\ & \langle \psi_u^{(o)}(c_{u,t}), \tilde{\phi}_i^{(o)}(c_{u,t}) \rangle \langle \phi_i^{(o)}(c_{u,t}), \tilde{\psi}_u^{(o)}(c_{u,t}) \rangle \mid \langle \phi_i^{(l)}(c_{u,t}), \psi_u^{(l)}(c_{u,t}) \rangle \end{aligned} \tag{5}$$

4.2 BPR with Multiple Feedbacks

Given that more than two types of feedbacks could be observed on e-commerce platforms, (namely *purchase, not purchase* as well as *refund*), the original BPR framework needs to be further extended. Let $I_{S_u}, I_{N_u}, I_{R_u}$ be the purchase, not purchase and refund item sets of user u. Compared to the items in *not purchase* set, the items in the *refund* set reflect explicit negative opinions of the consumers. Therefore, the partial relationship of consumer preference over these three feedbacks are $I_{S_u} \succeq I_{N_u} \succeq I_{R_u}$, which could be further expanded into three pairwise comparisons $I_S \succeq I_{N,u}, I_{S,u} \succeq I_{R_u}, I_N \succeq I_{R_u}$. Hence, the preference set for training in our scenario $D_{S'}$ could be defined as Eq. (6), where p, q are placeholders for the three feedbacks N, S, R. As suggested by Eq. (6), for any $(u, i, j) \in D_{S'}$, user u prefers item i to item j.

$$D_{S'} = \{(u, i, j) | i \in I_{p,u} \wedge j \in I_{q,u}, p \in \{S, N\}, p \succ q\} \tag{6}$$

Then the probability of consumers' preference pairs could be derived as shown in Eq. (7), where $\sigma(\cdot)$ is the sigmoid function, Θ is the experience related parameter set $\Theta = \{\Theta^e | 1 \leq e \leq e_{max}\}$, $\Theta^e = \{\psi^e; \phi^e; \alpha^e; \beta^e; \mathbf{b}^e\}$.

$$\prod_{u \in U} p(i >_u j | \Theta) = \prod_{(u,i,j) \in D_{S'}} \sigma(\hat{y}_{u,i}(\Theta^{e_{u,i}}) - \hat{y}_{u,j}(\Theta^{e_{u,j}})) \tag{7}$$

Combining the enhanced multiple feedback BPR framework with the feature-based latent factor model in Sect. 4.1, our exFBPR method that models the evolution of consumer preference as well as price sensitivity simultaneously could be developed.

4.3   exFBPR Method

The exFBPR method integrates the merits of the above mentioned feature-based LFM and BPR framework with multiple feedbacks. Concretely, the loss function of exFBPR consists of three parts, i.e., pairwise preference loss, regularization and experience constraint. The first part of loss is shown in Eq. (8), with ε represents the set of all experience parameters, L_{SN}, L_{NR}, L_{SR} are the loss of different preference pairs, e.g., $L_{SN} = -\sum_{u \in U, i \in I_{S_u}, j \in I_{N_u}} \ln \sigma(\hat{y}_{u,i,j})$, and $\lambda_{SN}, \lambda_{NR}, \lambda_{SR}$ are their weights respectively.

$$L(\varepsilon, \Theta) = \lambda_{SN} L_{SN} + \lambda_{NR} L_{NR} + \lambda_{SR} L_{SR} \tag{8}$$

The second part aims to regularize the parameters from overfitting. As the parameter set is experience-related, which should not change abruptly between adjacent experience levels. Hence, a regularizer $\pi(\Theta)$ is introduced to guarantee smooth transition between adjacent experience levels as shown in Eq. (9), where $\Theta^e = \{\Psi^e; \Phi^e; \alpha^e; \beta^e; \mathbf{b}^e\}$, $\Theta = \{\Theta^e | 1 \leq e \leq E\}$.

$$\pi(\Theta) = \sum_{e=2}^{E} \left\| \Theta^e - \Theta^{e-1} \right\|_2^2 \tag{9}$$

A third constraint that needs to be addressed in the loss function is the experience evolution of the users. As stated in Eq. (4), user experience gradually evolves as he/she consumes products at promotion. Altogether with these three parts above, the loss function could be derived as shown in Eq. (10).

$$L_{exFBPR} = L(\varepsilon, \Theta) + \lambda_\Theta \pi(\Theta), \text{where} \begin{cases} e_u(\tau_k) \geq e_u(\tau_{k-1}) \text{ if } \tau_k \in D \\ e_u(\tau_k) = e_u(\tau_{k-1}) \text{ otherwise} \end{cases} \tag{10}$$

4.4   Parameter Learning

In order to obtain the optimal parameter sets Θ and ε, we need to minimize the loss function shown in Eq. (10). As it is impractical to optimize Θ and ε simultaneously, coordinate descent method is used to optimize two sets of parameters alternatively.

Concretely, ε is held as constant when optimizing Θ, likewise Θ is held constantly when optimizing ε. The optimization is conducted alternatively until convergence. Note that as the loss function is non-convex, we have to settle for a local optimum. In practice, a bootstrapping based stochastic gradient descent procedure is used to optimize Θ while treating ε as constant. When optimizing ε, to tailor for the constraint in Eq. (10), a dynamic programming method is used with Bellman equation shown in Eq. (11). Therefore, the learning algorithm for exFBPR is shown in Algorithm 1.

$$f_u(k, e_{u,k}) = \min\{f_u(k+1, e_{u,k}) + \sum L_{u,i,j}(e_{u,k}, \hat{\Theta}), f_u(k, e_{u,k}+1)\} \quad (11)$$

5   Experiments

This section focuses on the performance evaluation of the proposed exFBPR method compared against the state-of-the-art methods. All experiments were implemented using Python 2.X and conducted on a PC with 64G RAM and i9 7900X CPU.

Algorithm 1. Learning algorithm for exFBPR

Input: T as transaction set, U as user set, I_{S_u} as bought item set, I_{N_u} as non purchased item set, I_{R_u} as refunded item set
Output: Parameters Θ, experience levels ε
1: Initialize parameters of Θ and ε
2: **repeat**
3:     **for all** $u \in U$ **do**
4:         Sample tuples $(u, i_S, t), (u, i_N, t), (u, i_R, t)$ uniformly from T, where $i_S \in I_{S_u}, i_N \in I_{N_u}, i_R \in I_{R_u}$
5:         Update $\hat{\Theta}_u^{e(u,i_S,t)} = \arg\min\{\ln\sigma(\hat{y}_{u,i_S,i_N}) + \ln\sigma(\hat{y}_{u,i_N,i_R}) + \ln\sigma(\hat{y}_{u,i_S,i_R}) + \lambda\pi(\Theta_u^{e(u,i_S,t)})\}$ ▷ Computed By Stochastic Gradient Descent
6:         Update $\hat{\varepsilon}_u = \arg\min\{L_u(\varepsilon_u, \hat{\Theta}_u)\}$ ▷ Computed By Dynamic Programming
7:     **end for**
8: **until** convergence

5.1   Dataset

The *Cosmetics* dataset used for the experiments was collected from an online store in Taobao, the largest e-commerce platform in China. Products sold at the store are mainly skin care and make-up products including cleanser, toner, lotion, etc. The dataset contains consumers' transactions and specific items the customers have bought in each transaction from 2010 to 2013. The number of valid orders exceeds 11 million. Through examining the retail price and promotion information it could be seen that 57% of the products in transaction logs were bought at discount. The descriptive statistics of the Cosmetics dataset is shown in Table 2.

5.2   Baselines and Metrics

In order to evaluate the effectiveness of our proposed method adequately and to directly compare with the closely related methods, we include three types state-of-the-art methods as baselines shown as follows.

Table 2. Descriptive statistics of Cosmetics dataset

Dataset	# items	# users	# transactions	# items bought	# items bought at discount	# items refunded
Cosmetics	27721	3094656	11454420	14311269	8154097 (57%)	214672

- *Basic recommendation methods*, including collaborative filtering (itemCF and userCF), matrix factorization (MF), Bayesian personalized ranking (BPR) [17];
- *Dynamic recommendation methods proposed in related research*, including feature-based BPR (FBPR) [20], experience-aware matrix factorization (exMF) [14], TimeSVD++ [12];
- *Variants of our method*, including experience-aware BPR (exBPR), experience-aware feature based matrix factorization (exFMF).

The baselines were optimized according to their objective functions and the optimized parameters were recorded for prediction. As it is widely accepted that it is more practical to rank items for each user instead of calculating the actual values [2], ranking metrics were adopted to compare the performance of our method with the baselines. In an overall sense, the area under curve(AUC) metric, which is suited for imbalanced binary prediction tasks [9] was used. As for the personalized ranking performance, Hit Ratio (HR) and Normalized Discounted Cumulative Gain ($NDCG$) were used to assess the performance of each method. HR measures whether the ground truth item is present on the ranked list, while NDCG accounts for the position of hit.

5.3   Evaluation Results

The leave-one-out evaluation method was adopted, i.e., the latest interaction of each user was held out for prediction and the model was trained with the data remained. In the testing phase, ranked lists of items were generated with different methods, the performances of which were evaluated according to the ground truth, i.e., the items that consumers actually bought.

As the two hyperparameters, i.e., number of latent factors and experience levels governed the performance of our method, they were first tuned as shown in Fig. 1. Likewise the baseline methods were also tuned to achieve their best performances.

The AUC, HR as well as NDCG scores of each method were shown in Table 3. It could be seen that the exFBPR method achieved the best performance compared with all three types of baselines. Through finer inspection, it could be seen that (a) the experience-aware method variants outperformed the plain ones without experience; (b) the feature-based method variants were generally superior to those without features; (c) the pairwise framework was more advantageous in providing accurate recommendations, which further validated the effectiveness of our modeling strategies.

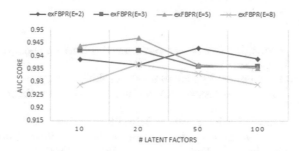

Fig. 1. Parameter tuning of exFBPR. Numbers of both latent factors and experience levels of exFBPR were tuned and the AUC score of each setting was recorded. It could be seen that exFBPR could achieve the best performance with 20 latent factors and 5 experience levels.

Table 3. The AUC, IIR, NDCG scores of different methods

	exFBPR	exBPR	exFMF	exMF	MF	FBPR	BPR	TimeSVD++	UserCF	ItemCF
AUC	**0.947**	0.945	0.906	0.900	0.886	0.916	0.911	0.867	0.782	0.786
HR@5	**0.223**	0.219	0.193	0.192	0.187	0.206	0.207	0.179	0.171	0.172
HR@10	**0.341**	0.337	0.316	0.317	0.304	0.328	0.328	0.299	0.289	0.288
HR@20	**0.447**	0.422	0.411	0.411	0.408	0.435	0.435	0.405	0.402	0.401
NDCG@5	**0.179**	0.172	0.162	0.161	0.156	0.163	0.164	0.153	0.151	0.151
NDCG@10	**0.202**	0.197	0.189	0.189	0.178	0.193	0.193	0.175	0.171	0.171
NDCG@20	**0.232**	0.231	0.227	0.227	0.201	0.229	0.229	0.189	0.187	0.187

6   Conclusion

In this study, a novel personalized promotion recommendation method, i.e., exF-BPR was proposed to model the temporal uncertainty of consumers' preference and price sensitivity simultaneously through the lens of consumer experience from transaction logs. Our proposed method has integrated the benefits of the feature-based latent factor model and the enhanced Bayesian personalized ranking framework. Furthermore, extensive empirical experiments have been conducted on a real-world dataset, with the results of revealing the effectiveness of the proposed method in providing satisfactory personalized recommendations. Finally, future research of this study can be carried out in an attempt to further explore more advanced fuzzy modeling accounting for the temporal uncertainty of consumer behavior in a wide variety of application contexts.

Acknowledgements. This study was partly supported by the National Natural Science Foundation of China (71490724/71772101), and the MOE Project of Key Research Institute of Humanities and Social Sciences at Universities of China (17JJD630006).

References

1. Bakos, Y.: The emerging role of electronic marketplaces on the internet. Commun. ACM **41**(8), 35–42 (1998)
2. Balakrishnan, S., Chopra, S.: Collaborative ranking. In: Proceedings of the Fifth ACM International Conference on Web Search and Data Mining, pp. 143–152. ACM (2012)
3. Bell, R.M., Koren, Y.: Scalable collaborative filtering with jointly derived neighborhood interpolation weights. In: Proceedings of Seventh IEEE International Conference on Data Mining (ICDM), pp. 43–52. IEEE (2007)
4. Chen, J., Jin, Q., Zhao, S., Bao, S., Zhang, L., Su, Z., Yu, Y.: Does product recommendation meet its Waterloo in unexplored categories? No, price comes to help. In: Proceedings of the 37th International ACM SIGIR Conference on Research & Development in Information Retrieval, pp. 667–676. ACM (2014)
5. Cunningham, P., Nowlan, N., Delany, S.J., Haahr, M.: A case-based approach to spam filtering that can track concept drift. In: International Conference on Case-Based Reasoning (ICCBR), vol. 3, pp. 03–2003 (2003)
6. Ding, J., Feng, F., He, X., Yu, G., Li, Y., Jin, D.: An improved sampler for Bayesian personalized ranking by leveraging view data. In: Companion of the The Web Conference 2018 on The Web Conference 2018, pp. 13–14. International World Wide Web Conferences Steering Committee (2018)
7. Ding, Y., Li, X.: Time weight collaborative filtering. In: Proceedings of the 14th ACM International Conference on Information and Knowledge Management, pp. 485–492. ACM (2005)
8. eMarketer: Retailers' brick-and-mortar worries out of whack (2017). https://retail.emarketer.com/article/retailers-brick-and-mortar-worries-of-whack-with-consumer-concerns/596fc8b6ebd40005284d5cd1. Accessed 19 July 2017
9. Fawcett, T.: An introduction to ROC analysis. Pattern Recog. Lett. **27**(8), 861–874 (2006)
10. Ge, Y., Xiong, H., Tuzhilin, A., Liu, Q.: Cost-aware collaborative filtering for travel tour recommendations. ACM Trans. Inf. Syst. (TOIS) **32**(1), 4 (2014)
11. Klinkenberg, R.: Learning drifting concepts: example selection vs. example weighting. Intell. Data Anal. **8**(3), 281–300 (2004)
12. Koren, Y.: Collaborative filtering with temporal dynamics. Commun. ACM **53**(4), 89–97 (2010)
13. Loni, B., Pagano, R., Larson, M., Hanjalic, A.: Bayesian personalized ranking with multi-channel user feedback. In: Proceedings of the 10th ACM Conference on Recommender Systems, pp. 361–364. ACM (2016)
14. McAuley, J.J., Leskovec, J.: From amateurs to connoisseurs: modeling the evolution of user expertise through online reviews. In: Proceedings of the 22nd International Conference on World Wide Web, pp. 897–908. ACM (2013)
15. Mukherjee, S., Lamba, H., Weikum, G.: Experience-aware item recommendation in evolving review communities. In: 2015 IEEE International Conference on Data Mining (ICDM), pp. 925–930. IEEE (2015)
16. Oard, D.W., Kim, J., et al.: Implicit feedback for recommender systems. In: Proceedings of the AAAI Workshop on Recommender Systems, WoUongong, vol. 83 (1998)
17. Rendle, S., Freudenthaler, C., Gantner, Z., Schmidt-Thieme, L.: BPR: Bayesian personalized ranking from implicit feedback. In: Proceedings of the Twenty-Fifth Conference on Uncertainty in Artificial Intelligence, pp. 452–461. AUAI Press (2009)

18. Tsymbal, A.: The problem of concept drift: definitions and related work. Comput. Sci. Dept. Trinity Coll. Dublin **106**(2), 58 (2004)
19. Umberto, P.: Developing a price-sensitive recommender system to improve accuracy and business performance of ecommerce applications. Int. J. Electron. Commer. Stud. **6**(1), 1 (2015)
20. Wan, M., Wang, D., Goldman, M., Taddy, M., Rao, J., Liu, J., Lymberopoulos, D., McAuley, J.: Modeling consumer preferences and price sensitivities from large-scale grocery shopping transaction logs. In: Proceedings of the 26th International Conference on World Wide Web, pp. 1103–1112. International World Wide Web Conferences Steering Committee (2017)
21. Widmer, G., Kubat, M.: Learning in the presence of concept drift and hidden contexts. Mach. Learn. **23**(1), 69–101 (1996)

Comparison Between Numerical Solutions of Fuzzy Initial-Value Problems via Interactive and Standard Arithmetics

Vinícius Francisco Wasques[1]([✉]), Estevão Esmi[1], Laécio C. Barros[1], and Barnabás Bede[2]

[1] Institute of Mathematics, Statistics and Scientific Computing,
University of Campinas, Campinas, Brazil
vwasques@outlook.com, eelaureano@gmail.com, laeciocb@ime.unicamp.br
[2] DigiPen Institute of Technology, Redmond, Washington, USA
bbede@digipen.edu

Abstract. In this work we propose a numerical solution for an n-dimensional initial-value problem where the initial conditions are given by interactive fuzzy numbers. The concept of interactivity is tied to the notion of joint possibility distribution. The numerical solutions are given by the fourth order Runge-Kutta method adapted for the arithmetic operations of interactive fuzzy numbers via sup-J extension, which is a generalization of the Zadeh's extension principle. We compare this method with the one based on the standard arithmetic. We show that the numerical solutions via interactive arithmetic are contained in the one via standard arithmetic. We provide an application to the SI epidemiological model to illustrate the results.

1 Introduction

Numerical methods such as Runge-Kutta, are used to approximate the analytical solution of ordinary differential equations (ODEs) with a given initial value. These methods can be used in fuzzy differential equations (FDEs) as well. In this case the arithmetic operations need to be adapted for fuzzy numbers, which means that an arithmetic on fuzzy numbers is necessary.

One can use the standard arithmetic which is obtained by the Zadeh's extension principle. The diameter of the solution, based on this approach, increases over time. Hence from the modeling point of view the standard arithmetic does not provide good results.

Another approach consists in extending these arithmetic operations by using the sup-J extension principle, which takes into account a special relationship between fuzzy numbers called *interactivity* [19]. The concept of interactivity arises from the notion of a joint possibility distribution [11], which resembles the definition of dependence. In particular we consider a specific joint possibility distribution namely J_0 [10]. In contrast to the standard arithmetic, the numerical methods based on J_0 produce good results with respect to diameter [16].

© Springer Nature Switzerland AG 2019
R. B. Kearfott et al. (Eds.): IFSA 2019/NAFIPS 2019, AISC 1000, pp. 704–715, 2019.
https://doi.org/10.1007/978-3-030-21920-8_62

This paper focuses on the SI epidemiological model [8]. This model describes the dynamics of a disease that affect the population whose is divided in two subpopulations, susceptible (S) and infected (I). A susceptible individual is at risk of becoming infected and subsequently, when this individual is infected, there is no cure for it. AIDS is an example of disease that behaves this way.

The initial values of susceptible and infected populations in the SI model may be uncertain, as well as the parameters involved. Classical models do not consider this fact, instead of that, the fuzzy numbers are being used to describe the uncertainties. One can observe that in several diseases, the risk of infection depends on factors such as the infectiousness of the source case, the closeness of contact and the host's immune status [1]. This type of correlation may be describe, in the context of fuzzy set theory, by interactive fuzzy numbers.

In this paper we provide a comparison of numerical solutions via Runge-Kutta method using the standard and interactive arithmetic emphasizing the use of interactivity. To this end this manuscript is organized as follows. Section 2 provides the main concepts used in this work. Section 3 presents the arithmetic necessary for this study and the novel results. Section 4 presents the proposed numerical solution with an application in the SI epidemiological model.

2   Mathematical Background

This section presents pertinent concepts and results about fuzzy set theory.

A fuzzy set A of a universe X is characterized by a function $\mu_A : X \to [0,1]$ called membership function, where $\mu_A(x)$ represents the membership degree of x in A for all $x \in X$ [18]. For notational convenience, we may simply use the symbol $A(x)$ instead of $\mu_A(x)$. The class of fuzzy subsets of X is denoted by $\mathscr{F}(X)$. Note that each classical subset of X can be uniquely identified with the fuzzy set whose membership function is given by its characteristic function.

The α-cuts of a fuzzy set $A \subseteq X$, denoted by $[A]^\alpha$, are defined as $[A]^\alpha = \{x \in X : A(x) \geq \alpha\}$, $\forall \alpha \in (0,1]$ [3]. In the case where X is also a topological space, we define the 0-cut of A by $[A]^0 = cl\{x \in X : A(x) > 0\}$, where cl Y, $Y \subseteq X$, denotes the closure of Y. An important subclass of $\mathscr{F}(\mathbb{R})$, denoted by $\mathbb{R}_\mathscr{F}$, is the class of fuzzy numbers which includes the sets of the real numbers as well as the set of the bounded closed intervals of $\mathbb{R}$. A fuzzy set A of $\mathbb{R}$ is said to be a fuzzy number if all α-cuts are bounded, closed and non-empty nested intervals for all $\alpha \in [0,1]$. The α-cuts of a fuzzy number A are denoted by $[A]^\alpha = [a_\alpha^-, a_\alpha^+]$. The subclass of $\mathbb{R}_\mathscr{F}$, denoted by $\mathbb{R}_{\mathscr{F}_\mathscr{C}}$, consists of fuzzy numbers A whose endpoints $a^-(\alpha) = a_\alpha^-$ and $a^+(\alpha) = a_\alpha^+$ are continuous functions with respect to α. Note that the class of triangular fuzzy number is contained in $\mathbb{R}_{\mathscr{F}_\mathscr{C}}$. Recall that a triangular fuzzy number A is denoted by the triple $(a;b;c)$, where $a \leq b \leq c$, and given by means of α-cuts as $[A]^\alpha = [a + \alpha(b-a), c - \alpha(c-b)]$, $\forall \alpha \in [0,1]$. The fuzzy numbers satisfy the following property $A \subseteq B \Leftrightarrow [A]^\alpha \subseteq [B]^\alpha$, for all $\alpha \in [0,1]$.

The Pompeiu-Hausdorff norm is defined by [7]

$$\|A\|_{\mathscr{F}} = \bigvee_{\alpha \in [0,1]} \max\{|a_\alpha^-|, |a_\alpha^+|\}.$$

where the symbol $\bigvee$ represents the supremum operator.

The *width* (or *diameter*) of $A \in \mathbb{R}_{\mathscr{F}}$ is defined by $width(A) = a_0^+ - a_0^-$.

The Zadeh's extension principle is a mathematical method to extend classical functions to functions that have fuzzy variables as arguments.

Definition 1. The Zadeh's extension of a given function $f : X_1 \times \cdots \times X_n \to Z$ is the fuzzy function $\widehat{f} : \mathscr{F}(X_1) \times \cdots \times \mathscr{F}(X_n) \to \mathscr{F}(Z)$ defined for each fuzzy set $A_i \in \mathscr{F}(X_i)$, for $i = 1, \ldots, n$, as the fuzzy number $\widehat{f}(A_1, \ldots, A_n)$, whose membership function is given by

$$\widehat{f}(A_1, \ldots, A_n)(z) = \bigvee_{(x_1, \ldots, x_n) \in f^{-1}(z)} A_1(x_1) \wedge \ldots \wedge A_n(x_n) \tag{1}$$

where $f^{-1}(z) = \{(x_1, \ldots, x_n) \in X_1 \times \cdots \times X_n \mid f(x_1, \ldots, x_n) = z\}$ and $\bigvee \emptyset = 0$.

An n-ary fuzzy relation R on $X = X_1 \times \ldots \times X_n$ is given by the mapping $R : X \to [0,1]$, where $R(x_1, \ldots, x_n) \in [0,1]$ is the degree of relationship among $x_1, \ldots, x_n$. A fuzzy relation $J \in \mathscr{F}(\mathbb{R}^n)$ is said to be a joint possibility distribution (JPD) among the fuzzy numbers $A_1, \ldots, A_n \in \mathbb{R}_{\mathscr{F}}$ if for all $i = 1, \ldots, n$, we have

$$A_i(y) = \bigvee_{x : x_i = y} J(x), \forall y \in \mathbb{R}, \tag{2}$$

One example of JPD is given as follows. Let t be a t-norm, that is, an associative, commutative and increasing operator $t : [0,1]^2 \to [0,1]$ that satisfies $t(x,1) = x$ for all $x \in [0,1]$. The fuzzy relation J_t given by $J_t(x_1, \ldots, x_n) = A_1(x_1) \ t \ldots t \ A_n(x_n)$ is called the t-norm-based joint possibility distribution of $A_1, \ldots, A_n \in \mathbb{R}_{\mathscr{F}}$.

For the case where the t-norm is given by the minimum operator ($t = \wedge$), that is,

$$J_\wedge(x_1, \ldots, x_n) = A_1(x_1) \wedge \ldots \wedge A_n(x_n), \tag{3}$$

the fuzzy numbers $A_1, \ldots, A_n$ are called *non-interactive*. This joint possibility distribution can be described by α-cuts as follows.

$$[J_\wedge]^\alpha = [A_1]^\alpha \times \ldots \times [A_n]^\alpha, \quad \forall \alpha \in [0,1]. \tag{4}$$

If a fuzzy relation J satisfies (2) and $J \neq J_\wedge$ then $A_1, \ldots, A_n$ are said to be J-*interactive* or simply *interactive*. Thus, the interactivity between fuzzy numbers arises from a given joint possibility distribution. The concept of interactivity resembles the relation of dependence in random variables.

Equation (2) guarantees that $J(x_1, \ldots, x_n) \leq A_i(x_i), \forall i = 1, \ldots, n$. This implies that $J(x_1, \ldots, x_n) \leq J_\wedge(x_1, \ldots, x_n), \forall (x_1, \ldots, x_n) \in \mathbb{R}^n$, which lead us to the following proposition.

Proposition 1. *Let $A_1, \ldots, A_n \in \mathbb{R}_{\mathscr{F}}$ and $J_{\wedge}$ be the joint possibility distribution given by (3). Then*

$$[J]^{\alpha} \subseteq [J_{\wedge}]^{\alpha}, \quad \forall \alpha \in [0, 1], \tag{5}$$

for every joint possibility distribution J of $A_1, \ldots, A_n$.

The definition of the sup-J extension principle is given as follows [11].

Definition 2 (Sup-J extension principle). Let $J \in \mathscr{F}(\mathbb{R}^n)$ be a joint possibility distribution of $(A_1, \ldots, A_n) \in \mathbb{R}_{\mathscr{F}}^n$ and $f : \mathbb{R}^n \to \mathbb{R}$. The sup-J extension of f at $(A_1, \ldots, A_n) \in \mathbb{R}_{\mathscr{F}}^n$, denoted by $f(J)$, is the fuzzy set defined by:

$$f(J)(y) = \bigvee_{(x_1, \ldots, x_n) \in f^{-1}(y)} J(x_1, \ldots, x_n), \tag{6}$$

where $f^{-1}(y) = \{(x_1, \ldots, x_n) \in \mathbb{R}^n : f(x_1, \ldots, x_n) = y\}$.

Note that the sup-J extension principle generalizes the Zadeh's extension principle, since Definition 2 boils down to Definition 1 for the case where $J = J_{\wedge}$, that is, when $A_1, \ldots, A_n$ are non-interactive. The sup-J extension gives rise to the arithmetic on interactive fuzzy numbers, as we see in Sect. 3. Theorem 1 provides a characterization of the extension principle by means of α-cuts [4,14].

Theorem 1. *Let $f : \mathbb{R}^n \to \mathbb{R}$ be a continuous function. For all $J \in \mathscr{F}(\mathbb{R}^n)$ and for all $\alpha \in [0, 1]$ we have that*

$$[f(J)]^{\alpha} = f([J]^{\alpha}) = \{f(x_1, \ldots, x_n) : (x_1, \ldots, x_n) \in [J]^{\alpha}\}.$$

Other type of interactivity between fuzzy numbers, which is not based on t-norm, is the one obtained by the concept of completely correlation. This concept was introduced by Fullér et al. [11] but only for two fuzzy numbers. Subsequently, the authors of [12,17] proposed an extension of this notion for n fuzzy numbers, $n > 2$. The fuzzy numbers $A_1, \ldots, A_n$ are said to be completely correlated if there exist $q = (q_2, \ldots, q_n), r = (r_2, \ldots, r_n)$ with $q_2 q_3 \ldots q_n \neq 0$ such that the corresponding joint possibility distribution $J = J_{\{q,r\}}$ is given by

$$J_{\{q,r\}}(x_1, \ldots, x_n) = A_i(x_i) \chi_U(x_1, \ldots, x_n), \quad \forall(x_1, \ldots, x_n) \in \mathbb{R}^n \tag{7}$$

for all $i = 1, \ldots, n$, where χ_U stands for the characteristic function of the set $U = \{(u, q_2 u + r_2, \ldots, q_n u + r_n) : \forall u \in \mathbb{R}\}$.

The JPD given by (7) can be used to provide solutions of fuzzy differential equations (FDEs) that consider interactivity [5,12,17]. However $J_{\{q,r\}}$ can only be applied to fuzzy numbers that have a co-linear relationship among their membership functions, which means that it can not be used to fuzzy numbers that do not have the same shape, triangular and trapezoidal fuzzy numbers for example.

Alternatively, Esmi et al. [10] employed a parametrized family of joint possibility distributions $\mathscr{J} = \{J_{\gamma} : \gamma \in [0, 1]\}$ to define interactive additions of fuzzy numbers in $\mathbb{R}_{\mathscr{F}_\mathscr{C}}$ such that the norm is increasing and continuous with respect to parameter γ. The JPDs of $\mathscr{J}$ can be applied to every pair of fuzzy numbers.

The authors of [16] considered a particular JPD of this family namely J_0 and used to produce a numerical solution of a fuzzy initial-value problem. Moreover, they provided several examples which verify that J_0 is more embracing than $J_{\{q,r\}}$.

Here we focus on distribution J_0 whose definition is given as follows. Let $A_1, A_2 \in \mathbb{R}_{\mathscr{F}_\mathscr{C}}$ and the functions $g_{1,2} : \mathbb{R} \times [0,1] \to \mathbb{R}$ given by

$$g_1(z,\alpha) = \bigwedge_{w \in [A_2]^\alpha} |w+z| \quad \text{and} \quad g_2(z,\alpha) = \bigwedge_{w \in [A_1]^\alpha} |w+z|. \tag{8}$$

Also, let us consider the classical sets

$$R_\alpha^i = \begin{cases} \{a_{i_\alpha}^-, a_{i_\alpha}^+\} & \text{if } \alpha \in [0,1) \\ [A_i]^1 & \text{if } \alpha = 1 \end{cases} \tag{9}$$

and

$$L^i(z,\alpha) = [A_{3-i}]^\alpha \cap [-g_i(z,\alpha) - z, g_i(z,\alpha) - z], \tag{10}$$

for all $z \in \mathbb{R}, \alpha \in [0,1]$ and $i = 1,2$.

Let J_0 be the fuzzy relation given by

$$J_0(x_1,x_2) = \begin{cases} A_1(x_1) \wedge A_2(x_2) & \text{, if } (x_1,x_2) \in P \\ 0 & \text{, otherwise} \end{cases}, \tag{11}$$

where $P = \bigcup_{i=1}^{2} \bigcup_{\alpha \in [0,1]} \{(x_1,x_2) : x_i \in R_\alpha^i \text{ and } x_{3-i} \in L^i(x_i,\alpha)\}, \forall \alpha \in [0,1]$.

Esmi et al. proved that the fuzzy relation J_0 is indeed a joint possibility distribution of A_1 and A_2. Moreover, they showed that the sum of two fuzzy numbers based on J_0 has the smallest norm than any other sum based on J.

Sussner et al. [15] employed shifts in order to define a new family of parametrized joint possibility distributions that can be used to control the width of the corresponding interactive addition as well.

Theorem 2. *Given $A_1, A_2 \in \mathbb{R}_{\mathscr{F}_\mathscr{C}}$ and $c = (c_1,c_2) \in \mathbb{R}^2$. Let $\tilde{A}_i \in \mathbb{R}_{\mathscr{F}_\mathscr{C}}$ be such that $\tilde{A}_i(x) = A_i(x+c_i), \forall x \in \mathbb{R}$ and $i = 1,2$. Let $\tilde{J}_0$ be the joint possibility distribution of fuzzy numbers $\tilde{A}_1, \tilde{A}_2 \in \mathbb{R}_{\mathscr{F}_\mathscr{C}}$ defined as in Eq. (11). The fuzzy relation J_0^c given by*

$$J_0^c(x_1,x_2) = \tilde{J}_0(x_1 - c_1, x_2 - c_2), \quad \forall (x_1,x_2) \in \mathbb{R}^2, \tag{12}$$

is a joint possibility distribution of A_1 and A_2.

The next section presents the standard and interactive arithmetic on fuzzy numbers.

3   Arithmetic on Fuzzy Numbers

The *standard arithmetic* on fuzzy numbers is obtained by the Zadeh's extension principle (see Definition 1) considering the arithmetic operations $f(x_1, x_2) = x_1 \otimes x_2$, where $\otimes \in \{+, -, \div, \cdot\}$. These operations are defined by

$$(A_1 \otimes_\wedge A_2)(y) = \bigvee_{x_1 \otimes x_2 = y} A_1(x_1) \wedge A_2(x_2), \quad \forall A_1, A_2 \in \mathbb{R}_{\mathscr{F}}. \tag{13}$$

The sup-J extension principle (see Definition 2) can be used to generate an arithmetic on fuzzy numbers as well. This arithmetic is called *interactive arithmetic* and is provided as follows

$$(A_1 \otimes_J A_2)(y) = \bigvee_{x_1 \otimes x_2 = y} J(x_1, x_2), \quad \forall A_1, A_2 \in \mathbb{R}_{\mathscr{F}}. \tag{14}$$

For this case the fuzzy numbers A_1 and A_2 are interactive and this arithmetic takes into account the interactivity between them.

Note that the standard arithmetic can be also given by the sup-J extension principle, considering $J = J_\wedge$. Hence the standard arithmetic is also called as *non-interactive arithmetic*. The next theorem establishes some connections between these two arithmetic.

Theorem 3. *Let $A, B, C, D \in \mathbb{R}_{\mathscr{F}}$ and $J_\wedge$ be the joint possibility distribution given by (3). Then for all joint possibility distribution J and $\otimes \in \{+, -, \div, \cdot\}$ we have*

(a) $A \otimes_J B \subseteq A \otimes_\wedge B$;
(b) $\lambda(A \otimes_J B) \subseteq \lambda(A \otimes_\wedge B)$, for all $\lambda \in \mathbb{R}$;
(c) If $A \subseteq B$ then $A \otimes_J C \subseteq B \otimes_\wedge C$;
(d) If $A \subseteq B$ and $C \subseteq D$ then $A \otimes_J C \subseteq B \otimes_\wedge D$;
(e) $(A \otimes_J (B \otimes_J C)) \subseteq (A \otimes_\wedge (B \otimes_\wedge C))$.

Proof.

(a) From Theorem 1 and Proposition 1 we obtain

$$[A \otimes_J B]^\alpha = \{a \otimes b : (a, b) \in [J]^\alpha\} \subseteq \{a \otimes b : (a, b) \in [J_\wedge]^\alpha\} = [A \otimes_\wedge B]^\alpha, \quad \forall \alpha \in [0, 1].$$

(b) It follows similarly from item (a).
(c) If $A \subseteq B$ then a combination of Theorem 1, Proposition 1 and Eq. (4) gives rise to the following

$$\begin{aligned}
[A \otimes_J C]^\alpha &= \{a \otimes c : (a, c) \in [J]^\alpha\} \subseteq \{a \otimes c : (a, c) \in [J_\wedge]^\alpha\} \\
&= \{a \otimes c : (a, c) \in [A]^\alpha \times [C]^\alpha\} \\
&\subseteq \{b \otimes c : (b, c) \in [B]^\alpha \times [C]^\alpha\} \\
&= [B \otimes_\wedge C]^\alpha, \quad \forall \alpha \in [0, 1].
\end{aligned}$$

(d) It follows similarly from item (c).
(e) A combination of items (a) and (c) proves this statement.

We use the results provided by Theorem 3 to compare the numerical solutions of fuzzy initial-value problems (FIVPs) using the interactive and the standard arithmetics. In the next section we present these solutions.

4   Fuzzy Numerical Solutions of FIVPs

This section presents a fuzzy numerical solution of FIVP with interactive fuzzy initial conditions. The solution is given by the Runge-Kutta method with the appropriate arithmetic operations for fuzzy numbers. In order to provide these solutions, let $y_i : \mathbb{R} \to \mathbb{R}^n$, with $i = 1, ..., n$, functions that depend on time t. Consider the following n-dimensional initial-value problem (IVP) composed of ODEs and initial conditions

$$\begin{cases} \frac{dy_1}{dt} = f_1(t, y_1, y_2, ..., y_n) \\ \quad \vdots \\ \frac{dy_n}{dt} = f_n(t, y_1, y_2, ..., y_n) \\ y_1(t_0) = y_1^{(0)}, \ldots, y_n(t_0) = y_n^{(0)} \in \mathbb{R}^n \end{cases} , \qquad (15)$$

where f_i is a function that depends on $y_1, y_2, ..., y_n$ and t, for each $i = 1, ..., n$.

The numerical solution of (15) based on the fourth order Runge-Kutta method is given by the following algorithm

$$y_i^{(k+1)} = y_i^{(k)} + \frac{h}{6}(k_1^{(k)} + 2k_2^{(k)} + 2k_3^{(k)} + k_4^{(k)}), \qquad (16)$$

where

$$k_1^{(k)} = f_i(t^{(k)}, y_1^{(k)}, ..., y_n^{(k)}),$$
$$k_2^{(k)} = f_i\left(t^{(k)} + \frac{h}{2}, y_1^{(k)} + \frac{h}{2}k_1^{(k)}, ..., y_n^{(k)} + \frac{h}{2}k_1^{(k)}\right),$$
$$k_3^{(k)} = f_i\left(t^{(k)} + \frac{h}{2}, y_1^{(k)} + \frac{h}{2}k_2^{(k)}, ..., y_n^{(k)} + \frac{h}{2}k_2^{(k)}\right),$$
$$k_4^{(k)} = f_i\left(t^{(k)} + h, y_1^{(k)} + hk_3^{(k)}, ..., y_n^{(k)} + hk_3^{(k)}\right), \qquad (17)$$

with $0 \leq k \leq N - 1$, where N is the number of partitions of the interval time divided in equally spaced intervals $[t^k, t^{k+1}]$ with size h and initial condition $(t^{(0)}, y_1^{(0)}, \ldots, y_n^{(0)})$.

The bidimensional SI model with vital dynamics is given by

$$\begin{cases} \frac{dS}{dt} = -\beta SI + (\eta - \mu)S, \\ \frac{dI}{dt} = \beta SI - \mu I, \\ S(0) = S^{(0)}, \quad I(0) = I^{(0)} \end{cases} , \qquad (18)$$

where η, μ, and, β are the rate of birth, death and infection of disease, respectively, and the initial conditions $S^{(0)}$ and $I^{(0)}$ are given by fuzzy numbers.

In the case where the initial conditions are considered as non-interactive, the fuzzy numerical solution has to be based on the standard arithmetic. Therefore the solution via Runge-Kutta method is given by (19) and is depicted in Fig. 1.

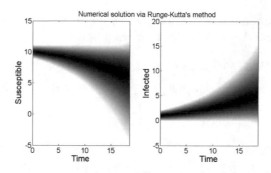

Fig. 1. The fuzzy solution given by Runge-Kutta method via standard arithmetic. The left and right figures present the susceptible and infected populations, respectively. The gray lines represent the α-cuts of the fuzzy solutions, where their endpoints for α varying from 0 to 1 are represented respectively from the gray-scale lines varying from white to black.

$$\begin{cases} S^{(k+1)} = S^{(k)} +_\wedge \frac{h}{6}(\hat{K}_{S_1}^{(k)} +_\wedge 2\hat{K}_{S_2}^{(k)} +_\wedge 2\hat{K}_{S_3}^{(k)} +_\wedge \hat{K}_{S_4}^{(k)}) \\ I^{(k+1)} = I^{(k)} +_\wedge \frac{h}{6}(\hat{K}_{I_1}^{(k)} +_\wedge 2\hat{K}_{I_2}^{(k)} +_\wedge 2\hat{K}_{I_3}^{(k)} +_\wedge \hat{K}_{I_4}^{(k)}) \end{cases}, \tag{19}$$

where

$$\hat{K}_{S_1}^{(k)} = f_S(t^{(k)}, S^{(k)}, I^{(k)}) = -\beta S^{(k)} \cdot_\wedge I^{(k)} +_\wedge (\eta - \mu)S^{(k)},$$

$$\hat{K}_{I_1}^{(k)} = f_I(t^{(k)}, S^{(k)}, I^{(k)}) = \beta S^{(k)} \cdot_\wedge I^{(k)} -_\wedge \mu S^{(k)},$$

$$\hat{K}_{S_i}^{(k)} = f_S\left(t^{(k)} + \frac{h}{2}, S^{(k)} +_\wedge \frac{h}{2}\hat{K}_{S_{i-1}}^{(k)}, I^{(k)} +_\wedge \frac{h}{2}\hat{K}_{S_{i-1}}^{(k)}\right), \text{for} \quad i = 2, 3$$

$$\hat{K}_{I_i}^{(k)} = f_I\left(t^{(k)} + \frac{h}{2}, S^{(k)} +_\wedge \frac{h}{2}\hat{K}_{I_{i-1}}^{(k)}, I^{(k)} +_\wedge \frac{h}{2}\hat{K}_{I_{i-1}}^{(k)}\right), \text{for} \quad i = 2, 3$$

$$\hat{K}_{S_4}^{(k)} = f_S\left(t^{(k)} + h, S^{(k)} +_\wedge h\hat{K}_{S_3}^{(k)}, I^{(k)} +_\wedge h\hat{K}_{S_3}^{(k)}\right),$$

$$\hat{K}_{I_4}^{(k)} = f_I\left(t^{(k)} + h, S^{(k)} +_\wedge h\hat{K}_{I_3}^{(k)}, I^{(k)} +_\wedge h\hat{K}_{I_3}^{(k)}\right).$$

The parameters used in all simulations are $h = 0.125$, $\eta = 0.2 \times 10^{-4}$, $\mu = 0.1 \times 10^{-4}$, $\beta = 0.01$, $S^{(0)} = (9; 10; 11)$ and $I^{(0)} = (0; 1; 2)$.

One can observe that in Fig. 1 the numerical method given by (19) yields a fuzzy solution which width increases over time. This fact holds due to the standard arithmetic for non-interactive fuzzy numbers. More precisely we have that $width(S^k +_\wedge A) \geq width(S^k)$, $\forall A \in \mathbb{R}_{\mathscr{F}}$, which implies that $width(S^{k+1}) \geq width(S^k)$ for all iteration k.

From the epidemiological point of view the numerical method based on the standard arithmetic is not a good approach to model the dynamic of the disease, since the uncertainty of the temporal evolution increases.

Now let us compare the solution given by (19) with the numerical solution produced by interactive arithmetic that we propose.

$$\begin{cases} S^{(k+1)} = S^{(k)} +_J \frac{h}{6}(K_{S_1}^{(k)} +_J 2K_{S_2}^{(k)} +_J 2K_{S_3}^{(k)} +_J K_{S_4}^{(k)}) \\ I^{(k+1)} = I^{(k)} +_J \frac{h}{6}(K_{I_1}^{(k)} +_J 2K_{I_2}^{(k)} +_J 2K_{I_3}^{(k)} +_J K_{I_4}^{(k)}) \end{cases}, \qquad (20)$$

where

$$K_{S_1}^{(k)} = f_S(t^{(k)}, S^{(k)}, I^{(k)}) = -\beta S^{(k)} \cdot_J I^{(k)} +_J (\eta - \mu)S^{(k)},$$

$$K_{I_1}^{(k)} = f_I(t^{(k)}, S^{(k)}, I^{(k)}) = \beta S^{(k)} \cdot_J I^{(k)} -_J \mu S^{(k)},$$

$$K_{S_i}^{(k)} = f_S\left(t^{(k)} + \frac{h}{2}, S^{(k)} +_J K_{S_{i-1}}^{(k)}\frac{h}{2}, I^{(k)} +_J K_{S_{i-1}}^{(k)}\frac{h}{2}\right), \text{ for } i = 2,3$$

$$K_{I_i}^{(k)} = f_I\left(t^{(k)} + \frac{h}{2}, S^{(k)} +_J K_{I_{i-1}}^{(k)}\frac{h}{2}, I^{(k)} +_J K_{I_{i-1}}^{(k)}\frac{h}{2}\right), \text{ for } i = 2,3$$

$$K_{S_4}^{(k)} = f_S\left(t^{(k)} + h, S^{(k)} +_J hK_{S_3}^{(k)}, I^{(k)} +_J hK_{S_3}^{(k)}\right),$$

$$K_{I_4}^{(k)} = f_I\left(t^{(k)} + h, S^{(k)} +_J hK_{I_3}^{(k)}, I^{(k)} +_J hK_{I_3}^{(k)}\right).$$

Remark 1. In general the arithmetic on fuzzy numbers does not satisfy the distributive and associative properties. Therefore, in order to compare the numerical solutions, the arithmetic operations in the methods need to be computed in the same order.

In view of Remark 1 let us assume that the arithmetical operations in (20) are computed in the same order as in (19). A combination of items (a), (b) and (c) of Theorem 3 ensures that $K_{I_i}^{(k)} \subseteq \hat{K}_{I_i}^{(k)}$ and $K_{S_i}^{(k)} \subseteq \hat{K}_{S_i}^{(k)}$, for all $i = 1, \dots, 4$ and for every iteration k. Furthermore the items (d) and (e) of Theorem 3 imply that the respective right side of the equations of the system (20) are contained in the right side of the equations of (19). Therefore we conclude that the fuzzy numerical solution via interactive arithmetic is contained in the numerical solution via standard arithmetic.

Recall that, in particular cases, the interactive arithmetic operations can not be computed. For example, the joint possibility distribution $J_{\{q,r\}}$ can not be used for the SI model with vital dynamic, since $S^{(k)}$ and $S^{(k+1)}$ are not completely correlated, as well as $I^{(k)}$ and $I^{(k+1)}$. Thus the numerical solution via Runge-Kutta method based on interactive arithmetic is contained in the numerical solution based on standard arithmetic if, the interactive arithmetic operations exist and they are calculated in the same order in both methods. To exemplify this result let us consider the joint possibility distribution $J = J_0^c$ (see (12)) which has no restrictions.

The fuzzy solution based on J_0^c is depicted in Fig. 2. One can observe that the interactive arithmetic operations based on J_0^c produce fuzzy numerical solutions with non-increasing width over time.

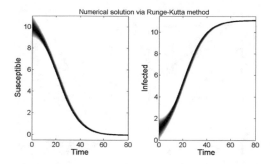

Fig. 2. The fuzzy solution given by Runge-Kutta method via interactive arithmetic. The left and right figures present the susceptible and infected populations, respectively. The gray lines represent the α-cuts of the fuzzy solutions, where their endpoints for α varying from 0 to 1 are represented respectively from the gray-scale lines varying from white to black.

Figure 3 illustrates that the numerical solution via interactive arithmetic is more specific than the solution given by (19). More precisely, the fuzzy solution given as in (20) is contained in the numerical solution via standard arithmetic, as we have shown.

The proposed numerical method can be applied to any n-dimensional initial-value problem. Here we focus on the SI model. The same methodology can be used to others epidemiological models. In fact for the SIS (Susceptible-Infected-Susceptible) and SIR (Susceptible-Infected-Recovered) models similar results are obtained.

Broadly speaking we can also deal with the n-dimensional FIVP by extending the functions f_i, given in (15), via sup-J extension principle. For this case we have

$$\left[K_1^{(k)}\right]^\alpha = \{f_1(t^{(k)}, y_1^{(k)}, ..., y_n^{(k)}) : (y_1^{(k)}, ..., y_n^{(k)}) \in [J]^\alpha\}$$

$$\subseteq \{f_1(t^{(k)}, y_1^{(k)}, ..., y_n^{(k)}) : (y_1^{(k)}, ..., y_n^{(k)}) \in [J_\wedge]]^\alpha\} = \left[\hat{K}_1^{(k)}\right]^\alpha . \ (21)$$

Analogously for $K_2^{(k)}, K_3^{(k)}$, and, $K_4^{(k)}$. Thus every $K_i^{(k)}$ is contained in $\hat{K}_i^{(k)}$ and by similar arguments used before, we conclude that the numerical solution obtained via interactive arithmetic is contained in the solution via standard arithmetic. However this approach is not equivalent to what we have done before, since in the SI model we extend every classical operation of the Runge-Kutta method, and in Eq. (21), we only extend the final result given by the function f_i.

It is important to observe that in the literature there are several fuzzy numerical solutions of FDEs using gH-difference [6] in conjunction with the standard arithmetic operations in the same method (see [2,13]). However from the point of view of joint possibility distributions concept these approaches are not well-defined, since the involved operations are not tied to the same JPD. In this sense our proposed method is more consistent.

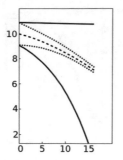 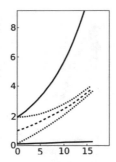

Fig. 3. The left and right figures present the susceptible and infected populations. The dotted and solid lines represent the 0-cut of the fuzzy solutions provided by the Runge-Kutta method based on interactive and standard arithmetic, respectively. The dashed line represents the classical solution of (18) considering the initial conditions given by $S^0 = 10$ and $I^0 = 1$.

5   Final Remarks

This paper presented a numerical method for solving n-dimensional FIVPs with initial conditions given by interactive fuzzy numbers. The numerical solution was obtained by the fourth order Runge-Kutta method, whose arithmetic operations were adopted for interactive fuzzy numbers via sup-J extension principle.

We focused on the SI epidemiological model with interactivity on state variables. We described this interactivity by the joint possibility distribution J_0 (cf. (11)) and used to produce an interactive arithmetic on fuzzy numbers.

We compared the fuzzy solutions via interactive and standard arithmetic. We verified that the solution given by the interactive arithmetic produces a more specific solution than the standard one. More precisely, the solution that considers interactivity is contained in the solution via standard arithmetic.

Finally, contrary to a fuzzy initial-value problem where the derivative is obtained by a fuzzy process [5,9], here we only used numerical methods for FIVPs, considering that initial conditions were given by interactive fuzzy numbers.

Acknowledgment. The authors would like to thank the support of CNPq under grants no. 142414/2017-4 and 306546/2017-5, FAPESP under grant no. 2016/26040-7 and CAPES - Finance Code 001.

References

1. Ahmad, S.: New approaches in the diagnosis and treatment of latent tuberculosis infection. Resp. Res. **11**, 1–169 (2010)
2. Ahmadian, A., Salahshour, S., Chan, C.S., Baleanu, D.: Numerical solutions of fuzzy differential equations by an efficient Runge-Kutta method with generalized differentiability. Fuzzy Sets Syst. **331**, 47–67 (2018)

3. Barros, L.C., Bassanezi, R.C., Lodwick, W.A.: A First Course in Fuzzy Logic, Fuzzy Dynamical Systems, and Biomathematics. Springer, Heidelberg (2017)
4. Barros, L.C., Bassanezi, R.C., Tonelli, P.A.: On the continuity of the Zadeh's extension. In: Proceedings of IFSA 1997 Congress, pp. 1–6 (1997)
5. Barros, L.C., Pedro, F.S.: Fuzzy differential equations with interactive derivative. Fuzzy Sets Syst. **309**, 64–80 (2017)
6. Bede, B.: Mathematics of Fuzzy Sets and Fuzzy Logic. Springer, Heidelberg (2013)
7. Diamond, P., Kloeden, P.: Metric Topology of Fuzzy Numbers and Fuzzy Analysis. Springer, Boston (2000)
8. Edelstein-Keshet, L.: Mathematical Models in Biology. SIAM, Bangkok (1988)
9. Esmi, E., Pedro, F.S., Barros, L.C., Lodwick, W.: Fréchet derivative for linearly correlated fuzzy function. Inf. Sci. **435**, 150–160 (2018)
10. Esmi, E., Sussner, P., Ignácio, G., Barros, L.C.: A parametrized sum of fuzzy numbers with applications to fuzzy initial value problems. Fuzzy Sets Syst. **331**, 85–104 (2018)
11. Fullér, R., Majlender, P.: On interactive fuzzy numbers. Fuzzy Sets Syst. **143**, 355–369 (2004)
12. Ibáñez, D.S., Esmi, E., Barros, L.C.: Linear ordinary differential equations with linearly correlated boundary values. In: IEEE International Conference on Fuzzy Systems, pp. 1–6 (2018)
13. Jayakumar, T., Maheskumar, D., Kanagarajan, K.: Numerical solution of fuzzy differential equations by Runge Kutta method of order five. Appl. Math. Sci. **6**, 2989–3002 (2012)
14. Nguyen, H.T.: A note on the extension principle for fuzzy sets. J. Math. Anal. Appl. **64**, 369–380 (1978)
15. Sussner, P., Esmi, E., Barros, L.C.: Controling the width of the sum of interactive fuzzy numbers with applications to fuzzy initial value problems. In: IEEE International Conference on Fuzzy Systems, pp. 85–104 (2016)
16. Wasques, V.F., Esmi, E., Barros, L.C., Sussner, P.: Numerical solutions for bidimensional initial value problem with interactive fuzzy numbers. In: Fuzzy Information Processing, pp. 84–95. Springer, Cham (2018)
17. Wasques, V.F., Esmi, E., Barros, L.C., Pedro, F.S., Sussner, P.: Higher order initial value problem with interactive fuzzy conditions. In: IEEE International Conference on Fuzzy Systems, pp. 1–8 (2018)
18. Zadeh, L.A.: Fuzzy sets. Inf. Cont. **8**, 338–353 (1965)
19. Zadeh, L.A.: The concept of a linguistic variable and its application to approximate reasoning-i, ii, iii. Inf. Sci. **8**, 301–357 (1975)

Fuzzy Octonion Numbers: Some Analytical Properties

Ricardo Augusto Watanabe[1]($\boxtimes$), Cibele Cristina Trinca Watanabe[2],
and Estevão Esmi[1]

[1] Institute of Mathematics, Statistics and Scientific Computing Campinas (IMECC),
University of Campinas (UNICAMP), Campinas, São Paulo 13083-872, Brazil
ricardoaw18@gmail.com, eelaureano@gmail.com
[2] Department of Biotechnology and Bioprocess Engineering,
Federal University of Tocantins (UFT), Gurupi, Tocantins 77402-970, Brazil
cibtrinca@yahoo.com.br

Abstract. An algebric result states that the sets of real, complex, quaternion and octonion numbers are the only normed division algebras. The mathematical analysis results concerning fuzzy and interval real, complex and quaternion numbers have been extensively studied in the literature. Thereby, it is natural to explore and study the fuzzy and interval extension for the last normed division algebra: the octonions. In this manuscript we show and discuss some important concepts and properties of mathematical analysis with respect to fuzzy and interval octonion numbers, namely, the existence of partial orders, metrics, supremum and infimun and limit of sequences of fuzzy octonion numbers. This work complete the picture of the study of the fuzzy mathematical analysis related to the normed division algebras since the sedenions are not an integral domain.

Keywords: Fuzzy octonion numbers · Interval octonion numbers ·
Fuzzy analytical properties

1 Introduction

The study of interval complex analysis was initialized by Boche [1] in 1966 and Henrici [2] in 1972. In the end of the decade of 1980 and beginning of the decade of 1990, the study of fuzzy complex numbers was developed by Renjun *et al.* [3], they had the first insight of the relation between the complex field and the fuzzy set in a two-dimensional Euclidean space; Buckley [4] developed the rudiment of an analytical theory for fuzzy complex numbers; Zhang [5] made a different formulation from Buckley for the fuzzy complex numbers and developed many results concerning Mathematical Analysis such as arithmetic operations, fuzzy distance and fuzzy limit of fuzzy complex numbers.

The work of Moura *et al.* in [6] was inspired by the work of [5] and it introduces the concepts of fuzzy quaternion numbers and interval quaternion

© Springer Nature Switzerland AG 2019
R. B. Kearfott et al. (Eds.): IFSA 2019/NAFIPS 2019, AISC 1000, pp. 716–726, 2019.
https://doi.org/10.1007/978-3-030-21920-8_63

numbers. Such a work furnishes us arithmetic operations, properties for fuzzy and interval quaternion numbers and limit of sequences of fuzzy quaternion numbers.

The sets $\mathbb{R}$ (real numbers), $\mathbb{C}$ (complex numbers), $\mathbb{H}$ (quaternion numbers) and $\mathbb{O}$ (octonion numbers) are the only normed division algebras [7]. It worth point out that there are many applications of the octonion numbers in physics [8], signal processing [9], neural networks [10] and other research fields. The mathematical analysis results concerning fuzzy real, complex and quaternion numbers have been studied in [11], [5] and [6], respectively. Thereby, it is natural to explore the extension for the last normed division algebra: the octonions.

Therefore, this work aims to provide some results related to the mathematical analysis for the set of fuzzy octonion numbers and it is organized as it follows: in Sect. 2 we define the basic representation of the classical octonions; in Sect. 3 we define the fuzzy octonion numbers with focus on their arithmetic operations; in Sect. 4 it is introduced the interval octonion numbers as well as their basic operations and properties; in Sect. 5 we introduce the concepts of supremun and infimum of fuzzy octonion numbers and, in Sect. 6, we exibit some results of limit of sequences of fuzzy octonion numbers.

2   Octonion Numbers

We utilize the references [8] and [7] for formalizing this section. The octonions are an 8-dimensional algebra with respect to the basis

$$\{1, e_1, e_2, e_3, e_4, e_5, e_6, e_7\}, \tag{1}$$

where e_0 is the unit element which can be identified with 1 and the multiplication of the basis elements is given by the following relations:

- $e_0^2 = 1$ and $e_1^2 = e_2^2 = \cdots = e_7^2 = -1$;
- e_i and e_j anticommute for all i, $j = 1, \ldots, 7$ and $i \neq j$:

$$e_i e_j = -e_j e_i; \tag{2}$$

- the *index cycling* identify holds:

$$e_i e_j = e_k \implies e_{i+1} e_{j+1} = e_{k+1}; \tag{3}$$

- the *index doubling* identify holds:

$$e_i e_j = e_k \implies e_{2i} e_{2j} = e_{2k}. \tag{4}$$

Together with a single nontrivial product, for instance, $e_1 e_2 = e_4$, these relations are enough to recover the whole multiplication of the basis elements. Such a multiplication can be provided by Table 1 which describes the result of multiplying the element of the i-th row by the element of the j-th column.

Table 1. Octonion multiplication table

	e_1	e_2	e_3	e_4	e_5	e_6	e_7
e_1	-1	e_4	e_7	$-e_2$	e_6	$-e_5$	$-e_3$
e_2	$-e_4$	-1	e_5	e_1	$-e_3$	e_7	$-e_6$
e_3	$-e_7$	$-e_5$	-1	e_6	e_2	$-e_4$	e_1
e_4	e_2	$-e_1$	$-e_6$	-1	e_7	e_3	$-e_5$
e_5	$-e_6$	e_3	$-e_2$	$-e_7$	-1	e_1	e_4
e_6	e_5	$-e_7$	e_4	$-e_3$	$-e_1$	-1	e_2
e_7	e_3	e_6	$-e_1$	$-e_5$	$-e_4$	$-e_2$	-1

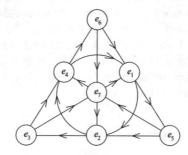

Fig. 1. Fano plane

The octonion product can be also represented by utilizing the **Fano plane** (see Fig. 1).

Fano plane is a mnemonic representation of the product of the elements of the octonion basis by consisting of 7 points and 7 lines, the *lines* are the sides of the triangle, its altitudes and the circle containing all the midpoints of the sides. Each pair of distinct points lies on a unique line. Each line contains three points and each of these triples has a cyclic ordering shown by the arrows. If e_i, e_j and e_k are cyclically ordered in this way, then

$$e_i e_j = e_k \quad \text{and} \quad e_j e_i = -e_k. \tag{5}$$

Together with the following rules:

- 1 is the multiplicative identity;
- $e_1, e_2, \ldots, e_7$ are square roots of -1;

the Fano plane completely describes the algebraic structure of the octonions. Index-doubling corresponds to rotating the picture a third of a turn.

The set of the octonion numbers is denoted by $\mathbb{O}$ and is referred as the classical set of the octonions. In Sect. 3 we explore the fuzzy case for $\mathbb{O}$.

3   Fuzzy Octonion Numbers

This section provides us the definition of fuzzy octonion numbers and some of their corresponding properties.

Definition 1 [12,13]. A fuzzy real set is a function $\bar{A} : \mathbb{R} \to [0,1]$.

Definition 2 [12,13]. A fuzzy real set $\bar{A}$ is a fuzzy real number if, and only if, it satisfies:

 (i) $\bar{A}$ is normal, i.e., there exists $x \in \mathbb{R}$ such that $\bar{A}(x) = 1$;
 (ii) $\bar{A}$ is fuzzy convex, i.e., $\bar{A}(tx+(1-t)y) \geq \min\{\bar{A}(x),\ \bar{A}(y)\}$, where $t \in [0,1]$ and $x,\ y \in \mathbb{R}$;
 (iii) $\bar{A}$ is upper semicontinuous on $\mathbb{R}$, i.e., given an arbitrary $x_0 \in \mathbb{R}$ and $\varepsilon > 0$, there exists a $\delta > 0$ such that if $|x - x_0| < \delta$, then $\bar{A}(x) - A(x_0) < \varepsilon$;
 (iv) $\bar{A}$ is compactly supported, i.e., $cl\{x \in \mathbb{R} \mid A(x) > 0\}$ is compact, where $cl(B)$ denotes the closure of the set B.

The set of the fuzzy real numbers is denoted by $\mathbb{R}_{\mathscr{F}}$.

Definition 3. A **fuzzy octonion number** is given by $o' : \mathbb{O} \to [0,1]$ such that

$$o'\left(\sum_{i=0}^{7} a_i e_i\right) = \bigwedge_{i=0}^{7} \bar{A}_i(a_i), \tag{6}$$

where $\bigwedge$ is the minimum operator, $a_i \in \mathbb{R}$ and $\bar{A}_i \in \mathbb{R}_{\mathscr{F}}$, for $i = 0, \ldots, 7$.

The fuzzy real part is represented by $Re_{\mathscr{F}}(o') = \bar{A}_0$ and the fuzzy imaginary part is represented by $Im_{\mathscr{F}}(o') = \bar{A}_i$, for $i = 1, \ldots, 7$.

The set of the fuzzy octonion numbers is denoted by $\mathbb{O}_{\mathscr{F}}$ and identified by $\mathbb{R}^8_{\mathscr{F}}$, where every element o' is associated with $(\bar{A}_i)$, for $i = 0, \ldots, 7$.

Zadeh's extension principle [12,13] applied to the octonion product furnishes us a way to define the fuzzy octonion product as it follows:

Definition 4. Let o', $f' \in \mathbb{O}_{\mathscr{F}}$, where $o' = (\bar{A}_i)$ and $f' = (\bar{B}_i)$, for $i = 0, \ldots, 7$. Then define

$$o' + f' = ((\bar{A}_i + \bar{B}_i)_i),\ \text{with } i = 0, \ldots, 7, \tag{7}$$

and

$$o'.f' = (\bar{C}_i),\ \text{with } i = 0, \ldots, 7,\ \text{where} \tag{8}$$

$$\bar{C}_0 = \bar{A}_0 \bar{B}_0 - \bar{A}_1 \bar{B}_1 - \bar{A}_2 \bar{B}_2 - \bar{A}_3 \bar{B}_3 - \bar{A}_4 \bar{B}_4 - \bar{A}_5 \bar{B}_5 - \bar{A}_6 \bar{B}_6 - \bar{A}_7 \bar{B}_7,$$

$$\bar{C}_1 = \bar{A}_0 \bar{B}_1 + \bar{A}_1 \bar{B}_0 + \bar{A}_2 \bar{B}_4 + \bar{A}_3 \bar{B}_7 - \bar{A}_4 \bar{B}_2 + \bar{A}_5 \bar{B}_6 - \bar{A}_6 \bar{B}_5 - \bar{A}_7 \bar{B}_3,$$

$$\bar{C}_2 = \bar{A}_0 \bar{B}_2 - \bar{A}_1 \bar{B}_4 + \bar{A}_2 \bar{B}_0 + \bar{A}_3 \bar{B}_5 + \bar{A}_4 \bar{B}_1 - \bar{A}_5 \bar{B}_3 + \bar{A}_6 \bar{B}_7 - \bar{A}_7 \bar{B}_6,$$

$$\bar{C}_3 = \bar{A}_0 \bar{B}_3 - \bar{A}_1 \bar{B}_7 - \bar{A}_2 \bar{B}_5 + \bar{A}_3 \bar{B}_0 + \bar{A}_4 \bar{B}_6 + \bar{A}_5 \bar{B}_2 - \bar{A}_6 \bar{B}_4 + \bar{A}_7 \bar{B}_1,$$

$$\bar{C}_4 = \bar{A}_0 \bar{B}_4 + \bar{A}_1 \bar{B}_2 - \bar{A}_2 \bar{B}_1 - \bar{A}_3 \bar{B}_6 + \bar{A}_4 \bar{B}_0 + \bar{A}_5 \bar{B}_7 + \bar{A}_6 \bar{B}_3 - \bar{A}_7 \bar{B}_5,$$

$$\bar{C}_5 = \bar{A}_0\bar{B}_5 - \bar{A}_1\bar{B}_6 + \bar{A}_2\bar{B}_3 - \bar{A}_3\bar{B}_2 - \bar{A}_4\bar{B}_7 + \bar{A}_5\bar{B}_0 + \bar{A}_6\bar{B}_1 + \bar{A}_7\bar{B}_4,$$
$$\bar{C}_6 = \bar{A}_0\bar{B}_6 + \bar{A}_1\bar{B}_5 - \bar{A}_2\bar{B}_7 + \bar{A}_3\bar{B}_4 - \bar{A}_4\bar{B}_3 - \bar{A}_5\bar{B}_1 + \bar{A}_6\bar{B}_0 + \bar{A}_7\bar{B}_2,$$

and

$$\bar{C}_7 = \bar{A}_0\bar{B}_7 + \bar{A}_1\bar{B}_3 + \bar{A}_2\bar{B}_6 - \bar{A}_3\bar{B}_1 + \bar{A}_4\bar{B}_5 - \bar{A}_5\bar{B}_4 - \bar{A}_6\bar{B}_2 + \bar{A}_7\bar{B}_0.$$

Since the fuzzy octonion numbers are an extension of the classical octonion numbers, then, as the reader might expect, the fuzzy octonion numbers are neither commutative nor associative in relation to the multiplication operation.

4   Interval Octonion Number

Let $\mathbb{I}(\mathbb{R}) = \{[a,b] : a, b \in \mathbb{R}\}$ be the set of the closed intervals endowed with the following arithmetic [14,15]:

- $[a,b] + [c,d] = [a + c, b + d]$;
- $[a,b] - [c,d] = [a - d, b - c]$, where $-[c,d] = [-d, -c]$;
- $[a,b] \cdot [c,d] = [\min\{a \cdot c,\ a \cdot d,\ b \cdot c,\ b \cdot d\},\ \max\{a \cdot c,\ a \cdot d,\ b \cdot c,\ b \cdot d\}]$;
- $[c,d]^{-1} = \dfrac{1}{[c,d]} = \left[\dfrac{1}{d}, \dfrac{1}{c}\right]$, if $0 \notin [c,d]$;

- $\dfrac{[a,b]}{[c,d]} = [a,b] \cdot [c,d]^{-1}$, whenever $0 \notin [c,d]$.

Definition 5. An **interval octonion number** O is an octuple $(A_i)_{i=0,\dots,7}$, where $A_i \in \mathbb{I}(\mathbb{R})$, with $i = 0, \dots, 7$. The set of the interval octonion numbers is denoted by $\mathbb{I}(\mathbb{O})$.

Definition 6. Let M, $N \in \mathbb{I}(\mathbb{O})$, where $M = (A_i)_{i=1,\dots,7}$ and $N = (B_i)_{i=0,\dots,7}$, then $M = N$ if, and only if, $\bigwedge_{i=0}^{7}(A_i = B_i)$.

Definition 7. Let M, $N \in \mathbb{I}(\mathbb{O})$, where $M = (A_i)_{i=0,\dots,7}$ and $N = (B_i)_{i=0,\dots,7}$, then

$$M + N = (A_i + B_i)_{i=0,\dots,7} \tag{9}$$

and

$$M \cdot N = (P_i)_{i=0,\dots,7}, \quad \text{where} \tag{10}$$

$$P_0 = A_0B_0 - A_1B_1 - A_2B_2 - A_3B_3 - A_4B_4 - A_5B_5 - A_6B_6 - A_7B_7,$$
$$P_1 = A_0B_1 + A_1B_0 + A_2B_4 + A_3B_7 - A_4B_2 + A_5B_6 - A_6B_5 - A_7B_3,$$
$$P_2 = A_0B_2 - A_1B_4 + A_2B_0 + A_3B_5 + A_4B_1 - A_5B_3 + A_6B_7 - A_7B_6,$$
$$P_3 = A_0B_3 - A_1B_7 - A_2B_5 + A_3B_0 + A_4B_6 + A_5B_2 - A_6B_4 + A_7B_1,$$
$$P_4 = A_0B_4 + A_1B_2 - A_2B_1 - A_3B_6 + A_4B_0 + A_5B_7 + A_6B_3 - A_7B_5,$$
$$P_5 = A_0B_5 - A_1B_6 + A_2B_3 - A_3B_2 - A_4B_7 + A_5B_0 + A_6B_1 + A_7B_4,$$
$$P_6 = A_0B_6 + A_1B_5 - A_2B_7 + A_3B_4 - A_4B_3 - A_5B_1 + A_6B_0 + A_7B_2,$$

and

$$P_7 = A_0B_7 + A_1B_3 + A_2B_6 - A_3B_1 + A_4B_5 - A_5B_4 - A_6B_2 + A_7B_0.$$

Proposition 1. *For all* M, N, $P \in \mathbb{I}(\mathbb{O})$:

- $M + N \in \mathbb{I}(\mathbb{O})$;
- $M + N = N + M$;
- $(M + N) + P = M + (N + P)$;
- *there is a* $0^* = (0_i)_{i=0,\dots,7}$, *where* $M + 0^* = M$ *and* $0 = [0, 0] \in \mathbb{I}(\mathbb{R})$.

Proof. It follows from Definition 7.

Proposition 2. *For all* M, $N \in \mathbb{I}(\mathbb{O})$:

- $M \cdot N \in \mathbb{I}(\mathbb{O})$;
- *there is a* $1^* = (1, 0, 0, 0, 0, 0, 0, 0) \in \mathbb{I}(\mathbb{O})$, *where* $M \cdot 1^* = M$ *and* $1 = [1, 1] \in \mathbb{I}(\mathbb{R})$.

Proof. It follows from Definition 7.

Observe that in $\mathbb{I}(\mathbb{R})$ we have neither the inverse of the addition operation nor the inverse of the multiplication operation. Also, $\mathbb{I}(\mathbb{O})$ does not present the properties of commutativity and associativity with respect to the multiplication operation.

Let $A = [a_1, a_2]$ and $B = [b_1, b_2]$ be elements from $\mathbb{I}(\mathbb{R})$. The set $\mathbb{I}(\mathbb{R})$ is endowed with the corresponding metric [14, 15]

$$d(A, B) = \max\{|a_1 - b_1|, |a_2 - b_2|\}. \tag{11}$$

The following proposition establishes a metric on the set $\mathbb{I}(\mathbb{O})$.

Proposition 3. *Let* $d : \mathbb{I}(\mathbb{R}) \times \mathbb{I}(\mathbb{R}) \to \mathbb{R}_+$ *be a metric. A function* $D : \mathbb{I}(\mathbb{O}) \times \mathbb{I}(\mathbb{O}) \to \mathbb{R}_+$ *defined as* $D(M, N) = \sum_{i=0}^{7} d(A_i, B_i)$ *is a metric.*

Proof. Let M, N, $P \in \mathbb{I}(\mathbb{O})$, where $M = (A_i)_{i=0,\dots,7}$, $N = (B_i)_{i=0,\dots,7}$ and $P = (C_i)_{i=0,\dots,7}$. Then

1. $D(M, N) = 0 \iff \sum_{i=0}^{7} d(A_i, B_i) = 0 \iff d(A_0, B_0) = 0, \dots, d(A_7, B_7)$
   $= 0 \iff A_i = B_i$, for $i = 0, \dots, 7 \iff M = N$;
2. if $d(A_i, B_i) \geq 0$, for $i = 0, \dots, 7$, then $D(M, N) \geq 0$;
3. $D(M, N) = \sum_{i=0}^{7} d(A_i, B_i) = \sum_{i=0}^{7} d(B_i, A_i) = D(N, M)$;
4. $D(M, N) = \sum_{i=0}^{7} d(A_i, B_i) \leq \sum_{i=0}^{7}(d(A_i, C_i) + d(C_i, B_i)) \leq D(M, P) + D(P, N)$.

Corollary 1. $(\mathbb{I}(\mathbb{O}), D)$ *is a metric space.*

Proof. It follows immediately from Proposition 3.

Next, it is possible to show that there exists a metric on $\mathbb{O}_{\mathscr{F}}$.

Proposition 4. *Let* $d : \mathbb{R}_{\mathscr{F}} \times \mathbb{R}_{\mathscr{F}} \to \mathbb{R}_+$ *be a metric. A function* $D : \mathbb{O}_{\mathscr{F}} \times \mathbb{O}_{\mathscr{F}} \to \mathbb{R}_+$ *defined as* $D(o_1', o_2') = \sum_{i=0}^{7} d(\bar{A}_i, \bar{B}_i)$ *is a metric.*

Proof. Analogous to the proof of Proposition 3.

Corollary 2. $(\mathbb{O}_{\mathscr{F}}, D)$ *is a metric space.*

Proof. It follows immediately from Proposition 4.

Definition 8. Let M, $N \in \mathbb{I}(\mathbb{O})$, where $M = (A_i)_{i=0,\dots,7}$ and $N = (B_i)_{i=0,\dots,7}$, then $M \leq N$ if, and only if, $\bigwedge_{i=0}^{7}(A_i \leq B_i)$ and $M < N$ if, and only if, $\bigwedge_{i=0}^{7}(A_i < B_i)$.

Note that Definition 8 establishes a partial order in $\mathbb{I}(\mathbb{O})$. The following definition is provided by the works [5] and [6].

Definition 9 [5,6]. A non-empty set C is said to be **dense** if, for every a, $b \in C$ and $a < b$, there exists a $c \in C$ with $a < c < b$.

Proposition 5 [5,6]. $\mathbb{I}(\mathbb{R})$ *is dense.*

Proposition 6. $\mathbb{I}(\mathbb{O})$ *is dense.*

Proof. Let M, $N \in \mathbb{I}(\mathbb{O})$, where $M = (A_i)_{i=0,\dots,7}$ and $N = (B_i)_{i=0,\dots,7}$ such that $M < N$, then $\bigwedge_{i=0}^{7}(A_i < B_i)$. By Proposition 5 $\mathbb{I}(\mathbb{R})$ is dense, therefore, there exists $T_i \in \mathbb{I}(\mathbb{R})$ such that $A_i < T_i < B_i$, for $i = 0,\dots,7$. Consequently, $\bigwedge_{i=0}^{7}(A_i < T_i < B_i)$ and, therefore, $M < T < N$, where $T = (T_i)$, with $i = 0,\dots,7$.

5   Supremum and Infimum for Fuzzy Octonion Numbers

In this section we define and establish results concerning least upper bound and greatest lower bound of fuzzy octonion numbers.

Definition 10 [16]. Let $\bar{A}$, $\bar{B} \in \mathbb{R}_{\mathscr{F}}$. We say that $\bar{A} \leq \bar{B}$ if, and only if, $\bar{A}[\alpha] \leq \bar{B}[\alpha]$, $\forall \alpha \in (0,1]$. Also, $\bar{A} < \bar{B}$ if, and only if, $\bar{A} \leq \bar{B}$ and there exists $\alpha_0 \in (0,1]$ such that $\bar{A}[\alpha_0] < \bar{B}[\alpha_0]$. In addition, $\bar{A} = \bar{B}$ if, and only if, $\bar{A} \leq \bar{B}$ and $\bar{B} \leq \bar{A}$.

Definition 11 [16]. We say that $\bar{A}$ is an **infinite fuzzy real number** if, and only if, for all $M \in \mathbb{R}^+$, there exists $\alpha_M \in (0,1]$ such that $[-M,M] \subseteq \bar{A}[\alpha_M]$. In this case, we denote $\bar{A}$ by $\bar{\infty}$.

Definition 12. Let $o' = (\bar{A}_i) \in \mathbb{O}_{\mathscr{F}}$. We say that o' is an **infinite fuzzy octonion number** if, and only if, $(\bar{A}_i) = \bar{\infty}$, for $i = 0,\dots,7$. In this case, we denote o' by ∞'.

Definition 13. Let $o'_1 = (\bar{A}_i)$ and $o'_2 = (\bar{B}_i)$ be elements from $\mathbb{O}_{\mathscr{F}}$. We say that $o'_1 \leq o'_2$ if, and only if, $\bigwedge_{i=0}^{7}(A_i \wedge B_i)$.

Definition 14. Let $R \subseteq \mathbb{O}_{\mathscr{F}}$. If there exists $M' \in \mathbb{O}_{\mathscr{F}}$, where $M' \neq \infty'$, such that $c' \leq M'$, for every $c' \in R$, then R is said to have an **upper bound** M'. Analogously, if there exists $m' \in \mathbb{O}_{\mathscr{F}}$, where $m' \neq \infty'$, such that $m' \leq c'$, for every $c' \in R$, then R is said to have a **lower bound** m'. A set with lower and upper bounds is said to be **bounded**.

Definition 15. We say that $s' \in \mathbb{O}_{\mathscr{F}}$ ($i' \in \mathbb{O}_{\mathscr{F}}$) is the **least upper bound** (**greatest lower bound**) for $R \subseteq \mathbb{O}_{\mathscr{F}}$ if s' (i') satisfies the following properties:

1. $c' \leq s'$ ($i' \leq c'$), for all $c' \in R$;
2. for any $\varepsilon > 0$, $\varepsilon \in \mathbb{R}$, there exists $c' \in R$ such that $s' < c' + \varepsilon$ ($c' - \varepsilon < i'$).

We denote $s' = \sup R$ ($i' = \inf R$).

Let $R \subseteq \mathbb{O}_{\mathscr{F}}$, we can define $Re_{(R)} = \{\mathbf{Re}(c') \in \mathbb{R}_{\mathscr{F}} : c' \in R\}$ and $Im_{(R)_i} = \{\mathbf{Im}(c') \in \mathbb{R}_{\mathscr{F}} : c' \in R\}$, for $i = 0, \ldots, 7$. Consequently, $R = Re_{(R)} \times_{i=0}^{7} Im_{(R)_i}$.

Proposition 7. *If $R \subseteq \mathbb{O}_{\mathscr{F}}$ has the least upper bound (greatest lower bound), then*

$$\sup(R) = (\sup Re_{(R)}, \; \sup_i Im_{(R)_i}) \tag{12}$$

and

$$\inf(R) = (\inf Re_{(R)}, \; \inf_i Im_{(R)_i}). \tag{13}$$

Proof. We prove the first part of the proposition, the second part is analogous. Since $R = \mathbf{Re}(R) \times \mathbf{Im}(R)$ and if R has the least upper bound, then there exist the least upper bounds for $\mathbf{Re}(R)$ and $\mathbf{Im}(R)$.

1. If $c' \in R$, then $\mathbf{Re}(c') \leq \sup Re_{(R)}$ and $\mathbf{Im}(c') \leq \sup_i Im_{(R)_i}$, for $i = 0, \ldots, 7$. Therefore, $c' \leq (\sup Re_{(R)}, \sup_i Im_{(R)_i})$;
2. Let $\varepsilon > 0$, so there exists $\bar{A}_0 \in Re_{(R)}$ and $\bar{A}_i \in Re_{(R)}$, for $i = 1, \ldots, 7$, where $\sup Re_{(R)} < \bar{A}_0 + \varepsilon$ and $\sup_i Im_{(R)_i} < \bar{A}_i + \varepsilon$, for $i = 1, \ldots, 7$. Consider $c' = (\bar{A}_0, \bar{A}_i)$, therefore, $(\sup Re_{(R)}, \sup_i Im_{(R)_i}) < c' + \varepsilon$.

6   Limit of a Sequence of Fuzzy Octonion Numbers

The results of Sect. 5 is employed in this section in order to obtain results concerning mathematical analysis.

Definition 16 [16]. Let d a metric on $\mathbb{R}_{\mathscr{F}}$, $\{\bar{A}_n\} \subset \mathbb{R}_{\mathscr{F}}$ and $\bar{A} \in \mathbb{R}_{\mathscr{F}}$. The sequence $\{\bar{A}_n\}$ is said to converge to $\bar{A}$ with respect to d if, for an arbitrary $\varepsilon > 0$, there exists $N > 0$ such that $d(\bar{A}_n, \bar{A}) < \varepsilon$, as $n \geq N$. We denote it by $\lim_{n \to \infty} \bar{A}_n = \bar{A}$.

Definition 17. Let d a metric on $\mathbb{R}_{\mathscr{F}}$, $o'_n \subset \mathbb{O}_{\mathscr{F}}$ and $\bar{o}' \in \mathbb{O}_{\mathscr{F}}$. Then o'_n is said to converge to o' if, for an arbitrary $\varepsilon > 0$, there exists an integer $N > 0$ such that $D(o'_n, o') < \varepsilon$, as $n \geq N$. We denote it by $\lim_{n \to \infty} o'_n = o'$.

Theorem 1. $\lim_{n \to \infty} o'_n = o'$ *if, and only if, the limits of the real and imaginary parts exist.*

Proof. It follows immediately from Definition 17.

Next, we review a result valid for $\mathbb{R}_{\mathscr{G}}$.

Theorem 2 [16]. *Let* $\bar{A}_n$, $\bar{B}_n \subset \mathbb{R}_{\mathscr{G}}$, $\bar{A}$, $\bar{B} \in \mathbb{R}_{\mathscr{G}}$ *and* $c \in \mathbb{R}$. *If* $\lim_{n \to \infty} \bar{A}_n = \bar{A}$ *and* $\lim_{n \to \infty} \bar{B}_n = \bar{B}$, *then*

1. $\lim_{n \to \infty} \bar{A}_n \pm \bar{B}_n = \bar{A} \pm \bar{B}$;
2. $\lim_{n \to \infty} c \cdot \bar{A}_n = c \cdot \bar{A}$.

Proof. See proof of the Theorem 3.2 in [16].

Theorem 3 *(Limit uniqueness theorem).* *If* $\lim_{n \to \infty} o'_n = o'$ *and* $\lim_{n \to \infty} o'_n = f'$, *then* $o' = f'$.

Proof. It follows immediately from Proposition 1.

Theorem 4 *(Sandwich theorem).* *Let* $\{o'_n\}$, $\{f'_n\} \subset \mathbb{O}_{\mathscr{G}}$ *and* o', $f' \in \mathbb{O}_{\mathscr{G}}$. *If* $\lim_{n \to \infty} o'_n = o'$ *and* $\lim_{n \to \infty} f'_n = f'$, *then*

1. $\lim_{n \to \infty} (o'_n \pm f'_n) = o' \pm f'$;
2. $\lim_{n \to \infty} c \cdot o'_n = c \cdot o'$.

Proof. 1. Let $o'_n = (\bar{A}_n)_i$ and $f'_n = (\bar{B}_n)_i$, for $i = 0, \ldots, 7$, where $Re(o'_n) = (\bar{A}_n)_0$, $Im(o'_n) = (\bar{A}_n)_i$, $Re(f'_n) = (\bar{B}_n)_0$ and $Im(f'_n) = (\bar{B}_n)_i$, for $i = 1, \ldots, 7$. Thus, by Theorem 1, it implies

$$\lim_{n \to \infty} (o'_n \pm f'_n) = \lim_{n \to \infty} ((\bar{A}_n \pm \bar{B}_n)_0, \ldots, (\bar{A}_n \pm \bar{B}_n)_7) =$$

$$= ((\bar{A} \pm \bar{B})_0, \ldots, (\bar{A} \pm \bar{B})_7) = o' \pm f'. \tag{14}$$

2. The proof is analogous to the first case.

Theorem 5 *(Boundedness theorem).* *Let* $\{o'_n\}$, $\{f'_n\}$, $\{p'_n\} \subset \mathbb{O}_{\mathscr{G}}$ *and* $o' \in \mathbb{O}_{\mathscr{G}}$. *If, for every* n, $f'_n \leq o'_n \leq p'_n$, *and* $\lim_{n \to \infty} f'_n = \lim_{n \to \infty} p'_n = o'$, *then* $\lim_{n \to \infty} o'_n = o'$.

Proof. It follows immediately from Definition 16.

Theorem 6. *Let* $h'_n \subset \mathbb{O}_{\mathscr{G}}$ *and* $h' \neq \infty'$. *If* h'_n *converges, then there exist* L', $l' \neq \infty'$ *such that* $l' \leq h'_n \leq L'$, *for every* n.

Proof. It follows immediately from Definition 16.

Theorem 7. *Let* $\lim_{n \to \infty} o'_n = o'$ *and* $\lim_{n \to \infty} f'_n = f'$. *Then* $\lim_{n \to \infty} D(o'_n, f'_n) = D(o', f')$.

Proof. It follows from Proposition 4 and Definition 17.

As the set formed by the sedenions is not a division algebra [7,8], then it is not possible to study some classical mathematical analysis properties for the sedenions since they are not an integral domain.

7   Conclusion

In the literature, the sets $\mathbb{R}$ (real numbers), $\mathbb{C}$ (complex numbers), $\mathbb{H}$ (quaternion numbers) and $\mathbb{O}$ (octonion numbers) are the only normed division algebras [7]. The mathematical analysis results concerning fuzzy real, complex and quaternion numbers have been studied in [11], [5] and [6], respectively. Therefore, it is natural to explore the extension for the last normed division algebra: the octonions. Then, in this work, we furnish properties of mathematical analysis for the set of fuzzy octonion numbers. Furthermore, since the sedenions are not an integral domain, then it is not possible to explore some classical mathematical analysis results for the set of sedenions.

Acknowledgements. The authors would like to thank the Brazilian Agency CAPES (Coordenação de Aperfeiçoamento de Pessoal de Nível Superior) for the financial support.

References

1. Boche, R.E.: Complex interval arithmetic with some applications. Technical report Number LMSC4-22-66-1, Lockheed General Research Program (1966)
2. Gargantini, I., Henrici, P.: Circular arithmetic and the determination of polynomial zeros. Numer. Math. **18**(4), 305–320 (1972)
3. Renjun, L., Shaoqing, Y., Baowen, L., Weihai, F.: Fuzzy complex numbers. BUSE-FAL **25**, 79–86 (1986)
4. Buckley, J.J.: Fuzzy complex numbers. Fuzzy Sets Syst. **33**(3), 333–345 (1989)
5. Zhang, G.-Q.: Fuzzy limit theory of fuzzy complex numbers. Fuzzy Sets Syst. **46**(2), 227–235 (1992)
6. Moura, R.P.A., Bergamashi, F.B., Santiago, R.H.N., Bedregal, B.R.C.: Fuzzy quaternion numbers. In: 2013 IEEE International Conference on Fuzzy Systems (FUZZ - IEEE). IEEE (2013). https://doi.org/10.1109/fuzz-ieee.2013.6622400
7. Schafer, R.D.: Introduction to Non-associative Algebras. Dover, New York (1995)
8. Baez, J.C.: The octonions. Bull. Am. Math. Soc. **39**, 145–205 (2002). Corrected Version: math.RA/0105155. https://doi.org/10.1090/S0273-0979-01-00934-X
9. Shen, M., Wang, R.: SA new singular value decomposition algorithm for octonion signal. In: 2018 24th International Conference on Pattern Recognition (ICPR), pp. 3233–3237. IEEE (2018)
10. Yasuaki, K., Hitoshi, L.: SA model of hopfield-type octonion neural networks and existing conditions of energy functions. In: 2016 International Joint Conference on Neural Networks (IJCNN), pp. 44264–44430. IEEE (2016)
11. Guang-Quan, Z.: Fuzzy limit theory of fuzzy numbers. Cybern. Syst. (World Scientific Publishing) **90**, 163–170 (1990)
12. Bede, B.: Mathematics of Fuzzy Sets and Fuzzy Logic. STUDFUZZ. Springer, Berlin (2013). https://doi.org/10.1007/978-3-642-35221-8
13. de Barros, L.C., Bassanezi, R.C., Lodwick, W.A.: A first course in fuzzy logic. In: Fuzzy Dynamical Systems, and Biomathematics: Theory and Applications. SFSC, vol. 347. Springer, Heidelberg (2017). https://doi.org/10.1007/978-3-662-53324-6
14. Moore, R.E.: Interval Analysis. Prentice Hall, Upper Saddle River (1966)

15. Moore, R.E., Kearfott, R.B., Cloud, M.J.: Introduction to Interval Analysis. Society for Industrial and Applied Mathematics (2009). https://doi.org/10.1137/1.9780898717716
16. Guangquan, Z.: Fuzzy distance and limit of fuzzy numbers. BUSEFAL **33**, 19–30 (1987)

Towards Hybrid Uncertain Data Modeling in Databases

Li Yan and Zongmin Ma[✉]

College of Computer Science and Technology,
Nanjing University of Aeronautics and Astronautics,
Nanjing 211106, Jiangsu, China
zongminma@nuaa.edu.cn

Abstract. Uncertain data extensively exists in many real-world applications. And uncertain data modeling has been investigated in various database models. This has resulted in numerous contributions. Actually, uncertainty in data has devise semantics, mainly including objective uncertainty and subjective uncertainty. Different types of uncertainty may occur together and we face with hybrid uncertain data with objective uncertainty and subjective uncertainty. This paper devotes to identify the semantics and expressive forms of hybrid uncertain data with both objective uncertainty and subjective uncertainty, and presents an up-to-date overview of the current state of the art in hybrid uncertain data modeling in relational databases and object-oriented databases.

1 Introduction

Uncertain information extensively exists in many real-world applications. As a result, we are facing more and more uncertain information which affect data and decision-making processing. Traditionally probability calculus, being a major and most commonly used tool to represent uncertainty, is applied to represent and deal with objective uncertainty. It should be noted that probability for objective uncertainty needs a huge workload of data preparation and there may not be sufficient volume of data enabling the performance statistical tests [19]. At this point, in practice, uncertainty may stem from subjective estimation made by experts because no information available at all is not always true and most case is some information available [26]. The subjective uncertainty is typically modeled by fuzzy sets [24]/possibility theory [25]. It is shown in some real-world applications (e.g., risk assessment [2,11]) that objective uncertainty and subjective uncertainty are not always separate and they are actually related each other [10], in which data are partially described by probability distributions and partially by possibility distributions. So, some efforts have devoted to investigate the interaction of objective uncertainty and subjective uncertainty [1,4,18,19, 27].

Database models are designed to represent an enormous wealth of data from various real-world applications, which provide an infrastructure of data management. Therefore, one of the major areas of database research has been the

© Springer Nature Switzerland AG 2019
R. B. Kearfott et al. (Eds.): IFSA 2019/NAFIPS 2019, AISC 1000, pp. 727–737, 2019.
https://doi.org/10.1007/978-3-030-21920-8_64

continuous effort to enrich existing database models with a more extensive collection of semantic concepts in order to satisfy the requirements in the real-world applications. One of some inadequacy of necessary semantics that traditional database models often suffer from is the inability to handle uncertain information. For this reason, uncertain data have been introduced into databases for imperfect information processing. In order to deal with objective uncertainty of data, probabilistic relational database model (e.g., [7,28,29]), and probabilistic object-oriented database model (e.g., [13]) have been proposed. In order to deal with subjective uncertainty of data, fuzzy relational database model (e.g., [15,16,23]) and fuzzy object-oriented database model (e.g., [20]) have been proposed also.

While probabilistic data models [12] and the fuzzy data models [15] have been developed separately to deal with objective uncertainty and subjective uncertainty, respectively, each of these two types of data models actually suffer from the inability to simultaneously handle fuzzy and probabilistic information (i.e., subjective uncertainty and objective uncertainty). So, there are several works on use database models to represent and process hybrid uncertain data which contain both probabilistic data and fuzzy data [5,6,21,22].

This paper summarizes the semantics and expressive forms of hybrid uncertain data with both probabilistic information and fuzzy information. On the basis, the paper presents a short review of up-to-date overview of the current state of the art in hybrid uncertain data modeling in relational databases and object-oriented databases. Note that the purpose of this paper is not to depict the details of hybrid uncertain data modeling techniques developed and used in the literature. Instead some basic concepts, adopted methods, and their relationships in hybrid uncertain data modeling are presented. More important, this paper tries to identify possible issues and directions of hybrid uncertain data modeling in the future.

The remainder of the paper is organized as follows. Section 2 presents some preliminaries, including imperfect data, probability distributions and fuzzy sets. In Sect. 3, the semantics and expressive forms of probabilistic data with fuzziness are investigated. Section 4 provides details of the different techniques in modeling hybrid uncertain data. Section 5 summarizes this paper and outlines possible issues for future research directions.

2   Preliminaries

This section provides preliminaries of imprecise data, fuzzy sets and probability theory.

2.1   Imperfect Data

Information in real-world applications is often imperfect. Efforts have been made to distinguish and classify different types and sources of imperfect information in the literature. Five basic kinds of imperfection are identified in [3], which are inconsistency, imprecision, vagueness, uncertainty, and ambiguity.

(1) Inconsistency is a type of semantic conflict, which indicates that a partic-
    ular data in the real world is recorded in one or more different databases
    or expressed as different semantics. Inconsistencies of these semantics are
    usually derived from the inheritance of information.
(2) Imprecision and vagueness are related to the content of a value, which indi-
    cates that a value is to be selected from a given range (interval or set), but
    the current choice is not known.
(3) Uncertainty is related to the true value of a value or a group of values, which
    indicates the degree of confidence in a given set of values or distributions.
(4) Ambiguity indicates that information lacks complete semantics, leading to
    a variety of possible explanations.

In general, a piece of information may contain several types of imperfection at
the same time. The age information of one person, for example, may be described
by a set of elements 21, 23, 24, 25 with possibility degrees of 70%, 84%, 80% and
79%, respectively. For example, the orientation of landing on for a coin being
tossed is described by a set of elements front, back with probability degrees of
50% and 50%, respectively. Many current approaches of representing imperfect
information are based on fuzzy sets [24]/possibility theory [25] and probability
theory. Among them, fuzzy sets and possibility theory are applied to represent
subjective uncertainty, and probability theory is applied to represent objective
uncertainty.

2.2 Stochastic Even and Probability Distributions

A stochastic phenomenon generally corresponds to multiple possible stochastic
events. To indicate the probability that a stochastic event will occur, the stochastic event is associated with a probability (i.e., probability measure). As a result,
the stochastic event is described by a probability distribution. Formally let e be
a stochastic event and its probability measure be $p_s(e)$ $(0 \leq p_s(e) \leq 1)$. Then,
for the stochastic phenomenon S with stochastic evens e_1, e_2,..., e_k, it can be
described by a probability distribution P as follows.

$$P = (e_1, p_s(e_1), (e_2, p_s(e_2), ..., (e_k, p_s(e_k)) \tag{1}$$

Probability distributions are applied to represent objective uncertainty.

As we may know, in the probability distribution $\{(e_1, p_s(e_1), (e_2, p_s(e_2),...,$
$(e_k, p_s(e_k))\}$ for the stochastic phenomenon P, we have $0 \leq p_s(e_i) \leq 1$ $(1 \leq i$
$\leq$ k). There is a probabilistic constraint on the stochastic events with respect
to the stochastic phenomenon. That is, the sum of probability measures of the
stochastic events must be less than or equal to one. Formally we have $\sum_{i=1,2,...,k}$
$p_s(e_i) \leq 1$.

2.3 Fuzzy Sets and Possibility Distribution

Fuzzy sets introduced by Zadeh [24] can be applied to represent subjective uncertainty. Let U be a universe of discourse and F be a fuzzy set in U. A membership

function $\mu_F\colon U \to [0,1]$ is defined for F, in which $\mu_F(u)$ for each $u \in U$ denotes the membership degree of u in the fuzzy set F. Then F is described by

$$F = \{(u_1, \mu_F(u_1)), (u_2, \mu_F(u_2)), ..., (u_n, \mu_F(u_n))\} \tag{2}$$

The set of the elements in fuzzy set F which membership degrees are non-zero is called the support of F, denoted by

$$\mathrm{supp}(F) = \{u \mid u \in U \quad and \quad \mu_F(u) > 0\} \tag{3}$$

With the fuzzy set $F = \{(u_1, \mu_F(u_1)), (u_2, \mu_F(u_2)), ..., (\mu_n, \mu_F(u_n))\}$, we can obtain a conventional set, which are consisted of the the elements in F with non-zero membership degrees. Such a set is called the support of F, denoted by

$$\mathrm{supp}(F) = \{u \mid u \in U \quad and \quad \mu_F(u) > 0\} \tag{4}$$

When the membership degree $\mu_F(u)$ above is explained as a measure of the possibility that a variable X has the value u, where X takes on values in U, a fuzzy value is described by the possibility distribution π_X [25].

$$\pi_X = \{(u_1, \pi_X(u_1)), (u_2, \pi_X(u_2)), ..., (u_n, \pi_X(u_n))\} \tag{5}$$

Here, $\pi_X(u_i)$ ($u_i \in U$ and $0 \le i \le n$) denotes the possibility that u_i is an actual value of variable X under consideration. Let π_X be the representation of possibility distribution for a variable X. This means that X may take on one of the possible values u_1, u_2, ..., and u_n, with each possible value (say u_i) associated with a possibility degree (say $\pi_X(u_i)$).

3   Probabilistic Data with Fuzziness

To deal with uncertain data, two major foundations have been developed, which are *probability theory* for objective uncertainty and *fuzzy* set theory for subjective uncertainty, respectively. Subjective uncertainty basically comes from subjective estimation and judgment made by human, for example, "*hot*" for temperature. Objective uncertainty mainly comes from objective statistics instead of subjective estimation and judgment, for example, possible orientation of landing on for a coin being tossed.

3.1   Subjective Probability Distributions

To deal with uncertain data, two major foundations have been developed, which are probability theory for objective uncertainty and fuzzy set theory for subjective uncertainty, respectively. Subjective uncertainty basically comes from subjective estimation and judgment made by human, for example, "hot" for temperature. Objective uncertainty mainly comes from objective statistics instead of subjective estimation and judgment, for example, possible orientation of landing on for a coin being tossed.

Traditionally. a stochastic event is associated with a probability measure and this probability measure is a crisp value in $[0, 1]$. Formally let e be a stochastic event and its probability measure be $p_s(e)$. Then $p_s(e)$ is a crisp value of $0 \leq p_s(e) \leq 1$. Here the crisp probability measure is generally obtained from statistical tests with sufficient volume of data. But it is possible that there is not sufficient volume of data enabling the performance statistical tests [19]. At this point, the probability measure is obtained by subjective estimation instead of statistical tests. As a result, the probability measure associated with a stochastic event is a subjective probability measure. Let us look at an example. In a real-world application of weather forecast, it is common to say "*the probability it will be mainly sunny tomorrow is 90%*". But "*the probability it will be mainly sunny tomorrow is very high*" may be easily understood by human beings. In this example, the probability measure of the stochastic event is a fuzzy linguistic term "*very high*" instead of a crisp value "*90%*".

Subjective probability measure is estimated by person who determine the probability of stochastic events occurrence according to her or his experience. Here subjective probability measure can be interpreted as the possibility of stochastic events occurrence, and is generally modeled by a linguistic term, which is represented by a fuzzy set or possibility distribution. Formally let e be a stochastic event and its subjective probability measure be $p_s(e)$. Then $p_s(e)$ is a fuzzy set with form as follows.

$$p_s(e) = \{(f_1, \mu_e(f_1)), (f_2, \mu_e(f_2)), ..., (f_m, \mu_e(f_m))\} \tag{6}$$

Here $0 \leq f_j \leq 1$ and $0 \leq \mu_e(f_j) \leq 1 (1 \leq j \leq m)$.

A stochastic phenomenon S with stochastic evens $e_1, e_2,..., e_k$ can be formally described by a subjective probability distribution P as follows.

$$p = \{(e_1, \{(f_{11}, \mu_{e1}(f_{11}), ...(f_{1m}, \mu_{e1}(f_{1m}))\}), ..., \\ (e_k, \{(f_{k1}, \mu_{ek}(f_{k1})), ..., (f_{kn}, \mu_{ek}(f_{kn}))\})\} \tag{7}$$

We know that there is a constraint on a traditional probability distribution: the sum of probability measures of all stochastic evens of a stochastic phenomenon is less than or equal to one. For a traditional probability distribution $\{(e_1, p_s(e_1)), (e_2, p_s(e_2)), ..., (e_k, p_s(e_k))\}$, we have $0 \leq \sum_{i=1,2,...,k} p_s(e_i) \leq 1$. Subjective probability distribution should satisfy such a constraint also. Formally, for a subjective probability distribution $\{(e_1, \{(f_{11}, \mu_{e1}(f_{11}), ...(f_{1m}, \mu_{e1}(f_{1m}))\}), ..., (e_k, \{(f_{k1}, \mu_{ek}(f_{k1})), ..., (f_{kn}, \mu_{ek}(f_{kn}))\})\}$, we have

$$0 \leq \max(\mathrm{supp}(p_s(e_1))), + \max(\mathrm{supp}(p_s(e_2))) + ... + \\ \max(\mathrm{supp}(p_s(e_k))) \leq 1 \tag{8}$$

Here supp $(p_s(e_i))$ means the support of $p_s(e_i)$ (i.e., fuzzy set $\{(f_{i1}, \mu_{ei}(f_{i1}), ...(f_{i2}, \mu_{ei}(f_{i2}))\}), ..., \{(f_{i1}, \mu_{ei}(f_{i1}))\})$, and max (supp $(p_s(e_i))$) means a value v_i with maximum $\mu_{ei}(v_i))$ in this support.

3.2   Subjective Status Descriptions of Stochastic Events

In a classical stochastic event associated with a probability measure, the probability measure is definitely known and represented by a crisp value in [0, 1], and the status of the stochastic event is definitely known and precisely described. But individuals in the real-world applications cannot be always known completely and their descriptions with some property values may be fuzzy. As a result, the stochastic event is one with subjective status descriptions.

Subjective status descriptions of stochastic event are presented by person who partially know the property values of stochastic event according to her or his knowledge. It should be noted that the probability measure associated with the stochastic event is crisp. Subjective status descriptions of stochastic event are generally modeled by linguistic terms, which are represented by fuzzy sets or possibility distributions.

A stochastic phenomenon S with stochastic evens e_1, e_2,..., e_k containing subjective status descriptions can be formally described by the probability distribution $P = \{(e_1, p_s(e_1)), (e_k, p_s(e_k))\}$. For each $e \in \{e_1, e_2,..., e_k\}$, let e have properties $A_1(e)$, $A_2(e)$, , $A_n(e)$. Then $A_i(e)$ ($1 \le i \le$ n) may be a fuzzy set with the following form.

$$A_i(e) = \{(v_1, \mu_e(v_1)), (v_2, \mu_e(v_2)), ..., (v_1, \mu_e(v_1))\} \tag{9}$$

Here $0 \le \mu_e(v_j) \le 1$ ($1 \le$ j $\le$ l) and $0 \le \sum_{i=1,2,...,k} p_s(e_i) \le 1$.

4   Modeling Probabilistic Data with Fuzziness in Databases

To represent and manipulate uncertain data in databases, two major foundations, which are probability theory and fuzzy sets, have been developed and applied to extend various database models [17]. This has resulted in numerous contributions, mainly with respect to the popular relational database (RDB) model. It should be noted that the fuzzy relational database model [15,23] or probabilistic relational database model [7,28,29] does not satisfy the requirement of modeling complex objects with inherent uncertainty. The object-oriented database (OODB) model can represent complex object structures without fragmenting the aggregate data and can also depict complex relationships among attributes. To deal with the problem of modeling complex structures and relationships with uncertainty, some recent efforts have concentrated on the fuzzy object-oriented databases [20] and probabilistic object-oriented databases [13].

Compared with many studies in the fuzzy databases and probabilistic databases, few efforts have been carried out to combine fuzzy sets/possibility theory and probability theory together in the context of databases. Imprecise probability measures are first introduced into the probabilistic relational databases in [14], in which a probability measure associated with a tuple is represented by an interval instead of a crisp value in (0, 1). To represent imprecise probability measures of tuples, i.e., interval probability measures, two additional attributes named *LB* and *UB* are introduced into the probabilistic relational scheme. Then we have the following probabilistic relational scheme.

$$R(A_1, A_2, ..., A_n, LB, UB) \tag{10}$$

Here $A_1, A_2, ..., A_n$ are the common attributes of R, and *LB* and *UB* are applied to represent the lower boundary and upper boundary of probability measures of tuples, respectively. Note that, being different from the common attributes in the probabilistic relational scheme, the domains of LB and UB are all $[0, 1]$. Interval probability measures in [11] are also applied to extend the object-oriented databases in [8].

To combine fuzzy sets and probability theory together in databases, a deductive probabilistic and fuzzy object-oriented database language is proposed in [6]. In this database model, a class property can contain fuzzy set values, and uncertain class membership and property applicability are measured by lower and upper bounds on probability. Similarly, in order to cope with fuzziness and probability in databases, an object base model is proposed in [5], which incorporates both fuzzy set values and probability degrees represented by interval probabilities. It is shown that, in [5,6], fuzzy sets are applied to represent the imprecise attribute values of objects (i.e., stochastic evens) and meanwhile the probability measures of objects are represented by interval values. Formally let o be an object with attributes

$$\{A_1, A_2, ..., A_k\} \tag{11}$$

and its probability measure be $p(o)(0 \leq p(o) \leq 1)$. Then o $[A_i]$ $(1 \leq i \leq k)$, which means the value of o on attribute A_i, may be a fuzzy value, and $LB \leq p(o) \leq UB$.

A very different approach to combining fuzzy sets and probability theory together in databases is investigated in [21,22], in which fuzzy sets are used to represent probability measures rather than attribute values. In [21], a probabilistic relational database model with fuzzy probability measures is proposed. The fuzzy probabilistic relational schema in [21] is defined as follows.

$$R(A_1, A_2, ..., A_n, A_{n+1}) \tag{12}$$

Here $A_1, A_2, ..., A_n$ are the common attributes of R, and $A_{n+1} \in R$ is the fuzzy probabilistic attribute of R. Let D_i $(1 \leq i \leq n+1)$ be the domain of attribute A_i. Then a fuzzy probabilistic relations instance r over R is a subset of the Cartesian product of $D_1 \times D_2 \times ... \times D_n \times D_n + 1$. Let $t \in r$ be a tuple of r. Then t $[A_i]$ $(1 \leq i \leq n)$ is a crisp value instead of fuzzy values in [5,6], and t $[A_{n+1}]$ is a crisp value in $[0, 1]$ or a fuzzy value in the universe of discourse of $[0, 1]$. Finally, let $t_1, t_2, ..., t_k$ be the tuples with the same primary key values in a fuzzy probabilistic relation and let t_i $[A_{n+1}] = \{\pi_{ti} (p_{i1})/p_{i1}, \pi_{ti}(p_{i2})/p_{i2}, ..., \pi_{ti}(p_{im})/p_{im}$ $(1 \leq i \leq n)$. Then we have the following.

$$\max(\text{supp}(t_1[A_{n+1}])) + \max(\text{supp}(t_2[A_{n+1}])) + ... + \max(\text{supp}(t_k[A_{n+1}])) \leq 1 \tag{13}$$

Also, in [21], four coalescence operations are defined to remove redundancy tuples, which are the coalescence-Plus operation ($\oplus$), the coalescence-Minus operation ($\ominus$), the coalescence-Times operation ($\otimes$) and the coalescence-Max operation (©). With these four coalescence operations, five primitive operations of the fuzzy probabilistic relational databases are defined, including union, *difference*, *selection*, *projection*, and *Cartesian product*. In addition, the natural join operation for the fuzzy probabilistic relations is defined because it is useful for the retrieval of relational data.

Following the similar idea in [21], a probabilistic object-oriented database model with fuzzy probability measures is proposed in [22]. Classes are the core of object-oriented databases. For the probabilistic object-oriented databases with fuzzy probability measures, let c be a probabilistic class with fuzzy probability measures, which contains attributes $\{a_1, a_2, ..., a_k, \Lambda\}$, in which $a_1, a_2, ..., a_k$ are common attributes of c, and Λ is the fuzzy probabilistic attribute of c. That is, $\Lambda(c) = \{a_1, a_2, , a_k, \Lambda\}$ and $\Lambda(c)$ is the fuzzy probabilistic attribute of c. Formally, the definition of a probabilistic class with fuzzy probability measure is presented as follows.

> CLASS *class-name* WITH PROBABILITY *degree*
>> INHERITS *superclass*$_1$ WITH PROBABILITY *degree*$_1$
>> ...
>> INHERITS *superclass*$_k$ WITH PROBABILITY *degree*$_k$
>> ATTRIBUTES
>>> *Attribute*$_1$: DOMAIN *dom*$_1$: TYPE OF *type*$_1$
>>> ...
>>> *Attribute*$_m$: DOMAIN *dom*$_m$: TYPE OF *type*$_m$
>>> *Fuzzy Probabilistic Attribute*: FUZZY DOMAIN: TYPE OF *real*
>> METHODS
>>> ...
> END

Let c be a probabilistic class with fuzzy probability measure, which contains attributes $\{a_1, a_2, ..., a_k, \Lambda\}$, and let o be an object on attribute set $\{a_1, a_2, ..., a_k, \Lambda\}$. Then o belongs to c with a probability. And let c_1 and c_2 be two probabilistic classes. Then c_1 is a subclass of c_2 with a probability. In addition, four coalescence operations in [21] are also applied in [22] in order to remove redundancy objects. With these four coalescence operations, two kinds of algebraic operations for the probabilistic classes with fuzzy probability measures are identified in [22]: one is for single class and another is for multiple classes. The latter contains *probabilistic product* ($\times$), *probabilistic join* ($\bowtie$), *probabilistic union* ($\cup$), *probabilistic difference* ($-$) and *probabilistic Intersection* ($\cap$) and the former contains *probabilistic selection* (σ) and the *probabilistic projection* (Π).

Table 1 presents a summary of major database models for probabilistic data with fuzziness.

Table 1. Major database models for probabilistic data with fuzziness

	Database models	Attribute values	Probability measures
[14]	RDB$_s$	*crisp values*	*interval values*
[8]	OODB$_s$	*crisp values*	*interval values*
[6]	*deductive* OODB$_s$	*fuzzy values*	*interval values*
[5]	OODB$_s$	*fuzzy values*	*interval values*
[21]	RDB$_s$	*crisp values*	*fuzzy values*
[22]	OODB$_s$	*crisp values*	*fuzzy values*

5   Summary and Discussion

Uncertain data extensively exists in many real-world applications and uncertain data modeling has been investigated in various database models. On the other hand, in practice, uncertain data generally have devise semantics. Typically, we have objective uncertainty based on probability theory and subjective uncertainty based on fuzzy sets/possibility theory. Two types of uncertainty can occur together and we face with modeling hybrid uncertain data with objective uncertainty and subjective uncertainty in database models. Actually, it is required in modeling real-world problems and constructing intelligent systems to integrate different methodologies and techniques, which has been the quest and focus of significant interdisciplinary research efforts. The advantages of such a hybrid system are that the strengths of its components are combined and the weaknesses of its components are complementary one to another. This paper identifies the semantics and expressive forms of hybrid uncertain data with both objective uncertainty and subjective uncertainty. An up-to-date overview of the current state of the art in hybrid uncertain data modeling in relational databases and object-oriented databases is presented.

Uncertain data in the real-world applications may have diverse semantics, and they need to be represented in database models for their management. Basically, the future work in hybrid uncertain data modeling and management will mainly focus on two aspects. First, the real-world applications are complex, and hybrid uncertain data with more complex structure and semantics should be considered and investigated. Typically, the characteristic values and probability measures of stochastic evens may be fuzzily unknown simultaneously. Second, diverse data models are expensively applied in various applications (e.g., relational databases, object-oriented databases, object-relational databases, XML, RDF and so on), and it is needed to extend and interoperate different data models for hybrid uncertain data management.

Acknowledgment. The work was supported in part by the National Natural Science Foundation of China (61772269 and 61370075).

References

1. Baldwin, J.M., Lawry, J., Martin, T.P.: A note on probability/possibility consistency for fuzzy events. In: Proceedings of the 6th International Conference on Information Processing and Management of Uncertainty in Knowledge-Based Systems, Granada, Spain, pp. 521–525, July 1996
2. Baudrit, C., Dubois, D., Guyonet, D.: Joint propagation and exploitation of probabilistic and possibilistic information in risk assessment. IEEE Trans. Fuzzy Syst. **14**, 593–607 (2006)
3. Bosc, P., Prade, H.: An introduction to fuzzy set and possibility theory based approaches to the treatment of uncertainty and imprecision in database management systems. In: Proceedings of the Second Workshop on Uncertainty Management in Information Systems: From Needs to Solutions (1993)
4. Buckley, J.J.: Fuzzy Probabilities: New Approach and Applications. Springer, Heidelberg (2005)
5. Cao, T.H., Nguyen, H.: Uncertain and fuzzy object bases: a data model and algebraic operations. Int. J. Uncertainty Fuzziness Knowl.-Based Syst. **19**(2), 275–305 (2011)
6. Cao, T.H., Rossiter, J.M.: A deductive probabilistic and fuzzy object-oriented database language. Fuzzy Sets Syst. **140**(1), 129–150 (2003)
7. Dey, D., Sarkar, S.A.: Probabilistic relational model and algebra. ACM Trans. Database Syst. **21**(3), 339–369 (1996)
8. Eiter, T., Lu, J.J., Lukasiewicz, T., Subrahmanian, V.S.: Probabilistic object bases. ACM Trans. Database Syst. **26**(3), 264–312 (2001)
9. Ferson, S.: What Monte Carlo method cannot do. Hum. Ecol. Risk Assess. **2**, 990–1007 (1996)
10. Gupta, C.P.: A note on transformation of possibilistic information into probabilistic information for investment decisions. Fuzzy Sets Syst. **56**, 175–182 (1993)
11. Guyonnet, D., Bourgine, B., Dubois, D., Fargier, H., Cme, B., Chils, P.J.: Hybrid approach for addressing uncertainty in risk assessment. J. Enviro. Eng. **126**, 68–76 (2003)
12. Haas, P.J., Suciu, D.: Special issue on uncertain and probabilistic databases. VLDB J. **18**(5), 987–988 (2009)
13. Kornatzky, Y., Shimony, S.E.: A probabilistic object-oriented data model. Data Knowl. Eng. **12**(2), 143–166 (1994)
14. Lakshmanan, L.V.S., Leone, N., Ross, R., Subrahmanian, V.S.: ProbView: a flexible probabilistic database system. ACM Trans. Database Syst. **22**(3), 419–469 (1997)
15. Ma, Z.M., Yan, L.: A literature overview of fuzzy database models. J. Inf. Sci. Eng. **24**(1), 189–202 (2008)
16. Ma, Z.M., Zhang, F., Yan, L.: Fuzzy information modeling in UML class diagram and relational database models. Appl. Soft Comput. **11**(6), 4236–4245 (2011)
17. Parsons, S.: Current approaches to handling imperfect information in data and knowledge bases. IEEE Trans. Knowl. Data Eng. **8**(2), 353–372 (1996)
18. Ralescu, A.: Fuzzy probabilities and their applications to statistical inference. In: Proceedings of the 5th International Conference on Processing and Management of Uncertainty in Knowledge-Based Systems, pp. 217–222 (1994)
19. Rebiasz, B.: New methods of probabilistic and possibilistic interactive data processing. J. Intell. Fuzzy Syst. **30**(5), 2639–2656 (2016)

20. Yan, L., Ma, Z.M.: Algebraic operations in fuzzy object-oriented databases. Inf. Syst. Front. **16**(4), 543–556 (2014)
21. Yan, L., Ma, Z.M.: A fuzzy probabilistic relational database model and algebra. Int. J. Fuzzy Syst. **15**(2), 244–253 (2013)
22. Yan, L., Ma, Z.: A probabilistic object-oriented database model with fuzzy measures and its algebraic operations. J. Intell. Fuzzy Syst. **28**(5), 1969–1984 (2015)
23. Yang, Q., Zhang, W.N., Liu, C.W., Wu, J., Yu, C.T., Nakajima, H., Rishe, N.: Efficient processing of nested fuzzy SQL queries in a fuzzy database. IEEE Trans. Knowl. Data Eng. **13**(6), 884–901 (2001)
24. Zadeh, L.A.: Fuzzy sets. Inf. Control **8**, 338–353 (1965)
25. Zadeh, L.A.: Fuzzy sets as a basis for a theory of possibility. Fuzzy Sets Syst. **1**, 3–28 (1978)
26. Zadeh, L.A.: The concept of a linguistic variable and its application to approximate reasoning (Parts 1, 2, and 3), Inf. Sci. **8**, 119–248 & 301–357; **9**, 43–80 (1975)
27. Zadeh, L.A.: Fuzzy probabilities. Inf. Process. Manag. **20**(3), 363–372 (1984)
28. Zhu, H., Zhang, C.C., Cao, Z.S., Tang, R.M.: On efficient conditioning of probabilistic relational databases. Knowl.-Based Syst. **92**, 112–126 (2016)
29. Zimanyi, E.: Query evaluation in probabilistic relational databases. Theor. Comput. Sci. **171**(1–2), 179–219 (1997)

Fuzzy Classifier with Convolution for Classification of Handwritten Digits

Rui Yin$^{(\boxtimes)}$ and Wei Lu

Dalian University of Technology, Dalian, China
1024183360@qq.com, luwei@dlut.edu.cn

Abstract. Traditional fuzzy classifier is an important part of artificial intelligence. It achieves classification based on membership function and fuzzy rules which can deal with the uncertainty of data and has semantics. However, the definition of fuzzy rules requires prior knowledge. And fuzzy rules is too sample to achieve high accuracy of classification for classification of handwritten digits. The classifier proposed in this paper combines convolution with fuzzy classifier to classify handwritten digits. The classifier can be divided into two parts: convolution feature extraction part and Gauss membership calculation part. Using back propagation algorithm, the classifier parameters are trained by a large number of labeled data. It can independently extract useful features of handwritten digits to build handwriting feature prototypes, and establish membership functions according to feature prototypes. Experiments on MNIST datasets show that, compared with traditional fuzzy classifiers, the proposed fuzzy classifier can greatly improve the accuracy with less raised time complexity. For MNIST datasets, the proposed fuzzy classifier with convolution can reach higher classification accuracy.

1 Introduction

In the big-data era, a lot of data about people's studying, working and life are generated in the real world. And, image is an important representation of data. Image classification is an important issue, and a lot of research works have been done. For example, convolutional neural network [1,10] is applied to scene understanding, and AFS algorithm is applied to multi-ethnic face recognition [15]. However, in practical applications, a large number of images are uncertain and contain a lot of noise, which is a big challenge for classification and recognition.

Fuzzy classifier is an intelligent method based on membership degree. Its classification principle is to calculate the probability of samples belonging to each category by establishing membership function, which is used as classification criterion. Its own characteristics determine its advantages in dealing with uncertain data and can eliminate noise to a certain extent. Therefore, the fuzzy classifier is widely used in image processing [16]. However, traditional fuzzy classifier is difficult to extract features from original images for classification, and it needs to extract image features by prior knowledge [3,14], which increases the workload.

© Springer Nature Switzerland AG 2019
R. B. Kearfott et al. (Eds.): IFSA 2019/NAFIPS 2019, AISC 1000, pp. 738–745, 2019.
https://doi.org/10.1007/978-3-030-21920-8_65

At the same time, traditional fuzzy classifier is difficult to achieve high accuracy on some issues, such as classification of handwritten digits.

Therefore, in order to take advantage of fuzzy classifier in dealing with uncertain data and solve the above problems, researchers try to combine deep learning with fuzzy algorithm [4,9], such as FDNN algorithm [5]. At the same time, convolution neural network has advantages in image classification [11]. Image features can be automatically extract by convolution operation and be used in classification algorithm. Literature [13] combines convolution with Gauss process, which proves that convolution can improve the image understanding ability of Gauss process. Therefore, in order to make the fuzzy classifier better applied to classification of handwritten digits, a classifier which combines convolution with fuzzy classifier is proposed in this paper. convolution operation which is a preprocessing to extracting the interesting features, is used in fuzzy classifier. Parameters are optimized by back propagation algorithm.

This paper is organized as follows. Section 2 provides a brief introduction to the convolution and Gauss membership function. In Sect. 3, a fuzzy classifier based on convolution operation is proposed. In Sect. 4, experiment results obtained by the proposed classifier are compared with other traditional fuzzy classifier when using MNIST datasets are given. In Sect. 5, we offer some conclusions.

2   Preliminaries

Before introducing the classifier in this paper, we need to know some relevant knowledge about convolution and Gauss membership function. Convolution is widely used in image classification because of its function of extracting local features of images. At the same time, the fuzzy algorithm is also applied to image semantic understanding because of its interpretability.

2.1   Convolution

Convolution is a mathematical operator that generates the third function by two different functions, which represents the area of the overlapped part of the function after flipping and translation. Its definition on continuous functions is shown in Eq. 1.

$$y(x) = \int_{-\infty}^{\infty} f(\tau)g(x - \tau)d\tau \tag{1}$$

$f(x)$ and $g(x)$ is an integrable function on the domain of definition. Convolution is widely used in signal processing and other fields. It can be considered to act on the input signal function through the system function, so as to filter signals which is similar to the characteristics of the system function. For discrete signals, the convolution formula is shown in Eq. 2.

$$y(n) = \sum_{i=-\infty}^{\infty} x(i)h(n - i) = x(n) * h(n) \tag{2}$$

For image data such as MNIST dataset, the image of handwritten digits is a two-dimensional numerical matrix, which can be regarded as discrete input signal. Therefore, convolution operation as shown in Fig. 1 is defined in depth learning [17].

The convolution operation of handwritten digital image is to use the convolution kernel W to slide on the matrix X. The gray value of the pixel on the image point and the corresponding value on the convolution core are multiplied, and then all the results are added as the gray values of the pixels on the image of handwritten digits corresponding to the intermediate pixels of the convolution core. Finally W slides the entire image of handwritten digits. After convolution operation, we can extract the local features. Different features can be obtained through multiple convolution kernels in practical applications.

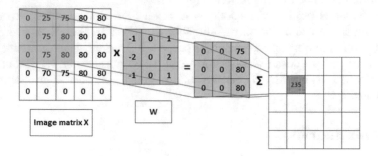

Fig. 1. Image convolution operation

2.2 Gauss Membership Function

The Gauss distribution shows that if a random variable x obeys a Gauss distribution with a mathematical expectation x_0 and a standard deviation σ_0 expressed as $x \sim N(x_0, \sigma_0^2)$,then its probability density function is shown in Eq. 3.

$$f(x) = ae^{-(x-x_0)^2/2\sigma_0^2} \tag{3}$$

The mathematical expectation x_0 determines the central position of its distribution. The parameter a determines the amplitude of the function. Gauss function is a commonly used membership function in the fuzzy classifier [7,12]. It can find out that the closer x is to x_0, the larger the value is. Therefore, in the fuzzy classifier, x_0 is often used as the prototype of a certain category. For a new sample, the greater the corresponding value, the closer it is to the prototype of the category. And the greater the possibility of belonging to the category. This indicates that the more likely the sample belong to this category.

3   The Proposed Classifier

The proposed classifier combines convolution operation with fuzzy classifier to realize MNIST classification. Among them, convolution operation part can be regarded as the preprocessing of dataset, including convolution operation, pooling operation and weighted summation of feature points at the same location, which can extract the features of the input original image of handwritten digits and reduce the dimension of the input data. The Gauss membership classification part calculates the membership degree of each category of the input features and makes decisions. In the process of building the model, the convolution kernel, the mean and variance of the Gauss membership function are optimized by the back propagation algorithm. The system diagram of this algorithm is shown in Fig. 2.

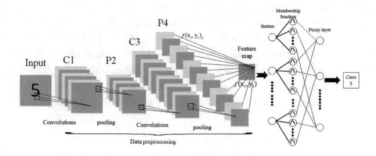

Fig. 2. Fuzzy classifier with convolution

3.1   Data Preprocessing

As shown in Fig. 2, data preprocessing includes the convolution layers $C1$ and $C3$, pooling layers $P2$ and $P4$, and a weighted accumulation layer of features. The convolution operation is used to extract the local features of the handwritten digital images, and the pooling operation can reduce the dimension of features while retaining useful information. Convolution and pooling operations preserve image location information while extracting features. In different result graphs, the pixel values of the same location represent the features extracted from the same location of the original handwritten digits through different convolution cores. Therefore, the most interesting feature map can be obtained by adding the weighted features of the same location in $P4$ layer.

For example, if the input sample is a gray image, the input date is a two-dimensional matrix which size is $M*N$. After randomly initializing k convolution cores, k feature images $C1$ can be obtained after one convolution operation. And the size of $C1$ is $M*N$. Then $P2$ can be obtained after the first pooling operation. The size of $P2$ is $(M/2)*(N/2)$. So, after two layers of convolution and pooling operations, the final multi-feature images $P4$ are obtained. Then, in order to get the most interesting feature map, all the features in $P4$ are calculated by Eq. 4.

The $f(x_i, y_i)_j$ shows the value of the (x_i, y_i) position in the j-th feature image, w_{ij} is the weight which is random initialization, and $f(x_i, y_i)$ is the combination characteristics value in feature map.

$$f(x_i, y_i) = \sum_j w_j f(x_i, y_i)_j \qquad (4)$$

3.2   Membership Calculation and Classification

For the fuzzy classifier, it is necessary to define the membership function of each category. For any input sample, the membership degree of each category is calculated. The largest membership degree is its category. As shown in Fig. 2, a two-dimensional feature map is obtained after data preprocessing, and it can be flattened into a vector $f_k (k \in [0, n])$, $f_k = f(x_i, y_i)$ when $k = i + j$. For a C classification problem, membership functions of C classes are randomly initialized. And membership degrees corresponding to n features of each class are calculated. Then the membership degree of each feature corresponding to the same class is summed as the feature vector's membership degree of each class. The calculation formula is shown in Eq. 5. After obtaining the membership degree, the classes corresponding to the maximum membership degree are taken as the classes to which the input belongs. The calculation formula is shown in Eq. 6.

$$pro_J = \sum_I e^{-(f_I - \mu_{IJ})^2 / 2\sigma_{IJ}^2} \qquad (5)$$

$$J = argmax(\sum_I e^{-(f_I - \mu_{IJ})^2 / 2\sigma_{IJ}^2}) \qquad (6)$$

f_I is the i-th feature value, I is belong to $[1, n]$, μ_{IJ} is the i-th feature value of the j-th class's feature protopyte, J is belong to $[1, C]$. σ_{IJ} is the standard deviation. By the Eq. 6, we can make a decision that the input data belongs to class J.

3.3   Training

The parameters in the classifier are all randomly initialized. In order to improve the accuracy of classification, a BP algorithm [6] Adam [8] is used. Loss function L is defined when the membership degree pro_J of each category is calculated.

$$L = -\sum_J ylab_J \times \log(pro_J) \qquad (7)$$

$ylab_J$ is the sample label corresponding to class J. The smaller the loss function L is, the closer the parameters are to the optimum. ε is a minimum value. When Eq. 8 is satisfied or the maximum number of iterations is reached, the parameters are considered to be optimal.

$$min(L) \leq \varepsilon \qquad (8)$$

4   Experiment

The proposed fuzzy classifier with convolution is test on MNIST data set which is a database of handwritten digits for 0 to 9 writing by U.S. Census Bureau employees and high school students. It contains a training set of 60,000 examples, a test set of 10000 examples and their labels. The digits have been size-normalized and centered in a fixed-size image. Each example is a $28 * 28$ pixel image which component of the vector is a value between 0 and 255 describing the intensity of the pixel. Using the algorithm introduced in Sect. 3, the maximum number of iterations is 70000. In each iteration, 100 samples are randomly selected from the training data set as input data. The initial learning rate is 0.0003. And the initial feature prototypes, convolution kernels and weights are all generated randomly. The training accuracy is shown in Fig. 3. Ten prototypes of handwritten digital features generated by training are shown in Fig. 4.

After the model was established, it is tested for 100 times. Each group randomly selects 100 samples from the test picture set as input data. The average accuracy of 100 rounds of testing is shown in the table, and the average consumption time is 45.5 ms. In order to illustrate the superiority of the proposed classifier compared with the traditional fuzzy classifier, this paper compares the proposed classifier with FCM classifier and K-means [2] classifier. For these two traditional fuzzy classifier, a sample of each category is randomly selected from the training data set as the initial clustering center. Then the clustering centers are optimized continuously until each of them is not changed. After that the clustering center is used to classify the test samples, and the accuracy of the comparison experiment is shown in the Table 1.

Table 1. The classification performance comparison on the MNIST database

Method	Classifier of this paper	FCM	K-means
Class accuracy	0.97	0.52	0.61
Time (ms)	45.5	1.96	0.91

Firstly, the training results are analyzed. From Fig. 3, we can see that the classification accuracy of the proposed classifier can quickly reach more than 0.6. After 70,000 training iterations, the accuracy can converge to a higher level, and the accuracy fluctuation range is between 0.93–1. As shown in Fig. 4, it is ten feature prototypes for all categories of handwritten digits established by training. (a)–(j) are the feature prototypes of the class of handwritten digits 0–9. It can be seen that the feature prototype of each category generated by training has its own characteristics. In the feature prototype, different shades of color represent different eigenvalues. So, Each feature prototype has a different color depth at the same location, which reflects the difference of each category. It can be used to make a classification. At the same time, we can see from the contrast experiment that the accuracy of this classifier is obviously improved, and the time loss increases less. On the other hand, there is no need for a

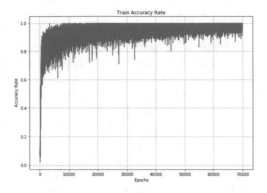

Fig. 3. Train accuracy

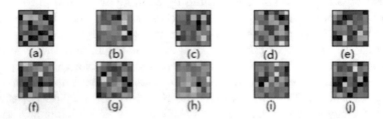

Fig. 4. Feature prototypes of ten handwritten digits

priori knowledge when initial the classification center for data sets with high dimension characteristics. Compared with the traditional fuzzy classifier, the proposed classifier has more advantages in classification of handwritten digits.

5   Conclusion

This classifier combines convolution operation with fuzzy classifier to realize classification of handwritten digits. The convolution operation is used to extract the local feature in the image of handwritten digits, and the convolution operation does not change the location information of the features. So the combination of the most grateful features corresponding to each part of the original image is obtained by weighted summation, which effectively reduces the input feature dimensions of the classification classifier. By introducing the BP algorithm of deep learning, the supervised learning method can be used to optimize the parameters, so that the proposed fuzzy classifier with convolution can achieve the desired accuracy. The classifier proposed in this paper automatically extracts the useful features of each category through training, and establishes the feature prototypes. It does not need to set up the feature set through prior knowledge. On the one hand, it reduces the workload, on the other hand, it reduces the interference of human factors. At the same time, the experimental results on the MNIST dataset show that compared with the traditional fuzzy classifier, the classification accuracy of the classifier has been greatly improved. For enhancing the understanding ability of the image data of the fuzzy classifier, it has certain research significance.

References

1. Alex, K., Ilya, S., Geoffrey, E.H.: Imagenet classification with deep convolutional neural networks. Adv. Neural Inf. Process. Syst. **25**, 1097–1105 (2012)
2. Arthur, D., Vassilvitskii, S.: k-means++: The advantages of careful seeding. In: Proceedings of the Eighteenth Annual ACM-SIAM Symposium on Discrete Algorithms, pp. 1027–1035. Society for Industrial and Applied Mathematics (2007)
3. Chuang, K.S., Tzeng, H.L., Chen, S., Wu, J., Chen, T.J.: Fuzzy c-means clustering with spatial information for image segmentation. Comput. Med. Imaging Graph. **30**(1), 9–15 (2006)
4. Duan, X., Wang, Y., Pedrycz, W., Liu, X., Wang, C., Li, Z.: AFSNN: a classification algorithm using axiomatic fuzzy sets and neural networks. IEEE Trans. Fuzzy Syst. **26**(5), 3151–3163 (2018)
5. Deng, Y., Ren, Z., Kong, Y., Bao, F., Dai, Q.: A hierarchical fused fuzzy deep neural network for data classification. IEEE Trans. Fuzzy Syst. **25**(4), 1006–1012 (2017)
6. Fan, H.W., Zhang, G.Y., Ding, A.L., Xie, C.R., Xu, T.: Improved BP algorithm and its application in detection of pavement crack. J. Chang'an Univ. **30**(1), 438–457 (2010)
7. Hameed, I.A.: Using gaussian membership functions for improving the reliability and robustness of students' evaluation systems. Expert Syst. Appl. **38**(6), 7135–7142 (2011)
8. Kingma, D.P., Ba, J.: Adam: a method for stochastic optimization. arXiv preprint arXiv:1412.6980 (2014)
9. Kulkarni, A.D., Lulla, K.: Fuzzy neural network models for supervised classification: multispectral image analysis. Geocarto Int. **14**(4), 42–51 (1999)
10. Lawrence, S., Giles, C.L., Tsoi, A.C., Back, A.D.: Face recognition: a convolutional neural-network approach. IEEE Trans. Neural Netw. **8**(1), 98–113 (2002)
11. Li, H., Lin, Z., Shen, X., Brandt, J., Hua, G.: A convolutional neural network cascade for face detection. In: Proceedings of the IEEE Conference on Computer Vision and Pattern Recognition, pp. 5325–5334 (2015)
12. Tay, K.M., Lim, C.P.: Optimization of Gaussian fuzzy membership functions and evaluation of the monotonicity property of fuzzy inference systems. In: IEEE International Conference on Fuzzy Systems, pp. 1219–1224 (2011)
13. Van der Wilk, M., Rasmussen, C.E., Hensman, J.: Convolutional Gaussian processes. In: Advances in Neural Information Processing Systems, pp. 2849–2858 (2017)
14. Winkler, R., Klawonn, F., Kruse, R.: Fuzzy c-means in high dimensional spaces. Int. J. Fuzzy Syst. Appl. (IJFSA) **1**(1), 1–16 (2011)
15. Xiaodong, D., Zedong, L., Cunrui, W., Back, A.D.: Research on multi-ethnic face semantic description and mining method based on AFS. Chin. J. Comput. **39**, 1435–1449 (2016)
16. Yongchuan, T., Yunsong, X.: Learning disjunctive concepts based on fuzzy semantic cell models through principles of justifiable granularity and maximum fuzzy entropy. Knowl.-Based Syst. **161**, 268–293 (2018)
17. Zeiler, M.D., Fergus, R.: Visualizing and understanding convolutional networks. In: European Conference on Computer Vision, pp. 818–833 (2014)

Logarithms Are Not Infinity: A Rational Physics-Related Explanation of the Mysterious Statement by Lev Landau

Francisco Zapata, Olga Kosheleva, and Vladik Kreinovich$^{(\boxtimes)}$

University of Texas at El Paso, El Paso, TX 79968, USA
fazg74@gmail.com, {olgak,vladik}@utep.edu

Abstract. Nobel-prize winning physicist Lev Landau liked to emphasize that logarithms are not infinity – meaning that from the physical
viewpoint, logarithms of infinite values are not really infinite. Of course,
from a literally mathematical viewpoint, this statement does not make
sense: one can easily prove that logarithm of infinity is infinite. However,
when a Nobel-prizing physicist makes a statement, you do not want to
dismiss it, you want to interpret it. In this paper, we propose a possible
physical explanation of this statement. Namely, in physics, nothing is
really infinite: according to modern physics, even the Universe is finite
in size. From this viewpoint, infinity simply means a very large value.
And here lies our explanation: while, e.g., the square of a very large value
is still very large, the logarithm of a very large value can be very reasonable – and for very large values from physics, logarithms are indeed very
reasonable.

1 Formulation of the Problem

Physicists use intuition. Physicists have been very successful in predicting
physical phenomena. Many fundamental physical phenomena can be predicted
with very high accuracy. The question is: how do physicists come up with the
corresponding models?

In this, physicists often use their intuition. This intuition is, however, difficult
to learn, because it is not formulated in precise terms – it is imprecise, it is
intuition, after all.

Can we formalize physicists' intuition – at least some of it? It would be
great to be able to emulate at least some of this intuition in a computer-based
systems, so that the same successful line of reasoning can be used to solve many
other problems.

Computers, however, only understand precise terms. So, to be able to emulate
physicists' intuition on a computer, we need describe it – or at least some aspects
of it – in precise terms.

**An example of physicists' intuition: Landau's statement about
logarithms.** Nobel-prize physicist Lev Landau often said that "logarithms are

© Springer Nature Switzerland AG 2019
R. B. Kearfott et al. (Eds.): IFSA 2019/NAFIPS 2019, AISC 1000, pp. 746–751, 2019.
https://doi.org/10.1007/978-3-030-21920-8_66

not infinity" – meaning that, in some sense, the logarithm of an infinite value is not really infinite; see, e.g., [4], p. 472; [10], p. 84; [12], p. 30.

Of course, this statement cannot be taken literally. From the purely mathematical viewpoint, this statement by Landau makes no sense: of course, the limit of $\ln(x)$ when x tends to infinity is infinite.

It is advisable to take this statement into account. This was a statement actively used by a Nobel-prize winning physicist, so we cannot just ignore it as a mathematically ignorant nonsense.

Formulation of the problem. But how can we make sense of this Landau's statement?

What we do in this paper. In this paper, we show how Landau's statement can be consistently formalized.

2   Why Infinities Are Important in Physics

Why are infinities important in the first place? At first glance, one may wonder why physicists are worried about infinities in the first place. In physics, everything is finite, infinities are mathematical abstractions, what is the big deal?

Alas, everything should be finite in physics, but infinities naturally appear. Yes, in physics, everything should be finite, but unfortunately, infinities creep in. Let us give a simple example of such a situation – the attempts to compute the overall mass m of an electron.

According to special relativity theory (see, e.g., [3,11]), this mass can be obtained by dividing the total energy E of the electron by the square of the speed of light c: $m = E/c^2$. This energy, in its turn, is equal to the sum of the rest energy $E_0 = m_0 \cdot c^2$ and the overall energy E_{el} of the electron's electric field.

According to the same relativity theory, the speed of all communications is limited by the speed of light. As a result, any elementary particle must be point-wise: otherwise, we would have different parts which – due to speed-of-light bound – would not be perfectly correlated and would, thus, constitute different sub-particles. The electric field E of a point-wise particle is well-known: it is determined by the usual Coulomb formula

$$E(x) = c_1 \cdot \frac{q}{r^2},$$

where c_1 is a constant, q is the electron's electric charge, and r is the distance from a given point x to the electron's location.

It is known that the field's energy density $\rho(x)$ is proportional to the square of the field: $\rho(x) = c_2 \cdot (E(x))^2$, i.e.,

$$\rho(x) = c_3 \cdot \frac{1}{r^4},$$

where $c_3 \overset{\text{def}}{=} c_2 \cdot (c_1 \cdot q)^2$. Thus, the overall energy of the electric field can be found if we integrate this density over the whole space:

$$E_{\text{el}} = \int \rho(x) \, dx = c_3 \cdot \int \frac{1}{r^4} \, dx.$$

Since the density function depends only on the distance r – i.e., is spherically symmetric – we can use the usual formulas of integrating spherically symmetric functions. Namely:

- First, for each radius r, we integrate over the sphere of this radius – whose area is $4\pi \cdot r^2$. On this sphere, the function is constant, so we simply multiply the expression by $4\pi \cdot r^2$.
- Then, we integrate the result over all possible values r.

In our case, the result is

$$E_{\text{el}} = c_3 \cdot \int_0^\infty \frac{4\pi \cdot r^2}{r^4} \, dr = c_4 \cdot \int \frac{1}{r^2} \, dr,$$

where $c_4 \overset{\text{def}}{=} c_3 \cdot 4\pi$. This integral is well know, so we get

$$E_{\text{el}} = -c_4 \cdot \left. \frac{1}{r} \right|_0^\infty.$$

For $r = \infty$, the expression $1/r$ is 0, but at the limit $r = 0$, we get a physically meaningless infinity!

This infinity problem is ubiquitous. The problem is not just in the specific formulas for the Coulomb law, the problem is much deeper: it can be traced to the fact that electromagnetic interactions – and many other physical interactions, e.g., gravitational ones – are *scale-invariant* in the sense that they have no physically preferable unit of length.

If we change from the original unit of length to a new one which is λ times smaller, then all numerical values of distance r will get multiplied by λ, so that the new values get the form $r' = \lambda \cdot r$. Scale-invariance means that all the physical equations - e.g., the equation that describes how the field energy density ρ depends on the distance r – remain the same after this change – provided, of course, that we appropriately change the unit for measuring energy density, to $\rho \to \rho' = c(\lambda) \cdot \rho$.

So, if in the original units, we have $\rho(r) = f(r)$ for some function f, then in the new units, we will have $\rho'(r') = f(r')$ for the exact same function $f(r)$. Here, $\rho' = c(\lambda) \cdot \rho$ and $r' = \lambda \cdot r$, so we conclude that $c(\lambda) \cdot \rho(r) = f(\lambda \cdot r)$. Since $\rho(r) = f(r)$, we thus conclude that

$$c(\lambda) \cdot f(r) = f(\lambda \cdot r).$$

It is known (see, e.g., [1]) that every measurable solution of this equation has the form $f(r) = c \cdot r^\alpha$ for some c and α. Thus, $\rho(r) = c \cdot r^\alpha$ and therefore, the overall energy of the corresponding field is equal to

$$\int \rho(x)\, dx = \int c \cdot r^\alpha \, dx = \int_0^\infty c \cdot r^\alpha \cdot 4\pi \cdot r^2 \, dr = c' \cdot \int_0^\infty r^{2+\alpha} \, dr,$$

where we denoted $c' \stackrel{\text{def}}{=} 4\pi \cdot c$.

When $\alpha \neq -3$, this integral is proportional to $r^{3+\alpha}|_0^\infty$:

- When $\alpha < -3$, this value is 0 at infinity, but infinite at $r = 0$.
- When $\alpha > -3$, this value is 0 for $r = 0$, but infinite for $r = \infty$.

In both cases, we get infinite energy.

When $\alpha = -3$, the integral is proportional to $\ln(x)|_0^\infty$. Logarithm is infinite both for $r = 0$ (when it is $-\infty$) and for $r = \infty$ (when it is $+\infty$), so the difference is infinite as well.

Comment. The situation is not limited to our 3-dimensional proper space (corresponding to 4-dimensional space-time), it can be observed in space-time of any dimension. Indeed, no matter what dimension d we assume for the proper space, the area of the sphere is proportional to r^{d-1}, thus the overall energy is proportional to the integral of $r^\alpha \cdot r^{d-1} = r^{\alpha+d-1}$. So:

- if $\alpha \neq -d$, this integral is proportional to $r^{\alpha+d}$ and is, thus, infinite either for $r = 0$ (when $\alpha < -d$) or for $r = \infty$ (when $\alpha > -d$);
- if $\alpha = -d$, the integral is proportional to $\ln(x)|_0^\infty$ and is, thus, infinite as well.

3   Towards Possible Physical Explanation of Landau's Statement

In reality, infinities are an idealization. In the above computations, we assumed that the distance r can take any value from 0 to infinity. In reality, the distance r cannot be too large: according to modern physics, a distance cannot be too large – it cannot exceed the current radius R of the Universe.

Similarly, the distance r cannot be too small: when the distance becomes too small, of order $r_0 \approx 10^{-33}$ cm, quantum effects become so relatively large that the notion of exact distance becomes impossible [3,11].

In physics, infinite usually means "very large", 0 often means "very small". In reality, when physicists talk about infinite value, what they mean is that in reality, the value is very large – so large that we can safely replace it with infinity. Indeed, the size of an electron is so small in comparison with the size R of the Universe that in most physical problems, we can safely assume that the Universe is infinite – just like when we measure short distances on Earth, we can safely ignore the fact that we are on a surface of a finite sphere, and use formulas of planar geometry – i.e., in effect, assume that the Earth is an infinite plane.

Similarly, when physicists talk about 0 values, what they mean is that the corresponding values are so small, that we can safely ignore this value. Indeed, in most physical problems, the quantum-effects distance 10^{-33} cm is so much smaller than anything we measure that we can safely take this distance to be 0.

What should we do. The notions "very large" and "very small" are clearly imprecise. So, to properly describe these notions – and to properly describe how physicists use them – it makes sense to use techniques specifically designed for dealing with such notions – namely, the techniques of fuzzy logic; see, e.g., [2,5–9,13].

This is something we will try to do, and this is something that we encourage interested readers to try. While such a formalization is still not done, what can we do?

Since there are no infinities, what is the problem? Why are mathematical infinities – which are not really infinite – still bothering physicists?

For example, if instead of using $r = 0$ as the lower bound on the integral, we use the quantum distance $r_0 = 10^{-33}$ cm, we will get a finite value proportional to $1/r_0$. The problem is that this value, while not infinite, is still too large to be physically meaningful. Indeed, the value r_0 is approximately 10^{-20} of the observed electron radius. Since the overall energy of the electric field is proportional to $1/r_0$, this means that the overall energy of the electron's electric field is 10^{20} times larger than we expected – too large.

Similarly, in all other cases: if we take a very large value, and raise it to a power, we still get a very large value.

But with logarithms it is different: a physical explanation of Landau's statement. Interestingly enough, the situation with logarithms is drastically different. Indeed, if we have a term proportional to $\ln(x)$, then, even if $x \approx 10^{20}$, this term is only proportional to $\ln(10^{20}) = 20 \cdot \ln(10) \approx 46$. If the coefficient of proportionality is 0.01 – as often happens in physics – the resulting term is smaller than 1!

This is probably what Landau had in mind when he made this statement:

- that when you have a power law like $y = r^\alpha$, then mathematical infinity usually means that the value of the quantity y is indeed too large to be meaningful;
- on the other hand, if we have a logarithmic dependence like $y = \ln(r)$, then, while mathematically we still have an infinity, in practice, even if we substitute a very large value r, we still get a very reasonable – and very finite – value of the corresponding quantity y.

Comment. Of course, this is just a qualitative explanation. To get a quantitative explanation, we need – as we have mentioned earlier – to further develop fuzzy (or similar) formalization of this idea.

Acknowledgments. This work was supported in part by the US National Science Foundation via grant HRD-1242122 (Cyber-ShARE Center of Excellence). The authors are thankful to the anonymous referees for valuable suggestions.

References

1. Aczél, J., Dhombres, J.: Functional Equations in Several Variables. Cambridge University Press, Cambridge (2008)
2. Belohlavek, R., Dauben, J.W., Klir, G.J.: Fuzzy Logic and Mathematics: A Historical Perspective. Oxford University Press, New York (2017)
3. Feynman, R., Leighton, R., Sands, M.: The Feynman Lectures on Physics. Addison Wesley, Boston (2005)
4. Gleick, J.: Genius: The Life and Science of Richard Feynman. Pantheon, New York (1992)
5. Klir, G., Yuan, B.: Fuzzy Sets and Fuzzy Logic. Prentice Hall, Upper Saddle River (1995)
6. Mendel, J.M.: Uncertain Rule-Based Fuzzy Systems: Introduction and New Directions. Springer, Cham (2017)
7. Nguyen, H.T., Kreinovich, V.: Nested intervals and sets: concepts, relations to fuzzy sets, and applications. In: Kearfott, R.B., Kreinovich, V. (eds.) Applications of Interval Computations, pp. 245–290. Kluwer, Dordrecht (1996)
8. Nguyen, H.T., Walker, C., Walker, E.A.: A First Course in Fuzzy Logic. Chapman and Hall/CRC, Boca Raton (2019)
9. Novák, V., Perfilieva, I., Močkoř, J.: Mathematical Principles of Fuzzy Logic. Kluwer, Boston (1999)
10. Sakharov, A.: Inventing and Solving Problems at the Frontier of Scientific Knowledge. Harvard University Press, Cambridge (1990)
11. Thorne, K.S., Blandford, R.D.: Modern Classical Physics: Optics, Fluids, Plasmas, Elasticity, Relativity, and Statistical Physics. Princeton University Press, Princeton (2017)
12. Weinberg, S.: The search for unity: notes on a history of quantum field theory. Daedalus **106**(4), 17–35 (1977)
13. Zadeh, L.A.: Fuzzy sets. Information and Control **8**, 338–353 (1965)

Fuzzy Transfer Learning in Heterogeneous Space Using Takagi-Sugeno Fuzzy Models

Hua Zuo[✉], Guangquan Zhang, and Jie Lu

Centre for Artificial Intelligence, University of Technology Sydney,
15 Broadway, Sydney, NSW 2007, Australia
{Hua.Zuo,Guangquan.Zhang,Jie.Lu}@uts.edu.au

Abstract. Transfer learning is gaining increasing attention due to its ability to leverage previously acquired knowledge (a source domain with a large amount of labeled data) to assist in completing a prediction task in a related domain (a target domain with little labeled data). Many transfer learning methods have been proposed, and especially the fuzzy transfer learning method, which is based on fuzzy systems, has been developed because of its capability to deal with the uncertainty. However, there is one issue with fuzzy transfer learning that has not yet resolved: the domain adaptation methods for regression tasks in heterogeneous space are still scarce, and the relation of features in two domains have not been explored to assist the construction of target model. In this work, we proposed a new fuzzy transfer learning method, which constructs the transformed mappings for the domain-independent and domain-dependent features, separately. The existing fuzzy rules of the source domain are transferred to the target domain through modifying the input space using the mappings, and the parameters of the mappings are optimized by the few labeled target data. The experiments on real-world datasets validate the effectiveness of the proposed method and discuss the impact of some important parameters to the performance of the constructed target model.

Keywords: Transfer learning · Fuzzy rules · Domain adaptation · Machine learning

1 Introduction

Transfer Learning [1], as a branch of machine learning, has gained growing attention due to its ability of transferring knowledge between the domains with different data distributions or features. Transfer learning addresses the problem of how to leverage the knowledge acquired previously (the source domain with a large amount of labeled data) to improve the efficiency and accuracy of learning in one domain (the target domain with few labeled data) that in some way relates to the original domain.

The techniques of transfer learning are applied mainly in two scenarios. In the first scenario, a well-performed model is built based on the historical data, but because of the rapidly changing environment, the existing model is outdated and cannot fit the new data [2]. In the second scenario, which usually happens in the new emerging area, few labeled or no labeled data is available, but there exists a related domain where a lot of

© Springer Nature Switzerland AG 2019
R. B. Kearfott et al. (Eds.): IFSA 2019/NAFIPS 2019, AISC 1000, pp. 752–763, 2019.
https://doi.org/10.1007/978-3-030-21920-8_67

labeled data is accessible [3]. Some examples of successful applications for the transfer learning methods include: classifying the French files leveraging the already-categorized English documents [4], predicting the status of the Australian banks, fail or survive, using the data of banks in America [5], and detecting the location of a user based on the previously collected WiFi data [6].

Since transfer learning belongs to a branch of machine learning, many methods in transfer learning are developed based on different prediction models, such as SVM [7], and neural networks [8]. There have been many survey papers that summarizes the techniques in transfer learning, especially in some specific areas, such as activity cognition [9], reinforcement learning [10], collaborative recommendation [11], and computational intelligence [12].

The methods of fuzzy transfer learning have been proposed to deal with the phenomenon of uncertainty in transfer learning problems. The integration of fuzzy logic with transfer learning has drawn considerable attention in the literatures. The researchers have applied fuzzy sets to represent linguistic variables when feature values cannot be precisely described in numerical values, and to describe fuzzy distance for the retrieval of similar cases. Transferring implicit and explicit knowledge from similar domains is hidden and uncertain by nature, thus using fuzzy logic and fuzzy rule theory to handle the associated vagueness and uncertainty is apt and can improve transfer accuracy. Thus, many scholars have turned to fuzzy systems as a solution for transfer learning problems with promising results. Deng et al. [13] proposed a series of transfer learning methods, using a Takagi-Sugeno-Kang (TSK) fuzzy model, and developed novel fuzzy logic systems algorithms by defining two new objective functions. Further, their methods were applied to deal with the insufficient scenarios, for example, recognizing electroencephalogram signals in environments with a data shortage. Behbood et al. [14] proposed a fuzzy-based transfer learning approach to long-term bank failure prediction models with source and target domains that have different data distributions. Liu et al. [15] focus on the unsupervised heterogeneous domain adaptation problem, and presented a novel transfer learning model via n-dimensional fuzzy geometry and fuzzy equivalence relations. A metric based on n- dimensional fuzzy geometry is defined to measure the features' similarity of a domain. Then the shared fuzzy equivalence relations are proposed and used to make the numbers of clustering categories the same under the same value of α, and knowledge can be transferred from the source domain to the target domain on heterogeneous space through the clustering categories.

Some work has been done to develop the domain adaptation ability of fuzzy rule-based models in regression tasks [16]. A set of algorithms are proposed for two different scenarios, where the datasets from the source domain and target domain are in homogeneous [17] and heterogeneous space [18], separately. In this paper, based on these works, we will explore the relation of features in two domains and apply it to assist the knowledge transfer process. The contribution of this work is using the the relation of the features between domains to assist the construction of target model.

The reminder of this paper is structured as followed. Section 2 presents some definitions of transfer learning, and the Takagi-Sugeno fuzzy model. Section 3 details the procedures of the proposed domain adaptation method in heterogeneous space.

Experiments in Sect. 4 validate the effectiveness of the presented transfer learning method and discuss the impact of some important parameters in the model. The final section concludes the paper and outlines further work.

2   Preliminaries

In this section, some important definitions in transfer learning, and the Takagi-Sugeno fuzzy model are introduced to make the readers have a clear cognition of the background knowledge of this work.

2.1   Definitions of Transfer Learning

A domain [1] is denoted by $D = \{F, P(X)\}$, where F is a feature space, and $P(X)$, $X = \{x_1, \cdots, x_n\}$, is the probability distribution of the instances.

A task [1] is denoted by $T = \{Y, f(\cdot)\}$, where $Y \in R$ is the label, and $f(\cdot)$ is an objective predictive function.

Definition 1 (Transfer Learning) [1]: Given a source domain D_s, a learning task T_s, a target domain D_t, and a learning task T_t, transfer learning aims to improve learning of the target predictive function $f_t(\cdot)$ in D_t using the knowledge in D_s and T_s where $D_s \neq D_t$ or $T_s \neq T_t$.

The process of transfer learning is illustrated in Fig. 1.

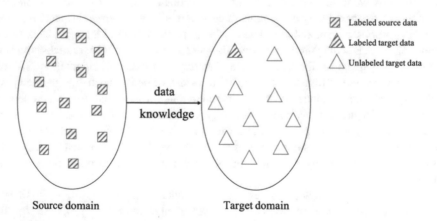

Fig. 1. The process of transfer learning

In Brief, transfer learning aims to utilize the previously acquired knowledge (or data) in the source domain to support the prediction model's construction in the target domain.

2.2 Takagi-Sugeno Fuzzy Model

The prediction model is a commonly used regression model in the fuzzy system area, the Takagi-Sugeno (TS) fuzzy model, which is consist of fuzzy rules in a nonlinear way [19]. A Takagi-Sugeno fuzzy model, which contains c fuzzy rules, is represented as:

$$\text{If } x \text{ is } A_i(x, v_i), \text{ then } y \text{ is } L_i(x, a_i) \; i = 1, \ldots, c \tag{1}$$

The TS fuzzy model could also be rewritten in the form of a neural network with the structure in Fig. 2. The first layer represents the input data, each neuron in the second layer represents a cluster, which also represents the condition of a fuzzy rule, and the third layer is the corresponding consequences of the fuzzy rules.

The construction of the TS model, a set of fuzzy rules, is based on a labeled dataset $\{(x_1, y_1), (x_2, y_2), \ldots, (x_N, y_N)\}$ using two procedures. In the first procedure, fuzzy C-means (FCM) [20] is applied to divide the data in an unsupervised learning process, so that clusters are learned, and the centers of the clusters are obtained. After getting the clusters, the coefficients of the linear functions, which are defined in each cluster, are calculated using the labeled datasets.

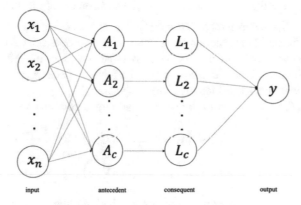

Fig. 2. TS model with the neural network structure

3 Knowledge Transfer Between Domains in Heterogeneous Space

This section presents the method of transferring fuzzy rules from the source domain to a target domain in heterogeneous space. In order to facilitate elaborating the proposed method, the heterogeneous domain adaptation problem is stated with formula, and the description of the used variables and datasets is given.

3.1   Problem Statement

Consider there are one source domain with a large amount of labeled data, and a target domain with very few labeled data. Suppose the dataset in the source domain is S and denoted as:

$$S = \left\{ \left(x_1^s, y_1^s\right), \cdots, \left(x_{N_s}^s, y_{N_s}^s\right) \right\} \tag{2}$$

where $\left(x_k^s, y_k^s\right)$ is the k th input-output data pair in the source domain. $x_k^s \in R^n$ is an input variable with n-dimension, the label $y_k^s \in R$ is the corresponding output, a continuous variable in a regression task, and N_s indicates the number of labeled data pairs in the source domain.

The target domain contains two subsets: T_U, the one with labels, and T_L, the one without labels:

$$T = \{T_L, T_U\} = \left\{ \left\{ \left(x_1^t, y_1^t\right), \cdots, \left(x_{N_{t1}}^t, y_{N_{t1}}^t\right) \right\}, \left\{ x_{N_{t1}+1}^t, \cdots, x_{N_t}^t \right\} \right\} \tag{3}$$

where $\left\{ \left(x_1^t, y_1^t\right), \cdots, \left(x_{N_{t1}}^t, y_{N_{t1}}^t\right) \right\}$ are the labeled data pairs in T_L, and $\left\{ x_{N_{t1}+1}^t, \cdots, x_{N_t}^t \right\}$ are the unlabeled data in T_U. The numbers of instances in T_L and T_U are N_{t1} and $N_t - N_{t1}$ respectively, and satisfy $N_{t1} \ll N_t$, $N_{t1} \ll N_s$.

Different with the homogeneous situation, in the heterogeneous domain adaptation problem, the feature space in the source and target domain are different. Here, we consider a special case, where dimensions of the feature space in two domains are the same, but the meanings of the features are not identical.

Suppose the feature spaces in the source and target domains are F_s and F_t:

$$F_s = \left(x_1^C, x_2^C, x_1^s, \ldots, x_{n-2}^s\right) \tag{4}$$

$$F_t = \left(x_1^C, x_2^C, x_1^t, \ldots, x_{n-2}^t\right) \tag{5}$$

where x_1^C and x_2^C are the two common features, which can be regarded as the domain-independent features, and the distributions of them are not identical in both source and target domains. $(x_1^s, \ldots, x_{n-2}^s)$, ..., and $(x_1^t, \ldots, x_{n-2}^t)$ are the domain-dependent features, which own different meaning in two domains. Here, we set the number of domain-independent features as two as an example, and it could be any number that doesn't exceed the dimension of the feature space.

Since the number of labeled data is sufficient in S, a well-performed model could be built for the source domain. Due to the different meanings of the features, however, the models for the source domain cannot be used directly to solve the regression tasks in the target domain. But the common features in the feature spaces of two domains provide the bridge that could transfer the shared knowledge between domains.

3.2   Knowledge Transfer in Heterogeneous Space Across Domains

The method of transferring knowledge from the source domain to the target domain in heterogeneous space could be summarized into three steps:

Step 1: Construct a TS model for the source domain.

Based on the labeled data in S, a TS fuzzy model M^s is built, and a set of fuzzy rules are obtained.

$$\text{If } x_k^s \text{ is } A_i\left(x_k^s, v_i^s\right), \text{ then } y_k^s \text{ is } L_i\left(x_k^s, a_i^s\right) \; i = 1, \ldots, cs \qquad (6)$$

Each rule, actually, is represented by the centers of the clusters v_i^s and the coefficients of the linear functions a_i^s. After the construction of the model M^s, a set of v_i^s and a_i^s are obtained, and the source data won't be used anymore, which could keep the privacy of the data, especially in some sensitive areas, for example the medical data.

The fuzzy rules in (6) have a high prediction accuracy on source data S, but a poor performance on target data T.

Step 2: Modify the existing rules in M^s to fit the target data.

The input space of the target data is changed through mappings to make the modified rules become compatible with target data. Since there are two types of features, domain-independent and domain-dependent, two different strategies are applied to change the features in the input space. The structures of mappings are shown in Fig. 3.

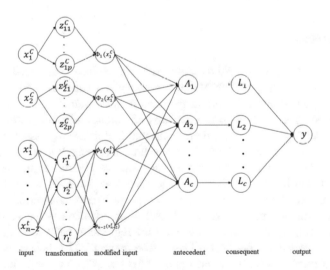

Fig. 3. Modification of input space for heterogeneous domain adaptation

For the domain-independent features, which are shared between the source and target domains, each input variable is changed using a three layers network. Although these features have the same meanings between domains, their distributions are quite different. Each domain-independent variable is assumed to be governed by some hidden features, so the different hidden features or the different weights of the features will lead to the difference of the distributions in two domains. Therefore, the idea of applying the mapping in Fig. 3 aims to adjust the hidden features in number or weight so that the modified distribution of the domain-independent features could fit the target data.

For the domain-dependent features, since there is no clear relation between the features in two domains, a method of fully connection with a hidden is applied to change the distributions of the domain-dependent features.

Step 3: Optimize the parameters of mappings using target data

The parameters in the mappings are obtained through a supervised learning way. The labeld target data set T_L is fed to the model and optimize the model to fit the target data. Denote $\left(\Phi\left(x_1^c, x_2^c\right), \phi\left(x_1^t, \ldots, x_{n-2}^t\right)\right) \equiv \Psi(x^t)$.

The following objective function is minimized:

$$Q = \sqrt{\frac{1}{N_{t1}} \sum_{k=1}^{N_{t1}} \left(\sum_{i=1}^{cs} A_i\left(\Psi\left(x_k^t\right), \Psi(v_i^s)\right) L_i\left(\Psi\left(x_k^t\right), a_i^s\right) - y_k^t\right)^2} + \frac{\lambda}{2} w^T w \quad (7)$$

The first term in (7) is the approximation error that aims to minimize the gap between the output of the modified model and the target data's real output. The second term introduces a structural risk term into the objective function. The parameter λ indicates the tradeoff between the quality of an approximation and the complexity of the approximation function; w is the vector of all the parameters optimized.

4 Experiments

The experiments using real-world datasets are implemented to validate the effectiveness of the proposed method in dealing with heterogeneous domain adaptation problem, and explore the impact of some important parameters in the transferring model.

Since the studies on regression problems of domain adaptation are scarce, especially in the heterogeneous space, there is no public datasets in these scenarios. In the work, therefore, two datasets from UCI Machine Learning Repository are used and modified to simulate the heterogeneous domain adaptation problems. A detailed description of modifying the datasets is provided to illustrate the datasets clearly. Further, all the models' construction apply the five-fold cross validation, the results are shown in the form of "mean ± variance". All the models are tested on the unlabeled target data to compare the ability of solving the regression tasks in target domain.

The first dataset concerns "Airfoil self-noise dataset". The dataset was split based on the "frequency" value. Five attributes, "frequency", "angle of attack", "chord length", "free-stream velocity", and "suction side displacement thickness", are used to predict the "scaled sound pressure level". Data with a "frequency" of greater than 800 hertzs formed the source domain, with 1000 instances, the remaining data, 450 instances, was used for the target domain. Further, the attributes "chord length", "free-stream velocity", and "suction side displacement thickness" in the source domain were perturbed with random numbers to simulate the domain-dependent features. All the instances in the source domain are labeled, but only 10 instances in the target domain are labeled.

The number of clusters, i.e. the amount of the fuzzy rules, is a crucial parameter that determines the structure of the model. Although some techniques, for example the infinite Gaussian mixture model and the heuristic algorithm, have been applied to explore the number of clusters, it is still not easy to determine in the high-dimension datasets. Therefore, the number of clusters is set as a hyper-parameter in the following experiments to explore the impact of it to the performance of transfer learning in fuzzy models.

With different number of clusters, the heterogeneous domain adaptation method is implemented, and the results are shown in Table 1. The first column in Table 1 indicates the number of clusters applied in the experiment. The results shown in the second to the fourth columns are the accuracies of three models on the unlabeled data in the target domain. The three models are: the source model built using source data, the target model built using only labeled target data, and the target data constructed using our proposed method. The results with the best performance are in bold.

Table 1. Results with different values of c in "Airfoil self-noise dataset"

Number of clusters	Performance of models on unlabeled target data		
	Source model	Target model using target data	Model using new method
2	0.5407 ± 0.0002	0.4279 ± 0.0088	$\mathbf{0.2013 \pm 0.0010}$
3	0.5524 ± 0.0001	0.2805 ± 0.0009	$\mathbf{0.1921 \pm 0.0009}$
4	0.5682 ± 0.0001	0.2550 ± 0.0052	$\mathbf{0.2184 \pm 0.0007}$
5	0.5672 ± 0.0001	0.2260 ± 0.0015	$\mathbf{0.1721 \pm 0.0001}$
6	0.5744 ± 0.0001	0.2175 ± 0.0008	$\mathbf{0.2165 \pm 0.0007}$
7	0.5806 ± 0.0001	0.2067 ± 0.0009	$\mathbf{0.1802 \pm 0.0011}$

Analyzing the results in Table 1, the high values in the first column indicate that the model built for the source domain does not fit to the target data, and the not small values in the third column manifest the poor performance of the target model only using the labeled target data, which also verify that the labeled target data is not sufficient to construct a well prediction model. Additionally, comparing the results in the second, third, and fourth columns, we can see that in each experiment with different number of clusters, the accuracies of the model applied the presented new method is higher than the accuracies of the other models, which indicates that our method is superior than other methods.

In the presented method, the number of neurons used to construct the mappings is an important parameter that determines the performance of the constructed target model. Since the input features are changed through two different ways, two parameters, p and q, are discussed here to explore the impact of them to the performance of the constructed model. The target model using the new method is built with different values of p and q. For comparison, the performance of the source model and target model using only labeled target data are shown in Table 2.

The performance of the target model using the proposed method is shown in Table 3. Both the values of p and q are changing from two to six, so twenty-five experiments are run to compare the results.

Table 2. Results of two compared models

Source model	0.5684 ± 0.0001
Target model using target data	0.2463 ± 0.0019

Table 3. Results with different values of p and q

p	q				
	2	3	4	5	6
2	0.1889 ± 0.0003	0.1984 ± 0.0001	0.2004 ± 0.0009	0.2035 ± 0.0004	0.2164 ± 0.0013
3	0.1812 ± 0.0001	0.1916 ± 0.0006	0.2062 ± 0.0017	0.2130 ± 0.0012	0.2263 ± 0.0023
4	0.1814 ± 0.0005	0.1977 ± 0.0009	0.2033 ± 0.0013	0.2264 ± 0.0031	0.2116 ± 0.0002
5	0.1790 ± 0.0003	0.2013 ± 0.0011	0.2134 ± 0.0007	0.1984 ± 0.0008	0.2112 ± 0.0019
6	0.1827 ± 0.0002	0.1936 ± 0.0008	0.1869 ± 0.0009	0.1965 ± 0.0009	0.2178 ± 0.0017

First, all the mean values in Table 3 is smaller than the mean values in Table 2, which validates the effectiveness of our method in deal with transfer learning in heterogeneous space. There is no obvious trend of the results with the change of p and q, and the best result appears when the p is equal to five and q is two.

In the second dataset, "Combined cycle power plant dataset (CCPP)", four attributes "temperature", "ambient pressure", "relative humidity", and "exhaust vacuum" were used to predict the "net hourly electrical energy output". The dataset was split into two source domains and a target domain using the attribute "temperature"; instances with a temperature of not greater than 25° were treated as source data, with 5000 instances, the remainder 2000 instances fell into the target domain. Also, the features "relative humidity", and "exhaust vacuum" in the source domain were perturbed with random numbers to simulate the domain-dependent features. All the instances in the source domain are labeled, but only 10 instances in the target domain are labeled.

Similarly, a set of experiments are designed to find out the impact of the changing number of clusters to the presented method. The results are shown in Table 4. In the first five experiments shown in Table 4, the performance of the model using the presented method is superior than the other two models. However, in the last experiment, where the number of clusters is equal to seven, the performance of the model applying only the target labeled target data is better than the proposed method, and the reason might derive from the quality of the labeled target data. Comparing the results in the third and fourth columns, although the new method shows better capability in dealing with the heterogeneous domain adaptation problem, it does not show a significant advantage. Although the labeled target data has small amount, they cover most of the clusters of target data, which lead to the good quality of the model using only target data.

Table 4. Results with different values of c in "CCPP"

Number of clusters	Performance of models on unlabeled target data		
	Source model	Target model using target data	Model using new method
2	0.5444 ± 0.0000	0.1021 ± 0.0006	**0.0720 ± 0.0000**
3	0.5422 ± 0.0000	0.0854 ± 0.0003	**0.0738 ± 0.0001**
4	0.5483 ± 0.0000	0.0857 ± 0.0001	**0.0728 ± 0.0000**
5	0.5511 ± 0.0000	0.0868 ± 0.0001	**0.0686 ± 0.0000**
6	0.5487 ± 0.0000	0.0817 ± 0.0000	**0.0803 ± 0.0002**
7	0.5492 ± 0.0000	**0.0815 ± 0.0000**	0.0872 ± 0.0002

The experimental results with varying values of p and q are shown in Tables 5 and 6. In the twenty-five experiments, the number of clusters is set to five, and the values of p and q go through the integers between two and six. Similarly, the proposed method shows its superiority in these experiments when comparing with the source model and the model built using target data only. The best result appears when the values of p and q are equal to six and four, separately. Additionally, the low variance of the results in all the experiments show the good generalization ability of the model constructed using the presented method.

Table 5. Results of two compared models

Source model	0.5492 ± 0.0001
Target model using target data	0.0803 ± 0.0001

Table 6. Results with different values of p and q

p	q				
	2	3	4	5	6
2	0.0735 ± 0.0001	0.0699 ± 0.0000	0.0717 ± 0.0000	0.0740 ± 0.0001	0.0750 ± 0.0000
3	0.0707 ± 0.0000	0.0717 ± 0.0001	0.0761 ± 0.0001	0.0746 ± 0.0000	0.0707 ± 0.0000
4	0.0706 ± 0.0000	0.0710 ± 0.0000	0.0718 ± 0.0000	0.0766 ± 0.0002	0.0720 ± 0.0001
5	0.0677 ± 0.0000	0.0695 ± 0.0000	0.0680 ± 0.0000	0.0759 ± 0.0000	0.0694 ± 0.0000
6	0.0687 ± 0.0001	0.0693 ± 0.0000	0.0674 ± 0.0000	0.0710 ± 0.0000	0.0741 ± 0.0000

5 Conclusion and Further Study

This work explores the transfer learning problems in heterogeneous space. The features in the source and target domains are identified by domain-independent features, which are shared between domains, and domain-dependent features, which only exist in an individual domain. Mappings with different structures are constructed for the two types of features, separately. For the domain-independent features, a shallow network is built for each feature to change the input distribution, and for the domain-dependent features, all the domain-dependent features are used to construct the transformation mapping. The parameters of the mappings are obtained through an optimization process based on the labeled target data. The experiments show that the accuracy of the model using our method is superior than the source model, and the model built with insufficient target labeled data. And an appropriate number of clusters is important to the performance of the target model. The change of values of p and q has slight impact to the final results.

The method presented in this paper focuses on the situation, where the dimensions of the feature space in two domains are the same, a special case in heterogeneous transfer learning. More general and challenging case, that the dimensions of the input data in two domains are different, will be considered and discussed in the future studies.

Acknowledgement. This work was supported by Australian Research Council under DP 170101623.

References

1. Pan, S.J., Yang, Q.: A survey on transfer learning. IEEE Trans. Knowl. Data Eng. **22**(10), 1345–1359 (2010)
2. Gong, B., Shi, Y., Sha, F., Grauman, K.: Geodesic flow kernel for unsupervised domain adaptation. In: 2012 IEEE Conference on Computer Vision and Pattern Recognition (CVPR), pp. 2066–2073 (2012)
3. Behbood, V., Lu, J., Zhang, G., Pedrycz, W.: Multistep fuzzy bridged refinement domain adaptation algorithm and its application to bank failure prediction. IEEE Trans. Fuzzy Syst. **23**(6), 1917–1935 (2015)
4. Xiao, M., Guo, Y.: Feature space independent semi-supervised domain adaptation via kernel matching. IEEE Trans. Pattern Anal. Mach. Intell. **37**(1), 54–66 (2015)
5. Behbood, V., Lu, J., Zhang, G.: Fuzzy bridged refinement domain adaptation: Long-term bank failure prediction. Int. J. Comput. Intell. Appl. **12**(01), 1350003 (2013)
6. Pan, S.J., Tsang, I.W., Kwok, J.T., Yang, Q.: Domain adaptation via transfer component analysis. IEEE Trans. Neural Netw. **22**(2), 199–210 (2011)
7. Tommasi, T., Orabona, F., Caputo, B.: Learning categories from few examples with multi model knowledge transfer. IEEE Trans. Pattern Anal. Mach. Intell. **36**(5), 928–941 (2014)
8. Hoo-Chang, S., et al.: Deep convolutional neural networks for computer-aided detection: CNN architectures, dataset characteristics and transfer learning. IEEE Trans. Med. Imaging **35**(5), 1285–1298 (2016)
9. Cook, D., Feuz, K.D., Krishnan, N.C.: Transfer learning for activity recognition: a survey. Knowl. Inf. Syst. **36**(3), 537–556 (2013)

10. Taylor, M.E., Stone, P.: Transfer learning for reinforcement learning domains: a survey. J. Mach. Learn. Res. **10**, 1633–1685 (2009)
11. Pan, W.: A survey of transfer learning for collaborative recommendation with auxiliary data. Neurocomputing **177**, 447–453 (2016)
12. Lu, J., Behbood, V., Hao, P., Zuo, H., Xue, S., Zhang, G.: Transfer learning using computational intelligence: a survey. Knowl.-Based Syst. **80**, 14–23 (2015)
13. Deng, Z., Jiang, Y., Ishibuchi, H., Choi, K.-S., Wang, S.: Enhanced knowledge-leverage-based TSK fuzzy system modeling for inductive transfer learning. ACM Trans. Intell. Syst. Technol. (TIST) **8**(1), 11 (2016)
14. Behbood, V., Lu, J., Zhang, G.: Fuzzy refinement domain adaptation for long term prediction in banking ecosystem. IEEE Trans. Ind. Inform. **10**(2), 1637–1646 (2014)
15. Liu, F., Zhang, G., Lu, J.: Unconstrained fuzzy feature fusion for heterogeneous unsupervised domain adaptation. In: 2018 IEEE International Conference on Fuzzy Systems (FUZZ-IEEE), pp. 1–8 (2018)
16. Zuo, H., Zhang, G., Pedrycz, W., Behbood, V., Lu, J.: Granular fuzzy regression domain adaptation in Takagi-Sugeno fuzzy models. IEEE Trans. Fuzzy Syst. **26**(2), 847–858 (2018)
17. Zuo, H., Zhang, G., Pedrycz, W., Behbood, V., Lu, J.: Fuzzy regression transfer learning in Takagi-Sugeno fuzzy models. IEEE Trans. Fuzzy Syst. **25**(6), 1795–1807 (2017)
18. Zuo, H., Lu, J., Zhang, G., Pedrycz, W.: Fuzzy rule-based domain adaptation in homogeneous and heterogeneous spaces. IEEE Trans. Fuzzy Syst. **27**(2), 348–361 (2018)
19. Hadjili, M.L., Wertz, V.: Takagi-Sugeno fuzzy modeling incorporating input variables selection. IEEE Trans. Fuzzy Syst. **10**(6), 728–742 (2002)
20. Bezdek, J.C., Ehrlich, R., Full, W.: FCM: The fuzzy c-means clustering algorithm. Comput. Geosci. **10**(2–3), 191–203 (1984)

Constraint Programming And Decision Making Papers from the 12th International Workshop CoProd'2019

Can We Improve the Standard Algorithm of Interval Computation by Taking Almost Monotonicity into Account?

Martine Ceberio[1], Olga Kosheleva[2], and Vladik Kreinovich[1]([⊠])

[1] Department of Computer Science, University of Texas at El Paso,
El Paso, TX 79968, USA
{mceberio,vladik}@utep.edu
[2] Department of Teacher Education, University of Texas at El Paso,
El Paso, TX 79968, USA
olgak@utep.edu

Abstract. In many practical situations, it is necessary to perform interval computations – i.e., to find the range of a given function $y = f(x_1, \ldots, x_n)$ on given intervals – e.g., when we want to find guaranteed bounds of a quantity that is computed based on measurements, and for these measurements, we only have upper bounds of the measurement error. The standard algorithm for interval computations first checks for monotonicity. However, when the function f is almost monotonic, this algorithm does not utilize this fact. In this paper, we show that such closeness-to-monotonicity can be efficiently utilized.

1 Formulation of the Problem

Need for Interval Computations. Most of the data comes from measurements. Measurements are never absolutely accurate, the measurement result $\widetilde{x}$ is, in general, different from from the actual (unknown) value of the corresponding quantity. In many cases, the only information that we have about the possible values of the measurement error $\Delta x \stackrel{\text{def}}{=} \widetilde{x} - x$ is the upper bounds Δ on its absolute value; see, e.g., [5].

Once we know this upper bound, then, based on the measurement result $\widetilde{x}$, the only thing that we can conclude about the actual value x is that this value belongs to the interval $\mathbf{x} = [\underline{x}, \overline{x}] = [\widetilde{x} - \Delta, \widetilde{x} + \Delta]$.

In many practical situations, we are interested in a quantity y which itself is difficult (or even impossible) to measure but which depends, in a known way, on the values of several easier-to-measure quantities $x_1, \ldots, x_n$: $y = f(x_1, \ldots, x_n)$ [5]. In such situations, to estimate y, we measure the quantities x_i. When we only have upper bounds on all the measurement errors, then the only information that we have about each of the quantities x_i is that it belongs to the interval

$$\mathbf{x}_i = [\underline{x}_i, \overline{x}_i] = [\widetilde{x}_i - \Delta_i, \widetilde{x}_i + \Delta_i].$$

© Springer Nature Switzerland AG 2019
R. B. Kearfott et al. (Eds.): IFSA 2019/NAFIPS 2019, AISC 1000, pp. 767–778, 2019.
https://doi.org/10.1007/978-3-030-21920-8_68

In this case, the set $\mathbf{y} = [\underline{y}, \overline{y}]$ of possible value of y is formed by the values $f(x_1, \ldots, x_n)$ corresponding to all possible combinations of values $x_i \in \mathbf{x}_i$:

$$\mathbf{y} = [\underline{y}, \overline{y}] = \{f(x_1, \ldots, x_n) : x_1 \in \mathbf{x}_1 \ \& \ \ldots \ \& \ x_n \in \mathbf{x}_n\}. \tag{1}$$

This range is also denoted by $f(\mathbf{x}_1, \ldots, \mathbf{x}_n)$.

Computing the values $\underline{y}$ and $\overline{y}$ based on the given algorithm $f(x_1, \ldots, x_n)$ and the given bounds $\underline{x}_i$ and $\overline{x}_i$ is known as *interval computations*; see, e.g., [1,3,4].

Need for Computing Enclosures. The above interval computations problem is, in general, NP-hard already for quadratic functions $f(x_1, \ldots, x_n)$; see, e.g., [2,6]. This means that, unless $P = NP$ (which most computer scientists believe to be false), no feasible algorithm is possible for solving all particular cases of the interval computation problem.

In many practically useful cases, we have efficient algorithms for interval computations. In other cases, all we can do is compute an *enclosure* $\mathbf{Y} \supseteq \mathbf{y}$ – and hope that this enclosure is close to the actual range $\mathbf{y}$.

The need for an enclosure comes from the fact that we often need to guarantee that the value y is within the given bounds $[y^-, y^+]$. If we produce an approximate solution $\mathbf{Y}$ and see that is it within the bounds, i.e., that $\mathbf{Y} \subseteq [y^-, y^+]$, then to guarantee that the actual value y is within the bounds, we need to be sure that all possible values of y are in the set $\mathbf{Y}$, i.e., that $\mathbf{y} \subseteq \mathbf{Y}$ – i.e., that $\mathbf{Y}$ is indeed an enclosure.

Case of Monotonic Functions. There are cases when computing the range (1) is easy: e.g., it is easy when the function $f(x_1, \ldots, x_n)$ is monotonic in each of its variables. For example, if the function $f(x_1, \ldots, x_n)$ is increasing in each of its variables, then its range is equal to $[f(\underline{x}_1, \ldots, \underline{x}_n), f(\overline{x}_1, \ldots, \overline{x}_n)]$.

An important particular case of monotonicity is the case when the function $f(x_1, \ldots, x_n)$ is linear, i.e., when $f(x_1, \ldots, x_n) = a_0 + \sum_{i=1}^{n} a_i \cdot x_i$. The range of the linear function is equal to $\mathbf{y} = [\widetilde{y} - \Delta, \widetilde{y} + \Delta]$, where $\widetilde{y} \stackrel{\text{def}}{=} f(\widetilde{x}_1, \ldots, \widetilde{x}_n)$ and $\Delta \stackrel{\text{def}}{=} \sum_{i=1}^{n} |a_i| \cdot \Delta_i$. In particular, when $f(x_1, x_2) = x_1 + x_2$, we get

$$[\underline{x}_1, \overline{x}_1] + [\underline{x}_2, \overline{x}_2] = [\underline{x}_1 + \underline{x}_2, \overline{x}_1 + \overline{x}_2], \tag{2}$$

and when $f(x_1, x_2) = x_1 - x_2$, we get

$$[\underline{x}_1, \overline{x}_1] - [\underline{x}_2, \overline{x}_2] = [\underline{x}_1 - \overline{x}_2, \overline{x}_1 - \underline{x}_2]. \tag{3}$$

Interval Arithmetic Operations. Monotonicity covers two arithmetic operations: addition and subtraction.

For multiplication, the situation is somewhat more complex, but still:

- For each x_1, the function $f(x_1, x_2) = x_1 \cdot x_2$ is either increasing or decreasing with respect to x_2 and thus, attains its minimum and maximum at one of the two endpoints of the interval $[\underline{x}_2, \overline{x}_2]$.

- Similarly, for each x_2, this function is either increasing or decreasing with respect to x_1, so to find its extreme values, it is sufficient to consider only the two endpoints $\underline{x}_1$ and $\overline{x}_1$.

Thus, to find the minimum and maximum of the product, it is sufficient to only consider 4 combinations of endpoints:

$$[\underline{x}_1, \overline{x}_1] \cdot [\underline{x}_2, \overline{x}_2] = [\min(\underline{x}_1 \cdot \underline{x}_2, \underline{x}_1 \cdot \overline{x}_2, \overline{x}_1 \cdot \underline{x}_2, \overline{x}_1 \cdot \overline{x}_2), \min(\underline{x}_1 \cdot \underline{x}_2, \underline{x}_1 \cdot \overline{x}_2, \overline{x}_1 \cdot \underline{x}_2, \overline{x}_1 \cdot \overline{x}_2)]. \quad (4)$$

The only remaining arithmetic operation is division x_1/x_2. In the computers, it is usually implemented as

$$x_1/x_2 = x_1 \cdot (1/x_2), \quad (5)$$

and the function $1/x_2$ is decreasing on any interval not containing 0, so

$$1/[\underline{x}_2, \overline{x}_2] = [1/\overline{x}_2, 1/\underline{x}_2] \text{ if } 0 \notin [\underline{x}_1, \overline{x}_2]. \quad (6)$$

Straightforward Interval Computations. In the computer, the only hardware supported operations are arithmetic operations, so any numerical algorithm is implemented as a sequence of such arithmetic operations; for example, computing $\sin(x)$ or $\exp(x)$ usually means computing the values of approximating polynomials (usually, the sum of the first few terms in the Taylor expansion).

Thus, one way to compute an enclosure is to replace each elementary arithmetic operation in the algorithm $f(x_1, \ldots, x_n)$ with the corresponding interval arithmetic operation (2)–(6).

The problem is that the resulting *straightforward* interval computations usually leads to a very wide enclosure.

Towards a Standard Algorithm for Interval Computations: First Stage. To get a narrower enclosure, a natural idea is as follows.

First, we check whether the given problem allows *simple* interval computation – i.e., whether the function is monotonic. According to calculus, a function is increasing with respect to x_i if and only if the corresponding partial derivative is non-negative, i.e., if $\dfrac{\partial f}{\partial x_i} \geq 0$ on the whole box $\mathbf{x} \stackrel{\text{def}}{=} \mathbf{x}_1 \times \ldots \times \mathbf{x}_n$. To check this inequality, we can use, e.g., straightforward interval computations to find the enclosure $[\underline{D}_i, \overline{D}_i]$ for the range of this partial derivative.

- If the whole enclosure $[\underline{D}_i, \overline{D}_i]$ is non-negative, i.e., if $\underline{D}_i \geq 0$, this means that the function $f(x_1, \ldots, x_n)$ is increasing in x_i.
- If the whole enclosure $[\underline{D}_i, \overline{D}_i]$ is non-positive, i.e., if $\overline{D}_i \leq 0$, then the function $f(x_1, \ldots, x_n)$ is decreasing in x_i.

If the function $f(x_1, \ldots, x_n)$ is increasing or decreasing with respect to each variable, then we can immediately compute its range. If it is increasing or decreasing with respect to only some of the variables, then we can reduce the problem to computing ranges of functions of fewer variables. For example, if the function is increasing with respect to x_1, then:

- to estimate $\underline{y}$, it is sufficient to consider the smallest value $x_1 = \underline{x}_1$, i.e., to consider the minimum of the function $\underline{F}(x_2, \ldots, x_n) \overset{\text{def}}{=} f(\underline{x}_1, x_2, \ldots, x_n)$ of $n - 1$ variables; and
- to estimate $\overline{y}$, it is sufficient to consider the largest value $x_1 = \overline{x}_1$, i.e., to consider the maximum of another function of $n - 1$ variables:

$$\overline{F}(x_2, \ldots, x_n) \overset{\text{def}}{=} f(\overline{x}_1, x_2, \ldots, x_n).$$

Towards a Standard Algorithm for Interval Computations: Second Stage. If the original problem is not exactly simple, a natural next idea is to find a *nearby simple* problem and use its solution.

In interval computations, we approximate the original function by a linear one – this can be naturally done by using the Mean Value Theorem, according to which, for each combination of values $x_i = \widetilde{x}_i - \Delta x_i \in \mathbf{x}_i$, we have

$$f(x_1, \ldots, x_n) = f(\widetilde{x}_1 - \Delta x_1, \ldots, \widetilde{x}_n - \Delta x_n) =$$

$$f(\widetilde{x}_1, \ldots, \widetilde{x}_n) - \sum_{i=1}^{n} \frac{\partial f}{\partial x_i}(\widetilde{x}_1 - \xi_1, \ldots, \widetilde{x}_n - \xi_n) \cdot \Delta x_i$$

for some $\xi_i \in [-\Delta_i, \Delta_i]$, leading to the following enclosure for the range $\mathbf{y} = f(\mathbf{x}_1, \ldots, \mathbf{x}_n)$:

$$\mathbf{y} \subseteq \widetilde{y} + \sum_{i=1}^{n}[\underline{D}_i, \overline{D}_i] \cdot [-\Delta_i, \Delta_i].$$

This formula is easy to compute – since the enclosures $[\underline{D}_i, \overline{D}_i]$ were estimated when we checked for monotonicity.

Towards a Standard Algorithm for Interval Computations: Third Stage. If the above method – of reducing a not-so-simple problem to a nearby simple one – does not lead to a sufficiently accurate enclosure, the next natural idea is to *divide the* original *problem into* several *simpler ones*. Usually:

- we divide one of the intervals $\mathbf{x}_i = [\underline{x}_i, \overline{x}_i]$ into two equal parts $\mathbf{x}'_i = [\underline{x}_i, \widetilde{x}_i]$ and $\mathbf{x}''_i = [\widetilde{x}_i, \overline{x}_i]$,
- we estimate the ranges $\mathbf{Y}'$ and $\mathbf{Y}''$ of the function $f(x_1, \ldots, x_n)$ on the corresponding sub-boxes $\mathbf{x}' = \mathbf{x}_1 \times \ldots \times \mathbf{x}_{i-1} \times \mathbf{x}'_i \times \mathbf{x}_{i+1} \times \ldots \times \mathbf{x}_n$ and

$$\mathbf{x}'' = \mathbf{x}_1 \times \ldots \times \mathbf{x}_{i-1} \times \mathbf{x}''_i \times \mathbf{x}_{i+1} \times \ldots \times \mathbf{x}_n;$$

- we take the union of these ranges: $\mathbf{Y} = \mathbf{Y}' \cup \mathbf{Y}''$.

This method works since when we approximate a function by a linear expression (as in mean value form), we thus ignore quadratic (and higher order) terms; thus, the accuracy of the mean valued form is of order Δ_i^2. When we divide the interval into two halves, each with half of the original value Δ_i, the corresponding error component decreases 4 times, to $(\Delta_i/2)^2 = \Delta_i^2/4$.

If we want a more accurate enclosure, we can bisect further, etc.

Thus, we arrive at the following algorithm – which is used in most interval computations software packages.

The Standard Algorithm for Interval Computations: Summary.

- first, we check if the function is monotonic with respect to (at least) some of the variables,
- then, we apply the mean value form,
- then, if needed, we bisect, and repeat the whole procedure for each sub-box.

A Problem with the Standard Algorithm – And What We Do in This Paper. The problem with the standard algorithm is that:

- when it checks whether a given problem is already simple, it checks for the general monotonicity property;
- however, when this algorithm takes into account a nearby simple problem, it only considers linear cases – but not more general monotonic ones.

As a result:

- when the function is linear, we get the exact range,
- and when the function is almost linear – i.e., it is very close to a linear one – we get an almost correct range, that tends to the exact one when the difference from a linear function tends to 0.

On the other hand, for non-linear functions:

- when the function is monotonic, we get the exact range,
- but when the function is almost monotonic – i.e., very close to a monotonic one – we do not utilize this closeness and get, in general, a lousy estimate.

It is therefore desirable to take this "almost monotonicity" into account, i.e., to come up with a method that would lead to the exact range when the difference from monotonicity tends to 0.

This is what we do in this paper.

2 How to Take Almost Monotonicity into Account: 1-D Case

Discussion. Let us see which bounds on the endpoints $\underline{y}$ and $\overline{y}$ can be extracted from the condition of almost monotonicity, and let us check whether these bounds can be better than the bounds coming from the usual mean value form. To make this comparison, we need to describe which bounds come from the mean value form.

Which Bounds Come from the Mean Value Form. In the 1-D case, the mean value form takes the form $f(\tilde{x}_1) + [\underline{D}_1, \overline{D}_1] \cdot [-\Delta_1, \Delta_1]$. For the product of the two intervals, the upper endpoint is equal to

$$\max(\underline{D}_1 \cdot \Delta_1, -\underline{D}_1 \cdot \Delta_1, \overline{D}_1 \cdot \Delta_1, -\overline{D}_1 \cdot \Delta_1) =$$

$$\max(\max(\underline{D}_1 \cdot \Delta_1, -\underline{D}_1 \cdot \Delta_1), \max(\overline{D}_1 \cdot \Delta_1, -\overline{D}_1 \cdot \Delta_1)).$$

In general, $\max(a, -a) = |a|$, so the upper endpoint is equal to $\Delta_1 \cdot \max(|\underline{D}_1|, |\overline{D}_1|)$. Thus, the upper bound for $f(x_1)$ takes the form

$$\overline{y} \le f(\tilde{x}_1) + \Delta_1 \cdot \max(|\underline{D}_1|, |\overline{D}_1|). \tag{7}$$

Similarly, the lower bound corresponding to the mean value form is of the type

$$\underline{y} \ge f(\tilde{x}_1) - \Delta_1 \cdot \max(|\underline{D}_1|, |\overline{D}_1|). \tag{8}$$

What does Almost Monotonicity Mean In 1-D Case. If the function $f(x_1)$ is increasing, then its range on the interval $[\underline{x}_1, \overline{x}_1]$ is equal to $[f(\underline{x}_1), f(\overline{x}_1)]$. The function is increasing if the range $[\underline{d}_1, \overline{d}_1]$ of its derivative contains only non-negative values, i.e., if $\underline{d}_1 \ge 0$.

In practice, usually, instead of the actual range $[\underline{d}_1, \overline{d}_1]$, we only know the enclosure $[\underline{D}_1, \overline{D}_1]$ for this range. Thus, we can utilize the monotonicity property if $\underline{D}_1 \ge 0$. In this terms, "almost monotonicity" means that the value $\underline{D}_1$ is negative but close to 0, i.e., has the form $\underline{D}_1 = -\varepsilon$, for some small $\varepsilon > 0$.

Similarly, since decreasing means that $\overline{D}_1 \le 0$, almost decreasing means that $\overline{D}_1 = \varepsilon > 0$ for some small ε.

What Bounds Can We Extract When the Function $f(x_1)$ is almost increasing: lower bounds. For an almost increasing function, for each $x_1 \in [\underline{x}_1, \overline{x}_1]$, we have

$$f(x_1) = f(\underline{x}_1) + \int_{\underline{x}_1}^{x_1} f'(x)\, dx.$$

Since $f'(x) \ge -\varepsilon$ for all $x_1 \in [\underline{x}_1, \overline{x}_1]$, we can thus conclude that

$$f(x_1) \ge f(\underline{x}_1) - \varepsilon \cdot (x_1 - \underline{x}_1).$$

Thus, for each x_1, we get $f(x_1) \ge \min_{x_1}(f(\underline{x}_1) - \varepsilon \cdot (x_1 - \underline{x}_1))$. The smallest value of the right-hand side is attained when x_1 is the largest, i.e., when $x_1 = \overline{x}_1$. In this case, $x_1 - \underline{x}_1 = \overline{x}_1 - \underline{x}_1 = 2 \cdot \Delta_1$, thus, we conclude that

$$f(x_1) \ge f(\underline{x}_1) - 2 \cdot \varepsilon \cdot \Delta_1.$$

Thus, when $\underline{D}_1 < 0$, we have a lower bound $\underline{y} \ge f(\underline{x}_1) + 2 \cdot \underline{D}_1 \cdot \Delta_1$. When $\underline{D}_1 \ge 0$, the lower bound is exactly $f(\underline{x}_1)$. Thus, in general, we get the following lower bound:

$$\underline{y} \ge f(\underline{x}_1) + 2 \cdot \min(0, \underline{D}_1) \cdot \Delta_1. \tag{9}$$

We can easily see that if the function $f(x_1)$ is increasing, we get the exact lower bound $\underline{y} = f(\underline{x}_1)$.

What Bounds Can We Extract When the Function $f(x_1)$ is Almost Increasing: Upper Bounds. Similarly, from the fact that

$$f(\overline{x}_1) - f(x_1) = \int_{x_1}^{\overline{x}_1} f'(x)\, dx \geq -\varepsilon \cdot (\overline{x}_1 - x_1),$$

it follows that $f(x_1) \leq f(\overline{x}_1) + \varepsilon \cdot (\overline{x}_1 - x_1)$ and thus, that

$$f(x_1) \leq \max_{x_1}(f(\overline{x}_1) + \varepsilon \cdot (\overline{x}_1 - x_1)).$$

The largest value is attained when x_1 is the smallest, i.e., when $x_1 = \underline{x}_1$; in this case, $\overline{x}_1 - x_1 - 2 \cdot \Delta_1$, so we conclude that $f(x_1) \leq f(\overline{x}_1) + 2 \cdot \varepsilon \cdot \Delta_1$. So, in general, we get an upper bound

$$\overline{y} \leq f(\overline{x}_1) - 2 \cdot \min(0, \underline{D}_1) \cdot \Delta_1. \tag{10}$$

If the function $f(x_1)$ is increasing, we get the exact upper bound $\overline{y} = f(\overline{x}_1)$.

What Bounds Can We Extract When the Function $f(x_1)$ is Almost Decreasing: Upper Bounds. For an almost decreasing function, for each $x_1 \in [\underline{x}_1, \overline{x}_1]$, from

$$f(x_1) = f(\underline{x}_1) + \int_{\underline{x}_1}^{x_1} f'(x)\, dx,$$

by using the fact that $f'(x) \leq \varepsilon$, we can thus conclude that

$$f(x_1) \leq f(\underline{x}_1) + \varepsilon \cdot (x_1 - \underline{x}_1).$$

Thus, for each x_1, we get

$$f(x_1) \geq \max_{x_1}(f(\underline{x}_1) + \varepsilon \cdot (x_1 - \underline{x}_1)).$$

The largest value of the right-hand side is attained when x_1 is the largest, i.e., when $x_1 = \overline{x}_1$. In this case, $x_1 - \underline{x}_1 = \overline{x}_1 - \underline{x}_1 = 2 \cdot \Delta_1$, thus, we conclude that

$$f(x_1) \leq f(\underline{x}_1) + 2 \cdot \varepsilon \cdot \Delta_1.$$

Thus, when $\overline{D}_1 > 0$, we have an upper bound $\overline{y} \leq f(\underline{x}_1) + 2 \cdot \overline{D}_1 \cdot \Delta_1$. When $\overline{D}_1 \leq 0$, the upper bound is exactly $f(\underline{x}_1)$. Thus, in general, we get the following upper bound:

$$\overline{y} \geq f(\underline{x}_1) + 2 \cdot \max(0, \overline{D}_1) \cdot \Delta_1. \tag{11}$$

If the function $f(x_1)$ is decreasing, we get the exact upper bound $\overline{y} = f(\underline{x}_1)$.

What Bounds Can We Extract When the Function $f(x_1)$ is almost decreasing: lower bounds. Similarly, from the fact that

$$f(\overline{x}_1) - f(x_1) = \int_{x_1}^{\overline{x}_1} f'(x)\, dx \geq -\varepsilon \cdot (\overline{x}_1 - x_1),$$

it follows that $f(x_1) \geq f(\overline{x}_1) - \varepsilon \cdot (\overline{x}_1 - x_1)$ and thus, that

$$f(x_1) \geq \min_{x_1}(f(\overline{x}_1) - \varepsilon \cdot (\overline{x}_1 - x_1)).$$

The smallest value is attained when x_1 is the smallest, i.e., when $x_1 = \underline{x}_1$; in this case, $\overline{x}_1 - x_1 = 2 \cdot \Delta_1$, so we conclude that $f(x_1) \geq f(\overline{x}_1) - 2 \cdot \varepsilon \cdot \Delta_1$. So, in general, we get a lower bound

$$\underline{y} \leq f(\overline{x}_1) - 2 \cdot \max(0, \overline{D}_1) \cdot \Delta_1. \tag{12}$$

If the function $f(x_1)$ is decreasing, we get the exact lower bound $\underline{y} = f(\overline{x}_1)$.

Summarizing. From the mean value form, we extract the bounds

$$\underline{y} \geq f(\widetilde{x}_1) - \Delta_1 \cdot \max(|\underline{D}_1|, |\overline{D}_1|) \text{ and } \overline{y} \leq f(\widetilde{x}_1) + \Delta_1 \cdot \max(|\underline{D}_1|, |\overline{D}_1|).$$

From the almost-increasing idea, we can conclude that

$$\underline{y} \geq f(\underline{x}_1) + 2 \cdot \min(0, \underline{D}_1) \cdot \Delta_1 \text{ and } \overline{y} \leq f(\overline{x}_1) - 2 \cdot \min(0, \underline{D}_1) \cdot \Delta_1.$$

From the almost-decreasing idea, we can conclude that

$$\underline{y} \leq f(\overline{x}_1) - 2 \cdot \max(0, \overline{D}_1) \cdot \Delta_1 \text{ and } \overline{y} \geq f(\underline{x}_1) + 2 \cdot \max(0, \overline{D}_1) \cdot \Delta_1.$$

Resulting Idea. In the non-monotonic case:

- instead of only computing the bounds $\underline{Y}$ and $\overline{Y}$ corresponding to the mean value form,
- why not also (or instead) compute the bounds corresponding to almost-monotonicity (at least some of them), and then take the largest of the lower bounds $\underline{Y}$ and the smallest of the upper bounds $\overline{Y}$?

To decide when to use the new bounds, let us analyze, on a simple example, when the next bounds are better.

What is the Simplest Case on Which We Can Compare the New Method with the Existing Ones. Linear functions are monotonic – and thus, for linear function, we do not need neither the mean value bounds, not any new bounds, we can easily compute the actual range.

Thus, to compare the two methods, we need to consider nonlinear functions. Let us consider the simplest nonlinear functions: the quadratic functions $f(x_1) = a_0 \cdot x_1^2 + a_1 \cdot x_1 + a_2$. Without losing generality, let us consider the case when $a_0 > 0$.

When we re-scale x_1 to $x_1' \stackrel{\text{def}}{=} \sqrt{a_0} \cdot x_1$, the interval changes, but, as one can see, the ranges produced by all three methods do not change. Thus, to compare the methods, it is sufficient to consider the case when $a_0 = 1$.

We can always represent the resulting expression $x_1^2 + a_1 \cdot x_1 + a_2$ as

$$(x_1 + a_1/2)^2 + \text{const}.$$

Replacing x_1 with the new variable $x_1' = x_1 + a_1/2$ also does not change the bounds, so we can safely assume that $a_1 = 0$, and thus, $f(x_1) = x_1^2 + a_2$.

If we subtract a constant from all the values of the function $f(x_1)$, then all the bounds will decrease by the same constant – and thus, this change will not affect which bound is smaller or larger (and hence, will not affect which method is better). So, for the purpose of comparing the estimates, we can subtract a_2 from all the values of the function $f(x_1)$ and thus, consider the function $f(x_1) = x_1^2$.

We consider the case when the function is not monotonic on the interval $[\underline{x}_1, \overline{x}_1]$, so we must have $\underline{x}_1 < 0 < \overline{x}_1$ – since on intervals not containing 0 the function $f(x_1) = x_1^2$ is clearly monotonic.

Without losing generality, let us assume that $|\underline{x}_1| \le \overline{x}_1$: if this is not true, we can switch from x_1 to $-x_1$ and get the above inequality – and this switch also does not affect the relative comparison of the above estimates.

Comparing the Estimates on the Simplest Case: General Discussion.
Let us consider this example of a function $f(x_1) = x_1^2$ on an interval $[\underline{x}_1, \overline{x}_1]$ for which $\underline{x}_1 < 0 < \overline{x}_1$ and $|\underline{x}_1| \le \overline{x}_1$. In this case, the size $\overline{x}_1$ of the increasing part is larger than (or equal to) the size $|\underline{x}_1|$ of the decreasing part. In this sense, the function is more on the increasing side, so it makes sense to compare the mean value estimate only with the almost increasing case.

Here, $\widetilde{x}_1 = \dfrac{\underline{x}_1 + \overline{x}_1}{2} = \dfrac{\overline{x}_1 - |\underline{x}_1|}{2}$ and $\Delta_1 = \dfrac{\overline{x}_1 - \underline{x}_1}{2} = \dfrac{\overline{x}_1 + |\underline{x}_1|}{2}$. In this example, $f'(x_1) = 2x_1$, so $\underline{D}_1 = 2 \cdot \underline{x}_1 = -2|\underline{x}_1|$ and $\overline{D}_1 = 2 \cdot \overline{x}_1$. Here,

$$[\underline{D}_1, \overline{D}_1] \cdot [-\Delta_1, \Delta_1] = [-\overline{D}_1 \cdot \Delta_1, \overline{D}_1 \cdot \Delta_1].$$

Mean Value Bounds for the Simplest Case. The mean value estimate has the following form

$$\underline{Y} = \left(\frac{\overline{x}_1 - |\underline{x}_1|}{2}\right)^2 - \overline{D}_1 \cdot \Delta_1 = \left(\frac{\overline{x}_1 - |\underline{x}_1|}{2}\right)^2 - 2 \cdot \overline{x}_1 \cdot \frac{\overline{x}_1 + |\underline{x}_1|}{2} =$$

$$\frac{1}{4} \cdot (\overline{x}_1)^2 - \frac{1}{2} \cdot \overline{x}_1 \cdot |\underline{x}_1| + \frac{1}{4} \cdot (\underline{x}_1)^2 - (\overline{x}_1)^2 - \overline{x}_1 \cdot |\underline{x}_1| =$$

$$-\frac{3}{4} \cdot (\overline{x}_1)^2 - \frac{3}{2} \cdot \overline{x}_1 \cdot |\underline{x}_1| + \frac{1}{4} \cdot (\underline{x}_1)^2; \tag{13}$$

$$\overline{Y} = \left(\frac{\overline{x}_1 - |\underline{x}_1|}{2}\right)^2 + \overline{D}_1 \cdot \Delta_1 = \left(\frac{\overline{x}_1 - |\underline{x}_1|}{2}\right)^2 + 2 \cdot \overline{x}_1 \cdot \frac{\overline{x}_1 + |\underline{x}_1|}{2} =$$

$$\frac{1}{4} \cdot (\overline{x}_1)^2 - \frac{1}{2} \cdot \overline{x}_1 \cdot |\underline{x}_1| + \frac{1}{4} \cdot (\underline{x}_1)^2 + (\overline{x}_1)^2 + \overline{x}_1 \cdot |\underline{x}_1| =$$

$$\frac{5}{4} \cdot (\overline{x}_1)^2 + \frac{1}{2} \cdot \overline{x}_1 \cdot |\underline{x}_1| + \frac{1}{4} \cdot (\underline{x}_1)^2. \tag{14}$$

Almost-Increasing Bounds for the Simplest Case. The almost-increasing estimates take the form

$$\underline{Y} = (\underline{x}_1)^2 - 2 \cdot |\underline{x}_1| \cdot (\overline{x}_1 + |\underline{x}_1|) = -2 \cdot |\underline{x}_1| \cdot \overline{x}_1 - (\underline{x}_1)^2; \tag{15}$$

$$\overline{Y} = (\overline{x}_1)^2 + 2 \cdot |\underline{x}_1| \cdot (\overline{x}_1 + |\underline{x}_1|) = (\overline{x}_1)^2 + 2 \cdot |\underline{x}_1| \cdot \overline{x}_1 + 2 \cdot (\underline{x}_1)^2. \qquad (16)$$

When does the New Method Lead to a Better Lower Bound? The lower bound produced by the new technique is better than the lower bound coming from the mean value form if

$$-\frac{3}{4} \cdot (\overline{x}_1)^2 - \frac{3}{2} \cdot \overline{x}_1 \cdot |\underline{x}_1| + \frac{1}{4} \cdot (\underline{x}_1)^2 < -2 \cdot |\underline{x}_1| \cdot \overline{x}_1 - (\underline{x}_1)^2,$$

i.e., equivalently, if

$$\frac{5}{4} \cdot (\underline{x}_1)^2 + \frac{1}{2} \cdot |\underline{x}_1| \cdot \overline{x}_1 - \frac{3}{4} \cdot (\overline{x}_1)^2 < 0.$$

Multiplying both sides of this inequality by 4 and dividing by $(\overline{x}_1)^2$, we conclude that $5z^2 + 2z - 3 < 0$, where we denoted $z \overset{\text{def}}{=} \dfrac{|\underline{x}_1|}{\overline{x}_1}$. This is equivalent to $z < 0.6$ – so it is better in most cases!

When does the New Method Lead to a Better Upper Bound? The upper bound produced by the new technique is better then the upper bound coming from the mean value if

$$(\overline{x}_1)^2 + 2 \cdot |\underline{x}_1| \cdot \overline{x}_1 + 2 \cdot (\underline{x}_1)^2 < \frac{5}{4} \cdot (\overline{x}_1)^2 + \frac{1}{2} \cdot \overline{x}_1 \cdot |\underline{x}_1| + \frac{1}{4} \cdot (\underline{x}_1)^2,$$

i.e., equivalently, that

$$\frac{7}{4} \cdot (\underline{x}_1)^2 + \frac{3}{2} \cdot \overline{x}_1 \cdot |\underline{x}_1| - \frac{1}{4} \cdot (\overline{x}_1)^2 < 0.$$

Multiplying both sides of this inequality by 4 and dividing by $(\overline{x}_1)^2$, we conclude that $7z^2 + 6z - 1 < 0$, i.e., that $z < 1/7$. Thus, the upper bound is better only if $|\underline{x}_1| < \dfrac{1}{7} \cdot \overline{x}_1$ – i.e., only when the function is indeed almost increasing.

Towards a General Conclusion. In the above simplest case, D_1 is simply proportional to x_1, with a positive coefficient (equal to 2). So, the comparison between the values $\underline{x}_1$ and $\overline{x}_1$ are equivalent to a similar comparison between $\underline{D}_1$ and $\overline{D}_1$. For example, the inequality $|\underline{x}_1| < 0.6 \cdot \overline{x}_1$ is equivalent to $|\underline{D}_1| < 0.6 \cdot \overline{D}_1$. Such inequalities can be easily formulated in the general case. Thus, we arrive at the following recommendations.

Comparison: Resulting Recommendation. When $\underline{D}_1 \geq 0$ or $\overline{D}_1 \leq 0$, the function is monotonic – so, computing its range is easy. The difficulties come when $\underline{D}_1 < 0 < \overline{D}_1$. Let us consider the case when $|\underline{D}_1| \leq \overline{D}_1$. Then, if we do not want to waste time on computing more than one pair of bounds:

- if $|\underline{D}_1| < \dfrac{1}{7} \cdot \overline{D}_1$, then we should only compute almost-increasing bounds;

- if $\dfrac{1}{7} \cdot \overline{D}_1 < |\underline{x}_1| \leq 0.6 \cdot \overline{D}_1$, then we should use the almost-increasing estimate to compute the lower bound and the mean value estimate for the upper bound;

- finally, if $0.6 \cdot \overline{D}_1 \leq |\underline{D}_1|$, we should only compute the mean value estimates.

Similarly, if $|\underline{D}_1| > \overline{D}_1$, then:

- if $\overline{D}_1 < \dfrac{1}{7} \cdot |\underline{D}_1|$, then we should only compute almost-decreasing bounds;
- if $\dfrac{1}{7} \cdot |\underline{D}_1| < \overline{D}_1 \leq 0.6 \cdot \overline{D}_1$, then we should use the mean value estimate for the lower bound and the almost-decreasing estimate to compute the upper bound;
- finally, if $0.6 \cdot |\underline{D}_1| \leq \overline{D}_1$, we should only compute the mean value estimates.

Comment. Of course, these recommendations are approximate, a future practical experience will hopefully provide better recommendations.

3   Multi-D Case

Almost Increasing Case: Formulas. In the general multi-D case, we can similarly conclude that when $\underline{D}_i < 0 < \overline{D}_i$ and $|\underline{D}_i| \leq \overline{D}_i$, then $\underline{y} \geq \underline{y}_{\underline{i}} - 2 \cdot |\underline{D}_i| \cdot \Delta_i$, where $\underline{y}_{\underline{i}}$ denotes the smallest possible value of the following auxiliary function of $n-1$ variables

$$f_{\underline{i}}(x_1, \ldots, x_{i-1}, x_{i+1}, \ldots, x_n) \overset{\text{def}}{=} f(x_1, \ldots, x_{i-1}, \underline{x}_i, x_{i+1}, \ldots, x_n).$$

Similarly, we have $\overline{y} \leq \overline{y}_{\overline{i}} + 2 \cdot |\underline{D}_i| \cdot \Delta_i$, where $\overline{y}_{\overline{i}}$ is the maximum of the function

$$f_{\overline{i}}(x_1, \ldots, x_{i-1}, x_{i+1}, \ldots, x_n) \overset{\text{def}}{=} f(x_1, \ldots, x_{i-1}, \overline{x}_i, x_{i+1}, \ldots, x_n).$$

Almost Increasing Case: Recommendation. Similar to the 1-D case, we can conclude that, if $|\underline{D}_i| < \dfrac{1}{7} \cdot \overline{D}_i$, then, instead of using the mean value bounds, we should reduce our original n-variable interval computations problem to two $(n-1)$-variable ones, for $f_{\underline{i}}$ and $f_{\overline{i}}$. Once we solve these fewer-variable problems and find the bounds $\underline{Y}_{\underline{i}} \leq \underline{y}_{\underline{i}}$ and $\overline{Y}_{\overline{i}} \geq \overline{y}_{\overline{i}}$, we can then use the above formulas to produce the bounds for the original problem:

$$\underline{Y} = \underline{Y}_{\underline{i}} - 2 \cdot |\underline{D}_i| \cdot \Delta_i \text{ and } \overline{Y} = \overline{Y}_{\overline{i}} + 2 \cdot |\underline{D}_i| \cdot \Delta_i.$$

When $\dfrac{1}{7} \cdot \overline{D}_i \leq |\underline{D}_i| < 0.6 \cdot \overline{D}_i$, then we should use the above reduction to compute the lower bound $\underline{Y}$, and the mean value technique for computing the upper bound $\overline{Y}$.

When $0.6 \cdot \overline{D}_i \leq |\underline{D}_i| \leq \overline{D}_i$, we should use mean value estimates for both bounds.

Almost Decreasing Case: Recommendation. Similarly, if $\overline{D}_i < \dfrac{1}{7} \cdot |\underline{D}_i|$, then we should reduce our original n-variable interval computations problem to two $(n-1)$-variable ones. Once we solve these fewer-variable problems and find the bounds $\underline{Y}_{\overline{i}} \leq \underline{y}_{\overline{i}}$ and $\overline{Y}_{\underline{i}} \geq \overline{y}_{\underline{i}}$, we can then produce the bounds for the original problem:

$$\underline{Y} = \underline{Y}_{\overline{i}} - 2 \cdot \overline{D}_i \cdot \Delta_i \text{ and } \overline{Y} = \overline{Y}_{\underline{i}} + 2 \cdot \overline{D}_i \cdot \Delta_i.$$

When $\frac{1}{7} \cdot |\underline{D}_i| \leq \overline{D}_i < 0.6 \cdot |\underline{D}_i|$, then we should use the above reduction to compute the lower bound $\underline{Y}$, and the mean value technique for computing the upper bound $\overline{Y}$.

When $0.6 \cdot \overline{D}_i \leq |\underline{D}_i| \leq \overline{D}_i$, we should use mean value estimates for both bounds.

Acknowledgements. This work was supported in part by the US National Science Foundation grant HRD-1242122 (Cyber-ShARE Center of Excellence).

References

1. Jaulin, L., Kiefer, M., Didrit, O., Walter, E.: Applied Interval Analysis, with Examples in Parameter and State Estimation, Robust Control, and Robotics. Springer, London (2001). https://doi.org/10.1007/978-1-4471-0249-6
2. Kreinovich, V., Lakeyev, A., Rohn, J., Kahl, P.: Computational Complexity and Feasibility of Data Processing and Interval Computations. Kluwer, Dordrecht (1998)
3. Mayer, G.: Interval Analysis and Automatic Result Verification. de Gruyter, Berlin (2017)
4. Moore, R.E., Kearfott, R.B., Cloud, M.J.: Introduction to Interval Analysis. SIAM, Philadelphia (2009)
5. Rabinovich, S.G.: Measurement Errors and Uncertainties: Theory and Practice. Springer, New York (2005)
6. Vavasis, S.A.: Nonlinear Optimization: Complexity Issues. Oxford University Press, New York (1991)

How Accurately Can We Determine the Coefficients: Case of Interval Uncertainty

Michal Cerny[1] and Vladik Kreinovich[2(✉)]

[1] Faculty of Informatics and Statistics, University of Economics,
nam. W. Churchilla 4, 13067 Prague, Czech Republic
`cernym@vse.cz`
[2] Department of Computer Science, University of Texas at El Paso,
El Paso, TX 79968, USA
`vladik@utep.edu`

Abstract. In many practical situations, we need to estimate the parameters of a linear (or more general) dependence based on measurement results. To do that, it is useful, before we start the actual measurements, to estimate how accurately we can, in principle, determine the desired coefficients: if the resulting accuracy is not sufficient, then we should not waste time trying and resources and instead, we should invest in more accurate measuring instruments. This is the problem that we analyze in this paper.

1 Formulation of the Problem

Need to Determine the Dependence Between Different Quantities. One of the main objectives of science is to find the dependencies $y = f(x_1, \ldots, x_n)$ between values of different quantities at different moments of time and at different locations.

Once we know such dependencies, we can then use them to predict the future values of different quantities.

For example, Newton's laws describe how the acceleration y of a celestial body depends on the current location and masses of this and other bodies $x_1, \ldots, x_n$ – and thus, these laws enable us to predict how these bodies will move.

Another important case is when we want to estimate the value of a quantity y which is difficult to directly measure. In such cases, it is often possible to find easier-to-measure quantities $x_1, \ldots, x_n$ knowing which we can determine y. For example, it is difficult to directly measure the distance y between two faraway locations on the Earth, but we can determine this distance if we use astronomical observations – or, nowadays, signals from the GPS satellites – to find the exact coordinates of each of the two locations.

How Can We Determine this Dependence. In some cases, we can use the known physical laws to derive the desired dependence. However, in most other cases, this dependence needs to be determined empirically:

© Springer Nature Switzerland AG 2019
R. B. Kearfott et al. (Eds.): IFSA 2019/NAFIPS 2019, AISC 1000, pp. 779–787, 2019.
https://doi.org/10.1007/978-3-030-21920-8_69

- we measure the values $x_1, \ldots, x_n$, and y in different situations, and then
- we use the measurement results to find the desired dependence.

Often, We Know the General Form of the Dependence, We Just Need to Find the Coefficients. In many cases, we know the general form of the desired dependence, i.e., we know that $y = F(x_1, \ldots, x_n, c_0, c_1, \ldots, c_m)$, where F is a known function, and the coefficients c_i need to be determined.

For example, we may know that the dependence is linear, i.e., that

$$y = c_0 + c_1 \cdot x_1 + \ldots + c_n \cdot x_n.$$

This is a typical situation when the values x_i have a narrow range $[\underline{X}_i, \overline{X}_i]$ and thus, we can expand the function $f(x_1, \ldots, x_n)$ in Taylor series over $x_i - \underline{X}_i$ and ignore quadratic (and higher order) terms in this expansion.

Need to Take Uncertainty into Account – In Particular, Interval Uncertainty. Measurements are never absolutely accurate: the measurement result $\widetilde{x}$ is, in general, different from the actual (unknown) value x. In many practical situations, the only information that we have about the measurement error $\Delta x \stackrel{\text{def}}{=} \widetilde{x} - x$ is the upper bound Δ on its absolute value: $|\Delta x| \leq \Delta$; see, e.g., [5].

In this case, after each measurement, the only information that we have about the actual value x is that this value is somewhere in the interval $[\widetilde{x} - \Delta, \widetilde{x} + \Delta]$. Because of this fact, this case is known as the case of *interval uncertainty*. There exist many algorithms for dealing with such uncertainty; see, e.g., [1,3,4].

Measurement Uncertainty Leads to Uncertainty in Coefficients. Since we can only measure the values x_i and y with some uncertainty, we can therefore only determine the coefficients c_i with some uncertainty.

It is therefore important to determine how accurate are the values c_i that we get as a result of these measurements.

Which Uncertainty Should Be Taken into Account. Strictly speaking, there are measurement uncertainties both when we measure easier-to-measure quantities $x_1, \ldots, x_n$, and when we measure the desired difficult-to-measure quantity y. However, usually, because of the very fact that y is much more difficult to measure than x_i, the measurement errors Δy corresponding to measuring y are much larger than the measurement errors of measuring x_i – so much larger that we can usually safely ignore the measurement errors of measuring x_i and assume that these values are known exactly.

Thus, in the linear case, we can safely assume that for each measurement k, we know the exact values $x_1^{(k)}, \ldots, x_n^{(k)}$, but we only know $y^{(k)}$ with uncertainty – i.e., based on the measurement result $\widetilde{y}^{(k)}$ and the known accuracy $\Delta > 0$, we know that the actual value $y^{(k)} = c_0 + \sum_{i=1}^{n} c_i \cdot x_i^{(k)}$ is between $\underline{y}^{(k)} = \widetilde{y}^{(k)} - \Delta$ and $\overline{y}^{(k)} = \widetilde{y}^{(k)} + \Delta$.

Once We Perform the Measurements, We Can Feasibly Find the Accuracy. One we have the measurement results, we can find the bounds on each of the coefficients c_i (and, similarly, the bounds on any linear combination of c_i) by solving the following linear programming problems (see, e.g., [7]): minimize (maximize) c_i under the constraints that

$$\underline{y}^{(k)} \leq c_0 + \sum_{i=1}^{n} c_i \cdot x_i^{(k)} \leq \overline{y}^{(k)}$$

for all the measurements $k = 1, \ldots, K$.

Remaining Question. But before we start spending our resources on measurements, it is desirable to check how accurately we can, in principle, determine the coefficients c_i.

This checking is important: If the resulting accuracy is not enough for us – then we should not waste time performing the measurements, and instead we should invest in a more accurate y-measuring instrument.

Of course, we can answer the above question by simulating measurement errors, but it would be great to have simple analytical expressions that would not require extensive simulation-related computations.

What We Do in This Paper. In this paper, we provide such expressions for the linear case.

2   Definitions and Results

Discussion. The range of each physical quantity is usually bounded:

- coordinates of Earth locations are bounded by the Earth's size,
- velocities are bounded by the speed of light, etc.

Thus, we can safely assume that for each variable x_i, we know the interval $[\underline{X}_i, \overline{X}_i]$ of its possible values.

Thus, we arrive at the following formulation of the problem.

Definition 1. *Let us assume that we are given the value $\Delta > 0$ and n intervals $[\underline{X}_i, \overline{X}_i]$, $i = 1, 2, \ldots, n$. We say that a tuple $(\Delta c_0, \Delta c_1, \ldots, \Delta c_n)$ is within the possible uncertainty if for each tuple $(c_0, c_1, \ldots, c_n)$ and for each combination of values $x_i \in [\underline{X}_i, \overline{X}_i]$, we have $|y' - y| \leq \Delta$, where:*

- $y \overset{\text{def}}{=} c_0 + \sum_{i=1}^{n} c_i \cdot x_i$ *and*
- $y' \overset{\text{def}}{=} c'_0 + \sum_{i=1}^{n} c'_i \cdot x_i$, *where $c'_i \overset{\text{def}}{=} c_i + \Delta c_i$.*

Comment. Because of the measurement uncertainty, after the measurement, the range of possible values of the corresponding quantity x is $[\tilde{x} - \Delta, \tilde{x} + \Delta]$. It may be therefore convenient to represent the intervals $[\underline{X}_i, \overline{X}_i]$ in the same form, as

$$[\underline{X}_i, \overline{X}_i] = [\tilde{X}_i - \Delta_i, \tilde{X}_i + \Delta_i].$$

For this, we need to take $\tilde{X}_i = \dfrac{\underline{X}_i + \overline{X}_i}{2}$ and $\Delta_i = \dfrac{\overline{X}_i - \underline{X}_i}{2}$.

Proposition 1. *For each Δ and $[\underline{X}_i, \overline{X}_i]$, a tuple $(\Delta c_0, \Delta c_1, \ldots, \Delta c_n)$ is within the possible uncertainty if and only if*

$$|\Delta c_0'| + \sum_{i=1}^{n} |\Delta c_i| \cdot \Delta_i \leq \Delta, \tag{1}$$

where $\Delta c_0' \overset{\text{def}}{=} \Delta c_0 + \sum_{i=1}^{n} \Delta c_i \cdot \widetilde{X}_i$.

Proof of Proposition 1. One can easily see that, since the dependence of y on c_i is linear, the difference $\Delta y \overset{\text{def}}{=} y' - y$ is equal to $\Delta y = \Delta c_0 + \sum_{i=1}^{n} \Delta c_i \cdot x_i$.

Each of the variables x_i independently runs over its own interval

$$[\underline{X}_i, \overline{X}_i] = [\widetilde{X}_i - \Delta_i, \widetilde{X}_i + \Delta_i].$$

Thus, each value x_i from this interval can be represented as $\widetilde{X}_i + \Delta x_i$, where $\Delta x_i \overset{\text{def}}{=} x_i - \widetilde{X}_i$ takes all possible values from the interval $[-\Delta_i, \Delta_i]$.

Substituting this expression for x_i into the above formula for Δy, we conclude that

$$\Delta y = \Delta c_0' + \sum_{i=1}^{n} \Delta c_i \cdot \widetilde{X}_i + \sum_{i=1}^{n} \Delta c_i \cdot \Delta x_i. \tag{2}$$

To make sure that always $|\Delta y| \leq \Delta$, i.e., that always $-\Delta \leq \Delta y \leq \Delta$, it is sufficient to make sure that

$$-\Delta \leq \underline{\Delta} \text{ and } \overline{\Delta} \leq \Delta,$$

where:

- $\underline{\Delta}$ is the smallest possible value of the expression (2), while
- $\overline{\Delta}$ is the largest possible value of the expression (2).

Let us find these smallest and largest values.

Each of the variables Δx_i independently runs over its own interval $[-\Delta_i, \Delta_i]$. Thus, the smallest possible value of (1) is attained when each of the terms in the sum (2) is the smallest.

- For $\Delta c_i \geq 0$, the term $\Delta c_i \cdot \Delta x_i$ is increasing with Δx_i, so its smallest value if when x_i is the largest: $\Delta x_i = -\Delta_i$. In this case, the value is equal to $-\Delta c_i \cdot \Delta_i$.
- For $\Delta c_i \leq 0$, the term $\Delta c_i \cdot \Delta x_i$ is decreasing with Δx_i, so its smallest value if when x_i is the largest: $\Delta x_i = \Delta_i$. In this case, the value is equal to $\Delta c_i \cdot \Delta_i$.

We can describe both terms by a single formula $-|\Delta c_i| \cdot \Delta_i$. Thus, the smallest possible value $\underline{\Delta}$ of Δy is equal to $\underline{\Delta} = \Delta c_0' - \sum_{i=1}^{n} |\Delta c_i| \cdot \Delta_i$, and the condition $-\Delta \leq \underline{\Delta}$ is equivalent to

$$- \Delta c_0' + \sum_{i=1}^{n} |\Delta c_i| \cdot \Delta_i \leq \Delta. \tag{3}$$

Similarly, the largest possible value of each term $\Delta c_i \cdot \Delta x_i$ is equal to $|\Delta c_i| \cdot \Delta_i$, thus

$$\overline{\Delta} = \Delta c_0' + \sum_{i=1}^{m} |\Delta c_i| \cdot \Delta_i,$$

and the condition $\overline{\Delta} \leq \Delta$ can be described as

$$\Delta c_0' + \sum_{i=1}^{n} |\Delta c_i| \cdot \Delta_i \leq \Delta. \tag{4}$$

Inequalities (3) and (4) are equivalent to requiring that the largest of the two left-hand sides is smaller than or equal to Δ, i.e., to the desired inequality. The proposition is proven.

Discussion. Based on Proposition 1, we can find bounds on each of the coefficient $\Delta c_1, \ldots, \Delta c_n$:

Proposition 2. *For each i from 1 to n, among all possible tuples which are within the possible uncertainty, the corresponding values of Δc_i form the interval*

$$\left[-\frac{\Delta}{\Delta_i}, \frac{\Delta}{\Delta_i} \right].$$

Comments. Thus, if we can measure y with accuracy Δ, and we can use any value x_i from the interval $[\tilde{X}_i - \Delta_i, \tilde{X}_i + \Delta_i]$, then we can determine the coefficient c_i that describes the dependence of y on x_i with accuracy $\dfrac{\Delta}{\Delta_i}$.

It is worth mentioning that the accuracy $\dfrac{\Delta}{\Delta_i}$ is what we can *guarantee* if we perform sufficiently many measurements. However, even with a primitive y-measuring device, for which the measurement accuracy Δ is high, we can get lucky and get much more accurate – even absolutely accurate – values of c_i.

Indeed, let us assume that for each tuple $\left(x_1^{(k)}, \ldots, x_n^{(k)} \right)$ of the x-values, for which the actual value of y is $y^{(k)} = c_0 + \sum_{i=1}^{n} c_i \cdot x_i^{(k)}$, we perform two y-measurements:

- in the first measurement, we get $\tilde{y}^{(k)} = y^{(k)} + \Delta$ and thus, based on this measurement result, we conclude that the actual value of $y^{(k)}$ belongs to the interval
$$[\tilde{y}^{(k)} - \Delta, \tilde{y}^{(k)} + \Delta] = [y^{(k)}, y^{(k)} + 2\Delta];$$

- in the second measurement, we get $\tilde{y}^{(k)} = y^{(k)} - \Delta$ and thus, based on this measurement result, we conclude that the actual value of $y^{(k)}$ belongs to the interval
$$[\tilde{y}^{(k)} - \Delta, \tilde{y}^{(k)} + \Delta] = [y^{(k)} - 2\Delta, y^{(k)}].$$

Since the value $y^{(k)}$ belongs to both intervals $[y^{(k)}, y^{(k)}+2\Delta]$ and $[y^{(k)}-2\Delta, y^{(k)}]$, it belongs to their intersection – and this intersection consists of the single point $y^{(k)}$. Thus, in this lucky case, we get the exact value of each y – and thus, after $n+1$ measurements, determine the exact values of all $n+1$ coefficients $c_0, c_1, \ldots, c_n$ by solving the corresponding system of linear equations

$$c_0 + \sum_{i=1}^{n} c_i \cdot x^{(k)} = y^{(k)}, \quad k = 1, \ldots, n+1.$$

Proof of Proposition 2. If Δc_i is a part of the tuple which is within the possible uncertainty, then from the inequality (1), we can conclude that $|\Delta c_i| \cdot \Delta_i \leq \Delta$, hence that

$$|\Delta c_i| \leq \frac{\Delta}{\Delta_i}. \tag{5}$$

Vice versa, for each value Δc_i that satisfies the inequality (5), we can take $\Delta c_1 = \ldots = \Delta c_{i-1} = \Delta c_{i+1} = \ldots \Delta c_n = 0$ and choose $\Delta c_0 = -\Delta x_i \cdot \widetilde{X}_i$, then $\Delta c_0' = 0$ and thus, the inequality (1) is satisfied.

The proposition is proven.

Proposition 3. *When 0 is a possible value of each variable x_i, then among all possible tuples which are within the possible uncertainty, the corresponding values of Δc_0 form the interval $[-\Delta, \Delta]$.*

Comment. Thus, if we can measure y with accuracy Δ, and we can use any value x_i from the interval $[\widetilde{X}_i - \Delta_i, \widetilde{X}_i + \Delta_i]$ containing 0, then we can determine the free term c_0 in the dependence of y on $x_1, \ldots, x_n$ with accuracy Δ.

Proof of Proposition 3. If a tuple $(\Delta c_0, \Delta c_1, \ldots, \Delta c_n)$ is within the possible uncertainty, then for possible value $x_1 = \ldots = x_n = 0$, we get $|\Delta c_0| \leq \Delta$.

Vice versa, if we have a value Δc_0 for which $|\Delta c_0| \leq \Delta$, then, by taking $\Delta c_1 = \ldots = \Delta c_n = 0$, we get a tuple that, as one can easily see, satisfies the desired inequality for all x_i and is, thus, within the possible uncertainty.

The proposition is proven.

Discussion. When for some i, $0 \notin [\underline{X}_i, \overline{X}_i]$, then all values $\Delta c_0 \in [-\Delta, \Delta]$ are still possible, but some values outside this interval are possible too.

Proposition 4. *For every value $\Delta c_0 \in [-\Delta, \Delta]$, there exists a tuple $(\Delta c_0, \Delta c_1, \ldots, \Delta c_n)$ which is within the possible uncertainty.*

Proof. This was, in effect, already proven in the proof of Proposition 3.

Proposition 5. *For $n = 1$, the range of possible values of Δc_0 is $[-\Delta', \Delta']$, where $\Delta' = \Delta + \dfrac{\Delta}{\Delta_1} \cdot m_1$ and:*

- $m_1 = 0$ *if $0 \in [\underline{X}_1, \overline{X}_1]$;*
- $m_1 = \underline{X}_1$ *if $\underline{X}_1 > 0$, and*
- $m_1 = |\overline{X}_1|$ *if $\overline{X}_1 < 0$.*

Proof. We have already proven this result for the case when $0 \in [\underline{X}_1, \overline{X}_1]$. Without losing generality, let us consider the case when $\underline{X}_1 > 0$; the case when $\overline{X}_1 < 0$ is proven similarly.

In this case, on the one hand, for $x_1 = \underline{X}_1$, we have $|\Delta c_0 + \Delta c_1 \cdot \underline{X}_1| \leq \Delta$, hence

$$|\Delta c_0| \leq |\Delta c_0 + \Delta c_1 \cdot \underline{X}_1| + |-\Delta c_1 \cdot \underline{x}_1| \leq \Delta + |\Delta c_1| \cdot \underline{X}_1.$$

By Proposition 1, we have $|\Delta c_1| \leq \dfrac{\Delta}{\Delta_1}$, hence indeed $|\Delta c_0| \leq \Delta'$.

On the other hand, let us prove that the value $\Delta c_0 = \Delta'$ is possible. Then, by swapping the signs of all Δc_i, we can prove that the value $-\Delta'$ is also possible. The inequalities $|\Delta c_0 + \Delta c_1 \cdot x_1| \leq \Delta$ that describe the set of possible tuples is an intersection of convex sets and is, thus, itself convex. So, with Δ' and $-\Delta'$, any convex combination of them is also possible – i.e., all the values from the interval $[-\Delta', \Delta']$.

Hence, it is sufficient to prove that the value $\Delta c_0 = \Delta'$ is possible. Indeed, we will prove that it is possible if we take $\Delta c_1 = -\dfrac{\Delta}{\Delta_1}$. We then need to prove that for these values Δc_i, we have $|\Delta c_0 + \Delta c_1 \cdot x_1| \leq \Delta$ for all $x_1 \in [\underline{X}_1, \overline{X}_1]$.

The left-hand side of the inequality is a convex function of x_1, so it is sufficient to check this inequality for the endpoints $x_1 = \underline{X}_1$ and $x_1 = \overline{X}_1$. For $x_1 = \underline{X}_1$, we have

$$\Delta c_0 + \Delta c_1 \cdot \underline{X}_1 = \Delta + \dfrac{\Delta}{\Delta_1} \cdot \underline{X}_1 - \dfrac{\Delta}{\Delta_1} \cdot \underline{X}_1 = \Delta,$$

and for $x_1 = \overline{X}_1$, we get

$$\Delta c_0 + \Delta c_1 \cdot \overline{X}_1 = \Delta + \dfrac{\Delta}{\Delta_1} \cdot \underline{X}_1 - \dfrac{\Delta}{\Delta_1} \cdot \overline{X}_1 =$$

$$\Delta - \dfrac{\Delta}{\Delta_1} \cdot (\overline{X}_1 - \underline{X}_1) = \Delta - \dfrac{\Delta}{\Delta_1} \cdot 2\Delta_1 = \Delta - 2\Delta = -\Delta.$$

In both cases, we have $|\Delta c_0 + \Delta c_1 \cdot x_1| \leq \Delta$. Thus, the proposition is proven.

3 Discussion

What if We Have Probabilistic Uncertainty. In the above text, we considered the case when we only know the upper bound on the measurement errors – i.e., when we only know the interval of possible values of the measurement error. In many practical situations, however, in addition to this upper bound, we also have some information about the probability of different values from this interval.

In such cases, it is convenient to represent the measurement error as the sum of two components:

- its mean, which is called *systematic error*, and
- the difference between the measurement error and its mean, which is called the *random error*.

Usually, we know the upper bound Δ_i on the absolute value of the systematic error, and we know some characteristics of the random error; see, e.g., [5]. With what accuracy can we then determine c_i?

Interestingly, we get the same answer as in the interval case. Indeed, if for the same example, we measure y several times, the arithmetic average of the measurement results tends to its mean value, i.e., to the actual value y plus the systematic error s_i; see, e.g., [6]. Thus, in measurement results obtained this way, the random error disappears and we get, in effect, the interval case.

What if We Consider Quadratic Dependencies. In the above text, we considered the case when we could ignore quadratic and higher order terms, and thus, safely assume that the dependence of y on x_i is linear. What if we want a more accurate description and thus, consider quadratic terms as well, i.e., consider the dependence

$$y = c_0 + \sum_{i=1}^{n} c_i \cdot x_i + \sum_{i=1}^{n} \sum_{j=1}^{n} c_{ij} \cdot x_i \cdot x_j.$$

In this case, even for a single tuple

$$(\Delta c_0, \Delta c_1, \ldots, \Delta c_n, \Delta c_{11}, \Delta c_{12}, \ldots, \Delta c_{nn}),$$

it is NP-hard (= intractable) to check whether this tuple is within the accuracy, i.e., whether

$$|\Delta y| = \left| \Delta c_0 + \sum_{i=1}^{n} \Delta c_i \cdot x_i + \sum_{i=1}^{n} \sum_{j=1}^{n} \Delta c_{ij} \cdot x_i \cdot x_j \right| \leq \Delta$$

for all values x_i from the corresponding intervals $[\underline{X}_i, \overline{X}_i]$: indeed, finding the maximum of a quadratic function under interval uncertainty is known to be NP-hard [2,8].

What if We Have an Ellipsoid. Instead of requiring that possible values of $(x_1, \ldots, x_n)$ form a box, we can consider the case when this set is an ellipsoid.

In this case, the range of a linear expression $\Delta c_0 + \sum_{i=1}^{n} \Delta c_i \cdot x_i$ can also be explicitly computed and thus, we also have an analytical expression describing tuples $(\Delta c_0, \Delta c_1, \ldots, \Delta c_n)$ which are within the possible uncertainty.

What if We also Have Relative Measurement Error. In our text, we assumed that the measurement accuracy Δ is the same for all y, i.e., in measurement terms, that we have an *absolute* error. In practice, we often also have *relative* error component, in which cases the upper bound $\Delta(y)$ on the y-measurement error depends on y as $\Delta(y) = \Delta_0 + c \cdot |y|$, for some $\Delta_0 > 0$ and $c > 0$.

Once we have measurement results, we can still use linear programming to find the accuracy with which we can determine the coefficients c_i, but it is not clear how to come up with an analytical expression for the tuples $(\Delta c_0, \Delta c_1, \ldots, \Delta c_n)$ which are within the possible uncertainty.

Acknowledgements. This work was supported in part by the US National Science Foundation grant HRD-1242122 (Cyber-ShARE Center of Excellence). M. Cerny acknowledges the support of the Czech Science Foundation (project 19-02773S).

References

1. Jaulin, L., Kiefer, M., Didrit, O., Walter, E.: Applied Interval Analysis, with Examples in Parameter and State Estimation, Robust Control, and Robotics. Springer, London (2001)
2. Kreinovich, V., Lakeyev, A., Rohn, J., Kahl, P.: Computational Complexity and Feasibility of Data Processing and Interval Computations. Kluwer, Dordrecht (1998)
3. Mayer, G.: Interval Analysis and Automatic Result Verification. de Gruyter, Berlin (2017)
4. Moore, R.E., Kearfott, R.B., Cloud, M.J.: Introduction to Interval Analysis. SIAM, Philadelphia (2009)
5. Rabinovich, S.G.: Measurement Errors and Uncertainties: Theory and Practice. Springer, New York (2005)
6. Sheskin, D.J.: Handbook of Parametric and Nonparametric Statistical Procedures. Chapman and Hall, Boca Raton (2011)
7. Vanderbei, R.J.: Linear Programming: Foundations and Extensions. Springer, New York (2014)
8. Vavasis, S.A.: Nonlinear Optimization: Complexity Issues. Oxford University Press, New York (1991)

Logical Differential Constraints Based on Interval Boolean Tests

Julien Alexandre dit Sandretto$^{(\boxtimes)}$ and Alexandre Chapoutot

ENSTA ParisTech, 828 bd des Maréchaux, 91120 Palaiseau, France
`julien.alexandre-dit-sandretto@ensta-paristech.fr`

Abstract. Continuous-time dynamical systems play a crucial role in the study or the design of systems in various domains. Checking the satisfaction of properties on these systems is important in particular in robotics or control-command systems. Constraint satisfaction problems is a well-suited framework for this purpose and recent papers extend this framework to deal with differential constraints. This article proposes an improvement of constraint differential satisfaction framework by providing a new solving algorithm based on interval Boolean functions.

1   Introduction

Continuous-time dynamical systems play a crucial role in the study or the design of systems in various domains. In particular, prediction of behaviors of systems or satisfaction of properties can be obtained using numerical simulation methods applied to these mathematical models.

Recently, an extension of constraint satisfaction problem (CSP) has been proposed in [2] to deal with dynamical systems which can be used to check temporal properties. This framework named *Set-based Constraint Satisfaction Differential Problems* (SCSDP) has as main feature the use of set-based constraints in order to have *specification robustness* against bounded uncertainties and also to have *model robustness* against model approximation. In consequence, this framework increases reliability of the computed solutions with respect to the real system. These two concepts are explained in the following paragraphs.

Properties of systems, or a specification, are usually given with margins in order to ensure safety or to increase robustness. For example, for autonomous vehicle reaching a particular point x in a map is given up to a given precision δ as sensors produce approximate information on the environment. In consequence, system properties are usually defined over sets of admissible behaviors, *i.e.*, a vehicle is considered to have reached x if its position p is such that $x - \delta \leqslant p \leqslant x + \delta$. Nonetheless, the combination of inequalities in classical CSP framework can be used to model such properties but may also lead to equality relation as in $x \geqslant 0 \wedge x \leqslant 0$. Usually, such equality constraints require relaxation techniques in CSP solver, *i.e.*, $x \in [-\varepsilon, \varepsilon]$. This relaxation should not be taken place inside solving algorithm, which usually applies the same relaxation for all equality

© Springer Nature Switzerland AG 2019
R. B. Kearfott et al. (Eds.): IFSA 2019/NAFIPS 2019, AISC 1000, pp. 788–792, 2019.
https://doi.org/10.1007/978-3-030-21920-8_70

constraints, but instead specification should be written in order to emphasize the margins that is important for the properties.

A system model M is usually only an approximation of the true system S, especially for continuous-time systems defined by ordinary differential equations (ODEs), where some unknown data have to be considered. For example, the position of a mobile robot is usually known up to a given precision depending on sensors. In the framework of bounded uncertainties, such data are represented by bounded sets and so the model M will be associated with a set of possible trajectories. Hence, properties on M should consider sets of values instead of a single value.

The main contribution of this article is to defined a new solving algorithm for SCSDP based on interval Boolean function. This improvement in regards to the solving algorithm presented in [2] allows for a simplification of the construction of complex constraints which may involve disjunctive logical operator.

2   Set-Based Constraint Satisfaction Differential Problems

In [2], a general class of differential equations are considered which can represent ODEs, Differential Algebraic Equations (DAEs) of index 1, and a mix of these equations with additional constraints, *e.g.*, to model energy preservation. More precisely, differential systems are of the form

$$\begin{cases} \dot{\mathbf{y}}(t) = \mathbf{F}(t, \mathbf{y}(t), \mathbf{x}(t), \mathbf{p}), \\ \quad 0 = \mathbf{G}(t, \mathbf{y}(t), \mathbf{x}(t)) \\ \quad 0 = \mathbf{H}(\mathbf{y}(t), \mathbf{x}(t)) \end{cases} \quad . \tag{1}$$

Non-linear functions $\mathbf{F} : \mathbb{R} \times \mathbb{R}^n \times \mathbb{R}^m \times \mathbb{R}^p \to \mathbb{R}^n$, $\mathbf{G} : \mathbb{R} \times \mathbb{R}^n \times \mathbb{R}^m \to \mathbb{R}^m$, $\mathbf{H} : \mathbb{R}^n \times \mathbb{R}^m \to \mathbb{R}$, $t \in [0, t_{\text{end}}]$, $\mathbf{y}(0) \in \mathscr{Y}_0$ and $\mathbf{p} \in \mathscr{P}$ are considered. More precisely, Initial Value Problems (IVP) for parametrized differential equations are considered over a finite time horizon $[0, t_{\text{end}}]$. Note that a bounded set of initial values and a bounded set of parameters are considered in this framework. This necessitates dealing with set of trajectories solution of Eq. (1). We assume classical hypothesis on $\mathbf{F}$, $\mathbf{Q}$, and $\mathbf{H}$ to ensure the existence and uniqueness of the solution of Eq. (1).

We denote by $\mathscr{Y}(\mathscr{T}, \mathscr{Y}_0, \mathscr{P})$ the solution set

$$\mathscr{Y}(\mathscr{T}, \mathscr{Y}_0, \mathscr{P}) = \{\mathbf{y}(t; \mathbf{y}_0, \mathbf{p}) \mid t \in \mathscr{T}, \mathbf{y}_0 \in \mathscr{Y}_0, \mathbf{p} \in \mathscr{P}\}. \tag{2}$$

Intuitively, $\mathscr{Y}(\mathscr{T}, \mathscr{Y}_0, \mathscr{P})$ gathers all the points reached by the solution $\mathbf{y}(t; \mathbf{y}_0, \mathbf{p})$ of Eq. (1) starting from all scalar initial values $\mathbf{y}_0$ and all scalar parameters $\mathbf{p}$. The proposed framework aims at checking if $\mathscr{Y}(\mathscr{T}, \mathscr{Y}_0, \mathscr{P})$ fulfills some specification defined in terms of set-based constraints.

To avoid problematic issue due to equality constraints set-based constraints are considered. More precisely, *inclusion* and *intersection* operators are considered. More precisely, constraints of the form

$$\mathbf{g}(\mathscr{A}) \subseteq \mathscr{B} \quad \text{and} \quad \mathbf{g}(\mathscr{A}) \cap \mathscr{B} = \emptyset,$$

where $\mathscr{A}$ and $\mathscr{B}$ are real compact sets and $\mathbf{g} : \mathbb{R}^n \to \mathbb{R}^m$ is a non-linear function. The lifting of $\mathbf{g}$ to sets is defined as usual by $\mathbf{g}(\mathscr{X}) = \{\mathbf{g}(x) : x \in \mathscr{X}\}$.

Note that these constraints can be seen as Boolean functions but while, from a mathematical formulation, the truth value always exist, it may not be the case when they can be computed. The contribution of this article is to give a formulation of theses constraints in terms of *interval test function* considering *interval Boolean values*, see Sect. 3.

The handling of differential constraints here follows the approach given in [3] in the exception of the solution operator of Eq. (1), which is here represented as a set of solution $\mathscr{Y}(\mathscr{T}, \mathscr{Y}_0, \mathscr{P})$ in order to unify the objects manipulated into constraints which are also sets. Set-based Constraints Satisfaction Differential Problems (SCSDP) based on a set-membership constraints and embedding differential constraints can now be defined.

Definition 1 (SCSDP). A SCSDP is a NCSP made of

- a finite set $\mathscr{S}$ of differential systems S_i as defined in Eq. (1).
- a finite set of variables $\mathscr{V}$ including the parameters of the differential systems S_i, *i.e.*, $(\mathbf{y}_0, \mathbf{p})$, a time variable t and some other algebraic variables $\mathbf{q}$;
- a domain $\mathscr{D}$ made of the domain of parameters $\mathbf{p} : \mathscr{D}_p$, of initial values $\mathbf{y}_0 : \mathscr{D}_{y_0}$, of the time horizon $t : \mathscr{D}_t$, and the domains of algebraic variables $\mathscr{D}_q$;
- a set of constraints $\mathscr{C}$ which may be defined by inclusion or disjunction constraints over variables of $\mathscr{V}$ and special variables $\mathscr{Y}_i(\mathscr{D}_t, \mathscr{D}_{y_0}, \mathscr{D}_p)$ representing the set of the solution of S_i in $\mathscr{S}$.

3 Interval Inclusion Test for SCSDP

Presented in [4], **inclusion tests** can be used to prove that all points in a box verify a property. These tests exploit the notion of interval Boolean values. The Boolean set is defined such that $\mathbb{B} = \{\texttt{false}, \texttt{true}\} = \{0, 1\}$. An interval Boolean is a subset of $\mathbb{B}$, *i.e.*, an element of the interval Boolean set $\mathbb{IB} = \{\emptyset, [0,0], [1,1], [0,1]\}$. The particular values $\emptyset$ and $[0,1]$, standing for the set $\{0,1\}$, mean, respectively, impossible and undetermined.

The operations on Booleans are extended to **interval Booleans**. If $[a] \in \mathbb{IB}$ and $[b] \in \mathbb{IB}$, the operations are defined such that:

- $[a] \wedge [b] = \{a \wedge b \mid a \in [a], b \in [b]\}$;
- $[a] \vee [b] = \{a \vee b \mid a \in [a], b \in [b]\}$;
- $\neg[a] = \{\neg a \mid a \in [a]\}$.

The behavior of undetermined interval of interval Boolean values is as follow: $0 \wedge [0,1] = 0$, $1 \wedge [0,1] = [0,1]$, $0 \vee [0,1] = [0,1]$, $1 \vee [0,1] = 1$.

A test function $\mathbf{t}$ maps $\mathbb{R}^n$ to $\mathbb{B}$. The interval extension of $\mathbf{t}$ is the inclusion test $[\mathbf{t}]$ mapping $\mathbb{IR}^n$ to $\mathbb{IB}$, such that for any $[\mathbf{x}] \in \mathbb{IR}^n$:

- $([\mathbf{t}]([\mathbf{x}]) = 1) \Rightarrow (\forall \mathbf{x} \in [\mathbf{x}], \mathbf{t}(\mathbf{x}) = 1)$
- $([\mathbf{t}]([\mathbf{x}]) = 0) \Rightarrow (\forall \mathbf{x} \in [\mathbf{x}], \mathbf{t}(\mathbf{x}) = 0)$

Example 1. Consider the simple test $t : \mathbb{R} \to \{0, 1\}$ such that

$$t : x \mapsto x < 2$$

such that $t(x) = 0$ if $x \geq 2$ and $t(x) = 1$ otherwise. The associated inclusion test is $[t] : \mathbb{IR} \to \mathbb{IB}$ such that

$$[t] : [x] \mapsto [x] < 2$$

$[t]([x]) = 0$ if $\underline{x} \geq 2$, $[t]([x]) = 1$ if $\overline{x} < 2$, and $[t]([x]) = [0, 1]$ otherwise. For example, $[t]([1, 3]) = [0, 1]$.

Inclusion tests for sets can also be defined. Let $\mathscr{A}$ be a subset of $\mathbb{R}^n$, an inclusion test $[\mathbf{t}_{\mathscr{A}}]$ for $\mathscr{A}$ is an inclusion test for the test $\mathbf{t}_{\mathscr{A}}(\mathbf{x}) \iff (\mathbf{x} \in \mathscr{A})$, *i.e.*, $[\mathbf{t}_{\mathscr{A}}]$ satisfies:

- $[\mathbf{t}_{\mathscr{A}}]([\mathbf{x}]) = 1 \Rightarrow (\forall \mathbf{x} \in [\mathbf{x}], \mathbf{t}_{\mathscr{A}}(\mathbf{x}) = 1) \iff ([\mathbf{x}] \subset \mathscr{A})$;
- $[\mathbf{t}_{\mathscr{A}}]([\mathbf{x}]) = 0 \Rightarrow (\forall \mathbf{x} \in [\mathbf{x}], \mathbf{t}_{\mathscr{A}}(\mathbf{x}) = 0) \iff ([\mathbf{x}] \cap \mathscr{A} = \emptyset)$;
- $[\mathbf{t}_{\mathscr{A}}]([\mathbf{x}]) = [0, 1]$ (nothing can be determined) about the inclusion of $[\mathbf{x}]$ in $\mathscr{A}$.

Contribution: We propose to transpose the approach based on interval Booleans for the inclusion test to the SCSDP. Operations 2 on sets used to define SCSDP have been transposed to intervals in a previous paper [2], with evaluation still in Boolean set $\mathbb{B}$. In order to obtain a unified formalism, with easier soundness understanding and larger expressivity, we propose here to use interval Booleans.

Operations we consider in our SCSDP formalism is as follow:

- $[\mathbf{t}]_{\mathbf{g}, \mathscr{A}, \mathscr{B}} = 1 \Rightarrow (\forall \mathbf{x} \in \mathscr{A}, \mathbf{g}(\mathbf{x}) \in \mathscr{B}) \iff \mathbf{g}(\mathscr{A}) \subset \mathscr{B}$;
- $[\mathbf{t}]_{\mathbf{g}, \mathscr{A}, \mathscr{B}} = 0 \Rightarrow (\forall \mathbf{x} \in \mathscr{A}, \mathbf{g}(\mathbf{x}) \notin \mathscr{B}) \iff \mathbf{g}(\mathscr{A}) \cap \mathscr{B} = \emptyset$;
- $[\mathbf{t}]_{\mathbf{g}, \mathscr{A}, \mathscr{B}} = [0, 1]$ nothing can be determined.

Example 2. The following Boolean formula ϕ can be defined

$$\phi \equiv ([\mathbf{t}]_{\mathbf{g}, \mathscr{A}, \mathscr{B}} \vee [\mathbf{t}]_{\mathbf{f}, \mathscr{A}, \mathscr{C}}) \wedge [\mathbf{t}]_{\mathbf{h}, \mathscr{C}, \mathscr{B}}$$

Remark in SCSDP, the set $\mathscr{A}$ can be the set $\mathscr{Y}(\mathscr{T}, \mathscr{Y}_0, \mathscr{P})$ to define dynamical constraints. In that case, $\mathscr{A}$ is parametrized by t. Hence, quantification over time can be considered, *e.g.*, to check that the set of trajectories is include in a particular safety set $\mathscr{B}$ for all time or that the solution reaches a particular set at a given time.

Solving a problem as given in Example 2 means determining sets $\mathscr{A}$, $\mathscr{B}$, and $\mathscr{C}$ such that the formula ϕ is satisfied. With interval methods, it is common to enclose a set by an inner and an outer paving [4] with a branching algorithm. The algorithm presented in Algorithm 1 is able to solve a SCSDP assuming that *Formula* is an interval Boolean function.

Algorithm 1. Branching algorithm for SCSDP

Require: $Stack = X$, $Stack_{in}$, $Stack_{out}$, $Formula$
  while $Stack \neq \emptyset$ **do**
    Pop $[x]$ in $Stack$
    Solve differential equation with $[x]$
    if $Formula([x]) = [1,1]$ **then**
      Push $[x]$ in $Stack_{in}$
    else if $Formula([x]) = [0,0]$ **then**
      Push $[x]$ in $Stack_{out}$
    else if $width([x]) > \epsilon$ **then**
      $[x] = [x_1] \cup [x_2]$
      Push $[x_1]$ and $[x_2]$ in $Stack$
    end if
  end while

---

4 Conclusion

An extension of the *Set-based Constraints Satisfaction Differential Problems* is proposed in this paper. The main idea is to equip the formalism with a full interval-based approach by considering interval of Boolean values. It provides a clearer soundness and a larger expressivity. Moreover, a basic branching algorithm can be used to solve a SCSDP, even if disjunctive logic operations appear in constraints. We are currently implementing the presented contribution in DynIbex [1]. As future work, experimentation will be performed.

References

1. dit Sandretto, J.A., Chapoutot, A.: DynIBEX: a differential constraint library for studying dynamical systems. In: Conference on Hybrid Systems: Computation and Control (2016). Poster
2. Dit Sandretto, J.A., Chapoutot, A., Mullier, O.: Constraint-based framework for reasoning with differential equations. In: Cyber-Physical Systems Security, pp. 23–41. Springer (2018)
3. Goldsztejn, A., Mullier, O., Eveillard, D., Hosobe, H.: Including ode based constraints in the standard CP framework. In: Principles and Practice of Constraint Programming, Volume 6308 of LNCS, pp. 221–235. Springer (2010)
4. Jaulin, L., Kieffer, M., Didrit, O., Walter, E.: Applied Interval Analysis. Springer (2001)

Fuzzy Approach to Optimal Placement of Health Centers

Juan Carlos Figueroa-García[1], Carlos Franco[2], and Vladik Kreinovich[3(✉)]

[1] Department of Industrial Engineering, Universidad Distrital,
Bogotá D.C., Colombia
filthed@gmail.com
[2] School of Management, Universidad del Rosario, Bogotá, Colombia
carlosa.franco@urosario.edu.co
[3] Department of Computer Science, University of Texas at El Paso,
El Paso, TX 79968, USA
vladik@utep.edu

Abstract. In countries with socialized medicine, it is important to decide how to distribute limited medical resources – in particular, where to place health centers. In this paper, we formulate and solve the corresponding constraint optimization problem. Once the locations are selected, it is necessary to decide which regions are served by each center. Traditionally, this decision is crisp, in the sense that each location is assigned to a single health center. We show that the medical service can be made more efficient if we allow fuzzy assignments, when some locations can be potentially served by two (or more) neighboring health centers.

1 Formulation of the Problem

Need for Health Centers. Many countries in the world have socialized medicine – in this sense, US is one of the few exceptions. In countries with socialized medicine, it is important to decide how to distribute the limited resources (and resources are always limited), so as to best serve the population.

In some case, all the patient needs is a regular general doctor. However, in many other cases, the patient also needs to undergo some tests – blood test, X-ray, etc., he/she may need to see a specialist, etc. From this viewpoint, it is more convenient for the patients if all the needed medical professionals are placed at a single location. This is the main idea behind health centers.

Where to Place Health Centers? Once we decided to place the doctors in health centers, the question is: where are the best locations for these centers? And, once we find these locations, what is the best way to assign each patient to one of these centers?

These are the problems that were raised in our previous paper [2]. These are the problems that we deal with in this paper as well.

© Springer Nature Switzerland AG 2019
R. B. Kearfott et al. (Eds.): IFSA 2019/NAFIPS 2019, AISC 1000, pp. 793–799, 2019.
https://doi.org/10.1007/978-3-030-21920-8_71

2   Where to Place the Health Centers: General Analysis

What We Know. Let X denote the area that we want to serve with health centers. Let $\rho(x)$ denote the population density at geographic location x, i.e., the number of people per unit area.

Once we know the population density, we can compute the overall number of people in any given area A as the integral $\int_A \rho(x)\, dx$. Let $P = \int_X \rho(x)\, dx$ denote the overall number of people in our area X.

What We Want. Based on knowing the population density, we need to decide how many health centers to place at different locations. Let $h(x)$ denote the number of health centers per unit area in the vicinity of a geographical location x.

Once we determine this density $h(x)$, we can compute the overall number of health centers in any given area A as the integral $\int_A h(x)\, dx$.

Main Limitation. Our resources are limited: we can only build so many health centers. Let N denote the overall number of health centers that we can build. This limitation means that

$$\int_X h(x)\, dx = N. \tag{1}$$

Objective Function: Informal Description. In the ideal world, every patient should be immediately seen by a doctor. In reality, it takes some time for a patient to reach the nearest health center. The smaller this time, the better.

Thus, a reasonable objective function that gauges the quality of health center placements is the average time that it takes for a patient to reach the nearest health center.

Let us describe this objective function in precise terms.

Objective Function: Towards a Formal Description. The time $t(x)$ that it takes for a patient at location x to reach the nearest health center can be computed as

$$t(x) = \frac{d(x)}{v(x)}, \tag{2}$$

where:

- $d(x)$ is the distance from location x to the nearest health center, and
- $v(x)$ is the average transportation speed in the vicinity of the location x.

The speed $v(x)$ is usually smaller in the city center, slightly larger in the suburbs, and even larger outside the city limits.

The maximum distance $m(x)$ that it takes for points in the vicinity of a location x to reach a health center corresponds to the case when the location is at the edge of the zone allocated to this center, i.e., at the edge of a disk of radius $m(x)$ served by this center. In this circle, there is exactly one health center.

On the other hand, based on the density $h(x)$ of health centers, we can estimate the number of health centers in this disk area as $\int h(x) \, dx \approx h(x) \cdot (\pi \cdot m(x)^2)$. From the condition that this value is 1 (meaning that there is only one health center in this disk area), we conclude that $h(x) \cdot (\pi \cdot m(x)^2) = 1$, i.e., that

$$m(x) = \frac{1}{\sqrt{\pi \cdot h(x)}}. \tag{3}$$

What is the average distance $d(x)$ from a center of the disk of radius $m(x)$ to a point on this disk? For all the points at distance r from the center, this distance is r, and the area of the small vicinity of this disk is $2\pi \cdot t \, dr$. Thus, the average distance can be computed as

$$\frac{1}{\pi \cdot (m(x))^2} \cdot \int_0^{m(x)} r \cdot (2\pi \cdot r \, dr) = \frac{1}{\pi \cdot (m(x))^2} \cdot \frac{2}{3} \cdot \pi \cdot (m(x))^2 = \frac{2}{3} \cdot m(x). \tag{4}$$

Substituting the expression (3) into this formula, we conclude that

$$d(x) = \frac{2}{3 \cdot \sqrt{\pi}} \cdot \frac{1}{\sqrt{h(x)}}. \tag{5}$$

Thus, from the formula (2), we conclude that

$$t(x) = \frac{d(x)}{v(x)} = \frac{2}{3 \cdot \sqrt{\pi}} \cdot \frac{1}{\sqrt{h(x)} \cdot v(x)}. \tag{6}$$

This is the time that it takes for each patient to reach the health center. The average time that it takes all the patients to reach the health center can be then computed as

$$\frac{1}{P} \cdot \int_X \rho(x) \cdot t(x) \, dx = \frac{2}{3 \cdot \sqrt{\pi} \cdot P} \cdot \int_X \frac{\rho(x)}{\sqrt{h(x)} \cdot v(x)}. \tag{7}$$

Now, we are ready to formulate the problem in precise terms.

Exact Formulation of the Problem. We know the functions $\rho(x)$ and $v(x)$. Based on this knowledge, we need to find the function $h(x)$ that minimizes the objective function (7) under the constraint (1).

Solving the Corresponding Problem. Multiplying all the value of the objective function by the same constant does not change which value is larger and which is smaller. Thus, minimizing the function (7) is equivalent to minimizing a simpler expression

$$\int_X \frac{\rho(x)}{\sqrt{h(x)} \cdot v(x)}. \tag{8}$$

To solve the problem of minimizing the expression (8) under the constraint (1), we can use the Lagrange multiplies method, according to which the above constraint optimization problem is equivalent, for an appropriate λ, to the unconstrained optimization problem of minimizing the expression

$$\int_X \frac{\rho(x)}{\sqrt{h(x) \cdot v(x)}} + \lambda \cdot \left(\int_X h(x)\, dx - N \right). \tag{9}$$

Differentiating this expression with respect to the unknown $h(x)$ and equating the derivative to 0, we conclude that

$$-\frac{1}{2} \cdot \frac{\rho(x)}{(h(x))^{2/3} \cdot v(x)} + \lambda = 0, \tag{10}$$

i.e., that

$$h(x) = c \cdot \left(\frac{\rho(x)}{v(x)} \right)^{2/3}, \tag{11}$$

for some constant c. This constant can be found if we substitute the expression (11) into the constraint (1). Then, we get the following solution.

Optimal Solution.

$$h(x) = N \cdot \frac{\left(\dfrac{\rho(x)}{v(x)} \right)^{2/3}}{\displaystyle\int_X \left(\dfrac{\rho(y)}{v(y)} \right)^{2/3} dy}. \tag{12}$$

Discussion. The density of health centers is proportional to the population density raised to the power 2/3. Thus, in the regions with higher population density $\rho(x)$, we place more health centers – but the number of health centers grows slower than the population density.

How Many Doctors Are Needed in Each Health Center. The number of medical personnel $M(x)$ needed for each health center is proportional to the number of people $N(x)$ served by each center:

$$M(x) = m_0 \cdot N(x). \tag{13}$$

The coefficient m_0 can be obtained if we know the overall number M of medical professionals, since for the whole population, the formula (13) implies that $M = m_0 \cdot P$ and thus, that $m_0 = \dfrac{M}{P}$ and hence,

$$M(x) = \frac{M}{P} \cdot N(x). \tag{14}$$

The number of people $N(x)$ served by a health center can be obtained by multiplying the population density $\rho(x)$ in the vicinity of a given location x by the area $1/h(x)$ covered by the center; thus,

$$M(x) = \frac{M}{P} \cdot \frac{\rho(x)}{h(x)}. \tag{15}$$

In view of the formula (12), we get

$$M(x) = \frac{M}{P \cdot N} \cdot \left(\int_X \left(\frac{\rho(y)}{v(y)} \right)^{2/3} dy \right) \cdot (\rho(x))^{1/3} \cdot (v(x))^{2/3}. \tag{16}$$

3   Where to Actually Place the Health Centers?

Formulation of the Problem. The above formulas describe how many health centers to place in the vicinity of each location x. But where exactly should we place them?

We need to find 2-D locations $p_1, \ldots, p_N$ so that the average distance from each place x to this location be the smallest possible. For each location x, let us denote, by $i(x)$, the number of the health center that will be associated with this location. Then, the distance from each location x to the corresponding health center is $d(x, p_{i(x)})$, and the corresponding travel time is equal to the ratio $\dfrac{d(x, p_{i(x)})}{v(x)}$. The average distance can be therefore computed as

$$\int \rho(x) \cdot \frac{d(x, p_{i(x)})}{v(x)} dx, \tag{17}$$

or, if we take into account the discrete character of the information, as the sum

$$\sum_x \rho(x) \cdot \frac{d(x, p_{i(x)})}{v(x)}. \tag{18}$$

We need to find the values $p_1, \ldots, p_N$ and the value $i(x)$ (for all $x \in X$) that minimize the expression (18).

Towards an Algorithm. It is difficult to immediately minimize the objective function (18) with respect to all the unknowns, so a natural idea is to minimize it iteratively. Namely, we start with some location of the centers. Then:

- First, we fix the locations p_i of the health centers and find the corresponding assignments $i(x)$. For each spatial location x, minimizing the expression (18) with respect to $i(x)$ means minimizing the distance $d(x, i(x))$, i.e., finding the health center which is the closest to this location.

- Then, we fix the assignments $i(x)$ and find the locations p_i that, for these assignments, minimize the expression (19). For each health center i, this is equivalent to finding the new location p_i for which the average distance

$$\sum_{x:i(x)=i} \rho(x) \cdot \frac{d(x, p_i)}{v(x)} \tag{19}$$

is the smallest possible. This can be done, e.g., by gradient descent.

Then, we repeat the procedure again and again until the process converges, i.e., until the distance between the locations of each health center on two consequent iteration is smaller than some pre-determined small value ε.

Comment. This is similar to the standard algorithm for computing fuzzy clusters (see, e.g., [1]), where we iteratively:

- first, assign each point to clusters depending on this point's distance to the cluster centers, and
- then find the new centers which are, on average, closest to all the points assigned to the corresponding cluster.

Resulting Algorithm. We first randomly place the centers in accordance with the center density $h(x)$ (that we computed in the previous section). Then, we iteratively do the following:

- First, for each spatial location x, we find the closest health center; we will denote the index of this health center by $i(x)$.
- Then, for each i from 1 to N, we find a new location p_i for which the average distance (19) is the smallest possible. This is done, e.g., by gradient descent.

Then, we repeat the procedure again and again until the process converges, i.e., until the distance between the locations p_i and p_i' of each health center on two consequent iteration is smaller than some pre-determined small value ε, i.e., until $d(p_i, p_i') \leq \varepsilon$ for all i.

4   Need for a Fuzzy Approach

Discussion. In the above description, we assumed that each location is assigned to exactly one health center. This assignment was based on the simplified assumption that the travel time (and waiting time) depends only on the distance from the location to the health center (and on the allowed traffic speed in the vicinity of this location).

In reality, as everyone who lives in a big city knows, travel time can change drastically. Sometimes there are traffic jams, sometimes there are accidents. Also, we only took into account travel time, but – again as everyone who ever went to a doctor knows – there is also waiting time.

Idea. From this viewpoint, if we have a patient who is slightly closer to one health center than to the other, it does not make sense to assign this patient always to the nearest health center: maybe there is a long waiting time in this nearest health center, but there is no waiting time in the other health center – as a result of which the patient will be served faster if he or she goes to this second health center this time.

In other words, instead of assigning each patient to a single health center, it is beneficial to make a "fuzzy" allocation: namely, to allow the patient to go to any health center in the nearest vicinity – namely, to the one for which the travel time + waiting time will be the smallest. There are many apps already for predicting travel time, there are similar apps for predicting the waiting time, and nowadays, when most medical records are electronic, it is not a problem to access the records from each of the health centers.

Acknowledgements. This work was supported in part by the US National Science Foundation grant HRD-1242122 (Cyber-ShARE Center of Excellence).

References

1. Bezdek, J.C.: A Primer on Cluster Analysis: 4 Basic Methods that (Usually) Work, 1st edn. Design Publishing (2017)
2. Franco, C., López-Santana, R.E., Figueroa-García, J.C.: A mathematical model under uncertainty for optimizing medicine logistics in hospitals. In: Proceedings of the 5th Workshop on Engineering Applications WEA 2018, Part II, Medellín, Colombia, 17–19 October 2018 (2018)

Constraint Programming Enabled Automated Ship Hull Geometric Design

Thomas Luke McCulloch[✉]

Bentley Systems Inc., Metairie, LA 70002, USA
`tlukemcculloch@gmail.com`

Abstract. Techniques for the generation of constrained and optimized geometry play a key role in the automated design of ships, particularly in the early stages. So called "form parameter design" systems are state-of-the-art for such ship hull geometry generation tasks. These systems solve a sequence of constrained nonlinear optimization problems to generate ship hull form geometry which conforms to constraints (form parameters). With such a system, a human expert, or search algorithm capable of discarding infeasible designs, is needed to pick constraint combinations that are at once feasible and also result in a high quality design. Simply finding a feasible set of form parameters is non-trivial enough that form parameter design of ship hulls has remained mostly an academic practice. Commercial tools exist, but their uptake by industry has been slow because this issue has given such tools a reputation for being hard to use.

In this paper a modified approach for hull shape generation is proposed, combining form parameter design with interval constraint logic programming. We use a constraint solver to pre-process the design space, ensuring all constraints are feasible. This ensures that the form parameter design tool always operates within the feasible sub-domain of the total design space. The new approach is briefly described. Its effectiveness is demonstrated by using a random number generator together with the constraint solver to generate feasible design candidates, and then valid ship hull geometry is generated by our form parameter design program.

1 Introduction

Design of a ship hull form is a complicated undertaking and the ship hull form effects all aspects of ship performance. Ships are often one of a kind designs so there is little chance to perfect a design by experience. Instead, the best way to mitigate design issues is to optimize the design in the early stages, by computational methods. This means ship designers need to generate and analyze many designs quickly. Thus, there is demand for ever improving methods to quickly generate and test candidate designs. State-of-the-art ship hull design generation tools are based on a technique called form parameter design.[1] These tools

---

[1] Tools such as Cases (Friendship Framework) [14] and PolyCAD [6].

© Springer Nature Switzerland AG 2019
R. B. Kearfott et al. (Eds.): IFSA 2019/NAFIPS 2019, AISC 1000, pp. 800–812, 2019.
https://doi.org/10.1007/978-3-030-21920-8_72

are intended to facilitate exactly this kind of early stage design optimization. These programs automate the generation of ship hull designs which conform to constraints. Despite commercial availability, these tools are not reliable without an engineer "in the loop." The engineer must intervene to specify a well posed system of constraints so that the tools can find valid geometry. This is the chief issue slowing the spread and usage of these design tools. The parametric design tools shift the burden on the designer from manipulating geometry directly to a task of guessing feasible constraint combinations. The former can be intuitive to naval architects. Unfortunately, the later is more often non-intuitive for a user. A natural question to ask, then, is "How do we ensure all design constraints are mutually attainable, in addition to encapsulating the individual geometric properties we desire?" We formalize the issues above with the following problem:

1.1 Problem Statement

Let an *interval valued design space* of dimension m be an m dimensional array S whose entries $S_j = [\underline{S}_j, \overline{S}_j]$ $(j = 1, \dots)$ are intervals. Each interval represents a particular form parameter design constraint, or possibly some design space parameter implicit in the ship hull design problem, such as gross ship dimensions. Note that the set of all such data which are input to a form parameter hull generation tool make up a single point in this design space. With these definitions, our problem may be stated as follows:

Find a form parameter, interval valued, thin, *design space*, S

$$S_j = [\underline{S}_j, \overline{S}_j] \tag{1}$$

such that all constraints and related properties are feasible,
and the form parameter design program returns a valid hull geometry.

Furthermore, we might add a wish that the new program should suggest feasible constraints when the user is not sure or has no opinion—as these constraints might still be needed for fully specifying the geometric design problem. A constraint solver can do this if there is enough data to propagate a solution to design variables which previously were uninitialized. For instance, if two variables of a ternary constraint are defined with interval data, this data can propagate to the third variable via consistency computations. See Listing Sect. 2.2 for an example of this with ternary constraints.

Now that we have posed our problem, and before we can give a solution, we must first give account of the prior art in this area of research.

1.2 Previous Work

Form parameter design (FPD) tools have been used as parametric modeling an optimization tools in the ship building industry and academia since the 1990s. FPD uses nonlinear constrained optimizing curve solvers together with surface generation techniques to produce hull form geometry from constraints. Nowaki used Bezier curves together with parametric input and variational techniques to describe the modern form parameter system [20]. Harries extended the technique to use B-splines in the late 1990s [13]. The method was extended to complex geometries by Kim [17]. B-splines are a standard representation for the ship hull geometry and thus also the standard computation mechanism for generating constraints and derivatives to facilitate the required nonlinear optimizations. NURBS are the standard for user manipulated geometry, but for efficiency purposes we restrict ourselves to B-splines for our optimization methods, as explained in more detail by Birk [5].

To date, form parameter design programs have never been combined with constraint solvers to ensure consistent inputs. In contrast, logical programming systems, under the guise of "expert systems," have been used for ship design, especially in the 1980s and 1990s. An excellent introduction to this body of work can be found in [3]. Expert systems have been used to determine gross ship parameters, but never in concert with FPD. Also of note, Harries has addressed aspects of the constraint feasibility problem with different tools, using local consistency estimates in his form parameter design programs [13]. It's worth mentioning that Harries' methods are designed to find suitable starting conditions for the nonlinear curve solver, given a set of constraints, rather than to ensure that the constraints themselves are feasible across the entire ship—as we do here. Abt et al. address the more holistic constraint handling issues in [1]. They suggest relaxing equality constraints and performing design of experiments in order to find feasible combinations. In contrast with this work, we focus on equality constraints without relaxation or search. Other programs have used, e.g., neural networks and fuzzy logic together with a design database for generating designs. See for instance Kim et al. [18]. These methods effectively search for good designs, using preexisting designs to aid the search, where our system instead rules out infeasible designs, without needing to learn from existing designs. Our algorithms could be used in concert with a meta-learner such as a neural net. Our program would ensure the meta-learner need only search for the best designs, avoiding the expenditure of searching infeasible design space.

Finally, our algorithms rely on interval constraint logic programming in order to efficiently reduce the design space in response to constraints. This processing capability draws on a half century of research involving both interval methods and constraint programming. Kearfott introduces interval arithmetic and details the wide array of uses of interval analysis in optimization in [16]. Algorithms for nonlinear interval optimization, including especially hull consistency are given by Walster and Hansen [12]. Implementation specific definitions of arc-consistency are provided by Collavizza et al. [10]. Our notion of local hull consistency used for arc-consistency corresponds to the notion of 2B-consistency defined there, and therefore we should note that our arc-consistency is only approximated locally. The use of a relational logic programming system to handle relational constraints was heavily inspired by the "miniKanren" and especially "microKanren" systems developed in papers by Friedman et al. [7,11]. Interval extensions to logic programming systems have been specified in the literature before. One of the first was by Cleary [9] A more recent review is given by Benhamou [2].

We implement our constraint solver in order to tie the entire system together with the greatest flexibility.

Outline

We give methodology for generating consistent form parameter design spaces for use in constraint based "form parameter ship hull design" tools in order to find consistent combinations of equality constraints across an entire ship hull design. This method acts as a design space preprocessor, pruning inconsistent constraint space configurations without the cost of a full attempt at generating the geometry. For verification of the idea, we generate valid ship hull geometry using a random number generator to select constraints. Details are given below. The remainder of this paper is structured as follows. Section 2 gives a sketch of the implementation, with particular emphasis on constraint solving and design rules. Section 3 explains form parameter design. Section 4 shows an example hull, and gives the random generation method. Section 5 gives our conclusions.

2 Program Development

For this project we had originally developed curve solver code here [5] and implemented a form parameter design tool based on this. Everything was written in Python [21]. By writing our own libraries for each step in the process, we could more easily integrate the needed subsystems. Accordingly, we implemented an interval, relational, constraint solver in python. The initial ideas were inspired by the logical, relational, constraint programming language miniKanren [11].

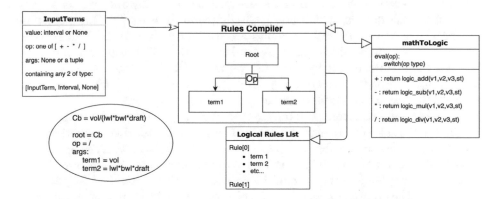

Fig. 1. Overview of the input system and rules compiler.

2.1   Python Internal Constraint Language

New code development starts with a binary unification algorithm extended to work with an interval data type. From there, we build up a constraint language for relational arithmetic by including functions to unify each binary case of the relation. For our purposes we need only addition, subtraction, multiplication, and division. These are ternary relations, and so it is relatively simple to perform unification across all terms using a generic function. See [19] for details. For general constraints, though, we need to support n-ary terms. We cannot write code for every term explicitly. Instead, we use operator overloading to build up a computational graph (binary tree) of the constraint input, and then walk the tree to compose ternary relational rules from each node.

Some implementations of reverse mode automatic differentiation follow a similar scheme. See that given by Christianson [8], Kearfott [16], for examples. Christenson states the idea as follows: "The operators from which the code for f is built are overloaded to append information about themselves and their arguments and results to the list dynamically as a side effect" [8]. In our case f is a rule formulated as an n-ary mathematical expression. Instead of a list, we assemble a computation tree. The `InputTerms` class at right in Fig. 1 implements this idea so that Python parses the incoming mathematical rules expressions and returns the syntax tree we want. An example term, "Cb," is shown circled at bottom left of Fig. 1. Our rules compiler walks the corresponding tree and assembles ternary relational rules from it. This is symbolized by the "Rules Compiler" shown in the center of Fig. 1. The logical rules list at bottom is the final output of the Rules "compiler." These rules are ready for processing by standard arc consistency algorithms. The `mathToLogic` compiler-helper-function takes the syntactic operation and operands of the rule as input. It returns a closure which can be used to filter the design space in a lazy and relational way. See [19] for the details, which are inspired by miniKanren [7]. Our algorithm will also make note of which variables are used in each rule. We can thereby connect all the rules which use a particular variable. This way if a variable's interval value is narrowed by one rule, we may call the related rules to see if this has some impact on the other variables in these relations. This provides the connectivity for arc consistency based constraint propagation.

An abstract algorithm for ternary constraint processing, ignoring many details for the sake of brevity, is given in Listing 1. To use this method of processing our constraints, subtraction $a - b = c$ is re-defined as addition $c + b = a$, and division of $a/b = c$ re-defined as $c * b = a$. For more complicated constraints, case by case implementations would be needed in order to handle the relational processing. The compiler generates lists of such rules which are equivalent to an entire n-ary expression, including automatically building the internal connective rules, so there is nothing required of the design engineer except to specify the rules and the design space of interest.

Programming in this manner allows a user to describe the design space with simple design rules of her choosing. These are input using basic mathematical

---

Procedure 1. Ternary Rules Processing

  Input operands a, b and result c
  Input An operator f and it's inverse, f^{-1}
  Input-Output env - design space mapping from variables to interval values
1: **procedure** TERNARY_RULES_PROCESSING(a,b,c,env)
2: **a,b,c** $\leftarrow env\left(a\right), env\left(b\right), env\left(c\right)$
3: env $\leftarrow$ unify($f\left(\mathbf{a}, \mathbf{b}\right), c,$)
4: env $\leftarrow$ unify($a, f^{-1}\left(\mathbf{c}, \mathbf{b}\right)$)
5: env $\leftarrow$ unify($b, f^{-1}\left(\mathbf{c}, \mathbf{a}\right)$)
6: return env

---

expressions common to any numerical computing language. Using this capability, we now give a method which can infer feasible designs from random input.

2.2 Feasibility and Design Space Narrowing

Starting with a design space, first we compute approximate arc consistency to ensure that the design space contains an estimate of the minimum of infeasible points. As mentioned above, we use the "2B" local consistency defined by Collavizza et al. [10]. Next we loop over all the parameters in the design space, assigning a random value to each parameter in sequence. Note that the random value is actually chosen by mapping a random number between 0 and 1 into the current feasible domain of that particular form parameter. After this parameter is assigned a random value from the existing design space, arc consistency is run in order to propagate the ramifications of this selection across the rest of the rules and to narrow the feasible domain for all other design parameters. The algorithm continues until all parameters are narrowed sufficiently that the design space is "thin"—in practice this means that the intervals are so narrow that they are below the tolerance of the nonlinear solvers to be used in curve generation. For purposes of geometry generation then, we have found a single design.

Note also that this since this system is generating designs in random fashion with only the dictates stipulated by the designer as constraints, this system is free to take the entire real line for an interval width where no information is provided for a particular variable. Also, if only two variables of a ternary constraint are defined with interval data, this data will propagate to the third variable via 2B-consistency. See code Listings Sect. 2.2 for an example of this property. Division by zero is handled by extended interval division [15]—and results in a design space splitting. Our system handles disjoint spaces by list processing techniques from miniKanren [7].

With our constraint programming system in hand, we now move to show a few of the design rules that are used in our system. An engineer is free to devise her own rules as needed. Here we state rules that are natural to a naval architect, and that proved helpful in computing valid design spaces for form parameter design.

```
a #exists  but  not  set
b = b == ia (1.,  5.)                    b   : ia (1.0 ,5.0)
c = c == ia (1.,2.)                      v1  : ia (1.0 ,2.0)
c = c == a/(b)                           c   : ia (1.0 ,2.0)
compute_arc_consistency ()               a   : ia (1.0 ,10.0)
```

Fig. 2. Python-pseudo-script showing instantiation of logical variables, assigning intervals to **b** and **c**, leaving the logical variable **a** "blank," and giving a relation between all 3 variables. Results shown after arc consistency at right. Local consistency has solved for **a**. A new term, **v1** has appeared—an automatically generated connective variable.

Table 1. A few hull definition terms. With our constraint system, the ship designer is free to define as many terms as needed for the problem.

Form parameters	
∇	submerged volume
L_{WL}	length of ship
B_{WL}	beam (horizontal extents)
D	depth (vertical extents of ship)
A_M	midship section area
LCB	location of the center of buoyancy of the ship

2.3   Design Rules

In contrast with heuristics often considered in optimization literature, the field of ship hull design comes equipped with a natural set of rules that designers use in standard practice. To name some of these rules, we first have to define a few simplified ship design form parameters. These terms are shown in Table 1. A few relational rules are then given in Table 2 (Fig. 2).

In the full program there can be as many relational rules as the hull designer sees fit. In our numerical experiments, we found that there should be enough rules to tie together the various pieces of hull geometry in gross form, i.e. at the level of overall dimension and cross-sectional area, and the like. Form parameters enter into the FPD tool as constraints. Hull coefficients are relations that hold between parameters. These are natural in naval architecture, in the sense that naval architects have been using them to think more clearly about hull shape for many years. Therefore, it seems natural to use them to pre-process the design space to ensure consistency between the form parameters.

Finally, note that there are typically many more form parameters than are shown here. Quite a number of constraints are used to generate a full ship, but the subset shown in this paper will serve to demonstrate the needed concepts. See [19] for details.

Table 2. Some hull-form relational rules. Each rule encodes a geometric relationship. The constraint solver must maintain the validity of these relationships by eliminating infeasible regions of the design space. We use relational interval arithmetic and constraint propagation to do so. These relationships ensure that the form parameters will work together to form a harmonious ship hull once processed by the FPD program.

Primary hull relational rules		
Form parameter	Symbol	Formula
Block coefficient	$C_B =$	$\frac{\nabla}{L_{WL} \times B_{WL} \times D}$
Prismatic coefficient	$C_P =$	$\frac{\nabla}{L_{WL} \times B_{WL} \times D \times C_M} = \frac{C_B}{C_M}$
Midship coefficient	$C_M =$	$\frac{A_M}{B \times D}$
Waterplane coefficient	$C_{WP} =$	$\frac{A_{WP}}{L_{WL} \times B_{WL}}$
Centerplane coefficient	$C_{CP} =$	$\frac{A_{CP}}{L_{WL} \times D}$
Length - beam ratio	$C_{LB} =$	$\frac{L_{WL}}{B_{WL}}$
Displacement - length ratio	$DLR =$	$\frac{\nabla}{(L_{WL})^3}$

3   Form Parmeter Design of Ship Hulls

Given a consistent design composed of form parameters and essential dimensions, FPD generates the corresponding ship hull by the following hierarchical process (Fig. 3):

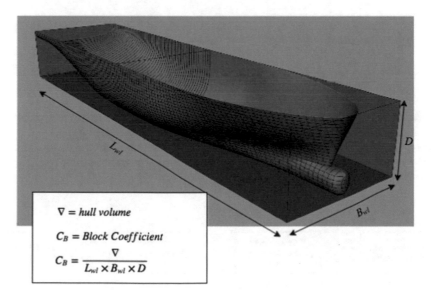

Fig. 3. Pictorial definition of block coefficient, C_B. Note we have not separated the underwater portion of the hull in this instance. This would be a minor change for the curve solvers and would leave the essentials of the constraint solver methodology unchanged.

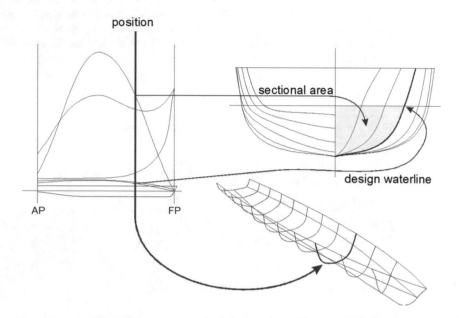

Fig. 4. FPD Process Example: using SAC to specify area at hull stations. Figure from Birk, [4].

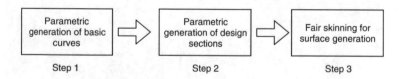

Fig. 5. FPD Essential steps of the shape definition process, as seen in, e.g. [14]. The critical problem is that initial set of feasible form parameters must be specified. Usually in form parameter design this task is offloaded to the user, or else an optimization algorithm must search and eliminate infeasible designs in order to find, construct and evaluate feasible candidates.

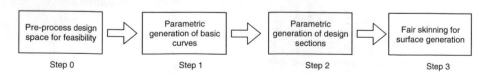

Fig. 6. Workflow for feasible form parameter design: we use our constraint solver and Python internal domain specific language to ensure form parameter consistency.

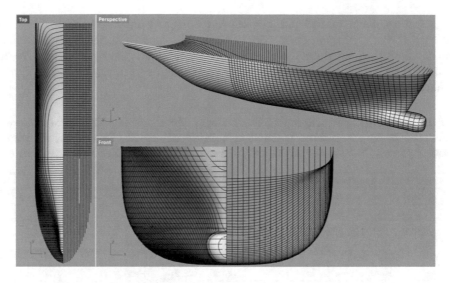

Fig. 7. Hull design for an offshore supply vessel; form parameters chosen at random, with the constraint solver ensuring feasibility.

- Initiate FPD with a set of "form parameter" constraints.
- Generate a set of "curves of form" which embody the incoming design parameters. For instance, the total displacement of the ship is a typical design parameter controlled by a curve called the sectional area curve (SAC). SAC stipulates the ship's transverse section area at any point on its length. SAC area is ship displacement.
- From the curves of form, a new set of curves will be designed. They will run in the transverse direction and take their constraint parameters from the curves of form. Figure 4 shows this process for sectional area in detail. In every case an optimizing curve solver generates a curve with the appropriate geometry.
- Once a set of transverse sectional curves are developed, a surface is generated which exactly interpolates these curves.

We modify FPD by preprocessing the design space—using the constraint solver to eliminate infeasible constraint values, thereby removing the need to have an expert pick feasible form parameters. Instead, even a random number generator can do so. The essential FPD work-flow is given in Fig. 5, and our modified work-flow is sketched in Fig. 6. Our modified work-flow adds a constraint solver to the beginning of this process (Fig. 7).

4   Results and Validation

Here we give a randomly generated set of form parameters and the resulting hull geometry created by the FPD solver. As detailed in Sect. 2.2, in order to randomly select designs, we select a thin interval for each design variable by mapping a unique random number between zero and one into each design space interval in turn, running constraint propagation each time. When this is done, we have a single, randomly selected design candidate. This geometry is representative of the type of hull shape that the system can robustly and randomly generate from a set of design space rules.

Table 3. Hull design parameters. The "Target Value" column shows thin intervals generated by the constraint solver and random algorithm. The "Actual Value" column shows how well the FPD solver was able to match this input. Values rounded for clarity.

Ship Hull Design Random Form parameters				
Symbol	Form Parameter	Target Value (1)	Actual Value (2)	%error 2
∇	SAC curve area	(33615.5,33615.5)	33615.539	0.0
L_{WL}	Waterline Length	(111.8,111.8)	111.8	0.0
B_{WL}	Ship Beam	(26.2,26.2)	26.2	0.0
D	Ship Draft	(17.1,17.1)	17.1	0.0
A_M	Midship Section Area	(428.2,428.2)	428.2	0.0
LCB	Longitudinal Center of Buoyancy	(55.2,55.2)	55.2	0.0
C_B	Block Coefficient	(0.6,0.6)	0.6	0.0
C_P	Prismatic Coefficient	(0.7,0.7)	0.7	0.0
C_M	Midship Area coefficient	(0.9,0.9)	0.9	0.0
CWL	Waterline area coefficient	(0.9,0.9)	0.9	0.0

This will serve as method validation in the following manner: Our goal is to allow a ship hull design program to operate with complete autonomy. To start off, a program will know nothing of the design space over which is searches. Therefore it is reasonable to approximate this design space search with random guessing. If our algorithm is robust with randomly generated inputs, then a machine learning, or other meta-learning algorithm can more efficiently focus on learning the feasible domain, without wasting efforts learning to avoid infeasible designs.

"Target" values are the initial consistent parameters generated by our methodology. "Actual" values come from final geometry. Percentage error is as follows:

$$\%error = \frac{\text{abs}\left(\text{mid}\, x_{clp} - x_{fpd}\right)}{\text{mid}\, x_{clp}} \times 100\% \qquad (2)$$

where x_{clp} denotes output from the constraint solver and x_{fpd} is the final value of the form parameter solved for by FPD. Such validation for our example hull, is given in Table 3. The FPD solver has no issue generating B-spline geometry with the constraints supplied by our system. For a larger tabulation of results, and usage details for generating an offshore supply vessel, see [19].

5   Conclusions

As shown in Table 3, the differences between target form parameters generated by the constraint solver and the resulting parameters of geometry generated by

FPD (the "Actual Values" given above) are negligible. This is a novel capability for form parameter design of ship hull forms. What we demonstrate here is the inherent capability to decide feasibility beforehand—as demonstrated by the fact that the constraint system eliminates infeasible choices from the design space and the FPD solution matches the target form parameters. Constraint logic programming allows a ship hull form parameter design program to operate in a newly robust and autonomous manner.

References

1. Abt, C., Harries, S., Hochkirch, K.: Constraint management for marine design applications. In: 8th International Symposium on Practical Design of Ships and Other Floating Structures (PRADS 2004), Lübeck-Travemünde, Germany (2004)
2. Benhamou, F.: Interval Constraint Logic Programming, pp. 1–21. Springer, Heidelberg (1995). https://doi.org/10.1007/3-540-59155-9_1
3. Bertram, V.: A survey on knowledge-based systems for ship design and ship operation. Int. J. Intell. Eng. Inform. 2(1), 71–90 (2013). https://doi.org/10.1504/IJIEI. 2013.056067
4. Birk, L., Harries, S.: Automated optimization – a complementing technique for the hydrodynamic design of ships and offshore structures. In: 1st International Conferecne on Computer Applications and Information Technology in the Maritime Industries (COMPIT), Potsdam, Germany (2000)
5. Birk, L., McCulloch, T.L.: Robust generation of constrained b-spline curves based on automatic differentiation and fairness optimization. Comput. Aided Geom. Des. 59, 49–67 (2018). https://doi.org/10.1016/j.cagd.2017.11. http://www.sciencedirect.com/science/article/pii/S0167839617301474
6. Bole, M., Lee, B.: Integrating parametric hull generation into early stage design. Ship Technol. Res./Schiffstechnik 53, 115–137 (2006)
7. Byrd, W.: Relational programming in minikanren: techniques, applications, and implementations. Ph.D. thesis, Indiana University (2009)
8. Christianson, B.: Automatic Hessians by Reverse Accumulation. Other Titles in Applied Mathematics, 2nd edn. SIAM, Philadelphia (2008)
9. Cleary, J.: Logical arithmetic. Future Comput. Syst. 2(2), 125–149 (1987)
10. Collavizza, H., Delobel, F., Rucher, M.: A note on partial consistencies over continuous domains. In: Proceedings of the 4th International Conference on Principles and Practice of Constraint Programming, CP 1998, pp. 147–161. Springer, London (1998)
11. Friedman, D., Byrd, W., Kiselyov, O.: The Reasoned Schemer. MIT Press, Cambridge (2005)
12. Hansen, E.,Walster, W.: Global Optimization Using Interval Analysis, 2nd edn. Marcel Dekker, New York City (2004)
13. Harries, S.: Parametric design and hydrodynamic optimization of ship hull forms. Ph.D. thesis, Technische Universität Berlin (D83), Berlin (1998)
14. Harries, S., Nowacki, H.: Form parameter approach to the design of fair hull shapes. In: 10th International Conference on Computer Applications in Shipbuilding (ICCAS 1999), Massachusetts Institute of Technology, Cambridge, MA (1999)
15. Kahan, W.M.: A more complete arithmetic, lecture notes for a summer course at the University of Michigan (1968)

16. Kearfott, B.: Rigorous Global Search: Continuous Problems. Nonconvex Optimization and its Applications, vol. 13. Kluwer Academic Publishers Group, Norwell/-Dordrecht (1996)
17. Kim, H.: Parametric design of ship hull forms. Ph.D. thesis, Technische Universität Berlin (D83), Berlin (2004)
18. Kim, S., H.C. K., Lee, Y.: Initial hull form design using fuzzy modeling. Ship Technol. Res./Schiffstechnik **43**, 175–180 (1996)
19. McCulloch, T.L.: Feasible form parameter design of complex ship hull form geometry. Ph.D. thesis, University of New Orleans (2018). https://scholarworks.uno.edu/td/2552
20. Nowacki, H., Bloor, M., Oleksiewicz, B. (eds.): Computational Geometry for Ships. World Scientific, Singapore (1995)
21. Rossum, G.: Python tutorial, technical report cs-r9526. Technical report, Centrum voor Wiskunde en Informatica, Amsterdam (1995)

Derivation of Louisville-Bratu-Gelfand Equation from Shift- or Scale-Invariance

Leobardo Valera, Martine Ceberio, and Vladik Kreinovich$^{(\boxtimes)}$

Department of Computer Science, University of Texas at El Paso,
El Paso, TX 79968, USA
leobardovalera@gmail.com, {mceberio,vladik}@utep.edu

Abstract. Louisville-Bratu-Gelfand equation appears in many different physical situations ranging from combustion to explosions to astrophysics. The fact that the same equation appears in many different situations seems to indicate that this equation should not depend on any specific physical process, that it should be possible to derive it from general principles. This is indeed what we show in this paper: that this equation can be naturally derived from basic symmetry requirements.

1 Formulation of the Problem

In Many Different Situations, We Have the Exact Same Louisville-Bratu-Gelfand Equation. In many different physical situations, we encounter the same differential equation

$$\nabla^2 \varphi = c \cdot \exp(a \cdot \varphi). \tag{1}$$

This equation – known as Louisville-Bratu-Gelfand equation – appears in the analysis of explosions, in the study of combustion, in astrophysics (to describe the matter distribution in a nebula), in electrodynamics – to describe the electric space charge around a glowing wire – and in many other applications areas; see, e.g., [2, 3, 5–8, 10, 11].

Challenge. The fact that the same equation appears in many different situations seems to indicate that this equation should not depend on any specific physical process, that it should be possible to derive it from general principles.

What We Do in This Paper. In this paper, we show that this equation can be naturally derived from basic symmetry requirements.

2 Laplace Equation – the Simplest Case of Louisville-Bratu-Gelfand Equation: Brief Reminder

Idea. The simplest form of Eq. (1) is when we take $c = 0$. In this case, we get a linear equation $\nabla^2 \varphi = 0$. This equation is known as the Laplace equation. So, in

R. B. Kearfott et al. (Eds.): IFSA 2019/NAFIPS 2019, AISC 1000, pp. 813–819, 2019.
https://doi.org/10.1007/978-3-030-21920-8_73

order to understand where the Eq. (1) comes from, let us first recall where the Laplace equation comes from.

Scalar Fields are Ubiquitous. To describe the state of the world, we need to describe the values of all the physical quantities at different locations. In physics, the dependence $\varphi(x)$ of a physical quantity φ on the location x is known as a *field*. Typical examples are components of an electric or magnetic fields, gravity field, etc.

In general, at each location x, there are many different physical fields. In some cases, we need to take all of them – or at least several of them – into account, since several fields are strong enough to affect the situation. However, in many practical situations, only one field is strong enough.

For example, when we analyze the motion of celestial bodies, we can safely ignore all the fields except for gravity. Similarly, if we analyze electric circuits, we can safely ignore all the fields but the electromagnetic field.

Case of Weak Fields. In general, equations describing fields are non-linear. However, in many real-life situations, fields are weak. In this case, we can safely ignore quadratic and higher order terms in terms of φ and consider linear equations.

General Case of Linear Equations. In physics, usually, we consider second order differential equations (see, e.g., [4,9]), i.e., equations that depend on the field φ, on its first order partial derivatives $\varphi_{,i} \stackrel{\text{def}}{=} \dfrac{\partial \varphi}{\partial x_i}$ and on its second order derivatives $\varphi_{,ij} \stackrel{\text{def}}{=} \dfrac{\partial^2 \varphi}{\partial x_i \partial x_j}$. The general linear equation containing these terms has the form

$$\sum_{i=1}^{3}\sum_{j=1}^{3} a_{ij} \cdot \varphi_{,ij} + \sum_{i=1}^{3} a_i \cdot \varphi_{,i} + a \cdot \varphi = 0. \tag{2}$$

Rotation-Invariance. In general, physics does not change if we simply rotate the coordinate system. Thus, it is reasonable to require that the system (2) be invariant with respect to arbitrary rotations. This requirement eliminates the terms proportional to the first derivatives $\varphi_{,i}$ – since otherwise, we have a selected vector a_i and thus, an expression which is not rotation-invariant.

Similarly, we cannot have different eigenvector of the matrix a_{ij} – this would violate rotation-invariance. Thus, this matrix must be proportional to the unit matrix with components δ_{ij} which are equal to 1 when $i = j$ and to 0 when $i \neq j$. So, $a_{ij} = a_0 \cdot \delta_{ij}$ for some a_0, and the Eq. (2) takes the form

$$a_0 \cdot \sum_{i=1}^{3} \varphi_{,ii} + a \cdot \varphi = 0. \tag{3}$$

Dividing both sides by a_0 and taking into account that $\sum_{i=1}^{3} \varphi_{,ii} = \nabla^2 \varphi$, we get the equation of the type

$$\nabla^2 \varphi + m \cdot \varphi = 0, \tag{4}$$

where we denoted $m \overset{\text{def}}{=} a/a_0$.

The Eq. (4) is indeed the general physics equation for a weak scalar field. The case of $m = 0$ corresponds to electromagnetic field or gravitational field – or, more generally, to any field whose quanta have zero rest mass, like photons or gravitons – quanta of the above fields. In the general case, when the quanta have non-zero rest mass, we get a more general Eq. (4) with $m \neq 0$ – e.g., for strong interactions whose quanta are π-mesons [4,9].

Additional Conditions are Needed to Pinpoint Laplace Equation. To explain why, out of all possible equations of type (4), Laplace equation – corresponding to $m = 0$ – is the most frequent, we need to use additional conditions. As such conditions, we will use the fundamental notions of scale- and shift-invariance.

Scale-Invariance: General Idea. Equations deal with numbers. To describe the value of a physical quantity as a number, we need to select a measuring unit. If we change the original unit to a one which is λ times small, then the same physical quantity which was previously described by the number x will now be described by a λ times larger number $x' = \lambda \cdot x$. For example, if we replace meters with a 100 times smaller unit – centimeter – all the length values are multiplied by 100: 1.7 m becomes $1.7 \cdot 100 = 170$ cm.

The choice of a measuring unit is a rather arbitrary procedure. It is therefore reasonable to require that the fundamental physical equations should not change if we simply change a measuring unit – i.e., if we replace all numerical values x of the corresponding quantity to re-scaled values $x' = \lambda \cdot x$.

Of course, different quantities may be related, so if we change the unit of one quantity, we may need to appropriate change units for measuring related quantities. For example, if we change the unit of time t, e.g., for hours to seconds, then, to preserve the relation $d = v \cdot t$ between the velocity v and the distance d, we need to also change the unit for measuring velocity – e.g., from kilometers per hour to kilometers per second.

Two Quantities. Equation (4) involves two physical quantities: the physical field φ and the coordinate (distance) x_i. Thus, we can consider scale-invariance with respect to both these quantities.

φ-Scale-Invariance. Since the Eq. (4) is linear in φ, it clearly does not change if we replace the original field $\varphi(x)$ with a φ-re-scaled field $\varphi'(x) = \lambda \cdot \varphi(x)$.

x-Scale-Invariance. If we change the unit of measuring x_i to a unit which is λ times smaller, then the numerical values will change from x_i to $\lambda \cdot x_i$. Thus, each derivative $\dfrac{\partial}{\partial x_i}$ gets divided by λ, and so, the second derivative is divided by λ^2, while the term $m \cdot \varphi$ remains unchanged. As a result, the Eq. (4) changes into

$$\frac{1}{\lambda^2} \cdot \nabla^2 \varphi + m \cdot \varphi = 0,$$

or, equivalently, into

$$\nabla^2 \varphi + m \cdot \lambda^2 \cdot \varphi = 0. \tag{5}$$

The only case when this equation is equivalent to the original Eq. (4) is when the coefficients at φ in the Eqs. (4) and (5) are the same, i.e., when $m = m \cdot \lambda^2$ and thus, $m = 0$.

It should be noticed that since re-scaling φ does not change the Eq. (4), if $m \neq 0$, the equations remains different no matter how we re-scale φ.

In other words, the only x-scale-invariant case of the general linear Eq. (4) is the Laplace equation.

Shift-Invariance: General Idea. For many physical quantities such as time or coordinate, the numerical value also depends on the selection of the starting point. If we change the starting point of measuring time to a new one which is s moments before, then, instead of the original measurement results t, we will new shifted numerical values $t' = t + s$.

The selection of a starting point is simply a matter of convenience, there is nothing fundamental about it. It is therefore reasonable to require that the fundamental physical equations do not change if we simply change the starting point. Of course, to preserve the equations, we may need to accordingly change something (measuring unit or a starting point) for some other quantities.

Let us see what we can conclude in our case by requiring shift-invariance for x_i and for φ.

x-Shift-Invariance. If we replace the original variables x_i with new variables $x_i' = x_i + s_i$, where s_i denote the corresponding shifts, then the derivatives do not change and thus, the Eq. (4) remains the same.

φ-Shift-Invariance. Let us now consider the consequences of requiring that the equation are invariant with respect to shifting the field φ, i.e., with respect to replacing all the values $\varphi(x)$ with the new values $\varphi'(x) = \varphi(x) + s$.

In many cases, such a shift makes perfect physical sense: indeed, e.g., the only way we measure electric potential $\varphi(x)$ is by measuring the difference $\varphi(x) - \varphi(x')$ between potentials at different locations. If we add the same value s to all the values of the field, then the differences remain the same – and thus, this addition will not affect any measurement results.

What happens if we apply this shift to the Eq. (4)? The derivatives do not change (since the derivative of a constant s is 0), but the term $m \cdot \varphi$ changes into $m \cdot (\varphi + s)$. Thus, instead of the original Eq. (4), we get a new equation

$$\nabla^2 \varphi + m \cdot \varphi + m \cdot s = 0. \tag{6}$$

The only possibility for the resulting Eq. (6) to be equivalent to (4) is when the additional term $m \cdot s$ is equal to 0, i.e., when $m = 0$.

In other words, the only φ-shift-invariant case of the general linear Eq. (4) is the Laplace equation.

Summary. To get Laplace equation $\nabla^2 \varphi = 0$ out of the general linear Eq. (4), we need to postulate either x-scale-invariance or φ-shift-invariance.

3   Derivation of Louisville-Bratu-Gelfand Equation

From Laplace Equation to Poisson Equation. The Laplace equation $\nabla^2\varphi = 0$ describes what happens in the absence of any external sources. If there is an external source for the field φ, then the expression $\nabla^2\varphi$ is, in general, not necessarily equal to 0. In other words, we have an equation of the type $\nabla^2\varphi = f$ for some external function f. This equation is known as the Poisson equation.

How to Describe Non-linearity. Non-linearity also means that the original linear equation is no longer exactly true, there are additional nonlinear terms in this equation. We can view these non-linear terms as a source for the field, i.e., in effect, we have the Poisson equation – with the only difference that now, the source term f is not an external term, it is a nonlinear function of the field itself:

$$\nabla^2\varphi = f(\varphi). \tag{7}$$

The question is: which function $f(\varphi)$ should we choose?

Which Function $f(\varphi)$ Should We Choose? Let Us use Symmetries. A natural idea for selecting the function $f(\varphi)$ is to use the same natural symmetries that we used to derive the Laplace equation in the first place: φ-shift-invariance and x-scale-invariance.

When is the Resulting Equation φ-Shift-Invariant? If we replace the original values of the field $\varphi(x)$ with the shifted values $\varphi'(x) = \varphi(x) + s$, then the derivatives will not change, so the Eq. (7) will take the form

$$\nabla^2\varphi = f(\varphi + s). \tag{8}$$

Literally speaking, these two equations coincide if $f(\varphi + s) = f(\varphi)$ for all φ and s, in which case, as we can easily see, the function f is simply a constant – so there is no nonlinearity.

However, as we have mentioned earlier, invariance does not mean that the equation remains the same without changing any other numerical values: sometimes, to preserve the equation, we need to accordingly make changes with other variables as well. In our case, this means that, to preserve the Eq. (7), in addition to a shift $\varphi \to \varphi + s$, we may also need to apply an appropriate re-scaling of the coordinates x_i: $x_i \to \lambda(s) \cdot x_i$. Under this re-scaling, the second derivatives are divided by λ^2, so, instead of the Eq. (8), we get a more complicated equation

$$\frac{1}{\lambda^2(s)} \cdot \nabla^2\varphi = f(\varphi + s),$$

or, equivalently,

$$\nabla^2\varphi = \lambda^2(s) \cdot f(\varphi + s). \tag{9}$$

For the new Eq. (9) to be equivalent to the original Eq. (7), we need to make sure that their right-hand sides coincide, i.e., that for all φ and s, we have

$$\lambda^2(s) \cdot f(\varphi + s) = f(\varphi),$$

or, equivalently,

$$f(\varphi + s) = C(s) \cdot f(\varphi), \tag{10}$$

where we denoted $C(s) \overset{\text{def}}{=} \dfrac{1}{\lambda^2(s)}$.

In physics, all dependencies are measurable, so the function $f(\varphi)$ is measurable. Thus, the function $C(s) = f(\varphi + s)/f(\varphi)$ is also measurable, as the ratio of two measurable functions. It is known (see, e.g., [1]) that for measurable functions, the only solutions to the functional Eq. (10) are functions

$$f(\varphi) = c \cdot \exp(a \cdot \varphi).$$

This is exactly Louisville-Bratu-Gelfand equation that we are trying to explain! Thus, we arrive at the following conclusion.

Preliminary Conclusion. The Louisville-Bratu-Gelfand equation can be uniquely determined if we require φ-shift-invariance.

What if We Require x-Scale-Invariance? What if now, instead of the φ-shift-invariance, we require x-scale-invariance?

If we replace the original values of the coordinates x_i with the re-scaled values $x'_i = \lambda \cdot x_i$, then the derivatives will divide by λ^2, while the term $f(\varphi)$ will not change. So the Eq. (7) will take the form

$$\frac{1}{\lambda^2} \cdot \nabla^2 \varphi = f(\varphi), \tag{11}$$

i.e., equivalently,

$$\nabla^2 \varphi = \lambda^2 \cdot f(\varphi). \tag{12}$$

Literally speaking, these two equations coincide if $f(\varphi) = \lambda^2 \cdot f(\varphi)$ for all φ and λ, in which case, as we can easily see, the function f is simply 0 – so there is no nonlinearity.

However, as we have mentioned earlier, invariance does not mean that the equation remains the same without changing any other numerical values: sometimes, to preserve the equation, we need to accordingly make changes with other variables as well. In our case, this means that, to preserve the Eq. (7), in addition to a re-scaling $x_i \to \lambda \cdot x_i$, we may also need to apply an appropriate shift $s(\lambda)$ of the φ-field: $\varphi(x) \to \varphi(x) + s(\lambda)$. Under this shift, the derivatives do not change, but the value $f(\varphi)$ is replaced by the value $f(\varphi + s(\lambda))$. So, instead of the Eq. (11), we get a more complicated equation

$$\frac{1}{\lambda^2} \cdot \nabla^2 \varphi = f(\varphi + s(\lambda)),$$

or, equivalently,

$$\nabla^2 \varphi = \lambda^2 \cdot f(\varphi + s(\lambda)). \tag{13}$$

For the new Eq. 12 to be equivalent to the original Eq. (7), we need to make sure that their right-hand sides coincide, i.e., that for all φ and λ, we have

$$\lambda^2 \cdot f(\varphi + s(\lambda)) = f(\varphi),$$

or, equivalently,

$$f(\varphi + s(\lambda)) = \lambda^{-2} \cdot f(\varphi). \tag{14}$$

For this equation, we also get $f(\varphi) = c \cdot \exp(a \cdot \varphi)$ [1], i.e., we also get the Louisville-Bratu-Gelfand equation.

General Conclusion. The simplest case of the Louisville-Bratu-Gelfand equation is the Laplace equation $\nabla^2 \varphi = 0$. To derive this equation from the general linear equation, we need to require either φ-shift-invariance or x-scale-invariance.

It turns out that in the nonlinear case, each of these two invariance requirements uniquely determines the Louisville-Bratu-Gelfand equation. The fact that this equation can be derived from natural symmetries explains why this same equation emerges in the description of many different physical phenomena.

Acknowledgements. This work was supported in part by the US National Science Foundation grant HRD-1242122 (Cyber-ShARE Center of Excellence).

References

1. Aczél, J., Dhombres, J.: Functional Equations in Several Variables. Cambridge University Press, Cambridge (2008)
2. Bateman, H.: Partial Differential Equations of Mathematical Physics. Cambridge University Press, Cambridge (1932)
3. Bratu, G.: Sur les équations intégrales non linéaires. Bulletin de la Société Mathématique de France **42**, 113–142 (1914)
4. Feynman, R., Leighton, R., Sands, M.: The Feynman Lectures on Physics. Addison Wesley, Boston (2005)
5. Gelfand, I.M.: Some problems in the theory of quailinear equations. Am. Math. Soc. Translations **29**(2), 295–381 (1963)
6. Joseph, D.D., Lundgren, T.S.: Quasilinear Dirichlet problems driven by positive sources. Arch. Ration. Mech. Anal. **49**(4), 241–269 (1973)
7. Liouville, J.: Sue l'équation aux différences partielles $\dfrac{d^2 \log \lambda}{dudv} \pm \dfrac{\lambda}{2a^2} = 0$. Journal de Mathématiques Pures et Appliquées, 71–72 (1853)
8. Richardson, O.W.: The Emission of Electricity from Hot Bodies. Longmans, Green, and Co., London (1921)
9. Thorne, K.S., Blandford, R.D.: Modern Classical Physics: Optics, Fluids, Plasmas, Elasticity, Relativity, and Statistical Physics. Princeton University Press, Princeton (2017)
10. Valera, L., Ceberio, M.: Model-order reduction using interval constraint solving techniques. J. Uncertain Syst. **11**(2), 84–103 (2017)
11. Walker, G.W.: Some problems illustrating the forms of nebulae. Proc. Roy. Soc. Lond. Ser. A **91**(631), 410–420 (1915)

Author Index